Ebook Set up and Use Instru

CW01083065

Use of the ebook is subject to the single user licence at the back of this book.

Electronic access to your price book is now provided as an ebook on the VitalSource® Bookshelf platform. You can access it online or offline on your PC/Mac, smartphone or tablet. You can browse and search the content across all the books you've got, make notes and highlights and share these notes with other users.

Setting up

1. Create a VitalSource Bookshelf account at https://online.vitalsource.com/user/new (or log into your existing account if you already have one).

2. Retrieve the code by scratching off the security-protected label inside the front cover of this book. Log in to Bookshelf and click the **Redeem** menu at the top right of the screen. and Enter the code in the **Redeem code** box and press **Redeem**. Once the code has been redeemed your Spon's Price Book will download and appear in your **library**. N.B. the code in the scratch-off panel can only be used once, and has to be redeemed before end December 2020.

When you have created a Bookshelf account and redeemed the code you will be able to access the ebook online or offline on your smartphone, tablet or PC/Mac. Your notes and highlights will automatically stay in sync no matter where you make them.

Use ONLINE

1. Log in to your Bookshelf account at https://online.vitalsource.com).
2. Double-click on the title in your **library** to open the ebook.

Use OFFLINE

Download BookShelf to your PC, Mac, iOS device, Android device or Kindle Fire, and log in to your Bookshelf account to access your ebook, as follows:

On your PC/Mac
Go to https://support.vitalsource.com/hc/en-us/ and follow the instructions to download the free VitalSource Bookshelf app to your PC or Mac. Double-click the VitalSource Bookshelf icon that appears on your desktop and log into your Bookshelf account. Select **All Titles** from the menu on the left – you should see your price book on your Bookshelf. If your Price Book does not appear, select **Update Booklist** from the **Account** menu. Double-click the price book to open it.

On your iPhone/iPod Touch/iPad
Download the free VitalSource Bookshelf App available via the iTunes App Store. Open the Bookshelf app and log into your Bookshelf account. Select **All Titles** – you should see your price book on your Bookshelf. Select the price book to open it. You can find more information at https://support.vitalsource.com/hc/en-us/categories/200134217-Bookshelf-for-iOS

On your Android™ smartphone or tablet
Download the free VitalSource Bookshelf App available via Google Play. Open the Bookshelf app and log into your Bookshelf account. You should see your price book on your Bookshelf. Select the price book to open it. You can find more information at https://support.vitalsource.com/hc/en-us/categories/200139976-Bookshelf-for-Android-and-Kindle-Fire

On your Kindle Fire
Download the free VitalSource Bookshelf App from Amazon. Open the Bookshelf app and log into your Bookshelf account. Select All Titles – you should see your price book on your Bookshelf. Select the price book to open it. You can find more information at https://support.vitalsource.com/hc/en-us/categories/200139976-Bookshelf-for-Android-and-Kindle-Fire

Support

If you have any questions about downloading Bookshelf, creating your account, or accessing and using your ebook edition, please visit https://support.vitalsource.com/

For questions or comments on content, please contact us on sponsonline@tandf.co.uk

Free Updates

with three easy steps…

1. Register today on www.pricebooks.co.uk/updates

2. We'll alert you by email when new updates are posted on our website

3. Then go to www.pricebooks.co.uk/updates
    and download the update.

All four Spon Price Books – *Architects' and Builders'*, *Civil Engineering and Highway Works*, *External Works and Landscape* and *Mechanical and Electrical Services* – are supported by an updating service. Two or three updates are loaded on our website during the year, typically in November, February and May. Each gives details of changes in prices of materials, wage rates and other significant items, with regional price level adjustments for Northern Ireland, Scotland and Wales and regions of England. The updates terminate with the publication of the next annual edition.

As a purchaser of a Spon Price Book you are entitled to this updating service for this 2020 edition – free of charge. Simply register via the website www.pricebooks.co.uk/updates and we will send you an email when each update becomes available.

If you haven't got internet access or if you've some questions about the updates please write to us at Spon Price Book Updates, Spon Press Marketing Department, 3 Park Square, Milton Park, Abingdon, Oxfordshire, OX14 4RN.

Find out more about Spon books
Visit www.pricebooks.co.uk for more details.

Spon's
Architects'
and Builders'
Price Book

2020

Spon's
Architects'
and Builders'
Price Book

Edited by

AECOM

2020

One hundred and forty-fifth edition

First edition 1873
One hundred and forty-fifth edition published 2020
by CRC Press
2 Park Square, Milton Park, Abingdon, Oxon, OX14 4RN

and by CRC Press
Taylor & Francis, 6000 Broken Sound Parkway, NW, Suite 300, Boca Raton, FL 33487

CRC Press is an imprint of the Taylor & Francis Group, an informa business

British Library Cataloguing in Publication Data
A catalogue record for this book is available from the British Library

ISBN: 978-0-367-26703-2
Ebook: 978-0-429-29470-9

ISSN: 0306-3046

Typeset in Arial by Taylor & Francis Books

MIX
Paper from
responsible sources
FSC® C013056

Printed and bound in Great Britain by
TJ International Ltd, Padstow, Cornwall

Contents

PART 1: GENERAL

PART 2: RATES OF WAGES AND LABOUR

PART 3: APPROXIMATE ESTIMATING

PART 4: PRICES FOR MEASURED WORKS

PART 5: FEES FOR PROFESSIONAL SERVICES

PART 6: DAYWORK AND PRIME COST

PART 7: USEFUL ADDRESSES FOR FURTHER INFORMATION

PART 8: TABLES AND MEMORANDA

Preface to the One Hundred and Forty–Fifth Edition

Recent Construction Activity Trends

Construction all work output recorded no discernible yearly change in 2018 Q4, according to the Office for National Statistics' (ONS) latest data. This extends a period through 2018 of variable year-on-year change rates, where no definitive trend was evident. A conclusion to draw from the most recent data point is that little change in the yearly rate implies no growth or expansion of industry output. Equally, the nominal output value also indicates still respectable levels of overall workload.

All major construction sectors recorded declines in the Q4 new orders data – housing (−7%), infrastructure (−38%) and commercial (−3%). Combined, these sectors account for well over 80% of all new work output, so the recorded falls will be felt eventually by the industry. Only the private industrial and public sectors offered positive yearly change in the 2018 Q4 data. However, these sectors account for less than 14% of all new work output, so these increases in work are not expected to move the national construction output dial materially.

AECOM's composite index for building costs – comprised of materials and labour inputs – continued to rise at 2019 Q1. The yearly change at 2018 Q4 was 4.1% and, although elevated, marks a slowdown in the rate of change experienced through 2018. Further increases on a yearly basis are expected in 2019, but at a slower rate that is somewhere nearer 3% than the 4–5% change rate seen for all of 2018. Input costs will likely remain higher overall than output prices in 2019, which maintains commercial pressure across the supply chain.

Materials

Construction materials experiencing the largest price increases in the 12 months to April 2019 in the UK are shown in the table below:

Construction materials	Change between April 2018 and April 2019
Insulating materials (thermal or acoustic)	+11.0%
Imported sawn or planed timber	+4.9%
Plastic pipes and fittings (rigid)	+4.5%
Precast concrete products	+3.7%
Fabricated structural steel	+3.5%
Cement	+3.0%

From April 2018 and April 2019 the average rise for the BEIS recorded materials show an increase of 2%.

Labour – Wage Agreements

Demand for labour and staff resources eased slightly at the start of 2019. That available capacity has opened up reflects a softening of workload and a corresponding need to secure resources. Nevertheless, wages for site trades still rose on a yearly change basis, which is consistent with respectable levels of overall workload creating demand for labour. Having shaken off some of the seasonal lull over the winter months, wage levels for most site classifications are returning to previous trend levels

Book Price Level

The price level of *Spon's A&B 2020* has been indexed at **124** (2015 = 100). Readers of *Spon's A&B* are reminded that Spon's is the only known price book in which key rates are checked against current tender prices. Users should note that this number is based on book prices arrived at by pricing our usual tender price models with prices taken from the book. It does not reflect a particular date which would appear in our normal published indices.

Profits and Overheads

The 2020 edition includes a 3% allowance main contractor's overheads and profit.

Preliminaries

There are continuing signs that main contractor preliminaries costs are levelling off, having been adjusted from the one-way downward flow of recent years. Indirect costs have steadily increased for the last few years. Whilst this is not across the board, and depends to a large degree on the project parameters, this is a likely response to industry output changes, lingering uncertainty for the medium-term and the need to retain a competitive edge. We have set our example provision for preliminaries this year at 12%, but the range can be over 20% for some specific projects and procurement routes.

- Preliminaries are ***not included*** within the main *Prices for Measured Works* or in the *Approximate Estimating Rates* sections of the book.
- Preliminaries are ***included*** in the rates within *Building Prices per Functional Unit* and *Building Prices per Square Metre* sections.

Prices included within this edition do not include for VAT, professional fees etc. which must be added if appropriate.

PARTS OF THIS BOOK

Part 1: General

This section contains advice on various construction specialism's; capital allowances; legislation; taxes; insurances; building cost and tender price indices and regional price variations.

Part 2 Rates of Wages and Labour

Shows current industry level wage agreement rates and how we have built up the gang labour rates being used in the book.

Part 3: Approximate Estimating

This section contains distinct areas:

- *Building Prices per Functional Unit; Building Prices per Square Metre and Building Cost Models*. It should be noted that these sections all *include site preliminaries*. The only occasion this happens within the book.
- *Approximate Estimating Rates* shows typical composite built-up rates organized by building elements. Please note these rates *do not include for any site preliminaries*.

There is also a section where we show typical preliminaries build up for a project valued at approximately £4,000,000. This is intended for guidance only and should not be used as part of any tender submission.

Part 4 Prices for Measured Work

These sections contain Prices for Measured Work organized using the NRM2 Work Sections for building works.

NOTE: All prices in Part 4 exclude the main contractor's preliminaries costs.

Part 5: Fees for Professional Services

This section contains guidance on fee levels for professional services; Quantity Surveyors; Architects and Consulting Engineers. Readers should always obtain fee proposals for their project prior to commencement as there are many factors that influence fee submissions.

NOTE: Professional fees are not included in any rates in the book.

Part 6: Daywork and Prime Cost

This section contains Daywork and Prime Cost allowances issued by the Royal Institute of Chartered Surveyors.

Part 7: Useful Addresses for Further Information

A list of useful trade associations' and professional bodies' contact details.

Part 8: Tables and Memoranda

This section contains general formulae, weights and quantities of materials, other design criteria and useful memoranda associated with each trade.

While every effort is made to ensure the accuracy of the information given in this publication, neither the Editors nor Publishers in any way accept liability for loss of any kind resulting from the use of such information

AECOM Ltd
Aldgate Tower
2 Leman Street
London
E1 8FA

Quality Auditing in Construction Projects

Abdul Razzak Rumane

This book provides construction professionals, designers, contractors and quality auditors involved in construction projects with the auditing skills and processes required to improve construction quality and make their projects more competitive and economical.

The processes within the book focus on auditing compliance to ISO, corporate quality management systems, project specific quality management systems, contract management, regulatory authorities' requirements, safety, and environmental considerations. The book is divided into seven chapters and each chapter is divided into numbered sections covering auditing-related topics that have importance or relevance for understanding quality auditing concepts for construction projects.

No other book covers construction quality auditing in such detail and with this level of practical application. It is an essential guide for construction and quality professionals, but also for students and academics interested in learning about quality auditing in construction projects.

June 2019: 234 × 156 mm: 600pp
Hb: 978-0-8153-8531-8 : £100.00

To Order: Tel: +44 (0) 1235 400524 Fax: +44 (0) 1235 400525
or Post: Taylor and Francis Customer Services,
Bookpoint Ltd, Unit T1, 200 Milton Park, Abingdon, Oxon, OX14 4TA UK
Email: book.orders@tandf.co.uk

For a complete listing of all our titles visit:
www.tandf.co.uk

Special Acknowledgements

Acodrain
Tel: 01462 816666
bulldIngdrainage@aco.co.uk
www.aco.co.uk
Drainage channels

Altro Floors
Tel: 0870 6065432/01462 707600
info@altro.co.uk
www.altro.co.uk
Floor coverings

ASSA ABLOY

ASSA ABLOY Entrance Systems Ltd
Tel: 0333 006 3443
sales.uk.entrance@assaabloy.com
www.assaabloyentrance.co.uk
Entrance systems

Building Innovation
Tel: 01926 888808
info@building-innovation.co.uk
www.building-innovation.co.uk
Tapered insulation

Halfen Ltd
Tel: 01582 470 300
brick@halfen.co.uk
www.halfen.co.uk
Brick accessories – channels and special products

Kingspan Insulated Panels
Tel: 01358 716100
info@kingspanpanels.com
www.kingspanels.co.uk
Insulated wall and roof panels

Parker & Highland Joinery Ltd
Tel: 01903 756 283
sales@parker-joinery.com
www.parker-joinery.com
Purpose-made joinery

Profile 22
Tel: 01952 290910
mail@profile22.co.uk
www.profile22.co.uk
uPVC windows

Severfield Ltd
Tel: 01845 577896
www.severfield.com
Structural steel

Acknowledgements

Acodrain
Tel: 01462 816666
buildingdrainage@aco.co.uk
www.aco.co.uk
Drainage channels

Allgood plc
Tel: 020 7387 9951
info@allgood.co.uk
www.allgood.co.uk
Ironmongery

Altro Floors
Tel: 0870 6065432 / 01462 707600
info@altro.co.uk
www.altro.co.uk
Floor coverings

Alumasc Exterior Building Products Ltd
Tel: 01744 648400
info@alumasc-exteriors.co.uk
www.alumasc-exteriors.co.uk
Aluminium rainwater goods

Alumasc Interior Building Products Ltd
Tel: 01952 580590
sales@pendock.co.uk
www.pendock.co.uk
Skirting trunking & casings

Amwell Systems Ltd
Tel: 01763 276200
sales@amwell-systems.com
www.amwell-systems.com
Toilet cubicles

Andrews Marble Tiles
Tel: 0113 262 4751
sales@andrews-tiles.co.uk
www.andrews-tiles.co.uk
Floor and wall tiles

Armitage Shanks Group
Tel: 01543 490253
arm-idealinfo@idealstandard.com
www.armitage-shanks.co.uk
Sanitary fittings

Armstrong Floor Products UK Ltd
Tel: 01642 768660
commercial_uk@armstrong.com
www.armstrong-flooring.co.uk
Flooring products

ASSA ABLOY Entrance Systems Ltd
Tel: 0333 006 3443
sales.uk.entrance@assaabloy.com
www.assaabloyentrance.co.uk
Entrance systems

Assa Abloy Ltd
Tel: 01902 364 648
info@assaabloy.com
www.assaabloy.co.uk
Ironmongery

Bison Concrete Products Ltd
Tel: 01283 495000
concrete@bison.co.uk
www.bison.co.uk
Precast concrete floors

Bolton Gate Co
Tel: 01204 871000
sales@boltongate.co.uk
www.boltongate.co.uk
Insulated shutter doors

Bradstone Structural Sales
Tel: 01285 646 884
bradstone-structural@aggregate.com
www.bradstone-structural.com
Walling, dressings and other products

British Gypsum Ltd
Tel: 01509 817 200
bgtechnical.enquiries@bpb.com
www.british-gypsum.com
Plasterboard and plaster products

Building Innovation
Tel: 01926 888808
info@building-innovation.co.uk
www.building-innovation.co.uk
Tapered insulation

Burlington Slate Ltd
Tel: 01229 889 681
sales@burlingtonstone.co.uk
www.burlingtonstone.co.uk
Westmoreland slating

Buttles
Tel: 01727 834 242
beeweb@buttle.co.uk
www.buttles.co.uk
Timber products

Catnic Ltd
Tel: 02920 337900
www.catnic.com
Steel lintels

Cavity Trays Ltd
Tel: 01935 474769
sales@cavitytrays.co.uk
www.cavitytrays.com
Cavity trays, closers and associated products

Cementation Foundations Skansa Ltd
Tel: 01923 423100
cementation.foundations@skanska.co.uk
www.skanska.co.uk
Piling systems

Colour Centre
Tel: 020 7609 1164
www.colourcentre.com
Paints, varnishes and stains

Concrete Canvas Ltd (UK)
Tel: 0845 680 1908
info@concretecanvas.co.uk
www.concretecanvas.co.uk
Concrete canvas

Cox Building Products
Tel: 01902 371800
sales@coxspan.co.uk
www.coxbp.com
Rooflights

Custom Metal Fabrications
Tel: 020 8844 0940
dgibbs@cmf.co.uk
www.cmf.co.uk
Metalwork and balustrading

Decra Roof Systems (UK) Ltd
Tel: 01293 545 058
sales@decra.co.uk
www.decra.co.uk
Roofing system and accessories

Dow Construction Products
Tel: 020 8917 5050
styrofoam-uk@dow.com
www.styrofoameurope.com
Insulation products

Dreadnought Clay Roof Tiles
Tel: 01384 77405
sales@dreadnought-tiles.co.uk
www.dreadnought-tiles.co.uk
Clay roof tiling

Dufaylite Developments Ltd
Tel: 01480 215000
enquiries@dufaylite.com
www.dufaylite.com
Clayboard void former

Envirodoor Ltd
Tel: 01482 659375
sales@envirodoor.com
www.envirodoor.com
Sliding and folding doors

Expamet Building Products
Tel: 01429 866611
sales@expamet.net
www.expamet.co.uk
Builder's metalwork products

Forbo Flooring UK Ltd
Tel: 01773 744 121
www.forbo-flooring.co.uk
Mats and matwells

Forterra Building Products Ltd
Tel: 01604 707600
www.forterra.co.uk
Facing bricks; aircrete blocks; aggregate blocks

Forticrete Ltd
Tel: 01525 244 900
masonry@forticrete.com
roofing@forticrete.com
nwblocksales@forticrete.com
www.forticrete.co.uk
Blocks and roof tiles

F P McCann Ltd
Tel: 028 7964 2558
www.fpmccann.co.uk
Precast concrete goods

Garador Ltd
Tel: 01935 443700
www.garador.co.uk
Garage doors

GCP
Tel: 01753 490000
SalesOrders.gcp-IRL@gcpat.com
https://gcpat.uk/en-gb
Expansion joint fillers and waterbars

Gradus Wall Protection
Tel: 01625 428 922
sales@gradusworld.com
www.gradusworld.com
Stair nosings

GRP Tanks
Tel: 0871 200 2082
www.grptanks.net
GRP water tanks

H+H Celcon UK Ltd
Tel: 01732 886333
info@hhcelcon.co.uk
www.hhcelcon.co.uk
Concrete blocks

Halfen Ltd
Tel: 01582 470 300
brick@halfen.co.uk
www.halfen.co.uk
Brick accessories – channels and special products

Hathaway Roofing Ltd
Tel: 01388 605 636
www.hathaway-roofing.co.uk
Sheet wall and roof claddings

Hillaldam Coburn Ltd
Tel: 020 8545 6680
sales@hillaldam.co.uk
www.coburn.co.uk
Sliding and folding door gear

HSS Hire Shops
Tel: 020 8260 3100
www.hss.co.uk
Tool hire

Hudevad
Tel: 02476 881200
sales@hudevad.co.uk
www.hudevad.co.uk
Radiators

Hunter Plastics Ltd
Tel: 01622 852798
info@hunterplastics.co.uk
www.hunterplastics.co.uks
Plastic rainwater goods

Ibstock Building Products
Tel: 01530 261999
marketing@ibstock.co.uk
www.ibstock.co.uk
Facing bricks; tilebricks

Icopal Ltd
Tel: 0843 224 7400
info.uksales@icopal.com
www.icopal.co.uk
Damp-proof products

James Latham
Tel: 01454 315 421
marketing@lathams.co.uk
www.lathamtimber.co.uk
Hardwood and panel products

Jeld-Wen UK Ltd
Tel: 0870 126 0000
marketing@jeld-wen.co.uk
www.jeld-wen.co.uk
Doors and windows

John Brash and Co Ltd
Tel: 01427 613858
info@johnbrash.co.uk
www.johnbrash.co.uk
Roofing shingles

John Guest Speedfit Ltd
Tel: 01895 449 233
www.speedfit.co.uk
Speedfit product range

Junkers Ltd
Tel: 01376 534 700
sales@junkers.co.uk
www.junkers.co.uk
Hardwood flooring

Kalzip Ltd
Tel: 01942 295500
kalzip-uk@corusgroup.com
www.kalzip.com
Kalzip roofing

KB Rebar Ltd
Tel: 0161 790 8635
www.kbrebar.co.uk
Reinforcement bar and mesh

Kingspan Access Floors
Tel: 01482 781 710
enquiries@kingspanaccessfloors.co.uk
www.kingspanaccessfloors.co.uk
Raised access floors

Kingspan Environmental
Tel: 01296 633000
pollutiongb@kingspanenv.com
www.kingspanenv.com
Interceptors and septic tanks

Kingspan Insulated Panels
Tel: 01358 716100
info@kingspanpanels.com
www.kingspanels.co.uk
Insulated wall and roof panels

Kingspan Insulation Ltd
Tel: 0870 850 8555
info.uk@insulation.kingspan.com
www.insulation.kingspan.com
Insulation products

Kingspan Structural Products
Tel: 01944 712 000
sales@kingspanstructural.com
www.kingspan.com
Multibeam purlins

Knauf Insulation Ltd
Tel: 01744 766 666
sales@knaufinsulation.com
www.knaufinsulation.co.uk
Insulation products

Landpro Ltd
Tel: 01252 795030
info@landpro.co.uk
www.landpro.co.uk
Landscaping consultants

Lignacite
Tel: 01992 464 441
info@lignacite.co.uk
www.lignacite.co.uk
Concrete blocks

Maccaferri
Tel: 01865 770 555
marketing@maccaferri.co.uk
www.maccaferri.co.uk
Gabions

Magrini Ltd
Tel: 01543 375311
sales@magrini.co.uk
www.magrini.co.uk
Baby equipment

Manhole Covers Ltd
Tel: 01296 668850
sales@manholecovers
www.manholecovers.com
Manhole covers

Marshalls CPM
Tel: 01179 814500
sales@cpm-group.co.uk
www.cpm-group.co.uk
Concrete pipes etc.

Marshalls Mono Ltd (drainage)
Tel: 01422 312000
drainage@marshalls.co.uk
www.marshalls.co.uk
Drainage channels

Marshalls Mono Ltd
Tel: 01422 312000
marshallspaving@marshalls.co.uk
www.marshalls.co.uk
Pavings

Metsec Lattice Beams Ltd
Tel: 01902 408011
technical@metseclatticebeams.com
www.metseclatticebeams.com
Lattice beams

Monier Ltd
Tel: 01293 618418
roofing.redland@monier.com
www.redland.co.uk
Redland roof tiles

NDM Metal Roofing & Cladding Ltd
Tel: 020 8991 7310
enquiries@ndmltd.com
www.ndmltd.com
Metal cladding and roofing

Parker & Highland Joinery Ltd
Tel: 01903 756 283
sales@parker-joinery.com
www.parker-joinery.com
Purpose-made joinery

Plumb Centre
Tel: 0870 1622 557
www.plumbcentre.co.uk
Cylinders & general plumbing

Polyflor Ltd
Tel: 0161 767 1111
www.polyflor.com
Polyfloor contract flooring

Polypipe Terrain
Tel: 01622 717811
commercialenquiries@polypipe.com
www.terraindrainage.com
Drainage goods

Premdor Ltd
Tel: 01793 708200
enquiries@premdor.com
www.premdor.co.uk
Doors and windows

Premier Loft Ladders
Tel: 0845 9000 195
sales@premierloftladders.com
www.premierloftladders.com
Loft ladders

Pressalit Care plc
Tel: 0844 880 6950
www.pressalitcare.com
Bathroom equipment

Profile 22
Tel: 01952 290910
mail@profile22.co.uk
www.profile22.co.uk
uPVC windows

Promat UK
Tel: 01344 381 301
promat@promat.co.uk
www.promat.co.uk
Fireproofing materials

Protim Solignum Ltd
Tel: 01628 486644
info@osmose.co.uk
www.osmose.co.uk
Paints and timber treatment

Radius Systems Ltd
Tel: 01773 811 112
www.radius-systems.co.uk
MDPE pipes and fittings

Rawlplug Ltd
Tel: 0141 638 225
info@rawlplug.co.uk
www.rawlplug.co.uk
Anchoring and fixing systems

Richard Potter Timber Merchants
Tel: 01270 625791
richardpotter@fortimber.demon.co.uk
www.fortimber.demon.co.uk
Carcasssing softwood

Rockwool Ltd
Tel: 01656 862 261
info@rockwool.co.uk
www.rockwool.co.uk
Pipe and other insulation products

Ryton's Building Products
Tel: 01536 511874
admin@rytons.com
www.vents.co.uk
Roof ventilation products

Safeguard Europe Ltd
Tel: 01403 212004
www.safeguardeurope.com
Damp-proofing and waterproofing

Saint Gobain PAM UK
Tel: 0115 930 5000
sales.uk.pam@saint-gobain.com
www.saint-gobain-pam.co.uk/
Cast iron soil, water and rainwater pipes and fittings

Sandtoft Roof Tiles Ltd
Tel: 01427 871200
info@sandtoft.co.uk
www.sandtoft.co.uk
Clay roof tiling

Schiedel Rite-Vent
Tel: 0191 416 1150
sales@schiedel.co.uk
www.isokern.co.uk
Flue pipes and gas blocks

Screeduct Ltd
Tel: 01789 459 211
sales@screeduct.com
www.screeduct.com
Trunking systems and conduits

Severfield Ltd
Tel: 01845 577896
www.severfield.com
Structural steel

Sheet Piling UK Ltd
Tel: 01772 794 141
enquiries@sheetpilinguk.com
www.sheetpilinguk.com
Sheet piling

Sheffield Insulation
Tel: 020 8477 9500
barking@sheffins.co.uk
www.SheffieldInsulation.co.uk
Insulation products

Siderise Insulation Ltd
Tel: 01656 730833
sales@siderise.com
www.siderise.com
Fire barriers

Slate UK David Wallace International Ltd
Tel: 015395 59289
www.slate.uk.com
Spanish roof slates

Stainless UK Ltd
Tel: 0114 244 1333
sales@stainless-uk.co.uk
www.stainless-uk.co.uk
Stainless steel rebar

Sterling Hydraulics (Huntley & Sparks) Ltd
Tel: 01460 722 22
Rigifix column guards

Stirling Lloyd Polychem Ltd
Tel: 01565 633 111
www.stirlinglloyd.com
Integritank products

Stressline Ltd
Tel: 0870 7503167
sales@stressline.ltd.uk
www.stressline.ltd.uk
Concrete lintels

Swish Building Products
Tel: 01827 317200
marketing@swishbp.co.uk
www.swishbp.co.uk
Swish celuka

Szerelmey Ltd
Tel: 0207 735 9995
info@szerelmey.com
www.szerelmey.com
Stonework

Tarkett-Marley Floors Ltd
Tel: 01622 854000
uksales@tarkett.com
www.tarkett-floors.com
Sheet and tile flooring

Tarmac Ltd
Tel: 020 8555 2415
enquiries@tarmac.co.uk
www.tarmac.co.uk
Ready-mixed concrete

Tarmac Mortar and Screeds
Tel: 08701 116 116
mortar@tarmac.co.uk
www.tarmac.co.uk
Readymix screeds and mortar

Tarmac Topblock Ltd
Tel: 01902 754 131
enquiries@tarmac.co.uk
www.topblock.co.uk
Concrete blocks

TATA Steel Ltd
Tel: 01724 405 060
www.tatasteeleurope.com
Steel

Timbmet
Tel: 01865 860351
marketing@timbmet.com
www.timbmet.com
Hardwood

Travis Perkins Trading Company
Tel: 01604 752484
www.travisperkins.co.uk
Builders merchant

VA Hutchison Flooring Ltd
Tel: 01243 841 175
www.hutchisonflooring.co.uk
Hardwood flooring

Vandex
Tel: 01403 210204
info@safeguardeurope.com
www.vandex.com
Vandex super and premix products

Velfac Ltd
Tel: 01223 897 100
post@velfac.co.uk
www.velfac.co.uk
Composite windows

Velux Company Ltd
Tel: 0870 166 7676
enquires@velux.co.uk
www.velux.co.uk
Velux roof windows & flashings

Visqueen Building Products
Tel: 01685 840 672
enquiries@visqueenbuilding.co.uk
www.visqueenbuilding.co.uk
Visqueen products

Wavin Plastics Ltd
Tel: 01249 766600
info@wavin.co.uk
www.wavin.co.uk
uPVC drainage goods

Web Dynamics Ltd
Tel: 01204 695666
www.webdynamics.co.uk
Breather membranes

Welco
Tel: 0121 4219000
enquiries@welconstruct.co.uk
www.welconstruct.co.uk
Lockers and shelving systems

Welsh Slate Ltd
Tel: 01248 600 656
enquiries@welshslate.com
www.welshslate.com
Natural Welsh slates

Wilde Contracts Ltd
Tel: 0161 624 6824
www.rogerwilde.com
Glass blocks

Yeoman Aggregates Ltd
Tel: 020 8896 6800
debra.ward@yeoman-aggregates.co.uk
www.yeoman-aggregates.co.uk
Hardcore, gravels, sand

Yorkshire Copper Tube Ltd
Tel: 0151 546 2700
info@yct.com
www.yorkshirecopper.com
Copper tube

How to use this Book

First time users of *Spon's Architects' and Builders' Price Book* (*Spon's A & B*) and others who may not be familiar with the way in which prices are compiled may find it helpful to read this section before starting to calculate the costs of building work. The level of information on a scheme and availability of detailed specifications will determine which section of the book and which level of prices users should refer to.

We have rebased our **TENDER PRICE INDEX to 2015 = 100** (from 1976 = 100) Users wishing to reference to or from the previous set of indices with a base date of 1976 = 100 need to apply a factor of **5.123**. Multiply the TPI 2015 series below by 5.123 to give the equivalent in the 1976 series.

Prices in the book do not necessarily, and are not intended to represent the lowest possible prices achievable, but are intended as a guide to expected price levels for the items described. AECOM cost a series of building models using current Spon's rates to calculate a book Tender Price Index (TPI).

For this edition of the book TPI has been calculated at 124 (2015 = 100).

Rates in the book include overhead and recovery margins but do not include main contractors' preliminaries: except for two sections:

- *Building Prices per Functional Unit* and
- *Building Prices per Square Metre* which are both within *Part 3 Approximating Estimating*.

New Rules of Measurement (NRM)

The NRM suite covers the life cycle of cost management and means that, at any point in a building's life, quantity surveyors will have a set of rules for measuring and capturing cost data. In addition, the BCIS *Standard Form of Cost Analysis* (SFCA) 4th edition has been updated so that it is aligned with the NRM suite. This book is intended for use with NRM1 and NRM2.

The three volumes of the NRM are as follows.

- NRM1 – Order of cost estimating and cost planning for capital building works.
- NRM2 – Detailed measurement for building works.
- NRM3 – Order of cost estimating and cost planning for building maintenance works.

APPROXIMATE ESTIMATING

For preliminary estimates/indicative costs before drawings are prepared or very little information is available, users are advised to refer to the average overall *Building Prices per Functional Unit* and multiply this by the proposed number of units to be contained within the building (i.e. number of bedrooms etc.) or *Building Prices per Square Metre* rates and multiply this by the gross internal floor area of the building (the sum of all floor areas measured within external walls) to arrive at an overall initial *Order of Cost Estimate*. These rates include preliminaries, but make no allowance for the cost of external works, VAT, or fees for provisional services.

Where preliminary drawings are available, one should be able to measure approximate quantities for all the major components of a building and multiply these by individual rates contained in the *Building Cost Models* or *Approximate Estimating Rates* sections to produce an *Elemental Cost Plan*. This should produce a more accurate estimate of cost than simply using overall prices per square metre. Labour and other incidental associated items, although normally measured separately within Bills of Quantities, are deemed included within approximate estimating rates. These rates do not include preliminaries or fees for professional services.

MEASURED WORKS

For more detailed estimates or documents such as Bills of Quantities (Quantities of supplied and fixed components in a building, measured from drawings), use rates from *Prices for Measured Works*. Depending upon the overall value of the contract reader may want to adjust the value and we have added a simple chart which shows typical adjustments that could be applied as shown later in this chapter. Items within the Measured Works sections are made up of many components: the cost of the material or product; any additional materials needed to carry out the work; plant required and the labour involved in unloading and fixing, etc.

Measured Works Rates

These components are usually broken down into:

Labour

This figure covers the cost of the operation and is calculated on the gang wage rate (skilled or unskilled) and the time needed for the job. A full explanation and build-up is provided. Large regular or continuous areas of work are more economical to install than smaller complex areas.

Plant

Plant covers the use of machinery ranging from excavators and dumpers to static plant and includes running costs such as fuel, water supply, electricity and waste disposal.

For this year's book we are showing Plant costs separately. Previously it has been included in the Materials column. We believe this will make the rates more transparent and understandable to users who did not always realize that Plant costs were included. Small handheld plant is not included.

Materials

Material prices include the cost of any ancillary materials, nails, screws, waste, etc., which may be needed in association with the main material product/s. If the material being priced varies from a standard measured rate, then identify the difference between the original PC price and the material price and add this to your alternative material price before adding to the labour cost to produce a new overall total rate. Alternative material prices, where given, are largely based upon list prices, before the deduction of quantity discounts etc., and therefore require discount adjustment before they can be substituted in place of PC figures given for Measured Work items.

Prime Cost

Commonly known as the PC; Prime Cost is the actual price of the material such as bricks, blocks, tiles or paint, as sold by suppliers. Prime Cost may be given as *per square metre*, *per 100 bags* or *each* according to the way the supplier sells the product. Unless otherwise stated, prices in *Spon's A & B* are deemed to be delivered to site, (in which case transport costs will be included) and also take account of trade and quantity discounts. Part loads generally cost more than whole loads but, unless otherwise stated, Prime Cost figures are based on average prices for full loads delivered to a hypothetical site in Acton, West London. Actual prices for live projects will vary depending on the contractor, supplier, the distance from the supplier to the site, the accessibility of the site, whether the whole quantity ordered is to be supplied in one delivery or at specified dates and market conditions prevailing at the time. Prime Cost figures for commonly used alternative materials are supplied in listed form at the beginning of some work sections.

Where a PC rate is entered alongside an item rate then the cost allowed for that item is in the overall material cost. For instance, bricks need mortar; paving needs sand bedding, so the PC cost is simply for the cost of bricks or paving, thus allowing the user to simply substitute an alternative product cost if desired.

Example:

Item *(NB example data only)*	PC £	Labour £	Material £	Unit	Total Rate £
Half brick wall in common bricks	380.00	18.05	27.14	m²	**45.19**
Replace with Forterra Brick; Brecken Grey @	476.40				
Calculation: (PC) £476.40 − £380.00 = £96.40 + 5% (OHP) = £101.22 add £1.69/m² to materials rate of £27.14 = £28.83					
Therefore, Brecken Grey bricks =	476.40	18.05	28.83	m²	**46.88**

Unit

The Unit is generally based upon measurement guidelines laid out in the New Rules of Measurement – Detailed measurement for building works (NRM2).

Total Rate

Prices in the Total Rate column generally include for the supply and fix of items, unless otherwise described.

Overheads and Profit

The general overheads of the Main Contractor's business – the head office overheads and any profit sought on capital and turnover employed, is usually covered under a general item of overheads and profit which is applied either to all measured rates as a percentage, or alternatively added to the tender summary or included within Preliminaries for site specific overhead costs.

Within this edition we are including an allowance of 3% for overheads and profit on built-up labour rates and material prices.

Preliminaries

Site specific Main Contractor's overheads on a contract, such as insurances, site accommodation, security, temporary roads and the statutory health and welfare of the labour force, are not directly assignable to individual items so they are generally added as a percentage or calculated allowances after all building component items have been costed and totalled. Preliminaries will vary from project to project according to the type of construction, difficulties of the site, labour shortage, or involvement with other contractors, etc. The overall addition for a scheme should be adjusted to allow for these factors. For this edition we have shown a calculated typical Preliminaries cost example of approximately 12%.

Sub/Specialist-Contractor's Costs

For the purpose of this book, these are deemed to include all the above costs, and assume a 2.5% main contractor's discount.

With the exclusion of main contractor's preliminaries, the above items combine to form item rates in the Prices for Measured Works sections. It should be appreciated that a variation in any one item in any group will affect the final measured work price. Any cost variation must be weighed against the total cost of the contract, and a small variation in Prime Cost where the items are ordered in thousands may have more effect on the total cost than a large variation on a few items, while a change in design which introduces the need to use, for example earth moving equipment, which must be brought to the site for that one task, will cause a dramatic rise in the contract cost. Similarly, a small saving on multiple items could provide a useful reserve to cover unforeseen extras.

COST PLANNING

Order of Cost Estimate

The purpose of an Order of Cost Estimate is to establish if the proposed building project is affordable and, if affordable, to establish a realistic cost limit for the project. The cost limit is the maximum expenditure that the employer is prepared to make in relation to the completed building project, which will be managed by the project team.

An Order of Cost Estimate is produced as an intrinsic part of RIBA Work Stages A: Appraisal and B: Design Brief or OGC Gateways 1 (Business Justification) and 2 (Delivery Strategy).

There are comprehensive guidelines within the NRM documentation and readers are recommended to read the relevant sections of the NRM where more detailed explanations and examples can be found.

At this early stage, in order for the estimate to be representative of the proposed design solution, the key variables that a designer needs to have developed to an appropriate degree of certainty are:

- The floor areas upon which the estimate is based
- Proposed elevations
- The implied level of specification

Rates will need to be updated to current estimate base date by the amount of inflation occurring from the base date of the cost data to the current estimate base date. The percentage addition can be calculated using published indices (i.e. tender price indices [TPI]).

Example 1:

New secondary school

Note: example data only

Gross Internal Floor Area (GIFA) = 15,000 m^2

Cost plan prepared with a TPI = 124

Start on site Q4 2022 TPI = 128.4

Location: North West adjustment = 0.89

From *Building Prices per Square Metre*

Assume rate of say £1,800 per m^2

		Cost (£)
School rate, say	£1,800 /m^2 × 15,000 m^2 =	27,000,000
Adjust for inflation to start date	(128.4/124) say +3.5%	945,000
	subtotal	27,945,000
Adjust for location	−10%, say	−2,795,000
	subtotal	25,150,000
Allow for contingencies	say 10%	250,000
Total Order of Cost Estimate		**25,400,000**

Main contractor's preliminaries, overheads and profit need not be added to the cost of building works as they are included within the Spon's building prices per square metre rates, but you will need to add on professional fees and other enabling works costs such as site clearance, demolition, external works, car parking, bringing services to site etc.

Elemental Cost Planning

Elemental cost plans are produced as an intrinsic part of RIBA Work Stages C: Concept, D: Design Development, E: Technical Design and F: Production Information; or when the OGC Gateway Process is used, Gateways 3A (Design Brief and Concept Approval) and 3B (Detailed Design Approval).

Cost Models can be used to quickly extract £/m^2 of GIFA:

Example 2:

Health Centre

Note: example data only

Gross Internal Floor Area = 1,000 m^2

Cost plan prepared with TPI = 124

Start on site Q4 2022 TPI = 128.4

Location: South West (adjustment = 0.90)

Element	Rate (£/m²)	Cost (£)
Substructures	106.53	106,530
Frame and Upper Floors	160	160,000
Roof	24	24,000
And so on for each element to give a total of	**1,300.56**	**1,300,560**
Contractors preliminaries, overheads and profit	say 15%	195,000
	subtotal	1,495,560
Adjust for inflation to start date	(128.4/124) say +3.5%	52,000
	subtotal	1,495,560
Adjust for location	say −10%	−155,000
	subtotal	1,392,560
Allow for contingencies	say 5%	70,000
	Total Elemental Cost Plan	**1,462,560**

Other allowances such as consultants fees, design fee, VAT, risk allowance, client costs, fixed price adjustment may need to be added to each of the examples above.

Formal Cost Planning Stages

The NRM schedules a number of formal cost planning stages, which are comparable with the RIBA Design and Pre-Construction Work Stages and OGC Gateways 3A (Design Brief and Concept Approval) and 3B (Detailed Design Approval) for a building project. The employer is required to 'approve' the cost plan on completion of each RIBA Work Stage before authorizing commencement of the next RIBA Work Stage.

Formal Cost Plan	RIBA Work Stage
1	2: Concept
2	3: Design Development
3	4: Technical Design

Formal Cost Plan 1 is prepared at a point where the scope of work is fully defined and key criteria are specified but no detailed design has begun. Formal Cost Plan 1 will provide the frame of reference for Formal Cost Plan 2. Likewise, Formal Cost Plan 2 will provide the frame of reference for Formal Cost Plan 3. Neither Formal Cost Plans 2 nor 3 involve the preparation of a completely new Elemental Cost Plan; they are progressions of the previous cost plans, which are developed through the cost checking of price significant components and cost targets as more design information and further information about the site becomes available.

Cost plans can be developed using from Elemental Cost Plans using both Approximate Estimating Rates and/or Prices for Measured Works depending upon the level of information available.

The cost targets within each formal cost plan approved by the employer will be used as the baseline for future cost comparisons. Each subsequent cost plan will require reconciliation with the preceding cost plan and explanations relating to changes made. In view of this, it is essential that records of any transfers made to or from the risk allowances and any adjustments made to cost targets are maintained, so that explanations concerning changes can be provided to both the employer and the project team.

Adjustment According to Contract Sum Cost

The construction costs for a project will depend on the size, type of building, standard of finish required, and location, the economic climate of the construction industry i.e. if there is a shortage of construction work available firms will reduce their tender in order to try and attract work. If the opposite is the case and there is a lot of work available, firms will increase their tenders, as they will not be too keen to obtain the contract which will stretch their resources unless it is worth their while financially. In a recession, construction firms can literally buy work in order to keep their workforce and to ensure some cash flow.

Building Costs can vary between builders/developers. This can be due to the size or purchasing abilities of a company or the discount that it receives from suppliers.

Adjustments can be made to reflect the value of a project by using the following table.

Contract Sum	% adjustment
£5,000,000	−1%
£3,500,000	0%
£2,750,000	1%
£2,000,000	2%
£1,250,000	3%
£1,000,000	4%

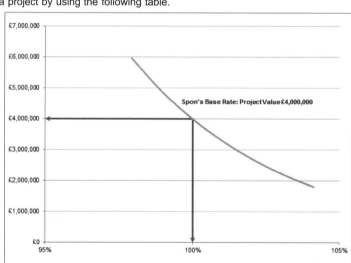

Spon's Architects' & Builders' Price Book is targeted at new build projects with a value range of approximately £3,000,000–£5,000,000.

Users should not simply apply percentage adjustments to any project regardless of size. We recommend project values between £2,000,000 and £5,000,000 for rates found in *Spon's A & B* could be adjusted according to the above table. This is given only as an indication and users should always remember that there are many factors that affect the overall project costs.

It is one thing to imagine a better world.

It's another to deliver it.

Creating iconic structures. Planning entire communities. Connecting people with transport systems. Rescuing historic buildings. Our team of circa 900 cost and project managers collaborate to turn imagination into reality, providing innovative solutions to complex challenges across the globe.

Imagine it. Delivered.

Estimator's Pocket Book, 2nd edition

Duncan Cartlidge

The *Estimator's Pocket Book, Second Edition* is a concise and practical reference covering the main pricing approaches, as well as useful information such as how to process sub-contractor quotations, tender settlement and adjudication. It is fully up to date with NRM2 throughout, features a look ahead to NRM3 and describes the implications of BIM for estimators.
It includes instructions on how to handle:

*   the NRM order of cost estimate;
*   unit-rate pricing for different trades;
*   pro-rata pricing and dayworks;
*   builders' quantities;
*   approximate quantities.

Worked examples show how each of these techniques should be carried out in clear, easy-to-follow steps. This is the indispensable estimating reference for all quantity surveyors, cost managers, project managers and anybody else with estimating responsibilities. Particular attention is given to NRM2, but the overall focus is on the core estimating skills needed in practice. Updates to this edition include a greater reference to BIM, an update on the current state of the construction industry as well as up-to-date wage rates, legislative changes and guidance notes.

Routledge Pocket Books

February 2019: 198 × 129 mm: 292pp
Pb: 978-1-138-36670-1 : £24.99

To Order: Tel: +44 (0) 1235 400524 Fax: +44 (0) 1235 400525
or Post: Taylor and Francis Customer Services,
Bookpoint Ltd, Unit T1, 200 Milton Park, Abingdon, Oxon, OX14 4TA UK
Email: book.orders@tandf.co.uk

For a complete listing of all our titles visit:
www.tandf.co.uk

PART 1

General

This part contains the following sections:

The Client Role in Successful Construction Projects

Jason Challender *et al.*

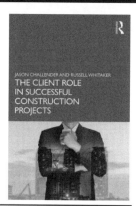

The Client Role in Successful Construction Projects is a practical guide for clients on how to initiate, procure and manage construction projects and developments. This book is written from the perspective of the client initiating a construction project as part of a business venture and differs from most available construction literature which can externalise the client as a risk to be managed by the design team. The book provides a practical framework for new and novice clients undertaking construction, giving them a voice and enabling them to:

- Understand the challenges that they and the project are likely to face.
- Communicate and interact effectively with key stakeholders and professionals within the industry.
- Understand in straightforward terms where they can have a positive impact on the project.
- Put in place a client-side due diligence process.
- Reduce their institutional risk and the risk of project failure.
- Discover how their standard models are able to co-exist and even transfer to a common client-side procedure for managing a construction project.

Written by clients, for clients, this book is highly recommended not only for clients, but for construction industry professionals who want to develop their own skills and enhance their working relationship with their clients. A supporting website for the book will be available, which will give practical examples of the points illustrated in the book and practical advice from specialists in the field.

May 2019: 234 × 156 mm: 318pp
Pb: 978-1-138-05821-7 : £29.99

To Order: Tel: +44 (0) 1235 400524 Fax: +44 (0) 1235 400525
or Post: Taylor and Francis Customer Services,
Bookpoint Ltd, Unit T1, 200 Milton Park, Abingdon, Oxon, OX14 4TA UK
Email: book.orders@tandf.co.uk

For a complete listing of all our titles visit:
www.tandf.co.uk

Capital Allowances

Introduction

Capital Allowances provide tax relief by prescribing a statutory rate of depreciation for tax purposes in place of that used for accounting purposes. They are utilized by government to provide an incentive to invest in capital equipment, including assets within commercial property, by allowing the majority of taxpayers a deduction from taxable profits for certain types of capital expenditure, thereby reducing or deferring tax liabilities.

The capital allowances most commonly applicable to real estate are those given for capital expenditure on existing commercial buildings in disadvantaged areas, and plant and machinery in all buildings other than residential dwellings, except for common areas. Relief for certain expenditure on industrial buildings and hotels was withdrawn from April 2011, although the ability to claim plant and machinery remains.

Enterprise Zone Allowances are also available for capital expenditure within designated areas only where there is a focus on high value manufacturing. Enhanced rates of allowances are also available on certain types of energy and water saving plant and machinery assets.

The Act

The primary legislation is contained in the Capital Allowances Act 2001. Major changes to the system were introduced in 2008, 2014 and 2018 affecting the treatment of tax relief to include plant and machinery allowances and the newly introduced Structures and Building Allowances (SBAs), which is awaiting legislative confirmation.

Plant and Machinery

Various legislative changes and case law precedents in recent years have introduced major changes to the availability of Capital Allowances on property expenditure. The Capital Allowances Act 2001 excludes expenditure on the provision of a building from qualifying for plant and machinery, with prescribed exceptions.

List A in Section 21 of the 2001 Act sets out those assets treated as parts of buildings:

- Walls, floors, ceilings, doors, gates, shutters, windows and stairs.
- Mains services, and systems, for water, electricity and gas.
- Waste disposal systems.
- Sewerage and drainage systems.
- Shafts, or other structures, in which lifts, hoists, escalators and moving walkways are installed.
- Fire safety systems.

Similarly, List B in Section 22 identifies excluded structures and other assets.

Both sections are, however, subject to Section 23. This section sets out expenditure, which, although being part of a building, may still be expenditure on the provision of Plant and Machinery.

Sections 21 and 22 do not affect the question whether expenditure on any item in List C is expenditure on the provision of Plant or Machinery.

List C in Section 23 is reproduced below:

1.　Machinery (including devices for providing motive power) not within any other item in this list.
2.　Gas and sewerage systems provided mainly –
　　a.　to meet the particular requirements of the qualifying activity, or
　　b.　to serve particular plant or machinery used for the purposes of the qualifying activity.
3.　Omitted.
4.　Manufacturing or processing equipment; storage equipment (including cold rooms); display equipment; and counters, checkouts and similar equipment.
5.　Cookers, washing machines, dishwashers, refrigerators and similar equipment; washbasins, sinks, baths, showers, sanitary ware and similar equipment; and furniture and furnishings.
6.　Hoists.
7.　Sound insulation provided mainly to meet the particular requirements of the qualifying activity.
8.　Computer, telecommunication and surveillance systems (including their wiring or other links).
9.　Refrigeration or cooling equipment.
10.　Fire alarm systems; sprinkler and other equipment for extinguishing or containing fires.
11.　Burglar alarm systems.
12.　Strong rooms in bank or building society premises; safes.
13.　Partition walls, where moveable and intended to be moved in the course of the qualifying activity.
14.　Decorative assets provided for the enjoyment of the public in hotel, restaurant or similar trades.
15.　Advertising hoardings; signs, displays and similar assets.
16.　Swimming pools (including diving boards, slides & structures on which such boards or slides are mounted).
17.　Any glasshouse constructed so that the required environment (namely, air, heat, light, irrigation and temperature) for the growing of plants is provided automatically by means of devices forming an integral part of its structure.
18.　Cold stores.
19.　Caravans provided mainly for holiday lettings.
20.　Buildings provided for testing aircraft engines run within the buildings.
21.　Moveable buildings intended to be moved in the course of the qualifying activity.
22.　The alteration of land for the purpose only of installing Plant or Machinery.
23.　The provision of dry docks.
24.　The provision of any jetty or similar structure provided mainly to carry Plant or Machinery.
25.　The provision of pipelines or underground ducts or tunnels with a primary purpose of carrying utility conduits.
26.　The provision of towers to support floodlights.
27.　The provision of –
　　a.　any reservoir incorporated into a water treatment works, or
　　b.　any service reservoir of treated water for supply within any housing estate or other particular locality.
28.　The provision of –
　　a.　silos provided for temporary storage, or
　　b.　storage tanks.
29.　The provision of slurry pits or silage clamps.
30.　The provision of fish tanks or fish ponds.
31.　The provision of rails, sleepers and ballast for a railway or tramway.
32.　The provision of structures and other assets for providing the setting for any ride at an amusement park or exhibition.
33.　The provision of fixed zoo cages.

Main Pool Plant and Machinery

Capital Allowances on main pool plant and machinery are currently given in the form of writing down allowances at the current rate of 18% per annum on a reducing balance basis. For every £100 of qualifying expenditure £18 is claimable in year 1, £14.76 in year 2 and so on until either all of the allowances have been claimed, or the asset is sold.

Integral Features

The category of qualifying expenditure on 'integral features' was introduced with effect from April 2008. The following items are integral features:

- An electrical system (including a lighting system)
- A cold water system
- A space or water heating system, a powered system of ventilation, air cooling or air purification, and any floor or ceiling comprised in such a system
- A lift, an escalator or a moving walkway
- External solar shading

Integral Features attract a writing down allowance of 8% per annum which reduces to 6% per annum from April 2019, on a reducing balance basis.

The Integral Features legislation introduced certain assets which were not previously allowable as plant and machinery. These assets include general power and lighting, thermal insulation to existing buildings, external solar shading and cold water systems.

Thermal Insulation

From April 2008, expenditure incurred on the installation of thermal insulation to existing buildings qualifies as Integral Feature plant and machinery which is available on a reducing balance basis at 8% per annum.

Long-Life Assets

A writing down allowance of 8% per annum is available on long-life assets.

A long-life asset is defined as plant and machinery that has an expected useful economic life of at least 25 years. The useful economic life is taken as the period from of first use until it is likely to cease to be used as a fixed asset of any business. It is important to note that this is likely to be a shorter period than an asset's physical life.

Plant and machinery provided for use in a building used wholly, or mainly, as a dwelling house, showroom, hotel, office, retail shop, or similar premises, or for purposes ancillary to such use, cannot be classified as a long-life asset.

In contrast certain plant and machinery assets in buildings such as factories, cinemas, hospitals etc. could potentially be treated as long-life assets.

Case Law

The fact that an item appears in List C does not automatically mean that it will qualify for capital allowances. It only means that it may potentially qualify.

Guidance about what can qualify as plant is found in case law dating back to 1887. The case of *Wimpy International Ltd and Associated Restaurants Ltd v Warland* in the late 1980s is one of the most important case law references for determining what can qualify as plant.

The Judge in that case applied three tests when considering whether, or not, an item is plant.

1. Is the item stock in trade? If the answer is yes, then the item is not plant.
2. Is the item used for carrying on the business? In order to pass the business use test the item must be employed in carrying on the business; it is not enough for the asset to be simply used in the business. For example, product display lighting in a retail store may be plant but general lighting in a warehouse would fail the test. *(Please note, this case law relates to the pre-Integral Feature Legislation, which introduced lighting as an eligible asset.)*
3. Is the item the business premises, or part of the business premises? An item cannot be plant if it fails the premises test, i.e. if the business use is the premises itself, or part of the premises, or a place in which the business is conducted. The meaning of 'part of the premises' in this context should not be confused with real property law. HMRC's internal manuals suggest there are four general factors to be considered, each of which is a question of fact and degree:

- Does the item appear visually to retain a separate identity
- With what degree of permanence has it been attached to the building
- To what extent is the structure complete without it
- To what extent is it intended to be permanent, or alternatively, is it likely to be replaced within a short period

Certain assets will qualify as plant in most cases. However, many others need to be considered on a case-by-case basis. For example, decorative assets in a hotel restaurant may be plant, but similar assets in an office reception area may be ineligible.

Refurbishment Schemes

Building refurbishment projects will typically be a mixture of capital costs and revenue expenses, unless the works are so extensive that they are more appropriately classified as a redevelopment. A straightforward repair or a 'like for like' replacement of part of an asset would be a revenue expense, meaning that the entire amount can be deducted from taxable profits in the same year.

Where capital expenditure is incurred which is incidental to the installation of plant or machinery then Section 25 of the Capital Allowances Act 2001 allows it to be treated as part of the expenditure on the qualifying item. Incidental expenditure will often include parts of the building that would be otherwise disallowed, as shown in the Lists reproduced above. For example, the cost of forming a lift shaft inside an existing building would be deemed to be part of the expenditure on the provision of the new lift.

The extent of the application of Section 25 was reviewed for the first time by the Special Commissioners in December 2007 and by the First Tier Tribunal (Tax Chamber) in December 2009, in the case of JD Wetherspoon. The key areas of expenditure considered were overheads and preliminaries where it was held that such costs could be allocated on a pro-rata basis; decorative timber panelling which was found to be part of the premises and so ineligible for allowances; toilet lighting which was considered to provide an attractive ambience and qualified for allowances; and incidental building alterations of which enclosing walls to toilets and kitchens and floor finishes did not qualify except for tiled splash backs, bespoke toilet cubicles and drainage did qualify, along with the related sanitary fittings and kitchen equipment.

The Enhanced Capital Allowances Scheme

The scheme is one of a series of measures introduced to ensure that the UK meets its target for reducing greenhouse gases under the Kyoto Protocol. 100% first year allowances are available on products included on the Energy Technology List published on the website at www.eca.gov.uk and other technologies supported by the scheme. All businesses will be able to claim the enhanced allowances, but only investments in new and unused Machinery and Plant can qualify.

There are currently 15 technologies with multiple sub-technologies currently covered by the scheme:

- Air-to-air energy recovery
- Automatic monitoring and targeting (AMT)
- Boiler equipment
- Combined heat and power (CHP)
- Compressed air equipment
- Heat pumps
- Heating ventilation and air conditioning (HVAC) equipment
- High speed hand air dryers
- Lighting
- Motors and drives
- Pipework insulation
- Refrigeration equipment
- Solar thermal systems
- Uninterruptible power supplies
- Waste heat to electricity conversion equipment

Finance Act 2003 introduced a new category of environmentally beneficial plant and machinery qualifying for 100% first-year allowances. The Water Technology List includes 14 technologies:

- Cleaning in place equipment
- Efficient showers

- Efficient taps
- Efficient toilets
- Efficient washing machines
- Flow controllers
- Greywater recovery and reuse equipment
- Leakage detection equipment
- Meters and monitoring equipment
- Rainwater harvesting equipment
- Small scale slurry and sludge dewatering equipment
- Vehicle wash water reclaim units
- Water efficient industrial cleaning equipment
- Water management equipment for mechanical seals

Buildings and structures and long–life assets, as defined above, cannot qualify under the scheme. However, following the introduction of the integral features rules, lighting in any non-residential building may potentially qualify for enhanced capital allowances, if it meets the relevant criteria. The same would apply to residential common areas.

A limited payable ECA tax credit equal to 19% of the loss surrendered was also introduced for UK companies in April 2008.

From April 2012, expenditure on plant and machinery for which tariff payments are received under the renewable energy schemes introduced by the Department of Energy and Climate Change (Feed-in Tariffs or Renewable Heat Incentives), will not be entitled to enhanced capital allowances.

The 2018 Budget announced that from April 2020, this accelerated ECA benefit will no longer be available.

Structure and Building Allowances

On 29 October 2018 the government published a Technical Note containing details of a new Structures and Buildings Allowance (SBA) for expenditure incurred on non-residential structures and buildings, following an announcement in the October 2018 budget. The legislation is still yet to be finalized, as at May 2019.

Whilst we await final confirmation of the legislation, the outline basic structure and timing of the relief is as follows:

The SBA will be available at a flat rate of 2% per annum over a 50-year period for expenditure incurred on new commercial structures and buildings, including costs for new conversions, or renovations, incurred on or after 29 October 2018, but only where all the contracts for the physical construction works were entered into on, or after 29 October 2018. The SBA extends to landlord capital contribution expenditure.

The relief will be available when the building, or structure, first comes into use and will be available for both UK and overseas assets, provided the business is within the charge to UK tax. Excluded from this relief is expenditure on dwelling houses.

The claimant must have an interest in the land on which the structure, or building is constructed.

The SBA expenditure will not qualify for the annual investment allowance (AIA).

The relief will cease to be available if the building, or structure, is brought into residential use, or if it is demolished. However, SBA continues to be available for periods of temporary disuse.

If the building, or structure is sold, the new owner takes over the remainder of the residue of the SBA over the remaining 50-year period.

Structures and buildings include offices, retail and wholesale premises, walls, bridges, tunnels, factories and warehouses. Capital expenditure on renovations, or conversions, of existing commercial structures, or buildings, will also qualify. Other qualifying uses include mines, transport undertakings, investment management businesses, and professions, or vocations, but UK, or EEA, furnished holiday lettings businesses will not qualify for SBA.

The costs of construction will include only the net direct costs related to physically constructing the building, or structure, after any discounts, refunds, or other adjustments. This will include demolition costs, or any land alterations necessary for construction and direct costs required to bring the building, or structure, into existence.

SBA qualifying expenditure will also be applicable to purchases of second-hand buildings, or structures. The basis of claim will be dependent upon the vendor's holding structure and whether the property has been used, or unused, at the time of acquisition.

Where a building, or structure, is acquired and subsequently altered, or renovated, this will trigger a new 50-year SBA qualification period, as will any further new streams of qualifying expenditure.

It is important to note that expenditure on plant and machinery does not qualify for SBA. At present, however, the draft legislation excludes the facility for a purchaser to reclassify expenditure on plant and machinery, which has been included within a prior claim for SBA.

Expenditure on qualifying land remediation will also not qualify for SBA.

Annual Investment Allowance

The Annual Investment Allowance (AIA) is available to all businesses of any size and allows a deduction for the whole AIA of qualifying expenditure on plant and machinery, including integral features and long-life assets. The AIA rates have fluctuated over the years:

- 1 April 2014 to 31 December 2015 – £500,000
- 1 January 2016 to 31 December 2018 – £200,000
- Currently, the AIA rate from 1 January 2019 to 31 December 2020 is £1,000,000.

For accounting periods less, or greater than, 12 months, or if claiming in periods where the rates have changed, time apportionment rules will apply to calculate hybrid rates applicable to the period of claim.

Enterprise Zones

The creation of 11 Enterprise Zones was announced in the 2011 Budget. Additional zones have since been added bringing the number to 24 in total. Originally introduced in the early 1980s as a stimulus to commercial development and investment, these zones had become virtually non-existent.

Enterprise zones benefit from a number of reliefs, including a 100% first year allowance for new and unused non-leased plant and machinery assets, where there is a focus on high-value manufacturing.

Flat Conversion Allowances

Tax relief was available on capital expenditure incurred on or after 11 May 2001 on the renovation or conversion of vacant, or underused space, above shops and other commercial premises to provide flats for rent.

This relief was abolished for expenditure incurred after April 2013.

Business Premises Renovation Allowance

The Business Premises Renovation Allowance (BPRA) was first announced in December 2003. The idea behind this scheme was to bring long-term vacant properties back into productive use by providing 100% capital allowances for the cost of renovating and converting unused premises in disadvantaged areas. The legislation was included in Finance Act 2005 and was finally implemented on 11 April 2007 following EU state aid approval.

The scheme will apply to properties within the areas specified in the Assisted Areas Order 2007 and Northern Ireland.

BPRA is available to both individuals and companies who own, or lease, business property which has been unused for 12 months, or more. Allowances will be available to a person who incurs qualifying capital expenditure on the renovation of business premises.

An announcement to extend the scheme by a further five years to 2017 was made within the 2011 Budget, along with a further 11 new designated Enterprise Zones.

Legislation was introduced in Finance Bill 2014 to clarify the scope of expenditure qualifying for relief to actual costs of construction and building work and for certain specified activities, such as architectural and surveying services. The changes came into effect for qualifying expenditure incurred on, or after, 1 April 2014 for businesses within the charge to corporation tax and 6 April 2014 for businesses within the charge to income tax.

Other Capital Allowances

Other types of allowances include those available for capital expenditure on Mineral Extraction, Research and Development, Know-How, Patents, Dredging and Assured Tenancy.

International Tax Depreciation

The UK is not the only tax regime offering investors, owners and occupiers valuable incentives to invest in plant and machinery and environmentally friendly equipment. Ireland, Australia, Malaysia and Singapore also have capital allowances regimes that are broadly similar to the UK and provide comparable levels of tax relief to businesses.

Many other overseas countries have tax depreciation regimes based on accounting treatment, instead of Capital Allowances. Some use a systematic basis over the useful life of the asset and others have prescribed methods spreading the cost over a statutory period, not always equating to the asset's useful life. Some regimes have prescribed statutory rates, whilst others have rates which have become acceptable to the tax authorities through practice.

BIM and Big Data for Construction Cost Management

Weisheng Lu *et al.*

This book is designed to help practitioners and students in a wide range of construction project management professions to understand what building information modelling (BIM) and big data could mean for them and how they should prepare to work successfully on BIM-compliant projects and maintain their competencies in this essential and expanding area.

In this book, the state-of-the-art information technologies that support high-profile BIM implementation are introduced, and case studies show how BIM has integrated core quantity surveying and cost management responsibilities and how big data can enable informed decision-making for cost control and cost planning. The authors' combined professional and academic experience demonstrates, with practical examples, the importance of using BIM and particularly the fusion of BIM and big data, to sharpen competitiveness in global and domestic markets.

This book is a highly valuable guide for people in a wide range of construction project management and quantity surveying roles. In addition, implications for project management, facilities management, contract administration, and dispute resolution are also explored through the case studies, making this book essential reading for built environment and engineering professionals.

October 2019: 234 × 156 mm: 178pp
Hb: 978-0-8153-9094-7 : £69.99

To Order: Tel: +44 (0) 1235 400524 Fax: +44 (0) 1235 400525
or Post: Taylor and Francis Customer Services,
Bookpoint Ltd, Unit T1, 200 Milton Park, Abingdon, Oxon, OX14 4TA UK
Email: book.orders@tandf.co.uk

For a complete listing of all our titles visit:
www.tandf.co.uk

Taylor & Francis
Taylor & Francis Group

Value Added Tax

Introduction

Value Added Tax (VAT) is a tax on the consumption of goods and services. The UK introduced a domestic VAT regime when it joined the European Community in 1973. The principal source of European law in relation to VAT is Council Directive 2006/112/EC, a recast of Directive 77/388/EEC, which is currently restated and consolidated in the UK through the VAT Act 1994 and various Statutory Instruments, as amended by subsequent Finance Acts.

VAT Notice 708: Buildings and construction (July 2018) provides HMRC's interpretation of the VAT law in connection with construction works, however, the UK VAT legislation should always be referred to in conjunction with the publication. Recent VAT tribunals and court decisions since the date of this publication will affect the application of the VAT law in certain instances. The Notice is available on HM Revenue & Customs website at www.hmrc.gov.uk.

The Scope of VAT

VAT is payable on:

- Supplies of goods and services made in the UK;
- By a taxable person;
- In the course or furtherance of business; and
- Which are not specifically exempted or zero-rated.

Rates of VAT

There are three rates of VAT:

- A standard rate, currently 20% since January 2011;
- A reduced rate, currently 5%; and
- A zero rate of 0%.

Additionally some supplies are exempt from VAT and others are considered outside the scope of VAT.

Recovery of VAT

When a taxpayer makes taxable supplies he must account for VAT, known as output VAT, at the appropriate rate of either 20%, 5% or 0%. Any VAT due then has to be declared and submitted on a VAT submission to HM Revenue & Customs and will normally be charged to the taxpayer's customers.

As a VAT registered person, the taxpayer is entitled to reclaim from HM Revenue & Customs, commonly referred to as input VAT, the VAT incurred on their purchases and expenses directly related to its business activities in respect of standard-rated, reduced-rated and zero-rated supplies. A taxable person cannot, however, reclaim VAT that relates to any non-business activities (but see below) or, depending on the amount of exempt supplies they made, input VAT may be restricted or not recoverable.

At predetermined intervals the taxpayer will pay to HM Revenue & Customs the excess of VAT collected over the VAT they can reclaim. However, if the VAT reclaimed is more than the VAT collected, the taxpayer, who will be in a net repayment position, can reclaim the difference from HM Revenue & Customs.

Example

X Ltd constructs a block of flats. It sells long leases to buyers for a premium. X Ltd has constructed a new building designed as a dwelling and will have granted a long lease. This first sale of a long lease is a VAT zero-rated supply. This means any VAT incurred in connection with the development, which X Ltd will have paid (e.g. payments for consultants and certain preliminary services) will be recoverable. For reasons detailed below, the contractor employed by X Ltd will have charged VAT on his construction services at the zero rate of VAT.

Use for Business and Non-Business Activities

Where a supply relates partly to business use and partly to non-business use, then the basic rule is that it must be apportioned on a fair and reasonable basis so that only the business element is potentially recoverable. In some cases, VAT on land, buildings and certain construction services, purchased for both business and non-business use, could be recovered in full by applying what is known as 'Lennartz' accounting, to reclaim VAT relating to the non-business use and account for VAT on the non-business use over a maximum period of 10 years. Following an ECJ case restricting the scope of this approach, its application to immovable property was removed completely in January 2011 by HMRC (business brief 53/10), when UK VAT law was amended to comply with EU Directive 2009/162/EU.

Taxable Persons

A taxable person is an individual, firm, company, etc., who is required to be registered for VAT. A person who makes taxable supplies above certain turnover limits is compulsorily required to be VAT registered. The current registration limit (since 1 April 2017), known as the VAT threshold, is £85,000 and will be maintained at this level for two years from 1 April 2018. If the threshold is exceeded in any 12 month rolling period, or there is an expectation that the value of the taxable supplies in a single 30 day period, or goods are received into the UK from the EU worth more than the £85,000, then registration for UK VAT is compulsory.

A person who makes taxable supplies below the limit is still entitled to be registered on a voluntary basis if they wish, for example, in order to recover input VAT incurred in relation to those taxable supplies, however output VAT will then become due on the sales and must be accounted for.

In addition, a person who is not registered for VAT in the UK but acquires goods from another EC member state, or make distance sales in the UK above certain value limits, may be required to register for VAT in the UK.

VAT Exempt Supplies

Where a supply is exempt from VAT this means that no output VAT is payable – but equally the person making the exempt supply cannot normally recover any of the input VAT on their own costs relating to that exempt supply.

Generally commercial property transactions such as leasing of land and buildings are exempt unless a landlord chooses to standard-rate its interest in the property by a applying for an option to tax. This means that VAT is added to rental income and also that VAT incurred, on say, an expensive refurbishment, is recoverable.

Supplies outside the scope of VAT

Supplies are outside the scope of VAT if they are:

- Made by someone who is not a taxable person;
- Made outside the UK; or
- Not made in the course or furtherance of business.

In course or furtherance of business

VAT must be accounted for on all taxable supplies made in the course or furtherance of business, with the corresponding recovery of VAT on expenditure incurred.

If a taxpayer also carries out non-business activities, then VAT incurred in relation to such supplies is generally not recoverable.

In VAT terms, business means any activity continuously performed which is mainly concerned with making supplies for a consideration. This includes:

- Anyone carrying on a trade, vocation or profession;
- The provision of membership benefits by clubs, associations and similar bodies in return for a subscription or other consideration; and
- Admission to premises for a charge.

It may also include the activities of other bodies including charities and non-profit making organizations.

Examples of non-business activities are:

- Providing free services or information;
- Maintaining some museums or particular historic sites;
- Publishing religious or political views.

Construction Services

In general the provision of construction services by a contractor will be VAT standard rated at 20%, however, there are a number of exceptions for construction services provided in relation to certain relevant residential properties and charitable buildings.

The supply of building materials is VAT standard rated at 20%, however, where these materials are supplied and installed as part of the construction services, the VAT liability of those materials follows that of the construction services supplied.

Zero-rated construction services

The following construction services are VAT zero-rated, including the supply of related building materials.

The construction of new dwellings

The supply of services in the course of the construction of a new building designed for use as a dwelling or number of dwellings is zero-rated, other than the services of an architect, surveyor or any other person acting as a consultant or in a supervisory capacity.

The following basic conditions must ALL be satisfied in order for the works to qualify for zero-rating:

1. A qualifying building has been, is being, or will be constructed;
2. Services are made 'in the course of the construction' of that building;
3. Where necessary, you hold a valid certificate; and
4. Your services are not specifically excluded from zero-rating.

The construction of a new building for 'relevant residential or charitable' use

The supply of services in the course of the construction of a building designed for use as a relevant residential purpose (RRP), or relevant charitable purpose (RCP), is zero-rated, other than the services of an architect, surveyor or any other person acting as a consultant or in a supervisory capacity.

A 'relevant residential' use building means:

1. A home or other institution providing residential accommodation for children;
2. A home or other institution providing residential accommodation with personal care for persons in need of personal care by reason of old age, disablement, past or present dependence on alcohol or drugs or past or present mental disorder;
3. A hospice;
4. Residential accommodation for students or school pupils;
5. Residential accommodation for members of any of the armed forces;
6. A monastery, nunnery, or similar establishment; or
7. An institution which is the sole or main residence of at least 90% of its residents.

A 'relevant residential' purpose building does not include use as a hospital, a prison, or similar institution, or as a hotel, inn, or similar establishment.

A 'relevant charitable' purpose means use by a charity in either, or both of the following ways:

1.   Otherwise than in the course or furtherance of a business; or
2.   As a village hall, or similarly in providing social or recreational facilities for a local community.

Non-qualifying use, which is not expected to exceed 10% of the time the building is normally available for use, can be ignored. The calculation of business use can be based on time, floor area, or head count, subject to approval being acquired from HM Revenue & Customs.

The construction services can only be zero-rated if a certificate is given by the end user to the contractor carrying out the works, confirming that the building is to be used for a qualifying purpose, i.e. for a 'relevant residential or charitable' purpose. It follows that such services can only be zero-rated when supplied to the end user and, unlike supplies relating to dwellings, supplies by subcontractors cannot be zero-rated.

The construction of an annex used for a 'relevant charitable' purpose

Construction services provided in the course of construction of an annexe for use entirely, or partly for a 'relevant charitable' purpose, can be zero-rated.

In order to qualify, the annexe must:

1.   Be capable of functioning independently from the existing building;
2.   Have its own main entrance; and
3.   Be covered by a qualifying use certificate.

The conversion of a non-residential building into dwellings, or the conversion of a building from non-residential use to 'relevant residential' use, where the supply is to a 'relevant' housing association

The supply to a 'relevant' housing association in the course of conversion of a non-residential building, or non-residential part of a building, into:

1.   A new eligible dwelling designed as a dwelling, or number of dwellings; or
2.   A building, or part of a building, for use solely for a relevant residential purpose.

Any services related to the conversion, other than the services of an architect, surveyor or any person acting as a consultant or in a supervisory capacity, are zero-rated.

A 'relevant' housing association is defined as:

1.   A private registered provider of social housing;
2.   A registered social landlord within the meaning of Part I of the Housing Act 1996 (Welsh registered social landlords);
3.   A registered social landlord within the meaning of the Housing (Scotland) Act 2001 (Scottish registered social landlords); or
4.   A registered housing association within the meaning of Part II of the Housing (Northern Ireland) Order 1992 (Northern Irish registered housing associations).

If the building is to be used for a 'relevant residential' purpose, the housing association should issue a qualifying use certificate to the contractor completing the works. Subcontractors' services that are not made directly to a relevant housing association are standard-rated.

The development of a residential caravan park

The supply in the course of the construction of any civil engineering work 'necessary for' the development of a permanent park for residential caravans, and any services related to the construction, other than the services of an architect, surveyor or any person acting as a consultant or in a supervisory capacity, are zero-rated when a new permanent park is being developed, the civil engineering works are necessary for the development of the park and the services are not specifically excluded from zero-rating. This includes access roads, paths, drainage, sewerage and the installation of mains water, power and gas supplies.

Certain building alterations for 'disabled' persons

Certain goods and services supplied to a 'disabled' person, or a charity making these items and services available to 'disabled' persons, can be zero-rated. The recipient of these goods or services needs to give the supplier an appropriate written declaration that they are entitled to benefit from zero rating.

The following services (amongst others) are zero-rated:

1. The installation of specialist lifts and hoists and their repair and maintenance;
2. The construction of ramps, widening doorways or passageways including any preparatory work and making good work;
3. The provision, extension and adaptation of a bathroom, washroom or lavatory; and
4. Emergency alarm call systems.

Sale of Reconstructed Buildings

A protected building is not to be regarded as substantially reconstructed unless, when the reconstruction is completed, the reconstructed building incorporates no more of the original building than the external walls, together with other external features of architectural or historical interest.

DIY Builders and Converters

Private individuals who decide to construct their own home are able to reclaim VAT they pay on goods they use to construct their home by use of a special refund mechanism made by way of an application to HM Revenue & Customs. This also applies to services provided in the conversion of an existing non-residential building to form a new dwelling.

The scheme is meant to ensure that private individuals do not suffer the burden of VAT if they decide to construct their own home.

Charities may also qualify for a refund on the purchase of materials incorporated into a building used for non-business purposes where they provide their own free labour for the construction of a 'relevant charitable' use building.

Reduced-Rated Construction Services

The following construction services are subject to the reduced rate of VAT of 5%, including the supply of related building materials.

Conversion – changing the number of dwellings

In order to qualify for the 5% rate, there must be a different number of 'single household dwellings' within a building than there were before commencement of the conversion works. A 'single household dwelling' is defined as a dwelling that is designed for occupation by a single household.

These conversions can be from 'relevant residential' purpose buildings, non-residential buildings and houses in multiple occupation.

A house in multiple occupation conversion

This relates to construction services provided in the course of converting a 'single household dwelling', a number of 'single household dwellings', a non-residential building or a 'relevant residential' purpose building into a house for multiple occupation, such as bed sit accommodation.

A special residential conversion

A special residential conversion involves the conversion of a 'single household dwelling', a house in multiple occupation, or a non-residential building into a 'relevant residential' purpose building. such as student accommodation or a care home.

Renovation of derelict dwellings

The provision of renovation services in connection with a dwelling or 'relevant residential' purpose building that has been empty for two or more years prior to the date of commencement of construction works can be carried out at a reduced rate of VAT of 5%.

Installation of energy saving materials

The supply and installation of certain energy saving materials including insulation, draught stripping, central heating, hot water controls and solar panels in a residential building, or a building used for a relevant charitable purpose.

Buildings that are used by charities for non-business purposes, and/or as village halls, are not eligible for the reduced rate for the supply of energy saving materials.

Grant-funded installation of heating equipment or connection of a gas supply

The grant-funded supply and installation of heating appliances, connection of a mains gas supply, supply, installation, maintenance and repair of central heating systems, and supply and installation of renewable source heating systems, to qualifying persons. A qualifying person is someone aged 60 or over or is in receipt of various specified benefits.

Grant-funded installation of security goods

The grant funded supply and installation of security goods to a qualifying person.

Housing alterations for the elderly

Certain home adaptations that support the needs of elderly people are reduced rated.

Building Contracts

Design and build contracts

If a contractor provides a design and build service relating to works to which the reduced or zero rate of VAT is applicable, then any design costs incurred by the contractor will follow the VAT liability of the principal supply of construction services.

Management contracts

A management contractor acts as a main contractor for VAT purposes and the VAT liability of his services will follow that of the construction services provided. If the management contractor only provides advice without engaging trade contractors, his services will be VAT standard rated.

Construction Management and Project Management

The project manager or construction manager is appointed by the client to plan, manage and coordinate a construction project. This will involve establishing competitive bids for all the elements of the work and the appointment of trade contractors. The trade contractors are engaged directly by the client for their services. The VAT liability of the trade contractors will be determined by the nature of the construction services they provide and the building being constructed.

The fees of the construction manager or project manager will be VAT standard rated. If the construction manager also provides some construction services, these works may be zero or reduced rated if the works qualify.

Liquidated and Ascertained Damages

Liquidated damages are outside of the scope of VAT as compensation. The employer should not reduce the VAT amount due on a payment under a building contract on account of a deduction of damages. In contrast, an agreed reduction in the contract price will reduce the VAT amount.

Similarly, in certain circumstances, HM Revenue & Customs may agree that a claim by a contractor under a JCT, or other form of contract, is also compensation payment and outside the scope of VAT.

Reverse Charge for Building and Construction Services

From 1 October 2019, a VAT reverse charge will be introduced for certain building and construction services. This measure has been introduced to combat missing trader fraud in the construction industry, removing the opportunity for fraudsters to charge VAT and then go missing, before paying it over to the HMRC.

For certain supplies of construction services, the customer will now be liable to account to HMRC for the VAT in respect of those purchases rather than the supplier. This reverse charge will apply through the supply chain where payments are required to be reported through the Construction Industry Scheme (CIS) up to the point where the customer receiving the supply is no longer a business that makes supplies of specified services (end users).

The introduction of a reverse charge does not change the liability of the supply of the specified services: it is the way in which the VAT on those supplies is accounted for. Rather than the supplier charging and accounting for the VAT, the recipient of those supplies accounts for the VAT its return instead of paying the VAT amount to its supplier. It will be able to reclaim that VAT amount as input tax, subject to the normal rules. The supplier will need to issue a VAT invoice that indicates the supplies are subject to the reverse charge.

The types of construction services covered by the reverse charge are based on the definition of 'construction operations' used in CIS under section 74 of the Finance Act 2004.

As well as excluding supplies of specified services to end users, the reverse charge does not capture supplies of specified services where the supplier and customer are connected in a particular way, and for supplies between landlords and tenants.

Project Risk and Opportunity Management

Agnar Johansen *et al.*

Effective risk and opportunity management is key to the successful delivery of any major engineering and construction project. This book looks at how all those involved can manage risk and capitalise on the opportunities that uncertainty present. The authors of this book highlight that uncertainties should be managed rather than avoided. This book will look at simple projects with a small team, to megaprojects where some hundreds of people are involved, and the consequences of delays or unforeseen costs. However, while the obvious risks can be planned for, the authors argue that it is often the opportunities in these situations that can have unexploited potential.

This book is about opportunity management seen from the owner's perspective. It will be an invaluable resource for those studying Engineering both undergraduate and postgraduate and set out ways in which projects should be managed from planning to completion. This book is also a great tool for those working in project management and the construction industry. While there are many books that demonstrate effective construction management, this book is the first of its kind to emphasise that there is opportunity in uncertainty, and possibility in the unexpected.

April 2019: 234 × 156 mm: 162pp
Pb: 978-1-138-36582-7 : £44.99

To Order: Tel: +44 (0) 1235 400524 Fax: +44 (0) 1235 400525
or Post: Taylor and Francis Customer Services,
Bookpoint Ltd, Unit T1, 200 Milton Park, Abingdon, Oxon, OX14 4TA UK
Email: book.orders@tandf.co.uk

For a complete listing of all our titles visit:
www.tandf.co.uk

The Aggregates Levy

The Aggregates Levy came into operation on 1 April 2002 in the UK, except for Northern Ireland where it was phased in over five years from 2003.

It was introduced to ensure that the external costs associated with the exploitation of aggregates are reflected in the price of aggregate, and to encourage the use of recycled aggregate. There continues to be strong evidence that the levy is achieving its environmental objectives, with sales of primary aggregate down and production of recycled aggregate up. The Government expects that the rates of the levy will at least keep pace with inflation over time, although it accepts that the levy is still bedding in.

The rate of the levy will continue to be £2.00 per tonne from 1 April 2019 and is levied on anyone considered to be responsible for commercially exploiting 'virgin' aggregates in the UK and should naturally be passed by price increase to the ultimate user.

All materials falling within the definition of 'Aggregates' are subject to the levy unless specifically exempted.

It does not apply to clay, soil, vegetable or other organic matter.

The intention is that it will:

- Encourage the use of alternative materials that would otherwise be disposed of to landfill sites
- Promote development of new recycling processes, such as using waste tyres and glass
- Promote greater efficiency in the use of virgin aggregates
- Reduce noise and vibration, dust and other emissions to air, visual intrusion, loss of amenity and damage to wildlife habitats

Definitions

'Aggregates' means any rock, gravel or sand which is extracted or dredged in the UK for aggregates use. It includes whatever substances are for the time being incorporated in it or naturally occur mixed with it.

'Exploitation' is defined as involving any one or a combination of any of the following:

- Being removed from its original site, a connected site which is registered under the same name as the originating site or a site where it had been intended to apply an exempt process to it, but this process was not applied
- Becoming subject to a contract or other agreement to supply to any person
- Being used for construction purposes
- Being mixed with any material or substance other than water, except in permitted circumstances

The definition of 'aggregate being used for construction purposes' is when it is:

- Used as material or support in the construction or improvement of any structure
- Mixed with anything as part of a process of producing mortar, concrete, tarmacadam, coated roadstone or any similar construction material

Incidence

It is a tax on primary aggregates production – i.e. 'virgin' aggregates won from a source and used in a location within the UK territorial boundaries (land or sea). The tax is not levied on aggregates which are exported or on aggregates imported from outside the UK territorial boundaries.

It is levied at the point of sale.

Exemption from Tax

An 'aggregate' is exempt from the levy if it is:

- Material which has previously been used for construction purposes
- Aggregate that has already been subject to a charge to the Aggregates Levy
- Aggregate which was previously removed from its originating site before the start date of the levy
- Aggregate which is moved between sites under the same Aggregates Levy Registration
- Aggregate which is removed to a registered site to have an exempt process applied to it
- Aggregate which is removed to any premises where china clay or ball clay will be extracted from the aggregate
- Aggregate which is being returned to the land from which it was won provided that it is not mixed with any material other than water
- Aggregate won from a farm land or forest where used on that farm or forest
- Rock which has not been subjected to an industrial crushing process
- Aggregate won by being removed from the ground on the site of any building or proposed building in the course of excavations carried out in connection with the modification or erection of the building and exclusively for the purpose of laying foundations or of laying any pipe or cable
- Aggregate won by being removed from the bed of any river, canal or watercourse or channel in or approach to any port or harbour (natural or artificial), in the course of carrying out any dredging exclusively for the purpose of creating, restoring, improving or maintaining that body of water
- Aggregate won by being removed from the ground along the line of any highway or proposed highway in the course of excavations for improving, maintaining or constructing the highway otherwise than purely to extract the aggregate
- Drill cuttings from petroleum operations on land and on the seabed
- Aggregate resulting from works carried out in exercise of powers under the New Road and Street Works Act 1991, the Roads (Northern Ireland) Order 1993 or the Street Works (Northern Ireland) Order 1995
- Aggregate removed for the purpose of cutting of rock to produce dimension stone, or the production of lime or cement from limestone
- Aggregate arising as a waste material during the processing of the following industrial minerals:
    - anhydrite
    - ball clay
    - barites
    - calcite
    - china clay
    - clay, coal, lignite and slate
    - feldspar
    - flint
    - fluorspar
    - fuller's earth
    - gems and semi-precious stones
    - gypsum
    - any metal or the ore of any metal
    - muscovite
    - perlite
    - potash
    - pumice
    - rock phosphates
    - sodium chloride
    - talc
    - vermiculite
    - spoil from the separation of the above industrial minerals from other rock after extraction
    - material that is mainly but not wholly the spoil, waste or other by-product of any industrial combustion process or the smelting or refining of metal

Anything that consists 'wholly or mainly' of the following is exempt from the levy (note that 'wholly' is defined as 100% but 'mainly' as more than 50%), thus exempting any contained aggregates amounting to less than 50% of the original volumes:

- clay, soil, vegetable or other organic matter
- drill cuttings from oil exploration in UK waters
- material arising from utility works, if carried out under the New Roads and Street Works Act 1991

However, when ground that is more than half clay is mixed with any substance (for example, cement or lime) for the purpose of creating a firm base for construction, the clay becomes liable to Aggregates Levy because it has been mixed with another substance for the purpose of construction.

Anything that consists completely of the following substances is exempt from the levy:

- Spoil, waste or other by-products from any industrial combustion process or the smelting or refining of metal – for example, industrial slag, pulverized fuel ash and used foundry sand. If the material consists completely of these substances at the time it is produced it is exempt from the levy, regardless of any subsequent mixing
- Aggregate necessarily arising from the footprint of any building for the purpose of laying its foundations, pipes or cables. It must be lawfully extracted within the terms of any planning consent
- Aggregate necessarily arising from navigation dredging
- Aggregate necessarily arising from the ground in the course of excavations to improve, maintain or construct a highway or a proposed highway
- Aggregate necessarily arising from the ground in the course of excavations to improve, maintain or construct a railway, monorail or tramway

Relief from the levy either in the form of credit or repayment is obtainable where:

- it is subsequently exported from the UK in the form of aggregate
- it is used in an exempt process
- where it is used in a prescribed industrial or agricultural process
- it is waste aggregate disposed of by dumping or otherwise, e.g. sent to landfill or returned to the originating site

An exemption for aggregate obtained as a by-product of railway, tramway and monorail improvement, maintenance and construction was introduced in 2007.

Discounts

Water which is added to the aggregate after the aggregate has been won (washing, dust dampening etc.) may be discounted from the tax calculations. There are two accepted options by which the added water content can be calculated.

The first is to use HMRC's standard added water percentage discounts listed below:

- washed sand 7%
- washed gravel 3.5%
- washed rock/aggregate 4%

Alternatively a more exact percentage can be agreed for dust dampening of aggregates.

Whichever option is adopted, it must be agreed in writing in advance with HMRC.

Impact

The British Aggregates Association suggested that the additional cost imposed by quarries is more likely to be in the order of £3.40 per tonne on mainstream products, applying an above average rate on these in order that by-products and low grade waste products can be held at competitive rates, as well as making some allowance for administration and increased finance charges.

With many gravel aggregates costing in the region of £20.00 per tonne, there is a significant impact on construction costs.

Avoidance

An alternative to using new aggregates in filling operations is to crush and screen rubble which may become available during the process of demolition and site clearance as well as removal of obstacles during the excavation processes.

Example: Assuming that the material would be suitable for fill material under buildings or roads, a simple cost comparison would be as follows (note that for the purpose of the exercise, the material is taken to be 1.80 tonnes per m³ and the total quantity involved less than 1,000 m³):

	£/m³	£/tonne
Disposing of site material:		
Cost of removing materials from site	30.92	17.18
Importing fill material:		
Cost of 'new' aggregates delivered to site	34.31	19.06
Addition for Aggregates Tax	3.60	2.00
Total cost of disposing waste aggregate and importing fill materials	68.83	38.24
Crushing site materials:		
Transportation of material from excavations or demolition to stockpiles	0.89	0.50
Transportation of material from temporary stockpiles to the crushing plant	2.39	1.33
Establishing plant and equipment on site; removing on completion	2.39	1.33
Maintain and operate plant	10.75	5.97
Crushing hard materials on site	15.53	8.62
Screening material on site	2.39	1.33
Total cost of crushing site materials ready for reuse	34.34	19.08

From the above it can be seen that potentially there is a great benefit in crushing site materials for filling rather than importing fill materials.

Setting the cost of crushing against the import price would produce a saving of £6.78 per m³. If the site materials were otherwise intended to be removed from the site, then the cost benefit increases by the saved disposal cost to £33.41 per m³.

Even if there is no call for any or all of the crushed material on site, it ought to be regarded as a useful asset and either sold on in crushed form or else sold with the prospects of crushing elsewhere.

Specimen Unit rates	Unit³	£
Establishing plant and equipment on site; removing on completion		
crushing plant	trip	1,450.00
screening plant	trip	750.00
Maintain and operate plant		
crushing plant	week	9,000.00
screening plant	week	2,200.00
Transportation of material from excavations or demolition places to temporary stockpiles	m³	2.00
Transportation of material from temporary stockpiles to the crushing plant	m³	2.00
Breaking up material on site using impact breakers		
mass concrete	m³	50.00
reinforced concrete	m³	71.00
brickwork	m³	36.00
Crushing material on site		
mass concrete not exceeding 1000 m³	m³	16.00
mass concrete 1000–5000 m³	m³	15.00
mass concrete over 5000 m³	m³	14.00
reinforced concrete not exceeding 1000 m³	m³	19.00
reinforced concrete 1000–5000 m³	m³	17.00
reinforced concrete over 5000 m³	m³	16.00
brickwork not exceeding 1000 m³	m³	15.00
brickwork 1000–5000 m³	m³	14.00
brickwork over 5000 m³	m³	13.00
Screening material on site	m³	3.00

More detailed information can be found on the HMRC website (www.hmrc.gov.uk) in Notice AGL1 Aggregates Levy published 1 April 2014 (updated 9 February 2017).

Accounting for Construction

Jim Meikle *et al.*

Accounting for Construction follows on from *Measuring Construction*, edited by the same team. It extends the coverage of some of the material in the first volume and expands the range of related topics to include, *inter alia,* shadow economies, accounting for informal construction and the treatment of the built environment sector in national accounts.

Taken together, the two volumes collate a range of topics that are only addressed, if addressed at all, in occasional academic papers and the publications of bodies such as national statistical offices and the World Bank. *Accounting for Construction* presents international examples from the UK, Australia and New Zealand and from both academic and professional contributors.

March 2019: 234 × 156 mm: 256pp
Hb: 978-1-138-29397-7 : £105.00

To Order: Tel: +44 (0) 1235 400524 Fax: +44 (0) 1235 400525
or Post: Taylor and Francis Customer Services,
Bookpoint Ltd, Unit T1, 200 Milton Park, Abingdon, Oxon, OX14 4TA UK
Email: book.orders@tandf.co.uk

For a complete listing of all our titles visit:
www.tandf.co.uk

Land Remediation

The purpose of this section is to review the general background of ground contamination, the cost implications of current legislation and to consider the various remedial measures and to present helpful guidance on the cost of Land Remediation.

It must be emphasized that the cost advice given is an average and that costs can vary considerably from contract to contract depending on individual Contractors, site conditions, type and extent of contamination, methods of working and various other factors as diverse as difficulty of site access and distance from approved tips.

We have structured this Unit Cost section to cover as many aspects of Land Remediation works as possible.

The introduction of the Landfill Directive in July 2004 has had a considerable impact on the cost of Remediation works in general and particularly on the practice of Dig and Dump. The number of Landfill sites licensed to accept Hazardous Waste has drastically reduced and inevitably this has led to increased costs.

Market forces will determine future increases in cost resulting from the introduction of the Landfill Directive and the cost guidance given within this section will require review in light of these factors.

Statutory Framework

In July 1999 new contaminated land provisions, contained in Part IIa of the Environmental Protection Act 1990 were introduced. Primary objectives of the measures included a legal definition of Contaminated Land and a framework for identifying liability, underpinned by a 'polluter pays' principle meaning that remediation should be paid for by the party (or parties) responsible for the contamination. A secondary, and indirect, objective of Part IIA is to provide the legislative context for remediation carried out as part of development activity which is controlled through the planning system. This is the domain where other related objectives, such as encouraging the recycling of brownfield land, are relevant.

Under the Act action to remediate land is required only where there are unacceptable actual or 'significant possibility of significant harm' to health, controlled waters or the environment. Only Local Authorities have the power to determine a site as Contaminated Land and enforce remediation. Sites that have been polluted from previous land use may not need remediating until the land use is changed; this is referred to as 'land affected by contamination'. This is a risk-based assessment on the site specifics in the context of future end uses. As part of planning controls, the aim is to ensure that a site is incapable of meeting the legal definition of Contaminated Land post-development activity. In addition, it may be necessary to take action only where there are appropriate, cost-effective remediation processes that take the use of the site into account.

The Environment Act 1995 amended the Environment Protection Act 1990 by introducing a new regime designed to deal with the remediation of sites which have been seriously contaminated by historic activities. The regime became operational on 1 April 2000. Local authorities and/or the Environment Agency regulate seriously contaminated sites which are known as 'special sites'. The risks involved in the purchase of potentially contaminated sites are high, particularly considering that a transaction can result in the transfer of liability for historic contamination from the vendor to the purchaser.

The contaminated land provisions of the Environmental Protection Act 1990 are only one element of a series of statutory measures dealing with pollution and land remediation that have been and are to be introduced. Others include:

- Groundwater regulations, including pollution prevention measures
- An integrated prevention and control regime for pollution
- Sections of the Water Resources Act 1991, which deals with works notices for site controls, restoration and clean up

April 2012 saw the first revision of the accompanying Part IIa Statutory Guidance. This has introduced a new categorization scheme for assessing sites under Part IIA. Category 1 is land which definitely is Contaminated Land and Category 4 is for land which definitely is not Contaminated Land. This is intended to assist prioritization of sites which pose the greatest risk. Still included in the statutory guidance are matters of inspection, definition, remediation, apportionment of liabilities and recovery of costs of remediation. The measures are to be applied in accordance with the following criteria:

- the planning system
- the standard of remediation should relate to the present use
- the costs of remediation should be reasonable in relation to the seriousness of the potential harm
- the proposals should be practical in relation to the availability of remediation technology, impact of site constraints and the effectiveness of the proposed clean-up method

Liability for the costs of remediation rests with either the party that 'caused or knowingly permitted' contamination, or with the current owners or occupiers of the land.

Apportionment of liability, where shared, is determined by the local authority. Although owners or occupiers become liable only if the polluter cannot be identified, the liability for contamination is commonly passed on when land is sold.

If neither the polluter nor owner can be found, the clean-up is funded from public resources.

The ability to forecast the extent and cost of remedial measures is essential for both parties, so that they can be accurately reflected in the price of the land.

At the end of March 2012, the National Planning Policy Framework replaced relevant planning guidance relating to remediation, most significantly PPS 23 Planning and Pollution Control. This has been replaced by a need to investigate and assess land contamination, which must be carried out by a competent person.

The EU Landfill Directive

The Landfill (England and Wales) Regulations 2002 came into force on 15 June 2002 followed by Amendments in 2004 and 2005. These new regulations implement the Landfill Directive (Council Directive 1999/31/EC), which aims to prevent, or to reduce as far as possible, the negative environmental effects of landfill. These regulations have had a major impact on waste regulation and the waste management industry in the UK.

The Scottish Executive implements the Landfill Directive through the Landfill (Scotland) Regulations 2003 and subsequent 2013 Amendments, and the Northern Ireland Assembly implements the Directive through the Landfill Regulations (Northern Ireland) 2003, the Waste and Emissions Trading Act 2003 and the Landfill Allowances Scheme (Northern Ireland) Regulations 2004 (as amended).

In summary, the Directive requires that:

- Sites are to be classified into one of three categories: hazardous, non-hazardous or inert, according to the type of waste they will receive
- Higher engineering and operating standards will be followed
- Biodegradable waste will be progressively diverted away from landfills
- Certain hazardous and other wastes, including liquids, explosive waste and tyres will be prohibited from landfills
- Pretreatment of wastes prior to landfilling will become a requirement

On 15 July 2004 the co-disposal of hazardous and non-hazardous waste in the same landfill site ended and in July 2005 new waste acceptance criteria (WAC) were introduced which also prevents the disposal of materials contaminated by coal tar.

The effect of this Directive has been to dramatically reduce the hazardous disposal capacity post July 2004, resulting in a SIGNIFICANT increase in remediating costs. This has significantly increased travelling distance and cost for disposal to landfill. The increase in operating expenses incurred by the landfill operators has also resulted in higher tipping costs.

However, there is now a growing number of opportunities to dispose of hazardous waste to other facilities such as soil treatment centres, often associated with registered landfills potentially eliminating landfill tax. Equally, improvements in on-site treatment technologies are helping to reduce the costs of disposal by reducing the hazardous properties of materials going offsite.

All hazardous materials designated for disposal off-site are subject to WAC tests. Samples of these materials are taken from site to laboratories in order to classify the nature of the contaminants. These tests, which cost approximately £200 each, have resulted in increased costs for site investigations and as the results may take up to 3 weeks this can have a detrimental effect on programme.

As from 1 July 2008 the WAC derogations which have allowed oil contaminated wastes to be disposed in landfills with other inert substances were withdrawn. As a result the cost of disposing oil contaminated solids has increased.

There has been a marked slowdown in brownfield development in the UK with higher remediation costs, longer clean-up programmes and a lack of viable treatment options for some wastes.

The UK Government established the Hazardous Waste Forum in December 2002 to bring together key stakeholders to advise on the way forward on the management of hazardous waste.

Effect on Disposal Costs

Although most landfills are reluctant to commit to future tipping prices, tipping costs have generally stabilized. However, there are significant geographical variances, with landfill tip costs in the North of England typically being less than their counterparts in the Southern regions.

For most projects to remain viable there is an increasing need to treat soil in situ by bioremediation, soil washing or other alternative long-term remediation measures. Waste untreatable on-site such as coal tar remains a problem. Development costs and programmes need to reflect this change in methodology.

Types of hazardous waste

- Sludges, acids and contaminated wastes from the oil and gas industry
- Acids and toxic chemicals from chemical and electronics industries
- Pesticides from the agrochemical industry
- Solvents, dyes and sludges from leather and textile industries
- Hazardous compounds from metal industries
- Oil, oil filters and brake fluids from vehicles and machines
- Mercury-contaminated waste from crematoria
- Explosives from old ammunition, fireworks and airbags
- Lead, nickel, cadmium and mercury from batteries
- Asbestos from the building industry
- Amalgam from dentists
- Veterinary medicines

[Source: Sepa]

Foam insulation materials containing ODP (Ozone Depletant Potential) are also considered as hazardous waste under the EC Regulation 2037/2000.

Land Remediation Techniques

There are two principal approaches to remediation – dealing with the contamination in situ or ex situ. The selection of the approach will be influenced by factors such as: initial and long term cost, timeframe for remediation, types of contamination present, depth and distribution of contamination, the existing and planned topography, adjacent land uses, patterns of surface drainage, the location of existing on-site services, depth of excavation necessary for foundations and below-ground services, environmental impact and safety, interaction with geotechnical performance, prospects for future changes in land use and long-term monitoring and maintenance of in situ treatment.

On most sites, contamination can be restricted to the top couple of metres, although gasholder foundations for example can go down 10 to 15 metres. Underground structures can interfere with the normal water regime and trap water pockets.

There could be a problem if contaminants get into fissures in bedrock.

In situ techniques

A range of in situ techniques is available for dealing with contaminants, including:

- Clean cover – a layer of clean soil is used to segregate contamination from receptor. This technique is best suited to sites with widely dispersed contamination. Costs will vary according to the need for barrier layers to prevent migration of the contaminant.
- On-site encapsulation – the physical containment of contaminants using barriers such as slurry trench cut-off walls. The cost of on-site encapsulation varies in relation to the type and extent of barriers required, the costs of which range from £50/m² to more than £175/m².

There are also in situ techniques for treating more specific contaminants, including:

- Bioremediation – for removal of oily, organic contaminants through natural digestion by microorganisms. Most bioremediation is ex situ, i.e. it is dug out and then treated on site in bio-piles. The process can be slow, taking up to three years depending upon the scale of the problem, but is particularly effective for the long-term improvement of a site, prior to a change of use.
- Phytoremediation – the use of plants that mitigate the environmental problem without the need to excavate the contaminant material and dispose of it elsewhere. Phytoremediation consists of mitigating pollutant concentrations in contaminated soils, water or air, with plants able to contain, degrade or eliminate metals, pesticides, solvents, explosives, crude oil and its derivatives and various other contaminants from the media that contain them.
- Vacuum extraction – involving the extraction of volatile organic compounds (e.g. benzene) from soil and groundwater by vacuum.
- Thermal treatment – the incineration of contaminated soils on site. Thermal processes use heat to increase the volatility to burn, decompose, destroy or melt the contaminants. Cleaning soil with thermal methods may take only a few months to several years.
- Stabilization – cement or lime, is used to physically or chemically bind oily or metal contaminants to prevent leaching or migration. Stabilization can be used in both in situ and ex situ conditions.
- Aeration – if the ground contamination is highly volatile, e.g. fuel oils, then the ground can be ploughed and rotovated to allow the substance to vaporize.
- Air sparging – the injection of contaminant-free air into the subsurface enabling a phase transfer of hydrocarbons from a dissolved state to a vapour phase.
- Chemical oxidization – the injection of reactive chemical oxidants directly into the soil for the rapid destruction of contaminants.
- Pumping – to remove liquid contaminants from boreholes or excavations. Contaminated water can be pumped into holding tanks and allowed to settle; testing may well prove it to be suitable for discharging into the foul sewer subject to payment of a discharge fee to the local authority. It may be necessary to process the water through an approved water treatment system to render it suitable for discharge.

Ex situ techniques

Removal for landfill disposal has, historically, been the most common and cost-effective approach to remediation in the UK, providing a broad spectrum solution by dealing with all contaminants. As mentioned above, the implementation of the Landfill Directive has resulted in other techniques becoming more competitive for the disposal of hazardous waste.

If used in combination with material-handling techniques such as soil washing, the volume of material disposed at landfill sites can be significantly reduced. The disadvantages of these techniques include the fact that the contamination is not destroyed, there are risks of pollution during excavation and transfer; road haulage may also cause a local nuisance. Ex situ techniques include:

- Soil washing – involving the separation of a contaminated soil fraction or oily residue through a washing process. This also involves the excavation of the material for washing ex situ. The dewatered contaminant still requires disposal to landfill. In order to be cost effective, 70–90% of soil mass needs to be recovered. It will involve constructing a hard area for the washing, intercepting the now-contaminated water and taking it away in tankers.
- Thermal treatment – the incineration of contaminated soils ex situ. The uncontaminated soil residue can be recycled. By-products of incineration can create air pollution and exhaust air treatment may be necessary.

Soil treatment centres are now beginning to be established. These use a combination of treatment technologies to maximize the potential recovery of soils and aggregates and render them suitable for disposal to the landfill. The technologies include:

- Physicochemical treatment – a method which uses the difference in grain size and density of the materials to separate the different fractions by means of screens, hydrocyclones and upstream classification.
- Bioremediation – the aerobic biodegradation of contaminants by naturally occurring microorganisms placed into stock-piles/windrows.
- Stabilization/solidification – a cement or chemical stabilization unit capable of immobilizing persistent leachable components.

Cost Considerations

Cost drivers

Cost drivers relate to the selected remediation technique, site conditions and the size and location of a project.

The wide variation of indicative costs of land remediation techniques shown below is largely because of differing site conditions.

Indicative costs of land remediation techniques for 2019 (excluding general items, testing, landfill tax and backfilling)		
Remediation technique	Unit	Rate (£/unit)
Removal – non-hazardous	disposed material (m³)	40–100
Removal – hazardous Note: excluding any pretreatment of material	disposed material (m³)	75–200
Clean cover	surface area of site (m²)	20–45
On-site encapsulation	encapsulated material (m³)	30–95
Bioremediation (in situ)	treated material (m³)	15–45
Bioremediation (ex situ)	treated material (m³)	20–45
Chemical oxidation	treated material (m³)	30–80
Stabilization/solidification	treated material (m³)	20–65
Vacuum extraction	treated material (m³)	25–75
Soil washing	treated material (m³)	40–95
Thermal treatment	treated material (m³)	100–400

Many other on-site techniques deal with the removal of the contaminant from the soil particles and not the wholesale treatment of bulk volumes. Costs for these alternative techniques are very much Engineer designed and site specific.

Factors that need to be considered include:

- waste classification of the material
- underground obstructions, pockets of contamination and live services
- ground water flows and the requirement for barriers to prevent the migration of contaminants
- health and safety requirements and environmental protection measures
- location, ownership and land use of adjoining sites
- distance from landfill tips, capacity of the tip to accept contaminated materials, and transport restrictions
- the cost of diesel fuel, currently approximately £1.30 per litre (at April 2019 prices)

Other project related variables include size, access to disposal sites and tipping charges; the interaction of these factors can have a substantial impact on overall unit rates.

The tables below set out the costs of remediation using dig-and-dump methods for different sizes of project, differentiated by the disposal of non-hazardous and hazardous material. Variation in site establishment and disposal cost accounts for 60–70% of the range in cost.

Variation in the costs of land remediation by removal: Non-hazardous Waste

Item	Disposal Volume (less than 3000 m³) (£/m³)	Disposal Volume (3000–10,000 m³) (£/m³)	Disposal Volume (more than 10,000 m³) (£/m³)
General items and site organizationcosts	55–90	25–40	7–20
Site investigation and testing	5–12	2–7	2–6
Excavation and backfill	18–35	12–25	10–20
Disposal costs (including tipping charges but not landfill tax)	20–35	20–35	20–35
Haulage	15–35	15–35	15–35
Total (£/m³)	**113–207**	**74–142**	**54–116**
Allowance for site abnormals	*0–10 +*	*0–15 +*	*0–10 +*

Variation in the costs of land remediation by removal: Hazardous Waste

Item	Disposal Volume (less than 3000 m³) (£/m³)	Disposal Volume (3000–10,000 m³) (£/m³)	Disposal Volume (more than 10,000 m³) (£/m³)
General items and site organizationcosts	55 – 90	25–40	7–20
Site investigation and testing	10–18	5–12	5–12
Excavation and backfill	18–35	12–25	10–20
Disposal costs (including tipping charges but not landfill tax)	80 – 170	80–170	80–170
Haulage	25–120	25–120	25–120
Total (£/m³)	**188–433**	**147–367**	**127–342**
Allowance for site abnormals	*0–10 +*	*0–15 +*	*0–10 +*

The strict health and safety requirements of remediation can push up the overall costs of site organization to as much as 50% of the overall project cost (see the above tables). A high proportion of these costs are fixed and, as a result, the unit costs of site organization increase disproportionally on smaller projects.

Haulage costs are largely determined by the distances to a licensed tip. Current average haulage rates, based on a return journey, range from £1.95 to £4.50 per mile. Short journeys to tips, which involve proportionally longer standing times, typically incur higher mileage rates, up to £9.00 per mile.

A further source of cost variation relates to tipping charges. The table below summarizes typical tipping charges for 2019, exclusive of landfill tax:

Typical 2019 tipping charges (excluding landfill tax)

Waste classification	Charges (£/tonne)
Non-hazardous wastes	15–45
Hazardous wastes	35–90
Contaminated liquid	40–75
Contaminated sludge	125–400

Tipping charges fluctuate in relation to the grades of material a tip can accept at any point in time. This fluctuation is a further source of cost risk. Furthermore, tipping charges in the North of England are generally less than other areas of the country.

Prices at licensed tips can vary by as much as 50%. In addition, landfill tips generally charge a tip administration fee of approximately £25 per load, equivalent to £1.25 per tonne. This charge does not apply to non-hazardous wastes.

Landfill Tax, increased on 1 April 2019 to £91.35 a tonne for active waste, is also payable. Exemptions are no longer available for the disposal of historically contaminated material (refer also to 'Landfill Tax' section).

Tax Relief for Remediation of Contaminated Land

The Finance Act 2001 included provisions that allow companies (but not individuals or partnerships) to claim tax relief on capital and revenue expenditure on the 'remediation of contaminated land' in the United Kingdom. The relief is available for expenditure incurred on or after 11 May 2001.

From 1 April 2009 there was an increase in the scope of costs that qualify for Land Remediation Relief where they are incurred on long-term derelict land. The list includes costs that the Treasury believe to be primarily responsible for causing dereliction, such as additional costs for removing building foundations and machine bases. However, while there is provision for the list to be extended, the additional condition for the site to have remained derelict since 1998 is likely to render this relief redundant in all but a handful of cases. The other positive change is the fact that Japanese Knotweed removal and treatment (on-site only) will now qualify for the relief under the existing legislation, thereby allowing companies to make retrospective claims for any costs incurred since May 2001 – provided all other entitlement conditions are met.

A company is able to claim an additional 50% deduction for 'qualifying land remediation expenditure' allowed as a deduction in computing taxable profits, and may elect for the same treatment to be applied to qualifying capital expenditure.

With Landfill Tax Exemption (LTE) now phased out, Land Remediation Relief (LRR) for contaminated and derelict land is the Government's primary tool to create incentives for brownfield development. LRR is available to companies engaged in land remediation that are not responsible for the original contamination.

Over 7 million tonnes of waste each year were being exempted from Landfill Tax in England alone, so this change could have a major impact on the remediation industry. The modified LRR scheme, which provides Corporation Tax relief on any costs incurred on qualifying land remediation expenditure, is in the long run designed to yield benefits roughly equal to those lost through the withdrawal of LTE, although there is some doubt about this stated equity in reality.

However, with much remediation undertaken by polluters or public authorities, who cannot benefit from tax relief benefits, the change could result in a net withdrawal of Treasury support to a vital sector. Lobbying and consultation continues to ensure the Treasury maintains its support for remediation.

While there are no financial penalties for not carrying out remediation, a steep escalator affected the rate of landfill tax for waste material other than inert or inactive wastes, which has increased to £91.35/tonne from 1 April 2019. This means that for schemes where there is no alternative to dig and dump and no pre-existing LTE, the cost of remediation has risen to prohibitive levels should contaminated material be disposed off-site to licensed landfills.

Looking forward, tax-relief benefits under LRR could provide a significant cash contribution to remediation. Careful planning is the key to ensure that maximum benefits are realized, with actions taken at the points of purchase, formation of JV arrangements, procurement of the works and formulation of the Final Account (including apportionment of risk premium) all influencing the final value of the claim agreed with HM Revenue & Customs.

The Relief

Qualifying expenditure may be deducted at 150% of the actual amount expended in computing profits for the year in which it is incurred.

For example, a property trading company may buy contaminated land for redevelopment and incurs £250,000 on qualifying land remediation expenditure that is an allowable for tax purposes. It can claim an additional deduction of £125,000, making a total deduction of £375,000. Similarly, a company incurring qualifying capital expenditure on a fixed asset of the business is able to claim the same deduction provided it makes the relevant election within 2 years.

What is Remediation?

Land remediation is defined as the doing of works including preparatory activities such as condition surveys, to the land in question, any controlled waters affected by the land, or adjoining or adjacent land for the purpose of:

- Preventing or minimizing, or remedying or mitigating the effects of, any relevant harm, or any pollution of controlled waters, by reason of which the land is in a contaminated state

Definitions

Contaminated land is defined as land that, because of substances on or under it, is in such a condition that relevant harm is or has the significant possibility of relevant harm being caused to:

- The health of living organisms
- Ecological systems
- Quality of controlled waters
- Property

Relevant harm is defined as meaning:

- Death of living organisms or significant injury or damage to living organisms.
- Significant pollution of controlled waters.
- A significant adverse impact on the ecosystem, or
- Structural or other significant damage to buildings or other structures or interference with buildings or other structures that significantly compromises their use.

Land includes buildings on the land, and expenditure on asbestos removal is expected to qualify for this tax relief. It should be noted that the definition is not the same as that used in the Environmental Protection Act Part IIA.

Sites with a nuclear license are specifically excluded.

Conditions

To be entitled to claim LRR, the general conditions for all sites, which must all be met, are:

- Must be a company
- Must be land in the United Kingdom
- Must acquire an interest in the land
- Must not be the polluter or have a relevant connection to the polluter
- Must not be in receipt of a subsidy.
- Must not also qualify for Capital Allowances (particular to capital expenditure only)

Additional conditions introduced since 1 April 2009:

- The interest in land must be major – freehold or leasehold longer than 7 years
- Must not be obligated to carry out remediation under a statutory notice

Additional conditions particular to derelict land:

- Must not be in or have been in productive use at any time since at least 1 April 1998
- Must not be able to be in productive use without the removal of buildings or other structures

In order for expenditure to become qualifying, it must relate to substances present at the point of acquisition.

Furthermore, it must be demonstrated that the expenditure would not have been incurred had those substances not been present.

New Aspects of Quantity Surveying Practice, 4th edition

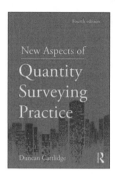

Duncan Cartlidge

In this fourth edition of New Aspects of Quantity Surveying Practice, renowned quantity surveying author Duncan Cartlidge reviews the history of the quantity surveyor, examines and reflects on the state of current practice with a concentration on new and innovative practice, and attempts to predict the future direction of quantity surveying practice in the UK and worldwide.

The book champions the adaptability and flexibility of the quantity surveyor, whilst covering the hot topics which have emerged since the previous edition's publication, including:

- the RICS 'Futures' publication;
- Building Information Modelling (BIM);
- mergers and acquisitions;
- a more informed and critical evaluation of the NRM;
- greater discussion of ethics to reflect on the renewed industry interest;
- and a new chapter on Dispute Resolution.

As these issues create waves throughout the industry whilst it continues its global growth in emerging markets, such reflections on QS practice are now more important than ever. The book is essential reading for all Quantity Surveying students, teachers and professionals. It is particularly suited to undergraduate professional skills courses and non-cognate postgraduate students looking for an up to date understanding of the industry and the role.

December 2017: 234 x 156 mm: 282 pp
Pb: 978-1-138-67376-2: £48.99

To Order: Tel: +44 (0) 1235 400524 Fax: +44 (0) 1235 400525
or Post: Taylor and Francis Customer Services,
Bookpoint Ltd, Unit T1, 200 Milton Park, Abingdon, Oxon, OX14 4TA UK
Email: book.orders@tandf.co.uk

For a complete listing of all our titles visit:
www.tandf.co.uk

The Landfill Tax

The Tax

The Landfill Tax came into operation on 1 October 1996. It is levied on operators of licensed landfill sites in England, Wales and Northern Ireland at the following rates with effect from 1 April 2019:

- Inactive or inert wastes [Lower Rate] £2.90 per tonne   Included are soil, stones, brick, plain and reinforced concrete, plaster and glass – lower rate

- All other taxable wastes [Higher Rate] £91.35 per tonne   Included are timber, paint and other organic wastes generally found in demolition work and builders skips – standard rate

The 2019 increase in the standard and lower rates of Landfill Tax are in line with the RPI, rounded to the nearest 5 pence and compare with previous rates as follows:

•	Lower Rate	From 1 April 2017	£2.70/tonne
•	Lower Rate	From 1 April 2018	£2.80/tonne
•	Lower Rate	From 1 April 2019	£2.90/tonne
•	Standard Rate	From 1 April 2017	£86.10/tonne
•	Standard Rate	From 1 April 2018	£88.95/tonne
•	Standard Rate	From 1 April 2019	£91.35/tonne

There has been no change in the loss on ignition threshold for fines which is a major eligibility factor for lower rate Landfill Tax since its introduction in April 2015. The threshold remains at 10%.

The Landfill Tax (Qualifying Material) Order 2011 came into force on 1 April 2011 and set out the qualifying criteria that are eligible for the lower rate of Landfill Tax. A waste will be lower rated for Landfill Tax only if it is listed as a qualifying material in the Landfill Tax (Qualifying Material) Order 2011, and is disposed of at an authorized waste site.

The principal qualifying materials are:

- Naturally occurring rocks and subsoils.
- Ceramic or concrete materials
- Processed or prepared minerals
- Furnace slags
- Ash arising from wood or waste combustion and from the burning of coal and petroleum coke (including when burnt with biomass
- Low activity inorganic compounds
- Calcium sulphate
- Calcium hydroxide and brine

From 1 April 2018, Landfill Tax is due on disposals of material at unauthorized sites in England and Northern Ireland. This also applies to disposals made prior to 1 April 2018 which are still on the site on 1 April 2018. Currently disposals on a site operated as a 'relevant regulated facility' which (as defined under the Environmental Permitting (England and Wales) Regulations 2016) includes sites which are 'a waste operation, such as a transfer station or treatment facility' are not considered to be taxable disposals. The Landfill Tax (Disposals of Material) Order 2018 provides more clarity on what constitutes a 'taxable disposal'. In particular, it clarifies which elements of a landfill cell are not taxable under the requirements of Part 3 of the Finance Act 1996, listing them as:

a)   Material that forms a layer which performs the function of drainage at the base of a landfill cell;
b)   Material that forms the impermeable layer that delimits the landfill cell;
c)   A pipe, pump or associated infrastructure inserted into a landfill cell for the purposes of the extraction or control of surplus liquid or gas from or within that cell;
d)   Material used for restoration of a landfill cell that only contains inert material; or
e)   Material placed in an information area within the meaning given by regulation 16 A(1) of the Landfill Tax Regulations 1996

Calculating the Weight of Waste

There are two options:

- If licensed sites have a weighbridge, tax will be levied on the actual weight of waste
- If licensed sites do not have a weighbridge, tax will be calculated by one of the three Specified Methods described in Excise Notice LFT1: a general guide to Landfill Tax (updated on 5 November 2018 and available via the HMRC website)

Water

If water greater than or equal to 25% of the material by weight has been added to the material to facilitate disposal, for the extraction of minerals or in the course of an industrial process, a landfill operator may apply to discount the water content of the material before tax is levied.

Effect on Prices

The tax is paid by landfill site operators only. Tipping charges reflect this additional cost.

As an example, *Spon's A & B* rates for mechanical disposal will be affected as follows:

- Inactive waste   *Spon's A & B 2019* net rate   £29.40/m³
                      Tax, 1.9 tonne per m³ (un-bulked) @ £2.90   £5.51/m³
                      Spon's rate including tax   £34.91/m³

- Active waste   Active waste will normally be disposed of by skip and will probably be mixed with inactive waste. The tax levied will depend on the weight of materials in the skip which can vary significantly.

Exemptions

The following disposals are exempt from Landfill Tax subject to meeting certain conditions:

- Dredgings which arise from the maintenance of inland waterways and harbours
- Naturally occurring materials arising from mining or quarrying operations
- Pet cemeteries
- Inert waste used to restore landfill sites and to fill working and old quarries where a planning condition or obligation is in existence
- Waste from visiting NATO forces

Devolution of Landfill Tax to Scotland from 1 April 2015

The Scotland Act 2012 provided for Landfill Tax to be devolved to Scotland. From 1 April 2015, operators of landfill sites in Scotland are no longer liable to pay UK Landfill Tax for waste disposed at their Scottish sites. Instead, they are liable to register and account for Scottish Landfill Tax (SLfT).

Operators of landfill sites only in Scotland were deregistered from UK Landfill Tax with effect from 31 March 2015.

The Scottish Government announced rates of Landfill Tax from 1 April 2019 to be as follows:

- Lower Rate – £2.90/tonne
- Standard Rate – £91.35/tonne

Both rates mirror those applicable in England and Northern Ireland.

Devolution of Landfill Tax to Wales from April 2018

The Wales Act 2014 provides for Landfill Tax to be devolved to Wales. This is took effect from 1 April 2018. Operators of landfill sites in Wales are liable to register and account for Landfill Disposals Tax (LDT); information is available on the Welsh Revenue Authority website, but in summary there are 3 rates of LDT:

- A lower rate of £2.90/tonne
- A standard rate of £91.35/tonne
- An unauthorized disposals rate of £137.00/tonne (set at 150% of the standard rate)

For further information contact the HMRC VAT and Excise Helpline, telephone 0300 200 3700.

The Devolution of Landfill Tax (Wales) (Consequential Transitional and Savings Provisions) Order 2018 provides for the closure of the Landfill Communities Fund (LCF) in Wales and the arrangements for a two year transitional period during which environmental bodies may continue to spend funds on projects in Wales.

Introduced in 1996 the LCF is a mechanism whereby landfill operators can receive a credit against their landfill tax liability by making contributions to environmental bodies involved in the scheme to improve the environment in the vicinity of landfill sites. On or after 1 April 2018 landfill tax credits provided under the existing landfill tax legislation will only be available to operators registered for and paying landfill tax in England and Northern Ireland, landfill tax having been devolved to Scotland in 2015 (see above). A transitional regime provides that landfill operators who have previously made contributions to environmental bodies carrying out work in Wales will remain eligible for tax credit provided that any unspent funds held by those environmental bodies and derived from contributions under the scheme are spent within 2 years of the date upon which devolution of landfill tax took place.

Architect's Legal Pocket Book, 3rd edition

Matthew Cousins

A little book that's big on information, the *Architect's Legal Pocket Book* is the definitive reference guide on legal issues for architects and architectural students. This handy pocket guide covers key legal principles which will help you to quickly understand the law and where to go for further information.

Now in its third edition, this bestselling book has been fully updated throughout to provide you with the most current information available. Subjects include contract administration, building legislation, planning, listed buildings, contract law, negligence, liability and dispute resolution. This edition also contains new cases, legislation, contract terms and certificates including the RIBA contract administration certificates, inspection duties, practical completion, the Hackitt review, the Report of the Independent Inquiry into the Construction of Edinburgh Schools and practical issues facing architects.

Illustrated with clear diagrams and featuring key cases, this is a comprehensive guide to current law for architects and an invaluable source of information. It is a book no architect should be without.

September 2019: 186 × 123 mm: 384pp
Pb: 978-1-138-50669-5 : £24.99

To Order: Tel: +44 (0) 1235 400524 Fax: +44 (0) 1235 400525
or Post: Taylor and Francis Customer Services,
Bookpoint Ltd, Unit T1, 200 Milton Park, Abingdon, Oxon, OX14 4TA UK
Email: book.orders@tandf.co.uk

For a complete listing of all our titles visit:
www.tandf.co.uk

Property Insurance

The problem of adequately covering by insurance the loss and damage caused to buildings by fire and other perils has been highlighted in recent years by the increasing rate of inflation.

There are a number of schemes available to the building owner wishing to insure his property against the usual risk. Traditionally the insured value must be sufficient to cover the actual cost of reinstating the building. This means that in addition to assessing the current value an estimate has also to be made of the increases likely to occur during the period of the policy and of rebuilding which, for a moderate size building, could amount to a total of three years. Obviously such an estimate is difficult to make with any degree of accuracy; if it is too low the insured may be penalized under the terms of the policy and if too high will result in the payment of unnecessary premiums.

There are variations on the traditional method of insuring which aim to reduce the effects of overestimating and details of these are available from the appropriate offices. For the convenience of readers who may wish to make use of the information contained in this publication in calculating insurance cover required the following may be of interest.

1 Present Cost

The current rebuilding costs may be ascertained in a number of ways:

- where the actual building cost is known this may be updated by reference to tender price changes
- by reference to average published prices per square metre of floor area. In this case it is important to understand clearly the method of measurement used to calculate the total floor area on which the rates have been based
- by professional valuation
- by comparison with the known cost of another recently built similar building

Whichever of these methods is adopted regard must be paid to any special conditions that may apply, e.g. a confined site; complexity of design; any demolition works and site clearance that may be required.

2 Allowances for Inflation

The Present Cost when established will usually, under the conditions of the policy, be the rebuilding cost on the first day of the policy period. To this must be added a sum to cover future price increases over the period of the policy. For this purpose, using the historical indices as a base and taking account of the likely change in future building costs and the tender climate, an anticipated re-build cost can be arrived at.

3 Fees

To the total of 1 and 2 above must be added an allowance for any design and other consultancy fees.

4 Value Added Tax (VAT)

To the total of 1 to 3 above must be added Value Added Tax.

Example:

An assessment for insurance cover is required from December 2020 for a property which cost £500,000 when initially built in 2010. This calculation uses AECOM Tender Price Indices. There are various methods for determining insurance cover needed. This is simply an example of one method and users would need to allow for any items that would apply to their own project. Indices would need to be amended/updated to reflect the most recent published data which is in the free Spon's *Updates* and in the *Market Forecast* articles, published in *Building* magazine

A) Update original build costs (2010Q4) to present day costs (2020Q4)

Known cost to build at end of 2010		£500,000
Calculation for present day build cost:		
AECOM tender price index 2010Q4 =	73	
Forecast tender index 2020Q4 =	120	
Increase in tender index (120/73)	65%	£322,400
Estimated present build cost (2020Q4)		**£822,400**

B) Calculate allowance for tender inflation to end of current policy: 31 December 2020

Estimated present build cost (2020Q4)		£822,400
Inflation calculation to end of policy date:		
Forecast tender index 2020Q4	120	
Forecast tender index 2021Q4	122	
Increase in tender index over period (122/120)	1.2%	£9,600
Estimated build cost at end of policy (2021Q4)		**£832,000**

Assuming that total damage is suffered on the last day of the policy: 31 December 2021, and that planning and documentation and general project set up would require a period of twelve months before re-building could commence on 31 December 2022, then a further allowance must be made to cover any inflation for that period. Again use the tender price index.

C) Calculate allowance for inflation to start of re-construction: 31 December 2022

Estimated build cost at end of policy (2021Q4)		£832,000
Inflation calculation to tender date:		
Forecast tender index 2021Q4	122	
Forecast tender index 2022Q4	124	
Increase in tender index over period (124/122)	1.8%	£15,000
Estimated rebuild cost		**£847,000**

D) Summary

You now need to add on any other costs such as fees, demolition, VAT as applicable etc.

Estimated cost of reinstatement		£847,000
Demolitions and site clearance		£50,000
Add professional fees	20%	£179,400
		£1,076,400
Add for VAT	20%	£215,280
		£1,291,680
Total insurance cover required	**say**	**£1,300,000**

Please note that this is just a simple example of one method to calculate the cost of re-building for insurance purposes. You always need to work with the most recent set of indices available if you are calculating the equivalent for your own property which are free Spon's *Updates* and also in the *Market Forecast* articles, published in *Building* magazine.

Metric Handbook:
Planning and Design Data,
6th edition

Pamela Buxton

- Significantly updated in reference to the latest construction standards and new building types
- Sustainable design integrated into chapters throughout
- Over half of the entire book has now been updated since 2015
- Over 100,000 copies sold to successive generations of architects and designers
- This book belongs in every design office.

The Metric Handbook is the major handbook of planning and design data for architects and architecture students. Covering basic design data for all the major building types it is the ideal starting point for any project. For each building type, the book gives the basic design requirements and all the principal dimensional data, and succinct guidance on how to use the information and what regulations the designer needs to be aware of.

As well as buildings, the Metric Handbook deals with broader aspects of design such as materials, acoustics and lighting, and general design data on human dimensions and space requirements. The Metric Handbook is the unique reference for solving everyday planning problems.

March 2018: 297 x 210 mm: 890 pp
Pb: 978-1-138-71468-7: £41.99

To Order: Tel: +44 (0) 1235 400524 Fax: +44 (0) 1235 400525
or Post: Taylor and Francis Customer Services,
Bookpoint Ltd, Unit T1, 200 Milton Park, Abingdon, Oxon, OX14 4TA UK
Email: book.orders@tandf.co.uk

For a complete listing of all our titles visit:
www.tandf.co.uk

Building Costs Indices, Tender Price Indices and Location Factors

Authors' Note

We have rebased our TENDER PRICE INDEX to 2015 = 100 (from 1976 = 100) Users wishing to reference to or from the previous set of indices with a base date of 1976 = 100 need to apply a factor of **5.123**. Multiply the TPI 2015 series below by 5.123 to give the equivalent in the 1976 series.

For this edition of the book TPI has been calculated at 124 (2015 = 100).

To avoid confusion, it is essential that the terms 'building costs' and 'tender prices' are clearly defined and understood.

- **Building Costs Indices** are the costs incurred by the contractor in the course of his business, the principal ones being those for labour and materials, i.e. cost to contractor
- **Tender Price Indices** represents the price for which the contractor offers to do the project for, i.e. cost to client

Readers are reminded that the following adjustments for time and location should be applied to the project as a whole and not be applied to individual elements or materials.

Building Cost Indices

Building costs are the costs incurred by the builder in the course of business i.e. wages, material prices, plant costs, rates, rents, overheads and taxes. AECOM's Building Cost Index measures movement of basic labour and materials costs to the builder. It is a composite index.

This table reflects the fluctuations since 2010 (base date 2015 = 100) in costs to the contractor. No allowance has been made for changes in productivity or for rates or hours worked which may occur in particular conditions and localities.

Year	First quarter	Second quarter	Third quarter	Fourth quarter	Annual average	% change year-on-year
2010	89.2	90.7	91.2	91.1	90.5	2.4%
2011	92.3	93.5	94.2	94.7	93.7	3.5%
2012	95.0	95.2	94.9	96.0	95.3	1.7%
2013	97.9	97.7	97.7	97.8	97.8	2.6%
2014	98.1	98.1	99.9	99.6	98.9	1.2%
2015	99.5	99.3	100.9	100.3	100.0	1.1%
2016	100.1	100.6	102.5	102.9	101.5	1.5%
2017	104.0	104.3	106.3	107.1	105.4	3.8%
2018	108.4	109.6	111.4	112.2	110.4	4.7%
2019	112.9 (P)	113.9 (F)	114.9	115.9	114.4 (F)	3.6%
2020	116.9	117.9	119.0	120.0	118.5	3.3%
2021	121.0	121.9	122.8	123.7	122.4	3.0%
2022	124.7	125.6	126.5	127.5	126.1	3.0%

Note: (P) Provisional, (F) = Forecast thereafter

Tender Price Indices

The extract tables below show the changes in building tender prices since 2010 (base date 2015 = 100). They include building costs but also take into account market considerations, the tendering climate, constructor sentiment and its assessment of prevailing market conditions amongst other things. They therefore include allowances for profits, risks and other oncosts such as preliminaries. The constructor also must anticipate cost changes during the lifetime of the contract. This means that, for example, in busier times tender prices may increase at a greater rate than building costs, whilst in a downturn the opposite may apply. The same concepts apply to trade contractors, and there is often a compounding of these trends up through the supply chain. AECOM's Tender Price Index is derived from analysis of project data and tender returns on AECOM's UK projects.

This year we have rebased our indices to 2015 = 100 (from 1976 = 100). Users wishing to use the old indices should multiply the relevant 2015 = 100 index × 5.123.

Tender prices are similar to building costs but also take into account market considerations such as the availability of labour and materials, and the prevailing economic situation. This means that in boom periods, when there is a surfeit of building work to be done, tender prices may increase at a greater rate than building costs, whilst in a period when work is scarce, tender prices may actually fall when building costs are rising.

Spon's A&B 2020 has been model calculated to give a book TPI of 124.

Year	First quarter	Second quarter	Third quarter	Fourth quarter	Annual average	% change year-on-year
2010	89.1	88.7	88.2	87.4	88.4	
2011	87.1	87.2	87.4	86.7	87.1	−1.4%
2012	87.0	87.0	86.5	85.7	86.6	−0.6%
2013	85.8	86.3	87.1	88.2	86.9	0.3%
2014	89.2	90.6	92.5	94.0	91.6	5.4%
2015	96.1	98.6	101.5	103.8	100.0	9.2%
2016	105.8	107.7	108.7	109.8	108.0	8.0%
2017	110.9	111.3	112.2	112.6	111.8	3.5%
2018	113.2	113.7	115.2	117.1	114.8	2.7%
2019	118.2 (P)	118.9 (F)	119.7	120.6	119.4 (F)	4.0%
2020	121.1	121.6	122.2	122.9	122.0	2.2%
2021	123.4	123.9	124.5	125.3	124.3	1.9%
2022	125.7	126.7	127.7	128.4	127.1	2.3%

Note: (P) = Provisional, (F) = Forecast thereafter

Readers can be kept abreast of tender price movements in the free Spon's *Updates* and also in the *Market Forecast* articles, published quarterly in *Building* magazine.

AECOM Tender Price, Building Cost and Consumer Price Indices chart

This chart shows the relative movement of the indices since 2010. This makes it much easier to see the relative movement between the indices.

Regional Variations

As well as being aware of inflationary trends when preparing an estimate, it is also important to establish the appropriate price level for the project location.

Prices throughout this book are at a price level index of 124 in Outer London (location factor = 1.00). Regional variations for certain inner London boroughs can be up to 10% higher, while prices in the Yorkshire and Humberside can be as much as 10% lower. Broad regional adjustment factors used to assist with the preparation of initial estimates are shown in the table below and map on the next page.

Over time, price differentials can change depending on regional workloads and local hot spots. *Spon's Updates* together with the *Market Forecast* features in *Building* magazine will keep readers informed of the latest regional developments and changes as they occur.

The regional variations shown in the following table are based on our forecast of price differentials in each of the economic planning regions at the time the book was published. Historically the adjustments vary little year-on-year; however there can be wider variations within any region, particularly between urban and rural locations. The following table shows suggested adjustments required to cost plans or estimates prepared using the *Spon's A&B 2020*. Please note: these adjustments need to be applied to the total cost plan value, not to individual material or element items.

Region	Adjustment to Measured Works section
Outer London (Spon's 2020)	1.00
Inner London	1.06
South East	0.96
South West	0.89
East of England	0.93
East Midlands	0.89
West Midlands	0.87
North East	0.84
North West	0.88
Yorkshire and Humberside	0.86
Wales	0.85
Scotland	0.87
Northern Ireland	0.76

Examples:

Adjust Location:

To adjust a cost plan derived using *Spon's A & B Price Book*. Location is East of England (0.92 in table)

Cost Plan = £3,200,000 × 0.92 = £2,944,000 (*adjusted*)

Spons Architects' and Builders' Price Book
2020 Edition
(Tender Index 124)

Scotland
0.87

Northern
Ireland
0.76

North East
0.84

Yorkshire and
Humberside
0.86

North West
0.88

East Midlands
0.89

East of England
0.93

West Midlands
0.87

Wales
0.85

South East
(excluding London) 0.96

South West 0.89

Inner London 1.06
Outer London 1.00

Quantity Surveyor's Pocket Book, 3rd Edition

D. Cartlidge

The third edition of the Quantity Surveyor's Pocket Book has been updated in line with NRM1, NRM2 and NRM3, and remains a must-have guide for students and qualified practitioners. Its focused coverage of the data, techniques and skills essential to the quantity surveying role makes it an invaluable companion for everything from initial cost advice to the final account stage.

Key features and updates included in this new edition:
- An up-to-date analysis of NRM1, 2 and 3;
- Measurement and estimating examples in NRM2 format;
- Changes in procurement practice;
- Changes in professional development, guidance notes and schemes of work;
- The increased use of NEC3 form of contract;
- The impact of BIM.

This text includes recommended formats for cost plans, developer's budgets, financial reports, financial statements and final accounts. This is the ideal concise reference for quantity surveyors, project and commercial managers, and students of any of the above.

March 2017; 186 × 123 mm, 466 pp
Pbk: 978-1-138-69836-9; £21.99

To Order: Tel: +44 (0) 1235 400524 Fax: +44 (0) 1235 400525
or Post: Taylor and Francis Customer Services,
Bookpoint Ltd, Unit T1, 200 Milton Park, Abingdon, Oxon, OX14 4TA UK
Email: book.orders@tandf.co.uk

For a complete listing of all our titles visit:
www.tandf.co.uk

PART 2

Rates of Wages and Labour

This part contains the following sections:

Contractual Procedures in the Construction Industry, 7th edition

Allan Ashworth and Srinath Perera

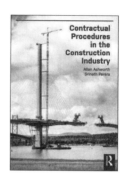

Contractual Procedures in the Construction Industry 7th edition aims to provide students with a comprehensive understanding of the subject, and reinforces the changes that are taking place within the construction industry. The book looks at contract law within the context of construction contracts, it examines the different procurement routes that have evolved over time and the particular aspects relating to design and construction, lean methods of construction and the advantages and disadvantages of PFI/PPP and its variants. It covers the development of partnering, supply chain management, design and build and the way that the clients and professions have adapted to change in the procurement of buildings and engineering projects.

Key features of the new edition include:

- A revised chapter covering the concept of value for money in line with the greater emphasis on added value throughout the industry today.
- A new chapter covering developments in information technology applications (building information modelling, blockchains, data analytics, smart contracts and others) and construction procurement.
- Deeper coverage of the strategies that need to be considered in respect of contract selection.
- Improved discussion of sustainability and the increasing importance of resilience in the built environment.
- Concise descriptions of some the more important construction case laws.

March 2018: 246 x 174 mm: 458 pp
Pb: 978-1-138-69393-7: £48.99

To Order: Tel: +44 (0) 1235 400524 Fax: +44 (0) 1235 400525
or Post: Taylor and Francis Customer Services,
Bookpoint Ltd, Unit T1, 200 Milton Park, Abingdon, Oxon, OX14 4TA UK
Email: book.orders@tandf.co.uk

For a complete listing of all our titles visit:
www.tandf.co.uk

Rates of Wages

CIJC Basic Rates of Pay

BUILDING INDUSTRY – ENGLAND, WALES AND SCOTLAND

Employers and Unions agreed a two year pay and conditions deal as part of the new Construction Industry Joint Council Working Rule Agreement. The first part became effective in June 2018 with a basic pay rise of 3.2%. The second part became effective in Jun 2019 with a 2.9% basic pay rise.

The following basic rates of pay effective from 24 June 2019	Rate per 39-hour week (£)	Rate per hour (£)
Craft Rate	495.18	12.70
Skill Rate 1	470.34	12.06
Skill Rate 2	453.18	11.62
Skill Rate 3	423.93	10.87
Skill Rate 4	400.53	10.27
General Operative	371.67	9.53

Apprentices/Trainees

The Construction Apprenticeship Scheme (CAS) is aimed mainly at school leavers entering the construction industry for the first time but is open to all ages and operates throughout England and Wales.

CAS is owned by the construction industry and administered by CITB, the Construction Industry Training Board.

For further information telephone CAS helpdesk – 0844 875 2274.

The following basic rates of pay effective from 24 June 2019	Rate per 39-hour week (£)	Rate per hour (£)
Year 1	222.30	5.70
Year 2	265.20	6.80
Year 3 without NVQ2	309.27	7.93
Year 3 with NVQ2	395.85	10.15
Year 3 with NVQ3	494.13	12.67
On Completion of Apprenticeship with NVQ2	494.13	12.67

Note: If an apprentice is 22 years and over, and in his/her second year of training, then the National Minimum Wage applies.

BUILDING AND ALLIED TRADES JOINT INDUSTRIAL COUNCIL

The Building and Allied Trades Joint Industrial Council (BATJIC), the partnership between the Federation of Master Builders (FMB) and the Transport and General Workers Union (TGWU).

Rates effective 24 June 2019	Rate per 39-hour week (£)	Rate per hour (£)
Craft operative (NQV3)	500.12	12.82
Craft operative (NQV2)	430.62	11.04
Adult general operative	382.42	9.81

For the latest wage/conditions information, go to www.fmb.org.uk.

BUILDING INDUSTRY – ENGLAND, WALES AND SCOTLAND

PLUMBING AND MECHANICAL ENGINEERING SERVICES INDUSTRY

Authorized rates of wages agreed by the Joint Industry Board for the Plumbing and Mechanical Engineering Services Industry in England and Wales

The Joint Industry Board for Plumbing and Mechanical Engineering Services in England and Wales

Lovell House
Sandpiper Court
Eaton Socon
St Neots
Huntingdon
Cambridgeshire
PE19 8EP
Tel: 01480 476 925
Fax: 01480 403 081
Email: info@jib-pmes.org.uk

Effective from 6 January 2020	Rate per 37.5-hour week (£)	Rate per hour (£)
Operatives		
Technical Plumber and Gas Service Technician	664.13	17.71
Advanced Plumber and Gas Service Engineer	597.75	15.94
Trained Plumber and Gas Service Fitter	513.38	13.69
Apprentices		
1st year of Training	248.25	6.62
2nd year of Training	285.38	7.61
3rd year of Training	321.75	8.58
3rd year of Training with NVQ Level 2	390.75	10.42
4th year of Training	395.63	10.55
4th year of Training with NVQ Level 3	449.63	11.99
4th year of Training with NVQ Level 4	496.50	13.24
Adult Trainees		
1st 6 months of Employment	400.50	10.68
2nd 6 months of Employment	429.38	11.45
3rd 6 months of Employment	447.00	11.92

The Joint Industry Board for the Plumbing Industry in Scotland and Northern Ireland

Bellevue House
22 Hopetoun Street
Edinburgh
EH7 4GH
Tel: 0131 556 0600
Email: info@snijib.org

BUILDING INDUSTRY – ENGLAND, WALES AND SCOTLAND

Effective from 1 July 2019	Rate per 37.5-hour week (£)	Rate per hour (£)
Operatives Plumbers & Gas Service Operatives		
Plumber and Gas Service Fitter	480.75	12.82
Advanced Plumber and Gas Service Engineer	547.13	14.59
Technician Plumber and Gas Service Technician	605.63	16.15
Plumbing Labourer	428.63	11.43
Apprentice Plumbers and Fitters		
1st Year Apprentice	157.50	4.20
2nd Year Apprentice	208.13	5.55
3rd Year Apprentice	252.00	6.72
4th Year Apprentice	325.13	8.67

Estimating and Tendering for Construction Work, 5th edition

Martin Brook

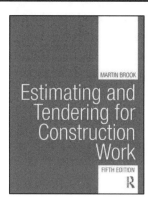

Estimators need to understand the consequences of entering into a contract, often defined by complex conditions and documents, as well as to appreciate the technical requirements of the project. Estimating and Tendering for Construction Work, 5th edition, explains the job of the estimator through every stage, from early cost studies to the creation of budgets for successful tenders.

This new edition reflects recent developments in the field and covers:

- new tendering and procurement methods
- the move from basic estimating to cost-planning and the greater emphasis placed on partnering and collaborative working
- the New Rules of Measurement (NRM1 and 2), and examines ways in which practicing estimators are implementing the guidance
- emerging technologies such as BIM (Building Information Modelling) and estimating systems which can interact with 3D design models

With the majority of projects procured using design-and-build contracts, this edition explains the contractor's role in setting costs, and design statements, to inform and control the development of a project's design.

Clearly-written and illustrated with examples, notes and technical documentation, this book is ideal for students on construction-related courses at HNC/HND and Degree levels. It is also an important source for associated professions and estimators at the outset of their careers.

July 2016; 246 × 189 mm, 334 pp
Pbk: 978-1-138-83806-2; £36.99

To Order: Tel: +44 (0) 1235 400524 Fax: +44 (0) 1235 400525
or Post: Taylor and Francis Customer Services,
Bookpoint Ltd, Unit T1, 200 Milton Park, Abingdon, Oxon, OX14 4TA UK
Email: book.orders@tandf.co.uk

For a complete listing of all our titles visit:
www.tandf.co.uk

Labour Rate Calculations

The format of this section is so arranged that, in the case of work normally undertaken by the Main Contractor, the constituent parts of the total rate are shown enabling the reader to make such adjustments as may be required. Similar details have also been given for work normally sublet although it has not been possible to provide this in all instances.

As explained in the Preface, there is a facility available to readers, which enables a comparison to be made between the level of prices in this section and current tenders by means of a tender index.

To adjust prices for other regions and times, the reader is recommended to refer to the explanations and examples on how to apply these tender indices.

There follow explanations and definitions of the basis of costs in the Prices for Measured Work section under the following headings:

- Overhead charges and profit
- Labour hours and Labour £ column
- Plant £ column
- Material £ column
- Total rate £ column

Overhead and profit charges

Rates checked against winning tenders include overhead charges and profit at current levels.

Labour hours and Labour £ columns

Labour rates are based upon typical gang costs divided by the number of primary working operatives for the trade concerned, and for general building work include an allowance for trade supervision (see below). Labour hours multiplied by Labour rate with the appropriate addition for overhead charges and profit gives Labour £. In some instances, due to variations in gangs used, Labour rate figures have not been indicated, but can be calculated by dividing Labour £ by Labour hours.

Building craft operatives and labourers

Effective from 24 June 2019 our calculated minimum weekly earnings for craft operatives and general operatives rates are £495.18 and £371.67 respectively; to these rates have been added allowances for the items below in accordance with the recommended procedure of the Chartered Institute of Building in its Code of Estimating Practice. The resultant hourly rates on which the Prices for Measured Work have generally been based are £16.89 for craft operative and £12.56 for general operatives.

We continue to use CIJC Working Rule Agreement rates to help ensure consistency between editions. Actual wage rates do vary, particularly between trades. It is relatively straightforward for users to apply their own rates.

Calculations take into account the following:

- Lost time
- Construction Industry Training Board Levy
- Holidays with pay
- Accidental injury, retirement and death benefits scheme
- Sick pay
- National insurance
- Severance pay and sundry costs
- Employer's liability and third-party insurance

The table which follows illustrates how the all-in hourly rates have been calculated based upon CIJC Working Rule Agreement. Productive time has been based on a total of 1727 hours (44.3 weeks) worked per year.

			Craft Operatives		General Operatives	
			£	£	£	£
Wages at standard basic rate						
Productive time	44.1	weeks	495.18	21,837.57	371.67	16,390.65
Lost time allowance, say	0.9	weeks	495.18	445.66	371.67	334.50
Overtime	0	weeks	0.00	0.00	0.00	0.00
				22,283.24		16,725.15
Extra payments under National Working Rules			0.00		0.00	
Holiday pay & Public holiday pay	6.0	weeks	495.18	2,971.10	371.67	2,230.02
Employer's contribution to:						
CITB Levy (0.50% of payroll)	0.50	year		126.27		94.78
EasyBuild Stakeholder Pension	52	weeks	7.50	390.00	7.50	390.00
National Insurance (average weekly payment)	52	weeks	44.11	2,293.88	27.40	1,424.60
				28,064.49		20,864.54
Severance pay and sundry costs, say	1.5%			420.97		312.97
				28,485.45		21,177.51
Employer's Liability and Third-Party Insurance	2%			569.71		423.55
Total cost per annum				**29,055.16**		**21,601.06**
Total cost per week				**£658.850**		**£489.82**
Total cost per hour				**£16.89**		**£12.56**

Similar calculations for other CIJC labour classifications result in the following cost per hour:

Based on basic rates of pay that came into effect 24 June 2019	Rate per hour (£)
Craft Operative	16.89
Skill Rate 1	16.02
Skill Rate 2	15.42
Skill Rate 3	14.39
Skill Rate 4	13.57
General Operative	12.56
Rates for responsibilities:	
Foreman	17.91
Ganger	13.57

Notes:

1.   Absence due to sickness has been assumed to be for periods not exceeding 3 days for which no payment is due.
2.   EasyBuild Stakeholder Pension effective from 1 July 2002. Death and accident benefit cover is provided free of charge.
3.   All NI Payments are at not-contracted out rates applicable from April 2019.
4.   Public holiday pay includes for 8 public holiday days.
5.   This table is an example showing a typical breakdown.

The labour rates used in the Measured Work sections have been based on the following gang calculations which generally include an allowance for supervision by a foreman or ganger. Alternative labour rates are given showing the effect of various degrees of bonus.

Typical Gang Build-up	Rate £/hour	Unit rate £/hour	Productive hours	Labour rates with bonus payment £/hour			
				Normal	+10%	+20%	+30%
Groundwork Gang							
1 Ganger	13.57	13.57					
6 Labourers	12.56	75.36					
		88.93	6.5 =	13.68	15.13	16.58	18.04
Concreting gang							
1 Foreman	17.91	17.91					
4 Skilled labourers	13.57	54.29					
		72.19	4.5 =	16.04	17.75	19.47	21.20
Steel Fixing Gang							
1 Foreman	17.91	17.91					
4 Steel fixers	16.89	67.57					
		85.48	4.5 =	18.30	20.23	22.17	24.12
Formwork Gang							
1 Foreman	17.91	17.91					
10 Carpenters	16.89	168.94					
1 Labourer	12.56	12.56					
		199.40	10.5 =	18.99	20.97	22.95	24.93
Bricklaying/Blockwork Gang							
1 Foreman	17.91	17.91					
6 Bricklayers	16.89	101.36					
1 Labourer	12.56	12.56					
		131.83	6.50 =	20.28	22.41	24.54	26.67
Carpentry/Joinery Gang							
1 Foreman	17.91	17.91					
5 Carpenters	16.89	84.47					
1 Labourer	12.56	12.56					
		114.93	5.5 =	20.90	23.09	25.29	27.50
Painter, Slater, etc.							
1 Craft operative	16.89	16.897	1 =	16.89	18.63	20.37	22.11
1 and 1 gang							
1 Craft operatives	16.89	16.89					
1 Skilled labourer	13.57	13.57					
		30.47	1 =	30.47	33.61	36.75	39.89
2 and 1 Gang							
2 Craft operatives	16.89	33.79					
1 Skilled labourer	13.57	13.57					
		47.36	2 =	23.68	26.12	28.56	31.00
Small Labouring Gang (making good)							
1 Foreman	17.91	17.91					
4 Skilled labourers	13.57	54.29					
		72.19	4.5 =	16.04	17.75	19.47	21.20
Drain Laying Gang/Clayware							
2 Skilled labourers	13.57	27.14	2 =	13.57	14.98	16.38	17.79

Subcontractor's operatives

Similar labour rates are shown in respect of subcontractor trades where applicable.

Plumbing operatives

The Joint Industry Board for Plumbing and Mechanical Engineering Services in England and Wales		
Effective from 6 January 2020	Rate per 37.5-hour week (£)	Rate per hour (£)
Operatives		
Technical Plumber and Gas Service Technician	664.13	17.71
Advanced Plumber and Gas Service Engineer	597.75	15.94
Trained Plumber and Gas Service Fitter	513.38	13.69
Apprentices		
1st year of Training	248.25	6.62
2nd year of Training	285.38	7.61
3rd year of Training	321.75	8.58
3rd year of Training with NVQ Level 2	390.75	10.42
4th year of Training	395.63	10.55
4th year of Training with NVQ Level 2	449.63	11.99
4th year of Training with NVQ Level 3	496.50	13.24
Adult Trainees		
1st 6 months of Employment	400.50	10.68
2nd 6 months of Employment	429.38	11.45
3rd 6 months of Employment	447.00	11.92

From 6 January 2020 the hourly earnings for technical and trained plumbers are £17.71 and £13.69 respectively; to these rates have been added allowances like those added for building operatives (see below). The resultant average hourly rate on which the Prices for Measured Work have been based is £25.72. The items referred to above for which allowance has been made are:

- Tool allowance
- Plumbers' welding supplement
- Holidays with pay
- Pension and welfare stamp
- National insurance
- Severance pay and sundry costs
- Employer's liability and third-party insurance

No allowance has been made for supervision as we have assumed the use of a team of technical or trained plumbers who are able to undertake such relatively straightforward plumbing works, e.g. on housing schemes, without supervision.

The table which follows shows how the average hourly rate referred to above has been calculated. Productive time has been based on a total of 1,680 hours worked per year.

Effective from 6 January 2020		Technical Plumber		Trained Plumber	
		Rate £	Total £	Rate £	Total £
Wages at standard basic rate					
Productive time	1680.0 hrs	17.71	29,752.80	13.69	22,999.20
Overtime (paid at basic rate)	336.0 hrs	17.71	5,950.56	13.69	4,599.84
Plumbers welding supplement	2016.0 hrs	0.50	1,008.00	–	–
			36,711.36		27,599.04
Employer's contribution to:					
Holiday credit/welfare					
Holiday pay for 31 days	31 days	132.83	4,117.58	102.68	3,182.93
Additional Holiday Pay (maximum)	60 credits	2.25	135.00	2.25	135.00
			40,963.94		30,916.97
Pension, say	7.5%		3,072.30		2,318.77
National Insurance	52 weeks	71.33	3,709.10	47.94	2,492.76
			47,745.33		35,728.50
Severance pay and sundry costs, say	1.5%		716.18		535.93
Liability and Third-Party Insurance	2.0%		969.23		725.29
Total cost per annum			49,430.74		36,989.72
Total cost per hour			29.42		22.02
Average all-in rate per hour				**25.72**	

Notes:

1. First 37.5 hours paid at normal rate.
2. 1.5 overtime rate after 37.5 hrs up to 8 pm, after which overtime rate is at double time.
3. National Insurance is payable on top-up funding as this is payable by the employer.
4. Absence due to sickness has been assumed to be for periods not exceeding 3 days for which no payment is due. The entitlement is payable from and including the fourth day of illness onwards.
5. This table is an example showing a typical breakdown.

LABOUR CATEGORIES

Schedule 1 to the WRA lists specified work establishing entitlement to the Skilled Operative Pay Rate 4, 3, 2, 1 or the Craft Rate as follows:

CIJC WORKING RULE AGREEMENT FOR THE CONSTRUCTION INDUSTRY – SCHEDULE 1	
Specified Work Establishing Entitlement to the Skilled Operative Pay Rate 4,3,2,1 or Craft Rate	
BAR BENDERS AND REINFORCEMENT FIXERS	
• Bender and fixer of Concrete Reinforcement capable of reading and understanding drawings and bending schedules and able to set out work	Craft Rate
CONCRETE	
• Concrete Leveller or Vibrator Operator	4
• Screeder and Concrete Surface Finisher working from datum such as road-form, edge beam or wire	4
• Operative required to use trowel or float (hand or powered) to produce high quality finished concrete	4
DRILLING AND BLASTING	
• Drills, rotary or percussive: mobile rigs, operator of	3
• Operative attending drill rig	4
• Shotfirer, operative in control of and responsible for explosives including placing, connecting and detonating charges	3
• Operatives attending on shotfirer, including stemming	4
DRYLINERS	
• Operatives undergoing approved training in drylining	4
• Operatives who can produce a certificate of training achievement indicating satisfactory completion of at least one unit of approved drylining training	3
• Dryliners who have successfully completed their training in drylining fixing and finishing	Craft Rate
FORMWORK CARPENTERS	
• 1st year trainee	4
• 2nd year trainee	3
• Formwork Carpenters	Craft Rate
GANGERS AND TRADE CHARGEHANDS	
(Higher grade payments may be made at the employer's discretion)	4
GAS NETWORK OPERATIONS	
Operatives who have successfully completed approved training to the standard of:	
• GNO Trainee	4
• GNO Assistant	3
• Team Leader – Services	2
• Team Leader – Mains	1
• Team Leader – Mains and Services	1
LINESMEN – ERECTORS	
• 1st Grade	2
Skilled in all works associated with the assembly, erection, maintenance and dismantling of Overhead Lines Transmission Lines on steel towers, concrete or wood poles, including all overhead lines construction elements	
• 2nd Grade	3
As above but lesser degree of skill – or competent and fully skilled to carry out some of the elements of construction listed above	
• Linesmen-erector's mate	4
Semi-skilled in works specified above and a general helper	
MASON PAVIORS	
• Operative assisting a Mason Pavior undertaking kerblaying, block and sett paving, flag laying, in natural stone and precast products	4
• Operative engaged in stone pitching or dry stone walling	3

LABOUR CATEGORIES

MECHANICS	
• **Maintenance Mechanic** capable of carrying out field service duties, maintenance activities and minor repairs	2
• **Plant Mechanic** capable of carrying out major repairs and overhauls including welding work, operating metal turning lathe or similar machine and using electronic diagnostic equipment	1
• **Maintenance/Plant Mechanics' Mate** on site or in depot	4
• **Tyre Fitter,** heavy equipment tyres	2
MECHANICAL PLANT DRIVERS AND OPERATORS	
Backhoe Loaders (with rear excavator bucket and front shovel and additional equipment such as blades, hydraulic hammers and patch planers)	
• Backhoe, up to and including 50 kW net engine power; driver of	4
• Backhoe, over 50 kW up to and including 100 kW net engine power; driver of	3
• Backhoe, over 100 kW net engine power; driver of	2
• **Compressors and Generators** Air compressor or generators over 10 kW; operator of	4
Concrete Mixers	
• Operative responsible for operating a concrete mixer or mortar pan up to and including 400 litre drum capacity	4
• Operative responsible for operating a concrete mixer over 400 litre and up to and including 1,500 litre drum capacity	3
• Operative responsible for operating a concrete mixer over 1,500 litre drum capacity	2
• Operative responsible for operating a mobile self-loading and batching concrete mixer up to 2,500 litre drum capacity	2
• Operative responsible for operating a mechanical drag-shovel	4
Concrete Placing Equipment	
• Trailer mounted or static concrete pumps: self-propelled concrete placers: concrete placing booms; operator of	3
• Self-propelled mobile concrete pump, with or without boom, mounted on lorry or lorry chassis; driver/operator of	2
Cranes/Mobile Cranes	
Self-propelled mobile crane on road wheels, rough terrain wheels or caterpillar tracks including lorry mounted:	
• Maximum lifting capacity at minimum radius, up to and including 5 tonne; driver of	4
• Maximum lifting capacity at minimum radius, over 5 tonne and up to and including 10 tonne; driver of	3
• Maximum lifting capacity at minimum radius, over 10 tonne	Craft Rate
Tower Cranes (including static or travelling: standard trolley or luffing jib)	
Up to and including 2 tonne max. Lifting capacity at min. radius; driver of	4
• Over 2 tonne up to and including 10 tonne max. Lifting capacity at min. radius; driver of	3
• Over 10 tonne up to and including 20 tonne max. Lifting capacity at minimum radius; driver of	2
• Over 20 tonne max. Lifting capacity at minimum radius; driver of	1
Miscellaneous Cranes and Hoists	
• Overhead bridge crane or gantry crane up to and including 10 tonne capacity; driver of	3
• Overhead bridge crane or gantry crane over 10 tonne up to and including 20 tonne capacity; driver of	2
• Power driven hoist or jib crane; operator of	4
• **Slinger/Signaller** appointed to attend Crane or hoist to be responsible for fastening or slinging loads and generally to direct lifting operations	4
Dozers	
• Crawler dozer with standard operating weight up to and including 10 tonne; driver of	3
• Crawler dozer with standard operating weight over 10 tonne and up to and including 50 tonne; driver of	2
• Crawler dozer with standard operating weight over 50 tonne; driver of	1

LABOUR CATEGORIES

Dumpers and Dump Trucks	
• Up to and including 10 tonne rated payload; driver of	4
• Over 10 tonne and up to and including 20 Tonne rated payload; driver of	3
• Over 20 tonne and up to and including 50 Tonne rated payload; driver of	2
• Over 50 tonne and up to and including 100 Tonne rated payload; driver of	1
• Over 100 tonne rated payload; driver of	Craft Rate
Excavators (360 degree slewing)	
• Excavators with standard operating weight up to and including 10 tonne; driver of	3
• Excavator with standard operating weight over 10 tonne and up to and including 50 tonne; driver of	2
• Excavator with standard operating weight over 50 tonne; driver of	1
• **Banksman** appointed to attend excavator or responsible for positioning vehicles during loading or tipping	4
Fork-Lifts Trucks and Telehandlers	
• Smooth or rough terrain fork lift trucks (including side loaders) and telehandlers up to and including 3 tonne lift capacity; driver of	4
• Over 3 tonne lift capacity; driver of	3
Motor Graders: driver of	2
Motorized Scrapers: driver of	2
Motor Vehicles (Road Licensed Vehicles) Driver and Vehicle Licensing Agency (DVLA)	
• Vehicles requiring a driving licence of category C1; driver of (Goods vehicle with maximum authorized mass (mam) exceeding 3.5 tonne but not exceeding 7.5 tonne and including such a vehicle drawing a trailer with a mam not over 750 kg)	4
• Vehicles requiring a driving licence of category C; driver of (Goods vehicle with a maximum authorized mass (mam) exceeding 3.5 tonne and including such a vehicle drawing a trailer with mam not over 750 kg)	2
• Vehicles requiring a driving licence of category C plus E; driver of (Combination of a vehicle in category C and a trailer with maximum authorized mass over 750 kg)	1
Power Driven Tools	
• Operatives using power-driven tools such as breakers, percussive drills, picks and spades, rammers and tamping machines	4
Power Rollers	
• Roller, up to and including 4 tonne operating weight; driver of	4
• Roller, over 4 tonne operating weight and upwards; driver of	3
Pumps, Power-driven pump(s); attendant of	4
Shovel Loaders, (Wheeled or tracked, including skid steer)	
• Up to and including 2 cubic metre shovel capacity; driver of	4
• Over 2 cubic metre and up to and including 5 cubic metre shovel capacity; driver of	3
• Over 5 cubic metre shovel capacity; driver of	2
Tractors (Wheeled or Tracked)	
• Tractor, when used to tow trailer and/or with mounted compressor, up to and including 100 kW rated engine power; driver of	4
• Tractor, ditto, over 100 kW up to and including 250 kW rated engine power; driver of	3
• Tractor, ditto, over 250 kW rated engine power; driver of	2
Trenchers (Type wheel, chain or saw)	
• Trenching Machine, up to and including 50 kW gross engine power; driver of	4
• Trenching Machine, over 50 kW and up to and including 100 kW gross engine power; driver of	3
• Trenching Machine, over 100 kW gross engine power; driver of	2
Winches, Power driven winch; driver of	4
PILING	
• General Skilled Piling Operative	4
• Piling Chargehand/Ganger	3
• Pile Tripod Frame Winch Driver	3

LABOUR CATEGORIES

Winches, Power driven winch; driver of	4
• CFA or Rotary or Driven Mobile Piling Rig Driver	2
• Concrete Pump Operator	3
PIPE JOINTERS	
• Jointers, pipes up to and including 300 mm dia.	4
• Jointers, pipes over 300 mm dia. and up to 900 mm dia.	3
• Jointers, pipes over 900 mm dia.	2
• **except** in HDPE mains when experienced in butt fusion and/or electrofusion jointing operations	2
PIPELAYERS	
• Operative preparing the bed and laying pipes up to and including 300 mm dia.	4
• Operative preparing the bed and laying pipes over 300 mm dia. and up to and including 900 mm dia.	3
• Operative preparing the bed and laying pipes over 900 mm dia.	2
PRE-STRESSING CONCRETE	
• Operative in control of and responsible for hydraulic jacks and other tensioning devices engaged in post-tensioning and/or pre-tensioning concrete elements	3
ROAD SURFACING WORK (includes rolled asphalt, dense bitumen macadam and surface dressings) Operatives employed in this category of work to be paid as follows:	
• Chipper	4
• Gritter Operator	4
• Raker	3
• Paver Operator	3
• Leveller on Paver	3
• Road Planer Operator	3
• Road Roller Driver, 4 tonne and upwards	3
• Spray Bar Operator	4
TIMBERMAN	
• Timberman, installing timber supports	3
• Highly skilled timber man working on complex supports using timbers of size 250 mm by 125 mm and above	2
• Operative attending	4
WELDERS	
• Grade 4 (Fabrication Assistant)	
• Welder able to tack weld using SMAW or MIG welding processes in accordance with verbal instructions and including mechanical preparation such as cutting and grinding	3
• Grade 3 (Basic Skill Level)	
• Welder able to weld carbon and stainless steel using at least one of the following processes SMAW, GTAW, GMAW for plate-plate fillet welding in all major welding positions, including mechanical preparation and complying with fabrication drawings	2
• Grade 2 (Intermediate Skill Level)	
• Welder able to weld carbon and stainless steel using manual SMAW, GTAW, semi-automatic MIG or MAG, and FCAW welding processes including mechanical preparation, and complying with welding procedures, specifications and fabrication drawings.	1
• Grade 1 (Highest Skill Level)	
• Welder able to weld carbon and stainless steel using manual SMAW, GTAW, semi-automatic GMAW or MIG or MAG, and FCA welding processes in all modes and directions in accordance with BSEN287–1 and/or 287–2 Aluminium Fabrications including mechanical preparation and complying with welding procedures, specifications and fabrication drawings.	Craft Rate
YOUNG WORKERS	
• Operatives below 18 years of age will receive payment 60% of the General Operative Basic Rate.	
• At 18 years of age or over the payment is 100% of the relevant rate.	

Architect's Pocket Book, 5th edition

Jonathan Hetreed, Ann Ross and
Charlotte Baden-Powell

This handy pocket book brings together a wealth of useful information that architects need on a daily basis – on-site or in the studio. It provides clear guidance and invaluable detail on a wide range of issues, from planning policy through environmental design to complying with Building Regulations, from structural and services matters to materials characteristics and detailing. This fifth edition includes the updating of regulations, standards and sources across a wide range of topics.

Compact and easy to use, the *Architect's Pocket Book* has sold well over 90,000 copies to the nation's architects, architecture students, designers and construction professionals who do not have an architectural background but need to understand the basics, fast.

This is the famous little blue book that you can't afford to be without.

April 2017: 186 × 123 mm: 412 pp
Pb: 978-1-138-64399-4: £25.99

To Order: Tel: +44 (0) 1235 400524 Fax: +44 (0) 1235 400525
or Post: Taylor and Francis Customer Services,
Bookpoint Ltd, Unit T1, 200 Milton Park, Abingdon, Oxon, OX14 4TA UK
Email: book.orders@tandf.co.uk

For a complete listing of all our titles visit:
www.tandf.co.uk

PART 3

Approximate Estimating

This part contains the following sections:

BIM and Quantity Surveying
S Pittard *et al.*

The sudden arrival of Building Information Modelling (BIM) as a key part of the building industry is redefining the roles and working practices of its stakeholders. Many clients, designers, contractors, quantity surveyors, and building managers are still finding their feet in an industry where BIM compliance can bring great rewards.

This guide is designed to help quantity surveying practitioners and students understand what BIM means for them, and how they should prepare to work successfully on BIM compliant projects. The case studies show how firms at the forefront of this technology have integrated core quantity surveying responsibilities like cost estimating, tendering, and development appraisal into high profile BIM projects. In addition to this, the implications for project management, facilities management, contract administration and dispute resolution are also explored through case studies, making this a highly valuable guide for those in a range of construction project management roles.

Featuring a chapter describing how the role of the quantity surveyor is likely to permanently shift as a result of this development, as well as descriptions of tools used, this covers both the organisational and practical aspects of a crucial topic.

December 2015: 234 x 156 mm: 258 pp
Pbk: 978-0-415-87043-6; £30.99

To Order: Tel: +44 (0) 1235 400524 Fax: +44 (0) 1235 400525
or Post: Taylor and Francis Customer Services,
Bookpoint Ltd, Unit T1, 200 Milton Park, Abingdon, Oxon, OX14 4TA UK
Email: book.orders@tandf.co.uk

For a complete listing of all our titles visit:
www.tandf.co.uk

Building Prices per Functional Unit

Prices given under this heading are given as a typical range of prices for new work. These should not be interpreted as a minimum or maximum range.

Prices include for Preliminaries, and Overheads and Profit recovery.

Unless otherwise stated, prices do not allow for external works, furniture or special equipment.

On certain types of buildings there exists a close relationship between its cost and the number of functional units that it accommodates. During the very early stages of a project an approximate estimate can be quickly derived by multiplying the proposed unit of accommodation (i.e. hotel bedrooms, car parking spaces etc.) by an appropriate rate.

The following indicative unit areas and costs have been derived from historical data and construction standards. It should be emphasized that the prices must be treated with reserve, as they represent the average of prices from our records and cannot provide more than a rough guide to the cost of a building. There are limitations when using this method of estimating, for example, the functional areas and costs of football stadia, hotels, restaurants etc. are strongly influenced by the extent of front and back of house facilities housed within it, and these areas can vary considerably from scheme to scheme.

The areas may also be used as a 'rule of thumb' in order to check on economy of designs. Where we have chosen not to show indicative areas, this is because either ranges are extensive or such figures may be misleading.

Costs have been expressed within a range, although this is not to suggest that figures outside this range will not be encountered, but simply that the calibre of such a type of building can itself vary significantly.

For assistance with the compilation of an *Order of Cost Estimate*, or of an *Elemental Cost Plan*, the reader is directed to the *Building Prices per Square Metre*, *Approximate Estimating Rates* and *Cost Models* sections of this book.

As elsewhere in this edition, prices do not include Value Added Tax or fees for professional services.

BUILDING PRICES PER FUNCTIONAL UNIT

Building type	Function	Range £		
Utilities, Civil Engineering facilities				
Car Parking				
surface car parking; 20 to 22 m²/car	car	2075.00	to	2900.00
surface car parking; landscaped 20 to 22 m²/car	car	2700.00	to	3700.00
multi-storey car parks flat slab; 24 to 28 m²/car	car	17000.00	to	23500.00
underground car parks partially underground under buildings; natural ventilation; 27 to 30 m² / car	car	20500.00	to	28500.00
underground car parks completely underground with landscaped roof; mechanical ventilation; 28 to 32 m²/car	car	36500.00	to	51000.00
underground car parks completely underground under buildings; mechanical ventilation; 28 to 32 m²/car	car	30000.00	to	42000.00
Offices				
business park; medium quality, non air-conditioned	person	13000.00	to	18000.00
low rise; air-conditioned; high quality speculative; 4–8 storeys	person	16000.00	to	22500.00
medium rise; air-conditioned; high quality speculative; 8–20 storeys	person	25000.00	to	34500.00
Health and Welfare facilities				
district hospitals; 65 to 85 m²/bed	bed	180000.00	to	255000.00
hospital teaching centres; 120+ m²/bed	bed	200000.00	to	280000.00
private hospitals; 75 to 100 m²/bed	bed	215000.00	to	300000.00
residential homes; 40 to 60 m²/bedroom	bed	70000.00	to	97000.00
nursing homes; 40 to 60 m²/bedroom	bed	93000.00	to	130000.00
Recreational facilities				
Theatres				
large – over 500 seats	seat	28500.00	to	39500.00
studio/workshop – less than 500 seats	seat	14500.00	to	20000.00
refurbishment	seat	11500.00	to	16000.00
Sports halls				
indoor bowls halls	rink	170000.00	to	240000.00
squash courts	court	98000.00	to	135000.00
Education facilities				
Schools				
nursery schools; 3 to 5 m²/pupil	pupil	7800.00	to	11000.00
secondary/middle schools; 6 to 10 m²/pupil	pupil	19000.00	to	27000.00
special schools; 18 to 20 m²/pupil	pupil	26000.00	to	36000.00
Universities				
arts buildings	student	15500.00	to	22000.00
computer buildings	student	21000.00	to	29000.00
science buildings	student	21000.00	to	29000.00
laboratories	student	41000.00	to	57000.00
Residential facilities				
Housing				
detached; four bedroom; 100 to 140 m² GIFA	house	175000.00	to	240000.00
semi-detached; three bedroom; 70 to 90 m² GIFA	house	105000.00	to	150000.00
terraced; 2 bedroom; 55 to 65 m² GIFA	house	68000.00	to	94000.00
low rise flats; two bedroom; 55 to 65 m² GIFA	flat	81000.00	to	110000.00
medium rise flats; two bedroom; 55 to 65 m² GIFA	flat	99000.00	to	140000.00

BUILDING PRICES PER FUNCTIONAL UNIT

Building type	Function	Range £		
Hotels				
luxury; city centre with conference and wet leisure facilities; 70 to 120 m²/bedroom	bed	440000.00	to	620000.00
business; town centre with conference and wet leisure facilities; 70 to 100 m²/bedroom	bed	270000.00	to	375000.00
mid-range; town centre with conference and leisure facilities; 50 to 60 m²/bedroom	bed	125000.00	to	175000.00
budget; city centre with dining and bar facilities; 35 to 45 m²/bedroom	bed	83000.00	to	115000.00
budget; roadside excluding dining facilities; 28 to 35 m²/bedroom	bed	56000.00	to	78000.00
Student residences				
large budget schemes; 200+ units with ensuite accommodation; 18 to 20 m²/bedroom	bed	28000.00	to	39000.00
smaller schemes (40–100 units) with mid range specifications; some with ensuite accommodation; 19 to 24 m²/bedroom	bed	38000.00	to	53000.00
smaller high quality courtyard schemes, college scheme; 24 to 28 m²/bedroom	bed	61000.00	to	84000.00

Building Regulations Pocket Book

Ray Tricker and Samantha Alford

This handy guide provides you with all the information you need to comply with the UK Building Regulations and Approved Documents. On site, in the van, in the office, wherever you are, this is the book you'll refer to time and time again to double check the regulations on your current job.

The *Building Regulations Pocket Book* is the must have reliable and portable guide to compliance with the Building Regulations.

*   Part 1 provides an overview of the Building Act
*   Part 2 offers a handy guide to the dos and don'ts of gaining the Local Council's approval for Planning Permission and Building Regulations Approval
*   Part 3 presents an overview of the requirements of the Approved Documents associated with the Building Regulations
*   Part 4 is an easy to read explanation of the essential requirements of the Building Regulations that any architect, builder or DIYer needs to know to keep their work safe and compliant on both domestic or non-domestic jobs

This book is essential reading for all building contractors and sub-contractors, site engineers, building engineers, building control officers, building surveyors, architects, construction site managers and DIYers. Homeowners will also find it useful to understand what they are responsible for when they have work done on their home (ignorance of the regulations is no defence when it comes to compliance!).

February 2018: 186 x 123 mm: 456 pp
Pb: 978-0-8153-6838-0 : £20.99

To Order: Tel: +44 (0) 1235 400524 Fax: +44 (0) 1235 400525
or Post: Taylor and Francis Customer Services,
Bookpoint Ltd, Unit T1, 200 Milton Park, Abingdon, Oxon, OX14 4TA UK
Email: book.orders@tandf.co.uk

For a complete listing of all our titles visit:
www.tandf.co.uk

Building Prices per Square Metre

Prices given under this heading are given as a typical range of prices for new work for a range of typical building types.

Prices in this include for Preliminaries and Overheads and Profit recovery.

Unless otherwise stated, prices do not allow for external works, furniture or special equipment.

Prices are based upon the total gross internal area (GIA) of all storeys measured in accordance with the 6th edition of the RICS *Code of Measuring Practice*, September 2007.

As in previous editions it is emphasized that the prices must be treated with a degree of reserve as they represent the average of prices from historic records and should be used to provide only an approximate order of cost for a building.

Costs have been expressed within a range, although this is not to suggest that figures outside this range will not be encountered, but simply that the calibre of such a type of building can itself vary significantly.

Prices can vary greatly between regions and even between individual sites. Professional interpretation will always need to be applied in the final analysis.

For assistance with the compilation of an *Order of Cost Estimate*, or of an *Elemental Cost Plan*, the reader is directed to the *Cost Models* and *Approximate Estimating Rates* sections in the book.

As elsewhere in this edition, prices do not include Value Added Tax or fees for professional services.

AUTHORS NOTE

RICS Property Measurement

Overview

The Royal Institution of Chartered Surveyors (RICS) has launched 'RICS Property Measurement', which in part, replaces the Code of Measuring Practice for office buildings. This came into effect on 1 January 2016.

The RICS Property Measurement incorporates the International Property Measurement Standards (IPMS) for offices, whilst the *Code of Measuring Practice* 6th edition remains in effect for all other building types.

IPMS was formed in 2013 at the World Bank, with the objective to provide global standardization for measurement of the built environment. It now comprises of more than 70 organizations including RICS. Each member organization is responsible for setting domestic measurement standards. Having successfully launched IPMS office buildings, they are currently working on IPMS residential buildings, followed by industrial, retail and mixed use.

Timetable

RICS Property Measurement was launched in May 2015. The use of IPMS office buildings becomes mandatory for RICS Members from 1 January 2016, unless clients instruct otherwise or local laws specify the use of a different standard.

Proposed changes from *Code of Measuring Practice* 6th edition: Office Buildings

The first key change is the use of new vocabulary. Gross External Area (GEA), Gross Internal Area (GIA) or Net Internal Area (NIA) are no longer used. This terminology has been replaced with IPMS 1, IPMS 2 and IPMS 3. Whilst these are not direct replacements they can be closely related to GEA, GIA and NIA respectively.

BUILDING PRICES PER SQUARE METRE

The second key change is the concept that whilst IPMS 1, IPMS 2 and IPMS 3 are more inclusive in measurement than GEA, GIA and NIA, these additional measures need to be stated separately. To ensure transparency, marked up plans should accompany measurement reports, which should be detailed to include the breakdown of the measure, this should provide benefit to the end user.

The third key change is the concept of limited use area. Whilst these areas are included in the measurement, they are then stated separately. They include areas with limited height (less than 1.5 metres), areas with limited natural light, internal structural walls/columns, area difference between internal dominant face and the wall/floor junction.

IPMS 1

The IPMS 1 measure includes the area of the building incorporating the external walls. The key differences between IPMS 1 and GEA are the inclusion of balconies under IPMS 1. However covered galleries and balconies are stated separately.

IPMS 2

IPMS 2 is measured to the internal dominant face. The internal dominant face is defined as the 'surface comprising more than 50% of the surface area for each vertical section forming an internal perimeter'. So for example, for an office building with unitized cladding, this would mean the glass surface.

Additional areas which would not be included in a GIA measure but is included under IPMS 2 are the balconies and covered canopies. These would then be stated separately.

IPMS 3

The most significant changes appear in the comparison between IPMS 3 and NIA. IPMS 3 include area taken by columns, piers/external columns and window reveals. All of these are excluded under NIA. Additionally space taken by balconies, covered galleries, roof terraces are all included but stated separately. Lift lobbies and area with headroom of less than 1.5 metres are also included but stated separately as limited use area.

It will take the wider market some time to wake up to this change, especially with the impact on end user market for valuation purposes, as it will have a direct impact on their comparable databases. There will be a period where the existing GEA, GIA and NIA will remain in favour and these are still being used in this book.

BUILDING PRICES PER SQUARE METRE

Item	Unit	Range £		
UNICLASS D1 UTILITIES, CIVIL ENGINEERING FACILITIES				
Car parking				
surface car parking	m^2	86.00	to	110.00
surface car parking; landscaped	m^2	110.00	to	140.00
Multi-storey car parks				
flat slab	m^2	580.00	to	740.00
Underground car parks				
partially underground under buildings; naturally ventilated	m^2	660.00	to	840.00
completely underground under buildings with mechanical ventilation	m^2	900.00	to	1150.00
completely underground with landscaped roof and mechanical ventilation	m^2	1075.00	to	1375.00
Transport facilities				
railway stations	m^2	3750.00	to	4800.00
bus and coach stations	m^2	2650.00	to	3400.00
bus garages	m^2	1000.00	to	1275.00
petrol stations	m^2	3300.00	to	4200.00
Vehicle showrooms with workshops, garages etc.				
up to 2,000 m^2	m^2	1275.00	to	1650.00
over 2,000 m^2	m^2	1100.00	to	1425.00
Vehicle showrooms without workshops, garages etc.				
up to 2,000 m^2	m^2	1275.00	to	1650.00
Vehicle repair and maintenance buildings				
up to 500 m^2	m^2	1900.00	to	2425.00
over 500 m^2 up to 2000 m^2	m^2	1275.00	to	1650.00
car wash buildings	m^2	1100.00	to	1425.00
Airport facilities (excluding aprons)				
airport terminals	m^2	3400.00	to	4300.00
airport piers/satellites	m^2	3550.00	to	4550.00
Airport campus facilities				
cargo handling bases	m^2	860.00	to	1075.00
distribution centres	m^2	480.00	to	610.00
hangars (type C and D facilities)	m^2	1475.00	to	1875.00
TV, radio and video studios	m^2	2450.00	to	3100.00
Telephone exchanges	m^2	1275.00	to	1650.00
Telephone engineering centres	m^2	870.00	to	1125.00
Branch post offices	m^2	1275.00	to	1650.00
Postal delivery offices/sorting offices	m^2	1275.00	to	1650.00
Electricity substations	m^2	1850.00	to	2350.00
Garages, domestic	m^2	790.00	to	1000.00
UNICLASS D2 INDUSTRIAL FACILITIES				
B1 Light industrial/offices buildings				
economical shell, and core with heating only	m^2	780.00	to	1000.00
medium shell and core with heating and ventilation	m^2	1100.00	to	1425.00
high quality shell and core with air conditioning	m^2	1725.00	to	2200.00
developers Category A fit-out	m^2	630.00	to	810.00
tenants Category B fit-out	m^2	420.00	to	530.00
Agricultural storage	m^2	660.00	to	850.00

BUILDING PRICES PER SQUARE METRE

Item	Unit	Range £		
UNICLASS D2 INDUSTRIAL FACILITIES – cont				
Factories				
for letting (incoming services only)	m^2	680.00	to	860.00
for letting (including lighting, power and heating)	m^2	860.00	to	1100.00
nursery units (including lighting, power and heating)	m^2	860.00	to	1100.00
workshops	m^2	800.00	to	1025.00
maintenance/motor transport workshops	m^2	1050.00	to	1325.00
owner occupation for light industrial use	m^2	1100.00	to	1425.00
owner occupation for heavy industrial use	m^2	1850.00	to	2350.00
Factory/office buildings high technology production				
for letting (shell and core only)	m^2	730.00	to	930.00
for owner occupation (controlled environment fully finished)	m^2	980.00	to	1250.00
High technology laboratory workshop centres, air-conditioned	m^2	3100.00	to	4000.00
Warehouse and distribution centres				
high bay (10–15 m high) for owner occupation (no heating) up to 10,000 m^2	m^2	350.00	to	450.00
high bay (10–15 m high) for owner occupation (no heating) over 10,000 m^2 up to 20,000 m^2	m^2	265.00	to	340.00
high bay (16–24 m high) for owner occupation (no heating) over 10,000 m^2 up to 20,000 m^2	m^2	390.00	to	495.00
high bay (16–24 m high) for owner occupation (no heating) over 20,000 m^2	m^2	320.00	to	410.00
Fit-out cold stores, refrigerated stores inside warehouse	m^2	700.00	to	900.00
Industrial buildings				
Shell with heating to office areas only				
500–1,000 m^2	m^2	570.00	to	730.00
1,000–2,000 m^2	m^2	485.00	to	620.00
greater than 2,000 m^2	m^2	530.00	to	680.00
Unit including services to production area				
500–1,000 m^2	m^2	790.00	to	1025.00
1,000–2,000 m^2	m^2	720.00	to	920.00
greater than 2,000 m^2	m^2	720.00	to	920.00
UNICLASS D3 ADMINISTRATIVE, COMMERCIAL, PROTECTIVE SERVICES FACILITIES				
Embassies	m^2	2350.00	to	3000.00
County courts	m^2	2800.00	to	3600.00
Magistrates courts	m^2	1550.00	to	1975.00
Civic offices				
non air-conditioned	m^2	1550.00	to	1975.00
fully air-conditioned	m^2	1850.00	to	2350.00
Probation/registrar offices	m^2	1175.00	to	1475.00
Offices for owner occupation				
low rise, medium quality; air-conditioned	m^2	1650.00	to	2125.00
medium rise, medium quality; air-conditioned	m^2	2075.00	to	2700.00
Offices – City and West End				
To developers Cat A finish only				
high quality, speculative 8–20 storeys, air-conditioned	m^2	2350.00	to	3000.00
high rise, air-conditioned, iconic speculative towers	m^2	3550.00	to	4550.00

BUILDING PRICES PER SQUARE METRE

Item	Unit	Range £		
Business park offices				
Generally up to 3 storeys only				
functional non air-conditioned less than 2,000 m^2	m^2	1350.00	to	1725.00
functional non air-conditioned more than 2,000 m^2	m^2	1425.00	to	1800.00
medium quality non air-conditioned less than 2,000 m^2	m^2	1475.00	to	1875.00
medium quality non air-conditioned more than 2,000 m^2	m^2	1575.00	to	2000.00
medium quality air-conditioned less than 2,000 m^2	m^2	1725.00	to	2200.00
medium quality air-conditioned more than 2,000 m^2	m^2	1850.00	to	2350.00
good quality – naturally ventilated to meet BCO specification (exposed soffits, solar shading) less than 2,000 m^2	m^2	1650.00	to	2125.00
good quality – naturally ventilated to meet BCO specification (exposed soffits, solar shading) more than 2,000 m^2	m^2	1725.00	to	2200.00
high quality air-conditioned less than 2,000 m^2	m^2	2075.00	to	2700.00
high quality air-conditioned more than 2,000 m^2	m^2	2150.00	to	2750.00
Large trading floors in medium sized offices	m^2	3400.00	to	4300.00
Two storey ancillary office accommodation to warehouses/factories	m^2	1350.00	to	1725.00
Fitting out offices (NIA – Net Internal Area)				
City and West End				
basic fitting out including carpets, decorations, partitions and services	m^2	620.00	to	790.00
good quality fitting out including carpets, decorations, partitions and services	m^2	680.00	to	860.00
high quality fitting out including carpets, decorations, partitions, ceilings, furniture, air conditioning and electrical services	m^2	740.00	to	940.00
reception areas	m^2	3100.00	to	3900.00
conference suites	m^2	2950.00	to	3800.00
meeting areas	m^2	3700.00	to	4700.00
back of house/storage	m^2	680.00	to	860.00
sub-equipment room	m^2	1975.00	to	2500.00
kitchen	m^2	4300.00	to	5500.00
Out-of-town (South East England)				
basic fitting out including carpets, decorations, partitions and services	m^2	490.00	to	630.00
good quality fitting out including carpets, decorations, partitions and services	m^2	550.00	to	710.00
high quality fitting out including carpets, decorations, partitions, ceilings, furniture, air conditioning and electrical services	m^2	680.00	to	860.00
reception areas	m^2	1850.00	to	2350.00
conference suites	m^2	2500.00	to	3200.00
meeting areas	m^2	2450.00	to	3150.00
back of house/storage	m^2	550.00	to	710.00
sub-equipment room	m^2	1975.00	to	2500.00
kitchen	m^2	2800.00	to	3600.00
Office refurbishment (including developers finish – (Gross Internal Floor Area – GIFA; Central London)				
minor refurbishment	m^2	580.00	to	740.00
medium refurbishment	m^2	1225.00	to	1575.00
major refurbishment	m^2	1900.00	to	2425.00
Banks and building societies				
banks; local	m^2	1775.00	to	2275.00
banks; city centre/head office	m^2	2500.00	to	3200.00
building society branches	m^2	1725.00	to	2200.00
building society; refurbishment	m^2	1100.00	to	1425.00

BUILDING PRICES PER SQUARE METRE

Item	Unit	Range £		
UNICLASS D3 ADMINISTRATIVE, COMMERCIAL, PROTECTIVE SERVICES FACILITIES – cont				
Shop shells				
small	m²	740.00	to	950.00
large, including department stores and supermarkets	m²	680.00	to	870.00
Fitting out shell for small shop (including shop fittings)				
simple store	m²	710.00	to	900.00
designer/fashion store	m²	1425.00	to	1800.00
Fitting out shell for department store or supermarket	m²	2025.00	to	2600.00
Retail warehouses				
shell	m²	520.00	to	660.00
fitting out, including all display and refrigeration units, check outs and IT systems	m²	310.00	to	390.00
Hypermarkets/Supermarkets				
shell	m²	620.00	to	790.00
supermarket fit-out	m²	1225.00	to	1575.00
hypermarket fit-out	m²	870.00	to	1100.00
Shopping centres				
Malls, including fit-out				
comfort cooled	m²	3900.00	to	5000.00
air-conditioned	m²	5000.00	to	6400.00
food court	m²	4350.00	to	5600.00
factory outlet centre – enclosed	m²	3550.00	to	4550.00
factory outlet centre – open	m²	710.00	to	910.00
anchor tenants; capped off services	m²	1175.00	to	1475.00
medium/small units; capped off services	m²	1100.00	to	1425.00
centre management	m²	2500.00	to	3200.00
enclosed surface level service yard	m²	1775.00	to	2275.00
landlords back of house and service corridors	m²	1775.00	to	2275.00
Refurbishment				
mall; limited scope	m²	1275.00	to	1650.00
mall; comprehensive	m²	1850.00	to	2350.00
Rescue/aid facilities				
ambulance stations	m²	1100.00	to	1425.00
ambulance control centres	m²	1725.00	to	2200.00
fire stations	m²	1850.00	to	2350.00
Police stations	m²	1775.00	to	2275.00
Prisons	m²	2225.00	to	2800.00
UNICLASS D4 MEDICAL, HEALTH AND WELFARE FACILITIES				
District hospitals	m²	2050.00	to	2600.00
refurbishment	m²	1475.00	to	1875.00
Hospices	m²	1650.00	to	2125.00
Private hospitals	m²	2050.00	to	2600.00
Pharmacies	m²	2225.00	to	2800.00
Hospital laboratories	m²	2450.00	to	3150.00
Ward blocks	m²	1850.00	to	2350.00
refurbishment	m²	1100.00	to	1425.00
Geriatric units	m²	1850.00	to	2350.00
Psychiatric units	m²	1725.00	to	2200.00

BUILDING PRICES PER SQUARE METRE

Item	Unit	Range £		
Psycho-geriatric units	m²	1650.00	to	2125.00
Maternity units	m²	1850.00	to	2350.00
Operating theatres	m²	2450.00	to	3150.00
Outpatients/casualty units	m²	1850.00	to	2350.00
Hospital teaching centres	m²	1575.00	to	2050.00
Health centres	m²	1475.00	to	1875.00
Welfare centres	m²	1575.00	to	2050.00
Day centres	m²	1475.00	to	1875.00
Group practice surgeries	m²	1650.00	to	2125.00
Homes for the mentally handicapped	m²	1350.00	to	1725.00
Homes for the physically handicapped	m²	1550.00	to	1975.00
Geriatric day hospital	m²	1725.00	to	2200.00
Accommodation for the elderly				
residential homes	m²	1100.00	to	1425.00
nursing homes	m²	1475.00	to	1875.00
Homes for the aged	m²	1725.00	to	2200.00
refurbishment	m²	920.00	to	1175.00
Observation and assessment units	m²	1225.00	to	1575.00
Primary health care				
doctors surgery – basic	m²	1275.00	to	1650.00
doctors surgery/medical centre	m²	1550.00	to	1975.00
Hospitals				
diagnostic and treatment centres	m²	3100.00	to	4000.00
acute services hospitals	m²	3000.00	to	3850.00
radiotherapy and ocology units	m²	3100.00	to	4000.00
community hospitals	m²	2500.00	to	3200.00
trauma unit	m²	2350.00	to	3000.00
UNICLASS D5 RECREATIONAL FACILITIES				
Public houses	m²	1850.00	to	2350.00
Dining blocks and canteens	m²	1475.00	to	1875.00
Restaurants	m²	1650.00	to	2125.00
Community centres	m²	1575.00	to	2050.00
General purpose halls	m²	1350.00	to	1725.00
Visitors' centres	m²	2225.00	to	2800.00
Youth clubs	m²	1850.00	to	2350.00
Arts and drama centres	m²	1975.00	to	2500.00
Theatres, including seating and stage equipment				
large – over 500 seats	m²	4200.00	to	5300.00
studio/workshop – less than 500 seats	m²	2750.00	to	3500.00
refurbishment	m²	2225.00	to	2800.00
Concert halls, including seats and stage equipment	m²	3100.00	to	4000.00
Cinema				
shell	m²	980.00	to	1250.00
multiplex; shell only	m²	1975.00	to	2500.00
fitting out, including all equipment, air-conditioned	m²	1225.00	to	1575.00
Exhibition centres	m²	1850.00	to	2350.00

BUILDING PRICES PER SQUARE METRE

Item	Unit	Range £		
UNICLASS D5 RECREATIONAL FACILITIES – cont				
Swimming pools				
international standard	m²	4200.00	to	5300.00
local authority standard	m²	2750.00	to	3500.00
school standard	m²	1275.00	to	1650.00
leisure pools, including wave making equipment	m²	3250.00	to	4150.00
Ice rinks	m²	1475.00	to	1875.00
Rifle ranges	m²	1225.00	to	1575.00
Leisure centres				
dry	m²	1725.00	to	2200.00
extension to hotels; shell and fit-out, including small pool	m²	2450.00	to	3150.00
wet and dry	m²	2600.00	to	3300.00
Sports halls including changing rooms	m²	1175.00	to	1475.00
School gymnasiums	m²	1100.00	to	1425.00
Squash courts	m²	1175.00	to	1475.00
Indoor bowls halls	m²	820.00	to	1050.00
Bowls pavilions	m²	1025.00	to	1275.00
Health and fitness clubs	m²	1725.00	to	2200.00
Sports pavilions	m²	1225.00	to	1575.00
changing only	m²	1425.00	to	1800.00
social and changing	m²	1425.00	to	1800.00
Clubhouses	m²	1175.00	to	1475.00
UNICLASS D6 RELIGIOUS FACILITIES				
Temples, mosques, synagogues	m²	1575.00	to	2050.00
Churches	m²	1475.00	to	1875.00
Mission halls, meeting houses	m²	1575.00	to	2050.00
Convents	m²	1725.00	to	2200.00
Crematoria	m²	1975.00	to	2500.00
Mortuaries	m²	2700.00	to	3450.00
UNICLASS D7 EDUCATION, SCIENTIFIC AND INFORMATION FACILITIES				
Nursery schools	m²	1475.00	to	1875.00
Primary/junior schools	m²	1550.00	to	1975.00
Secondary/middle schools	m²	1850.00	to	2350.00
Secondary schools and further education colleges				
classrooms	m²	1225.00	to	1575.00
laboratories	m²	1350.00	to	1725.00
craft design and technology	m²	1350.00	to	1725.00
music	m²	1550.00	to	1975.00
Extensions to schools				
classrooms	m²	1475.00	to	1875.00
laboratories	m²	1550.00	to	1975.00
Sixth form colleges	m²	1475.00	to	1875.00
Special schools	m²	1225.00	to	1575.00
Training colleges	m²	1225.00	to	1575.00
Management training centres	m²	1575.00	to	2050.00

BUILDING PRICES PER SQUARE METRE

Item	Unit	Range £		
Universities				
arts buildings	m^2	1350.00	to	1725.00
science buildings	m^2	1650.00	to	2125.00
College/University libraries	m^2	1350.00	to	1725.00
Laboratories and offices, low level servicing	m^2	2600.00	to	3300.00
Computer buildings	m^2	2225.00	to	2800.00
Museums and art galleries				
national standard museum	m^2	6000.00	to	7600.00
national standard independent specialist museum, excluding fit-out	m^2	4200.00	to	5300.00
regional, including full air conditioning	m^2	3550.00	to	4550.00
local, including full air conditioning	m^2	2650.00	to	3400.00
conversion of existing warehouse to regional standard museum	m^2	1650.00	to	2125.00
conversion of existing warehouse to local standard museum	m^2	1425.00	to	1800.00
Galleries				
international standard art gallery	m^2	3600.00	to	4600.00
national standard art gallery	m^2	2750.00	to	3500.00
independent commercial art gallery	m^2	1550.00	to	1975.00
arts and drama centre	m^2	1475.00	to	1875.00
Learning resource centre				
economical	m^2	1350.00	to	1725.00
high quality	m^2	1725.00	to	2200.00
Libraries				
regional; 5,000 m^2	m^2	3100.00	to	3900.00
national; 15,000 m^2	m^2	4000.00	to	5100.00
international; 20,000 m^2	m^2	5200.00	to	6700.00
Conference centres	m^2	2150.00	to	2750.00
UNICLASS D8 RESIDENTIAL FACILITIES				
Local Authority and Housing Association schemes				
Housing Asociation Developments (Code for Sustainable Homes Level 3)				
Bungalows				
semi-detached	m^2	1175.00	to	1525.00
terraced	m^2	1125.00	to	1425.00
Two storey housing				
detached	m^2	1175.00	to	1525.00
semi-detached	m^2	1125.00	to	1425.00
terraced	m^2	990.00	to	1275.00
Three storey housing				
semi-detached	m^2	1175.00	to	1525.00
terraced	m^2	990.00	to	1275.00
Apartments/flats				
low rise	m^2	1175.00	to	1525.00
medium rise	m^2	1450.00	to	1850.00
Sheltered housing with wardens accommodation	m^2	1125.00	to	1425.00
Private Developments				
single detached houses	m^2	1575.00	to	2025.00
two and three storey houses	m^2	1650.00	to	2100.00

BUILDING PRICES PER SQUARE METRE

Item	Unit	Range £		
UNICLASS D8 RESIDENTIAL FACILITIES – cont				
Private Developments – cont				
Apartments/flats generally				
standard quality; 3–5 storeys	m^2	1975.00	to	2500.00
warehouse/office conversion to apartments	m^2	2175.00	to	2800.00
high quality apartments in residential tower – Inner London	m^2	2650.00	to	3400.00
Hotels (including fittings, furniture and equipment)				
luxury city-centre with conference and wet leisure facilities	m^2	3500.00	to	4500.00
business town centre with conference and wet leisure facilities	m^2	2600.00	to	3300.00
mid-range with conference and leisure facilities	m^2	1975.00	to	2550.00
budget city-centre with dining and bar facilities	m^2	1750.00	to	2250.00
budget roadside excluding dining facilities	m^2	1525.00	to	1950.00
Hotel accommodation facilities (excluding fittings, furniture and equipment)				
bedroom areas	m^2	1050.00	to	1350.00
front of house and reception	m^2	1350.00	to	1725.00
restaurant areas	m^2	1575.00	to	2025.00
bar areas	m^2	1400.00	to	1775.00
function rooms/conference facilities	m^2	1350.00	to	1725.00
Students residences				
large budget schemes with en-suite accommodation	m^2	1350.00	to	1725.00
smaller schemes (40–100 units) with mid range specifications	m^2	1650.00	to	2075.00
smaller high quality courtyard schemes, college style	m^2	2225.00	to	2850.00

Building Cost Models

The authors have been publishing cost models in *Building* magazine for many years. Although the scope and coverage of the cost models has expanded considerably, the objectives have always remained constant. They are:

- To provide detailed elemental cost information derived from a generic building that can be applied to other projects
- To provide a commentary on cost drivers and other design and specification issues
- To compare suitable procurement routes that secure the clients objectives

Models are based on actual projects and are not compiled from current book rates. Users should remember that rates are always intended purely as a guide and are not a substitute for prices received for actual projects, with specific location, specifications, conditions etc. which can all affect the cost.

For this edition we have published updated elemental cost data for several building types. Locations do vary for each model, so please note the location and location factor for any adjustments that may need to be made. The TPI for each individual model reflects the location of that model, which may not be the same as the book TPI which has a location factor of 1.00. This is important if users are adjusting model costs to a different location or time.

A summary of UK location factors can be found in *Building Costs Indices, Tender Price Indices and Location Factors* section of this book, together with examples showing how to apply those factors.

Readers should note that these models are based around actual projects and are intended as an indication to build costs only. There are many factors which need to be considered when putting together an early cost plan or estimate and all need to be considered when arriving at a Cost Plan for any particular project. Rates and totals are rounded.

Readers may refer to the *Approximate Estimating Rates* section of this book to make adjustments to the models for alternative specifications within any of the elements.

As elsewhere in this edition prices do not include Value Added Tax or professional services fees.

Building Cost Models

RETAIL DISTRIBUTION UNIT

This cost model is based on a single-storey, rectangular-shaped warehouse new building (12 m height to haunch) with a two-storey office, with gross internal floor areas of 29,640 m² and 1,680 m² respectively. The development is on a brownfield site, which has previously been cleared. Any land remediation has been undertaken as part of a separate contract. The building has a reinforced in situ concrete ground-bearing slab, a steel portal frame with reinforced concrete pads, and an aluminium built-up cladding system to roof and walls. External works consist of access and circulation road, car parking, service yard, pedestrian areas, landscaping areas and gatehouse.

Warehouse: Gross internal floor area	29,640 m²
Office Shell: Gross internal floor area	1,680 m²
Model location is based on West Midland location	Model TPI = 108; LF = 0.87

This cost model is copyright of AECOM Ltd

Retail Distribution Unit	Quantity	Unit	Rate (£)	Total (£)	Cost (£/m²)
Substructure				**1,939,400**	**65.43**
Pad foundations; excavations; concrete; reinforcement; holding down bolts; averaged size 2.50 × 2.50 × 0.75 m	45	nr	1,350		
Reinforced column casings below slab finished level; including formwork	45	nr	100		
Reinforced precast concrete perimeter upstand/ground beam below cladding; 425 × 300 mm wide	608	m	130		
Reinforced concrete perimeter retaining wall 1400 mm high	120	m	390		
Foundations to dock levellers: 2975 mm × 325 mm deep reinforced concrete	24	m	1,025		
Strip foundations to retaining walls; concrete; formwork; reinforcement	120	m	280		
Backfilling retaining walls	203	m³	30		
Imported fill to make up levels – 250 mm thick	7,410	m³	26		
Ground slab 200 mm thick minimum. Slab to have approved sealant/surface hardener, FM2 special finish	29,640	m²	44		
150 mm type 1 under slab	4,446	m²	39		
Design calculations; laser level; floor survey; sacrificial sealant to joints (maybe included with slab quote)		item	5,500		
Inspections by a specialist flooring consultant		item	8,300		
Frame				**1,514,700**	**51.10**
Structural steel portal frame 12 m; hit and miss valley columns, painted, including cold rolled wall and roof sections; service loading 0.25 kn/m²	29,640	m²	48		
Site treatment/decoration	29,640	m²	1		
Roof				**1,932,500**	**65.20**
LPCB approved trapezoidal roof panels, standard colour	29,640	m²			
GRP triple skin roof lights to 15% of the ambient	4,446	m²			
Valley gutters; membrane lined system, including stop end and the like	741	m²			

RETAIL DISTRIBUTION UNIT

Retail Distribution Unit	Quantity	Unit	Rate (£)	Total (£)	Cost (£/m²)
Eaves gutter abutment; steel outer skin, insulation board, membrane lined system, including stop ends and the like	494	m			
Ridge/hip flashing	988	m			
Access hatches, including flashings	2	nr			
Access ladder to roof from warehouse floor	2	nr			
Allow for safe travel across roof by means of safety ropes; Top-fix fall arrest cable system, includes user equipment	734	m			
Allow access equipment (hoisting, distribution and power)		item			
Penetrations and air testing		item			
Syphonic drainage HDPE pipework; including primary and secondary system and external concrete splash pads	1,235	m			
Hydrabrake at first manhole		item			
External Walls, Windows and Doors				**875,600**	**29.54**
Micro-rib composite panels LPC Grade B approved; 1000 mm cover width panels, standard colour; including vertical top hat, joint flashing and the like; laid horizontally	3,910	m²	90		
Extra over; 1-hour fire-rated	3,910	m²	4		
Trapezoidal composite panels LPC Grade B approved; standard colour, including end laps, flashing and the like; laid vertically	4,928	m²	49		
Profiled cladding to rear of parapet; laid vertically	704	m²	55		
Sill closure cill flashing	704	m	33		
Parapet/coping flashing detail 750 mm girth	704	m	48		
Single personnel fire escape doors with internal panic bars	16	nr	950		
Louvres in external wall cladding to plant rooms	200	m²	330		
Allowance for access equipment	1	nr	82,500		
Inspections by a specialist cladding consultant		item	5,500		
Dock Leveller Installations				**271,500**	**9.16**
Insulated sectional overhead type dock door (2.75 m × 3 m) finished in a standard colour range; including two vision panels 1500 mm above FFL	24	nr	1,550		
Dock shelters heavy duty scissor retracting frame type with 650 mm wide curtain and 1200 mm deep top curtain in high frequency welded PVC	24	nr	1,250		
Electrohydraulic dock levellers with 480, lip and platform size 2500 mm long × 2000 mm wide; dynamic capacity 6000 kg gross	24	nr	2,600		

RETAIL DISTRIBUTION UNIT

Retail Distribution Unit	Quantity	Unit	Rate (£)	Total (£)	Cost (£/m²)
2 Nr internal bollards 138 mm × 11000 mm high to protect door track	24	nr	170		
Heavy duty rubber dock buffers; 175 mm thick × 250 mm wide 4450 mm deep	24	nr	360		
Internal dock loading lights	24	nr	240		
External red/green traffic lights	24	nr	160		
Precast concrete pits to levellers	24	nr	3,050		
External escape stairs – (allowance only)	14	nr	2,750		
Bollards to external stairs 225 mmx 1500 mm	28	nr	280		
Level Access Doors				**4,500**	**0.15**
Insulated sectional overhead type level access doors finished in a colour from a standard range with emergency hand chain; including two vision panels; Size 4000 mm × 5000 mm high	1	nr	3,850		
2 Nr internal bollards 225 × 1100 mm to protect door track	1	pair	280		
2 Nr external bollards 225 × 1500 mm to protect door jambs	1	pair	330		
Preliminaries and Contingency				**1,307,600**	**44.12**
Overheads and profit, site establishment and supervision @		15%			
Contingency @		5%			
Construction cost (Warehouse shell only, rate based on GIFA)				**7,845,800**	**264.70**

Office Shell and Fit-Out	Quantity	Unit	Rate (£)	Total (£)	Cost (£/m²)
Substructure				**104,500**	**62.20**
Imported fill (type 1) making up levels; 250 mm thick	210	m³	33		
Pad foundations; excavations; concrete; reinforcement; holding down bolts; average size 1.75 × 1.75 × 0.75 m	24	nr	825		
Reinforced column casings below slab finished level; including formwork	24	m	105		
Reinforced precast concrete perimeter upstand/ground beam below cladding; 425 × 300 mm wide	86	m	85		
Lift Pit	2	n	3,300		
Ground floor slab reinforced concrete 150 mm thick; 250 mm type 1 granular fill material	840	m²	33		
Reinforced concrete perimeter retaining wall 1500 mm high	86	m	390		
Frame, Upper Floors and Stairs				**421,300**	**250.77**
Fabricated steel; painted, including cold rolled sections	1,680	m²	80		
Site treatment/decoration	1,680	m²	1		
Board fire protection to compartment wall		item	28,000		
1 hr intumescent fire protection to columns and beams		item	39,000		
Office bridges structure supports		item	33,000		

RETAIL DISTRIBUTION UNIT

Office Shell and Fit-Out	Quantity	Unit	Rate (£)	Total (£)	Cost (£/m²)
Composite concrete floor on structural metal decking	830	m²	55		
Accommodation staircase to serve upper office levels; including finishes, balustrades and hand railing	2	nr	33,000		
Accommodation staircase; ground to plant deck including finishes and handrails	2	nr	16,500		
Precast concrete/metal stairs, ground to second floor plant deck; including finishes, balustrading and hand railing	1	nr	22,000		
Fire stopping to perimeter of offices	116	m	28		
Handrail to plant deck	116	m	130		
Roof				**56,500**	**33.63**
Either Kingspan composite KS1000RW LPC Grade B approved trapezoidal roof panels or decked/insulated single ply membrane system	840	m²	33		
Eaves gutter abutment; 3 mm thick post-galvanized steel outer, 45 mm thick insulation board, 0.7 mm inner skin; including stop ends	116	m	140		
Access hatches, 900 x 900 mm, including	2	nr	2,200		
Allow for safe travel across roof by means of safety ropes; Latchways Mansafe	116	m	51		
Allow access equipment (hoisting, distribution and power)		item	2,200		
External Walls, Windows and Doors				**250,100**	**148.87**
Micro-rib composite panels LPC grade B approved, laid horizontally	1,022	m²	65		
Sill closure drip flashing	116	m	28		
Profiled cladding to rear of parapet; vertically laid (height 1.00 m)	116	m²	33		
Parapet/coping flashing detail 750 mm girth	116	m	65		
Powder coated, thermally broken curtain walling by Schuco/Kawneer	100	m²	390		
Automatic entrance doors to match curtain walling (Only required if no lobby)	1	nr	2,750		
Powder coated, thermally broken double-glazed windows; glazing to be anti-sun outer pane with low-e coating	270	m²	390		
Glass entrance canopy supported on powder coated ground bearing structure; minimum size to be 15 m²	1	nr	22,000		
Internal Partitions and Doors				**537,600**	**320.00**
Office fit-out ground & upper floors (including wall, floor, ceiling finishes, doors etc.)	1,680	m²	320		

RETAIL DISTRIBUTION UNIT

Office Shell and Fit-Out	Quantity	Unit	Rate (£)	Total (£)	Cost (£/m²)
Mechanical Installations				340,800	202.86
Heating/Cooling	1,680	m²	130		
Ventilation	1,680	m²	55		
Rainwater harvesting tank; 20,000 litres		item	16,500		
Solar Thermal		item	13,500		
Electrical Installations				404,900	241.01
Power to doors	18	nr	825		
LV installations	1,680	m²	22		
Submain cabling	1,680	m²	9		
Containment installations	1,680	m²	11		
Lighting installations	1,680	m²	55		
Small power installations	1,680	m²	28		
Earthing and bonding installations	1,680	m²	1		
Gatehouse		item	16,500		
Testing and commissioning		3%			
Subcontractor preliminaries		12%			
Lift installation; 8 persons electrohydraulic	2	nr	43,000		
Allowance for builder's work in connection with services		5%			
Preliminaries and Contingency				419,200	249.52
Overheads and profit, site establishment and supervision @		15%			
Contingency @		5%			
Construction cost (Office shell and fit-out only, rate based on Office GIA)				2,534,900	1,508.86

External Works	Quantity	Unit	Rate (£)	Total (£)	Cost (£/m²)
Site Works				2,712,400	86.60
Site clearance including removal of trees, shrubs etc.	80,098	m²	0.40		
Disposal of excavated material off site	3,231	m³	20		
Excavate to reduce levels and stockpile on site	16,020	m²	3		
Filling to make to levels with excavated material	8,010	m³	3		
Level and compact surfaces of excavation or filling	80,098	m²	0.50		
Geotextile membrane	80,098	m²	0.60		
Disposal of excavated material to Landscaped areas	4,779	m²	3		
SW drainage to building	30,480	m²	9		
SW drainage to HGV yard and road	16,562	m²	9		
SW drainage to car parking	8,135	m²	9		
Paving	8,163	m²	9		
FW drainage to foul drain runs	387	m	170		
FW drainage to manholes	10	nr	1,350		

RETAIL DISTRIBUTION UNIT

External Works	Quantity	Unit	Rate (£)	Total (£)	Cost (£/m²)
Petrol interceptors	3	nr	6,100		
Connection to main sewer	1	nr	1,650		
Heavy duty access road and service yard; 175 mm brushed concrete on and including 150 type 1 subbase with 2800 g polythene membrane; sealed construction joints	16,562	m²	39		
Trief kerbs on and including concrete foundations/ haunching	182	m	105		
Standard kerbs	2,157	m	17		
Road marking/linings, signage		item	11,000		
Car parking/light duty road; comprising 30 mm dense bitumen wearing course; 80 mm dense bitumen base course; 150 mm type 1 subbase	8,135	m²	25		
Standard kerbs	532	m	17		
Road marking/linings, signage		item	5,500		
Car parking/light duty road; comprising 30 mm dense bitumen wearing course; 80 mm dense bitumen base course; 150 mm type 1 subbase	8,135	m²	25		
Standard kerbs	532	m	17		
Road marking/linings, signage		item	5,500		
Public footpaths to be 600 mm × 600 mm × 50 mm precast concrete paving slabs to provide a minimum 1200 mm wide path; on and including 50 mm sand bed on 150 mm type 1 granular material	8,163	m²	20		
Perimeter fencing; 2.4 m high Paladin fence with concrete posts	1,142	m	80		
Entrance barriers	1	nr	22,000		
Entrance gates	1	nr	28,000		
Grassed/planted areas	13,655	m²	11		
Tree planting allowance	51	nr	550		
12 months maintenance to begin from completion of planting		item	11,000		
Perimeter fencing; 2.4 m high Paladin fence with concrete posts	1,142	m	80		
Entrance barriers	1	nr	22,000		
Entrance gates	1	nr	28,000		
External Services				**371,700**	**11.87**
External lighting		item	110,000		
Gas; including housing		item	44,000		
Electricity; including hosing		item	55,000		
Water; including hosing		item	39,000		
Telecoms		item	2,200		

RETAIL DISTRIBUTION UNIT

External Works	Quantity	Unit	Rate (£)	Total (£)	Cost (£/m²)
External hydrant main and hydrants		item	82,500		
Service trenches to electricity, gas, water, hydrant mains distribution		item	110,000		
Ancillary Buildings				89,900	2.87
Prefabricated gatehouse 20 m² (215 sq ft)		item	77,500		
Allowance for foundations, base, drainage and services required		item	4,700		
Vehicle barriers and BWIC (1 nr barrier/lane)	2	nr	3,850		
Preliminaries and Contingency				634,800	20.27
Overheads; profit, site establishment and site supervision @		15%			
Contingency @		5%			
Construction cost (External works only, rate based on GIFA)				3,808,800	121.61
TOTAL DISTRIBUTION CENTRE CONSTRUCTION COST (rate based on GIFA)				14,189,500	453.05

SMALL INDUSTRIAL UNIT

A single storey new building with a gross internal floor area of 900 m², subdivided into five industrial units. Reinforced concrete ground bearing slab and pads to receive a steel portal frame. Wall and roof cladding is aluminium built up system, with internal blockwork division walls. Each of the five units has a separate entrance door and one roller shutter door, together with a single WC. Units vary in size from 150 m² to 360 m².

Gross internal floor area	900 m²
Model location is South East England	Model TPI = 120; LF = 0.96

This cost model is copyright of AECOM Ltd

Small Industrial Unit	Quantity	Unit	Rate (£)	Total (£)	Cost (£/m²)
Substructure				**86,000**	**95.56**
Excavation and disposal off site	190	m³	30		
Reinforced concrete ground slab, including ground beams and column bases	900	m²	65		
Power floated and hardener	900	m²	9		
Strip foundations for party walls	80	m	170		
Frame and Upper Floors				**71,000**	**78.89**
Steel propped portal frame, cold rolled purlins, surface treatments (@ 40 kg/m²)	36	tonne	1,600		
Intumescent paint fire protection to steelwork	36	tonne	320		
Allowance for miscellaneous works, protecting columns		item	1,800		
Roof				**67,700**	**75.22**
Built up aluminium roof cladding with 180 mm thick insulation, including all labours	950	m²	46		
Extra over for Rooflights (10% of total roof area)	95	m²	80		
Mansafe system	80	m	100		
Rainwater drainage, aluminium gutters and downpipes	120	m	70		
External Wall, Windows and Doors				**149,400**	**166.00**
Built up aluminium wall cladding with 130 mm thick insulation	520	m²	50		
2.5 m high inner leaf of 140 mm thick fairface blockwork	380	m²	41		
3000 mm × 4600 mm high steel sectional overhead doors	5	nr	4,100		
Aluminium single entrance doors	5	nr	1,450		
Coated aluminium double-glazed window system	150	m²	490		
Polycarbonate canopy entrance – 1500 mm × 1000 mm	5	nr	1,300		
Internal Walls and Partitions and Doors				**47,800**	**53.11**
2-hour fire-resistant blockwork party walls	450	m²	85		
Fireproofing between blockwork and roof		item	1,800		
Metal stud partitions	50	m²	65		
Laminated faced internal doorset with softwood frames and ironmongery	5	nr	900		

SMALL INDUSTRIAL UNIT

Small Industrial Unit	Quantity	Unit	Rate (£)	Total (£)	Cost (£/m²)
Wall Finishes				5,700	6.33
Emulsion paint to blockwork wall surfaces generally	1,370	m²	4		
Ceramic wall tiles splashbacks to WC area	5	m²	59		
Floor Finishes				900	1.00
Screed and non-slip vinyl sheeting to WC areas	15	m²	59		
Ceiling Finishes				800	0.89
Moisture-resistant plasterboard to WC with ceiling grid and paint finish	15	m²	47		
Sanitary Appliances				7,000	7.78
Disabled WC suite including all sanitary and fittings	5	nr	1,400		
Disposal Installations				2,700	3.00
Waste, soil and vent installation; uPVC pipework and fittings	900	m²	3		
Hot and Cold Water Installations				4,200	4.67
Hot and cold-water supplies to WCs	5	nr	825		
Electrical Installations				33,600	37.33
Small power, basic and emergency lighting	900	m²	26		
Supply to WC for ventilation, heater etc.	5	nr	1,900		
External lighting generally item		item	680		
Incoming Services				13,500	15.00
Allowance for incoming, electrical, gas and water services		item	13,500		
Protective Installations				1,200	1.33
Lightning protection	900	m²	1		
Communication Installations				11,700	13.00
Fire and intruder alarm	900	m²	13		
Builders Work in Connection				800	0.89
Forming holes and chases etc.	1	nr	800		
Preliminaries and Contingency				104,600	116.22
Overheads; profit, site establishment and site supervision @		15%			
Contingency @		5%			
Construction cost (shell only; rate based on GIFA)				**608,600**	**676.22**

INNER LONDON OFFICES

This cost model is based on a typical central London office scheme arranged over 10 floors above ground and one basement floor. It has a gross internal floor area of 25,000 m² with a net-to-gross efficiency of 72% and a wall-to-floor ratio of 0.45. It Achieves a BREEAM rating of 'Excellent'. It assumes a steel frame building with four-pipe fan coil unit system and a metal tile suspended ceiling. Included are all main contractor preliminaries, fees and risk allowance to complete the building to Category A standard. Demolitions, site preparation, external works and services, fit-out beyond Category A standard, tenant enhancements, professional fees and VAT are all excluded.

Gross internal floor area	25,000 m²
Net internal floor area	18,000 m²
Model location is Inner London	Model TPI = 131; LF = 1.06

This cost model is copyright of AECOM Ltd

OFFICE: Shell and Core Works	Quantity	Unit	Rate (£)	Total (£)	Cost (£/m²)
Substructure				5,148,300	205.93
Dewatering		item	295,000		
Break out existing slabs, piles, obstructions and allowance for probing/testing		item	590,000		
Foundations; bored piles with under-ream; ground beams; pile caps	2,150	m²	480		
Piling platform; mini piles and other works to boundary walls		item	355,000		
RC basement slab 300 mm thick, including waterproofing, excavation and disposal	2,150	m²	300		
RC mat slab 1200 mm thick, including waterproofing, excavation and disposal	250	m²	725		
Reinforced concrete retaining walls 300 mm thick	1,100	m²	380		
Reinforced concrete ground floor slab 130 mm thick on profiled metal sheet decking	2,150	m²	180		
Allowance for car park ramp, slab thickenings to stair foundations, lift/escalator pits, drainage channels, concrete transfer walls etc.		item	415,000		
Allowance for crane base including base piles		item	60,000		
Attendance on archaeologists and movement monitoring		item	120,000		
Below slab drainage; other items and sundries		item	650,000		
Frame and Upper Floors				9,804,000	392.16
Structural steel frame including fittings based on 80 kg/m² overall GIFA including fittings	1,828	tonne	2,400		
Secondary steelwork, based on an extra 5 kg/m²	114	tonne	2,950		
Concrete encased beams at ground floor		item	90,000		
Fire protection to steel frame (90 min intumescent paint)	1,942	tonne	650		
Reinforced concrete core walls average	3,300	m²	360		
Allowance for other structures (e.g. within plant rooms etc.)		item	180,000		

Building Cost Models

INNER LONDON OFFICES

OFFICE: Shell and Core Works	Quantity	Unit	Rate (£)	Total (£)	Cost (£/m²)
Allowance for expansion joints and other sundries		item	60,000		
Lightweight reinforced concrete 130 mm thick on profiled steel decking	17,430	m²	115		
Upstands plinths; walkways etc.		item	295,000		
Roof				**1,303,300**	**52.13**
Profiled steel decking with 200 mm lightweight concrete with mesh reinforcement; insulation and acoustics to soffit	2,200	m²	170		
Proprietary roof; paving slabs; insulation and ballast	2,200	m²	240		
Green roof	440	m²	240		
Insulation to exposed soffits and acoustic treatment		item	72,500		
Allowance for upstands to perimeter/terraces	222	m	600		
Upstands/plinths, hatches/ladders, safety hooks and latchways		item	90,000		
Stairs				**259,000**	**10.36**
Steel pan staircases; concrete infills to stair treads; painted mild steel balustrades and handrails (basement to roof)		item	43,000		
Steel pan staircases; concrete infills to stair treads; painted mild steel balustrades and handrails (basement to ground floor – two flights)	2	nr	18,000		
Feature entrance stairs		item	120,000		
Allowance for stairs/cat ladders and safety rails to plant rooms		item	60,000		
External Works				**12,467,900**	**498.72**
Unitized curtain walling; solid spandrel panels; selective high performance glass and some solar control	9,633	m²	1,125		
External structural glazed wall to entrance lobby	200	m²	1,450		
Glazed screen to retail shopfronts	70	m²	1,450		
Aluminium screening to plant enclosures	450	m²	725		
Glass entrance canopies; cantilevered from building	250	m²	1,800		
Extra for louvres		item	90,000		
Blockwork walls at roof level, including windposts	100	m²	180		
Allowance for visual mock-ups and performance tests		item	355,000		
External Windows and Doors				**276,000**	**11.04**
Single and double doors, including disabled pass doors		item	30,000		
Extra over cladding for revolving doors	2	nr	90,000		
Steel roller shutter to loading bay and car park	1	nr	30,000		
Metal doors in service areas		item	36,000		

INNER LONDON OFFICES

OFFICE: Shell and Core Works	Quantity	Unit	Rate (£)	Total (£)	Cost (£/m²)
Internal Walls and Internal Partitions				2,517,500	100.70
Fairfaced blockwork walls at basement, ground and roof levels	4,000	m²	180		
Curved blockwork entrance feature wall	300	m²	240		
Drylined core walls	6,950	m²	130		
Extra for double thickness drylined core walls	1,000	m²	130		
Other walls/partitions to plant areas, additional walls		item	215,000		
Veneer faced WC cubicles/doors; access panelling	90	nr	5,300		
Internal Doors				629,000	25.16
Single timber doors	160	nr	2,250		
Double timber doors	30	nr	3,550		
Profilex riser doors	50	nr	1,450		
Other doors: plantrooms; additional access door hatches		item	90,000		
Wall Finishes				1,451,600	58.06
Stone cladding to main entrance lobby	880	m²	480		
Feature wall panelling on steel frame in main entrance lobby	150	m²	1,300		
Paint to fair face block walls	2,150	m²	12		
Plaster and paint to blockwork/concrete	3,820	m²	24		
Skim coat and paint to drylined walls	1,700	m²	14		
Stone cladding to toilets	450	m²	420		
Granite cladding to lift lobbies	800	m²	480		
Lift architraves		item	120,000		
Floor Finishes				1,150,200	46.01
Granite/stone tiles to main entrance lobby and lift lobbies	1,250	m²	480		
Stone tiles to toilets including, waterproofing, screed; skirtings	500	m²	480		
Lightweight screed to circulation and core areas	1,280	m²	42		
Sealant/hardener to car park, loading bay and plant rooms	1,140	m²	110		
Vinyl flooring to security areas		item	12,000		
Entrance mats and matwells		item	53,000		
Allowance for white lining to carpark and loading bay		item	30,000		
Allowance for other floor finishes		item	36,000		
Ceiling Finishes				877,200	35.09
GRG feature ceiling to main entrance lobby	870	m²	480		
Feature drylined ceiling to lift lobbies	380	m²	240		
Metal tile suspended ceilings to toilets	500	m²	120		

INNER LONDON OFFICES

OFFICE: Shell and Core Works	Quantity	Unit	Rate (£)	Total (£)	Cost (£/m²)
Painted plasterboard on metal framing to corridors etc.	840	m²	110		
Insulation to car park/loading bay soffits	1,000	m²	36		
Access panels, bulkheads; detailing; sundry ceiling finishes		item	180,000		
Fittings/Fitting Out (excludes loose furniture)				833,500	33.34
Main entrance reception desk and security desks		item	120,000		
Stone vanity tops in toilets for basins/taps with mirrors behind	80	m	2,400		
Soap dispensers/tanks, roll holders, paper towels etc.	90	nr	650		
Fittings to disabled toilets	10	nr	1,800		
Rubbish compactor		item	30,000		
Column guards, bollards/crash rails to loading bay/ car park, cycle racks, traffic management, statutory signage		item	415,000		
Sanitary Appliances				194,400	7.78
WCs, basins, cleaners sinks, urinals (average rate per point)	300	nr	600		
Disabled toilets	12	nr	1,200		
Disposal Installations				439,000	17.56
Rainwater disposal system	25,000	m²	4		
Soil waste and vent installation	25,000	m²	11		
Drainage to retail areas		item	24,000		
Condensate drainage	25,000	m²	2		
Water Installations				712,500	28.50
Cold water services: incoming, storage, pumps, distribution	25,000	m²	12		
Hot water heaters and distribution	25,000	m²	10		
Miscellaneous water supplies	25,000	m²	5		
Water services for retail\kitchenettes	25,000	m²	2		
Space Heating and Air Treatment				3,838,000	153.52
Gas installation		item	48,000		
Boilers, flues and primary pipework distribution		item	180,000		
Air handling units and ductwork distribution	25,000	m²	39		
Water cooled chillers	25,000	m²	20		
Heat rejection plant	25,000	m²	15		
LTHW heating including pumps	25,000	m²	20		
CHW installation including pumps and riser pipework	25,000	m²	26		
Condenser water installation	25,000	m²	14		
Metering LTHW/CHW installations	4	nr	30,000		
Reception air conditioning	25,000	m²	5		

INNER LONDON OFFICES

OFFICE: Shell and Core Works	Quantity	Unit	Rate (£)	Total (£)	Cost (£/m²)
Ventilation Installations				712,500	28.50
Toilet and smoke extract ventilation	25,000	m²	8		
Ventilation to plant room, lift motor rooms, refuse area, etc.	25,000	m²	4		
Car park and basement ventilation	25,000	m²	7		
Stair and lobby pressurization	25,000	m²	10		
Electrical Installations				2,925,000	117.00
HV switchgear and transformer	25,000	m²	7		
LV distribution and rising busbars	25,000	m²	50		
Power to mechanical plant	25,000	m²	4		
Small power installation	25,000	m²	7		
Lighting, emergency lighting	25,000	m²	24		
Lighting, emergency lighting to car park and basement	25,000	m²	4		
Enhanced lighting in lobby and other areas		item	60,000		
External building lighting		item	180,000		
Life safety only standby generator installation, including flues, acoustics, fuel installation		item	210,000		
Miscellaneous wireways	25,000	m²	1		
Earthing and bonding	25,000	m²	2		
Lifts				2,530,000	101.20
Passenger lifts, 2.5 m/s, 21 person serving 10 floors with enhanced car finishes	6	nr	255,000		
Goods lift serving 10 floors	1	nr	240,000		
Vehicle park lift	1	nr	190,000		
Firefighting lift	2	nr	190,000		
Builders work, control system etc.		item	190,000		
Protective Installations				720,000	28.80
Sprinkler installations; tanks, pumps, risers etc.	25,000	m²	24		
Dry riser installation	25,000	m²	2		
Lightning protection	25,000	m²	2		
Communication Installations				610,500	24.42
Fire alarm installations	25,000	m²	18		
Landlord security provisions	25,000	m²	5		
Disabled alarms, refuge alarms, induction loops		item	48,000		
Special Installations				1,370,000	54.80
Building management system	25,000	m²	30		
Leak detection system	25,000	m²	1		
Allowance for facade cleaning equipment		item	590,000		

Building Cost Models

INNER LONDON OFFICES

OFFICE: Shell and Core Works	Quantity	Unit	Rate (£)	Total (£)	Cost (£/m²)
Builders Work				650,300	26.01
Builder's work in connection with services installations, including machine bases,	25,000	m²	26		
Preliminaries and Contingencies				16,887,000	675.48
Contractor's preliminaries and overheads and profit		23.0%			
Risk transfer and Contingency @		8.0%			
Construction cost (shell and core work only)				68,306,700	2,732.27

Office: Category A Works	Quantity	Unit	Rate (£)	Total (£)	Cost (£/m²)
Wall Finishes				165,000	6.60
Emulsion paint finish to office side of core walls	2,150	m²	6		
Column casings, including paint, subframe, etc.	1,450	m²	105		
Floor Finishes				975,600	39.02
Dust sealer to concrete slabs	18,000	m²	1		
Medium grade fully accessible raised floor, metal faced ply core; 150 mm nominal depth; including fire barriers	18,000	m²	53		
Ceiling Finishes				1,170,000	46.80
Concealed grid metal tray suspended ceiling to office areas; acoustic quilt and fire breaks	18,000	m²	65		
Fittings/Fitting Out				32,400	1.30
Statutory signage	18,000	m²	2		
Space Heating and Air Treatment				3,657,600	146.30
Four pipe fan coil units	18,000	m²	32		
Distribution ductwork, grilles etc.	18,000	m²	75		
CHW installation; insulation	18,000	m²	46		
LTHW installation; insulation	18,000	m²	36		
Condensate installation; insulation	18,000	m²	14		
Electrical Installations				1,949,400	77.98
Lighting and emergency lighting installation	18,000	m²	75		
Distribution boards	18,000	m²	6		
Small power: supply only floor boxes	18,000	m²	5		
Earthing and bonding	18,000	m²	4		
Mechanical power supplies	18,000	m²	5		
Lighting control	18,000	m²	14		
Protective Installations				432,000	17.28
Sprinkler protection to offices	18,000	m²	24		
Special Installations				468,000	18.72
Building management system	18,000	m²	26		

INNER LONDON OFFICES

Office: Category A Works	Quantity	Unit	Rate (£)	Total (£)	Cost (£/m²)
Builders Work in Connection				158,400	6.34
Builders work in connection with Category A services	18,000	m²	9		
Preliminaries and Contingency				2,342,300	93.69
Contractor's overheads and profit, site establishment and supervision @		20%			
Contingency @		5%			
Construction cost (Category A only, rate based on GIFA)				11,350,700	454.03
TOTAL OFFICE CONSTRUCTION COST (rate based on GIFA)				79,657,400	3,186.30

Building Cost Models

BUSINESS PARK OFFICE

The cost model features a standard office building located in the South-East. The specification includes: GIFA of 6,000 m² (65,000 ft²), two storeys arranged around a central street, steel frame and standard FCUs fitted. The model covers shell-and-core works only. Fit-out costs to Category A specification and external works are excluded.

Gross internal floor area   6,000 m²

Model location is South East England   Model TPI = 120; LF = 0.96

This cost model is copyright of AECOM Ltd

Business Park Office	Quantity	Unit	Rate (£)	Total (£)	Cost (£/m²)
Substructure				496,000	82.15
Excavation and disposal off site; reinforced concrete ground slab, including ground beams and column bases	3,100	m³	160		
Frame, Upper Floors and Stairs				924,700	153.14
Steel propped portal frame, cold rolled purlins, surface treatments (@ 400 kg/m²)	300	tonne	1,700		
Composite slab, metal deck and in situ concrete	2,900	m²	75		
Fire protection to steelwork		item	50,000		
Allowance for miscellaneous works, protecting columns etc.		item	17,500		
Main steel internal staircases; handrails and balustrades (rate per flight)	4	nr	9,600		
Steel external staircases; handrails, balustrades, framing (rate per flight)	4	nr	12,000		
Atrium balustrading	45	m	850		
Miscellaneous metalwork in shafts and risers		item	5,000		
Roof				754,900	125.02
In situ concrete roof slab; metal deck and in situ concrete	3,080	m²	75		
In situ concrete rooflight upstand walls	120	m²	200		
Timber roof trusses		item	60,000		
Monopitch roof, coated aluminium roof cladding, including purlins, trims, cappings, insulation	1,780	m²	120		
Flat roof, asphalt, including insulation, trims, skirtings, flashings, linings to gutters, paving slabs	1,500	m²	110		
Mansafe system		item	1,250		
Rooflights	100	m²	600		
External Wall, Windows and Doors				966,300	160.03
Coated aluminium double glazed curtain walling	150	m²	450		
Fire escape doors	8	nr	1,200		
Coated aluminium double glazed window system	850	m²	420		
Solar shading; fixed aluminium louvres	300	m	360		
Glazed main double entrance doors		item	6,000		
Aluminium wall cladding to roof level plant etc.	610	m²	54		

BUSINESS PARK OFFICE

Business Park Office	Quantity	Unit	Rate (£)	Total (£)	Cost (£/m²)
Render on blockwork, insulation and boarding	550	m²	200		
Facing brickwork in cavity walls	690	m²	180		
Facing brick cladding to columns	675	m²	120		
Miscellaneous masonry		item	70,000		
Internal Walls and Partitions and Doors				**375,800**	**62.24**
Blockwork	1,200	m²	60		
Drywall partition; 1 hr fire-resistant	570	m²	75		
Hardwood glazed screens; 1 hr fire-resistant	180	m²	600		
Hardwood doors and frames; stainless steel ironmongery	80	nr	1,350		
Allowance for fire containment in common areas		item	10,000		
WC cubicles		item	35,000		
Wall Finishes				**71,700**	**11.87**
Emulsion paint to blockwork wall surfaces generally		item	40,000		
Plasterboard to cores and common areas	1,650	m²	19		
Floor Finishes				**90,200**	**14.49**
Screeds to cores and common areas; 150 mm thick to make up levels	640	m²	24		
Carpet tiling to common areas	170	m²	36		
Ceramic tiling to cores and common areas	470	m²	110		
Floor finishes to internal stairs		item	5,000		
Entrance matting	2	nr	6,000		
Ceiling Finishes				**63,000**	**10.43**
Plasterboard and skim to core and common areas	1,050	m²	60		
Furniture and Fittings				**38,000**	**6.29**
Granite vanity units		item	27,000		
Glass surrounds to vanity units		item	5,000		
Miscellaneous architectural metalwork		item	6,000		
Sanitary Appliances				**29,100**	**4.82**
WC suite, including disabled facility including all sanitary and fittings	6,050	m²	5		
Disposal Installations				**38,200**	**6.33**
Waste, soil and vent installation; uPVC pipework and fittings	6,050	m²	3		
Rainwater installation	6,050	m²	4		
Water Installations				**3,900**	**0.65**
Hot and cold water supplies to WCs	5	nr	775		

Building Cost Models

BUSINESS PARK OFFICE

Business Park Office	Quantity	Unit	Rate (£)	Total (£)	Cost (£/m²)
Space Heating and Air Treatment and Ventilation				488,800	80.95
Air handling units, chillers; ductwork, pipework, insulation	6,050	m²	75		
Toilet extract ventilation		item	15,000		
Cooling and ventilation installations to common areas		item	20,000		
Electrical and Gas Installations				312,800	51.80
Mains and sub-mains distribution	6,050	m²	24		
Small power to landlord's areas	975	m²	18		
Electrical supplies to mechanical plant		item	17,500		
Lighting to landlord's areas	975	m²	60		
External feature lighting		item	5,000		
External lighting generally		item	65,000		
Electric trench heating to common areas		item	4,000		
Incoming Services				10,000	1.66
Allowance for incoming electrical, gas and water services		item	10,000		
Protective Installations				8,500	1.41
Lightning protection		item	6,000		
Earthing and bonding		item	2,500		
Communication Installations				42,300	7.01
Fire alarm and smoke detection	6,050	m²	6		
Disabled WC alarm		item	2,000		
Allowance for containment		item	4,000		
Specialist Installations				108,900	18.04
BMS controls	6,050	m²	18		
Builders Work in Connection				10,000	1.66
Forming holes and chases etc. @		item	10,000		
Preliminaries and Contingency				991,300	164.17
Overheads, profit, site establishment and site supervision @		17%			
Contingency @		3%			
Construction cost (shell and core only; rate based on GIFA)				5,824,400	964.61

DATA CENTRE

This cost model is based on a tier 3 two-storey, new-build data centre for loads up to 1,500 W/m². The scheme comprises two technical spaces, each with a gross internal area of 1,350 m², 850 m² of associated internal plant areas, 600 m² of external plant deck and 950 m² of support facilities. Electrical infrastructure to the site is via A and B supplies with on-site dedicated substations. Further resilience is provided by uninterruptible power supplies and diesel generators at N+1. The technical areas are completed including full fit-out and are capable of delivering up to 20 kW per rack. The m² rate in the cost breakdown is based on the technical area, not gross floor area. Costs of site preparation, external works and external services are included.

Technical Floor area	2,700 m²
Model location is South East	Model TPI = 120; LF = 0.96

This cost model is copyright of AECOM Ltd

Data Centre	Quantity	Unit	Rate (£)	Total (£)	Cost (£/m²)
Substructure				**1,640,800**	**607.70**
Excavation and disposal off site; reinforced concrete ground slab, including ground beams and column bases	3,500	m²	31		
Reinforced concrete ground slab, hardcore, dpm, ground beams	675	m²	530		
Piled foundations and pile caps 450 mm dia., 20 m long	490	nr	1,400		
Pile caps	490	nr	700		
Concrete slab to energy centre; 250 mm thick including boundary walls	225	m²	460		
1 m high concrete bund to external plant		item	42,000		
Frame, Upper Floors and Stairs				**1,581,700**	**585.81**
Main building steel frame, cold rolled purlins, surface treatments (@ 85 kg/m²):	420	tonne	2,300		
Energy centre steel frame, cold rolled purlins, surface treatments (@ 85 kg/m²):	28	tonne	2,300		
Additional fire protection	448	tonne	480		
200 mm in situ reinforced concrete upper floors	2,250	m²	115		
Internal steel staircase and balustrade with concrete finish		item	34,000		
Main entrance reinforced concrete steps		item	9,400		
Galvanized steel external escape stairs and balustrades	4	nr	3,800		
Galvanized steel external stairs and balustrades	3	nr	6,300		
Roof				**225,000**	**83.33**
Kalzip embossed aluminium standing seam roof	2,250	m²	100		
External Wall, Windows and Doors				**736,400**	**272.74**
Kingspan MR panels, 45 mm thick with Fire Safe LC PB approved core including windows, louvres, internal blockwork and Metsec framing	1,700	m²	290		
Main entrance security revolving doors		item	75,000		
Coated aluminium double glazed curtain walling		item	18,000		

DATA CENTRE

Data Centre	Quantity	Unit	Rate (£)	Total (£)	Cost (£/m²)
Energy centre compound frame, cladding and louvres	335	nr	160		
Mechanical centre compound frame and cladding	220	m²	440		
Internal Walls and Partitions and Doors				452,000	167.41
140 mm thick blockwork walls to plant areas	2,100	m²	50		
Lift shaft walls		item	57,000		
Internal partitions 60/120 min fire-rating	1,200	m²	90		
Column encasements and bulkheads		item	56,000		
Bump protection rails to corridors		item	25,000		
Cubicles and vanity units		item	5,000		
Internal single doors 30 min fire-rating	40	nr	2,400		
Wall Finishes				96,400	35.70
Emulsion paint to wall surfaces generally	4,100	m²	12		
Hardwood glazed screens; 1 hr fire-resistant	4,200	m²	11		
Wall tiling upstand to kitchen and toilet areas including shower		item	2,600		
Floor Finishes				471,900	174.78
Raised access floor 850 mm overall height to technical spaces; Kingspan 600 mm square galvanized-steel encapsulated particle board panels	2,700	m²	120		
Raised access floors to other areas including plantrooms	800	m²	115		
Dust sealant to plant and technical areas	3,550	m²	7		
Floor finishes to offices and meeting rooms	400	m²	19		
Floor tiles to reception area	20	m²	190		
Vinyl to stairs		item	1,450		
Epoxy floor paint to plant areas	850	m²	10		
Allowance for protection to raised access floors	2,700	m²	4		
Ceiling Finishes				138,500	51.30
Suspended 600 × 600 mm grid ceiling to office and technical areas	3,550	m²	39		
Furniture and Fittings				15,700	5.81
Reception desk and furniture		item	3,150		
Tea kitchens		item	12,500		
Sanitary Appliances				11,100	4.11
WCs, urinals, wash hand basins; generally	14	nr	500		
Hand dryers	5	nr	440		
Shower installations		item	1,900		

DATA CENTRE

Data Centre	Quantity	Unit	Rate (£)	Total (£)	Cost (£/m²)
Disposal Installations				**12,400**	**4.59**
Above ground drainage installation to support areas	950	m²	13		
Water Installations				**120,700**	**44.70**
Hot and cold water supply to WC, kitchen	950	m²	65		
Hot and cold water installations to plant areas	850		23		
External water installations		Item	9,400		
Water storage installation including ancillary pipework		Item	30,000		
Space Heating and Air Treatment and Ventilation				**3,862,500**	**1,430.56**
Indirect fresh air handling plant (with DX coil) and associated ductwork		item	3,380,000		
water treatment and pipework for adiabatic cooling to air handling plant:		item	110,000		
Supply and extract ventilation system to technical areas	2,700	m²	65		
Ventilation to switch rooms and ancillary areas	850	m²	25		
Office supply and extract ventilation	950	m²	105		
VRF heating and cooling installation to office, meeting rooms and support areas	400	m²	190		
Electrical Installations				**14,091,100**	**5,218.93**
High voltage switchgear and distribution based on a 7 MVA dual supply requirement		item	515,000		
Standby generator installation based on 4 nr 2,500 kVA gen sets, providing n+2 complete with flues and 48 hr bulk storage fuel tanks		item	3,440,000		
Low voltage switchgear including, mechanical panels, UPS panels, utility panels, 4 nr 2.5 MVA package transformers including delivery and installation		item	3,020,000		
Static UPS installation consisting of 4 nr 1100 kVA modules per system installed in n+1 configuration (IT systems) and 2 nr 550 kVA modules installed in n+1 configuration mechanical systems), including distribution		item	3,730,000		
14 nr dual input STS PDUs and 14 nr single input PDUs including installation		item	1,300,000		
LV distribution including cabling to PDUs, CRACs, IT and mechanical panels including busbar installation		item	1,350,000		
LV distribution equipment		item	50,000		
General lighting and emergency lighting to all areas complete with PIR controls	4,500	m²	57		
External lighting		item	32,000		

DATA CENTRE

Data Centre	Quantity	Unit	Rate (£)	Total (£)	Cost (£/m²)
Small power installation to plant and support areas	1,800	m²	32		
Primary and secondary containment installation to plant, support and technical areas	4,500	m²	70		
Allowance for trace heating		item	25,000		
Protective and Special Installations				**1,873,600**	**693.93**
Earthing and bonding installation including clean earth to plant	4,500	m²	65		
Lightning protection installation	4,500	m²	8		
Water mist installation to part of plant area and technical areas	3,200	m²	140		
Leak detection installation to plant and technical areas	3,550	m²	21		
Dry riser installation		item	7,600		
Energy management systems/SCADA installation	4,500	m²	140		
Building management system	4,500	m²	85		
Communications Installations				**763,500**	**282.78**
L1 fire and smoke detection and alarm system to plant, support and technical areas	4,500	m²	31		
High sensitivity smoke detection installation to technical areas	2,700	m²	26		
Intruder detection, based on monitored doors, windows and PIR detection	4,500	m²	10		
Access control to technical spaces, plant and support areas including entrances and exits, integrated with intruder detection	4,500	m²	50		
CCTV installation to technical spaces, plant and support areas including entrances and exits	4,500	m²	50		
Interphone video telephony to entrance area		item	22,500		
Public address installation	4,500	m²	5		
Disabled toilet alarm system	2	nr	3,450		
Data installation to office and meeting room areas	400	m²	25		
Lift Installations				**309,000**	**114.44**
Passenger/goods lift; hydraulic 2.5 tonnes; serving two stops	1	nr	165,000		
Passenger lift; hydraulic eight person; serving two stops	1	nr	70,000		
Scissor lifts	2	nr	37,000		

DATA CENTRE

Data Centre	Quantity	Unit	Rate (£)	Total (£)	Cost (£/m²)
Testing and Commissioning				297,000	110.00
Testing and commissioning of mechanical services		item	36,000		
Testing and commissioning of electrical services		item	26,000		
IST commissioning		item	175,000		
Allow for electrical heat load test		item	60,000		
Builders Work in Connection				640,300	237.15
Forming holes and chases etc. @		3%			
External Works				566,000	209.63
Allowance for vehicle crash barriers		item	36,000		
Allowance for external security fence		item	125,000		
Allowance for soft landscaping works		item	405,000		
Preliminaries and Contingency				5,790,400	2,144.59
Overheads, profit, site establishment and site supervision @		15%			
Contingency @		5%			
Construction cost (shell and core only; rate based on Technical Floor Area)				33,696,000	12,479.99

OFFICE FIT-OUT

The cost model is based on a notional scheme that is typical for a West End based professional firm. The design is of a high specification, with a mixture of open-planned and cellular office use. It should be noted that the cost of internal finishes can vary considerably and is dependent on a client's aspirations. The cost model assumes that the floor space has been taken from shell and core with a negotiated landlord contribution for the Category A installation.

The office fit-out model is presented as a cost breakdown by functional area rather than in a traditional elemental format. The fit-out is designed to provide the following areas:

Open plan/circulation areas	1,300 m²	Tea points	40 m²
Client meeting rooms	280 m²	Staff dining	150 m²
Client reception	80 m²	Kitchen	60 m²
Cellular offices	1,000 m²	Communications/data rooms	60 m²
Staff meeting rooms	500 m²	Ancillary areas	175 m²
		Total usable area	3,645 m²

Model location is Inner London   Model TPI = 131; LF = 1.06

This cost model is copyright of AECOM Ltd

Office Fit-Out	Quantity	Unit	Rate (£)	Total (£)	Cost (£/m² of area)
Open Plan/Circulation (1,300 m²)				**1,072,200**	**825**
Medium grade raised access floor 150 mm high	1,300	m²	56		
Carpet tiles	1,300	m²	35		
Suspended perforated metal ceiling	1,300	m²	100		
Emulsion paint to core walls		item	14,000		
Mechanical and electrical services; including heating, air treatment, ventilation, power, lighting, protective installations, communications, special installations	1,300	m²	530		
Structured cabling	1,300	m²	49		
Roller blinds to perimeter windows	1,300	m²	28		
Statutory signage		item	2,800		
Builders work in connection with services	1,300	m²	14		
Client Meeting Rooms (280 m²)				**1,127,600**	**4,027**
Acoustic rated metal stud partitions, 60-minute fire-rated, hardwood skirting and ply strengthening	170	m	370		
Double glazed partitions, manifestation	50	m	1,250		
Folding wall; acoustic finish	15	nr	2,800		
American Walnut doors	15	nr	3,150		
Medium grade raised access floor 150 mm high	280	m²	56		
Carpet tiles	280	m²	56		
Suspended perforated metal ceiling with feature raft	280	m²	140		
Fabric wall coverings	900	m²	380		

OFFICE FIT-OUT

Office Fit-Out	Quantity	Unit	Rate (£)	Total (£)	Cost (£/m² of area)
Mechanical and electrical services; including heating, air treatment, ventilation, power, lighting, protective installations, communications, special installations	280	m²	850		
Structured cabling	280	m²	42		
Black out blinds to perimeter windows	280	m²	70		
Statutory signage		item	3,150		
Branding and corporate signage		item	14,000		
Specialist joinery including credenzas		item	210,000		
Builders work in connection with services	280	m²	13.80		
Client Reception (80 m²)				**299,300**	**3,741**
Double glazed door	1	nr	5,550		
Natural stone flooring on boarded substrate	80	nr	380		
Painted plasterboard suspended ceilings including bulkheads, access panels and ceiling feature	80	m²	180		
Timber panel and polished plaster wall finishes		item	42,000		
Mechanical and electrical services; including heating, air treatment, ventilation, power, lighting, protective installations, communications, special installations	80	m²	850		
Structured cabling	80	m²	42		
Statutory signage		item	1,400		
Branding and corporate signage		item	28,000		
Specialist joinery including reception desk		item	105,000		
Builders work in connection with services	80	m²	13.80		
Cellular Offices (1,000 m²)				**1,644,300**	**1,644**
Acoustic rated metal stud partitions, 60-minute fire-rated, MDF skirting and ply strengthening	600	m	350		
Single glazed partitions, manifestation, blinds	150	m	975		
Solid core timber veneer doors	120	nr	2,800		
Medium grade raised access floor 150 mm high	1,000	m²	56		
Carpet tiles	1,000	m²	35		
Suspended perforated metal tiled ceiling	1,000	m²	100		
Emulsion paint wall finish	3,000	m²	8		
Mechanical and electrical services; including heating, air treatment, ventilation, power, lighting, protective installations, communications, special installations	1,000	m²	630		
Structured cabling	1,000	m²	49		
Roller blinds to perimeter windows	1,000	m²	35		
Statutory and corporate signage		item	8,300		
Builders work in connection with services	1,000	m²	14		

Building Cost Models

OFFICE FIT-OUT

Office Fit-Out	Quantity	Unit	Rate (£)	Total (£)	Cost (£/m² of area)
Staff Meeting Rooms (500 m²)				790,300	1,581
Acoustic rated metal stud partitions, 60-minute fire-rated, MDF skirting and ply strengthening	320	m	350		
Double glazed partitions, manifestation	80	m	1,250		
Solid core timber veneer doors	15	nr	2,800		
Medium grade raised access floor 150 mm high	500	m²	56		
Carpet tiles	500	m²	35		
Suspended perforated metal tiled ceiling	500	m²	100		
Emulsion paint wall finish	1,750	m²	8		
Mechanical and electrical services; including heating, air treatment, ventilation, power, lighting, protective installations, communications, special installations	500	m²	630		
Structured cabling	500	m²	42		
Roller blinds to perimeter windows	500	m²	35		
Statutory and corporate signage		item	2,800		
Specialist joinery		item	63,000		
Builders work in connection with services	500	m²	14		
Tea Points (40 m²)				111,600	2,790
Standard metal stud partitions, 60-minute fire-rated, MDF skirting and ply strengthening	40	m	280		
Medium grade raised access floor 150 mm high including tanking to slab	40	m²	85		
Slip-resistant floor finish on two layers of boarding	40	m²	85		
Suspended perforated metal tiled ceiling	40	m²	100		
Emulsion paint wall finish and ceramic splashbacks		item	4,850		
Mechanical and electrical services; including disposal, water installations, zip taps, heating, air treatment, transfer ventilation, power, lighting, protective installations, communications, leak detection, special installations	40	m²	700		
Statutory and signage		item	140		
Joinery units, work surfaces and equipment		item	56,000		
Builders work in connection with services	40	m²	14		
Staff Dining (150 m²)				265,800	1,772
Standard metal stud partitions, 60-minute fire-rated, MDF skirting and ply strengthening	40	m	280		
Medium grade raised access floor 150 mm high including tanking to slab	150	m²	56		
Porcelain floor tiles on two layers of boarding and ditra matting	150	m²	210		
Suspended perforated metal tiled ceiling	150	m²	140		

OFFICE FIT-OUT

Office Fit-Out	Quantity	Unit	Rate (£)	Total (£)	Cost (£/m² of area)
Emulsion paint wall finish and ceramic splash backs		item	17,000		
Mechanical and electrical services; including disposal, water installations, zip taps, heating, air treatment, transfer ventilation, power, lighting, protective installations, communications, leak detection, special installations	150	m²	660		
Statutory and signage		item	5,550		
Servery and specialist joinery item		item	70,000		
Builders work in connection with services	150	m²	14		
Kitchen (60 m²)				**280,800**	**4,680**
Standard metal stud partitions, 60-minute fire-rated, MDF skirting and ply strengthening	50	m	280		
Metal pivot door	1	nr	4,150		
Medium grade raised access floor 150 mm high including tanking to slab	60	m²	85		
Slip-resistant floor finish on two layers of boarding and ditra matting	60	m²	85		
Hygienic finished ceiling	60	m²	140		
Altro Whiterock wall finish	150	m²	180		
Mechanical and electrical services; including disposal, water installations, zip taps, heating, air treatment, ventilation, dedicated extract, power, lighting, protective installations, communications, leak detection, special installations	60	m²	1,225		
Statutory signage		item	1,400		
Stainless steel kitchen units, worktops and equipment		item	140,000		
Builders work in connection with services	60	m²	35		
Communications/Data Rooms (60 m²)				**238,600**	**3,977**
Standard metal stud partitions, 90-minute fire-rated, MDF skirting and ply strengthening	40	m	320		
Solid timber painted doors; 90 mins fire-rated	1	nr	2,450		
Heavy duty raised access floor 150 mm high with anti-static floor finish	60	m²	170		
Dust sealant to concrete soffit	60	m²	21		
Emulsion paint wall finish	120	m²	8		
Mechanical & electrical services; including air treatment, ventilation, stand-alone air conditioning units, electrical, power, lighting, protective installations, communications, special installations, leak detection	60	m²	3,450		
Structured cabling	60	m²	49		
Statutory signage		item	140		
Builders work in connection with services	60	m²	14		

OFFICE FIT-OUT

Office Fit-Out	Quantity	Unit	Rate (£)	Total (£)	Cost (£/m² of area)
Ancillary Areas (175 m²)				276,300	1,579
(including stores, reprographics, mail room etc.)					
Standard metal stud partitions, 60-minute fire-rated, MDF skirting and ply strengthening	150	m	280		
Solid timber painted doors	10	nr	2,100		
Medium grade raised access floor 150 mm high	175	m²	49		
Sheet vinyl flooring including boarding substrate	175	m²	56		
Suspended perforated metal tiled ceiling	175	m²	100		
Emulsion paint wall finish	800	m²	8		
Mechanical and electrical services; including disposal, water installations, heating, air treatment, ventilation, power, lighting, protective installations, communications, special installations	175	m²	560		
Statutory signage		item	350		
Fixed fittings and shelving		item	70,000		
Builders work in connection with services	175	m²	14		
Preliminaries and contingencies				1,282,400	352
Main Contractor's management, site establishment and site supervision @		10%			
Main Contractor's overhead & profit @		5%			
Design reserve @		5%			
Construction cost (rate based on total usable area)				7,389,200	2,027

OUT OF TOWN RETAIL UNIT

A 6,500 m² non-food retail unit on an existing retail park in the South East. The unit is comprised of two full height levels which will give the operator the flexibility to trade over both floors without having to compromise on a reduced head height that is often encountered on mezzanine levels of retail park units. The cost model is based on a shell only specification and allows for the construction of the retail unit whilst maintaining continuous trade for all other tenants within the retail park. The overall programme duration for is based on 18 weeks.

Gross internal floor area   6,500 m²

Model location is South East   Model TPI = 120; LF = 0.96

This cost model is copyright of AECOM Ltd

Out of Town Retail Unit	Quantity	Unit	Rate (£)	Total (£)	Cost (£/m²)
Substructure				**702,600**	**108.09**
Excavation, disposal, underground drainage, backfill		item	80,000		
Concrete pad foundations	64	nr	2,000		
Reinforced concrete ground floor slab, 250 mm thick including gas and vapour barrier, edge beams	3,250	m²	140		
Lift and escalator pits	4	nr	9,900		
Frame and Upper Floors				**1,114,000**	**171.38**
Steel propped portal frame, cold rolled purlins, surface treatments (@ 60 kg/m²)	390	tonne	2,000		
Secondary steelwork and signage framework		item	26,500		
Holorib and concrete upper floor slab 150 mm thick	3,250	m²	70		
Precast concrete stairs and metal balustrades to cores	4	nr	20,000		
Roof				**598,700**	**92.11**
Standing seam metal roof, insulated roof	2,844	m²	90		
Feature rooflight, clerestory glazed		item	80,000		
Concrete plant deck, inverted roof including waterproofing and ballast	540	m²	105		
Sundry roof works, flashings, forming openings etc.		item	120,000		
Roof parapet screens		item	40,000		
Handrails and GRP designated maintenance walkways		item	46,000		
External Walls, Windows and Doors				**1,791,500**	**275.62**
Structural glazing system to front and side elevations, planar type system, low 'G' value, low iron glass, glazed fins and capless mastic joints, stainless steel trims to perimeter framing	830	m²	850		
Acrylic render system, including insulation and blockwork, to stair cores and soffits of colonnade	740	m²	200		
Punched windows to office areas	9	nr	1,350		
Framed curtain walling to stair cores	130	m²	400		
Half round profiled composite cladding, including block work inner skin	1,270	m²	150		

OUT OF TOWN RETAIL UNIT

Out of Town Retail Unit	Quantity	Unit	Rate (£)	Total (£)	Cost (£/m²)
Granite cladding system to columns	336	m²	530		
Granite cladding system to feature entrance columns	166	m²	750		
Brise soleil, aluminium fins to glazed facades		item	200,000		
Entrance lobby, glazed sides and 2 nr sets of double plus single door and security screens		item	92,500		
Engineering brick base to facades	105	m²	140		
Screens to sprinkler tanks and plant		item	40,000		
Roller shutter doors to service yard		item	13,500		
Fire escape doors		item	20,000		
Internal Walls and Partitions				**40,000**	**6.15**
Sound reducing board and metal stud partitions; generally 2 layers of board each side; various levels of fire and sound insulations; part glazed as appropriate:	500	m²	80		
Internal Doors				**16,000**	**2.46**
Internal doorsets to stair cores; solid core; including vision panels and overpanels; painted finish; ironmongery		item	16,000		
Disposal Installations				**40,000**	**6.15**
Rainwater disposal; downpipes and fittings		item	40,000		
Water Installations				**40,000**	**6.15**
Hot and cold water supply to WC, kitchen		item	40,000		
Electrical and Gas Installations				**111,600**	**17.17**
Incoming electrical services including new substation and termination cubicle		item	105,000		
Incoming gas supply		item	6,600		
Protective, Communications and Special Installations				**151,600**	**23.32**
Lightning protection, earthing and bonding:		item	6,600		
Sprinkler tanks and pump set (internal pipework and sprinkler heads part of fit-out)		item	145,000		
Builders Work				**33,000**	**5.08**
Builders work in connection with services @		item	33,000		
Preliminaries and Contingency				**816,500**	**125.62**
Management costs; site establishment; site supervision @		12.0%			
Contingency @		5.0%			
Construction cost (rate based on GIFA)				**5,455,500**	**839.30**

CAR DEALERSHIP

A purpose built new dealership on an automotive retail park, comprising showroom and workshop of 900 m² and 1,000 m² respectively, with additional office and ancillary space, giving a gross internal floor area of 3,500 m². It has a reinforced in situ concrete ground-bearing slab, a steel frame with piled foundations, reinforced concrete pile caps and ground beams. There is a full height glazed front, aluminium cladding on the side walls and single ply roofing and rooftop car parking. There is full comfort cooling in all areas and the cost of fitting out this and the display lighting is included. External works consist of car parking, forecourt display areas, landscaping.

Gross internal floor area   3,500 m²

Model location is South East England   Model TPI = 120; LF = 0.96

This cost model is copyright of AECOM Ltd

Car Dealership	Quantity	Unit	Rate (£)	Total (£	Cost (£/m²)
Substructure				838,900	239.69
Excavation and disposal off site	7,000	m³	16		
CFA piled foundations; excavations; concrete; reinforcement; 600 mm dia.; 15 m deep	277	nr	1,100		
Reinforced concrete pile caps	94	nr	650		
Reinforced concrete ground beams 500 mm × 600 mm	500	m	170		
Reinforced concrete ground slab, hardcore, dpm, ground beams and column bases for steel frame (0.25 W/m².K):	2,200	m²	105		
Lift pit	1	nr	3,300		
Recessed slab for specialist equipment		item	39,000		
Frame, Upper Floors and Stairs				884,000	252.57
Steel propped portal frame, cold rolled purlins, surface treatments	220	tonne	2,050		
Intumescent paint to give 60 minute fire protection to steelwork:		item	44,000		
Decorations to exposed steelwork		item	11,000		
150 mm reinforced concrete suspended slab	2,900	m²	100		
Feature customer staircase; steel with cement screed infill; including finishes, balustrading and hand railing	1	nr	44,000		
Internal precast concrete staircase to serve upper office levels; including finishes, balustrades and hand railing	2	nr	22,000		
Envelope				1,230,300	351.51
Engineering brick plinth	30	m²	280		
Composite wall cladding – front of house	600	m²	190		
Composite wall cladding – back of house	1,300	m²	85		
Glazed curtain walling	400	m²	500		
Vehicle access doors in curtain walling	3	nr	11,000		
Revolving access doors	3	nr	44,000		
Entrance feature		item	22,000		
Personnel doors	3	nr	3,300		

CAR DEALERSHIP

Car Dealership	Quantity	Unit	Rate (£)	Total (£	Cost (£/m²)
Workshop doors	7	nr	8,800		
Roof to dealership; single ply membrane with tapered insulation	780	m²	100		
Reinforced concrete slab to rooftop car park	1,300	m²	110		
Kingspan Styrozone car park system to rooftop parking; including screed, waterproofing, asphalt, etc.	1,300	m²	140		
Steel ramp to rooftop parking, single vehicle width		item	55,000		
Vehicle barrier to rooftop parking	270	m	170		
Syphonic roof drainage	2,080	m²	13		
Escape stairs enclosure	10	m²	775		
Internal Walls and Partitions and Doors				**298,500**	**85.29**
Plasterboard and metal stud partitions; generally 2 layers of plasterboard each side; various levels of fire and sound insulations	3,000	m²	80		
Glazed partition system	80	m	575		
Workshop lining system (Whiterock)	100	m²	70		
Column casings		item	5,500		
Internal Finishes				**438,200**	**125.20**
Barrier matting	90	m²	330		
Floor tiles	2,200	m²	75		
Tiled skirting	80	m	33		
Carpet	450	m²	33		
Timber skirting	500	m	16		
Timber (MDF) skirting	200	m	11		
Vinyl flooring	320	m²	44		
Floor paint	300	m²	22		
Painting wall and timber	5,000	m²	13		
Wall tiles	300	m²	70		
Suspended ceilings	1,300	m²	33		
Metal framed ceiling and bulkheads	1,200	m²	50		
Fire barriers		item	5,500		
Space Heating and Air Treatment Ventilation				**754,500**	**215.57**
Incoming services and metering		item	20,000		
Ventilation	3,500	m²	85		
Heat recovery system	3,500	m²	85		
Hot and cold water services		item	46,000		
Gas distribution		item	11,000		
Workshop heating		item	33,000		
Above ground drainage		item	11,000		

CAR DEALERSHIP

Car Dealership	Quantity	Unit	Rate (£)	Total (£	Cost (£/m²)
Photovoltaic system		item	17,500		
Testing and commissioning		item	21,000		
Electrical and Gas Installations				713,100	203.74
Mains switchgear and distribution		item	24,000		
Sub-mains cabling		item	10,400		
Lighting and emergency lighting installations	3,500	m²	85		
Small power installations	3,500	m²	85		
Power to doors and workshop equipment		item	29,000		
External lighting		item	39,000		
Earthing and bonding		item	2,200		
Testing and commissioning		item	13,500		
Lift Installation				55,000	15.71
8 person electrohydraulic serving 3 floors	1	nr	55,000		
General Fixtures, furnishings and Equipment				126,500	36.14
Feature car display lighting trusses		item	22,000		
Benches and seating		item	11,000		
Signage		item	5,500		
WC sanitary ware and cubicles		item	60,000		
Kitchen units and coffee station		item	28,000		
Builders Work in Connection				2,800	0.80
Forming holes and chases etc. @		item	2,800		
Preliminaries and Contingency				985,600	281.60
Overheads, profit, site establishment and site supervision @		15%			
Contingency @		3%			
Construction cost for shell only (rate based on GIFA)				6,327,400	1,808

Car Dealership – External Works	Quantity	Unit	Rate (£)	Total (£)	Cost (£/m²)
Hard and Soft Landscaping				244,800	69.94
Grassed/planted areas	650	m²	11		
Car parking and access road	3,700	m²	50		
Standard kerbs	610	nr	24		
Road markings		item	2,200		
Paving to pedestrian areas	200	m²	39		
Stainless steel bollards	100	nr	280		

Building Cost Models

CAR DEALERSHIP

Car Dealership – External Works	Quantity	Unit	Rate (£)	Total (£)	Cost (£/m²)
Perimeter Fencing				**17,400**	**4.97**
Perimeter fencing 2.4 m high palading welded mesh fence	102	m²	50		
Entrance vehicle barrier		item	8,200		
Entrance gates	3	nr	1,350		
Drainage and External Services				**205,600**	**58.74**
Storm water drainage	720	m	90		
Storm water manholes	17	nr	1,350		
Gullies, rainwater outlets		item	4,400		
Connection to main sewer		item	5,500		
Foul water drainage	500	m	90		
Foul water manholes	11	nr	1,350		
Petrol interceptor	1	nr	8,800		
Connection to main sewer		item	3,300		
Utilities (gas, water, electricity)		item	36,000		
Preliminaries and Contingency				**86,300**	**24.66**
Contractors preliminaries, overhead and profit		15%			
Contingency @		3%			
Construction cost for external works only (rate based on GIFA)				**554,100**	**158.31**
TOTAL CONSTRUCTION COST (rate based on GIFA)				**6,881,500**	**1,966**

PRIVATE PATIENT WARD

The cost model is for a stand-alone new-build 20-bed private patient ward unit with associated ancillary support services; both internal and external finishes to a good quality. Theatres are provided elsewhere within the hospital estate.

Gross internal floor area, including tiers	1,700 m²
Model location is Outer London	Model TPI = 124; LF = 1.00

This cost model is copyright of AECOM Ltd

Private Patient Ward	Quantity	Unit	Rate (£)	Total (£)	Cost (£/m²)
Substructure				**509,400**	**299.65**
Strip and level site		item	67,500		
Perimeter footings, 0.6 m deep and 0.6 m wide, mass concrete, reinforcement 200 kg/m³	110	m³	700		
Pad foundations, 0.8 m deep and 0.75 × 0.75 m, mass concrete, reinforcement @ 250 kg/m³,	32	m²	825		
150 mm concrete slab; hardcore, dpm, insulation, mesh reinforcement	1,750	m²	140		
Under slab drainage, including excavation		item	80,000		
Allow for trenching of services		item	13,500		
Frame				**96,100**	**56.53**
Precast concrete pad stones bedded on mortar	120	nr	70		
Structural steel beams	20	tonne	2,650		
Structural steel columns	8	tonne	2,650		
Sundry items, additional secondary steel		item	13,500		
Roof				**510,100**	**300.06**
Kalzip aluminium standing seam, 250 mm mineral wool insulation; waterproof membrane and roof structure	700	m²	460		
Single-ply roof membrane, rigid insulation, vapour control layer and roof structure	1,050	m²	160		
Solar shading timber louvered canopy		item	6,600		
Rainwater goods; gutters and downpipes		item	13,500		
External Walls, Windows and Doors				**529,600**	**311.53**
Cavity wall construction, brick external finish, insulation, 100 mm medium-density load-bearing blockwork inner leaf	550	m²	210		
Cavity wall construction, timber cladding system, insulation, 100 mm medium-density load-bearing blockwork inner leaf	450	m²	330		
Cavity wall construction, aluminium cladding panels to external finish, insulation, 100 mm medium-density load-bearing blockwork inner leaf	20	m²	200		
External liquid oxygen enclosure		item	13,500		
Generator enclosure		item	6,600		

Building Cost Models

PRIVATE PATIENT WARD

Private Patient Ward	Quantity	Unit	Rate (£)	Total (£)	Cost (£/m²)
Curtain walling, double-glazed units, powder coated aluminium, including all necessary ironmongery, fixings, mastic pointing	150	m²	750		
External doors, powder-coated aluminium double-glazed units, including ironmongery and integral cills		item	34,000		
Windows, powder-coated aluminium double-glazed units, including all necessary ironmongery and integral cills,		item	95,000		
Internal Walls, Partitions, Windows and Doors				**319,800**	**188.12**
Load-bearing blockwork partitions, 140 mm thick	1,250	m²	80		
Internal partition, 70 mm studs; 30 min fire-rating, 2 nr wall board 12.5 mm plasterboard to both sides,	80	m	120		
Internal partition, 70 mm studs with mineral fibre, 90 min fire-rating, insulation, 2 nr 15 mm plasterboard to both sides	130	m	160		
Internal partition, 70 mm studs, 30 min fire-rating, 12.5 mm wallboard plaster board & 1 nr 12.5 mm moisture board	180	m	120		
Internal glazed screens, double-glazed partitions	70	m²	900		
IPS panelling, including all necessary fixtures and fittings		item	16,000		
Privacy screens, including all necessary fixtures and fittings	2	nr	675		
Single leaf timber doors and ironmongery	50	nr	1,125		
One and half leaf timber doors and ironmongery	12	nr	1,350		
Double leaf timber doors and ironmongery	8	nr	1,450		
Cupboard/riser timber doors	4	nr	850		
Wall Finishes				**128,100**	**75.35**
2 coat plaster to blockwork; 13 mm thick	2,150	m²	24		
1 coat plaster to blockwork; 3 mm thick	500	m²	7		
Painting to plastered surface, 1 mist and 2 top coats	2,650	m²	11		
Hygiene wall cladding	280	m²	75		
Feature glass, impact resistant, on proprietary carrier system	35	m²	530		
PVC wall protection system		item	6,600		
Floor Finishes				**103,700**	**61.00**
Levelling screed, to insulated floor areas	1,620	m²	7		
Carpet, heavy duty carpet tiles, including timber skirting	120	m²	40		
Barrier matting, including aluminium frame	30	m²	360		

PRIVATE PATIENT WARD

Private Patient Ward	Quantity	Unit	Rate (£)	Total (£)	Cost (£/m²)
Standard vinyl floor finish, including coved skirting	1,050	m²	46		
Non-slip vinyl floor finish, including coved skirting	330	m²	53		
Ceramic floor tiles to entrance and reception	120	m²	85		
Epoxy paint finish	50	m²	27		
Ceiling Finishes				**165,400**	**97.29**
Suspended ceiling, 600 mm × 600 mm, T grid, including suspension system and perimeter edge trim	1,090	m²	53		
Suspended hygiene ceiling, 600 mm × 600 mm, T grid, including suspension system and perimeter edge trim	80	m²	70		
Suspended plasterboard ceiling; MF plasterboard system, including suspension system and perimeter edge trim	300	m²	53		
Slatted timber ceiling, western red cedar, prefabricated timber panels with support profiles and concealed fixings and edge trims	230	m²	200		
Plaster skim coat to plasterboard, 3 mm thick applied in 1 coat	1,700	m²	13		
Painting to plasterboard surfaces, 1 mist coat and 2 top coats	1,700	m²	11		
Fittings and Furnishings				**162,200**	**95.41**
Bespoke reception desk, with built-in storage		item	10,500		
Bespoke reception desk, 2 workstations, with built-in storage		item	5,250		
Touchdown base, I workstation, with built-in storage		item	6,600		
Shelving generally		item	6,600		
Mirrors		item	3,950		
Grab rails		item	6,600		
Window blinds and curtains		item	6,600		
Worktops and base units		item	42,000		
Ceiling hoists, including track		item	49,000		
Wall protection generally		item	16,000		
Client and statutory signage		item	6,600		
Baby change unit		item	1,350		
Door stops	50	nr	13		
Coat hooks		item	400		
Mechanical Installations				**690,000**	**405.88**
Clinical wash hand basin	20	nr	590		
Wash hand basins	20	nr	400		
Stainless steel sink	4	nr	200		

PRIVATE PATIENT WARD

Private Patient Ward	Quantity	Unit	Rate (£)	Total (£)	Cost (£/m²)
WC suite and concealed cistern	7	nr	460		
WC suite, Doc M pack	7	nr	1,125		
Macerator pump	1	nr	460		
Bath, to be HTM64 compliant	1	nr	2,000		
Shower, with thermostatic mixer	8	nr	1,350		
Disposal installations		item	34,000		
Water installations – cold water systems		item	42,000		
Water installations – hot water systems		item	140,000		
Heat source		item	95,000		
Space heating and air treatment		item	300,000		
Ventilating systems		item	34,000		
Electrical Installations				**1,230,900**	**724.06**
LV installation including 200 kVA standby generator		item	210,000		
Lighting		item	145,000		
Small power		item	82,500		
Gas installations		item	7,900		
Protective installations		item	13,500		
Communications installations		item	135,000		
Special installations		item	240,000		
Testing and commissioning of building services installations		item	42,000		
Builders work in connection with services		item	90,000		
Incoming infrastructure, gas, statutory water, sewer and telecoms connection		item	265,000		
Preliminaries and Contingency				**890,000**	**523.53**
Management costs, site establishment and site supervision. Contractor's preliminaries, overheads and profit and contingencies @		20%			
Construction cost (rate based on GIFA)				**5,335,300**	**3,138**

EXTRA CARE HOME

The cost model is for a 77 one and two bedroom apartments housed in a three-storey building of 6,855 m² gross internal floor area, traditional construction.

Gross internal floor area, including tiers   6,855 m²

Model location is North West   Model TPI = 109; LF = 0.88

This cost model is copyright of AECOM Ltd

Extra Care Home	Quantity	Unit	Rate (£)	Total (£)	Cost (£/m²)
Substructure				**741,500**	**108.17**
Site clearance, excavation & disposal	6,210	m²	8		
Site remediation, includes: removal of contaminated material, vaporized membrane to footprint of the building and 600 mm cover to all soft landscaped areas		item	160,000		
Trench fill foundations, beam and block floor, floor insulation and screed	1,970	m²	190		
Piled foundations, beam and block floor, insulation and screed		item	160,000		
Frame and Upper Floors				**539,900**	**78.76**
Steel framing and supporting columns to ground floor main entrance		item	27,000		
Concrete beam and block upper floors	4,884	m²	105		
Roof				**362,200**	**52.84**
Concrete pitched tiled roof, timber structure batterns and felt incl. overhang	1,380	m²	160		
Single ply membrane roof, timber structure	750	m²	180		
Front entrance steel and roof structure		item	6,400		
Stairs				**54,000**	**7.88**
Precast concrete stairs over four storeys including balustrade/handrail; over four storeys; half landing	3	nr	18,000		
External Walls, Windows and Doors				**1,016,500**	**148.29**
Traditional external wall construction comprising facing brick, cavity and internal skin of external wall block work, including DPC, cavity trays and insulation, joints and lintels	916	m²	170		
Traditional external wall construction comprising block work and through colour render, cavity and internal skin of external wall block work including DPC, cavity trays and insulation, joints and lintels	1,131	m²	170		
Traditional external wall construction comprising blockwork and through dressed metal colour cladding, cavity and internal skin of external wall blockwork including DPC, cavity trays and insulation, joints and lintels	39	m²	250		

Building Cost Models

EXTRA CARE HOME

Extra Care Home	Quantity	Unit	Rate (£)	Total (£)	Cost (£/m²)
Traditional external wall construction comprising blockwork and through accent colour render, cavity and internal skin of external wall blockwork including DPC, cavity trays and insulation, joints and lintels	32	m²	170		
Juliette balconies including steel structure posts	32	m²	470		
Allowance for balconies including steel structure posts, foundations and non-slip decking	26	nr	4,650		
Composite windows	708	m²	400		
Curtain walling and cladding to resident stairwells	110	m²	470		
External doors, single and double	89	nr	2,050		
Internal Walls, Partitions, Windows and Doors				**789,600**	**115.19**
Metal stud partition to internal walls	3,767	m²	41		
Blockwork internal walls – 300 mm	4,249	m²	85		
Internal timber six panel painted apartment entrance door, 60 min fire-rated, ironmongery	77	nr	800		
Internal apartment door 30 min fire-rated; ironmongery	287	nr	320		
Internal apartment store door, single 30 min fire-rated; ironmongery	111	nr	270		
Internal apartment store door; double 30 min fire-rated; ironmongery	9	nr	400		
Internal lobby entrance door, powder coated aluminium	2	nr	2,000		
Ancillary communal doors; 30 min fire-rated where required, ironmongery	62	nr	420		
Communal circulation door; double veneered 60 min fire-rated including vision panel	23	nr	900		
Internal windows to apartments, 60 min fire-rated	77	nr	470		
Wall Finishes				**481,000**	**70.17**
Plasterboard and skim finish to all internal walls and the internal face of the external wall face incl. emulsion	18,152	m²	23		
Ceramic wall tiles to apartment bathrooms & kitchens, craft room kitchenette, staff room kitchen, communal WCs, assisted bathrooms, including trim and sealant	1,427	m²	47		
Floor Finishes				**160,000**	**23.34**
Dust sealer to apartments, halls, bedrooms, lounges and stores	3,375	m²	6		
Safety vinyl flooring with coved skirtings	1,269	m²	45		
Non-slip ceramic tiles	203	m²	80		
Carpet to communal areas	1,206	m²	51		

EXTRA CARE HOME

Extra Care Home	Quantity	Unit	Rate (£)	Total (£)	Cost (£/m²)
High quality carpet	54	m²	57		
Matting (grime buster)	4	m²	105		
Ceiling Finishes				**227,700**	**33.22**
Plasterboard and skim suspended ceiling to apartments and communal areas with emulsion finish	5,610	m²	36		
Plasterboard and skim suspended ceiling to wet apartments and communal areas with emulsion finish	501	m²	38		
Allowance for access hatches and reflected suspended ceiling details		item	6,700		
Fittings and Furnishings				**438,500**	**63.97**
Standard kitchen units including oven, hob and extract	77	nr	2,700		
Bespoke reception desk allowance		item	6,700		
Allowance for main kitchen and servery including store and cold store		item	100,000		
Shelving generally		item	1,350		
Recycle bins	77	nr	41		
Hair dressers salon and nail bar		item	6,700		
Staff laundry		item	27,000		
Painted curtain battens, window boards, handrails, pipe boxing, etc.		item	27,000		
Internal and external signage		item	13,500		
Lockers and benches to staff room		item	4,000		
Kitchenettes	2	nr	1,025		
Bathroom cupboards and shelves	77	nr	470		
Stores and meeting room	4	nr	725		
Sanitary Fittings				**196,200**	**28.62**
Apartment ensuite incl. wash basin, WC, shower and curtain rail	77	nr	2,250		
Assisted bathroom incl. argo huntleigh malibu automatic disinfectant bath, wash basin, WC	1	nr	13,000		
Other WC and wash basin	5	nr	1,600		
Staff WC and shower	1	nr	1,900		
Mechanical Installations and Lifts				**2,898,600**	**422.84**
Mechanical and electrical installation comprising hot and cold water installation; CHP installation; sprinkler system; lighting and power; TV and telephone; fire/smoke alarm; CCTV/intruder alarm; door entry system; external lighting; incoming supply and distribution; heating and ventilation; allowance for BWIC with services	6,855	m²	410		

EXTRA CARE HOME

Extra Care Home	Quantity	Unit	Rate (£)	Total (£)	Cost (£/m²)
Lift installation for 13 persons		item	54,000		
Lift installation for 8 persons		item	34,000		
Preliminaries and Contingency				**1,755,800**	**256.13**
Management costs; site establishment; site supervision @		13%			
Main contractor's overheads and profit @		5%			
Contingency @		3%			
Construction cost (rate based on GIFA)				**9,661,500**	**1,409.42**

MENTAL HEALTH FACILITY

The cost model is based on a new build 49 bed adult acute mental health facility in Greater London. The building has a gross internal floor area of 4,258 m² and is aiming to achieve BREEAM Excellent. Professional fees, external works excluded.

Gross internal floor area   4,258 m²

Model location is Greater London   Model TPI = 124; LF = 1.00

This cost model is copyright of AECOM Ltd

Mental Health Facility	Quantity	Unit	Rate (£)	Total (£)	Cost (£/m²)
Substructure				**1,012,600**	**237.81**
Allowance for site clearance		item	40,000		
Excavate and disposal to formation level		item	60,000		
Plain strip foundations to ground-floor external walls, including concrete, brickwork, DPC etc.	609	m	300		
Plain strip foundations to ground-floor internal walls, including concrete, brickwork, DPC etc.	554	m	270		
Ground-floor slab; assumed 225 mm deep beam and block floor; structural screed; damp-proof membrane; insulation	3,802	m²	150		
Concrete lift pit and walls; excavate and disposal		item	10,000		
Frame and Upper Floors				**558,400**	**131.14**
Allowance for lower ground floor steelwork		item	490,000		
Upper floors and roof deck; 250 mm thick reinforced concrete	4,275	m²	16		
Roof				**752,400**	**176.70**
Single-ply membrane roof, insulation, vapour barrier etc.	3,802	m²	150		
Rainwater goods including anti-scale downpipes and connection to drains	3,802	m²	20		
Rooflights; 2.00 m × 1.00 m	24	nr	1,500		
Mansafe system		item	20,000		
Allowance for screening to rooftop plant		item	50,000		
Stairs				**20,000**	**4.70**
Main feature staircase incl. balustrades, handrails and half landings		item	20,000		
External Walls, Windows and Doors				**1,776,500**	**417.21**
Traditional external wall construction of medium quality facing bricks, cavity, inner blockwork skin and insulation	3,434	m²	250		
Extra over for timber cladding on brickwork	343	m²	150		
Automatic main entrance aluminium double door including all ironmongery double door including all ironmongery	1	nr	15,000		
Automatic FM entrance aluminium double door including all ironmongery	1	nr	10,000		

MENTAL HEALTH FACILITY

Mental Health Facility	Quantity	Unit	Rate (£)	Total (£)	Cost (£/m²)
Automatic rear entrance aluminium double door including all ironmongery	1	nr	10,000		
External aluminium single door including all ironmongery	4	nr	2,500		
External aluminium double door including all ironmongery	14	nr	3,500		
Security windows including all ironmongery	1,030	m²	750		
Internal Walls, Partitions, Windows and Doors				**1,106,700**	**259.91**
Plasterboard partitions with ply backing, 4.00 m high	5,448	m²	90		
IPS panelling to ensuites and WCs	63	nr	1,000		
Internal security-glazed screens	272	m²	360		
Single-leaf timber doorset with anti-ligature ironmongery and vistamatic to bedrooms	49	nr	2,400		
Single-leaf timber doorset with anti-ligature ironmongery to ensuites	49	nr	1,400		
Single-leaf timber doorset with anti-ligature ironmongery to other areas	133	nr	1,400		
Double-leaf timber doorset with anti-ligature ironmongery to other areas	15	nr	1,900		
Airlock double doors between wards	3	nr	7,500		
Airlock double door to main entrance	1	nr	10,000		
Wall Finishes				**338,000**	**79.38**
2 coats emulsion to inner leaf of external wall	3,434	m²	8		
2 coats emulsion to internal partitions	10,896	m²	8		
Extra over for PVC Altro to ensuites/WCs/shower areas	1,500	m²	120		
Sheet-cushioned vinyl to seclusion room and de-escalation room	105	m²	100		
Window boards; 25 mm primed softwood; 1 coat underseal and 2 coats of gloss paint	256	m²	50		
Allowance for enhanced finishes		item	20,000		
Floor Finishes				**338,300**	**79.45**
Self-levelling screed and dpm	4,258	m²	20		
Extra over for screed to falls in showers	50	m²	150		
Vinyl sheet flooring with 100 mm high coved skirting	4,258	m²	55		
Extra for non-slip vinyl flooring to ensuites and bathrooms	177	m²	10		
Extra for resin flooring to seclusion room and de-escalation room	35	m²	80		
Extra barrier matting	15	m²	230		

MENTAL HEALTH FACILITY

Mental Health Facility	Quantity	Unit	Rate (£)	Total (£)	Cost (£/m²)
Extra for timber-sprung floor to fitness suite	40	m²	85		
Ceiling Finishes				**266,500**	**62.59**
Taped and jointed plasterboard ceiling with ply backing	4,258	m²	45		
Extra over for moisture-resistant to wet areas	177	m²	10		
Decoration to the above	4,258	m²	10		
Extra over for tamper-proof access panels to services	30	nr	300		
Allowance for bulkheads		item	21,500		
Fittings and Furnishings				**395,000**	**92.77**
FF&E; Group 1; supply and fix		item	365,000		
FF&E; Group 2; fix only		item	30,000		
Mechanical Installations				**377,500**	**88.66**
Above-ground drainage to saniteryware, utility rooms etc.	256	nr	350		
Above-ground drainage to central kitchen		item	3,500		
Incoming water supply, storage, water treatment and plantroom pipework and equipment		item	43,000		
Cold water distribution	256	outlet	450		
Hot water calorifiers including heat exchanger		item	21,500		
Hot water distribution, 174 nr outlets 95,700	174	outlet	550		
Allowance for providing hot and cold water supply to main kitchen		item	4,000		
Allowance for CAT 5 cold water supply to refuse areas		item	5,000		
Sanitaryware				**244,100**	**57.33**
Clinical anti-ligature wash hand basins	55	nr	750		
Clinical anti-ligature WCs	67	nr	600		
Clinical anti-ligature showers 48 nr 21,600	48	nr	450		
Clinical anti-ligature assisted WCs 14 nr 14,000	14	nr	1,000		
Clinical anti-ligature assisted wash hand basins	14	nr	750		
Clinical anti-ligature assisted showers	10	nr	450		
Patient assisted bathroom complete with WC and wash hand basins	4	nr	13,500		
WCs	9	nr	500		
Wash hand basins	13	nr	400		
Showers	4	nr	250		
Multi-faith foot wash	1	nr	800		
Kitchen sinks	5	nr	500		
Combined sink and slop hoppers	4	nr	4,000		

Building Cost Models

MENTAL HEALTH FACILITY

Mental Health Facility	Quantity	Unit	Rate (£)	Total (£)	Cost (£/m²)
Cleaner's sinks	4	nr	1,250		
Installation of the above	256	nr	90		
Heat Source, Space Heating, Air Treatment and Ventilation Systems				**1,333,000**	**313.06**
Natural gas supplies to primary plant		item	12,500		
Natural gas to therapy kitchen/servery etc.		item	13,000		
Provisional allowance for providing 3 nr 500 kW gas-fired condensing boilers, including gas train, controls and flue	3	nr	20,000		
Plantroom pipework, valves and fittings etc.		item	87,500		
CHP including flue		item	130,000		
Plantroom pipework, valves and fittings etc.		item	43,000		
LTHW plantroom pipework, valves and fittings (secondary) including equipment etc.		item	43,000		
LTHW heating distribution including insulation		item	130,000		
Heat emitters (radiators, air curtains)		item	11,000		
Underfloor heating		item	150,000		
Chilled water plant and distribution		item	110,000		
Local cooling		item	20,000		
General supply & extract AHUs and ventilation ductwork distribution including grilles, dampers etc.		item	405,000		
Dirty extract fans and associated ductwork, grilles, dampers etc.		item	110,000		
Kitchen extract system		item	8,000		
Electrical Installations				**1,011,500**	**237.55**
Main LV switch panel and incoming supply		item	65,000		
Containment 51,096		item	52,500		
LV sub-mains distribution including distribution boards		item	77,500		
Small power (1,250 pts)		item	215,000		
Lighting (combination of non-anti-ligature and anti-ligature)		item	300,000		
Emergency lighting (combination of non-anti-ligature and anti-ligature)		item	110,000		
Lighting control including wiring		item	110,000		
Earthing and bonding		item	21,500		
Supply and installation of 13-person machine room less passenger lift to 2 floors, DDA compliant		item	60,000		

MENTAL HEALTH FACILITY

Mental Health Facility	Quantity	Unit	Rate (£)	Total (£)	Cost (£/m²)
Protective, Communication and Specialist Installation				**2,532,300**	**594.72**
Lighting protection		item	13,000		
Gas suppression		item	13,000		
Fire alarm including void detection		item	130,000		
CCTV installation (internally only)		item	110,000		
Nurse call alarms including WCs		item	130,000		
Staff attack system		item	175,000		
Data and communication (circa 1,300 points)		item	150,000		
TV/radio cabling (wire-ways only)		item	17,500		
Induction loop		item	4,300		
Intruder alarm		item	13,000		
Access control based on circa 170 nr doors		item	260,000		
Video entry		item	10,000		
Public address/voice alarm		item	10,000		
Containment		item	43,000		
Controls/BMS		item	215,000		
Leak detection		item	215,000		
Testing and commissioning		7.5%			
Mechanical and Electrical subcontractor preliminaries		12.0%			
Builders work in connection with services		item	110,000		
Preliminaries and Contingency				**2,533,200**	**594.93**
Contractors preliminaries, overheads and profit		21.0%			
Construction cost (rate based on GIFA)				**14,596,000**	**3,428**

PALLIATIVE CARE

The cost model is for a three storey extension to an existing palliative care unit, creating inpatient and office accommodation with associated ancillary support services; both internal and external finishes to a good quality.

Gross internal floor area, including tiers　　　　　　　1,444 m²

Model location is Greater London　　　　Model TPI = 124; LF = 1.00

This cost model is copyright of AECOM Ltd

Palliative Care	Quantity	Unit	Rate (£)	Total (£	Cost (£/m²)
Substructure				**294,000**	**203.60**
Allowance for site clearance, excavation into gradient site & disposal		item	87,500		
Excavate and dispose of slab and foundations		item	33,000		
450 mm wide × 500 mm deep strip foundation with 140 mm load-bearing blockwork to DPC	25	m³	280		
650 mm wide × 500 mm deep strip foundation with 140 mm load-bearing blockwork inner skin and 103 mm facing brickwork to DPC	40	m³	280		
175 mm thick reinforced concrete ground bearing slab with 2 layers of A142 mesh	630	m²	110		
Concrete lift pit and walls; excavate and disposal		item	8,700		
500 mm thick reinforced slab to underside of PCC retaining wall	87	m²	220		
Allowance for 75 mm structural screed to Ground Floor slab	630	m²	33		
Installation of new drainage system, including connection and integration within new substructure works	630	m²	33		
Waterproofing details		item	16,500		
Frame and Upper Floors				**221,900**	**153.67**
Allowance for lower ground floor steelwork		item	8,700		
Upper floors – 150 mm precast concrete hollow core floor units (5.00 kN/m²)	500	m²	80		
Allowance for 75 mm structural screed	500	m²	33		
Padstones – allowance for lower ground floor		item	1,650		
Allowance for timber framing to ground floor, first floor & roof		item	155,000		
Roof				**132,700**	**91.90**
Concrete roof tiles, including roof coverings, ridges, troughs, hips etc., 370 m² @ £80	370	m²	90		
Double glazed rooflights and frames, including ironmongery, restrictors and vents	12	nr	1,100		
Rainwater goods	700	m²	22		
Man safe system		item	8,200		
Parapet wall detail; aluminium capping	40	m	180		

PALLIATIVE CARE

Palliative Care	Quantity	Unit	Rate (£)	Total (£	Cost (£/m²)
Single ply membrane roof, insulation, vapour barrier etc.	340	m²	150		
Allowance for roof access hatch		item	4,350		
Stairs				28,000	19.39
Main feature staircase incl. balustrades, handrails and half landings		item	28,000		
External Walls, Windows and Doors				286,600	198.48
Traditional external wall construction of medium quality facing bricks, cavity, inner blockwork skin and insulation	740	m²	200		
Extra over wall tiles; including battens	190	m²	90		
Extra over render; assumed 2 coats at 15 mm thick	180	m²	55		
Aluminium framed glazed system, incl. all ironmongery	50	m²	490		
Double glazed windows and frames, incl. all ironmongery, lintels, restrictors and vents	70	m²	420		
Window boards; 25 mm primed softwood; 1 coat underseal and 2 coat gloss	40	m	55		
Main entrance aluminium double doors incl. all ironmongery	1	nr	11,000		
Aluminium external doorsets, incl. all ironmongery, double	4	nr	2,650		
Aluminium external doorsets, incl. all ironmongery, single leaf and a half	12	nr	2,000		
Aluminium louvre double plant room door and frame, incl. all ironmongery; fire-rated FD60	3	nr	3,300		
Internal Walls, Partitions, Windows and Doors				430,600	298.20
140 mm thick blockwork partitions/metsec stud	870	m²	70		
Metal stud internal walls; 2 layers of plasterboard to each side	960	m²	95		
Internal glazed screens in softwood framing	60	m²	550		
IPS panelling		item	22,000		
Allowance for bed head trunking		item	49,000		
Extra over for pattressing walls		item	3,300		
Allowance for lintels to door openings		item	5,500		
Allowance for single leaf timber doors	40	nr	925		
Allowance for one and a half leaf timber doors	55	nr	1,550		
Allowance for double leaf timber doors	9	nr	2,000		
Allowance for single leaf glazed doors	5	nr	1,750		
Allowance for double leaf glazed doors	2	nr	2,850		
Extra over for fire protection (FD30)		item	11,000		

Building Cost Models

PALLIATIVE CARE

Palliative Care	Quantity	Unit	Rate (£)	Total (£	Cost (£/m²)
Wall Finishes				128,100	88.71
Plaster to blockwork and skim finish to stud walls	1,240	m²	16		
Allowance for splashback to wet areas		item	2,200		
Whiterock to bathrooms	780	m²	75		
100 mm skirting; 1 coat of underseal and 2 coats of high gloss paint		item	2,200		
Allowance for sundry items [corner protection, handrails]		item	11,000		
Allowance to paint all walls with two coats of emulsion paint	3,110	m²	11		
Floor Finishes				68,900	47.71
Self-levelling screed	1,380	m²	11		
Heavy duty carpet, incl. underlay etc.	368	m²	33		
Altro vinyl sheet flooring with 100 mm high coved skirting	1,000	m	38		
E/O wet areas	110	m²	11		
Epoxy resin floor paint; Watco Dustop or equivalent	12	m²	22		
Barrier matting – generally		item	2,200		
Ceiling Finishes				97,200	67.31
Taped and jointed plasterboard ceiling	1,380	m²	38		
E/O for access panels to services		item	11,000		
E/O for moisture-resistant		item	2,200		
E/O for fire protection		item	16,500		
Decoration to the above	1,380	m²	11		
Fittings and Furnishings				165,700	114.75
Nurse base station, with built-in storage	2	nr	8,700		
Shelving; generally		item	5,500		
Mirrors; generally		item	3,300		
Hand/Grab rail; generally		item	22,000		
Window blinds/curtain		item	5,500		
Worktops, base units and the like		item	35,000		
Ceiling hoist track		item	55,000		
Staff lockers		item	5,500		
Client and statutory signage; generally		item	5,500		
Sundry and fitting of Group 2 items		item	11,000		
Mechanical Installations				447,600	309.97
WC	29	nr	725		
Clinical wash hand basin	32	nr	410		
Wash hand basins	29	nr	460		

PALLIATIVE CARE

Palliative Care	Quantity	Unit	Rate (£)	Total (£	Cost (£/m²)
Stainless steel sink kitchen sink with drainer	3	nr	380		
Tea point	3	nr	380		
Shower, with thermostatic mixer	29	nr	1,100		
Sluice	3	nr	3,300		
Installation of the above		item	11,000		
Disposal installations		item	87,500		
Water installations		item	135,000		
Heat source		item	33,000		
Space heating and air conditioning		item	67,500		
Ventilating systems		item	22,000		
Electrical Installations				**670,000**	**463.99**
Electrical mains and sub-mains distributions		item	77,500		
Lighting		item	135,000		
Allowance for external lighting		item	38,000		
Small power		item	49,000		
Fuel installations/systems		item	5,500		
Lift and conveyor installations/systems		item	87,500		
Fire and lightning protection		item	5,500		
Communication, security and control systems		item	185,000		
External services		item	11,000		
Testing and commissioning of building services installations		item	29,000		
Builders work in connection		item	47,000		
Preliminaries and Contingency				**594,200**	**411.50**
Management costs, site establishment and site supervision. Contractor's preliminaries, overheads and profit @		18%			
Construction cost (rate based on GIFA)				**3,565,500**	**2,469**

Building Cost Models

PRIMARY CARE HEALTH CENTRE

A three-storey new build primary care health centre for GPs, community nurses, midwifery services, mental health services. The building in the cost study contains a gross floor area of 2,500 m² located in Greater London. Costs include for group 1 and fitting of group 2 furniture, fixtures and equipment, but exclude external works, utilities, design/construction/client contingency, professional fees.

Gross internal floor area 2,500 m²

Model location is Greater London Model TPI = 124; LF = 1.00

This cost model is copyright of AECOM Ltd

Primary Care Health Centre	Quantity	Unit	Rate (£)	Total (£)	Cost (£/m²)
Substructure				**457,000**	**182.80**
Allowance for site clearance		item	21,500		
Excavate and disposal to formation		item	48,000		
Perimeter footings	115	m³	550		
Pad foundations	220	m³	500		
Groundbearing slab	1,150	m²	140		
Concrete lift pit		item	11,000		
Drainage	1,150	m²	32		
Allowance for soft spots and below ground obstructions		item	5,400		
Frame and Upper Floors				**621,300**	**248.52**
Steel frame, including fittings, brackets and supports		item	330,000		
Allowance for intumescent paint, 60 min fire-rating		item	87,500		
150 mm precast concrete floor	1,350	m²	80		
Allowance for 75 mm structural screed	1,350	m²	55		
Blockwork to lifts and stairwell, 200 mm thick		item	21,500		
Roof				**519,900**	**207.96**
150 mm deep precast concrete floor	1,200	m²	80		
Bitumen felt roof waterproofing system; including underlay, vapour control layer, insulation, flashings, edge trims,	1,200	m²	140		
Double-glazed rooflights and frames, including ironmongery, restrictors and vents	80	m²	925		
Rainwater goods	1,200	m²	22		
Mansafe system		item	21,500		
Parapet capping	230	m	105		
Plant screening		item	38,000		
Main entrance canopy		item	27,000		
Roof terrace and balustrade	40	m²	320		
Roof sundries: walkways, access hatches, plant bases		item	32,000		

PRIMARY CARE HEALTH CENTRE

Primary Care Health Centre	Quantity	Unit	Rate (£)	Total (£)	Cost (£/m²)
Stairs				**65,000**	**26.00**
Precast concrete staircases, 2 nr from ground to second floor, including handrails and finishes		item	60,000		
External Walls, Windows and Doors				**731,200**	**292.48**
External insulated render and cavity wall	1,460	m²	240		
Extra for timber cladding	380	m²	190		
Aluminium framed glazed system	220	m²	675		
Aluminium framed double-glazed windows	250	m²	410		
Main entrance aluminium automatic double doors and ironmongery		item	11,000		
Aluminium external double doorsets	4	nr	2,700		
Aluminium louvre double plant room door, FD 60	2	nr	3,200		
Insulated render soffit	80	m²	300		
Building signage		item	5,400		
Internal Walls and Partitions				**456,900**	**182.76**
Metal stud internal walls; 2 layers plasterboard to each side	965	m	280		
Metal wall lining to rear of external wall	1,460	m²	95		
Allow for column boxing		item	48,000		
Internal Doors				**305,400**	**122.16**
Internal single leaf timber doors	85	nr	1,025		
Internal 1.5 leaf timber doors	45	nr	1,750		
Internal double leaf timber doors	6	nr	2,000		
Additional acoustic and fire protection to walls and floors		item	55,000		
Internal glazing		item	16,000		
IPS panelling		item	27,000		
Glazed balustrade	15	m	700		
Sliding partition	5	m	1,600		
Shutter to reception	1	nr	11,000		
Wall Finishes				**105,200**	**42.08**
Emulsion paint to walls	7,250	m²	11		
Coloured Perspex screens to reception		item	11,000		
Splashback tiling to wet areas		item	2,700		
Whiterock finish to wet areas	100	m²	85		
Sundry items, handrails, protection barriers		item	5,400		
Floor Finishes				**138,800**	**55.52**
Self-levelling screed	2,500	m²	11		
Vinyl sheet flooring	2,125	m²	41		

PRIMARY CARE HEALTH CENTRE

Primary Care Health Centre	Quantity	Unit	Rate (£)	Total (£)	Cost (£/m²)
Extra for wet areas	100	m²	11		
Carpet to office areas	150	m²	32		
Ceramic tiling to entrance areas	150	m²	80		
Epoxy resin floor paint to plant rooms	75	m²	22		
Barrier matting		item	5,400		
Ceiling Finishes				**134,000**	**53.60**
Mineral fibre suspended ceilings	2,275	m²	38		
Acoustic timber slatted ceiling	150	m²	270		
Bulkheads and access panels to services		item	5,400		
Moisture-resistant ceiling tiles	100	m²	16		
Fittings and Furnishings				**191,000**	**76.40**
Reception area desk		item	15,000		
Secondary welcome points with storage	2	nr	6,600		
General storage, worktops and room cabinetry	2,500	m²	36		
Statutory signage and wayfinding		item	16,000		
Built-in specimen and prescription-off boxing		item	6,900		
Window blinds	470	m²	38		
Curtain tracks		item	16,000		
Mirrors		item	5,400		
Staff lockers		item	27		
Sundry items and fitting of Group 2 FF&E items		item	8,000		
Sanitary Fittings and Disposal Installations				**161,000**	**64.40**
WCs including pan, concealed cistern and waste; WHBs and clinical WHBs including taps, waste and trap; disabled WCs including pan, concealed cistern, waste and grab rails; showers including tray, thermostatic valves and head; dirty utility disposal units, cleaners' sinks including bib taps, waste and trap, tea point sink units including taps, waste, trap and boiling/ chilled water dispensers,	2,500	m²	43		
Rainwater disposal installation including gullies and downpipes	2,500	m²	5		
Above ground SVP disposal and condensate installation	2,500	m²	16		
Water Installations				**154,300**	**61.72**
Cold water system	2,500	m²	28		
HWS distribution including incoming storage calorifiers, circulating pumpset and connections to sanitaryware,	2,500	m²	32		
Miscellaneous Cat 5 water supplies		item	4,250		
Cold water system	2,500	m²	28		

PRIMARY CARE HEALTH CENTRE

Primary Care Health Centre	Quantity	Unit	Rate (£)	Total (£)	Cost (£/m²)
Space Heating and Air Treatment				653,500	261.40
Low-nox condensing boilers, primary pumpset, primary distribution and flue		item	27,000		
LTHW distribution including heat emitters/ underfloor heating and over door heaters	2,500	m²	45		
Supply and extract AHUs, DX condenser unit, supply and extract ductwork distribution, grilles and diffusers	2,350	m²	190		
DX cooling systems to serve reception/waiting areas/office/IT hub rooms	2,500	m²	27		
Ventilation Installations				16,000	6.40
Combined extract systems linked to general extract distribution serving WC areas, dirty utility rooms, baby change	2,500	m²	6		
Gas Installations				6,400	2.56
Gas Installations, all costs associated with the supply and installation of gas		item	6,400		
Electrical Installations				611,800	244.72
Incoming LV supply, main distribution panel, containment, sub-mains cabling, distribution boards, PV panels, mechanical supplies, earthing and bonding, leak detection to IT hub rooms	2,500	m²	80		
Photovoltaics	50	m²	625		
Luminaires including emergency luminaires with 3 hr battery packs, final circuits, lighting control, medical lighting, enhanced lighting to reception	2,500	m²	105		
Medical lighting		item	38,000		
Small power distribution	2,500	m²	32		
Protective Installations				5,500	2.20
Lightning protection; earthing installations:	2,500	m²	2		
Communications Installations				387,800	155.12
Data outlets, cabling and wireways	2,500	m²	18		
Fire detection and alarm system and disabled refuge	2,500	m²	34		
Intruder detection, CCTV, access control, nurse call, disabled WC alarm systems, intercoms	2,500	m²	55		
BMS controls and occupancy monitoring	2,500	m²	48		
Lifts				130,000	52.00
1 nr MRL stretcher evacuation lift; 1600 kg, 21 person serving 2 levels; 0.63 m/s minimum speed; 1 nr MRL passenger lift; 1,275 kg, 13 person serving 3 levels		item	130,000		

PRIMARY CARE HEALTH CENTRE

Primary Care Health Centre	Quantity	Unit	Rate (£)	Total (£)	Cost (£/m²)
Builders Work in Connection with Services				63,800	25.52
Builder's work in connection with services		3%			
Preliminaries and Contingency				1,183,100	473.24
Contractors preliminaries, overheads and profit		20%			
Construction cost (rate based on GIFA)				7,098,900	2,839.56

PRIMARY SCHOOL EXTENSION

A single storey, three-classroom extension to a primary school. Constructed using traditional masonry cavity walls on concrete strip foundations, with a pitched tiled roof over. Individual classrooms are formed by load-bearing blockwork partitions.

Gross internal floor area including tiers	310 m²
Model location is South East England	Model TPI = 120; LF = 0.96

This cost model is copyright of AECOM Ltd

Primary School: Three Classroom Extension	Quantity	Unit	Rate (£)	Total (£)	Cost (£/m²)
Substructure				**49,900**	**160.97**
Excavation & disposal	140	m³	35		
Concrete strip foundations, masonry work below DPC; blockwork and facing brickwork	90	m	190		
Reinforced in situ concrete ground slab, including service trench; vapour barrier; hardcore, excavation and disposal	310	m²	90		
Roof				**65,900**	**212.58**
Softwood roof trusses	380	m²	52		
Board insulation to roof	310	m²	23		
Cement slate roofing including all eaves, ridge tiles and labours; measured on plan	380	m²	70		
Aluminium rainwater down pipes	40	m	95		
Aluminium gutters	90	m	85		
Fire barriers	40	m	23		
External Walls				**41,800**	**134.84**
Brick cavity wall, facing brick outer skin with cavity, 140 mm inner blockwork leaf	220	m²	190		
Windows and External Doors				**39,100**	**126.13**
Proprietary aluminium framed, double glazed windows, doors and solid aluminium faced panels; powder coated finish	55	m²	490		
Double door steel security doors, including all ironmongery	1	nr	2,000		
Double door aluminium framed glazed doors and screens to paved areas, including all ironmongery	15	m²	675		
Internal Walls and Partitions				**15,800**	**50.97**
Partitions; 100/140 blockwork	285	m²	46		
WC cubicle partitions; laminated plastics, including all ironmongery	4	nr	650		
Internal Doors				**12,600**	**40.65**
Internal fire doors, Georgian wired glass vision panel, stainless steel ironmongery to classrooms	5	nr	1,150		

PRIMARY SCHOOL EXTENSION

Primary School: Three Classroom Extension	Quantity	Unit	Rate (£)	Total (£)	Cost (£/m²)
Double fire door, Georgian wired glass, stainless steel ironmongery to corridor	1	nr	1,650		
Wrot softwood storage cupboard door, stainless steel ironmongery	5	nr	575		
Non fire-rated doors, stainless steel ironmongery to WC	3	nr	750		
Wall Finishes				**18,600**	**60.00**
Plaster and 1 mist coat, 3 coats of emulsion paint	760	m²	21		
Ceramic wall tiles, full height in toilets and selected classroom areas	40	m²	70		
Floor Finishes				**20,000**	**64.52**
Sand cement screed	310	m²	23		
Carpet tiles	150	m²	29		
Safety vinyl to WC areas & practical areas, including skirtings	90	m²	46		
Heavy duty vinyl to circulation areas, including skirtings	68	m²	52		
Entrance matting with aluminium matwell	2	m²	410		
Ceiling Finishes				**12,900**	**41.61**
Plasterboard ceiling, plaster skim and emulsion paint finish	280	m²	41		
Moisture resistance plasterboard ceiling, plaster skim and emulsion paint finish	30	m²	46		
Furniture and Fittings				**25,000**	**80.65**
Storage trays containers	3	nr	3,150		
Storage units – allowance of 1 double cupboard per classroom	3	nr	650		
Worktops, including cut out for sink, 3000 × 600 mm	3	nr	170		
Coat hooks, fixed to masonry	100	nr	35		
Pinboards, 1000 × 2000 mm	6	nr	95		
Pinboards, 1000 × 1200 mm	2	nr	65		
Whiteboards: Interactive	3	nr	1,850		
Whiteboards: Magnetic	3	nr	130		
Signage	1	item	2,550		
Mirrors 640 × 460 mm	7	nr	52		
Sanitary Fittings				**8,300**	**26.77**
WCs	6	nr	400		
Urinals, including side panel	2	nr	270		
Hand basins	8	nr	250		
Disabled toilet, including WC, wash hand basin, grab rails and other fittings	1	nr	1,500		

PRIMARY SCHOOL EXTENSION

Primary School: Three Classroom Extension	Quantity	Unit	Rate (£)	Total (£)	Cost (£/m²)
Single stainless steel sinks to classrooms, 1200 × 600 mm	3	nr	290		
Cleaners sink, 510 × 380 mm	1	nr	925		
Disposal Installations				7,200	23.23
Waste, soil and vent installation; uPVC pipework and fittings	310	m²	23		
Water Installations				12,700	40.97
Cold water points	21	nr	350		
Hot water points	13	nr	410		
Space Heating and Air Treatment				46,500	150.00
Space heating, all costs associated with the supply & installation of the heating system, temperature control and distribution pipework	310	m²	150		
Ventilation Installations				5,800	18.71
Ventilation extraction to toilet areas		item	5,800		
Electrical and Gas Installations				45,600	147.10
Mains and sub-mains installation	310	m²	29		
Small power installation	310	m²	41		
Lighting and general luminaires; emergency lighting	310	m²	70		
Gas installations, all costs associated with the supply and installation of gas	310	m²	7		
Protective, Communications and Special Installations				17,400	56.13
Lightning protection		item	1,400		
Fire alarm installation; smoke detectors; call points		item	5,200		
Telephone and data wireways; internal telephone system		item	2,900		
Security installation; intruder detection, CCTV to existing control unit etc.		item	7,900		
Builders Work				6,900	22.26
Forming holes, chases etc.		item	6,900		
Preliminaries and Contingency				108,000	348.39
Management costs; site establishment; site supervision @		18%			
Contractors contingency @		5%			
Construction cost (rate based on GIFA)				**560,000**	**1,806.48**

Building Cost Models

SECONDARY SCHOOL BLOCK

A small secondary school extension to provide a new science block with two teaching rooms attached to an existing building.

Gross internal floor area including tiers	200 m²
Model location is South East England	Model TPI = 120; LF = 0.96

This cost model is copyright of AECOM Ltd

Science Block and Classrooms	Quantity	Unit	Rate (£)	Total (£)	Cost (£/m²)
Substructure				17,800	89.00
Excavation and disposal	45	m³	37		
Concrete strip foundations, masonry work below DPC; blockwork and facing brickwork	45	m	180		
Reinforced in situ concrete ground slab, vapour barrier; granular fill, insulation (0.25 W/m².K)	100	m²	80		
Upper Floors and Stairs				14,800	74.00
Prestressed precast concrete plank floor 200 mm thick	100	m²	70		
Precast concrete staircase with half landing	1	nr	5,900		
Nylon coated steel handrail	20	m	95		
Roof				19,800	99.00
Steel profiled sheet roof decking with single layer waterproof membrane, insulation and vapour control layer	100	m²	190		
Aluminium rainwater pipework and hopper heads	14	m	53		
External Walls				56,200	281.00
Facing brickwork, insulated cavity, 100 mm concrete blockwork inner leaf	85	m²	150		
Timber cladding on battens with breather membrane on block cavity wall	170	m²	230		
Aluminium flashings to timber boarding	115	m	37		
Windows and External Doors				33,000	165.00
Polyester powder coated double glazed aluminium windows, steel lintels, painted window boards	44	m²	510		
Polyester powder coated double glazed aluminium doors, steel lintels					
Single leaf external doors	3	nr	2,250		
Leaf and a half entrance doors	1	nr	3,800		
Internal Walls and Partitions				6,000	30.00
Lightweight aerated concrete blockwork 140 mm thick	130	m²	46		
Internal Doors				5,000	25.00
Painted solid core timber doors with vision panels, linings and ironmongery	6	nr	825		

SECONDARY SCHOOL BLOCK

Science Block and Classrooms	Quantity	Unit	Rate (£)	Total (£)	Cost (£/m²)
Wall Finishes				11,900	59.50
Plaster and eggshell paint	600	m²	20		
Floor Finishes				17,800	89.00
Screed, latex and slip-resistant vinyl sheet flooring and skirtings to toilets	9	m²	75		
Screed, latex and vinyl safety flooring and skirtings to classrooms	170	m²	75		
Screed, latex and heavy contract polypropylene carpet to lobbies and landings	50	m²	53		
Aluminium stair edging	30	m	40		
Painted timber skirting	50	m	11		
Ceiling Finishes				6,600	33.00
Demountable suspended ceiling, sound-absorbing tiles	200	m²	33		
Furniture and Fittings				18,900	94.50
Black out blinds	9	nr	160		
Science laboratory furniture	2	nr	8,600		
Mirrors	3	nr	70		
Sanitary Fittings				5,500	27.50
Plastic laminate wall linings	3	nr	430		
WC suite	4	nr	460		
Wash hand basin	3	nr	330		
Warm air hand dryers	4	nr	330		
Disposal Installations				5,000	25.00
Waste, soil and vent installation; uPVC pipework and fittings	200	m²	25		
Water Installations				5,800	29.00
Hot and cold water supply	200	m²	29		
Space Heating and Air Treatment				36,000	180.00
Space heating, all costs associated with the supply and installation of the heating system, temperature control and distribution pipework	200	m²	180		
Ventilation Installations				9,000	45.00
Ventilation extraction to toilet areas	200	m²	43		
Electrical and Gas Installations				45,600	228.00
Distribution boards and sub-mains	200	m²	20		
General power installation	200	m²	59		
General lighting installation	200	m²	120		
Gas installation	200	m²	29		

SECONDARY SCHOOL BLOCK

Science Block and Classrooms	Quantity	Unit	Rate (£)	Total (£)	Cost (£/m²)
Protective, Communications and Special Installations				**26,600**	**133.00**
Lightning protection	200	m²	11		
Data installation	200	m²	22		
Fire alarm installation	200	m²	22		
Intruder alarm installation	200	m²	21		
CCTV installation	200	m²	24		
Door access installation	200	m²	33		
Builders Work				**17,700**	**88.50**
Forming holes, chases etc.	200	m²	16		
Alterations in connecting to existing building		item	14,500		
Preliminaries and Contingency				**96,300**	**481.50**
Management costs; site establishment; site supervision @		18%			
Contingency @		7.5%			
Construction cost (rate based on GIFA)				**455,300**	**2,276.50**

LABORATORY

This cost model is based on a university-funded research laboratory. The laboratories are primarily containment level 2 and are constructed to VC-A criteria. All floors have a 4.6 m storey height and the building achieves an overall BREEAM 'Excellent' rating. The building has a gross internal floor area of 11,000 m², of which 3,800 m² is net laboratory and associated technical space, and 3,000 m² is office, write-up and teaching space. Services are distributed vertically to the laboratory areas from a rooftop plant room. The offices are naturally ventilated along with under-floor displacement ventilation. The costs of site preparation, external works and external services are not included. Loose fittings, equipment and specialist laboratory equipment are also not included.

Gross internal floor area including tiers	11,000 m²
Model location is South East	Model TPI = 120; LF = 0.96

This cost model is copyright of AECOM Ltd

Laboratory	Quantity	Unit	Rate (£)	Total (£)	Cost (£/m²)
Substructure				**3,176,500**	**288.77**
CFA piles, 450 mm and 600 mm, including piling mat	3,400	m²	320		
Excavation, including lift pits, disposal off site	2,300	m³	90		
Formation of ground-floor slab, 350 mm thick, including pile caps and beams	3,400	m²	390		
Below slab drainage	3,400	m²	70		
Incoming service trenches; water, gas, electric and data		item	87,500		
Allowance for sundry groundwork items		item	230,000		
Upper Floors and Stairs				**3,111,000**	**282.82**
Reinforced concrete core/shear walls, 250–450 mm thick	1,350	m²	320		
Reinforced concrete columns and beams, various sizes	10,000	m²	75		
Reinforced concrete floor slabs; 275–400 mm thick	6,600	m²	190		
Structural steel columns and beams to roof structures, fire protection	175	tonne	3,200		
Allowance for sundry items, fire stopping		item	115,000		
Roof				**1,707,500**	**155.23**
Reinforced concrete roof slab; 300 mm thick	3,400	m²	180		
Single-ply membrane roof including insulation, upstands, parapets and openings for services	2,400	m²	210		
Proprietary metal standing seam roof covering, including insulation	1,000	m²	290		
Feature glazed roof light		item	175,000		
Allowance for rainwater goods, gutters and downpipes		item	115,000		
Fall arrest		item	11,500		

LABORATORY

Laboratory	Quantity	Unit	Rate (£)	Total (£)	Cost (£/m²)
Stairs and Ramps				**597,500**	**54.32**
Reinforced in situ concrete stair, general circulation, including handrails	11	nr	15,000		
Feature atrium stairs and balustrades	2,400	item	230,000		
Allowance for access ladders, steps and plant room walkways		item	115,000		
Balustrades and handrails to external balcony and courtyard		item	87,500		
External Walls and Doors				**5,690,300**	**517.30**
Metal rainscreen cladding system, two finishes, including Metsec, concealed fixings, flashings trims	2,780	m²	825		
Proprietary metal cassette cladding panels, concealed fixings, flashings, trims	1,320	m²	650		
Aluminium mesh screens to plant areas	720	m	480		
Allowance for sundry brickwork		item	87,500		
Curtain walling, powder coated, capped system, including opening windows to offices	2,840	m²	700		
Allowance for brise-soleil	3	nr	875		
External doors; powder coated aluminium glazed doorsets including ironmongery		item	115,000		
Internal Walls and Partitions				**1,718,600**	**156.24**
Metal stud partitions; dB ratings 40 dB–65 dB	9,600	m²	100		
Glazed internal partitions	680	m²	520		
Allowance for forming holes for laboratory services through walls		item	175,000		
Allowance for fire protection through penetrations		item	230,000		
Internal Doors				**854,000**	**77.64**
Solid core doors, door frames, ironmongery, vision panels, PVC coated, to laboratories	160	nr	3,200		
Solid core doors, door frames, ironmongery, vision panels, veneered, to offices	180	nr	1,900		
Wall Finishes				**643,500**	**58.50**
Painting to walls, emulsion	22,700	m²	7		
Timber acoustic wall finishes to offices and atrium	1,560	m²	260		
Allowance for enhanced finishes to specialist laboratories		item	57,500		
Ceramic tiles to WCs and shower cores	250	m²	95		
Floor Finishes				**1,196,000**	**108.73**
Levelling screed and dpm	4,900	m²	40		
Medium duty raised access floor	3,650	m²	65		
Non-slip vinyl floor finish, including coved skirting	4,900	m²	90		

LABORATORY

Laboratory	Quantity	Unit	Rate (£)	Total (£)	Cost (£/m²)
Carpet, heavy duty carpet tiles, including timber skirting	3,650	m²	65		
Ceramic floor tiles	300	m	90		
Allowance for sundry floor finishes		item	57,500		
Ceiling Finishes				1,223,500	111.23
Suspended metal plank system	4,900	m²	100		
Suspended perforated plasterboard systems, including acoustic treatment and painting	3,650	m²	140		
Suspended plasterboard system, including painting	300	m²	70		
Allowance for painting concrete soffits		item	29,000		
Allowance for fire protection above ceilings		item	115,000		
Allowance for sundry ceiling finishes		item	57,500		
Furniture and Fittings				6,198,000	563.45
Cold room fit-out	2	nr	46,000		
Autoclaves; single sided and pass through, various sizes	4	nr	145,000		
Allowance for fixed laboratory benching, shelving, service spines, under bench and over bench cupboards, lab sinks, wash hand basins and eye wash stations	3,800	m²	750		
Fume cupboards and safety cabinets	90	nr	12,000		
Allowance for fit-out to specialist laboratories		item	570,000		
WC fittings, IPS panelling, cubicles, mirrors, hand dryers etc.	30	nr	3,450		
Allowance for miscellaneous fixtures and fittings, reception desk, kitchenettes, copy stations, whiteboards etc.		item	175,000		
Window blinds	1,500	m²	140		
Lecture theatre seating, raked	250	nr	550		
Audio visual installation, to lecture theatre, meeting rooms, communal spaces		item	285,000		
Allowance for signage, statutory, directional and external building signage		item	115,000		
Sanitary Fittings and Disposal Installations				410,200	37.29
Allowances for sanitary appliances including WCs, urinals, wash hand basins, disabled access fittings, showers	11,000	m²	11		
Rainwater disposal	11,000	m²	5		
Above ground drainage and condensate	11,000	m²	9		
Laboratory drainage	3,800	m²	35		

LABORATORY

Laboratory	Quantity	Unit	Rate (£)	Total (£)	Cost (£/m²)
Water Installations				1,161,900	105.63
Incoming mains and domestic cold water system	11,000	m²	28		
Domestic hot water system	11,000	m²	14		
Laboratory cold water installation	3,800	m²	29		
Purified water system to laboratory spaces	3,800	m²	60		
Allowance for irrigation and WC flushing system		item	23,000		
Allowance for centralized steam installation, serving autoclaves and laboratory appliances	3,800	m²	90		
Heat Source				902,800	82.07
Ground source heat pump installation, including boreholes, pipework and all associated plant		item	740,000		
Gas fired boilers including flues	11,000	m²	15		
Space Heating and Air Treatment				4,255,000	386.82
Low temperature hot water heating installation	11,000	m²	90		
Supply and extract ventilation to laboratory spaces including air handling units, heat recovery and ductwork	3,800	m²	450		
Supply and extract ventilation to office/write-up spaces including air handling units, heat recovery and ductwork	3,000	m²	40		
Extra for supply and extract ventilation to lecture theatre		item	57,500		
Chilled water system including air cooled chillers, plantroom installation, distribution pipework, fan coil units and chilled beams	11,000	m²	120		
DX cooling to IT rooms		item	57,500		
Ventilation Installations				1,017,000	92.45
Ventilation extraction to toilet areas		item	29,000		
Fume cupboard and safety equipment extract systems	3,800	m²	260		
Electrical and Gas Installations				3,811,000	346.45
HV installation including incoming main		item	52,000		
LV panel, sub-mains installation and distribution boards	11,000	m²	60		
Containment generally	11,000		52		
Lighting generally, including feature lighting and external lighting	11,000		150		
Small power generally including power provision to laboratory benches	11,000		52		
Standby generator and fuel supply		item	130,000		
Photovoltaic installation including inverters		item	175,000		

LABORATORY

Laboratory	Quantity	Unit	Rate (£)	Total (£)	Cost (£/m²)
Fuel Installation				98,900	8.99
Gas installation generally including external supply		item	11,000		
Laboratory gas installation	3,800	m²	21		
Lift Installation				490,000	44.55
Goods lift	1	nr	200,000		
Passenger lift	2	nr	145,000		
Fire and Lightning Protection				48,300	4.39
Dry riser installation		item	23,000		
Lightning protection	11,000	m²	2		
Communications, Security and Control Systems				2,292,000	208.36
Fire alarm and detection system	11,000	m²	23		
Data installation, category 6 wiring	11,000	m²	35		
Allowance for disabled induction loops, disabled refuge and alarm system		item	17,500		
Security system; access control		item	145,000		
Security system; intruder alarm system	11,000	m²	4		
Security system; CCTV system		item	23,000		
Building management system	11,000	m²	130		
Specialist Installations				1,281,000	116.45
Compressed air system including packaged plantroom	3,800	m²	100		
Allowance for laboratory gases, including four reticulated, four directly fed to various rooms and gas detection system	3,800	m²	220		
Allowance for process chilled water system, complete with distribution pipework	3,800	m²	17		
Builders Work				473,100	43.01
Forming holes, chases etc.		3%			
Preliminaries and Contingency				14,384,400	1,307.67
Mechanical and electrical services contractor preliminaries and commissioning management @		15%			
Main contractor's preliminaries, overheads and profit @		20%			
Main contractor's pre-construction stage services	25	week			
Design reserve @		5%			
Construction cost (rate based on GIFA)				56,742,000	5,158

Building Cost Models

AFFORDABLE HOUSING

This cost model is based on a new build scheme of a mix of two and three bedroom houses. The buildings comply with Code for Sustainable Homes Level 3, are of a timber frame construction with a reinforced in situ concrete ground slab, facing brick external wall and concrete tiles on timber roof trusses. The softwood doors and windows are painted and internal subdivision is by load-bearing timber stud partition. Features for the Lifetimes Homes Standard, National Housing Federation Standards and Housing Quality Indicators are also included.

Gross internal floor area	820 m²
Model location is South East England	Model TPI = 120; LF = 0.96

This updated cost model is copyright of AECOM Ltd

Affordable Housing	Quantity	Unit	Rate (£)	Total (£)	Cost (£/m²)
Substructure				**98,100**	**119.63**
Excavation and disposal	144	m²	40		
Strip foundation with cavity masonry wall to DPC level	244	m	200		
Ground floor slab with thickenings, including insulation (0.13 W/m².K)	414	m²	105		
Frame, Upper Floors and Stairs				**155,700**	**189.88**
Structural timber frame	820	m²	140		
Timber cassette upper floor, including any lintels required	402	m²	70		
Wooden, straight flight stairs	10	nr	1,275		
Roof				**94,700**	**115.49**
Timber roof trusses and associated timbers	432	m²	100		
Insulation to roof at rafter level (0.13 W/m².K)	402	m²	43		
Concrete interlocking roof tiles	432	m²	50		
Entrance canopy, approx 3 m²	10	nr	460		
uPVC rainwater down pipes	108	m	21		
uPVC gutters	220	m	26		
External Walls, Windows and Doors				**228,900**	**279.15**
Facing brick outer skin to structural timber frame	950	m²	95		
Forming cavity and fit insulation (0.22 W/m².K)	950	m²	22		
Application of enhanced construction details to reduce air tightness to 8.0 m³/(h.m²)	950	m²	14		
Party wall, 240 mm thick timber stud frame with Isowool quilt, enhanced insulation and plasterboard plank (0.2 W/m².K)	300	m²	100		
Softwood double glazed windows to meet secured by design requirements (0.15 W/m².K)	88	m²	580		
Softwood window boards	55	nr	62		
Single external doors (2.0 W/m².K)	20	nr	1,000		

AFFORDABLE HOUSING

Affordable Housing	Quantity	Unit	Rate (£)	Total (£)	Cost (£/m²)
Internal Walls and Partitions				**41,500**	**50.61**
Load-bearing timber stud partition faced with plasterboard, skim finish	250	m²	85		
Non load-bearing timber stud partition, faced with plasterboard, skim finish	270	m²	75		
Internal Doors				**46,100**	**56.22**
Internal doors, wood veneer, ironmongery	70	nr	430		
Wrot softwood storage cupboard door, ironmongery	20	nr	320		
Softwood door linings, architrave, painted	800	m	12		
Wall Finishes				**25,200**	**30.73**
Plaster skim with emulsion paint finish	2,200	m²	10		
Ceramic wall tiles	50	m²	50		
Floor Finishes				**40,400**	**49.27**
Screed	432	m²	24		
Carpet	520	m²	31		
Vinyl tiling including skirting	150	m²	31		
Softwood skirtings with gloss paint finish	670	m	14		
Ceiling Finishes				**29,200**	**35.61**
Plasterboard ceiling, plaster skim and emulsion paint finish	656	m²	34		
Moisture resistance plasterboard ceiling, plaster skim and emulsion paint finish	164	m²	42		
Fittings and Furnishings				**25,500**	**31.10**
General fittings – kitchen units, worktops, sink	10	nr	2,550		
Sanitary Fittings				**22,500**	**27.44**
Sanitary fittings, full fitted bathroom, wash hand basin, bath with shower over	10	nr	2,250		
Disposal Installations				**8,500**	**10.37**
Waste, soil and vent installation; uPVC pipework and fittings	820	m²	10		
Water Installations				**60,600**	**73.90**
Hot water	60	nr	550		
Cold water	60	nr	460		
Space Heating and Ventilation				**53,300**	**65.00**
Electric high efficiency condensing boiler, distribution pipework and panels with temperature control zones	820	m²	65		
Electrical and Gas Installations				**40,400**	**49.27**
Mains and sub-mains connection, small power distribution, cooker point, lighting distribution	820	m²	43		
Gas Installations, all costs associated with the supply and installation of gas supply	10	nr	510		

AFFORDABLE HOUSING

Affordable Housing	Quantity	Unit	Rate (£)	Total (£)	Cost (£/m²)
Protective, Communications and Special Installations				**8,600**	**10.49**
Earthing and bonding	860	m²	2		
Smoke detectors; telephone points and TV aerial, including additional data and power sockets and wiring	820	m²	9		
Special Installations				**52,600**	**64.15**
Solar thermal hot water installation	10	nr	3,850		
Mechanical ventilation with heat recovery	820	m²	17		
Builders Work				**20,000**	**24.39**
Builders work in connection with services installation	10	nr	1,225		
Additional works to comply with Lifetime Homes requirements	10	nr	775		
Preliminaries and Contingency				**249,400**	**304.15**
Management costs; site establishment; site supervision @		18%			
Contingency @		5%			
Construction cost (rates based on GIFA)				**1,301,200**	**1,586.85**

APARTMENTS – PRIVATE RENTED SECTOR

The apartments are being built for PRS and are arranged in a single eleven-storey building with two cores. Each core has its own apartment entrance lobby. The building also includes a residents' gym (GIA 185 m²), a private screening room and cinema (92 m²) and a residents' lounge and multifunction area (185 m²), as well as a dedicated estate management suite (GIA of 139 m²) and two retail units (GIA 280 m²).

Model location is Central London (zones 2/3)   Model TPI = 131; LF = 1.06

This cost model is copyright of AECOM Ltd

Apartment Shell and Core	Quantity	Unit	Rate (£)	Total (£)	Cost (£/m²)
Substructure				**1,296,100**	**84.35**
Works to existing site; site clearance and removal of existing vegetation, etc.	1,280	m²	11		
Excavation and disposal of material off site; assumed average 1.5 m reduce level dig	2,200	m³	60		
Allow for 15% contaminated soil	330	m³	120		
Allow for removal of unknown obstructions		item	29,000		
Allowance for piling mat to footprint of building; say 400 mm deep; including excavation and disposal and compacting	1,280	m²	46		
Allowance for mobilization and demobilization of piling rig; include for setting up piling rig at pile positions		item	85,000		
CFA piling; approx. 600 mm dia. piling 35 m deep; cut off top of pile; allow for pile cap; disposal of spoil; etc.	2,625	m	170		
Allowance for sundries: integrity tests, NDT, etc.		item	23,000		
Allowance for 500 mm thick reinforced concrete slab; excavation and disposal; 350 mm thick hardcore bed, 50 mm sand blinding, dpm, in situ concrete slab, reinforcement (150 kg/m³), etc.	1,280	m²	290		
Allowance for under slab drainage	1,280	m²	40		
Allowance for plant bases		item	11,500		
Allowance for lift pits	4	nr	8,500		
Frame, Upper Floors and Stairs				**5,071,900**	**330.09**
Allowance for reinforced concrete frame; comprising columns and core walls to ground floor	1,280	m²	230		
Allowance for reinforced concrete frame; including columns, beams and core walls to upper floors	14,085	m²	150		
Upper floor; post-tensioned concrete slab; 225 mm thick slab; 100 kg/m³ reinforcement; 6 kg/m² post-tensioned strands, formwork, etc.	14,085	m²	160		
Precast reinforced concrete stairs; with powder coated metal balustrading and handrails; ground to 11th floor	22	m²	9,100		
Allowance for sundry platforms to plantrooms		item	8,500		
Access ladders to roof	4	m²	2,850		

APARTMENTS – PRIVATE RENTED SECTOR

Apartment Shell and Core	Quantity	Unit	Rate (£)	Total (£)	Cost (£/m²)
Extra over allowance for enhancements to podium slab; transfer structures over retail areas, etc.	1,280	m²	120		
Allowance for steel support to lifts	44	nr	850		
Roof				**567,400**	**36.93**
Roof slab; post-tensioned concrete slab; 225 mm thick slab; 100 kg/m³ reinforcement; 6 kg/m² post-tensioned strands, formwork etc.	1,280	m²	160		
Allowance for lift overruns, AOVs, etc.		item	17,000		
Allowance for membrane roof covering to concrete slab including insulation, waterproofing, drainage, etc.	1,280	m²	150		
Extra over allowance for green roof coverings to podium roof	1,280	nr	120		
External Walls, Windows and Doors				**6,896,600**	**448.85**
Structurally glazed wall to principle facade, apartment entrance lobbies and retail	878	m²	925		
Rainscreen cladding system: comprising timber rainscreen and brickwork panels on insulation and Metsec/SFS backing; 60% of facade area	3,834	m²	675		
Extra over for material interfaces and feature detailing		item	285,000		
Allowance for double glazed thermally broken aluminium windows; powder coated; 40% of facade area	2,556	m²	625		
Extra over for glazed sliding double doors to retail units, apartment entrance lobbies, gym and estate management suite	6	nr	4,000		
Extra over for glazed double door to residents lounge	1	nr	2,300		
External louvered doorsets to residential refuse stores, retail, plant area and roof	7	nr	5,700		
Allowance for canopies to entrances	6	nr	11,500		
Allowance for scaffolding to external wall areas generally	7,268	m²	70		
Allowance for single doors to balconies	200	nr	1,450		
Allowance for balconies to apartments; bolt on steel balconies; including balustrades; cold bridging details, etc.	200	nr	3,400		
Internal Walls, Partitions and Doors				**1,682,400**	**109.50**
Blockwork walls; 300 mm overall thickness to ground floor; including windposts	1,300	m²	150		
Blockwork walls; 100 mm overall thickness to ground floor; including windposts	380	m²	105		
Partitions to residential corridor; 260 mm overall thickness, 15 mm soundblock plasterboard, 12.5 mm WBP plywood lining, vapour control layer, 2 nr 50 mm metal studs, 100 mm air gap, 2 nr 15 mm soundblock plasterboard	2,376	m²	140		

APARTMENTS – PRIVATE RENTED SECTOR

Apartment Shell and Core	Quantity	Unit	Rate (£)	Total (£)	Cost (£/m²)
Partitions between apartments; 260 mm overall thickness, 15 mm soundblock plasterboard, 12.5 mm WBP plywood lining, vapour control layer, 2 nr 50 mm metal studs, 100 mm air gap, 2 nr 15 mm soundblock plasterboard	4,800	m²	140		
Double leaf timber doors, to ground floor storage	2	nr	2,050		
Single leaf, timber doors to ground floor storage and WCs	5	nr	975		
Single leaf, timber entrance doors to apartments	200	nr	1,025		
Single leaf, timber doors to amenity areas	72	nr	975		
Single leaf, timber doors to risers	96	nr	575		
Allowance for WC and shower cubicles to amenity areas, etc.	8	nr	2,300		
Allowance for temporary protection to doors, etc.		item	85,000		
Wall Finishes				**583,900**	**38.00**
Plasterboard lining to core walls, service cupboards, etc.	3,496	m²	46		
Plasterboard linings to blockwork walls	2,843	m²	46		
Paint to internal partitions; residential corridors, linings and back of house areas	8,895	m²	9		
MDF skirtings with painted finish to lift lobbies, residential corridors, gym, estate management suite	2,905	m²	17		
Enhanced finishes to apartment entrances and lift lobbies; at ground level		item	57,500		
Timber panelling to residents lounge	150	m²	290		
Acoustic panelling to cinema room	115	m²	170		
Extra over for forming recesses for projection screen and other detailing generally; to cinema room		item	5,700		
Service panelling to WCs and showers	75	m²	340		
Porcelain tiles to WCs and showers	120	m²	85		
Floor Finishes				**327,100**	**21.29**
75 mm levelling screed to corridors and lift lobbies, apartment entrance lobbies, cinema room, residents lounge, estate management suite and gym	3,118	m²	34		
Sealant to plant areas	418	m²	17		
Power float finish to retail areas	279	m²	3		
Carpet finish to residential corridors and lift lobbies	2,282	m²	46		
Luxury Carpet to cinema room	88	m²	85		
Stone floor finish to apartment entrance lobbies	186	m²	290		
Timber batten floor to residents lounge and estate management suite	300	m²	40		
Engineered timber flooring to gym area	162	m²	150		

APARTMENTS – PRIVATE RENTED SECTOR

Apartment Shell and Core	Quantity	Unit	Rate (£)	Total (£)	Cost (£/m²)
Porcelain tiles to WCs and showers	53	m²	120		
Epoxy resin to storage areas, refuse areas, etc.	46	m²	23		
Allowance for drainage channels including stainless steel grilles to changing facilities to gym	10	m	290		
Ceiling Finishes				194,000	12.63
Painted plasterboard ceiling to residential corridors, cinema room, residents lounge and back of house areas	2,561	m²	43		
Acoustic rated demountable suspended ceiling, including paint finish to estate management suite and gym	293	m²	70		
Allowance for specialist acoustic ceiling to cinema room	88	m²	85		
Enhanced finishes to reception and lift lobby areas at ground level		item	29,000		
Painted moisture-resistant plasterboard ceiling to WCs and showers	53	m²	51		
Sealer to exposed ceiling soffit to retail units, plant and storage areas	743	m²	17		
Allowance for additional bulkheads and level changes generally		item	11,500		
Furniture and Fittings				342,000	22.26
Statutory signage		item	34,000		
Mail boxes	200	nr	150		
Allowance for mirrors to gym		item	11,500		
Allowance for vanity units to changing rooms		item	8,500		
Allowance for cubicles, showers and fittings to gym changing areas	6	nr	2,850		
Allowance for fittings to changing rooms and WCs areas; hand driers, soap dispensers, towel racks, etc.		item	5,700		
Allowance for lockers to changing rooms	20	nr	575		
Allowance for benches to changing rooms		item	2,850		
Allowance for FF&E to cinema room; including projection screen and equipment; joinery cabinetry and millwork, etc.		item	23,000		
Allowance for cinema room seating	12	nr	1,700		
Allowance for back of house storage for estate management suite and maintenance facilities		item	23,000		
Allowance for FF&E to gym areas; vanity units, mirrors, etc.		item	17,000		
Allowance for servery joinery and equipment to residents lounge		item	5,700		
Study desks and panelling to residents lounge	10	nr	1,150		

APARTMENTS – PRIVATE RENTED SECTOR

Apartment Shell and Core	Quantity	Unit	Rate (£)	Total (£)	Cost (£/m²)
Residents lounge bookcases	5	m	1,700		
Sundry joinery to residents lounge; window seating, bars, etc.		item	23,000		
Timber screens and panelling to residents lounge		item	11,500		
Fireplace and wall feature to residents lounge		item	8,500		
Allowance for general storage and joinery to estate management suite		item	17,000		
Sundries; notice boards, signage, etc.		item	5,700		
Reception desks to apartment entrance areas and estate management suite		item	46,000		
Exclusions: meeting room furniture, desks & seats, gym equipment					
Sanitary Fittings				**32,400**	**2.11**
Cleaners sinks	3	nr	510		
Sanitary ware to estate management suite	4	nr	750		
Sanitary ware including showers to gym	12	nr	900		
Sanitary ware to cinema room	2	nr	750		
Sanitary ware to work residents lounge	8	nr	750		
Extra over disabled provision		item	5,700		
Associated subcontractor preliminaries		item	3,800		
Disposal Installations				**350,200**	**22.79**
Rainwater disposal from roof outlets connection to underground drainage	15,365	m²	3		
Foul disposal to amenity areas and apartment complete with associated vent stacks, connection to underground drainage	15,365	m²	15		
Drainage from plant rooms, bin stores and the like via floor gullies		item	16,000		
Capped connection to retail		item	2,850		
Associated subcontractor preliminaries		item	44,000		
Water Installations				**403,100**	**26.23**
Cold water installation incoming main, storage tank, water treatment, multi-booster pump set, distribution to apartments including tenant sets complete with water meter	15,365	m²	19		
Cold water distribution to sanitary ware including estate management suite, cleaners cupboards and gym		item	21,000		
CAT 5 water points & plant to wash down points and to roof BMU		item	17,000		
Hot water installation to sanitary ware including estate management suite, cleaners cupboards and gym via a plate heat exchanger, including pumps, pipe work etc.		item	21,000		

APARTMENTS – PRIVATE RENTED SECTOR

Apartment Shell and Core	Quantity	Unit	Rate (£)	Total (£)	Cost (£/m²)
Capped connections to retail		item	5,700		
Associated subcontractor preliminaries		item	51,000		
Heat Source, Space Heating and Air Treatment				**785,000**	**51.09**
Gas fired boiler installation	3	nr	30,000		
Flue to atmosphere		item	90,000		
Plant room pipework, pumps sets, controls etc.		item	44,000		
Associated subcontractor preliminaries		item	32,000		
Distribution to apartments, valves, etc.; terminating in HIUs (HIUs part of fit-out)	15,365	m²	18		
Landlords heating via radiators to stair cores and back of house areas		item	13,000		
Underfloor heating to reception areas		item	8,000		
Air treatment to reception, gym, office areas; VRF installation and mini AHU		item	170,000		
Associated subcontractor preliminaries		item	67,500		
Ventilation Installations				**302,900**	**19.71**
Plant room ventilation	418	m²	105		
Stairwell make up via AoV	2	nr	8,500		
Smoke clearance to corridors	24	vents	8,500		
Associated subcontractor preliminaries (lot @ £33,150)		item	38,000		
Electrical and Gas Installations				**1,250,600**	**81.39**
Primary distribution board, landlords power and lighting boards, cabling and containment, sub-metering, ryefield installation to apartments	15,365	m²	28		
Life safety standby generator and flue		item	130,000		
Life safety cabling & equipment to firefighting lifts, smoke extract and sprinkler installations	15,365	m²	3		
Small power to landlords areas and circulation space		item	9,100		
Power to mechanical installations	15,365	m²	3		
Lighting to landlord areas and circulation areas including lighting control and emergency lighting	2,978	m²	70		
Back of house lighting	15,365	m²	1		
Power and lighting to office and gym		item	107,500		
Feature lighting internal/external		item	46,000		
Earthing and bonding	15,365	m²	2		
Capped connections to retail		item	5,700		
Associated subcontractor preliminaries		item	150,000		
Gas installation to boilers	15,365	m²	2		
Associated subcontractor preliminaries		item	3,750		

APARTMENTS – PRIVATE RENTED SECTOR

Apartment Shell and Core	Quantity	Unit	Rate (£)	Total (£)	Cost (£/m²)
Protective, Communications and Special Installations				**889,900**	**57.92**
Sprinkler distribution to apartments, monitored floor valves	15,365	m²	11		
Dry risers	24	outlet	2,300		
Lightning & surge protection	15,365	m²	1		
Associated subcontractor preliminaries		item	36,000		
Fire alarm to landlord areas, interlink to apartments	15,365	m²	7		
Disabled refuge and WC alarm installation		item	75,000		
Containment for data/telephone installations	15,365	m²	4		
Satellite farm and aerial installation to roof		item	23,000		
Telephone/TV/Satellite to apartments	15,365	m²	8		
CCTV and access control to ground floor areas and plant rooms (12 nr CCTV cameras, 6 nr access control points)		item	72,500		
Data outlets to landlord areas and Wi-Fi installation to residents lounge		item	29,000		
Door entry and sub-door entry installation		item	34,000		
Associated subcontractor preliminaries		item	75,000		
Special Installations and Lifts				**1,166,800**	**75.94**
Remote metering to apartments and billing system for LTHW heating only	200	nr	340		
BMS	15,365	m²	10		
Cinema room installation		item	115,000		
Associated subcontractor preliminaries		item	49,000		
13 person, 1.6 m/s	2	nr	150,000		
17 person, 1.6 m/s	2	nr	170,000		
Enhanced finishes to lift car and 2 sets of drapes	2	nr	29,000		
Associated subcontractor preliminaries		item	80,000		
Builders Work				**470,000**	**30.59**
Builders work in connection with services installation:		item	470,000		
Preliminaries and Contingency				**7,066,100**	**459.88**
Management costs; site establishment; site supervision @		25%			
Contingency @		5%			
Construction cost shell and core (rate based on GIFA)				**29,678,400**	**1,932**
Residential Fit-Out	Quantity	Unit	Rate (£)	Total (£)	Cost (£/m²)
Internal Walls and Partitions and Doors				**1,213,300**	**78.97**
Plasterboard stud partitions within apartments	8,700	m²	60		
EO allowance for boxing out and acoustic treatment to SVP's including crossovers		item	57,500		

APARTMENTS – PRIVATE RENTED SECTOR

Residential Fit-Out	Quantity	Unit	Rate (£)	Total (£)	Cost (£/m²)
Internal apartment doors: flush painted, solid core, single leaf doorset including ironmongery	475	nr	850		
Utility cupboard doors; flush paint, solid core, double leaf doorset; including ironmongery	200	nr	1,150		
Wall Finishes				899,800	58.56
Painted plasterboard lining to internal face of external wall, core walls, external wall of bathroom pod, etc.	8,500	m²	46		
Paint finish to all internal partitions	17,400	m²	9		
Plywood behind kitchen walls	1,800	m²	17		
Allowance for glass splashbacks to kitchens generally	500	m²	230		
Allowance for mastic	200	nr	290		
MDF skirting with paint finish	8,633	m²	17		
Floor Finishes				1,246,900	81.15
Acoustic timber batten subfloor to general areas	10,072	m²	40		
Carpet finishes to bedrooms; approximately 30% of net residential area	3,022	m²	46		
Timber flooring to all other areas	7,050	m²	100		
Ceiling Finishes				557,100	36.26
Painted plasterboard ceiling	10,072	m²	43		
Allowance for blind boxes/recess	1,400	m	60		
Access hatches	200	nr	200		
Furniture and Fittings				5,363,100	349.05
Allowance for kitchen including sink, taps and white goods (oven, hob, fridge freezer, washer/dryer, extract hood); studio	15	item	8,000		
Allowance for kitchen including sink, taps and white goods (oven, hob, fridge freezer, washer/dryer, extract hood)	80	nr	9,100		
Allowance for kitchen including sink, tap and white goods (oven, hob, fridge freezer, washer/dryer, extract hood); 2 bed	105	nr	10,200		
Allowance for enhanced warranty provision	200	nr	290		
Roller blinds	2,556	m²	46		
Wardrobes to bedrooms generally	654	m	925		
Bathroom pod incorporating; sanitary ware, finishes and fittings	305	nr	9,100		
Sanitary Fittings and Disposal Installations				119,000	7.74
Soil and waste to white goods, connection to bathroom pods	200	nr	520		
Associated subcontractor preliminaries		item	15,000		

APARTMENTS – PRIVATE RENTED SECTOR

Residential Fit-Out	Quantity	Unit	Rate (£)	Total (£)	Cost (£/m²)
Water Installations				**246,000**	**16.01**
Hot and cold water pipework to white goods, connection to bathroom pods	200	m²	1,050		
Associated subcontractor preliminaries		item	36,000		
Heat Source, Space Heating and Air Treatment				**948,200**	**61.71**
Twin plate heat exchanger for heat and hot water generation with heat meter	200	nr	1,650		
LTHW underfloor heating	11,585	m²	43		
Associated subcontractor preliminaries		item	120,000		
Ventilation Installations				**490,000**	**31.89**
Whole house ventilation with heat recovery and boost function	200	nr	2,150		
Associated subcontractor preliminaries		item	60,000		
Electrical Installation				**1,255,000**	**81.68**
Tenant's distribution board, meter	200	nr	460		
Small power distribution points, connection to pods	200	nr	1,500		
Lighting, wiring, installation and switches, connections to pods	200	nr	3,350		
Earthing and bonding	200	nr	190		
Associated subcontractor preliminaries		item	155,000		
Protective, Communications and Special Installations				**683,000**	**44.45**
Sprinkler protection	200	nr	1,200		
Fire alarm	200	nr	300		
Data/voice/TV outlets, wiring	200	nr	700		
Entry system	200	nr	575		
Local control to heating	200	nr	410		
Associated subcontractor preliminaries		item	46,000		
Builders Work				**190,000**	**12.37**
Builders work in connection with services installation:		item	190,000		
Preliminaries and Contingency				**4,128,500**	**268.70**
Management costs; site establishment; site supervision @		25%			
Contingency @		5%			
Construction cost fit-out (rate based on GIFA)				**17,339,900**	**1,129**
External Works and Utilities	Quantity	Unit	Rate (£)	Total (£)	Cost (£/m²)
External Works				**285,000**	**18.55**
Allowance for external works		item	285,000		
Utilities				**529,000**	**34.43**
Water Infrastructure charges	200	nr	400		

Building Cost Models

APARTMENTS – PRIVATE RENTED SECTOR

External Works and Utilities	Quantity	Unit	Rate (£)	Total (£)	Cost (£/m²)
Plus connection charge		item	29,000		
Foul infrastructure charges	200	nr	400		
Plus connection charge		item	46,000		
Electrical connection – no reinforcement	1	nr	145,000		
Fibre installation		item	115,000		
Gas		item	34,000		
Construction cost external works (rate based on GIFA)				**814,000**	**52.98**
TOTAL BUILDING COST (Shell and Core, Residential Fit-out and External Works)				**47,832,300**	**3,113**

COMMUNITY CENTRE

An 860 m² building built to achieve BREEAM Very Good level of performance. Main structure is a concrete ground bearing slab and steel frame on concrete pads. External walls are a mixture of facing bricks at ground level and render above. Roof construction is a single ply warm roof construction, with a small amount of patent glazing.

Gross internal floor area	860 m²
Model location is South East England	Model TPI = 120; LF = 0.96

This cost model is copyright of AECOM Ltd

Community Centre	Quantity	Unit	Rate (£)	Total (£)	Cost (£/m²)
Substructure				**100,500**	**116.86**
Excavation & disposal off site	430	m³	24		
Reinforced concrete ground slab, hardcore, dpm, ground beams and column bases for steel frame (0.25 W/m².K)	860	m²	105		
Frame				**127,200**	**147.91**
Steel propped portal frame, cold rolled purlins, surface treatments (@ 60 kg/m²)	52	tonne	2,000		
Intumescent paint to give 30 minute fire protection to steelwork	52	tonne	400		
Allowance for miscellaneous works, protecting columns		item	3,300		
Roof				**158,300**	**184.07**
Ply roofing with rigid board insulation and single ply polymeric roof cladding (0.18 W/m².K)	887	m²	160		
Roof glazing, aluminium, double glazed (2.2 W/m².K)	16	m²	800		
Rainwater drainage, aluminium gutters and downpipes	50	m	70		
External Walls, Windows and Doors				**225,700**	**262.44**
Light gauge steel framing; board; natural insulation; clay facing brick externally (0.20 W/m².K)	47	m²	240		
Light gauge steel framing; board; natural insulation; rendered finish on cement profile substrate (0.20 W/m².K)	497	m²	200		
Curtain walling with powder coated aluminium frame	125	m²	800		
Aluminium triple glazed windows	25	m²	400		
Steel fire escape doors	2	nr	1,175		
Canopy over entrance	1	nr	2,650		
Internal Walls and Partitions				**66,100**	**76.86**
Sound reducing board and metal stud partitions; generally 2 layers of board each side; various levels of fire and sound insulations; part glazed as appropriate	700	m²	85		
Fireproofing between partitions and roof		item	6,600		

COMMUNITY CENTRE

Community Centre	Quantity	Unit	Rate (£)	Total (£)	Cost (£/m²)
Internal Doors				43,100	50.12
Timber internal doorset with softwood frames and ironmongery; vision panels; including fire-resisting where necessary	38	nr	925		
Timber acoustic sliding stacking panel partition; 6 m × 2.6 m high	1	nr	7,900		
Wall Finishes				27,800	32.33
Plaster blocks and board; emulsion paint generally	1,100	m²	17		
Ceramic wall tiles splashbacks to WC areas	130	m²	70		
Floor Finishes				60,600	70.47
Demountable stage	20	m²	200		
Vinyl flooring	300	m²	100		
Carpet tiling	275	m²	46		
Laminate timber	250	m²	53		
Entrance matwell	2	m²	330		
Ceiling Finishes				36,000	41.86
Concealed grid plasterboard suspended ceiling	500	m²	53		
Plasterboard fixed directly to roof purlins	350	m²	27		
Fittings and Furnishings				33,000	38.37
Kitchen fitting; worktops; shelving		item	33,000		
Sanitary Fittings				25,600	29.77
WC suite; urinals, including disabled facility including all sanitary and fittings	15	nr	1,300		
Equipment for shower rooms; including disabled person equipment	2	nr	2,550		
Disposal Installations				5,000	4.52
Waste, soil and vent installation; uPVC pipework and fittings	15	nr	320		
Water Installations				39,600	46.05
Hot and cold water supply to WC, kitchen	860	m²	45		
Space Heating and Air Treatment and Ventilation				85,800	99.77
Boiler		item	32,500		
Underfloor heating system	860	m²	58		
Toilet and kitchen extract		item	1,900		
Electrical and Gas Installations				66,700	77.56
Small power, basic and emergency lighting	860	m²	19		
General lighting	860	m²	26		
External building lighting generally		item	6,400		
Incoming services		item	19,000		

COMMUNITY CENTRE

Community Centre	Quantity	Unit	Rate (£)	Total (£)	Cost (£/m²)
Protective, Communications and Special Installations				55,900	65.00
Lightning protection, earthing and bonding	860	m²	5		
Fire and intruder alarms, panic alarm buttons	860	m²	15		
Data installations and containment	860	m²	13		
Basic audio visual installation, projector, screen, satellite TV installation		item	26,000		
Builders Work				9,800	11.40
Forming holes, chases etc.		item	9,600		
Preliminaries and Contingency				242,000	281.40
Management costs; site establishment; site supervision @		15%			
Contingency @		5%			
Construction cost (rate based on GIFA)				1,408,700	1,636.76

Building Cost Models

CAR PARK

This cost model looks at a six-storey car park that forms part of a high-end multi-use development. It incorporates external cladding and an open roof, with external d-ramps and fully natural ventilation with BREEAM equal to very good. The multi-storey car park has a gross floor area of 40,000 m² has spaces for 1,300 cars.

Gross internal floor area	40,000 m²
Car parking spaces	1,300 nr
Model location is South East England	Model TPI = 120; LF = 0.96

This cost model is copyright of AECOM Ltd

Car Park	Quantity	Unit	Rate (£)	Total (£)	Cost (£/m²)
Substructures				**2,058,000**	**51.45**
400 mm dia. CFA in situ concrete piles, average 18 m long, 335 nr	6,000	m	53		
Pile caps including excavation, disposal, concrete, reinforcement, formwork	4,800	m²	150		
Ground floor construction including ground beams/post-tensioned beams	4,250	m²	180		
Tarmacadam surfacing to car park including type 1, base layer, binder and top coat	4,250	m²	60		
Frame, Upper Floors and Stairs				**10,507,500**	**262.69**
In situ reinforced concrete frame	40,000	m²	100		
160 mm thick post-tensioned in situ reinforced concrete fat slab flooring	40,000	m²	160		
Precast concrete stairs including balustrades and handrails	10	nr	8,900		
Allowance for upstands and movement joints		item	18,500		
Roof				**609,200**	**15.23**
Top finish including waterproofing to stair cores and plant rooms	280	m²	160		
Asphalt finish to top deck of car park	6,780	m²	80		
Metal decking TATA Comflor CF80 including concrete, reinforcement, edge lipping to cores and plant rooms	280	m	65		
TATA D100 decking including insulation, waterproofing and top finish coat to cores	50	m	75		
External Walls, Windows and Doors				**2,352,200**	**58.81**
Timber cladding on 140 mm blockwork substrate to stair cores	200	m²	380		
Single skin 100 mm facing blockwork to party wall, stretcher bond	2,400	m²	95		
Timber louvres/blades on a metal carrier frame	5,000	m²	330		
140 mm dense fair faced blockwork to form staircase lobbies	830	m²	65		

CAR PARK

Car Park	Quantity	Unit	Rate (£)	Total (£)	Cost (£/m²)
Capless stick system built external curtain walling system with glazed insulated clear glass units	100	m²	650		
Aluminium louvered screen	450	m²	360		
Double leaf external doorsets including ironmongery	5	nr	2,650		
Flush steel faced external single doorsets including ironmongery	37	nr	2,350		
Double leaf external doorsets to risers and plant areas including ironmongery	4	nr	2,350		
Single leaf external doorsets to risers including ironmongery	4	nr	1,900		
Internal Walls and Doors				**200,800**	**5.02**
Blockwork walls 100 mm thick	85	m²	60		
Blockwork walls 140 mm thick	450	m²	65		
Blockwork walls 200 mm thick	630	m²	100		
Single doorsets including frame and ironmongery	23	nr	1,125		
Double doorsets including ironmongery	50	nr	1,550		
Wall Finishes				**38,100**	**0.95**
Paint finish to exposed columns in car park	250	m²	6		
Decorations to walls/upstands	6,200	m²	6		
Floor Finishes				**252,800**	**6.32**
Asphalt finish to car park ramps and top level	360	m²	53		
Paint finish to ramps	1,440	m²	6		
screed to plant rooms	245	m²	42		
Paint finish to floor of plant rooms	245	m²	18		
Paint finish to pedestrian walkways and around disabled parking bays	3,164	m²	18		
Paint to floor stairs and landings	850	m²	8		
Parking lines including zebra crossing, arrows, parking spaces etc.	25,000	m²	6		
Ceiling Finishes				**232,500**	**5.81**
Paint to soffit of slabs and edges of exposed beams to car park	36,600	m²	6		
Paint to soffits of ramps	1,440	m²	6		
Paint to soffit of stairs and landings	850	m²	9		
Fittings and Furnishings				**178,000**	**4.45**
Statutory signage		item	12,000		
Directional signage		item	36,000		
Bollards		item	100,000		
Trolley bays tubular stainless steel guarding		item	30,000		

CAR PARK

Car Park	Quantity	Unit	Rate (£)	Total (£)	Cost (£/m²)
Sanitary Fittings				**8,300**	**0.21**
Unisex disabled toilet suite including WC, wash hand basin, grab rails, and other fittings	2	nr	1,650		
Cleaners sink	6	nr	825		
Disposal Installation				**311,000**	**7.78**
Car park drainage/rainwater installation	40,000	m²	7		
Petrol interceptor		item	13,500		
Foul drainage installation to sanitary appliances		item	13,500		
Water Installation				**21,600**	**0.54**
Water installation for car washing	3	nr	2,700		
Hot and cold water supplies to sanitary appliances		item	13,500		
Space Heating and Air Treatment and Ventilation				**68,600**	**1.72**
Parking attendant heating		item	1,450		
Trace heating		item	6,700		
Staircase ventilation for firefighting purposes		item	47,000		
Plant room ventilation		item	6,700		
Toilet ventilation		item	6,700		
Electrical Installation				**2,192,000**	**54.80**
LV distribution	40,000	m²	13		
General lighting	40,000	m²	24		
Emergency lighting (self-contained)	40,000	m²	7		
Lighting control system	40,000	m²	7		
Small power installation	40,000	m²	2		
General earthing	40,000	m²	1		
Protective, Communications and Special Installations				**1,730,100**	**43.25**
Dry riser installations	12	landing	3,450		
Lightning protection	40,000	m²	1		
Fire alarm installation	40,000	m²	20		
Fire telephones and disabled refuge alarms		item	20,000		
CCTV installation	50	nr	4,500		
Car park barriers	4	nr	20,000		
VMS signage (one per floor)	6	nr	27,000		
Pay-on-foot machines	12	nr	10,000		
Electric charging points – active (20%)	260	nr	310		
Electric charging points – passive (20%)	26	nr	310		
Solar PV panels	200	m²	725		

CAR PARK

Car Park	Quantity	Unit	Rate (£)	Total (£)	Cost (£/m²)
Lift Installations				**460,000**	**11.50**
13-person passenger/goods lifts 1.0 m/s	2	nr	95,000		
13-person passenger/firefighting lifts 1.0 m/s	2	nr	135,000		
Builders Work				**120,000**	**3.00**
Forming holes, chases etc.		item	120,000		
Preliminaries and Contingency				**3,224,600**	**80.62**
Management costs; site establishment; site supervision @		15%			
Testing and commissioning		item	23,500		
Construction cost (rate based on GIFA)				**24,565,300**	**614.15**
Cost per parking space				**19,000**	

MULTIPLEX CINEMA

This model is based on a 12-screen multiplex cinema situated above retail units within a retail-led development in the southeast UK. The cinema occupies the entire first floor of the block and includes mezzanine accommodation for projection rooms and administration space. The ground floor of the development consists of retail shells and escape routes from the cinema complex. The breakdown separates the costs of the shell and fit-out construction.

Gross internal floor area   18,000 m²

Model location is South East England   Model TPI = 120; LF = 0.96

This cost model is copyright of AECOM Ltd

Multiplex Shell and Core	Quantity	Unit	Rate (£)	Total (£)	Cost (£/m²)
Frame, Upper Floors and Stairs				**1,921,800**	**106.77**
Structural steel frame including fire protection (based on 80 kg/m²)	422	tonne	1,950		
Allowance for secondary steel – bracings, support to signage & spiral stairs etc.		item	82,500		
Acoustic pads to cinema columns and support stubs	169	nr	180		
100 mm thick Bison solid precast concrete plank with concrete topping	4,136	m²	65		
Acoustic floor slab and insulation to ramps, lower seating in auditoria	1,400	m²	180		
Forming ramps/concrete steps into auditoria		item	110,000		
Elastomeric strip to top of steel where precast plank rests		item	22,000		
Composite metal slab	3,090	m²	90		
Form steps for seating		item	55,000		
Roof				**1,195,100**	**66.39**
Structural steel to roof (based on 48 kg/m²)	158	tonne	1,950		
200 mm thick hollowcore PCC planks with structural topping	972	m²	90		
450 mm thick hollowcore PCC planks with structural topping	2,330	m²	130		
Reinforced laminated PVC single layer polymeric warm roof with closed cell rigid insulation tapered to falls	2,354	2.0%	70		
Reinforced laminated PVC single layer polymeric inverted roof system	1,964	m²	110		
Extra over for maintenance walkway		item	11,000		
Allowance for sundry metalworks, cat ladder, access hatches etc.		item	27,000		
Glazed roof canopy	70	m²	1,100		
External Walls				**1,717,700**	**95.43**
Proprietary metal cladding panel	3,690	m²	350		

MULTIPLEX CINEMA

Multiplex Shell and Core	Quantity	Unit	Rate (£)	Total (£)	Cost (£/m²)
Profiled aluminium channel	1,019	m	270		
Clay facing brickwork	116	m²	240		
Aluminium copings	356	m	170		
Aluminium framed hinged double door to cinema entrance	3	nr	8,100		
Aluminium framed pivot door	1	nr	16,500		
Steel single doors	6	nr	2,200		
Steel double doors	2	nr	4,350		
Internal Walls and Doors				673,700	37.43
Blockwork at ground/first floor	1,681	m²	65		
Additional layer of wall to inside face of external walls	3,000	m²	170		
Timber doors; single	40	nr	380		
Steel doors; single	8	nr	2,200		
Steel door; double	8	nr	2,700		
Internal Finishes				98,900	5.49
Top coat sealer to walls	4,145	m²	3		
Dry lining to internal face of external walls	1,762	m²	33		
Allowance for secondary ceiling grid to support services		item	27,000		
Floor finishes to cinema (excluded)			–		
Space Heating and Air Treatment and Ventilation				1,118,000	62.11
Sanitary appliances (see fit-out)			–		
Disposal installations		item	87,500		
Water installations		item	92,500		
Space heating and air conditioning		item	33,000		
Ventilation systems		item	120,000		
Electrical installations		item	495,000		
Gas and other fuel installations		item	33,000		
Lift installations (by tenant)			–		
Escalators (by tenant)			–		
Fire and lightning protection		item	27,000		
Communication, security and control Systems		item	120,000		
Specialist installations		item	55,000		
Builders work in connection with services		item	55,000		
Preliminaries and Contingency				1,681,300	93.41
Management costs; site establishment; site supervision @		25%			
Construction cost (rate based on GIFA)				8,406,500	467.03

MULTIPLEX CINEMA

Multiplex Fit-Out	Quantity	Unit	Rate (£)	Total (£)	Cost (£/m²)
Internal Walls and Doors				**1,952,100**	**108.45**
Glazed screen at ground level	60	m²	875		
Standard 150 mm thick partitions	794	m²	110		
150 mm thick acoustic wall	246	m²	220		
300 mm thick acoustic wall	4,323	m²	250		
550 mm thick acoustic wall	1,665	m²	300		
Wall mounted handrails to perimeter corridor, steps/stairs	196	m	200		
Balustrades to auditoria and steps in slab to projection level	205	m	200		
Allowance for shaftwall construction for hoist enclosures		item	5,400		
Allowance for free-standing screens to auditoria and WC entrances		item	27,000		
Melamine-faced solid laminate access panelling to WCs etc.	16	nr	550		
Allowance for additional partitions to administration areas		item	5,400		
Acoustic doors; fire-rated; steel-faced with insulated cores; steel goalpost frames; ironmongery	12	nr	3,800		
Security shutters; to concession unit entrances; including allowance for subframe		item	5,400		
Internal Finishes				**768,200**	**42.68**
13 mm two-coat plaster to blockwork	1,681	m²	13		
Feature plaster finish to entrance lobby	100	m²	130		
Two coats emulsion paint; to plaster/plasterboard generally	1,762	m²	9		
Ceramic tiling	200	m²	70		
Ceramic floor tiling including screed	100	m²	80		
Vinyl sheet flooring including screed; to part auditoria	1,000	m²	55		
Carpet; on underlay; including screed; to circulation, part auditoria and office spaces	2,000	m²	65		
Allowance for enhanced foyer finishes		item	27,000		
Mass barrier ceiling to soffit of auditorium	3,600	m²	110		
MF ceiling to cinema corridors	1,040	m²	44		
Allowance for ceiling to back of house	460	m²	44		
Allowance for access hatches and fire barriers		item	22,000		
Fittings and Furnishings				**53,400**	**2.97**
Allowance for additional joinery		item	11,000		
Allowance for internal signage		item	16,500		

MULTIPLEX CINEMA

Multiplex Fit-Out	Quantity	Unit	Rate (£)	Total (£)	Cost (£/m²)
Allowance for fixing tenant's supply-only fixtures and fittings		item	16,500		
WCs, urinals, wash-hand basins	23	nr	380		
Allowance for fittings for disabled WCs	2	nr	330		
Preliminaries and Contingency				693,400	38.52
Management costs; site establishment; site supervision @		25%			
Construction cost for fit-out only (rate based on GIFA)				3,467,100	192.62
TOTAL CONSTRUCTION COST (rate based on GIFA)				11,873,600	659.65

CONCERT HALL

A new, international standard 2,000 seat concert hall, built on a greenfield site in London with a gross floor area of 18,000 m². Excluded are demolitions, services diversions or reinforcement and highways alterations.

Gross internal floor area	18,000 m²
Model location is Inner London	Model TPI = 131; LF = 1.06

This cost model is copyright of AECOM Ltd

Concert Hall	Quantity	Unit	Rate (£)	Total (£)	Cost (£/m²)
Substructure				8,820,000	490.00
Piled foundations; RC ground bearing slab; lift pits & sumps. Partial basement 1,000 m² assumed secant piled with RC lining walls and cavity drain	18,000	m²	490		
Frame				14,400,000	800.00
Steel frame and composite upper floors; large span trusses over auditorium, rehearsal room and to create balconies in auditorium. 'Box in box' construction to main auditorium and rehearsal room.	18,000	m²	800		
Roof				3,060,000	170.00
Composite roof with inverted roof finish; 20% roof lights and 20% terraces including associated balustrading. Including roof mounted BMU; acoustic louvres to hide plant and smoke extract hatches	18,000	m²	170		
Stairs				2,160,000	120.00
Precast concrete escape stairs; feature foyer stairs; including balustrading and handrails	18,000	m²	120		
External Walls, Windows and External Doors				14,850,000	825.00
Solid cladding (70%) and glazing (30%); entrance doors; terrace doors; plantroom and substation doors and fire escape doors; canopy over entrance	18,000	m²	825		
Internal Walls and Partitions				2,700,000	150.00
Blockwork and dry lined internal walls; balustrading to internal voids and balcony fronts	18,000	m²	150		
Internal Doors				1,530,000	85.00
Internal fire doors, Georgian wired glass vision panel, stainless steel ironmongery to classrooms	18,000	m²	85		
Internal Finishes				9,900,000	550.00
Wall, floor and ceiling finishes throughout. Specialist panelling in auditorium and enhanced finishes in public foyer areas; provision for acoustic treatment	18,000	m²	550		

CONCERT HALL

Concert Hall	Quantity	Unit	Rate (£)	Total (£)	Cost (£/m²)
Fixed Furniture and Equipment				**6,120,000**	**340.00**
Includes main auditorium seating (2,000 nr @ £400), fixed fittings in box office, dressing rooms; cloakroom; workshops; front and back bars; servery counters; lockers and main information point	18,000	m²	340		
Specialist Installations				**9,360,000**	**520.00**
Stage engineering including orchestra pit lift, choir wagon, adjustable acoustic reflectors over the orchestra platform, lighting bridges, equipment bars and variable acoustics systems in audience areas; sound & communications including AV installation; specialist lighting	18,000	m²	520		
Mechanical Services				**13,950,000**	**775.00**
Sanitary ware, hot and cold water services, above ground drainage, sprinklers, BMS control systems	18,000	m²	775		
Electrical Services				**11,700,000**	**650.00**
Mains and sub-mains distribution, standby generation, protective, communications and special Installations	18,000	m²	650		
Lift Installation				**2,160,000**	**120.00**
Passenger and goods lifts including one number heavy duty scenery lift and disabled platform lifts	18,000	m²	120		
Builders Work				**1,350,000**	**75.00**
Forming holes, chases etc.	18,000	m²	75		
Landscaping				**1,530,000**	**85.00**
Allowance for landscaping to 5 m perimeter strip around building	18,000	m²	85		
Utilities Connections				**522,000**	**29.00**
Allowance for utilities connections including electric, gas, telecoms and drainage	18,000	m²	29		
External Drainage				**322,200**	**17.90**
Allowance for below ground drainage	18,000	m²	18		
Preliminaries and Contingency				**49,582,400**	**2,754.58**
Management costs; site establishment; site supervision @		20%			
Contractors contingency @		7%			
Design reserve and construction contingency @		15%			
Total construction cost (rate based on GIFA)				**154,016,600**	**8,556.48**

Building Measurement, 2nd Edition

A. Packer

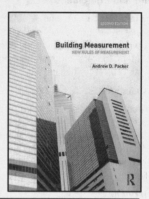

Measurement of buildings is the core skill of the quantity surveyor. It underpins the procurement, management, delivery and subsequent commissioning of a completed building, and must now be completed using New Rules of Measurement 2 (NRM 2).

In this much-needed new measurement textbook, the measurement of the most common building elements is described using NRM2. Extensive worked examples including fully up to date hand-drawn diagrams and supporting take-off lists ensure that the reader develops a confidence in their ability to measure using NRM2 in practice.

A practical step-by-step approach is used to explain and interpret the detail of the specific Work Sections of NRM2, covering a broad range of different trades, including mechanical and electrical systems; external works; groundwork; masonry; joinery; and internal finishes.

Presuming no prior knowledge of measurement or NRM2, and fully up to date with current practice, including consideration of Building Information Modelling, this is the ideal text for students of measurement at HND or BSc level, as well as practitioners needing a crash course in how to apply NRM2.

October 2016; 246 × 189 mm, 358 pp
Pbk: 978-1-138-83814-7; £38.99

To Order: Tel: +44 (0) 1235 400524 Fax: +44 (0) 1235 400525
or Post: Taylor and Francis Customer Services,
Bookpoint Ltd, Unit T1, 200 Milton Park, Abingdon, Oxon, OX14 4TA UK
Email: book.orders@tandf.co.uk

For a complete listing of all our titles visit:
www.tandf.co.uk

Taylor & Francis
Taylor & Francis Group

Preliminaries Build-up Example

The number of items priced in the preliminaries section of Bills of Quantities and the manner in which they are priced vary considerably between contractors. Some contractors, by modifying their percentage factor for overheads and profit, attempt to cover the costs of preliminary items in their Prices for Measured Work. However, the cost of Preliminaries will vary widely according to job size and complexity, site location, accessibility, degree to which mechanization is practicable, position of the contractor's head office and relationships with local labour/domestic subcontractors. It is therefore usually far safer to price preliminary items separately on their merits according to the project.

It is not possible for the quantity surveyor/cost manager to quantify the main contractors' preliminaries and it is left for the contractor to interpret the tender documentation and ascertain his resources and method of working they require to complete the works.

This preliminaries bill can therefore be a relatively simple pricing schedule of headings under which the contractor prices their preliminaries items. Allowances should be determined using predicted time durations and quantities for resources.

An example in pricing preliminaries follows, and this assumes the form of contract used is the JCT Standard Building Contract With Quantities (SBC/Q) and the value, including preliminaries, is approximately £4,400,000. The contract is estimated to take 52 weeks to complete and the value is built up as follows:

	£
Contractors' Labour, Plant and Materials value	3,150,000
Subcontracts and Provisional Sums	750,000
Preliminaries, say	£500,000
Project value	**£4,400,000**

At the end of the section the data are summarized to give a total value of preliminaries for the project example.

NOTE: The term 'Not priced', where used throughout this section means either that the cost implication is negligible or that it is usually included elsewhere in the tender.

Preliminaries

Pricing Schedule

Employer's requirements
Site accommodation – not priced
Site records – not priced
Completion and post-completion requirements

Defects after completion	Based on 0.2% of the contract sum	**£9,000**

Main Contractor's cost items

Management and staff – main contractor's project specific management and staff

Management and staff	Based on 6% of the tender (excluding preliminaries)	**£234,000**

Site establishment – main contractors and common user temporary site accommodation

Management and staff	Based on 0.35% of the tender (excluding preliminaries)	**£18,000**

Temporary services

Lighting and power	Based on 1% of the tender (excluding preliminaries)	**£39,000**
Water for the works (provided from mains)	Based on 0.1% of the tender (excluding preliminaries)	**£4,000**
Temporary telephones etc.	Based on 0.15% of the tender (excluding preliminaries)	**£6,000**

Security – allow for staff and security equipment

Security	Based on 0.15% of the tender (excluding preliminaries)	**£6,000**
Temporary hoarding, fencing etc.	2.4 m high fencing of OSB boarding one side; pair of gates	**£15,000**

Safety and environmental protection – compliance with all welfare facilities, first aid etc.

Safety, health and welfare	Based on 0.1% of the tender (excluding preliminaries)	**£8,000**

Control and protection – allowance for setting out, protection of the works, sampling

Control and protection of the works	Based on 0.1% of the tender (excluding preliminaries)	**£4,000**
Drying the works	Based on 0.03% of the tender (excluding preliminaries)	**£2,000**

Mechanical plant – common user mechanical plant and equipment

Plant/Transport	Based on 2% of the tender (excluding preliminaries)	**£88,000**
Small plant and tools	Based on 0.2% of the tender (excluding preliminaries)	**£9,000**

Temporary works – common user access scaffolding

Temporary roads and walkways	Say 250 × 3.5 m hardcore/recycled concrete	**£10,000**
Access scaffolding	Based on 0.75% of the tender (excluding preliminaries)	**£30,000**

Site records – photographs, progress reporting etc. – not priced

Completion and post-completion requirements – testing and commissioning, handover plan etc. – not priced

Cleaning – the cost would normally represent an allowance for final clearing of the works on completion with the residue for cleaning throughout the contract period. This amount being reduced with the introduction of waste management plans

Removing rubbish and cleaning	Based on 0.15% of the tender (excluding preliminaries)	**£7,000**

Fees and charges – any miscellaneous fees charges (rates etc.) – not priced

Site services – temporary works not specific to an element

Traffic regulations	Based on 0.05% of the tender (excluding preliminaries)	**£3,000**
Additional temporary items	Based on 0.25% of the tender (excluding preliminaries)	**£11,000**

Insurances, bonds, guarantees and warranties

Contract value (including preliminaries), say	£4,488,000
Estimated increased costs during contract period, say 2%	£88.0900
	£4,715,400
Estimated increased costs incurred during period of reinstatement, say 5%	£224,400
	£4,712,400
Professional fees, say 15.0%	£706,860
	£5,419,260
Insurance of the works	Based on 0.15% of project value	£9,000

If at the Contractor's risk, the insurance cover must be sufficient to include the full cost of reinstatement, all increases in cost, professional fees and any consequential costs such as demolition. The average provision for fire risk is 0.15% of the value of the work after adding for increased costs and professional fees.

NOTE: Insurance premiums are liable to considerable variation, depending on the contractor, the nature of the work and the market in which the insurance is placed.

Summary of Preliminaries Costs Example (NB rounded to nearest £1000)	
Defects after completion	£9,000
Management and staff	£234,000
Site accommodation	£18,000
Lighting and power for the works	£39,000
Water for the works	£4,000
Temporary telephones	£6,000
Security	£6,000
Hoardings, fans, fencing, etc.	£15,000
Safety, health and welfare	£8,000
Protection of the works	£4,000
Drying the works	£2,000
Large plant/transport	£88,000
Small plant and tools	£9,000
Temporary roads and walkways	£10,000
Access scaffolding	£30,000
Removing rubbish, etc., and cleaning	£7,000
Traffic regulations	£3,000
Additional temporary works	£11,000
Insurances	£9,000
TOTAL PRELIMINARIES	£512,000

It is emphasized that the above is an example only of the way in which Preliminaries may be priced and it is essential that for any particular contract or project the items set out in Preliminaries should be assessed on their respective values. The value of the Preliminaries items in recent tenders received by the editors varies from a 10% to 18% addition to all other costs. The above example represents approximately a 13% addition to the value of measured work.

Approximate Estimating Rates

Estimating by means of priced approximate quantities is always more accurate than by using overall building prices per square metre. Prices given in this section, which is arranged in elemental order, are derived from *Prices for Measured Works* section, but also include for all the incidental items and labours which are normally measured separately in Bills of Quantities. They have been established with a tender price level as indicated on the book cover. They include overheads and profit but *do not include for main contractors preliminaries*, for which an example build-up is given in an earlier section.

Whilst every effort is made to ensure the accuracy of these figures, they have been prepared for approximate estimating purposes and guidance only and on no account should they be used for the preparation of tenders.

Unless otherwise described units denoted as m² refer to appropriate unit areas (rather than gross internal floor areas).

As elsewhere in this edition prices do not include Value Added Tax or fees for professional services.

Approximate Estimating Rates are exclusive of any main contractor's preliminaries.

Cost plans can be developed from Elemental Cost Plans using both *Approximate Estimating Rates* and/or *Prices for Measured Works* depending upon the level of information available.

The cost targets within each formal cost plan approved by the employer will be used as the baseline for future cost comparisons. Each subsequent cost plan will require reconciliation with the preceding cost plan and explanations relating to changes made. In view of this, it is essential that records of any transfers made to or from the risk allowances and any adjustments made to cost targets are maintained, so that explanations concerning changes can be provided to both the employer and the project team

1.1 SUBSTRUCTURE

Item	Unit	Range £		
1.1 SUBSTRUCTURE				
1.1.1 Standard Foundations				
Foundations and floor slab composite				
Strip or trenchfill foundations with masonry up to 150 mm above floor level only; blinded hardcore bed; slab insulation; reinforced ground bearing slab 200 mm thick. To suit residential and small commercial developments with good ground bearing capacity				
shallow foundations up to 1.00 m deep	m^2	190.00	to	230.00
shallow foundations up to 1.50 m deep	m^2	220.00	to	270.00
Foundations in poor ground; mini piles; typically 300 mm dia.; 15 m long; 175 mm thick reinforced concrete slab; for single storey commercial type development				
minipiles to building columns only; 1 per column	m^2	150.00	to	185.00
minipiles to columns and perimeter ground floor beam; piles at 2.00 m centres to ground beams	m^2	300.00	to	360.00
minipiles to entire building at 2.00 m × 2.00 m grid	m^2	330.00	to	400.00
Raft foundations				
simple reinforced concrete raft on poorer ground for development up to two storey high	m^2	170.00	to	205.00
Trench fill foundations				
Machine excavation, disposal, plain in situ concrete 20.00 N/mm^2 – 20 mm aggregate (1:2:4) trench fill, 300 mm high cavity masonry in cement mortar (1:3), pitch polymer damp roof course				
With 3 courses of common brick outer skin up to DPC level (PC bricks @ £380/1000)				
600 mm × 1000 mm deep	m	125.00	to	150.00
600 mm × 1500 mm deep	m	170.00	to	205.00
With 3 courses of facing bricks brick outer skin up to DPC level (PC bricks @ £500/1000)				
600 mm × 1000 mm deep	m	130.00	to	155.00
600 mm × 1500 mm deep	m	170.00	to	210.00
With 3 courses of facing bricks brick outer skin up to DPC level (PC bricks @ £700/1000)				
600 mm × 1000 mm deep	m	130.00	to	160.00
600 mm × 1500 mm deep	m	175.00	to	210.00

1.1 SUBSTRUCTURE

Item	Unit	Range £		
Strip foundations				
Excavate trench 600 mm wide, partial backfill, partial disposal, earthwork support (risk item), compact base of trench, plain in situ concrete 20.00 N/mm² – 20 mm aggregate (1:2:4), 250 mm thick, 3 courses of common brick up to DPC level, cavity brickwork/blockwork in cement mortar (1:3), pitch polymer damp-proof course machine excavation				
With 3 courses of common brick outer skin up to DPC level				
(PC bricks @ £380/1000)				
600 mm × 1000 mm deep	m	135.00	to	165.00
600 mm × 1500 mm deep	m	180.00	to	220.00
With 3 courses of facing brick outer skin up to DPC level				
(PC bricks @ £500/1000)				
600 mm × 1000 mm deep	m	140.00	to	170.00
600 mm × 1500 mm deep	m	190.00	to	230.00
With 3 courses of facing brick outer skin up to DPC level				
(PC bricks @ £700/1000)				
600 mm × 1000 mm deep	m	150.00	to	180.00
600 mm × 1500 mm deep	m	190.00	to	235.00
Column bases				
Excavate pit in firm ground by machine, partial backfill, partial disposal, support, compact base of pit; in situ concrete 25.00 N/mm²; reinforced at 50 kg/m³				
rates per base for base size				
600 mm × 600 mm × 300 mm; 1000 mm deep pit	nr	425.00	to	510.00
900 mm × 900 mm × 450 mm; 1250 mm deep pit	nr	510.00	to	620.00
1500 mm × 1500 mm × 600 mm; 1500 mm deep pit	nr	850.00	to	1025.00
2700 mm × 2700 mm × 1000 mm; 1500 mm deep pit	nr	2450.00	to	3000.00
rates per m³ for base size				
600 mm × 600 mm × 300 mm; 1000 mm deep pit	m³	3800.00	to	4600.00
900 mm × 900 mm × 450 mm; 1250 mm deep pit	m³	1425.00	to	1725.00
1500 mm × 1500 mm × 600 mm; 1500 mm deep pit	m³	640.00	to	780.00
2700 mm × 2700 mm × 1000 mm; 1500 mm deep pit	m³	350.00	to	420.00
extra for reinforcement at 75 kg/m³ concrete, base size	m³	83.00	to	100.00
extra for reinforcement at 100 kg/m³ concrete, base size	m³	100.00	to	125.00
1.1.2 Specialist Foundations				
Piling				
Enabling works				
excavate to form piling mat; supply and lay imported hardcore – recycled brick and similar to form piling mat	m³	11.30	to	13.70
provision of plant (1 nr rig); including bringing to and removing from site; maintenance, erection and dismantling at each pile position	item	19000.00	to	23000.00
Supply and install concrete Continuous Flight Auger (CFA) piles; cart away inactive spoil; measure total length of pile				
450 mm dia. reinforced concrete CFA piles	m	52.00	to	63.00
600 mm dia. reinforced concrete CFA piles	m	83.00	to	100.00
750 mm dia. reinforced concrete CFA piles	m	130.00	to	155.00
900 mm dia. reinforced concrete CFA piles	m	185.00	to	225.00

1.1 SUBSTRUCTURE

Item	Unit	Range £		
1.1 SUBSTRUCTURE – cont				
Piling – cont				
Pile testing				
using tension piles as reaction; typical loading test 1000 kN to 2000 kN	item	7900.00	to	9500.00
integrity testing; minimum 20 per visit	nr	16.00	to	19.30
Secant wall piling, 750 mm dia. piles including site mobilization and demobilization; Measure total area of piling (total length × total depth)				
supply and install secant wall piling	m²	270.00	to	325.00
Steel sheet piling; Measure total area of piling (total length × total depth)				
interlocking steel sheet piling to excavation perimeter; Corus LX or similar; extraction on completion	m²	200.00	to	240.00
Pile caps				
Excavate pit in firm ground by machine, partial backfill, partial disposal, support, compaction, cut off pile and prepare reinforcement, formwork; reinforced in situ concrete cap 25 N/mm²; reinforcement at 50 kg/m³; rates per cap for cap size				
900 mm × 900 mm × 1000 mm; 1–2 piles	nr	750.00	to	910.00
2100 mm × 2100 mm × 1000 mm; 2–3 piles	nr	1675.00	to	2050.00
2700 mm × 2700 mm × 1500 mm; 3–5 piles	nr	3400.00	to	4100.00
Excavate pit in firm ground by machine, partial backfill, partial disposal, support, compaction, cut off pile and prepare reinforcement, formwork; reinforced in situ concrete cap 25 N/mm²; reinforcement at 50 kg/m³; rates per m³ for typical cap size				
up to 900 mm × 900 mm × 1000 mm; 1–2 piles	m³	900.00	to	1075.00
up to 2100 mm × 2100 mm × 1000 mm; 2–3 piles	m³	390.00	to	470.00
up to 2700 mm × 2700 mm × 1500 mm; 3–5 piles	m³	310.00	to	380.00
extra for reinforcement at 75 kg/m³ concrete	m³	82.00	to	99.00
extra for reinforcement at 100 kg/m³ concrete	m³	100.00	to	125.00
extra for alternative strength concrete C30 (30.00 N/mm²)	m³	1.40	to	1.70
extra for alternative strength concrete C40 (40.00 N/mm²)	m³	7.10	to	8.60
Concrete ground beams				
Reinforced in situ concrete ground beams; bar reinforcement; formwork				
300 mm × 300 mm, reinforcement at 180 kg/m³	m	53.00	to	64.00
450 mm × 450 mm, reinforcement at 200 kg/m³	m	100.00	to	120.00
450 mm × 600 mm, reinforcement at 270 kg/m³	m	150.00	to	180.00
Precast concrete ground beams; square ends and dowelled. Supplied and installed in lengths to span between stanchion bases; concrete strength C50; maximum UDL of 4.5 kN/m; in situ work at beam ends not included.				
350 mm × 425 mm	m	93.00	to	110.00
350 mm × 875 mm	m	220.00	to	270.00
350 mm × 1075 mm	m	300.00	to	365.00
450 mm × 400 mm	m	110.00	to	130.00
Concrete lift pits				
Excavate and disposal; reinforced concrete floor and walls; bitumen tanking as necessary				
1.65 m × 1.81 m × 1.60 m deep pit – 8 person lift / 630 kg	nr	2650.00	to	3200.00
1.80 m × 2.50 m × 1.60 m deep pit –13 person lift / 1000 kg	nr	3300.00	to	4000.00
3.00 m × 2.50 m × 1.60 m deep pit – 21 person lift / 1700 kg	nr	3850.00	to	4650.00

1.1 SUBSTRUCTURE

Item	Unit	Range £		
Underpinning				
In stages not exceeding 1500 mm long from one side of existing wall and foundation, excavate preliminary trench by machine and underpinning pit by hand, partial backfill, partial disposal, earthwork support (open boarded), cutting away projecting foundations, prepare underside of existing, compact base of pit, plain in situ concrete 20.00 N/mm² – 20 mm aggregate (1:2:4), formwork, brickwork in cement mortar (1:3), pitch polymer damp-proof course, wedge and pin to underside of existing with slates Commencing at 1.00 m below ground level with common bricks, depth of underpinning				
900 mm high, one brick wall	m²	395.00	to	480.00
1500 mm high, one brick wall	m²	570.00	to	690.00
extra for commencing 2.00 m below ground level	m²	78.00	to	94.00
extra for commencing 3.00 m below ground level	m²	150.00	to	185.00
extra for commencing 4.00 m below ground level	m²	210.00	to	260.00
1.1.3 Lowest Floor Construction				
Ground floor construction				
Mechanical excavation to reduce levels, disposal, level and compact, hardcore bed blinded with sand, 1200 gauge polythene damp-proof membrane, in situ concrete 20.00 N/mm² – 20 mm aggregate (1:2:4)				
150 mm thick concrete slab with 1 layer of A195 fabric reinforcement	m²	96.00	to	115.00
200 mm thick concrete slab with 1 layer of A252 fabric reinforcement	m²	105.00	to	125.00
250 mm thick concrete slab with 1 layer of A393 fabric reinforcement	m²	105.00	to	130.00
extra for every additional 50 mm thick concrete	m²	7.45	to	9.00
extra for alternative strength concrete C30 (30.00 N/mm²)	m³	1.40	to	1.70
extra for alternative strength concrete C40 (40.00 N/mm²)	m³	7.10	to	8.60
extra for reinforcement at 25 kg/m³ concrete	m³	31.00	to	37.00
extra for reinforcement at 50 kg/m³ concrete	m³	51.00	to	62.00
extra for reinforcement at 75 kg/m³ concrete	m³	82.00	to	99.00
extra for reinforcement at 100 kg/m³ concrete	m³	100.00	to	125.00
Warehouse ground floor				
Steel fibre reinforced floor slab placed using large pour construction techniques providing a finish floor flatness complying with FM2 special +/- 15 mm from datum. NOTE: Excavation, subbase and damp-proof membrane not included				
nominal 200 mm thick in situ concrete floor slab, concrete grade C40, reinforced with steel fibres, surface power floated and cured with a spray application of curing and hardening agent	m²	32.00	to	39.00
Suspended ground floor				
Beam and block flooring				
suspended floor with 150 mm deep precast concrete beams and infill blocks	m²	58.00	to	70.00
suspended floor with 225 mm deep precast concrete beams and infill blocks	m²	100.00	to	120.00

2.1 FRAME

Item	Unit	Range £		

1.1 SUBSTRUCTURE – cont

Board or slab insulation
Kingspan Thermafloor TF70 (Thermal conductivity 0.022 W/mK) rigid urethane floor insulation for solid concrete and suspended ground floors

50 mm thick	m²	11.90	to	14.40
75 mm thick	m²	15.80	to	19.10
100 mm thick	m²	18.80	to	23.00
125 mm thick	m²	24.50	to	29.50
150 mm thick	m²	28.50	to	34.50

Styrofoam Floormate 500 (Thermal conductivity 0.033 W/mK) extruded polystyrene foam or other equal and approved

50 mm thick	m²	18.70	to	22.50
75 mm thick	m²	19.70	to	24.00
100 mm thick	m²	23.00	to	28.00

1.1.5 Basements

Reinforced concrete basement floors; NOTE: excluding bulk excavation costs and hardcore fill

200 mm thick waterproof concrete with 2 × layers of A252 mesh reinforcement	m²	74.00	to	90.00

Reinforced concrete basement walls; NOTE: excluding bulk excavation costs and hardcore fill

200 mm thick waterproof reinforced concrete	m²	380.00	to	460.00

2.1 FRAME

2.1.1 Frame and Floor Composite

Concrete frame; flat slab reinforced concrete floors up to 250 mm thick
Suspended slab; no coverings or finishes

up to 6 storeys	m²	150.00	to	180.00
7-12 storeys	m²	160.00	to	195.00
13-18 storeys	m²	185.00	to	225.00

Steel frame; composite beam and slab floors
Suspended slab; permanent steel shuttering with 130 mm thick concrete; no coverings or finishes

up to 6 storeys	m²	150.00	to	180.00
7-12 storeys	m²	170.00	to	210.00
13-18 storeys	m²	185.00	to	225.00

Concrete frame

Reinforced in situ concrete columns, bar reinforcement, formwork. Generally all formwork assumes four uses
Reinforcement rate 180 kg/m³; column size

225 mm × 225 mm	m	89.00	to	110.00
300 mm × 600 mm	m	190.00	to	235.00
450 mm × 900 mm	m	350.00	to	425.00

2.1 FRAME

Item	Unit	Range £		
Reinforcement rate 240 kg/m³; column size				
225 mm × 225 mm	m	98.00	to	120.00
300 mm × 600 mm	m	210.00	to	255.00
450 mm × 900 mm	m	390.00	to	470.00
In situ concrete casing to steel column, formwork; column size				
225 mm × 225 mm	m	79.00	to	96.00
300 mm × 600 mm	m	185.00	to	220.00
450 mm × 900 mm	m	315.00	to	380.00
Reinforced in situ concrete beams, bar reinforcement, formwork. Generally all formwork assumes four uses				
Reinforcement rate 200 kg/m³; beam size				
225 mm × 450 mm	m	110.00	to	140.00
300 mm × 600 mm	m	180.00	to	220.00
450 mm × 600 mm	m	230.00	to	280.00
600 mm × 600 mm	m	280.00	to	340.00
Reinforcement rate 240 kg/m³; beam size				
225 mm × 450 mm	m	120.00	to	150.00
300 mm × 600 mm	m	190.00	to	230.00
450 mm × 600 mm	m	240.00	to	290.00
600 mm × 600 mm	m	300.00	to	365.00
In situ concrete casing to steel beams, formwork; beam size				
225 mm × 450 mm	m	105.00	to	125.00
300 mm × 600 mm	m	155.00	to	190.00
450 mm × 600 mm	m	205.00	to	250.00
600 mm × 600 mm	m	245.00	to	300.00
Other floor and frame constructions				
reinforced concrete cantilevered balcony; up to 1.50 m wide × 1.20 m deep	nr	2800.00	to	3400.00
reinforced concrete cantilevered walkways; up to 1.00 m wide	m²	195.00	to	235.00
reinforced concrete walkways and supporting frame; up to 1.00 m wide × 2.50 m high	m²	255.00	to	310.00
Steel frame				
Fabricated steelwork erected on site with bolted connections, primed				
smaller sections ne 40 kg/m	tonne	2100.00	to	2550.00
universal beams; grade S275	tonne	2000.00	to	2425.00
universal beams; grade S355	tonne	2075.00	to	2500.00
universal columns; grade S275	tonne	2000.00	to	2425.00
universal columns; grade S355	tonne	2075.00	to	2500.00
composite columns	tonne	1950.00	to	2375.00
hollow section circular	tonne	2250.00	to	2700.00
hollow section square or rectangular	tonne	2250.00	to	2700.00
cellular beams (FABSEC)	tonne	2250.00	to	2700.00
lattice beams	tonne	2750.00	to	3350.00
roof trusses	tonne	2900.00	to	3500.00
Steel finishes				
grit blast and one coat zinc chromate primer	m²	8.95	to	10.80
touch up primer and one coat of two pack epoxy zinc phosphate primer	m²	5.60	to	6.75
blast cleaning	m²	2.95	to	3.60
galvanizing	m²	20.00	to	24.00
galvanizing	tonne	295.00	to	360.00

2.1 FRAME

Item	Unit	Range £		
2.1 FRAME – cont				
Steel frame – cont				
Other floor and frame constructions				
steel space deck on steel frame, unprotected	m²	51.00	to	62.00
18.00 m high bay warehouse; steel propped portal frame, cold rolled purlin				
sections, primed surface treatment only, excluding decorations and protection	m²	110.00	to	130.00
columns and beams to mansard roof, 60 min fire protection	m²	175.00	to	210.00
feature columns and beams to glazed atrium roof unprotected	m²	180.00	to	220.00
Fire protection to steelwork				
Sprayed mineral fibre; gross surface area				
60 minute protection	m²	17.40	to	21.00
90 minute protection	m²	31.00	to	37.50
Sprayed vermiculite cement; gross surface area				
60 minute protection	m²	19.90	to	24.00
90 minute protection	m²	29.00	to	35.00
Supply and fit fire-resistant boarding to steel columns and beams; noggins, brackets and angles, intumescent paste. Beamclad or similar; measure board area				
30 minute protection for concealed applications	m²	28.00	to	34.00
60 minute protection for concealed applications	m²	62.00	to	75.00
30 minute protection left exposed for decoration	m²	31.00	to	37.50
60 minute protection left exposed for decoration	m²	74.00	to	90.00
Intumescent fire protection coating/decoration to exposed steelwork; Gross surface area (m²) or per tonne; On site application, spray applied				
30 minute protection per m²	m²	12.80	to	15.50
30 minute protection per tonne	tonne	325.00	to	395.00
60 minute protection per m²	m²	16.20	to	19.70
60 minute protection per tonne	tonne	410.00	to	500.00
Intumescent fire protection coating/decoration to exposed steelwork; Gross surface area (m²) or per tonne; Off site application, spray applied				
60 minute protection per m²	m²	24.00	to	29.00
60 minute protection per tonne	tonne	620.00	to	750.00
90 minute protection per m²	m²	48.00	to	59.00
90 minute protection per tonne	tonne	1225.00	to	1475.00
120 minute protection per m²	m²	73.00	to	88.00
120 minute protection per tonne	tonne	1850.00	to	2225.00

2.3 ROOF

Item	Unit	Range £		
2.2 UPPER FLOORS				
2.2.1 Floors				
Composite steel deck flooring				
Composite steel and concrete upper floors (Note: all floor thicknesses are nominal); A142 mesh reinforcement				
TATA Slimdek SD225 steel decking 1.25 mm thick; 130 mm reinforced concrete topping	m^2	95.00	to	115.00
Re-entrant type steel deck 0.90 mm thick; 150 mm reinforced concrete	m^2	81.00	to	98.00
Re-entrant type steel deck 60 mm deep, 1.20 mm thick; 200 mm reinforced concrete	m^2	87.00	to	105.00
Trapezoidal steel decking 0.90 mm thick; 150 mm reinforced concrete topping	m^2	83.00	to	100.00
Trapezoidal steel decking 1.20 mm thick; 150 mm reinforced concrete topping	m^2	87.00	to	105.00
Composite precast concrete and in situ concrete upper floors				
Omnidec (Hanson Building Products) flooring system comprising of 50 mm thick precast concrete deck (as permanent shuttering); 210 mm polystyrene void formers; concrete topping and reinforcement	m^2	115.00	to	140.00
Precast concrete suspended floors; 1200 mm wide suspended slab; 75 mm thick screed; no coverings or finishes				
3.00 m span; 150 mm thick planks; 5.00 kN/m^2 loading	m^2	76.00	to	88.00
6.00 m span; 150 mm thick planks; 5.00 kN/m^2 loading	m^2	78.00	to	90.00
7.50 m span; 200 mm thick planks; 5.00 kN/m^2 loading	m^2	82.00	to	95.00
9.00 m span; 250 mm thick planks; 5.00 kN/m^2 loading	m^2	86.00	to	99.00
12.00 m span; 350 mm thick planks; 5.00 kN/m^2 loading	m^2	90.00	to	105.00
3.00 m span; 150 mm thick planks; 8.50 kN/m^2 loading	m^2	75.00	to	87.00
6.00 m span; 200 mm thick planks; 8.50 kN/m^2 loading	m^2	77.00	to	89.00
7.50 m span; 250 mm thick planks; 8.50 kN/m^2 loading	m^2	86.00	to	99.00
3.00 m span; 150 mm thick planks; 12.50 kN/m^2 loading	m^2	78.00	to	90.00
6.00 m span; 250 mm thick planks; 12.50 kN/m^2 loading	m^2	86.00	to	99.00
Post-tensioned concrete upper floors				
reinforced post-tensioned suspended concrete slab 150–225 mm thick, 40 N/mm^2, reinforcement 60 kg/m^3, formwork	m^2	110.00	to	135.00
Softwood floors				
Joisted floor; plasterboard ceiling; skim; emulsion; t&g chipboard, sheet vinyl flooring and painted softwood skirtings	m^2	68.00	to	87.00
2.3 ROOF				
2.3.1 Roof Structure				
Flat roof decking; structure only				
Softwood				
comprising roof joists; 100 mm × 50 mm wall plates; herringbone strutting; 18 mm thick external quality plywood boarding	m^2	52.00	to	63.00
Metal decking				
galvanized steel roof decking; insulation; three layer felt roofing and chippings; 0.70 mm thick steel decking (U-value = 0.25 W/m^{2}K)	m^2	77.00	to	93.00
aluminium roof decking; three layer felt roofing and chippings; 0.90 mm thick aluminium decking (U-value = 0.25 W/m^{2}K)	m^2	8.80	to	10.60

2.3 ROOF

Item	Unit	Range £		
2.3 ROOF – cont				
Flat roof decking – cont				
Concrete				
precast concrete suspended slab with sand: cement screed over reinforced concrete slabs; on steel permanent steel shuttering; 150 mm	m²	82.00	to	99.00
reinforced concrete topping	m²	83.00	to	100.00
Screeds/Decks to receive roof coverings				
50 mm thick cement and sand screed	m²	18.10	to	22.00
75 mm thick lightweight bituminous screed and vapour barrier	m²	64.00	to	77.00
Softwood trussed pitched roofs; structure only				
Timber roof trusses and bracings only; plan area				
comprising 75 mm × 50 mm roof trusses at 600 mm centres (measured on plan)	m²	35.50	to	43.00
comprising 100 mm × 38 mm roof trusses at 600 mm centres (measured on plan)	m²	42.00	to	51.00
Mansard type roof comprising 100 mm × 50 mm roof trusses at 600 mm centres; 70° pitch	m²	43.00	to	52.00
forming small dormers up to 1.50 m²	m²	720.00	to	870.00
2.3.2 Roof Coverings				
Timber trusses with tile coverings				
Timber roof trusses; insulation; roof coverings; PVC rainwater goods; plasterboard; skim and emulsion to ceilings (U-value = 0.25 W/m² K); roof plan area				
concrete interlocking tile coverings	m²	185.00	to	225.00
clay plain tile coverings	m²	195.00	to	235.00
clay pan tile coverings	m²	210.00	to	250.00
natural Welsh slate coverings	m²	250.00	to	305.00
reconstructed stone coverings	m²	240.00	to	295.00
Timber dormer roof trusses; insulation; roof coverings; PVC rainwater goods; plasterboard; skim and emulsion to ceilings (U-value = 0.25 W/m² K)				
concrete interlocking tile coverings	m²	220.00	to	260.00
clay pantile coverings	m²	200.00	to	245.00
plain clay tile coverings	m²	215.00	to	260.00
natural slate coverings	m²	310.00	to	375.00
reconstructed stone coverings	m²	270.00	to	330.00
Comparative tiling and slating finishes; including underfelt, battening, eaves courses and ridges				
concrete troughed or bold roll interlocking tiles; sloping	m²	67.00	to	81.00
Tudor clay pantiles; sloping	m²	81.00	to	98.00
blue composition (cement fibre) slates; sloping	m²	100.00	to	125.00
Spanish slates; sloping	m²	130.00	to	155.00
Welsh slates; sloping	m²	200.00	to	245.00
reconstructed stone slates; random slates; sloping	m²	165.00	to	200.00

2.3 ROOF

Item	Unit	Range £		
Steel trussed with metal sheet cladding				
Steel roof trusses and beams; thermal and acoustic insulation (U-value = $0.25\,W/m^2\,K$)				
aluminium profiled composite cladding	m^2	290.00	to	350.00
copper roofing on boarding	m^2	320.00	to	385.00
Sheet roof claddings				
Fibre cement sheet profiled cladding				
Profile 6; single skin; natural grey finish	m^2	25.50	to	31.00
P61 Insulated System; natural grey finish; metal inner lining panel (U-value = $0.25\,W/m^2\,K$)	m^2	49.50	to	60.00
extra for coloured fibre cement sheeting	m^2	1.70	to	2.10
double skin GRP translucent roof sheets	m^2	56.00	to	68.00
triple skin GRP translucent roof sheets	m^2	62.00	to	75.00
Steel PVF2 coated galvanized trapezoidal profile cladding fixed to steel purlins (not included); for roof pitches greater than 4°				
single skin trapezoidal sheeting only	m^2	16.80	to	20.00
composite insulated roofing system; 80 mm overall panel thickness (U-value = $0.25\,W/m^2\,K$)	m^2	44.50	to	54.00
composite insulated roofing system; 115 mm overall panel thickness (U-value = $0.18\,W/m^2\,K$)	m^2	47.00	to	57.00
composite insulated roofing system; 150 mm overall panel thickness (U-value = $0.14\,W/m^2\,K$)	m^2	63.00	to	76.00
For very shallow roof pitches				
standing seam joints composite insulated roofing system; 90 mm overall panel thickness (U-value = $0.25\,W/m^2\,K$)	m^2	72.00	to	87.00
standing seam joints composite insulated roofing system; 110 mm overall panel thickness (U-value = $0.20\,W/m^2\,K$)	m^2	78.00	to	94.00
standing seam joints composite insulated roofing system; 125 mm overall panel thickness (U-value = $0.18\,W/m^2\,K$)	m^2	81.00	to	98.00
Aluminium roofing; standing seam				
Kalzip standard natural aluminium; 0.9 mm thick; 180 mm glassfibre insulation; vapour control layer; liner sheets; (U-value = $0.25\,W/m^2\,K$)	m^2	66.00	to	80.00
Copper roofing				
copper roofing with standing seam joints; insulation breather membrane or vapour barrier (U-value = $0.25\,W/m^2\,K$)	m^2	190.00	to	230.00
Stainless steel				
Terne-coated stainless steel roofing on and including Metmatt underlay (U-value = $0.25\,W/m^2\,K$)	m^2	160.00	to	195.00
Zinc; standing seam				
Terne-coated zinc roofing on and including Metmatt underlay (U-value = $0.25\,W/m^2\,K$)	m^2	115.00	to	140.00
Lead				
Roof coverings in welded seam construction including Geotec underlay fixed to prevent lifting and distortion	m^2	160.00	to	195.00
Flat roofing systems				
Includes insulation and vapour control barrier; excludes decking (U-value = $0.25\,W/m^2\,K$)				
single layer polymer roofing membrane	m^2	74.00	to	89.00
single layer polymer roofing membrane with tapered insulation	m^2	145.00	to	175.00
20 mm thick polymer modified asphalt roofing including felt underlay	m^2	100.00	to	125.00

2.3 ROOF

Item	Unit	Range £		
2.3 ROOF – cont				
Flat roofing systems – cont				
Includes insulation and vapour control barrier – cont				
high performance bitumen felt roofing system	m²	98.00	to	120.00
high performance polymer modified bitumen membrane	m²	110.00	to	140.00
Kingspan KS1000TD composite Single Ply roof panels for roof pitches greater than 0.7° (after deflection); 1.5 mm Single Ply External Covering, Internal Coating Bright White Polyester (steel)				
71 mm thick (U-value = 0.25)	m²	59.00	to	71.00
100 mm thick (U-value = 0.18)	m²	59.00	to	72.00
120 mm thick (U-value = 0.15)	m²	68.00	to	83.00
Edges to felt flat roofs; softwood splayed fillet;				
280 mm × 25 mm painted softwood fascia; no gutter aluminium edge trim	m	45.00	to	55.00
Edges to flat roofs; code 4 lead drip dresses into gutter; 230 mm × 25 mm painted softwood fascia;				
100 mm uPVC gutter	m	78.00	to	94.00
Edges to felt flat roofs; softwood splayed fillet; buildup				
280 mm × 25 mm painted softwood fascia; no gutter aluminium edge trim	m	45.50	to	55.00
Edges to flat roofs; code 4 lead drip dresses into gutter; 230 mm × 25 mm painted softwood fascia; buildup				
100 mm uPVC gutter	m	78.00	to	94.00
100 mm cast iron gutter; decorated	m	68.00	to	83.00
2.3.3 Specialist Roof Systems				
Landscaped roofs				
Polyester-based elastomeric bitumen waterproofing and vapour equalization layer, copper lined bitumen membrane root barrier and waterproofing layer, separation and slip layers, protection layer, 50 mm thick drainage board, filter fleece, insulation board, Sedum vegetation blanket				
intensive (high maintenance – may include trees and shrubs require deeper substrate layers, are generally limited to flat roofs)	m²	190.00	to	235.00
extensive (low maintenance – herbs, grasses, mosses and drought tolerant succulents such as Sedum)	m²	165.00	to	200.00
Ethylene tetrafluoroethylene (ETFE) systems				
Multiple layered ETFE inflated cushions supported by a lightweight aluminium or steel structure	m²	820.00	to	990.00
2.3.4 Roof Drainage				
Gutters				
100 mm uPVC gutter	m	28.50	to	34.50
170 mm uPVC gutter	m	47.00	to	57.00
100 mm cast iron gutter; decorated	m	47.00	to	57.00
150 mm cast iron gutter; decorated	m	76.00	to	92.00
Rainwater downpipes pipes; fixed to backgrounds; including offsets and shoes				
68 mm dia. uPVC	m	26.00	to	31.00
110 mm dia. uPVC	m	47.00	to	57.00
75 mm dia. cast iron; decorated	m	71.00	to	86.00
100 mm dia. cats iron; decorated	m	92.00	to	110.00

2.3 ROOF

Item	Unit	Range £		
2.3.5 Rooflights/Patent Glazing and Glazed Roofs				
Rooflights				
individual polycarbonate rooflights; rectangular; fixed light	m^2	520.00	to	630.00
individual polycarbonate rooflights; rectangular; manual opening	m^2	720.00	to	870.00
individual polycarbonate rooflights; rectangular; electric opening	m^2	1075.00	to	1325.00
Velux style rooflights to traditional roof construction (tiles/slates)	m^2	570.00	to	690.00
Patent glazing; including flashings, standard aluminium alloy bars; Georgian wired or laminated glazing; fixed lights				
single glazed	m^2	200.00	to	240.00
low-e clear toughened and laminated double glazed units	m^2	520.00	to	630.00
2.3.6 Roof Features				
Flat roof feature				
roof walkways 600 mm × 600 mm × 50 mm precast concrete slabs on support system; pedestrian access	m^2	52.00	to	63.00
solar reflective paint	m^2	5.30	to	6.40
limestone chipping finish	m^2	5.70	to	6.90
grip tiles in hot bitumen	m^2	70.00	to	85.00
Insulation				
Glass fibre roll; Crown Loft Roll 44 (Thermal conductivity 0.044 W/mK) or other equal; laid loose				
100 mm thick	m^2	2.80	to	3.40
150 mm thick	m^2	3.70	to	4.45
200 mm thick	m^2	4.60	to	5.60
Glass fibre quilt; Isover Modular roll (Thermal conductivity 0.043 W/mK) or other equal and approved; laid loose				
100 mm thick	m^2	2.75	to	3.30
150 mm thick	m^2	3.60	to	4.35
200 mm thick	m^2	5.35	to	6.50
Crown Rafter Roll 32 (Thermal conductivity 0.032 W/mK) glass fibre flanged building roll; pinned vertically or to slope between timber framing				
50 mm thick	m^2	5.80	to	7.00
75 mm thick	m^2	7.55	to	9.15
100 mm thick	m^2	9.20	to	11.20
Thermafleece EcoRoll (0.039 W/mK)				
50 mm thick	m^2	7.10	to	8.65
75 mm thick	m^2	9.15	to	11.10
100 mm thick	m^2	11.40	to	13.80
140 mm thick	m^2	15.40	to	18.70

2.4 STAIRS AND RAMPS

Item	Unit	Range £		
2.4 STAIRS AND RAMPS				
2.4.1 Stair/Ramp Structures				
Reinforced concrete construction				
Escape staircase; granolithic finish; mild steel balustrades and handrails				
3.00 m rise; dogleg	nr	6400.00	to	7800.00
plus or minus for each 300 mm variation in storey height	nr	630.00	to	760.00
Staircase; terrazzo finish; mild steel balustrades and handrails; plastered soffit; balustrades and staircase soffit decorated				
3.00 m rise; dogleg	nr	9700.00	to	12000.00
plus or minus for each 300 mm variation in storey height	nr	950.00	to	1150.00
Staircase; terrazzo finish; stainless steel balustrades and handrails; plastered and decorated soffit				
3.00 m rise; dogleg	nr	11500.00	to	14000.00
plus or minus for each 300 mm variation in storey height	nr	1150.00	to	1375.00
Staircase; high quality finishes; stainless steel and glass balustrades; plastered and decorated soffit				
3.00 m rise; dogleg	nr	18000.00	to	21500.00
plus or minus for each 300 mm variation in storey height	nr	2075.00	to	2500.00
Metal construction				
Steel access/fire ladder				
3.00 m high	nr	750.00	to	900.00
4.00 m high; epoxide finished	nr	1150.00	to	1400.00
Light duty metal staircase; galvanized finish; perforated treads; no risers; balustrades and handrails; decorated				
3.00 m rise; straight; 900 mm wide	nr	4300.00	to	5200.00
plus or minus for each 300 mm variation in storey height	nr	350.00	to	430.00
Light duty circular metal staircase; galvanized finish; perforated treads; no risers; balustrades and handrails; decorated				
3.00 m rise; straight; 1548 mm dia.	nr	4750.00	to	5700.00
plus or minus for each 300 mm variation in storey height	nr	415.00	to	500.00
Heavy duty cast iron staircase; perforated treads; no risers; balustrades and hand rails; decorated				
3.00 m rise; straight	nr	5400.00	to	6500.00
plus or minus for each 300 mm variation in storey height	nr	550.00	to	670.00
3.00 m rise; spiral; 1548 mm dia.	nr	6100.00	to	7400.00
plus or minus for each 300 mm variation in storey height	nr	610.00	to	740.00
Feature metal staircase; galvanized finish perforated treads; no risers; decorated				
3.00 m rise; spiral balustrades and handrails	nr	7100.00	to	8500.00
3.00 m rise; dogleg; hardwood balustrades and handrails	nr	8400.00	to	10000.00
3.00 m rise; dogleg; stainless steel balustrades and handrails	nr	11000.00	to	13000.00
plus or minus for each 300 mm variation in storey height	nr	720.00	to	870.00
galvanized steel catwalk; nylon coated balustrading 450 mm wide	m	400.00	to	480.00

2.5 EXTERNAL WALLS

Item	Unit	Range £		

Timber construction
Softwood staircase; softwood balustrades and hardwood handrail; plasterboard; skim and emulsion to soffit

2.60 m rise; standard; straight flight	nr	1100.00	to	1350.00
2.60 m rise; standard; top three treads winding	nr	1275.00	to	1550.00
2.60 m rise; standard; dogleg	nr	1400.00	to	1700.00

Oak staircase; balustrades and handrails; plasterboard; skim and emulsion to soffit

2.60 m rise; purpose-made; dogleg	nr	9400.00	to	11500.00
plus or minus for each 300 mm variation in storey height	nr	1250.00	to	1525.00

2.4.2 Stair/Ramp Finishes
Finishes to treads and risers; including nosings etc.

vinyl or rubber	m	58.00	to	70.00
carpet (PC sum £22/m²)	m	88.00	to	105.00

2.4.3 Stair/Ramp Balustrades and Handrails
Wall handrails

softwood handrail and brackets	m	84.00	to	100.00
hardwood handrail and brackets	m	120.00	to	140.00
mild steel handrail and brackets	m	155.00	to	190.00
stainless steel handrail and brackets	m	195.00	to	240.00

Balustrading and handrails

mild steel balustrade and steel or timber handrail	m	305.00	to	370.00
balustrade and handrail with metal infill panels	m	390.00	to	470.00
balustrade and handrail with glass infill panels	m	430.00	to	520.00
stainless steel balustrade and handrail	m	510.00	to	620.00
stainless steel and structural glass balustrade	m	890.00	to	1075.00

2.5 EXTERNAL WALLS

2.5.1 External Enclosing Walls above ground level

Brick/block walling; solid walls
Common brick solid walls; bricks PC £380.00/1000

half brick thick	m²	58.00	to	70.00
one brick thick	m²	110.00	to	130.00
one and a half brick thick	m²	160.00	to	190.00
two brick thick	m²	200.00	to	245.00
extra for fair face per side	m²	2.55	to	3.10

Engineering brick walls; class B; bricks PC £590.00/1000

half brick thick	m²	64.00	to	78.00
one brick thick	m²	120.00	to	145.00
one and a half brick thick	m²	170.00	to	210.00
two brick thick	m²	225.00	to	275.00

Facing brick walls; sand faced facings; bricks PC £500.00/1000

half brick thick; pointed one side	m²	74.00	to	89.00
one brick thick; pointed both sides	m²	140.00	to	170.00

2.5 EXTERNAL WALLS

Item	Unit	Range £		
2.5 EXTERNAL WALLS – cont				
Brick/block walling – cont				
Facing bricks solid walls; handmade facings; bricks PC £700.00/1000				
half brick thick; fair face one side	m²	89.00	to	110.00
one brick thick; fair face both sides	m²	175.00	to	210.00
add or deduct for each variation of £10.00/1000 in PC value				
half brick thick	m²	0.75	to	0.90
one brick thick	m²	1.50	to	1.80
one and a half brick thick	m²	2.20	to	2.70
two brick thick	m²	3.00	to	3.60
Aerated lightweight block walls				
100 mm thick	m²	32.00	to	39.00
140 mm thick	m²	41.50	to	50.00
215 mm thick	m²	78.00	to	94.00
Dense aggregate block walls				
100 mm thick	m²	42.00	to	51.00
140 mm thick	m²	55.00	to	67.00
Coloured dense aggregate masonry block walls; Lignacite or similar				
100 mm thick single skin, weathered face, self-coloured	m²	84.00	to	100.00
100 mm thick single skin, polished face	m²	130.00	to	160.00
Brick/block walling; cavity walls				
Cavity wall; facing brick outer skin; insulation; with plaster on standard weight block inner skin; emulsion (U-value = 0.30 W/m² K)				
machine-made facings; PC £550.00/1000	m²	135.00	to	165.00
machine-made facings; PC £700.00/1000	m²	150.00	to	185.00
self-finished masonry block, weathered finish outer skin and paint grade block inner skins; fair faced both sides	m²	135.00	to	165.00
self-finished masonry block, polished finish outer skin and paint grade block inner skins; fair faced both sides	m²	180.00	to	220.00
Cavity wall; rendered block outer skin; insulation; with plaster on standard weight block inner skin; emulsion (U-value = 0.30 W/m² K)				
block outer skin; insulation; lightweight block inner skin outer block rendered	m²	165.00	to	200.00
Cavity wall; facing brick outer skin; insulation; plasterboard on stud inner skin; emulsion (U-value = 0.30 W/m² K)				
machine-made facings; PC £550.00/1000	m²	110.00	to	130.00
machine-made facings; PC £700.00/1000	m²	125.00	to	150.00
Reinforced concrete walling				
In situ reinforced concrete 25.00 N/mm²; 15 kg/m² reinforcement; formwork both sides				
150 mm thick	m²	135.00	to	165.00
225 mm thick	m²	150.00	to	185.00
300 mm thick	m²	170.00	to	205.00
Panelled walling				
Precast concrete panels; including insulation; lining and fixings generally 7.5 m × 0.15 m thick × storey height (U-value = 0.30 W/m² K)				
standard panels	m²	265.00	to	320.00
standard panels; exposed aggregate finish	m²	290.00	to	350.00
reconstructed stone faced panels	m²	330.00	to	400.00

2.5 EXTERNAL WALLS

Item	Unit	Range £		
brick clad panels (PC £350.00/1000 for bricks)	m²	435.00	to	530.00
natural stone faced panels (Portland Stone or similar)	m²	560.00	to	680.00
marble or granite faced panels	m²	740.00	to	900.00
Kingspan TEK cladding panel; 142 mm thick (15 mm thick OSB board and 112 mm thick rigid eurathane core) fixed to frame; metal edge trim flashings	m²	360.00	to	435.00
Sheet claddings				
Cement fibre profiled cladding				
Profile 6; single skin; natural grey finish	m²	26.50	to	32.00
P61 Insulated System; natural grey finish; metal inner lining panel (U-value = 0.30 W/m² K)	m²	49.50	to	60.00
P61 Insulated System; natural grey finish; 100 mm lightweight concrete block (U-value = 0.30 W/m² K)	m²	89.00	to	110.00
P61 Insulated System; natural grey finish; 12 mm self-finished plasterboard lining (U-value = 0.30 W/m² K)	m²	68.00	to	82.00
extra for coloured fibre cement sheeting	m²	1.50	to	1.85
Metal profiled cladding				
Standard trapazoidal profile				
composite insulated roofing system; 60 mm overall panel thickness (U-value = 0.35 W/m² K)	m²	52.00	to	63.00
composite insulated roofing system; 80 mm overall panel thickness (U-value = 0.26 W/m² K)	m²	54.00	to	66.00
composite insulated roofing system; 100 mm overall panel thickness (U-value = 0.20 W/m² K)	m²	58.00	to	70.00
Micro-rib profile				
composite insulated roofing system; micro-rib panel; 60 mm overall panel thickness (U-value = 0.35 W/m² K)	m²	140.00	to	170.00
composite insulated roofing system; micro-rib panel; 80 mm overall panel thickness (U-value = 0.26 W/m² K)	m²	155.00	to	190.00
composite insulated roofing system; micro-rib panel; 100 mm overall panel thickness (U-value = 0.20 W/m² K)	m²	170.00	to	200.00
Flat panel				
composite insulated roofing system; micro-rib panel; 60 mm overall panel thickness (U-value = 0.35 W/m² K)	m²	230.00	to	280.00
composite insulated roofing system; micro-rib panel; 80 mm overall panel thickness (U-value = 0.26 W/m² K)	m²	220.00	to	270.00
composite insulated roofing system; micro-rib panel; 100 mm overall panel thickness (U-value = 0.20 W/m² K)	m²	260.00	to	310.00
Glazed walling				
Curtain walling				
stick curtain walling with double glazed units, aluminium structural framing and spandrel rails. Standard colour powder coated capped	m²	540.00	to	650.00
unitized curtain walling system with double glazed units, aluminium structural framing and spandrel rails. Standard colour powder coated	m²	1100.00	to	1350.00
unitized curtain walling bespoke project specific system with double glazed units, aluminium structural framing and spandrel rails. Standard colour powder coated	m²	1100.00	to	1350.00
visual mock-ups for project specific curtain walling solutions	item	36000.00	to	43000.00

2.5 EXTERNAL WALLS

Item	Unit	Range £		
2.5 EXTERNAL WALLS – cont				
Glazed walling – cont				
Other systems				
lift surround of double glazed or laminated glass with aluminium or stainless steel framing	m²	980.00	to	1175.00
Other cladding systems				
Tiles (clay/slate/glass/ceramic)				
machine-made clay tiles; including battens	m²	32.50	to	39.50
best handmade sand faced tiles; including battens	m²	45.00	to	54.00
concrete plain tiles; including battens	m²	46.50	to	56.00
natural slates; including battens	m²	87.00	to	105.00
20 mm × 20 mm thick mosaic glass or ceramic; in common colours; fixed on prepared surface	m²	120.00	to	150.00
Steel				
vitreous enamelled insulated steel sandwich panel system; with insulation board on inner face	m²	210.00	to	250.00
Formalux sandwich panel system; with coloured lining tray; on steel cladding rails	m²	240.00	to	290.00
Comparative external finishes				
Comparative concrete wall finishes				
wrought formwork one side including rubbing down	m²	6.00	to	7.30
shotblasting to expose aggregate	m²	7.30	to	8.90
bush hammering to expose aggregate	m²	18.80	to	23.00
Comparative in situ finishes				
two coats Sandtex Matt cement paint to render	m²	6.50	to	7.90
13 mm thick cement and sand plain face rendering	m²	24.50	to	29.50
ready-mixed self-coloured acrylic resin render on blockwork	m²	64.00	to	78.00
three coat Tyrolean rendering; including backing	m²	46.00	to	55.00
Stotherm Lamella; 6 mm thick work in 2 coats	m²	94.00	to	110.00
Insulation				
Crown Dritherm Cavity Slab 37 (Thermal conductivity 0.037 W/mK) glass fibre batt or other equal; as full or partial cavity fill; including cutting and fitting around wall ties and retaining discs				
50 mm thick	m²	4.35	to	5.30
75 mm thick	m²	6.00	to	7.30
100 mm thick	m²	6.90	to	8.35
Crown Dritherm Cavity Slab 34 (Thermal conductivity 0.034 W/mK) glass fibre batt or other equal; as full or partial cavity fill; including cutting and fitting around wall ties and retaining discs				
65 mm thick	m²	4.55	to	5.50
75 mm thick	m²	5.10	to	6.10
85 mm thick	m²	6.50	to	7.85
100 mm thick	m²	6.65	to	8.05

2.5 EXTERNAL WALLS

Item	Unit	Range £		
Crown Dritherm Cavity Slab 32 (Thermal conductivity 0.032 W/mK) glass fibre batt or other equal; as full or partial cavity fill; including cutting and fitting around wall ties and retaining discs				
65 mm thick	m^2	5.25	to	6.35
75 mm thick	m^2	5.90	to	7.20
85 mm thick	m^2	6.40	to	7.75
100 mm thick	m^2	7.20	to	8.70
Crown Frametherm Roll 40 (Thermal conductivity 0.040 W/mK) glass fibre semi-rigid or rigid batt or other equal; pinned vertically in timber frame construction				
90 mm thick	m^2	6.10	to	7.35
140 mm thick	m^2	8.10	to	9.85
Kay-Cel (Thermal conductivity 0.033 W/mK) expanded polystyrene board standard grade SD/N or other equal; fixed with adhesive				
25 mm thick	m^2	5.60	to	6.75
50 mm thick	m^2	7.60	to	9.20
75 mm thick	m^2	9.60	to	11.60
100 mm thick	m^2	11.40	to	13.80
Kingspan Kooltherm K8 (Thermal conductivity 0.022 W/mK) zero ODP rigid urethene insulation board or other equal; as partial cavity fill; including cutting and fitting around wall ties and retaining discs				
40 mm thick	m^2	15.70	to	19.00
50 mm thick	m^2	15.70	to	19.00
60 mm thick	m^2	18.10	to	22.00
75 mm thick	m^2	17.90	to	22.00
Kingspan Thermawall TW50 (Thermal conductivity 0.022 W/mK) zero ODP rigid urethene insulation board or other equal and approved; as partial cavity fill; including cutting and fitting around wall ties and retaining discs				
25 mm thick	m^2	12.10	to	14.70
50 mm thick	m^2	14.70	to	17.80
75 mm thick	m^2	19.90	to	24.00
100 mm thick	m^2	25.00	to	30.00
Thermafleece TF35 high density wool insulating batts (0.035 W/mK); 60% British wool, 30% recycled polyester and 10% polyester binder with a high recycled content				
50 mm thick	m^2	15.00	to	18.10
75 mm thick	m^2	19.20	to	23.00
2.5.3 Solar/Rain Screen				
Rainscreen				
25 mm thick tongued and grooved tanalized softwood boarding; including timber battens	m^2	48.50	to	59.00
timber shingles, Western Red Cedar, preservative treated in random widths; not including any subframe or battens	m^2	62.00	to	76.00
25 mm thick tongued and grooved Western Red Cedar boarding including timber battens	m^2	55.00	to	67.00

2.6 WINDOWS AND EXTERNAL DOORS

Item	Unit	Range £		

2.5 EXTERNAL WALLS – cont

2.5.3 Solar/Rain Screen – cont
Rainscreen – cont

Item	Unit		Range £	
25 mm thick Western Red Cedar tongued and grooved wall cladding on and including treated softwood battens on breather membrane, 10 mm Eternit Blueclad board and 50 mm insulation board; the whole fixed to Metsec frame system; including sealing all joints etc.	m²	135.00	to	165.00
Sucupira Preta timber boardin 19 mm thick × 75 mm wide slats with chamfered open joints fixed to 38 mm × 50 mm softwood battens with stainless steel screws on 100 mm Kingspan K2 insulations and 28 mm WBP plywood fixed to 50 mm × 1000 softwood studs; complete with breather membrane, aluminium cills and Sucupira Petra batten at external joints	m²	305.00	to	370.00
Trespa single skin rainscreen cladding, 8 mm thick panel, with secondary support/frame system; adhesive fixed panels; open joints; 100 mm Kingspan K15 insulation; aluminium subframe fixed to masonry/concrete	m²	250.00	to	305.00
Terracotta rainscreen cladding; aluminium support rails; anti-graffiti coating	m²	440.00	to	530.00
Corium brick tiles in metal tray system; standard colour; including all necessary angle trim on main support system; comprising of polythene vapour check; 75 mm × 50 mm timber studs; 50 mm thick rigid insulation with taped joints; 38 mm × 50 mm timber counter battens	m²	330.00	to	400.00

Solar shading

Item	Unit		Range £	
fixed aluminium Brise Soleil including uni-strut supports; 300 mm deep	m	180.00	to	215.00

2.6 WINDOWS AND EXTERNAL DOORS

2.6.1 External Windows

Softwood windows (U-value = 1.6 W/m² K)
Standard windows

Item	Unit		Range £	
painted; double glazed; up to 1.50 m²	m²	470.00	to	570.00
painted; double glazed; over 1.50 m², up to 3.20 m²	m²	350.00	to	425.00

Purpose-made windows

Item	Unit		Range £	
painted; double glazed; up to 1.50 m²	m²	650.00	to	780.00
painted; double glazed; over 1.50 m²	m²	580.00	to	700.00

Hardwood windows (U-value = 1.4 W/m² K)
Standard windows; stained

Item	Unit		Range £	
double glazed	m²	1250.00	to	1525.00

Purpose-made windows; stained

Item	Unit		Range £	
double glazed	m²	740.00	to	900.00

Steel windows (U-value = 1.6 W/m² K)
Standard windows

Item	Unit		Range £	
double glazed; powder coated	m²	490.00	to	590.00

Purpose-made windows

Item	Unit		Range £	
double glazed; powder coated	m²	540.00	to	660.00

2.6 WINDOWS AND EXTERNAL DOORS

Item	Unit	Range £		
uPVC windows				
Windows; standard ironmongery; sills and factory glazed with low-e 24 mm double glazing				
standard with low-e 24 mm double glazing	m^2	220.00	to	265.00
WER A rating	m^2	220.00	to	270.00
WER C rating	m^2	220.00	to	265.00
Secured by Design accreditation	m^2	220.00	to	270.00
extra for colour finish to uPVC	m^2	56.00	to	68.00
Composite aluminium/timber windows; U value = 1.5 W/m^2 K				
Purpose-made windows; stainless steel ironmongery; Velfac System 200 or similar				
fixed windows up to 1.50 m^2	m^2	290.00	to	350.00
fixed windows over 1.50 m^2 up to 4.00 m^2	m^2	260.00	to	310.00
outward opening pivot windows up to 1.50 m^2	m^2	700.00	to	850.00
outward opening pivot windows over 1.50 m^2 up to 4.00 m^2	m^2	310.00	to	375.00
round porthole window 1.40 m dia.	nr	950.00	to	1150.00
round porthole window 1.80 m dia.	nr	145.00	to	175.00
round porthole window 2.60 m dia.	nr	2800.00	to	3350.00
purpose-made entrance screens and doors double glazed	m^2	1225.00	to	1475.00
2.6.2 External Doors				
Softwood external doors				
Standard external softwood doors and hardwood frames; doors painted; including ironmongery				
Matchboarded, framed, ledged and braced door, 838 mm × 1981 mm	nr	570.00	to	690.00
flush door; cellular core; plywood faced; 838 mm × 1981 mm	nr	580.00	to	710.00
Heavy duty solid flush door				
single leaf	nr	1225.00	to	1475.00
double leaf	nr	1975.00	to	2425.00
single leaf; emergency fire exit	nr	1675.00	to	2025.00
double leaf; emergency fire exit	nr	2450.00	to	2950.00
Steel external doors				
Standard doors				
single external steel door, including frame, ironmongery, powder coated finish	nr	1025.00	to	1225.00
single external steel security door, including frame, ironmongery, powder coated finish	nr	2125.00	to	2600.00
double external steel security door, including frame, ironmongery, powder coated finish	nr	3150.00	to	3800.00
Overhead doors				
single skin; manual	m^2	160.00	to	200.00
single skin; electric	m^2	300.00	to	360.00
electric operation standard lift, 42 mm thick insulated sandwich panels	m^2	230.00	to	275.00
rapid lift fabric door, external, electric operation	m^2	980.00	to	1175.00

2.6 WINDOWS AND EXTERNAL DOORS

Item	Unit	Range £		
2.6 WINDOWS AND EXTERNAL DOORS – cont				
Steel external doors – cont				
Dock shelters				
Curtain mechanical shelter; extruded aluminium frame; spring loaded-frame connected with parallel bracing arms; 2 side curtains, one top curtain; double-layered high quality polyester, coated on both sides, longitudinal strength: approx. 750 N, transverse strength	n	1275.00	to	1475.00
Inflatable mechanical shelter; hot dipped galvanized surface treatment, polyester painted, top bag with polyester fabric panels; side bags with polyester fabric panels, strong blower for straight movement of top and side bags; side section with steel guards; colour from standard range	n	4000.00	to	4600.00
uPVC external doors				
Entrance doors; residential standard; PVC-u frame; brass furniture (spyhole/security chain/letter plate/draught excluder/multipoint locking)				
overall 900 × 2100 mm half glazed	nr	465.00	to	560.00
overall 900 × 2100 mm half glazed; WER A rated	nr	470.00	to	570.00
overall 900 × 2100 mm half glazed; WER C rated	nr	465.00	to	560.00
overall 900 × 2100 mm half glazed; coloured	nr	530.00	to	640.00
Composite aluminium/timber entrance screens and doors; U-value = 1.5 W/m^2 K				
Purpose-made doors				
glazed single personnel door; stainless steel ironmongery; Velfac System100 or similar	nr	2075.00	to	2500.00
glazed double personnel door; stainless steel ironmongery; Velfac System 100 or similar	pair	3300.00	to	4000.00
automatic revolving door; 2000 mm dia.; clear laminated glazing; 4 nr wings; glazed curved walls	nr	38000.00	to	46000.00
automatic sliding door; bi-parting	m^2	7300.00	to	8900.00
Stainless steel entrance screens and doors				
Purpose-made screen; double glazed				
with manual doors	m^2	2400.00	to	2900.00
with automatic doors	m^2	2700.00	to	3300.00
Shop fronts, shutters and grilles				
Purpose-made screen				
temporary timber shop fronts	m^2	84.00	to	100.00
hardwood and glass; including high enclosed window beds	m^2	930.00	to	1125.00
flat facade; glass in aluminium framing; manual centre doors only	m^2	790.00	to	950.00
grilles or shutters	m^2	430.00	to	520.00
fire shutters; powers operated	m^2	460.00	to	560.00

2.7 INTERNAL WALLS AND PARTITIONS

Item	Unit	Range £		
Automatic glazed entrance doors				
Automatic glazed sliding doors in aluminium; polyester powder coated; linear sliding doors; inner pivoted pocket screens; glazed with safety units in accordance with BS. ASSA ABLOY SL510 P RC2 Anti-Burglar Sliding Door System complete with 100 mm operator providing automated operation of the sliding door leaves				
Double bi-parting; opening 2.00 m × 2.30 m	nr	9100.00	to	11500.00
Automatic revolving door, framed. ASSA ABLOY automatic revolving door system providing hands free operation and escape functionality. Modern modular design comprising four collapsible door leaves with electromechanical breakout, framed curved glazed outer walls, radially segmented ceiling and semi-circular dustcovers, complete with operator and imbedded digital control				
3 or 4 wing 2100 mm dia., 2200 mm internal height	nr	27000.00	to	34000.00
3 or 4 wing 3000 mm dia., 2200 mm internal height	nr	29000.00	to	36500.00
Automatic revolving door, structural glass. ASSA ABLOY automatic glass revolving door system providing hands free operation and Hi-Tech functionality. Modern modular design comprising four door leaves, curved glazed outer walls, glass ceiling/canopy, complete with in-floor mounted operator and imbedded digital control				
3 or 4 wing 1800 mm dia., 2200 mm internal height	nr	40000.00	to	50000.00
3 or 4 wing 3000 mm dia., 2200 mm internal height	nr	43000.00	to	54000.00

2.7 INTERNAL WALLS AND PARTITIONS

2.7.1 Internal Walls and Partitions

Item	Unit	Range £		
Frame and panel partitions				
Timber stud partitions				
structure only comprising 100 mm × 38 mm softwood studs at 600 mm centres; head and sole plates	m²	24.00	to	29.50
softwood stud comprising 100 mm × 38 mm softwood studs at 600 mm centres; head and sole plates; 12.5 mm thick plasterboard each side; tape and fill joints; emulsion finish	m²	61.00	to	73.00
Metal stud and board partitions; height range from 2.40 m to 3.30 m				
73 mm partition; 48 mm studs and channels; one layer of 12.5 mm Gyproc Wallboard each side; joints filled with joint filler and joint tape; emulsion paint finish; softwood skirtings with gloss finish	m²	46.50	to	56.00
102 mm partition; 70 mm studs and channels; one layer of 15 mm Gyproc Wallboard each side; joints filled with joint filler and joint tape; emulsion paint finish; softwood skirtings with gloss finish	m²	49.50	to	60.00
102 mm thick partition; 70 mm steel studs at 600 mm centres generally; 1 layer 15 mm Fireline board each side; joints filled with joint filler and joint tape; emulsion paint finish; softwood skirtings with gloss finish	m²	58.00	to	71.00
130 mm thick partition; 70 mm steel studs at 600 mm centres generally; 2 layers 15 mm Fireline board each side; joints filled with joint filler and joint tape; emulsion paint finish; softwood skirtings with gloss finish	m²	69.00	to	84.00
102 mm thick partition; 70 mm steel studs at 600 mm centres generally; 1 layer 15 mm Soundbloc board each side; joints filled with joint filler and joint tape; emulsion paint finish; softwood skirtings with gloss finish	m²	58.00	to	70.00
130 mm thick partition; 70 mm steel studs at 600 mm centres generally; 2 layers 15 mm Soundbloc board each side; joints filled with joint filler and joint tape; emulsion paint finish; softwood skirtings with gloss finish	m²	77.00	to	93.00

2.7 INTERNAL WALLS AND PARTITIONS

Item	Unit	Range £		
2.7 INTERNAL WALLS AND PARTITIONS – cont				
Frame and panel partitions – cont				
Alternative board finishes				
12 mm plywood boarding	m²	11.00	to	13.30
one coat Thistle board finish 3 mm thick work to walls; plasterboard base	m²	7.15	to	8.65
extra for curved work	%	15.00	to	20.00
Acoustic insulation to partitions				
Mineral fibre quilt; Isover Acoustic Partition Roll (APR 1200) or other equal;				
pinned vertically to timber or plasterboard				
25 mm thick	m²	2.70	to	3.20
50 mm thick	m²	3.50	to	4.20
Brick partitions				
common brick half brick thick wall; bricks PC £380.00/1000	m²	55.00	to	67.00
Block partitions				
aerated/lightweight block partitions				
100 mm thick	m²	33.00	to	40.00
140 mm thick	m²	45.00	to	54.00
200/215 mm thick	m²	57.00	to	69.00
dense aggregate block walls				
100 mm thick	m²	42.00	to	51.00
140 mm thick	m²	55.00	to	67.00
extra for fair face (rate per side)	m²	1.00	to	1.20
extra for plaster and emulsion (rate per side)	m²	21.00	to	25.00
extra for curved work	%	15.00	to	20.00
Concrete partitions				
Reinforced concrete walls; C25 strength; standard finish; reinforced at 100 kg/m³				
150 mm thick	m²	165.00	to	200.00
225 mm thick	m²	190.00	to	230.00
300 mm thick	m²	210.00	to	255.00
extra for plaster and emulsion (rate per side)	m²	21.00	to	25.00
extra for curved work	%	20.00	to	30.00
Glass				
Hollow glass block walling; cement mortar joints; reinforced with 6 mm dia.				
stainless steel rods; pointed both sides with mastic or other equal and approved				
240 mm × 240 mm × 80 mm clear blocks	m²	360.00	to	430.00
240 mm × 240 mm × 80 mm craft blocks	m²	360.00	to	430.00
190 mm × 190 mm × 100 mm glass blocks; 30 minute fire-rated	m²	390.00	to	475.00
190 mm × 190 mm × 100 mm glass blocks; 60 minute fire-rated	m²	650.00	to	780.00
2.7.2 Balustrades and Handrails				
Wall handrails				
softwood handrail and brackets	m	84.00	to	100.00
hardwood handrail and brackets	m	120.00	to	140.00
mild steel handrail and brackets	m	155.00	to	190.00
stainless steel handrail and brackets	m	195.00	to	240.00

2.7 INTERNAL WALLS AND PARTITIONS

Item	Unit	Range £		
Balustrading and handrails				
mild steel balustrade and steel or timber handrail	m	305.00	to	370.00
balustrade and handrail with metal infill panels	m	390.00	to	470.00
balustrade and handrail with glass infill panels	m	430.00	to	520.00
stainless steel balustrade and handrail	m	510.00	to	620.00
stainless steel and structural glass balustrade	m	890.00	to	1075.00
2.7.3 Moveable Room Dividers				
Demountable/Folding partitions				
Demountable partitioning; aluminium framing; veneer finish doors				
medium quality; 46 mm thick panels factory finish vinyl faced	m²	370.00	to	450.00
high quality; 46 mm thick panels factory finish vinyl faced	m²	550.00	to	660.00
Demountable aluminium/steel partitioning and doors				
high quality; folding	m²	435.00	to	530.00
high quality; sliding	m²	930.00	to	1125.00
Demountable fire partitions				
enamelled steel; half hour	m²	670.00	to	810.00
stainless steel; half hour	m²	1050.00	to	1275.00
soundproof partitions; hardwood doors luxury veneered	m²	330.00	to	395.00
Aluminium internal patent glazing				
single glazed laminated	m²	200.00	to	245.00
double glazed; 1 layer toughened and 1 layer laminated glass	m²	270.00	to	330.00
Stainless steel glazed manual doors and screens				
high quality; to inner lobby of malls	m²	880.00	to	1075.00
Acoustic folding partition; headtrack suspension and bracing; aluminium framed with high density particle board panel with additional acoustic insulation; melamine laminate finish; acoustic seals. Nominal weight approximately 55 kg per m²				
sound reduction 48 db (Rw)	m²	620.00	to	750.00
sound reduction 55 db (Rw)	m²	750.00	to	910.00
2.7.4 Cubicles				
Framed panel cubicles				
Changing and WC cubicles; high pressure laminate faced MDF; proprietary system				
standard quality WC cubicle partition sets; aluminium framing; melamine face chipboard dividing panels and doors; ironmongery; small range (up to 5 cubicles); standard cubicle set (rate per cubicle)	nr	485.00	to	590.00
medium quality WC cubicle partition sets; stainless steel framing; real wood veneer face chipboard diiding panels and doors; ironmongery; small range (up to 6 cubicles); standard cubicle set (rate per cubicle)	nr	930.00	to	1125.00
high end specification flush fronted system floor to ceiling 44 mm doors, no visible fixings; real wood veneer or HPL or high Gloss paint and lacquer; cubicle set; 800 mm × 1500 mm × 2400 mm high per cubicle, with satin finished stainless steel ironmongery; 30 mm high pressure laminated (HPL) chipboard divisions and 44 mm solid cored real wood veneered doors and pilasters; small range (up to 6 cubicles); standard cubicle set (rate per cubicle)	nr	2475.00	to	3000.00

2.8 INTERNAL DOORS

Item	Unit	Range £	
2.7 INTERNAL WALLS AND PARTITIONS – cont			
Framed panel cubicles – cont			
IPS duct panel systems; melamine finish chipboard; softwood timber subframe 2.70 m high IPS back panelling system; to accommodate wash hand basins or urinals; access hatch; frame support	m²	150.00 to	180.00
2.8 INTERNAL DOORS			
2.8.1 Internal Doors			
The following rates include for the supply and hang of doors, complete with all frames, architrave, typical medium standard ironmongery set and appropriate finish			
Standard doors			
Standard doors; cellular core; softwood; softwood architrave; aluminium ironmongery (latch only)			
single leaf; moulded panel; gloss paint finish	nr	310.00 to	375.00
single leaf; Sapele veneered finish	nr	350.00 to	420.00
Purpose-made doors			
Softwood panelled; softwood lining; softwood architrave; aluminium ironmongery (latch only); brass or stainless ironmongery (latch only); painting and polishing			
single leaf; four panels; mouldings	nr	475.00 to	570.00
double leaf; four panels; mouldings	nr	930.00 to	1125.00
Hardwood panelled; hardwood lining; hardwood architrave; aluminium ironmongery (latch only); brass or stainless ironmongery (latch only); painting and polishing			
single leaf; four panelled doors; mouldings	nr	960.00 to	1175.00
double leaf; four panelled doors; mouldings	nr	1875.00 to	2250.00
Fire doors			
Standard fire doors; cellular core; softwood lining; softwood architrave; aluminuim ironmongery (lockable, self-closure); painting or polishing;			
single leaf; Oak veneered; 30 min fire resistance; polished	nr	500.00 to	610.00
double leaf; Oak veneered; 30 min fire resistance; polished	nr	1150.00 to	1400.00
single leaf; Oak veneered; 60 min fire resistance; polished	nr	780.00 to	950.00
double leaf; Oak veneered; 60 min fire resistance; polished	nr	1450.00 to	1750.00
Ironmongery sets			
Stainless steel ironmongery; euro locks; push plates; kick plates; signage; closures; standard sets			
office door; non-locking; fire-rated	nr	330.00 to	400.00
office/store; lockable; fire-rated	nr	380.00 to	455.00
classroom door; lockable; fire-rated	nr	495.00 to	600.00
maintenance/plant room door; lockable; fire-rated	nr	355.00 to	430.00
standard bathroom door (unisex)	nr	290.00 to	355.00
accessible toilet door	nr	170.00 to	205.00
fire escape door	nr	1825.00 to	2200.00

3.1 WALL FINISHES

Item	Unit	Range £		
Overhead doors				
single skin; manual	m²	160.00	to	200.00
single skin; electric	m²	300.00	to	360.00
electric operation standard lift, 42 mm thick insulated sandwich panels	m²	230.00	to	275.00
rapid lift fabric door, internal, electric operation	m²	980.00	to	1175.00
3.1 WALL FINISHES				
3.1.1 Wall Finishes				
Internal wall area (unless otherwise described)				
In situ wall finishes				
Comparative finishes				
one mist and two coats emulsion paint	m²	4.20	to	5.10
two coats of lightweight plaster	m²	14.80	to	17.90
two coats of lightweight plaster with emulsion finish	m²	19.00	to	23.00
two coat sand cement render and emulsion finish	m²	26.50	to	32.00
plaster and vinyl wallpaper coverings	m²	24.00	to	29.00
polished plaster system; Armourcoat or similar; 11 mm thick first coat and 2 mm thick finishing coat with a polished finish	m²	120.00	to	145.00
Rigid tile/panel/board finishes				
Timber boarding/panelling; on and including battens; plugged to wall				
12 mm thick softwood boarding	m²	41.00	to	50.00
hardwood panelling; t&g & v-jointed	m²	125.00	to	150.00
Ceramic wall tiles; including backing				
economical quality	m²	43.00	to	52.00
medium to high quality	m²	49.00	to	60.00
Porcelain; including backing				
Porcelain mosaic tiling; walls and floors	m²	82.00	to	99.00
Marble				
Roman Travertine marble wall linings 20 mm thick; polished	m²	310.00	to	375.00
Granite				
Dakota mahogany granite cladding 20 mm thick; polished finish; jointed and pointed in coloured mortar	m²	410.00	to	495.00
Dakota mahogany granite cladding 40 mm thick; polished finish; jointed and pointed in coloured mortar	m²	680.00	to	820.00
Comparative woodwork finishes				
Knot; one coat primer; two undercoats; gloss on wood surfaces; number of coats:				
two coats gloss	m²	10.60	to	12.80
three coats gloss	m²	11.60	to	14.10
three coats gloss; small girth n.e. 300 mm	m	7.80	to	9.45
Polyurethane lacquer				
two coats	m²	7.90	to	9.60
three coats	m²	8.70	to	10.50

3.2 FLOOR FINISHES

Item	Unit	Range £	
3.1 WALL FINISHES – cont			
Comparative woodwork finishes – cont			
Flame-retardant paint			
Unitherm or similar; two coats	m²	19.20 to	23.00
Polish			
wax polish; seal	m²	15.30 to	18.50
wax polish; stain and body in	m²	18.90 to	23.00
French polish; stain and body in	m²	28.00 to	34.00
Other wall finishes			
Wall coverings			
decorated paper backed vinyl wall paper	m²	8.90 to	10.80
PVC wall linings – Altro or similar; standard satins finish	m²	75.00 to	91.00
3.2 FLOOR FINISHES			
3.2.1 Floor Finishes			
Internal floor area (unless otherwise described)			
In situ screed and floor finishes; laid level; over 300 mm wide			
Cement and sand (1:3) screeds; steel trowelled			
50 mm thick	m²	18.10 to	22.00
75 mm thick	m²	23.00 to	28.00
100 mm thick	m²	25.50 to	31.00
Flowing screeds			
Latex screeds; self-colour; 3 mm to 5 mm thick	m²	10.80 to	13.00
Lafarge Gyvlon flowing screed 50 mm thick	m²	11.90 to	14.40
Treads, steps and the like (small areas)	m²	33.00 to	40.00
Granolithic; laid on green concrete			
20 mm thick	m²	25.00 to	30.50
38 mm thick	m²	29.00 to	35.50
Resin floor finish			
Altrotect; 2 coat application nominally 350–500 micron thick	m²	16.70 to	20.00
AltroFlow EP; 3 part solvent-system; up to 3 mm thick	m²	36.00 to	43.50
Atro Screed (TB Screed; Quartz; Multiscreed), 3 mm to 4 mm thick	m²	47.00 to	57.00
Altrocrete heavy duty polyurethane screed, 6 mm to 8 mm screed	m²	52.00 to	63.00
Sheet/board flooring			
Chipboard			
18 mm to 22 mm thick chipboard flooring; t&g joints	m²	17.70 to	21.00
Softwood			
22 mm thick wrought softwood flooring; 150 mm wide; t&g joints;	m²	30.00 to	36.00
softwood skirting, gloss paint finish	m	17.30 to	21.00
MDF skirting, gloss paint finish	m	16.50 to	20.00
Hardwood			
Wrought hardwood t&g strip flooring; polished; including fillets	m²	115.00 to	140.00
Hardwood skirting, stained finish	m	22.00 to	26.50

3.2 FLOOR FINISHES

Item	Unit	Range £		
Sprung floors				
Taraflex Combisport 85 System with Taraflex Sports M Plus; t&g plywood 22 mm thick on softwood battens and crumb rubber cradles	m²	170.00	to	210.00
Sprung composition block flooring (sports), court markings, sanding and sealing	m²	110.00	to	140.00
Rigid tile/slab finishes (includes skirtings; excludes screeds)				
Quarry tile flooring	m²	73.00	to	88.00
Glazed ceramic tiled flooring				
standard plain tiles	m²	47.50	to	58.00
anti-slip tiles	m²	52.00	to	63.00
designer tiles	m²	110.00	to	130.00
Terrazzo tile flooring 28 mm thick polished	m²	54.00	to	65.00
York stone 50 mm thick paving	m²	170.00	to	210.00
Slate tiles, smooth; straight cut	m²	78.00	to	95.00
Portland stone paving	m²	280.00	to	340.00
Roman Travertine marble paving; polished	m²	275.00	to	330.00
Granite paving 20 mm thick paving	m²	420.00	to	510.00
Parquet/wood block wrought hardwood block floorings; 25 mm thick; polished; t&g joints	m²	150.00	to	180.00
Flexible tiling; welded sheet or butt joint tiles; adhesive fixing				
vinyl floor tiling; 2.00 mm thick	m²	15.50	to	18.80
vinyl safety flooring; 2.00–2.50 mm thick	m²	48.00	to	59.00
vinyl safety flooring; 3.5 mm thick heavy duty	m²	56.00	to	68.00
linoleum tile flooring; 333 mm × 333 mm × 3.20 mm tiles	m²	41.00	to	50.00
linoleum sheet flooring; 2.00 mm thick	m²	32.00	to	39.00
rubber studded tile flooring; 500 mm × 500 mm × 2.50 mm thick	m²	43.50	to	53.00
Carpet tiles; including underlay, edge grippers				
heavy domestic duty; to floors; PC Sum £38/m²	m²	62.00	to	75.00
heavy domestic duty; to treads and risers; PC Sum £38/m²	m	49.00	to	60.00
heavy contract duty; Forbo Flotex HD; PC Sum £25/m²	m²	48.50	to	59.00
Entrance matting				
door entrance and circulation matting; soil and water removal	m²	58.00	to	70.00
Entrance matting and matwell				
18 mm thick Gradus Topguard barrier matting with aluminium frame	m²	375.00	to	455.00
18 mm thick Gradus Topguard barrier matting with stainless steel frame	m²	320.00	to	385.00
18 mm thick Gradus Topguard barrier matting with polished brass frame	m²	445.00	to	540.00
3.2.2 Raised Access Floors				
Raised access floors: including 600 mm × 600 mm steel encased particle boards on height adjustable pedestals < 300 mm				
light grade duty	m²	50.00	to	60.00
medium grade duty	m²	50.00	to	60.00
heavy grade duty	m²	67.00	to	81.00
battened raft chipboard floor with sound insulation fixed to battens; medium quality carpeting	m²	100.00	to	120.00
Common floor coverings bonded to access floor panels				
anti-static vinyl	m²	31.00	to	38.00
heavy duty fully flexible vinyl tiles	m²	25.00	to	30.00
needle punch carpet	m²	19.00	to	23.00

3.3 CEILING FINISHES

Item	Unit	Range £		
3.3 CEILING FINISHES				
3.3.1 Finishes to Ceiling				
In situ finishes				
Decoration only to soffits; one mist and two coats emulsion paint				
to exposed steelwork (surface area)	m²	6.40	to	7.75
to concrete soffits (surface area)	m²	4.65	to	5.65
to plaster/plasterboard	m²	4.25	to	5.15
Plaster to soffits				
plaster skim coat to plasterboard ceilings	m²	7.40	to	8.95
lightweight plaster to concrete	m²	19.60	to	24.00
Plasterboard to soffits				
9 mm Gyproc board and skim coat	m²	30.00	to	36.50
12.50 mm Gyproc board and skim coat	m²	30.50	to	37.00
Other board finishes; with fire-resisting properties; excluding decoration				
12.50 mm thick Gyproc Fireline board	m²	23.00	to	28.00
15 mm thick Gyproc Fireline board	m²	24.00	to	29.00
12 mm thick Supalux	m²	64.00	to	78.00
Specialist plasters; to soffits				
sprayed acoustic plaster; self-finished	m²	83.00	to	100.00
rendering; Tyrolean finish	m²	83.00	to	100.00
Other ceiling finishes				
softwood timber t&g boarding	m	51.00	to	67.00
Comparative ceiling finishes				
Emulsion paint to plaster				
two coats	m²	4.80	to	5.80
one mist and two coats	m²	5.20	to	6.30
Gloss paint to timber or metal				
primer and two coats	m²	7.10	to	8.60
primer and three coats	m²	10.20	to	12.40
Other coatings				
Artex plastic compound one coat; textured to plasterboard	m²	8.10	to	9.80
3.3.3 Demountable Suspended Ceilings				
Armstrong suspended ceiling; assume large rooms over 250 m²				
mineral fibre; basic range; Cortega, Tatra, Academy; exposed grid	m²	21.50	to	28.00
mineral fibre; medium quality; Corline; exposed grid	m²	31.00	to	41.00
mineral fibre; medium quality; Corline; concealed grid	m²	46.00	to	61.00
mineral fibre; medium quality; Corline; silhouette grid	m²	58.00	to	76.00
mineral fibre; Specific; Bioguard; exposed grid	m²	26.00	to	34.00
mineral fibre; Specific; Bioguard; clean room grid	m²	41.00	to	54.00
mineral fibre; Specific; Hygiene, Cleanroom; exposed grid	m²	56.00	to	73.00
metal open cell; Cellio Global White; exposed grid	m²	52.00	to	69.00
metal open cell; Cellio Black; exposed grid	m²	63.00	to	83.00
wood veneer board; Microlook 8; plain; exposed grid	m²	69.00	to	90.00
wood veneer; plain; exposed grid	m²	94.00	to	125.00
wood veneers; perforated; exposed grid	m²	160.00	to	215.00
wood veneers; plain; concealed grid	m²	155.00	to	205.00
wood veneers; perforated; concealed grid	m²	215.00	to	280.00

4.1 FITTINGS, FURNISHINGS AND EQUIPMENT

Item	Unit	Range £		
Other suspended ceilings				
perforated aluminium ceiling tiles 600 mm × 600 mm	m²	36.00	to	43.00
perforated aluminium ceiling tiles 1200 mm × 600 mm	m²	43.00	to	52.00
metal linear strip; Dampa/Luxalon	m²	55.00	to	66.00
metal linear strip micro-perforated acoustic ceiling with Rockwool acoustic infill	m²	67.00	to	81.00
metal tray	m²	57.00	to	69.00
egg-crate	m²	87.00	to	105.00
open grid; Formalux/Dimension	m²	115.00	to	140.00
Integrated ceilings				
coffered; with steel surfaces	m²	150.00	to	180.00
acoustic suspended ceilings on anti-vibration mountings	m²	69.00	to	84.00

4.1 FITTINGS, FURNISHINGS AND EQUIPMENT

4.1.1 General Fittings, Furnishings and Equipment

Office furniture and equipment

There is a large quality variation for office furniture and we have assumed a medium level, even so prices will vary between suppliers

Item	Unit	Range £		
Reception desk				
straight counter; 3500 mm long; 2 person	nr	2350.00	to	2950.00
curved counter; 3500 mm long; 2 person	nr	5600.00	to	7100.00
curved counter; 3500 mm long; 2 person; real wood veneer finish	nr	11000.00	to	14000.00
Furniture and equipment to general office area; standard off the shelf specification				
workstation; 2000 mm long desk; drawer unit; task chair	nr	900.00	to	1125.00

4.1.2 Domestic Kitchen Fittings and Equipment

Residential fittings (volume housing)

Kitchen fittings for residential units (not including white goods). NB quality and quantity of units can varies enormously. Always obtain costs from your preferred supplier. The following assume medium standard units from a large manufacturer and includes all units, worktops, stainless steel sink and taps, not including white goods

Item	Unit	Range £		
one person flat	nr	2275.00	to	2900.00
two person flat/house	nr	2500.00	to	3200.00
three person house	nr	4200.00	to	5300.00
four person house; includes utility area	nr	7800.00	to	9900.00
five person house; includes utility area	nr	10500.00	to	13500.00

Hotel bathroom pods

Fully fitted out, finished and furnished bathroom pods; installed; suitable for business class hotels

Item	Unit	Range £		
standard pod (4.50 m plan area)	nr	5600.00	to	7100.00
accessible pod (4.50 m plan area)	nr	7600.00	to	9500.00

5.1 SANITARY INSTALLATIONS

Item	Unit	Range £		
4.1 FITTINGS, FURNISHINGS AND EQUIPMENT – cont				
4.1.3 Special Purpose Fittings, Furnishings and Equipment				
Window blinds				
Louvre blind				
89 mm louvres, manual chain operation, fixed to masonry	m²	61.00	to	77.00
127 mm louvres, manual chain operation, fixed to masonry	m²	53.00	to	67.00
Roller blind				
fabric blinds, roller type, manual chain operation, fixed to masonry	m²	47.00	to	59.00
solar black out blinds, roller type, manual chain operation, fixed to masonry	m²	81.00	to	100.00
Fire curtains				
Electrically operated automatic 2 hr fire curtains to form a virtually continuous barrier against both fire and smoke; size in excess of 4 m² (minimum size 800 mm)	m²	1675.00	to	2125.00
Dock levellers and shelters				
ASSA ABLOY Entrance Systems				
curtain mechanical shelter; 2 side curtains, one top curtain	nr	1325.00	to	1675.00
teledock leveller; electrohydraulic operation with two lifting cylinders; one lip ram cylinder; movable telescopic lip; Control for door and leveller from one communal control panel, including interlocking of leveller/door	nr	3900.00	to	5000.00
inflatable mechanical shelter; top bag with polyester fabric panels, 1000 mm extension; side bags with polyester fabric panels, 650 mm extension, strong blower for straight movement of top and side bags; side section with steel guards; colour from standard range	nr	4200.00	to	5300.00
Stand-alone Load House; a complete loading unit attached externally to building; modular steel skin/insulated cladding; two wall elements, one roof element	nr	6600.00	to	8400.00
5.1 SANITARY INSTALLATIONS				
5.1.1 Sanitary Appliances				
Please refer to *Spon's Mechanical and Electrical Services Price Book* for a more comprehensive range of rates				
Rates are for gross internal floor area – GIFA unless otherwise described				
shopping malls and the like (not tenant fit-out works)	m²	1.05	to	1.50
office building – multi-storey	m²	9.60	to	13.40
office building – business park	m²	8.50	to	11.80
performing arts building (medium specification)	m²	15.60	to	21.50
sports hall	m²	12.30	to	17.10
hotel	m²	41.00	to	57.00
hospital; private	m²	24.50	to	34.00
school; secondary	m²	11.80	to	16.40
residential; multi-storey tower	m²	0.60	to	0.90
supermarket	m²	1.60	to	2.20

5.6 SPACE HEATING AND AIR CONDITIONING

Item	Unit	Range £		
Comparative sanitary fittings/sundries				
Note: Material prices vary considerably, the following composite rates are based on average prices for mid priced fittings:				
Individual appliances (including fittings)				
low level WC's; vitreous china pan and cistern; black plastic seat; low pressure ball valve; plastic flush pipe; fixing brackets	nr	395.00	to	500.00
bowl type wall urinal; white glazed vitreous china flushing cistern; chromium plated flush pipes and spreaders; fixing brackets	nr	395.00	to	510.00
sink; glazed fireclay; chromium plated waste; plug and chain	nr	600.00	to	760.00
sink; stainless steel; chromium plated waste; plug and chain				
single drainer; double bowl (bowl and half)	nr	365.00	to	465.00
double drainer; double bowl (bowl and half)	nr	390.00	to	495.00
bath; reinforced acrylic; chromium plated taps; overflow; waste; chain and plug; P trap and overflow connections	nr	360.00	to	460.00
bath; enamelled steel; chromium plated taps; overflow; waste; chain and plug; P trap and overflow connections	nr	760.00	to	960.00
shower tray; glazed fireclay; chromium plated waste; riser pipe; rose and mixing valve	nr	640.00	to	820.00
5.4 WATER INSTALLATIONS				
5.4.1 Mains Water Supply				
Please refer to *Spon's Mechanical and Electrical Services Price Book* for a more comprehensive range of rates				
Hot and cold water installations; mains supply; hot and cold water distribution.				
Gross internal floor area (unless described otherwise)				
shopping malls and the like (not tenant fit-out works)	m²	13.10	to	18.10
office building – multi-storey	m²	19.10	to	26.50
office building – business park	m²	14.30	to	19.90
performing arts building (medium specification)	m²	22.50	to	31.00
sports hall	m²	17.00	to	23.50
hotel	m²	48.00	to	67.00
hospital; private	m²	58.00	to	80.00
school; secondary; potable and non potable to labs, art rooms	m²	44.00	to	62.00
residential; multi-storey tower	m²	54.00	to	75.00
supermarket	m²	30.00	to	41.50
distribution centre	m²	2.05	to	2.85
5.6 SPACE HEATING AND AIR CONDITIONING				
5.6.1 Space Heating and Air Conditioning				
Please refer to *Spon's Mechanical and Electrical Services Price Book* for a more comprehensive range of rates				
Gross internal floor area (unless described otherwise)				
shopping malls and the like (not tenant fit-out works); LTHW, air conditioning, ventilation	m²	86.00	to	110.00
office building – multi-storey; LTHW, ductwork, chilled water, ductwork	m²	120.00	to	160.00
office building – business park; LTHW, ductwork, chilled water, ductwork	m²	135.00	to	180.00

5.8 ELECTRICAL INSTALLATIONS

Item	Unit	Range £		
5.6 SPACE HEATING AND AIR CONDITIONING – cont				
5.6.1 Space Heating and Air Conditioning – cont				
Gross internal floor area (unless described otherwise) – cont				
performing arts building (medium specification); LTHW, ductwork, chilled water, ductwork	m²	240.00	to	315.00
sports hall; warm air to main hall, LTHW radiators to ancillary	m²	42.00	to	56.00
hotel; air conditioning	m²	160.00	to	210.00
hospital; private; LPHW	m²	240.00	to	315.00
school; secondary; LTHW, cooling to ICT server room	m²	130.00	to	170.00
residential; multi-storey tower; LTHW to each apartment	m²	72.00	to	95.00
supermarket	m²	18.00	to	23.50
distribution centre; LTHW to offices, displacement system to warehouse	m²	25.00	to	33.00
5.7 VENTILATION SYSTEMS				
5.7.1 Ventilation Systems				
Please refer to *Spon's Mechanical and Electrical Services Price Book* for a more comprehensive range of rates				
Gross internal area (unless otherwise described)				
shopping malls and the like (not tenant fit-out works)	m²	41.00	to	54.00
office building – multi-storey; toilet and kitchen areas only	m²	7.10	to	9.30
office building – business park; toilet and kitchen areas only	m²	8.25	to	10.80
performing arts building (medium specification); toilet and kitchen areas, workshop	m²	16.40	to	21.50
sports hall	m²	15.30	to	20.00
hotel; general toilet extraction, bathrooms, kitchens	m²	48.00	to	63.00
hospital; private; toilet and kitchen areas only	m²	8.80	to	11.60
school; secondary; toilet and kitchen areas, science labs	m²	20.00	to	26.50
residential; multi-storey tower	m²	41.00	to	54.00
supermarket	m²	5.95	to	7.85
distribution centre; smoke extract system	m²	5.95	to	7.85
5.8 ELECTRICAL INSTALLATIONS				
5.8.1 Electric Mains and Sub-Mains Distrubition				
Please refer to *Spon's Mechanical and Electrical Services Price Book* for a more comprehensive range of rates				
Including LV distribution, HV distribution, lighting, small power. Gross internal floor area (unless described otherwise)				
shopping malls and the like (not tenant fit-out works)	m²	130.00	to	180.00
office building – multi-storey	m²	110.00	to	150.00
office building – business park	m²	58.00	to	81.00
performing arts building (medium specification)	m²	210.00	to	290.00
sports hall	m²	64.00	to	89.00
hotel	m²	200.00	to	280.00
hospital; private	m²	235.00	to	325.00
school; secondary	m²	140.00	to	195.00
residential; multi-storey tower	m²	93.00	to	130.00
supermarket	m²	75.00	to	105.00
distribution centre	m²	70.00	to	97.00

5.10 LIFT AND CONVEYOR INSTALLATIONS

Item	Unit	Range £		
Comparative fittings/rates per point				
Consumer control unit; 63–100 amp 230 volt; switched and insulated; RCDB protection. Gross internal floor area (unless described otherwise)	nr	340.00	to	425.00
Fittings; excluding lamps or light fittings				
lighting point; PVC cables	nr	50.00	to	64.00
lighting point; PVC cables in screwed conduits	nr	57.00	to	72.00
lighting point; MICC cables	nr	75.00	to	95.00
Switch socket outlet; PVC cables				
single	nr	63.00	to	80.00
double	nr	83.00	to	105.00
Switch socket outlet; PVC cables in screwed conduit				
single	nr	94.00	to	120.00
double	nr	110.00	to	140.00
Switch socket outlet; MICC cables				
single	nr	91.00	to	115.00
double	nr	105.00	to	130.00
Other power outlets				
Immersion heater point (excluding heater)	nr	115.00	to	145.00
Cooker point; including control unit	nr	190.00	to	240.00
5.9 FUEL INSTALLATIONS				
5.9.2 Fuel Distribution Systems				
Please refer to *Spon's Mechanical and Electrical Services Price Book* for a more comprehensive range of rates				
Gas mains service to plantroom. Gross internal floor area				
shopping mall/supermarket	m²	3.50	to	4.40
warehouse/distribution centre	m²	1.00	to	1.25
office/hotel	m²	1.60	to	2.05
5.10 LIFT AND CONVEYOR INSTALLATIONS				
5.10.1 Lifts and Enclosed Hoists				
Please refer to *Spon's Mechanical and Electrical Services Price Book* for a more comprehensive range of rates				
Passenger lifts				
Passenger lifts (standard brushed stainless steel finish; 2 panel centre opening doors)				
8-person; 4 stops; 1.0 m/s speed	nr	84000.00	to	100000.00
8-person; 4 stops; 1.6 m/s speed	nr	89000.00	to	110000.00
13-person; 6 stops; 1.0 m/s speed	nr	105000.00	to	130000.00
13-person; 6 stops; 1.6 m/s speed	nr	110000.00	to	140000.00
13-person; 6 stops; 2.0 m/s speed	nr	115000.00	to	140000.00
21-person; 8 stops; 1.0 m/s speed	nr	130000.00	to	160000.00
21-person; 8 stops; 1.6 m/s speed	nr	150000.00	to	180000.00
21-person; 8 stops; 2.0 m/s speed	nr	150000.00	to	180000.00
extra for enhanced finishes; mirror; carpet	nr	4150.00	to	5000.00
extra for lift car LCD TV	nr	9000.00	to	11000.00
extra for intelligent group control; 5 cars; 11 stops	nr	43000.00	to	52000.00

5.11 FIRE AND LIGHTNING PROTECTION

Item	Unit	Range £		

5.10 LIFT AND CONVEYOR INSTALLATIONS – cont

Passenger lifts – cont

Special installations				
wall climber lift; 10-person; 0.50 m/s; 10 levels	nr	420000.00	to	510000.00
disabled platform lift single wheelchair; 400 kg; 4 stops; 0.16 m/s	nr	10000.00	to	12500.00

Non-passenger lifts

Goods lifts; prime coated internal finish				
2000 kg load; 4 stops; 1.0 m/s speed	nr	105000.00	to	130000.00
2000 kg load; 4 stops; 1.6 m/s speed	nr	115000.00	to	140000.00
2500 kg load; 4 stops; 1.0 m/s speed	nr	110000.00	to	140000.00
2000 kg load; 8 stops; 1.0 m/s speed	nr	145000.00	to	175000.00
2000 kg load; 8 stops; 1.6 m/s speed	nr	150000.00	to	180000.00
Other goods lifts				
Hoist	nr	29000.00	to	35000.00
Kitchen service hoist 50 kg; 2 levels	nr	12000.00	to	15000.00

5.10.2 Escalators

30° escalator; 0.50 m/s; enamelled steel glass balustrades				
3.50 m rise; 800 mm step width	nr	92000.00	to	110000.00
4.60 m rise; 800 mm step width	nr	99000.00	to	120000.00
5.20 m rise; 800 mm step width	nr	100000.00	to	125000.00
6.00 m rise; 800 mm step width	nr	110000.00	to	140000.00
extra for enhanced finish; enamelled finish; glass balustrade	nr	12000.00	to	14500.00

5.10.6 Dock Levellers and Scissor Lifts

Electrohydraulic operation with two lifting cylinders; integrated frame to suit various pit applications; Control for door and leveller from one communal control pane

2500 mm long dock leveller with swing lip	nr	3400.00	to	4150.00
2500 mm long dock leveller with telescope lip	nr	3650.00	to	4400.00

5.11 FIRE AND LIGHTNING PROTECTION

5.11.1 Fire Fighting Systems

Please refer to *Spon's Mechanical and Electrical Services Price Book* for a more comprehensive range of rates

Gross internal floor area (unless described otherwise); Including lightning protection, sprinklers unless stated otherwise

shopping malls and the like (not tenant fit-out works)	m²	16.30	to	23.00
office building – multi-storey	m²	35.00	to	49.00
office building – business park	m²	5.80	to	8.05
performing arts building (medium specification); lightning protection only	m²	1.75	to	2.45
sports hall; lightning protection only	m²	3.50	to	4.85
hospital; private; lightning protection only	m²	2.30	to	3.20
school; secondary; lightning protection only	m²	2.20	to	3.05
hotel	m²	40.00	to	56.00
residential; multi-storey tower	m²	58.00	to	81.00
supermarket; lightning protection only	m²	1.15	to	1.60
distribution centre	m²	45.00	to	63.00

5.14 BUILDER'S WORK IN CONNECTION WITH SERVICES

Item	Unit	Range £		

5.12 COMMUNICATION AND SECURITY INSTALLATIONS

5.12.1 Communication, Security and Control Systems
Please refer to *Spon's Mechanical and Electrical Services Price Book* for a more comprehensive range of rates
Gross internal floor area (unless described otherwise); Including fire alarm, public address system, security installation unless stated otherwise

Item	Unit		Range £	
shopping malls and the like (not tenant fit-out works)	m²	45.50	to	60.00
office building – multi-storey	m²	33.00	to	43.00
office building – business park; fire alarm and wireways only	m²	15.90	to	21.00
performing arts building (medium specification)	m²	100.00	to	130.00
sports hall	m²	66.00	to	86.00
hotel	m²	110.00	to	145.00
hospital; private	m²	85.00	to	110.00
school; secondary	m²	63.00	to	83.00
residential; multi-storey tower	m²	63.00	to	83.00
supermarket	m²	19.60	to	26.00
distribution centre	m²	23.00	to	30.50

5.13 SPECIAL INSTALLATIONS

5.13.3 Specialist Mechanical Installations
Window cleaning equipment

Item	Unit		Range £	
twin track	m	185.00	to	250.00
manual trolley/cradle	nr	13000.00	to	17000.00
automatic trolley/cradle	nr	26000.00	to	35000.00

Sauna

Item	Unit		Range £	
2.20 m × 2.20 m internal Finnish sauna; benching; heater; made of Sauna grade Aspen or similar	nr	10000.00	to	13500.00

Jacuzzi Installation

Item	Unit		Range £	
2.20 m × 2.20 m × 0.95 m; 5 adults; 99 jets; lights; pump	nr	13000.00	to	18000.00

5.13.4 Specialist Electrical/Electronic Installations
Solar PV systems

Item	Unit		Range £	
Residential solar water heating including collectors; dual coil cylinders; pump; controller (NB excludes any grant allowance)	m²	920.00	to	1225.00
Solar power including 4 kWp monocrystalline solar modules (12 × 185 Wp) on roof mounting kit; certified inverter; DC and AC isolation switches and connection to the grid and certification	nr	8000.00	to	11000.00

5.14 BUILDER'S WORK IN CONNECTION WITH SERVICES

5.14.1 Builder's Work in Connection with Services
Please refer to *Spon's Mechanical and Electrical Services Price Book* for a more comprehensive range of rates
Warehouses, sports halls and shopping malls. Gross internal floor area

Item	Unit		Range £	
main supplies, lighting and power to landlord areas	m²	3.75	to	4.25
central heating and electrical installation	m²	11.40	to	12.90
central heating, electrical and lift installation	m²	12.90	to	14.60
air conditioning, electrical and ventilation installations	m²	29.00	to	33.00

8.2 ROADS, PATHS, PAVINGS AND SURFACINGS

Item	Unit	Range £		
5.14 BUILDER'S WORK IN CONNECTION WITH SERVICES – cont				
5.14.1 Builder's Work in Connection with Services – cont				
Offices and hotels. Gross internal floor area				
main supplies, lighting and power to landlord areas	m²	11.00	to	14.60
central heating and electrical installation	m²	16.00	to	21.00
central heating, electrical and lift installation	m²	18.90	to	25.00
air conditioning, electrical and ventilation installations	m²	35.00	to	46.50
8.2 ROADS, PATHS, PAVINGS AND SURFACINGS				
8.2.1 Roads, Paths and Pavings				
Paved areas				
Gravel paving rolled to falls and chambers paving on subbase; including excavation	m²	17.20	to	22.00
Resin bound paving				
16 mm–24 mm deep of natural gravel	m²	71.00	to	90.00
16 mm–24 mm deep of crushed rock	m²	72.00	to	91.00
16 mm–24 mm deep of marble chips	m²	78.00	to	98.00
Tarmacadam paving				
two layers; limestone or igneous chipping finish paving on subbase; including excavation and type 1 subbase	m²	90.00	to	110.00
Slab paving				
precast concrete paving slabs on subbase; including excavation	m²	63.00	to	79.00
precast concrete tactile paving slabs on subbase; including excavation	m²	130.00	to	165.00
York stone slab paving on subbase; including excavation	m²	150.00	to	190.00
imitation York stone slab paving on subbase; including excavation	m²	94.00	to	120.00
Brick/Block/Setts paving				
brick paviors on subbase; including excavation	m²	82.00	to	100.00
precast concrete block paviors to footways including excavation; 150 mm hardcore subbase with dry sand joints	m²	100.00	to	130.00
granite setts on 100 mm thick concrete subbase; including excavation; pointing mortar joints	m²	180.00	to	230.00
cobblestone paving cobblestones on 100 mm thick concrete subbase; including excavation; pointing mortar joints	m²	185.00	to	230.00
Reinforced grass construction				
Grasscrete or similar on 200 mm type 1 subbase; topsoil spread across units and grass seeded upon completion	m²	79.00	to	99.00
Car parking alternatives				
Surface parking; include drains, kerbs, lighting				
surface level parking	m²	100.00	to	125.00
surface car parking with landscape areas	m²	130.00	to	160.00
Rates per car				
surface level parking	car	2175.00	to	2800.00
surface car parking with landscape areas	car	2800.00	to	3550.00

8.2 ROADS, PATHS, PAVINGS AND SURFACINGS

Item	Unit	Range £		
Other surfacing options				
permeable concrete block paving; 80 mm thick Formpave Aquaflow or equal; maximum gradient 5%; laid on 50 mm sand/gravel; subbase 365 mm thick clean crushed angular non plastic aggregate, geotextiles filter layer to top and bottom of subbases	m²	58.00	to	73.00
All purpose roads				
Tarmacadam or reinforced concrete roads, including all earthworks, drainage, pavements, lighting, signs, fencing and safety barriers				
7.30 m wide two lane carriageway	m	3550.00	to	4500.00
10.00 m wide two lane carriageway	m	4400.00	to	5500.00
dual two lane road 7.30 m wide carriageway	m	7100.00	to	9000.00
dual three lane road 11.00 m wide carriageway	m	11000.00	to	14000.00
Road crossings				
NOTE: Costs include road markings, beacons, lights, signs, advance danger signs etc.				
Zebra crossing	nr	21000.00	to	27000.00
Pelican crossing	nr	37000.00	to	47000.00
Underpass				
Provision of underpasses to new roads, constructed as part of a road building programme				
Precast concrete pedestrian underpass				
3.00 m wide × 2.50 m high	m	6200.00	to	7800.00
Precast concrete vehicle underpass				
7.00 m wide × 5.00 m high	m	27000.00	to	34500.00
14.00 m wide × 5.00 m high	m	64000.00	to	81000.00
8.2.2 Special Surfacing and Pavings Sports Grounds				
Pitch plateau construction and pitch drainage not included				
Hockey and support soccer training (no studs)				
Sand filled polypropelene synthetic artificial pitch to FIH national standard. Notts Pad XC FIH shockpad; sport needlepunched polypropelene sand filled synthetic multi sport carpet; white lining; 3 m high perimeter fence	m²	32.00	to	42.00
FIFA one star football and IRB reg 22 rugby union				
Lano Rugby Max 60 mm monofilament laid on shockpad on a dynamic base infill with 32 kg/m² sand and 11 kg/m² SBR rubber – IRB Clause 22 compliant; supply and spread to required infill rates, 2EW sand and 0.5–2 mm SBR Rubber. Sand (32 kg/m²)and rubber (11 kg/m²) spread and brushed into surface; white lining; 3.5 m high perimeter fence	m²	26.50	to	35.00
Tennis and netball court				
Lay final layer on existing subbase of 50 mm porous stone followed by 2 coats of porous bitumen macadam. 40 mm binder course; 14 mm aggregate and 20 mm surface course with 6 mm aggregate. Laid to fine level tolerances and laser leveled, +/- 6 mm tolerance over 3 m; white lining; 3 m high perimeter fencing	m²	42.50	to	56.00
Natural grass winter sports				
Strip topsoil and level existing surface (cut and fill); laser grade formation and proof roll. Load and spread topsoil, grade and consolidate; cultivate and grade topsoil. Supply and spread 25 mm approved sand and mix into topsoil by power harrow; apply pre-seeding fertilizer and sow winter rye sports grass	m²	9.60	to	12.60

8.4 FENCING, RAILINGS AND WALLS

Item	Unit	Range £		

8.3 SOFT LANDSCAPING, PLANTING AND IRRIGATION SYSTEMS

8.3.1 Seeding and Turfing
Please refer to *Spon's External Works and Landscape Price Book* for a more comprehensive range of rates
Surface treatment

Item	Unit		Range £	
spread and lightly consolidate top soil from spoil heap 150 mm thick; by machine	m²	2.35	to	2.85
spread and lightly consolidate top soil from spoil heap 150 mm thick; by hand	m²	8.10	to	9.80
Plant supply, planting, maintenance and 12 months guarantee				
seeded areas	m²	5.20	to	6.30
turfed areas	m²	6.75	to	8.20

8.3.2 External Planting
Planted areas (per m² of planted area)

Item	Unit			
herbaceous plants	m²	6.10	to	7.40
climbing plants	m²	9.70	to	11.70
general planting	m²	23.50	to	28.00
woodland	m²	35.00	to	42.00
shrubbed planting	m²	59.00	to	72.00
dense planting	m²	60.00	to	73.00
shrubbed area including allowance for small trees	m²	77.00	to	93.00
Trees				
light standard bare root tree (PC £10)	nr	66.00	to	80.00
standard root balled tree (PC £25)	nr	78.00	to	94.00
heavy standard root ball tree (PC £40)	nr	145.00	to	175.00
semi-mature root balled tree (PC £130)	nr	460.00	to	560.00

Parklands
NOTE: Work on parklands will involve different techniques of earth shifting and cultivation. The following rates include for normal surface excavation, they include for the provision of any land drainage.

Item	Unit			
parklands, including cultivating ground, applying fertilizer, etc. and seeding with parks type grass	ha	23000.00	to	28000.00

Lakes including excavation average 1.0 m deep, laying 1000 micron sheet with welded joints; spreading top soil evenly on top 200 mm deep

Item	Unit			
regular shaped lake	1000 m²	26000.00	to	32000.00

8.4 FENCING, RAILINGS AND WALLS

8.4.1 Fencing and Rails
Chain link fencing; plastic coated

Item	Unit			
1.20 m high	m	23.00	to	28.00
1.80 m high	m	30.00	to	37.00
Timber fencing				
1.20 m high chestnut pale facing	m	23.00	to	28.00
1.80 m high cross-boarded fencing	m	31.50	to	38.00

8.4 FENCING, RAILINGS AND WALLS

Item	Unit	Range £		
8.4.2 Walls and Screens				
Screen walls; one brick thick; including foundations etc.				
1.80 m high facing brick screen wall	m	330.00	to	395.00
1.80 m high smooth faced masonry block boundary wall	m	200.00	to	240.00
8.4.3 Retaining Walls				
Crib retaining walls				
Permacrib timber crib retaining walls on concrete foundation; granular infill material; measure face area m². Excludes backfill material to the rear of the wall				
wall heights up to 2.0 m	m²	200.00	to	245.00
wall heights up to 3.0 m	m²	215.00	to	260.00
wall heights up to 4.0 m	m²	230.00	to	280.00
wall heights up to 5.0 m	m²	240.00	to	290.00
wall heights up to 6.0 m	m²	250.00	to	305.00
Andacrib concrete crib retaining walls on concrete foundation; granular infill material; measure face area m². Excludes backfill material to the rear of the wall				
wall heights up to 2.0 m	m²	215.00	to	260.00
wall heights up to 3.0 m	m²	230.00	to	280.00
wall heights up to 4.0 m	m²	240.00	to	290.00
wall heights up to 5.0 m	m²	260.00	to	310.00
wall heights up to 6.0 m	m²	270.00	to	325.00
Textomur green faced reinforced soil system; geogrid/geotextile reinforcement and compaction of reinforced soil mass.				
wall heights up to 3.0 m	m²	145.00	to	175.00
wall heights up to 4.0 m	m²	150.00	to	180.00
wall heights up to 5.0 m	m²	160.00	to	190.00
wall heights up to 6.0 m	m²	165.00	to	200.00
Titan/Geolock vertical modular block reinforced soil system; steel ladder. Geogrid reinforcement and soil fill material				
wall heights up to 3.0 m	m²	240.00	to	290.00
wall heights up to 4.0 m	m²	255.00	to	310.00
wall heights up to 5.0 m	m²	265.00	to	320.00
wall heights up to 6.0 m	m²	275.00	to	335.00
8.4.4 Barriers and Guardrails				
Post and rail fencing				
open metal post and rail fencing 1.00 m high	m	165.00	to	200.00
galvanized steel post and rail fencing 2.00 m high	m	190.00	to	235.00
Guard rails				
steel guard rails and vehicle barriers	m	66.00	to	80.00
Bollards and barriers				
parking bollards precast concrete or steel	nr	180.00	to	215.00
vehicle control barrier; manual pole	nr	1000.00	to	1200.00

8.6 EXTERNAL DRAINAGE

Item	Unit	Range £		
8.5 EXTERNAL FIXTURES				
8.5.1 Site/Street Furniture and Equipment				
Please refer to *Spon's External Works and Landscape Price Book* for a more comprehensive range of rates				
Street Furniture				
Roadsigns				
reflected traffic signs 0.25 m² area on steel post	nr	140.00	to	165.00
internally illuminated traffic signs; dependent on area	nr	220.00	to	270.00
externally illuminated traffic signs; dependent on area	nr	860.00	to	1050.00
Benches; bolted to ground				
benches – hardwood or precast concrete	nr	1125.00	to	1375.00
Litter bins; bolted to ground				
precast concrete	nr	205.00	to	250.00
hardwood slatted	nr	210.00	to	255.00
cast iron	nr	420.00	to	510.00
large aluminium	nr	640.00	to	770.00
Bus stops including basic shelter	nr	2650.00	to	3200.00
Pillar box	nr	640.00	to	770.00
Galvanized steel cycle stand	nr	52.00	to	63.00
Galvanized steel flag staff	nr	1250.00	to	1525.00
Playground equipment				
Modern swings with flat rubber safety seats: four seats; two bays	nr	1575.00	to	1925.00
Stainless steel slide, 3.40 m long	nr	1850.00	to	2250.00
Climbing frame – igloo type 3.20 m × 3.75 m on plan × 2.00 m high	nr	1225.00	to	1475.00
See-saw comprising timber plank on sealed ball bearings				
3960 mm × 230 mm × 70 mm thick	nr	105.00	to	130.00
Wicksteed Tumbleguard type safety surfacing around play equipment	m²	13.80	to	16.70
Bark particles type safety surfacing 150 mm thick on hardcore bed	nr	13.80	to	16.70
8.6 EXTERNAL DRAINAGE				
8.6.1 Surface Water and Foul Water Drainage				
Overall £/m² of drained area allowances				
site drainage (per m² of paved area)	m²	24.00	to	30.50
building storm water drainage (per m² of gross internal floor area)	m²	19.80	to	25.00
Machine excavation, grade bottom, earthwork support, laying and jointing pipes and accessories, backfill and compact, disposal of surplus soil.				
uPVC pipes and fittings, lip seal coupling joints				
up to 1.50 m deep; nominal size				
100 mm dia. pipe	m	53.00	to	67.00
160 mm dia. pipe	m	64.00	to	80.00
over 1.50 m not exceeding 3.00 m deep; nominal size				
100 mm dia. pipe	m	93.00	to	120.00
160 mm dia. pipe	m	100.00	to	130.00

8.6 EXTERNAL DRAINAGE

Item	Unit	Range £		
Machine excavation, grade bottom, earthwork support, laying and jointing pipes and accessories, backfill and compact, disposal of surplus soil. uPVC Ultra-Rib ribbed pipes and fittings, sealed ring push fit joints				
up to 1.50 m deep; nominal size				
150 mm dia. pipe	m	54.00	to	68.00
225 mm dia. pipe	m	69.00	to	87.00
300 mm dia. pipe	m	80.00	to	100.00
over 1.50 m not exceeding 3.00 m deep; nominal size				
150 mm dia. pipe	m	94.00	to	120.00
225 mm dia. pipe	m	115.00	to	145.00
300 mm dia. pipe	m	130.00	to	160.00
Machine excavation, grade bottom, earthwork support, laying and jointing pipes and accessories, backfill and compact, disposal of surplus soil. Vitrified clay pipes and fittings, Hepsleve; shingle bed and surround				
up to 1.00 m deep; nominal size				
150 mm dia. pipe	m	72.00	to	91.00
225 mm dia. pipe	m	97.00	to	120.00
300 mm dia. pipe	m	170.00	to	210.00
over 1.50 m not exceeding 3.00 m deep; nominal size				
150 mm dia. pipe	m	110.00	to	140.00
225 mm dia. pipe	m	135.00	to	170.00
300 mm dia. pipe	m	205.00	to	260.00
Machine excavation, grade bottom, earthwork support, laying and jointing pipes and accessories, backfill and compact, disposal of surplus soil. Class M tested concrete centrifugally spun pipes and fittings, flexible joints; concrete bed and surround				
up to 3.00 m deep; nominal size				
300 mm dia. pipe	m	180.00	to	225.00
525 mm dia. pipe	m	275.00	to	350.00
900 mm dia. pipe	m	490.00	to	620.00
1200 mm dia. pipe	m	710.00	to	890.00
Machine excavation, grade bottom, earthwork support, laying and jointing pipes and accessories, backfill and compact, disposal of surplus soil. Cast iron Timesaver drain pipes and fittings, mechanical coupling joints				
up to 1.50 m deep; nominal size				
100 mm dia. pipe	m	110.00	to	140.00
150 mm dia. pipe	m	160.00	to	200.00
over 1.50 m not exceeding 3.00 m deep; nominal size				
100 mm dia. pipe	m	150.00	to	190.00
150 mm dia. pipe	m	195.00	to	250.00

8.6 EXTERNAL DRAINAGE

Item	Unit	Range £		
8.6 EXTERNAL DRAINAGE – cont				
Brick manholes				
Excavate pit in firm ground, partial backfill, partial disposal, earthwork support, compact base of pit, 150 mm plain in situ concrete 20.00 N/mm² – 20 mm aggregate (1:2:4) base, formwork, one brick wall of engineering bricks in cement mortar (1:3) finished fair face, vitrified clay channels, plain in situ concrete 25.00 N/mm² – 20 mm aggregate (1:2:4) cover and reducing slabs, fabric reinforcement, formwork step irons, medium duty cover and frame; Internal size of manhole				
600 mm × 450 mm; cover to invert				
not exceeding 1.00 m deep	nr	480.00	to	580.00
over 1.00 m not exceeding 1.50 m deep	nr	640.00	to	770.00
over 1.50 m not exceeding 2.00 m deep	nr	790.00	to	960.00
900 mm × 600 mm; cover to invert				
not exceeding 1.00 m deep	nr	630.00	to	760.00
over 1.00 m not exceeding 1.50 m	nr	860.00	to	1025.00
over 1.50 m not exceeding 2.00 m	nr	1075.00	to	1300.00
900 mm × 900 mm; cover to invert				
not exceeding 1.00 m deep	nr	750.00	to	900.00
over 1.00 m not exceeding 1.50 m	nr	1000.00	to	1225.00
over 1.50 m not exceeding 2.00 m	nr	1275.00	to	1525.00
1200 × 1800 mm; cover to invert				
not exceeding 1.00 m deep	nr	1475.00	to	1775.00
over 1.00 m not exceeding 1.50 m deep	nr	1925.00	to	2325.00
over 1.50 m not exceeding 2.00 m deep	nr	2375.00	to	2850.00
Concrete manholes				
Excavate pit in firm ground, disposal, earthwork support, compact base of pit, plain in situ concrete 20.00 N/mm² – 20 mm aggregate (1:2:4) base, formwork, reinforced precast concrete chamber and shaft rings, taper pieces and cover slabs bedded jointed and pointed in cement; mortar (1:3) weak mix concrete filling to working space, vitrified clay channels, plain in situ concrete 25.00 N/mm² – 20 mm aggregate (1:1:5:3) benchings, step irons, medium duty cover and frame; depth from cover to invert; Internal dia. of manhole				
900 mm dia.; cover to invert				
over 1.00 m not exceeding 1.50 m deep	nr	750.00	to	910.00
1050 mm dia.; cover to invert				
over 1.00 m not exceeding 1.50 m deep	nr	820.00	to	1000.00
over 1.50 m not exceeding 2.00 m deep	nr	940.00	to	1125.00
over 2.00 m not exceeding 3.00 m deep	nr	1150.00	to	1375.00
1500 mm dia.; cover to invert				
over 1.50 m not exceeding 2.00 m deep	nr	1550.00	to	1875.00
over 2.00 m not exceeding 3.00 m deep	nr	1975.00	to	2400.00
over 3.00 m not exceeding 4.00 m deep	nr	2425.00	to	2950.00
1800 mm dia.; cover to invert				
over 2.00 m not exceeding 3.00 m deep	nr	2700.00	to	3200.00
over 3.00 m not exceeding 4.00 m deep	nr	3300.00	to	4000.00

8.6 EXTERNAL DRAINAGE

Item	Unit	Range £		
2100 mm dia.; cover to invert				
over 2.00 m not exceeding 3.00 m deep	nr	4100.00	to	4950.00
over 3.00 m not exceeding 4.00 m deep	nr	5100.00	to	6100.00
Polypropylene inspection chambers				
475 mm dia. PPIC inspection chamber including all excavations; earthwork support; cart away surplus spoil; concrete bed and surround; lightweight cover and frame				
600 mm deep	nr	355.00	to	430.00
900 mm deep	nr	410.00	to	495.00
8.6.2 Ancillary Drainage Systems				
Septic tanks				
Excavate for, supply and install Klargester glass fibre septic tank, complete with lockable cover				
2800 litre capacity	nr	3700.00	to	3700.00
3800 litre capacity	nr	4600.00	to	4600.00
4600 litre capacity	nr	5400.00	to	5400.00
Urban and landscape drainage				
Exacavate for and lay oil separators, complete with lockable cover				
Excavate for, supply and lay 1000 litre polyethelyne by pass oil separator	nr	3750.00	to	3750.00
Excavate for, supply and lay 1000 litre polyethelyne full retention oil separator	nr	3900.00	to	3900.00
8.6.4 Land Drainage				
NOTE: If land drainage is required on a project, the propensity of the land to flood will decide the spacing of the land drains. Costs include for excavation and backfilling of trenches and laying agricultural clay drain pipes with 75 mm dia. lateral runs average 600 mm deep, and 100 mm dia. mains runs average 750 mm deep.				
land drainage to parkland with laterals at 30 m centres and main runs at 100 m centres	ha	8800.00	to	10000.00
Rainwater harvesting				
FP McCann Ltd Easi-Rain domestic and commercial rainwater harvesting tank system designed to reduce demands on the mains network by providing a sustainable water collection and distribution system ideal for non-potable uses such as toilet flushing, washing and landscape or garden watering. Including excavations				
1500 litre tank	nr	170.00	to	170.00
7500 litre tank	nr	3400.00	to	3400.00
Soakaways: stormwater management				
Soakaway crates; heavyweight (60 tonne) polypropylene high-void box units; including excavation and backfilling as required. Note final surfacing not included	m^3	165.00	to	200.00
Soakaway crates; lightweight (20 tonne) polypropylene high-void box units; including excavation and backfilling as required. Not efinal surfacing not included	m^3	160.00	to	195.00

8.7 EXTERNAL SERVICES

Item	Unit	Range £	
8.7 EXTERNAL SERVICES			
8.7.1 Water Mains Supply			
Water main; all laid in trenches including excavation and backfill with excavated material			
up to 50 mm dia. MDPE pipe	m	110.00 to	130.00
Service connection charges			
The privatization of telephone, water, gas and electricity has complicated the assessment of service connection charges. Typically, service connection charges will include the actual cost of the direct connection plus an assessment of distribution costs from the main. The latter cost is difficult to estimate as it depends on the type of scheme and the distance from the mains. In addition, service charges are complicated by discounts that maybe offered. For instance, the electricity boards will charge less for housing connections if the house is all electric. However, typical charges for a reasonably sized housing estate might be as follows:			
Water and sewerage connections			
water connections; water main up to 2 m from property	house	540.00 to	650.00
water infrastructure charges for new properties	house	340.00 to	415.00
waste and water sewerage infrastructure charges for new properties	house	690.00 to	830.00
8.7.2 Electricity Mains Supply			
Electric main; all laid in trenches including excavation and backfill with excavated material			
600/1000 volt cables. Two core 25 mm dia. cable including 100 mm dia. clayware duct	m	120.00 to	140.00
Service connection charges			
The privatization of telephone, water, gas and electricity has complicated the assessment of service connection charges. Typically, service connection charges will include the actual cost of the direct connection plus an assessment of distribution costs from the main. The latter cost is difficult to estimate as it depends on the type of scheme and the distance from the mains. In addition, service charges are complicated by discounts that maybe offered. For instance, the electricity boards will charge less for housing connections if the house is all electric. However, typical charges for a reasonably sized housing estate might be as follows:			
all electric housing	house	2450.00 to	2950.00
pre-packaged substation housing	nr	34000.00 to	41500.00
8.7.5 Gas Mains Supply			
Gas main; all laid in trenches including excavation and backfill with excavated material			
150 mm dia. gas pipe	m	320.00 to	390.00

8.8 ANCILLARY BUILDINGS AND STRUCTURES

Item	Unit	Range £		

Service connection charges

The privatization of telephone, water, gas and electricity has complicated the assessment of service connection charges. Typically, service connection charges will include the actual cost of the direct connection plus an assessment of distributioncosts from the main. The latter cost is difficult to estimate as it depends on the type of scheme and the distance from the mains. In addition, service charges are complicated by discounts that maybe offered. For instance, the electricity boards will charge less for housing connections if the house is all electric. However, typical charges for a reasonably sized housing estate might be as follows

Item	Unit		Range £	
Gas				
gas connection to house	house	320.00	to	385.00
governing station	nr	18500.00	to	22500.00

8.7.6 Telecommunication and other Communication System Connections

Telephone duct; all laid in trenches including excavation and backfill with excavated material

Item	Unit		Range £	
100 mm dia. uPVC duct	m	76.00	to	92.00

BT approved communication boxes, including excavations and bedding, precast reinforced concrete box with cast iron cover (medium grade)

Item	Unit		Range £	
communication box 1200 mm long × 600 mm wide × 895 mm high	nr	400.00	to	460.00
communication box 910 mm long × 890 mm wide × 1000 mm high	nr	465.00	to	540.00
telephone connections charges	house	195.00	to	240.00

8.7.9 External/Street Lighting Systems

Lighting

Item	Unit		Range £	
lighting to pedestrian areas an estate roads on 4.00 m–6.00 m columns with up to 70 W lamps	nr	270.00	to	325.00
lighting to main roads;10.00 m–12.00 m columns with 250 W lamps	nr	590.00	to	710.00
lighting to main roads; 12.00 m–15.00 m columns with 400 W high pressure sodium lighting	nr	730.00	to	890.00

8.8 ANCILLARY BUILDINGS AND STRUCTURES

8.8.2 Ancillary Buildings and Structures

Footbridges

Footbridge of either precast concrete or steel construction up to 6.00 m wide, 6.00 m high including deck, access stairs and ramp, parapets etc.

Item	Unit		Range £	
5 m span between piers or abutments	m²	4450.00	to	5600.00
20 m span between piers or abutments	m²	4150.00	to	5200.00
Footbridge of timber (stress graded with concrete piers)				
12 m span between piers or abutments	m²	1475.00	to	1875.00

9.2 MAIN CONTRACTOR'S PRELIMINARIES

Item	Unit	Range £		
8.8 ANCILLARY BUILDINGS AND STRUCTURES – cont				
Roadbridges				
Reinforced concrete bridge with precast beams; including all excavation, reinforcement, formwork, concrete, bearings, expansion joints, deck waterproofing and finishing's, parapets etc. deck area				
10.00 m span	m²	3450.00	to	4350.00
15.00 m span	m²	3350.00	to	4250.00
Reinforced concrete bridge with prefabricated steel beams; including all excavation, reinforcement, formwork, concrete, bearings, expansion joints, deck waterproofing and finishing's, parapets etc. deck area				
20.00 m span	m²	3650.00	to	4600.00
30.00 m span	m²	3450.00	to	4350.00
Multiparking systems/stack parkers				
Fully automatic systems				
integrated robotic parking system using robotic car transporter to store vehicles	car	24000.00	to	30000.00
Semi-automatic systems				
integrated parking system, transverse and vertical positioning; semi-automatic parking achieving 17 spaces in a 6 car width × 3 car height grid	car	13000.00	to	17000.00
Integrated stacker systems				
integrated parking system, vertical positioning only; double width, double height pit stacker achieving 4 spaces with each car stacker	car	6900.00	to	8700.00
integrated parking system, vertical positioning only; triple stacker achieving 3 spaces with each car stacker, generally 1 below ground and 2 above ground	car	14000.00	to	17500.00
integrated parking system, vertical positioning only; triple height double width stacker, achieving 6 spaces with each car stacker, generally 2 below ground and 4 above ground	car	10000.00	to	13000.00
9.2 MAIN CONTRACTOR'S PRELIMINARIES				
9.2.2 Site Establishment				
Typical rates for site accommodation; Minimum hire periods generally apply				
Office cabins – 24 ft × 9 ft (20 m²)	week	28.00	to	35.00
Office cabins – 32 ft × 10 ft (30 m²)	week	35.00	to	44.50
Fire-rated cabins – 24 ft × 9 ft (20 m²)	week	59.00	to	74.00
Fire-rated cabins – 32 ft × 10 ft (30 m²)	week	80.00	to	100.00
Meeting room–24 ft × 9 ft (20 m²)	week	28.00	to	35.00
Meeting room–32 ft × 9 ft (30 m²)	week	35.00	to	44.50
Mess cabins (incl. furniture)	week	33.00	to	42.00
Drying rooms (incl. furniture)	week	37.50	to	47.00
Safestore – 20ft × 8ft (15 m²)	week	19.30	to	24.00
Container with padlock – 20ft × 8ft (15 m²)	week	10.70	to	13.50
Toilets – 13 ft × 9 ft	week	43.00	to	54.00
Signs and notices	week	54.00	to	68.00
Fire extinguishers	week	8.55	to	10.80
Kitchen with cooker, fridge, sink, water heater; 32 ft × 10 ft	week	160.00	to	200.00
Mess room with wash basin, water heater, seating; 16 ft × 7 ft	week	96.00	to	120.00
Toilets – fitted to mains; three pan unit	week	170.00	to	215.00
Toilets – fitted to mains; four pan unit	week	200.00	to	260.00
Haulage to and from site; Site offices, Storage sheds, Toilets	item	4700.00	to	6000.00

9.2 MAIN CONTRACTOR'S PRELIMINARIES

Item	Unit	Range £		

9.2.7 Mechanical Plant

Power and lighting; minimum hire periods generally apply

Item	Unit		Range £	
3 kVA transformer	week	10.60	to	13.30
5 kVA transformer	week	17.50	to	22.00
10 kVA transformer	week	35.00	to	44.50
4 Way distribution Box; 110 V	week	9.00	to	11.30
14 m extension cable	week	4.30	to	5.45
25 m extension cable	week	11.30	to	14.30
2 × 500 W floodlights	week	13.20	to	16.70
2 × 500 W floodlights on stands	week	16.50	to	21.00
Generators; minimum hire periods generally apply				
1 kVA 240 V portable (low noise)	week	41.00	to	52.00
2 kVA dual volt, portable petrol	week	28.50	to	36.00
5 kVA silenced	week	80.00	to	100.00
7.5 kVA silenced	week	120.00	to	150.00
15 kVA silenced	week	170.00	to	215.00
40 kVA silenced	week	255.00	to	320.00
60 kVA silenced	week	270.00	to	340.00
108 kVA silenced	week	330.00	to	420.00
165 kVA silenced	week	410.00	to	520.00
Mixers and small plant				
small mixer; 5–3½ diesel	week	38.50	to	48.50
mixer; 7–5 diesel	week	48.00	to	61.00
compresser	week	100.00	to	130.00
Small dehumidifers	week	26.00	to	33.00
Large dehumidifers	week	70.00	to	89.00
Turbo dryer	week	44.00	to	55.00
Gas jet air heater; 260,000 Btu	week	53.00	to	66.00

9.2.8 Temporary Works

Roadways

Item	Unit		Range £	
formation of temporary roads to building perimeter comprising of geotextile membrane and 300 mm MOT type 1; reduce level; spoil to heap on site within 25 m of excavation	m²	22.00	to	27.00
installation of wheel wash facility and maintenance	nr	3000.00	to	3600.00

A Strategic Approach for the UK Construction Industry

Stephen Gruneberg

The construction industry is a microcosm of the economy as a whole, and as such the economics of the sector contains many of the aspects of the economy in general, albeit with fascinating and unique features. What are the implications of economic theory for the future of UK construction? How does the industry ensure innovation, quality and efficiency? What priorities might best serve the construction industry, those working in it and their customers?

In seeking answers to these and other questions, the UK government has commissioned a number of reports on the construction industry including the Latham and Egan reports and more recently *Construction 2025*. These have invariably proposed time and cost targets for the construction industry. In this new book, Stephen Gruneberg stands in stark contrast to those reports and presents the relevant theoretical aspects of construction economics to account for the behaviour of construction firms and suggest a strategy for future growth and sustainability. He discusses the theory and data relating to the output of firms in relation to the type of firm, the market and how these firms behave as a result. The purpose of this book is to advocate the measures needed to create the kind of industry that must be fostered to ensure the quality of its output, sustainability and the fair terms and conditions of employment for its workforce.

December 2018: 216 × 138 mm: 92pp
Hb: 978-1-138-08977-8 : £45.00

To Order: Tel: +44 (0) 1235 400524 Fax: +44 (0) 1235 400525
or Post: Taylor and Francis Customer Services,
Bookpoint Ltd, Unit T1, 200 Milton Park, Abingdon, Oxon, OX14 4TA UK
Email: book.orders@tandf.co.uk

For a complete listing of all our titles visit:
www.tandf.co.uk

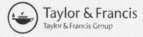

PART 4

Prices for Measured Works

This part contains the following NRM2 work sections:

NEC4: 100 Questions and Answers

Kelvin Hughes

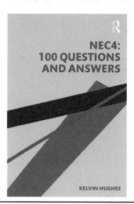

This book details some of the most important and interesting questions raised about the NEC4 family of contracts and provides clear, comprehensive answers to those questions.

Written by an NEC expert with over 20 years' experience using, advising and training others, the book has several distinctive features:

*   It covers the *whole* NEC4 family
*   It is written by a very experienced NEC author who explains sometimes complex issues in a simple and accessible style
*   The questions and answers range from beginner level up to a masterclass level
*   The questions are real life questions asked by actual NEC practitioners on real projects.

The book includes questions and answers relating to tendering, early warnings, programme issues, quality management, payment provisions, compensation events, liabilities, insurances, adjudication, termination and much more. It is essential reading for anyone working with the NEC4 family of contracts, whether professionals or students in construction, architecture, project management and engineering.

May 2019: 234 × 156 mm: 246pp
Pb: 978-1-138-36525-4 : £45.99

To Order: Tel: +44 (0) 1235 400524 Fax: +44 (0) 1235 400525
or Post: Taylor and Francis Customer Services,
Bookpoint Ltd, Unit T1, 200 Milton Park, Abingdon, Oxon, OX14 4TA UK
Email: book.orders@tandf.co.uk

For a complete listing of all our titles visit:
www.tandf.co.uk

INTRODUCTION

The rates contained in *Prices for Measured Works* are intended to apply to a project in the Outer London area costing about £4,250,000 including Preliminaries, and assume that reasonable quantities of all types of work are required. Similarly it has been necessary to assume that the size of the project warrants the subletting of all types of work normally sublet. Adjustments should be made to standard rates for time, location, local conditions, site constraints and all the other factors likely to affect costs of any individual project.

Adjustments may need to be made for different project values. In *How to use this Book* you will find some guidance on adjustments according to contract cost.

The distinction between builders' work and work normally sublet is stressed because prices for work which can be sublet may well be inadequate for the contractor who is called upon to carry out relatively small quantities of such work themselves.

Measured Works prices are generally based upon wage rates and known material costs from May 2019. Built up prices and subcontractor rates include an allowance of 3% for overheads and profit.

As elsewhere in this edition, prices do not include Value Added Tax or professional services fees.

Measured Work rates are exclusive of any main contractor's preliminaries.

Prices for Measured Works section is structured to reflect NRM2.

To aid readers with the transition we have scheduled the SMM7 Work Sections and shown the NRM2 Work Sections alongside:

SMM7 Work Section	NRM2 Work Section
	2 Off-site manufactured materials, components and buildings:
C Demolition/Alteration/Renovation	3 Demolitions
	4 Alterations, repairs and conservation
D Groundwork	5 Excavate and filling
D20 Excavation and Filling	7 Piling
D30 Cast In Place Piling	8 Underpinning
D32 Steel Piling	9 Diaphragm walls and embedded retaining walls
D40 Embedded Retaining Walling	10 Crib walls, gabions and reinforced earth
D41 Crib Walls/Gabions/Reinforced Earthworks	
D50 Underpinning	
E In Situ Concrete/Large Precast Concrete	11 In situ concrete works
E10 In Situ Concrete Construction	12 Precast/composite concrete
E20 Formwork For In Situ Concrete	13 Precast concrete
E30 Reinforcement For In Situ Concrete	
E40 Designed Joints In Situ Concrete	
E41 Worked Finishes/Cutting To In Situ Concrete	
E42 Accessories Cast Into In Situ Concrete	
E50 Precast Concrete Large Units	
E60 Precast/Composite Concrete Decking	

SMM7 Work Section	NRM2 Work Section
F Masonry	14 Masonry
F10 Brick/Block Walling	
F11 Glass block Walling	
F20 Natural Stone Rubble Walling	
F22 Cast Stone Walling/Dressings	
F30 Accessories/Sundry Items For Brick/Block/Stone Walling	
F31 Precast Concrete Sills/Lintels/Coping features	
G Structural/Carcassing Metal/Timber	15 Structural metalwork
G10 Structural Steel Framing	16 Carpentry
G12 Isolated Structural Metal Members	
G20 Carpentry/Timber Framing/First Fixing	
H Cladding/Covering	17 Sheet roof coverings
H10 Patent Glazing	18 Tile and slate roof and wall coverings
H11 Curtain Walling	21 Cladding and covering
H20 Rigid Sheet Cladding	
H30 Fibre Cement Profile Sheet Cladding	
H31 Metal Profiled/Flat Sheet Cladding/Covering/Siding	
H32 Plastic Profiled Sheet Cladding/Covering/Siding	
H41 Glass Reinforced Plastic Panel Cladding Features	
H51 Natural Stone Cladding Features	
H53 Clay Slab/Cladding/Features	
H60 Plain Roof Tiling	
H61 Fibre Cement Slating	
H62 Natural Slating	
H63 Reconstructed Stone Slating/Tiling	
H64 Timber Shingling	
H71 Lead Sheet Coverings/Flashes	
H72 Aluminium Sheet Coverings/Flashings	
H73 Copper Strip Sheet Coverings/Flashings	
H74 Zinc Strip Sheet Coverings/Flashings	
H75 Stainless Steel Sheet Coverings/Flashings	
H76 Fibre Bitumen Thermoplastic Sheet Coverings/Flashings	
H92 Rainscreen Cladding	
J Waterproofing	19 Waterproofing
J10 Specialist Waterproof Rendering	
J20 Mastic Asphalt Tanking/Damp-Proof Membranes	
J21 Mastic Asphalt Roofing/Insulation/Finishes	
J30 Liquid Applied Tanking/Damp-Proof Membranes	

SMM7 Work Section	NRM2 Work Section
J40 Flexible Sheet Tanking/Damp-Proof Membranes	
J41 Built Up Felt Roof Coverings	
J42 Single Layer Plastic Roof Coverings	
J43 Proprietary Roof Decking With Felt Finish	
K Linings/Sheathing/Dry partitioning	20 Proprietary linings and partitions
K10 Plasterboard Dry Lining/Partitions/Ceilings	30 Suspended ceilings
K11 Rigid Sheet Flooring/Sheathing/Linings/Casings	
K13 Rigid Sheet Fine Linings/Panelling	
K14 Glass Reinforced Gypsum Linings/Panelling	
K20 Timber Board Flooring/Sheathing/Linings/Casings	
K30 Demountable Partitions	
K32 Framed Panel Cubicle Partitions	
K33 Concrete/Terrazzo Partitions	
K40 Demountable Suspended Ceilings	
K41 Raised Access Floors	
L Windows/Doors/Stairs	23 Windows, screens and lights
L10 Windows/Rooflights/Screens/Louvres	24 Doors, shutters and hatches
L20 Doors/Shutters/Hatches	25 Stairs, walkways and balustrades
L30 Stairs/Walkways/Balustrades	27 Glazing
L40 General Glazing	
M Surface Finishes	28 Floor, wall, ceiling and roof finishes
M10 Cement: Sand/Concrete Screeds/Granolithic Screeds/Topping	29 Decoration
M11 Mastic Asphalt Flooring/Floor Underlays	
M12 Trowelled Bitumen/Resin/Rubber Latex	
M20 Plastered/Rendered/Roughcast Coating	
M21 Insulation With Rendered Finish	
M22 Sprayed Mineral Fibre Coatings	
M30 Metal Mesh Lathing/Anchored Reinforcement For Plastered Ceilings	
M31 Fibrous Plaster	
M40 Stone/Concrete/Quarry/Ceramic Tiling	
M41 Terrazzo Tiling/In Situ Terrazzo	
M42 Wood Block/Composition Block/Parquet	
M50 Rubber/Plastic/Cork/Lino/Carpet Tiling/Sheeting	
M51 Edge Fixed Carpeting	
M52 Decorative Papers/Fabrics	
M60 Painting/Clear Finishing	

SMM7 Work Section	NRM2 Work Section
N Furniture/Equipment	32 Furniture, fittings and equipment
N10/11 General Fixtures/Kitchen Fittings	
N13 Sanitary Appliances/Fittings	
N15 Signs/Notices	
P Building Fabric Sundries	22 General joinery
P10 Sundry Insulation/Proofing Work/Fire Stops	31 Insulation, fire stopping and fire protection
P20 Unframed Isolated Trims/Skirtings/Sundry Items	
P21 Ironmongery	
P30 Trenches/Pipeways/Pits For Buried Engineering Services	
P31 Holes/Chases/Covers/Supports for Services	
Q Paving/Planting/Fencing/Site Furniture	35 Site work
Q10 Kerbs/Edgings/Channels/Paving Access	36 Fencing
Q20 Hardcore/Granular/Cement Bound Bases	37 Soft landscaping
Q21 In Situ Concrete Roads/Pavings	
Q22 Coated Macadam/Asphalt Roads/Pavings	
Q23 Gravel/Hoggin/Woodchip Roads/Pavings	
Q25 Slab/Brick/Block/Sett/Cobble Pavings	
Q26 Special Surfacings/Pavings For Sport	
Q30 Seeding/Turfing	
Q31 Planting	
Q40 Fencing	
R Disposal Systems	33 Drainage above ground
R10 Rainwater Pipework/Gutters	34 Drainage below ground
R11 Foul Drainage Above Ground	
R12 Drainage Below Ground	
R13 Land Drainage	
S Piped Supply Systems	
S10/S11 Hot And Cold Water	
S13 Pressurized Water	
T Mechanical Heating/Cooling/Refrigeration Systems	
T10 Gas/Oil Fired Boilers	
T31 Low Temperature Hot Water	
V Electrical Systems	39 Electrical services
V21/V22 General Lighting and Low Voltage Power	
W Security systems	
W20 Lightning Protection	
X Transport Systems	41 Builder's work in connection with services

03 DEMOLITIONS

Item	PC £	Labour hours	Labour £	Plant £	Material £	Unit	Total rate £
3.01 DEMOLITIONS							
NOTE: Demolition rates vary enormously from project to project, depending upon access, type of construction, method of demolition, redundant or recycleable materials etc. Always obtain specific quotations for each project under consideration. The following rates for simple demolition works may be of some assistance for comparative purposes. Generally scaffold and other access equipment is not included in the rates.							
Demolishing structures; by machine; note disposal not included							
Demolishing to ground level; single storey brick out-building; timber flat roof; grub up shallow foundations							
single storey out building approximately 50 m³	–	–	–	–	–	m³	**88.20**
single storey out building approximately 200 m³	–	–	–	–	–	m³	**60.48**
single storey out building approximately 500 m³	–	–	–	–	–	m³	**28.98**
Demolishing to ground level; light steel framed and sheet roofed cycle shelter; grub up shallow foundations	–	–	–	–	–	m³	**63.00**
Concrete framed multi-storey car park	–	–	–	–	–	m³	**42.84**
Demolishing to ground level; multi-storey masonry building; flat roof; 1–4 storeys; volume	–	–	–	–	–	m³	**31.50**
Warehouse to slab level; 7 m eaves	–	–	–	–	–	m²	**17.64**
Warehouse to below slab; 7 m eaves; grub up shallow foundations (pads, ground beams and ground slab)	–	–	–	–	–	m²	**28.98**
Reinforced concrete frame building; easy access	–	–	–	–	–	m³	**50.40**
Reinforced concrete frame building; restricted access	–	–	–	–	–	m³	**94.50**
Concrete encased steel frame building; easy access	–	–	–	–	–	m³	**35.28**
Concrete encased steel frame building; restricted access	–	–	–	–	–	m³	**84.42**
Demolishing parts of structures; by hand; cost of skip included							
Breaking up plain concrete bed; load into skip							
100 mm thick	–	0.45	15.91	6.31	–	m²	**22.22**
150 mm thick	–	0.67	23.56	12.68	–	m²	**36.24**
200 mm thick	–	0.90	31.82	12.60	–	m²	**44.42**
300 mm thick	–	1.33	47.10	18.49	–	m²	**65.59**
Breaking up reinforced concrete bed; load into skip							
100 mm thick	–	0.50	17.82	7.01	–	m²	**24.83**
150 mm thick	–	0.75	26.41	13.37	–	m²	**39.78**
200 mm thick	–	1.00	35.33	14.02	–	m²	**49.35**
300 mm thick	–	1.50	53.15	20.88	–	m²	**74.03**
Demolishing reinforced concrete column or cutting away concrete casing to steel column; load into skip	–	9.99	353.26	96.56	–	m³	**449.82**
Demolishing reinforced concrete beam or cutting away concrete casing to steel beam; load into skip	–	11.47	405.77	105.49	–	m³	**511.26**

03 DEMOLITIONS

Item	PC £	Labour hours	Labour £	Plant £	Material £	Unit	Total rate £
3.01 DEMOLITIONS – cont							
Demolishing parts of structures – cont							
Demolishing reinforced concrete wall; load into skip							
100 mm thick	–	0.99	35.00	9.60	–	m²	**44.60**
150 mm thick	–	1.50	53.15	14.34	–	m²	**67.49**
225 mm thick	–	2.25	79.56	21.58	–	m²	**101.14**
300 mm thick	–	3.00	105.97	28.91	–	m²	**134.88**
Demolishing reinforced concrete suspended slabs; load into skip							
100 mm thick	–	0.84	29.59	8.92	–	m²	**38.51**
150 mm thick	–	1.25	44.23	13.02	–	m²	**57.25**
225 mm thick	–	1.87	66.19	19.51	–	m²	**85.70**
300 mm thick	–	2.50	88.48	26.35	–	m²	**114.83**
Breaking up small concrete plinth; make good structures; load into skip	–	3.83	135.58	46.55	7.40	m³	**189.53**
Breaking up small precast concrete kerb; make good structures; load into skip	–	0.41	14.64	3.92	7.40	m	**25.96**
Remove precast concrete window sill; set aside for reuse	–	1.33	47.10	–	–	m	**47.10**
Remove brick on-edge-coping; prepare walls for raising; load into skip							
one brick thick	–	0.38	12.52	0.68	–	m	**13.20**
one and a half brick thick	–	0.50	16.68	1.02	–	m	**17.70**
Demolishing brick chimney to 300 mm below roof level; sealing off flues with slates; piecing in treated sawn softwood rafters and making good roof coverings; load into skip (excluding scaffold access)							
680 mm × 680 mm × 900 mm high above roof	–	11.40	316.46	19.61	85.58	nr	**421.65**
add for each additional 300 mm height	–	2.08	57.76	5.23	–	nr	**62.99**
680 mm × 1030 mm × 900 mm high above roof	–	16.86	467.99	20.92	122.64	nr	**611.55**
add for each additional 300 mm height	–	3.11	86.50	7.85	–	nr	**94.35**
1030 mm × 1030 mm × 900 mm high above roof	–	25.52	708.59	32.69	166.44	nr	**907.72**
add for each additional 300 mm height	–	4.71	130.82	10.46	–	nr	**141.28**
Demolishing defective brick chimney to roof level; re-build using 25% new facing bricks; new lead flashings; core flues; reset chimney pot; load into skip (excluding scaffold access)							
680 mm × 680 mm × 900 mm high above roof	–	10.73	299.81	19.61	129.14	nr	**448.56**
add for each additional 300 mm height	–	2.08	57.76	5.23	13.56	nr	**76.55**
680 mm × 1030 mm × 900 mm high above roof	–	15.66	435.08	20.92	155.84	nr	**611.84**
add for each additional 300 mm height	–	3.11	86.50	7.85	16.70	nr	**111.05**
1030 mm × 1030 mm × 900 mm high above roof	–	24.02	667.48	32.69	181.56	nr	**881.73**
add for each additional 300 mm height	–	4.71	130.82	10.46	21.29	nr	**162.57**
Demolishing external brick walls; load into skip							
half brick thick	–	0.87	23.93	5.14	–	m²	**29.07**
two half brick thick skins if cavity wall	–	1.46	39.97	9.94	–	m²	**49.91**
one brick thick	–	1.50	41.21	9.94	–	m²	**51.15**
one and a half brick thick	–	2.04	56.00	14.90	–	m²	**70.90**
two brick thick	–	2.62	71.80	19.87	–	m²	**91.67**
extra for plaster, render or pebbledash finish per side	–	0.08	2.22	1.02	–	m²	**3.24**

03 DEMOLITIONS

Item	PC £	Labour hours	Labour £	Plant £	Material £	Unit	Total rate £
Demolish external stone walls; load into skip							
300 mm thick	–	1.00	27.41	10.46	–	m²	**37.87**
400 mm thick	–	1.33	36.52	13.60	–	m²	**50.12**
600 mm thick	–	2.00	54.83	20.53	–	m²	**75.36**
Demolish external stone walls; clean off and set aside for reuse							
300 mm thick	–	1.50	41.12	2.62	–	m²	**43.74**
400 mm thick	–	2.00	54.83	3.92	–	m²	**58.75**
600 mm thick	–	3.00	82.24	5.23	–	m²	**87.47**
Remove fireplace surround and hearth							
breaking up 500 mm wide concrete hearth; load into skip	–	1.50	41.21	3.42	–	m	**44.63**
remove fire surround; tiled interior; load into skip	–	1.54	42.19	7.85	–	nr	**50.04**
cast iron surround; set aside for reuse	–	2.58	70.81	–	–	nr	**70.81**
stone surround; set aside for reuse	–	6.74	184.80	–	–	nr	**184.80**
fill in opening with common bricks; air brick; 2 coat plaster; timber skirting; re-screed floor	–	0.95	31.34	–	77.86	m²	**109.20**
Removing roof timbers complete (NB not coverings); including rafters, purlins, ceiling joists, plates etc.; load into skip; measured flat on plan	–	0.28	8.05	4.02	–	m²	**12.07**
Remove softwood floor structure; load into skip							
joists at ground floor level	–	0.20	5.62	1.37	–	m²	**6.99**
joists at first floor level	–	0.41	11.35	1.37	–	m²	**12.72**
joists at second floor	–	0.58	16.03	1.37	–	m²	**17.40**
individual timber members	–	0.23	6.17	0.03	–	m	**6.20**
Remove boarding; withdraw nails; set aside for reuse							
softwood flooring at ground floor level	–	0.30	8.22	–	–	m²	**8.22**
softwood flooring at first floor level	–	0.51	14.04	–	–	m²	**14.04**
softwood flooring at second floor level	–	0.61	16.78	–	–	m²	**16.78**
fascia/barge board/gutter board second floor level; load into skip	–	0.52	14.12	0.03	–	m	**14.15**
Remove doors and windows; set aside for reuse							
solid single door only (frame left in place)	–	0.33	9.13	1.02	–	nr	**10.15**
solid single door; frame to skip	–	0.55	15.07	1.72	–	nr	**16.79**
solid double door only (frame left in place)	–	0.66	18.10	2.05	–	nr	**20.15**
solid double door; frame to skip	–	1.00	27.38	3.42	–	nr	**30.80**
glazed screen and doors	–	1.00	27.41	5.14	–	m²	**32.55**
casement window frame; set aside frame; glass into skip; up to 1.00 m²	–	0.75	20.56	0.26	–	m²	**20.82**
casement window frame; set aside frame; glass into skip; 1.00 to 2.00 m²	–	1.00	27.41	0.26	–	m²	**27.67**
casement window frame; set aside frame; glass into skip; 2.00 to 3.00 m²	–	1.10	30.16	0.26	–	m²	**30.42**
pair French windows and frame; upto 1200 mm × 1200 mm	–	3.33	91.28	5.14	–	nr	**96.42**
removing double hung sash window and frame; store for reuse	–	2.50	68.47	–	–	nr	**68.47**

03 DEMOLITIONS

Item	PC £	Labour hours	Labour £	Plant £	Material £	Unit	Total rate £
3.01 DEMOLITIONS – cont							
Demolishing parts of structures – cont							
Remove doors and windows; load into skip							
remove solid timber door	–	0.30	8.22	1.72	–	nr	**9.94**
remove door frame	–	0.10	2.74	1.72	–	nr	**4.46**
casement window frame; up to 1.00 m²	–	0.60	16.45	1.72	–	m²	**18.17**
casement window frame; 1.00 to 2.00 m²	–	0.80	21.94	1.72	–	m²	**23.66**
casement window frame; 2.00 to 3.00 m²	–	0.90	24.67	1.72	–	m²	**26.39**
Demolishing internal partitions; load into skip							
half brick thick brickwork	–	0.87	23.93	5.14	–	m²	**29.07**
one brick thick brickwork	–	1.50	41.21	9.94	–	m²	**51.15**
one and a half brick thick brickwork	–	2.04	56.00	14.90	–	m²	**70.90**
75 mm blockwork	–	0.58	16.03	3.77	–	m²	**19.80**
90 mm blockwork	–	0.62	17.03	4.44	–	m²	**21.47**
100 mm blockwork	–	0.67	18.25	5.14	–	m²	**23.39**
115 mm blockwork	–	0.71	19.49	5.14	–	m²	**24.63**
125 mm blockwork	–	0.75	20.48	5.47	–	m²	**25.95**
140 mm blockwork	–	0.79	21.72	5.82	–	m²	**27.54**
150 mm blockwork	–	0.84	22.94	6.49	–	m²	**29.43**
190 mm blockwork	–	0.98	26.89	8.21	–	m²	**35.10**
215 mm blockwork	–	1.08	29.60	8.89	–	m²	**38.49**
255 mm blockwork	–	1.25	34.27	10.61	–	m²	**44.88**
extra for plaster finish per side	–	0.08	2.22	1.02	–	m²	**3.24**
breaking up brick plinths	–	3.33	91.28	34.20	–	m³	**125.48**
stud walls (metal or timber); solid board each side	–	0.38	10.37	3.42	–	m²	**13.79**
stud walls (metal or timber); glazed panels	–	0.50	13.81	3.42	–	m²	**17.23**
lightweight steel mesh security screen	–	0.41	11.35	1.70	–	m²	**13.05**
solid steel demountable partition	–	0.62	17.03	2.40	–	m²	**19.43**
glazed demountable partition including removal of glass	–	0.84	22.94	3.42	–	m²	**26.36**
glazed screen including any doors incorporated	–	1.00	27.41	5.14	–	m²	**32.55**
remove large folding partition and track	–	9.90	271.40	52.03	–	m²	**323.43**
remove timber skirting, architrave, dado rail; load into skip	–	0.10	2.74	0.03	–	m	**2.77**
Removing timber staircase and balustrades							
single straight flight	–	2.92	79.94	34.20	–	nr	**114.14**
single dogleg flight	–	4.17	114.24	51.30	–	nr	**165.54**
Perimeter hoarding							
security fencing; 2400 mm high painted plywood hoarding	–	–	–	–	–	m	**157.50**

03 DEMOLITIONS

Item	PC £	Labour hours	Labour £	Plant £	Material £	Unit	Total rate £
3.02 TEMPORARY SUPPORT OF STRUCTURES, ROADS AND THE LIKE							
Temporary support							
NOTE: The requirement for shoring and strutting for the formation of large openings are dependent upon a number of factors; for example the weight of the superimposed structure to be supported; the number of windows; number of floors; roof type; whether raking shores are required; depth to load-bearing surface; duration support is to be left in place. Prices therefore are best built-up by assessing the use and waste of materials and the labours involved. The follow guide needs to be tested against each particular project.							
Support of structures not to be demolished							
strutting to window openings over proposed new openings	–	0.56	18.54	–	8.57	nr	**27.11**
plates, struts, braces and hardwood wedges in supports to floors and roof of openings	–	1.11	36.74	–	25.31	nr	**62.05**
dead shore and needle using die square timber with sole plates, braces, hardwood wedges and steel dogs	–	27.75	918.58	–	126.19	nr	**1044.77**
set of two raking shores using die square timber with 50 mm thick wall pieces, hardwood wedges and steel dogs, including forming holes for needles and making good	–	33.30	1102.30	–	128.14	nr	**1230.44**
cut holes through one brick wall for die square needle and make good on completion of works, including facings externally and plaster internally	–	5.36	177.43	–	27.61	nr	**205.04**

04 ALTERATIONS, REPAIRS AND CONSERVATION

Item	PC £	Labour hours	Labour £	Plant £	Material £	Unit	Total rate £
4.01 REMOVING							
NOTE: It is highly unlikely for exactly the same composite items of alteration works will be encountered on different projects. The following spot items have been included to allow the reader to build up the composite rates for estimating purposes. Rates include for the removal of debris from site, but do not include for any temporary works, shoring, scaffolding etc. or for any redecorations.							
Fixtures and fittings; remove and load into skip							
Handrails and balustrades							
timber or metal handrail and brackets	–	0.27	7.44	0.35	–	m	**7.79**
timber balustrades	–	0.30	8.24	1.02	–	m	**9.26**
metal balustrades	–	0.45	12.23	1.02	–	m	**13.25**
Kitchen fittings							
wall units (up to 500 mm wide)	–	0.41	11.17	5.14	–	nr	**16.31**
floor units (up to 500 mm wide)	–	0.27	7.44	7.52	–	nr	**14.96**
larder units (up to 500 mm wide)	–	0.36	9.84	17.10	–	nr	**26.94**
Bathroom fittings							
bath panels and bearers	–	0.36	9.84	0.35	–	nr	**10.19**
toilet roll holders or soap dispensers; 2 or 3 screw fixings	–	0.27	8.26	0.03	–	nr	**8.29**
towel rails; 4 or 6 screw fixings	–	0.54	14.89	0.03	–	nr	**14.92**
mirror; up to 600 mm × 600 mm; 4 screw fixings	–	0.58	17.69	0.05	–	nr	**17.74**
pipe casings	–	0.27	7.44	1.37	–	m	**8.81**
Other fixtures and fittings							
shelves, window boards and the like; screw fixings	–	0.30	9.14	0.35	–	m	**9.49**
curtain track, plastic; screw fixing at 600 mm centres	–	0.10	3.04	–	–	m	**3.04**
small nameplates or individual numerals/letters; screw fixings	–	0.25	7.60	–	–	nr	**7.60**
small notice board and frame from walls; up to 600 mm × 900 mm; screw fixings	–	0.50	15.20	–	–	nr	**15.20**
Finishes; removing and load into skip unless stated otherwise							
Breaking up floor screeds							
asphalt paving	–	0.54	14.89	0.79	–	m²	**15.68**
floor screed up to 100 mm thick unreinforced	–	0.33	9.05	0.79	–	m²	**9.84**
granolithic floor screed up to 100 mm thick unreinforced	–	0.58	15.96	0.79	–	m²	**16.75**
Floor finishes							
carpet and underfelt	–	0.11	2.93	1.31	–	m²	**4.24**
carpetgrip edge fixing strip	–	0.01	0.30	–	–	m	**0.30**
vinyl or similar sheet flooring	–	0.09	2.39	0.94	–	m²	**3.33**
vinyl or similar tile flooring with tile remover	–	0.17	4.52	1.09	–	m²	**5.61**
ceramic floor tiles with tile remover	–	0.40	10.97	1.15	–	m²	**12.12**
woodblock flooring; set aside for reuse	–	0.67	18.35	–	–	m²	**18.35**

04 ALTERATIONS, REPAIRS AND CONSERVATION

Item	PC £	Labour hours	Labour £	Plant £	Material £	Unit	Total rate £
Hack off wall finishes with chipping hammer							
plaster to walls	–	0.33	9.05	1.06	–	m²	10.11
cement rendering, pebbledash or similar	–	0.25	6.85	0.72	–	m²	7.57
ceramic wall tiles	–	0.40	10.97	1.06	–	m²	12.03
Strip off wallpaper use steam paper stripper							
1 layer	–	0.10	2.74	0.13	–	m²	2.87
2 layers	–	0.14	3.84	0.13	–	m²	3.97
3 layers	–	0.17	4.80	0.13	–	m²	4.93
Remove wall linings including battening behind							
plain sheet boarding	–	0.27	7.44	1.37	–	m²	8.81
matchboarding	–	0.36	9.84	2.05	–	m²	11.89
Ceiling finishes							
plasterboard and skim; removes nails or screws	–	0.20	5.48	1.02	–	m²	6.50
wood lathe an plaster; removes nails or screws	–	0.22	6.04	1.72	–	m²	7.76
suspended ceilings; lay in grid; hangers and tee sections	–	0.45	12.23	1.72	–	m²	13.95
plain sheet boarding; including battening	–	0.41	11.17	1.37	–	m²	12.54
matchboarding; including battening	–	0.54	14.89	2.05	–	m²	16.94
Wall panelling							
remove timber wall panelling; clean off and set aside for reuse	–	0.58	15.96	–	–	m²	15.96
Removing coverings load into skip							
Roof coverings							
slates	–	0.45	12.23	0.68	–	m²	12.91
slates; set aside for reuse	–	0.54	14.89	–	–	m²	14.89
nibbed tiles	–	0.36	9.84	0.68	–	m²	10.52
nibbed tiles; set aside for reuse	–	0.45	12.23	–	–	m²	12.23
underfelt and nails	–	0.04	1.07	0.35	–	m²	1.42
felt or polymer membrane	–	0.22	6.12	0.68	–	m²	6.80
profiled metal sheeting; insulation; steel liner	–	0.75	20.62	1.31	–	m²	21.93
sheet metal coverings	–	0.45	12.23	0.68	–	m²	12.91
remove tiling battens	–	0.07	1.86	0.35	–	m²	2.21
Remove roof coverings; select and refix including providing 25% new tiles including nails							
natural slates; Welsh Blue	–	1.08	29.52	0.35	54.50	m²	84.37
clay plain tiles	–	0.99	27.12	0.35	8.09	m²	35.56
asbestos free artificial blue/black slates	–	0.99	27.12	0.35	7.76	m²	35.23
concrete interlocking tiles	–	0.63	17.28	0.35	3.26	m²	20.89
removing ridge or hip tile; provide and fit new	–	0.45	12.23	–	11.57	m	23.80
4.02 CUTTING OR FORMING OPENINGS							
Cutting openings or recesses; spoil into adjacent skip							
Through reinforced concrete walls; make good; not including any props or supports							
150 mm thick wall	–	5.02	172.63	7.42	11.82	m²	191.87
225 mm thick wall	–	6.95	236.21	10.74	16.03	m²	262.98
300 mm thick wall	–	8.75	302.27	10.25	19.84	m²	332.36

04 ALTERATIONS, REPAIRS AND CONSERVATION

Item	PC £	Labour hours	Labour £	Plant £	Material £	Unit	Total rate £
4.02 CUTTING OR FORMING OPENINGS – cont							
Cutting openings or recesses – cont							
Through reinforced concrete suspended slab; make good; not including any props or supports							
150 mm thick wall	–	3.81	131.51	6.96	17.74	m²	156.21
225 mm thick wall	–	5.66	195.04	10.74	16.03	m²	221.81
300 mm thick wall	–	7.04	243.43	10.25	19.84	m²	273.52
Through brick or block walls or partitions							
half brick thick	–	2.38	65.15	5.82	–	m²	70.97
one brick thick	–	3.95	108.23	11.63	–	m²	119.86
one and a half brick thick	–	5.52	151.31	17.45	–	m²	168.76
two brick thick	–	7.09	194.39	23.26	–	m²	217.65
100 mm thick blockwork	–	1.75	47.87	5.14	–	m²	53.01
140 mm thick blockwork	–	2.10	57.71	7.19	–	m²	64.90
215 mm thick blockwork	–	2.60	71.27	10.94	–	m²	82.21
Cut opening through wall for new window (not included); 1200 mm × 1200 mm; quoining up jambs; cut and pin galvanized steel lintel; new soldier course in facing bricks							
one brick thick wall or two half brick thick skins	8.00	7.18	191.28	16.74	164.74	nr	372.76
one and a half brick thick wall	8.00	9.44	253.32	25.13	164.74	nr	443.19
two brick thick wall	8.00	11.70	315.36	33.49	164.74	nr	513.59
Cut opening through wall for new door (not included); 1200 mm × 2100 mm; quoining up jambs; cut and pin galvanized steel lintel; new soldier course in facing bricks							
one brick thick wall or two half brick thick skins	8.00	17.11	495.61	29.30	240.34	nr	765.25
one and a half brick thick wall	8.00	22.75	659.46	43.97	278.18	nr	981.61
two brick thick wall	8.00	28.39	824.82	58.61	315.98	nr	1199.41
Diamond cutting							
Cutting to create openings in block wall							
up to 1.00 m girth; up to 150 mm thick	–	–	–	–	–	m	28.33
up to 1.00 m girth; 200 mm thick	–	–	–	–	–	m	33.80
up to 1.00 m girth; 250 mm thick	–	–	–	–	–	m	39.28
up to 1.00 m girth; 300 mm thick	–	–	–	–	–	m	44.74
3 m–3.5 m girth; up to 150 mm thick	–	–	–	–	–	m	127.50
3 m–3.5 m girth; 200 mm thick	–	–	–	–	–	m	152.09
3 m–3.5 m girth; 250 mm thick	–	–	–	–	–	m	176.71
3 m–3.5 m girth; 300 mm thick	–	–	–	–	–	m	201.30
9 m–10 m girth; up to 150 mm thick	–	–	–	–	–	m	423.17
9 m–10 m girth; 200 mm thick	–	–	–	–	–	m	564.23
9 m–10 m girth; 250 mm thick	–	–	–	–	–	m	705.28
9 m–10 m girth; 300 mm thick	–	–	–	–	–	m	846.34
Cutting to create openings in brick wall							
up to 1.00 m girth; up to 150 mm thick	–	–	–	–	–	m	35.41
up to 1.00 m girth; 200 mm thick	–	–	–	–	–	m	42.25
up to 1.00 m girth; 250 mm thick	–	–	–	–	–	m	49.08

04 ALTERATIONS, REPAIRS AND CONSERVATION

Item	PC £	Labour hours	Labour £	Plant £	Material £	Unit	Total rate £
up to 1.00 m girth; 300 mm thick	–	–	–	–	–	m	55.90
3 m–3.5 m girth; up to 150 mm thick	–	–	–	–	–	m	255.04
3 m–3.5 m girth; 200 mm thick	–	–	–	–	–	m	304.20
3 m–3.5 m girth; 250 mm thick	–	–	–	–	–	m	353.45
3 m–3.5 m girth; 300 mm thick	–	–	–	–	–	m	402.59
9 m–10 m girth; up to 150 mm thick	–	–	–	–	–	m	846.34
9 m–10 m girth; 200 mm thick	–	–	–	–	–	m	1128.47
9 m–10 m girth; 250 mm thick	–	–	–	–	–	m	1410.59
9 m–10 m girth; 300 mm thick	–	–	–	–	–	m	1692.68
Cutting to create openings in concrete wall							
up to 1.00 m girth; up to 150 mm thick	–	–	–	–	–	m	141.70
up to 1.00 m girth; 200 mm thick	–	–	–	–	–	m	169.00
up to 1.00 m girth; 250 mm thick	–	–	–	–	–	m	196.36
up to 1.00 m girth; 300 mm thick	–	–	–	–	–	m	223.66
3 m–3.5 m girth; up to 150 mm thick	–	–	–	–	–	m	1275.20
3 m–3.5 m girth; 200 mm thick	–	–	–	–	–	m	1520.99
3 m–3.5 m girth; 250 mm thick	–	–	–	–	–	m	1767.14
3 m–3.5 m girth; 300 mm thick	–	–	–	–	–	m	2012.96
9 m–10 m girth; up to 150 mm thick	–	–	–	–	–	m	4231.73
9 m–10 m girth; 200 mm thick	–	–	–	–	–	m	5642.29
9 m–10 m girth; 250 mm thick	–	–	–	–	–	m	7052.87
9 m–10 m girth; 300 mm thick	–	–	–	–	–	m	8463.43
Cutting to create openings in timber wall							
up to 1.00 m girth; up to 150 mm thick	–	–	–	–	–	m	42.50
up to 1.00 m girth; 200 mm thick	–	–	–	–	–	m	50.70
up to 1.00 m girth; 250 mm thick	–	–	–	–	–	m	58.90
up to 1.00 m girth; 300 mm thick	–	–	–	–	–	m	67.10
3 m–3.5 m girth; up to 150 mm thick	–	–	–	–	–	m	382.58
3 m–3.5 m girth; 200 mm thick	–	–	–	–	–	m	456.28
3 m–3.5 m girth; 250 mm thick	–	–	–	–	–	m	530.16
3 m–3.5 m girth; 300 mm thick	–	–	–	–	–	m	603.90
9 m–10 m girth; up to 150 mm thick	–	–	–	–	–	m	1269.52
9 m–10 m girth; 200 mm thick	–	–	–	–	–	m	1692.68
9 m–10 m girth; 250 mm thick	–	–	–	–	–	m	2115.86
9 m–10 m girth; 300 mm thick	–	–	–	–	–	m	2539.03
Quoining up jambs to openings							
Common bricks							
half brick thick wall	–	0.90	29.86	–	9.00	m	38.86
one brick thick wall	–	1.35	44.63	–	18.00	m	62.63
one and a half brick thick wall	–	1.75	57.79	–	27.01	m	84.80
two brick thick wall	PC	2.15	71.32	–	36.01	m	107.33
100 mm blockwork wall	–	0.63	20.87	–	13.56	m	34.43
140 mm blockwork wall	–	0.78	25.69	–	20.16	m	45.85
215 mm blockwork wall	–	0.97	32.11	–	25.39	m	57.50
Cutting back projections; spoil into skip							
Brick projection flush with adjacent wall							
225 mm × 112 mm	–	0.27	8.99	0.35	–	m	9.34
225 mm × 225 mm	–	0.45	14.77	0.68	–	m	15.45
337 mm × 112 mm	–	0.63	20.87	1.02	–	m	21.89
450 mm × 225 mm	–	0.81	26.65	1.37	–	m	28.02
chimney breast	–	1.57	52.02	10.94	–	m²	62.96

04 ALTERATIONS, REPAIRS AND CONSERVATION

Item	PC £	Labour hours	Labour £	Plant £	Material £	Unit	Total rate £
4.02 CUTTING OR FORMING OPENINGS – cont							
Temporary screens							
Providing and erecting; maintaining temporary dust proof screens; 50 mm × 75 mm softwood framing and 12 mm thick plywood covering to one side; single layer of polythene sheet to the other; clear away on completion	–	0.90	29.86	1.31	15.92	m²	**47.09**
4.03 FILLING IN OPENINGS							
Fill existing openings							
Fill in small holes							
make good hole where small (up to 12 mm dia.) pipe removed; cement mortar	–	0.19	6.29	–	0.28	nr	**6.57**
Larger holes fill in with common brickwork or blockwork							
half brick thick	–	1.66	54.91	–	34.06	m²	**88.97**
one brick thick	–	2.74	90.55	–	72.07	m²	**162.62**
one and a half brick thick	–	3.77	124.91	–	108.11	m²	**233.02**
two brick thick	–	4.71	156.05	–	144.14	m²	**300.19**
100 mm blockwork	–	0.94	31.14	–	22.78	m²	**53.92**
140 mm blockwork	–	1.13	37.25	–	31.18	m²	**68.43**
215 mm blockwork	–	1.44	47.52	–	46.63	m²	**94.15**
cavity wall; facing bricks; 100 mm cavity fill insulation; 140 mm block	–	4.13	136.79	–	83.71	m²	**220.50**
4.04 REMOVE EXISTING AND REPLACING							
Remove existing material for replacement							
NB: Access equipment is not included							
Cutting out decayed, defective or cracked work and replacing with new common bricks; gauged mortar							
small areas; half brick thick walling	–	4.56	150.95	3.26	39.46	m²	**193.67**
small areas; one brick thick walling	–	8.88	293.94	5.23	81.53	m²	**380.70**
small areas; one and a half brick thick walling	–	12.58	416.42	9.16	122.29	m²	**547.87**
small areas; two brick thick walling	–	16.10	532.94	11.77	163.06	m²	**707.77**
individual bricks; half brick thick walling	–	0.28	9.26	–	0.84	nr	**10.10**
Cutting out decayed, defective or cracked work and replacing with new facing bricks; gauged mortar; facing to one side							
small areas; half brick thick walling (PC £ per 1000)	500.00	6.75	223.44	3.26	50.26	m²	**276.96**
small areas; half brick thick walling (PC £ per 1000)	700.00	6.75	223.44	3.26	68.26	m²	**294.96**
individual bricks; half brick thick walling (PC £ per 1000)	500.00	0.42	13.91	–	1.02	m²	**14.93**

04 ALTERATIONS, REPAIRS AND CONSERVATION

Item	PC £	Labour hours	Labour £	Plant £	Material £	Unit	Total rate £
Cutting out decayed, defective or cracked soldier course arch and replace with new							
facing bricks (PC £ per 1000)	500.00	1.80	59.58	0.35	13.57	m²	**73.50**
facing bricks (PC £ per 1000)	700.00	1.80	59.58	0.35	18.37	m²	**78.30**
Cutting out raking cracks in brickwork; stitching in new common bricks							
half brick thick (PC £ per 1000)	380.00	2.96	97.98	2.74	18.01	m²	**118.73**
one brick thick (PC £ per 1000)	380.00	5.41	179.08	5.14	38.65	m²	**222.87**
one and a half brick thick (PC £ per 1000)	380.00	8.09	267.79	7.86	56.68	m²	**332.33**
Cutting out raking cracks in brickwork; stitching in new facing bricks; half brick thick; facing to one side							
facing bricks (PC £ per 1000)	500.00	4.40	145.64	2.74	25.13	m²	**173.51**
facing bricks (PC £ per 1000)	700.00	4.44	146.98	2.74	34.13	m²	**183.85**
Cutting out raking cracks in cavity brickwork; stitching in new common bricks one side; facing bricks the other side; both skins half brick thick; facing to one side							
facing bricks (PC £ per 1000)	500.00	7.59	251.24	5.47	41.42	m²	**298.13**
facing bricks (PC £ per 1000)	700.00	7.59	251.24	5.47	53.86	m²	**310.57**
Cutting away old angle fillets and replacing with new; cement mortar; 50 mm face width	–	0.23	7.61	–	7.87	m	**15.48**
Cutting out ends of joists and plates from walls; making good in common bricks in cement mortar							
175 mm joists; 400 mm centres	–	0.60	19.86	–	15.17	m	**35.03**
225 mm joists; 400 mm centres	–	0.74	24.49	–	17.45	m	**41.94**
Cutting and pinning to existing brickwork ends of joists	–	0.37	12.25	–	–	nr	**12.25**
Remove defective wall and rebuild with new materials							
defective parapet wall; 600 mm high; with two courses of tiles and brick-on-edge coping over; rebuilding with new facing bricks, tiles and coping	–	6.16	203.90	9.58	90.41	m	**303.89**
defective capping stones and haunching; replace stones and re-haunch in cement mortar	–	1.39	46.01	–	15.76	m	**61.77**

4.05 PREPARING EXISTING STRUCTURES FOR CONNECTION

Prepare existing structures for connection

Making good where intersecting wall has been removed							
half brick thick	–	0.28	9.26	–	2.63	m	**11.89**
one brick thick	–	0.37	12.25	–	5.26	m	**17.51**
100 mm blockwork	–	0.23	7.61	–	2.63	m	**10.24**
150 mm blockwork	–	0.27	8.94	–	2.63	m	**11.57**
215 mm blockwork	–	0.32	10.60	–	5.26	m	**15.86**
225 mm blockwork	–	0.36	11.92	–	5.26	m	**17.18**

04 ALTERATIONS, REPAIRS AND CONSERVATION

Item	PC £	Labour hours	Labour £	Plant £	Material £	Unit	Total rate £
4.06 REPAIRING							
Repairing concrete							
Reinstating plain concrete bed with site mixed in situ concrete; mix 20.00 N, where opening no longer required							
100 mm thick	–	0.44	15.16	–	18.05	m^2	**33.21**
150 mm thick	–	0.72	25.06	–	27.08	m^2	**52.14**
Reinstating reinforced concrete bed with site mixed in situ concrete; mix 20.00 N; mesh reinforcement, where opening no longer required							
100 mm thick	–	0.66	22.93	–	24.06	m^2	**46.99**
150 mm thick	–	0.91	31.78	–	33.08	m^2	**64.86**
Reinstating reinforced concrete suspended floor with site mixed in situ concrete; mix 25.00 N; mesh reinforcement; formwork, where opening no longer required							
150 mm thick	–	2.96	102.18	–	47.42	m^2	**149.60**
225 mm thick	–	3.12	107.84	–	61.32	m^2	**169.16**
300 mm thick	–	3.60	124.81	–	75.19	m^2	**200.00**
Reinstating small hole through concrete suspended slab with site mixed in situ concrete; mix 25.00 N, where opening no longer required							
50 mm dia.	–	0.39	13.79	–	0.07	nr	**13.86**
100 mm dia.	–	0.51	18.04	–	0.29	nr	**18.33**
150 mm dia.	–	0.65	22.98	–	0.64	nr	**23.62**
Clean out minor crack and fill with cement: mortar mixed with bonding agent	–	0.31	8.50	–	1.79	m	**10.29**
Clean out crack to form a 20 mm × 20 mm groove and fill with cement: mortar mixed with bonding agent	–	0.61	16.73	0.13	6.82	m	**23.68**
Repairing metal							
Overhauling and repairing metal casement windows adjust and oil ironmongery; prepare affected parts for redecoration	–	1.39	46.01	–	–	nr	**46.01**
Repairing timber							
Removing or punching projecting nails; refixing timber flooring							
loose boards	–	0.14	4.63	–	–	m^2	**4.63**
refix floorboards previously set aside	–	0.74	24.49	–	1.10	m^2	**25.59**
Remove damaged softwood flooring and fix new plain edge softwood flooring							
small areas	–	1.06	35.09	–	38.94	m^2	**74.03**
individual boards 150 mm wide	–	0.28	9.26	–	3.96	m	**13.22**
Sanding down and resurfacing existing flooring; preparing; body in with wax polish							
softwood	–	0.70	19.88	–	2.00	m^2	**21.88**
hardwood	–	0.85	24.14	–	2.00	m^2	**26.14**

04 ALTERATIONS, REPAIRS AND CONSERVATION

Item	PC £	Labour hours	Labour £	Plant £	Material £	Unit	Total rate £
Fitting existing softwood skirting or architrave to new frames							
75 mm high	–	0.09	2.98	–	0.02	m	**3.00**
150 mm high	–	0.12	3.97	–	0.02	m	**3.99**
225 mm high	–	0.15	4.97	–	0.02	m	**4.99**
Piecing in new 25 mm × 150 mm moulded softwood skirtings to match existing where old has been removed; prepare for decoration	–	0.35	11.12	–	5.78	m	**16.90**
Repair doors							
Easing and adjusting softwood doors; oil ironmongery	–	0.59	19.54	–	–	nr	**19.54**
Remove softwood doors; easing and adjust; oil ironmongery; rehang	–	0.75	24.83	–	–	nr	**24.83**
Remove softwood door; plane 12 mm from bottom edge; rehang	–	1.11	36.74	–	–	nr	**36.74**
Take off existing softwood doorstops; supply and fit new 25 mm × 38 mm doorstop	–	0.10	3.31	–	1.82	m	**5.13**
Cutting out infected or decayed structural members; supply and fix new treated sawn softwood members							
Floors and flat roofs							
50 mm × 125 mm	–	0.37	12.25	–	3.78	m	**16.03**
50 mm × 150 mm	–	0.41	13.57	–	4.54	m	**18.11**
50 mm × 175 mm	–	0.44	14.57	–	4.24	m	**18.81**
Pitched roofs							
38 mm × 100 mm	–	0.33	10.92	–	4.31	m	**15.23**
50 mm × 100 mm	–	0.42	13.91	–	3.02	m	**16.93**
50 mm × 125 mm	–	0.46	15.23	–	3.78	m	**19.01**
50 mm × 150 mm	–	0.51	16.88	–	4.54	m	**21.42**
Kerb bearers							
50 mm × 75 mm	–	0.42	13.91	–	1.81	m	**15.72**
50 mm × 100 mm	–	0.52	17.21	–	2.42	m	**19.63**
75 mm × 100 mm	–	0.63	20.86	–	3.62	m	**24.48**
75 mm × 150 mm	–	0.66	21.85	–	5.45	m	**27.30**
75 mm × 175 mm	–	0.70	23.17	–	6.35	m	**29.52**
4.07 REPOINTING							
Repointing masonry							
Raking out decayed masonry joints and repointing in cement mortar							
brickwork walls generally	–	0.69	22.84	–	2.63	m²	**25.47**
cutting out staggered cracks and repointing to match existing along brick joints	–	0.37	12.25	–	–	m	**12.25**
brickwork and re-wedge horizontal flashing	–	0.23	7.61	–	1.32	m	**8.93**
brickwork and re-wedge stepped flashing	–	0.34	11.26	–	1.32	m	**12.58**
brickwork in chimney stacks	–	1.11	36.74	–	2.63	m²	**39.37**
uncoursed stonework	–	1.11	36.74	–	1.32	m	**38.06**

Prices for Measured Works

04 ALTERATIONS, REPAIRS AND CONSERVATION

Item	PC £	Labour hours	Labour £	Plant £	Material £	Unit	Total rate £
4.09 CLEANING SURFACES							
Cleaning surfaces by hand using handheld manual tools only (brushes, scrapers etc.)							
Masonry/Concrete							
cleaning surface of moss and lichen from walls	–	0.28	7.68	–	–	m²	**7.68**
cleaning bricks of mortar, sort and stack for reuse	–	9.25	253.58	–	–	1000	**253.58**
clean surface of concrete to receive new damp-proof membrane	–	0.14	3.84	–	–	m²	**3.84**
4.10 DECONTAMINATION							
Insecticide or fungicidal treatments							
Treating individual timbers with two coats of proprietary insecticide and fungicide by brush or spray application as appropriate							
general boarding	–	–	–	–	–	m²	**13.73**
structural members	–	–	–	–	–	m²	**11.86**
skirtings, architraves etc.	–	–	–	–	–	m	**12.23**
remove cobwebs, dust and roof insulation; treat exposed timbers with proprietary insecticide and fungicide by spray application	–	–	–	–	–	m²	**22.94**
lifting necessary floorboards; treating flooring with two coats of proprietary insecticide and fungicide by spray application; refix boards	–	–	–	–	–	m²	**21.96**
treating surfaces of adjoining masonry or concrete with two coats of proprietary dry rot by spray application	–	–	–	–	–	m²	**13.48**

05 EXCAVATE AND FILLING

Item	PC £	Labour hours	Labour £	Plant £	Material £	Unit	Total rate £
5.01 SITE CLEARANCE AND PREPARATION							
Site clearance							
Prices are applicable to excavation in firm soil							
Fell and removing trees							
girth 500–1500 mm	–	18.50	260.70	15.68	–	nr	**276.38**
girth 1500–3000 mm	–	32.50	457.99	15.68	–	nr	**473.67**
girth exceeding 3000 mm	–	46.50	655.28	19.59	–	nr	**674.87**
Removing tree stumps							
girth 500 mm–1.50 m	–	2.00	28.18	22.04	–	nr	**50.22**
girth 1.50–3.00 m	–	3.00	42.27	31.51	–	nr	**73.78**
girth exceeding 3.00 m	–	6.00	84.55	44.18	–	nr	**128.73**
Clearing site vegetation							
bushes, scrub, undergrowth, hedges and trees and tree stumps not exceeding 500 mm girth	–	0.03	0.42	1.23	–	m²	**1.65**
Lifting turf for preservation, by hand							
stacking	–	0.32	4.51	–	–	m²	**4.51**
Topsoil for preservation; to spoil heap less than 50 m from excavations							
average depth 150 mm	–	0.02	0.28	0.88	–	m²	**1.16**
add or deduct for each 25 mm variation in average depth	–	0.01	0.15	0.21	–	m²	**0.36**
5.02 EXCAVATION							
Excavating by machine							
To reduce levels							
maximum depth not exceeding 0.25 m	–	0.03	0.47	0.67	–	m³	**1.14**
maximum depth not exceeding 1.00 m	–	0.03	0.47	0.67	–	m³	**1.14**
maximum depth not exceeding 2.00 m	–	0.04	0.52	0.74	–	m³	**1.26**
maximum depth not exceeding 4.00 m	–	0.04	0.57	0.80	–	m³	**1.37**
Basements and the like; commencing level exceeding 0.25 m below existing ground level							
maximum depth not exceeding 1.00 m	–	0.06	0.84	1.23	–	m³	**2.07**
maximum depth not exceeding 2.00 m	–	0.07	0.99	1.23	–	m³	**2.22**
maximum depth not exceeding 4.00 m	–	0.08	1.12	1.51	–	m³	**2.63**
maximum depth not exceeding 6.00 m	–	0.09	1.27	1.90	–	m³	**3.17**
maximum depth not exceeding 8.00 m	–	0.12	1.69	2.17	–	m³	**3.86**
Pits							
maximum depth not exceeding 0.25 m	–	0.31	4.37	4.45	–	m³	**8.82**
maximum depth not exceeding 1.00 m	–	0.33	4.65	4.45	–	m³	**9.10**
maximum depth not exceeding 2.00 m	–	0.39	5.50	5.02	–	m³	**10.52**
maximum depth not exceeding 4.00 m	–	0.47	6.62	5.68	–	m³	**12.30**
maximum depth not exceeding 6.00 m	–	0.49	6.90	5.96	–	m³	**12.86**
Extra over pit excavating for commencing level exceeding 0.25 m below existing ground level							
1.00 m below	–	0.03	0.42	0.66	–	m³	**1.08**
2.00 m below	–	0.05	0.70	0.95	–	m³	**1.65**
3.00 m below	–	0.06	0.84	1.23	–	m³	**2.07**
4.00 m below	–	0.09	1.27	1.61	–	m³	**2.88**

05 EXCAVATE AND FILLING

Item	PC £	Labour hours	Labour £	Plant £	Material £	Unit	Total rate £
5.02 EXCAVATION – cont							
Excavating by machine – cont							
Trenches; width not exceeding 0.30 m							
maximum depth not exceeding 0.25 m	–	0.26	3.67	3.50	–	m³	**7.17**
maximum depth not exceeding 1.00 m	–	0.28	3.94	3.50	–	m³	**7.44**
maximum depth not exceeding 2.00 m	–	0.33	4.65	4.07	–	m³	**8.72**
maximum depth not exceeding 4.00 m	–	0.40	5.63	5.02	–	m³	**10.65**
maximum depth not exceeding 6.00 m	–	0.46	6.48	5.96	–	m³	**12.44**
Trenches; width exceeding 0.30 m							
maximum depth not exceeding 0.25 m	–	0.23	3.24	3.12	–	m³	**6.36**
maximum depth not exceeding 1.00 m	–	0.25	3.52	3.12	–	m³	**6.64**
maximum depth not exceeding 2.00 m	–	0.30	4.22	3.78	–	m³	**8.00**
maximum depth not exceeding 4.00 m	–	0.35	4.93	4.45	–	m³	**9.38**
maximum depth not exceeding 6.00 m	–	0.43	6.06	5.68	–	m³	**11.74**
Extra over trench excavating for commencing level exceeding 0.25 m below existing ground level							
1.00 m below	–	0.03	0.42	0.66	–	m³	**1.08**
2.00 m below	–	0.05	0.70	0.95	–	m³	**1.65**
3.00 m below	–	0.06	0.84	1.23	–	m³	**2.07**
4.00 m below	–	0.09	1.27	1.61	–	m³	**2.88**
For pile caps and ground beams between piles							
maximum depth not exceeding 0.25 m	–	0.35	4.93	6.06	–	m³	**10.99**
maximum depth not exceeding 1.00 m	–	0.39	5.50	5.96	–	m³	**11.46**
maximum depth not exceeding 2.00 m	–	0.39	5.50	6.62	–	m³	**12.12**
To bench sloping ground to receive filling							
maximum depth not exceeding 0.25 m	–	0.07	0.99	1.61	–	m³	**2.60**
maximum depth not exceeding 1.00 m	–	0.09	1.27	1.61	–	m³	**2.88**
maximum depth not exceeding 2.00 m	–	0.09	1.27	1.90	–	m³	**3.17**
Extra over any types of excavating irrespective of depth							
excavating below ground water level	–	0.13	1.83	2.17	–	m³	**4.00**
next to existing services	–	0.35	4.93	1.23	–	m³	**6.16**
around existing services crossing excavation	–	0.60	8.46	3.50	–	m³	**11.96**
Extra over any types of excavating irrespective of depth for breaking out existing materials							
rock	–	2.95	41.57	19.65	–	m³	**61.22**
concrete	–	2.55	35.94	15.49	–	m³	**51.43**
reinforced concrete	–	3.60	50.73	22.59	–	m³	**73.32**
brickwork, blockwork or stonework	–	1.85	26.07	11.40	–	m³	**37.47**
Extra over any types of excavating irrespective of depth for breaking out existing hard pavings, 75 mm thick							
coated macadam or asphalt	–	0.19	2.68	0.97	–	m²	**3.65**
Extra over any types of excavating irrespective of depth for breaking out existing hard pavings, 150 mm thick							
concrete	–	0.39	5.50	2.35	–	m²	**7.85**
reinforced concrete	–	0.58	8.18	3.21	–	m²	**11.39**
coated macadam or asphalt and hardcore	–	0.26	3.67	1.09	–	m²	**4.76**

05 EXCAVATE AND FILLING

Item	PC £	Labour hours	Labour £	Plant £	Material £	Unit	Total rate £
Working space allowance to excavations 600 mm wide							
reduce levels, basements and the like	–	0.07	0.99	1.23	–	m²	2.22
pits	–	0.19	2.68	3.50	–	m²	6.18
trenches	–	0.18	2.53	3.12	–	m²	5.65
pile caps and ground beams between piles	–	0.20	2.82	3.50	–	m²	6.32
Extra over excavating for working space for backfilling in with special materials							
hardcore	–	0.13	1.83	1.55	20.04	m²	23.42
sand	–	0.13	1.83	1.55	26.80	m²	30.18
40 mm–20 mm gravel	–	0.13	1.83	1.55	37.38	m²	40.76
plain in situ ready mixed designated concrete C7.5–40 mm aggregate	–	0.93	15.37	1.62	59.99	m²	76.98
Excavating by hand							
Topsoil for preservation							
average depth 150 mm	–	0.23	3.24	–	–	m²	3.24
add or deduct for each 25 mm variation in average depth	–	0.03	0.42	–	–	m²	0.42
To reduce levels							
maximum depth not exceeding 0.25 m	–	1.44	20.29	–	–	m³	20.29
maximum depth not exceeding 1.00 m	–	1.63	22.97	–	–	m³	22.97
maximum depth not exceeding 2.00 m	–	1.80	25.37	–	–	m³	25.37
maximum depth not exceeding 4.00 m	–	1.99	28.05	–	–	m³	28.05
Basements and the like; commencing level exceeding 0.25 m below existing ground level							
maximum depth not exceeding 1.00 m	–	1.90	26.77	–	–	m³	26.77
maximum depth not exceeding 2.00 m	–	2.04	28.75	–	–	m³	28.75
maximum depth not exceeding 4.00 m	–	2.73	38.47	–	–	m³	38.47
maximum depth not exceeding 6.00 m	–	3.33	46.93	–	–	m³	46.93
maximum depth not exceeding 8.00 m	–	4.02	56.65	–	–	m³	56.65
Pits							
maximum depth not exceeding 0.25 m	–	2.13	30.01	–	–	m³	30.01
maximum depth not exceeding 1.00 m	–	2.75	38.75	–	–	m³	38.75
maximum depth not exceeding 2.00 m	–	3.30	46.50	–	–	m³	46.50
maximum depth not exceeding 4.00 m	–	4.18	58.91	–	–	m³	58.91
maximum depth not exceeding 6.00 m	–	5.17	72.85	–	–	m³	72.85
Extra over pit excavating for commencing level exceeding 0.25 m below existing ground level							
1.00 m below	–	0.42	5.92	–	–	m³	5.92
2.00 m below	–	0.88	12.40	–	–	m³	12.40
3.00 m below	–	1.30	18.32	–	–	m³	18.32
4.00 m below	–	1.71	24.10	–	–	m³	24.10
Trenches; width not exceeding 0.30 m							
maximum depth not exceeding 0.25 m	–	1.85	26.07	–	–	m³	26.07
maximum depth not exceeding 1.00 m	–	2.76	38.89	–	–	m³	38.89
Trenches; width exceeding 0.30 m							
maximum depth not exceeding 0.25 m	–	1.80	25.37	–	–	m³	25.37
maximum depth not exceeding 1.00 m	–	2.46	34.67	–	–	m³	34.67
maximum depth not exceeding 2.00 m	–	2.88	40.58	–	–	m³	40.58
maximum depth not exceeding 4.00 m	–	3.66	51.57	–	–	m³	51.57
maximum depth not exceeding 6.00 m	–	4.68	65.95	–	–	m³	65.95

05 EXCAVATE AND FILLING

Item	PC £	Labour hours	Labour £	Plant £	Material £	Unit	Total rate £
5.02 EXCAVATION – cont							
Excavating by hand – cont							
Extra over trench excavating for commencing level exceeding 0.25 m below existing ground level							
1.00 m below	–	0.42	5.92	–	–	m³	5.92
2.00 m below	–	0.88	12.40	–	–	m³	12.40
3.00 m below	–	1.30	18.32	–	–	m³	18.32
4.00 m below	–	1.71	24.10	–	–	m³	24.10
For pile caps and ground beams between piles							
maximum depth not exceeding 0.25 m	–	2.78	39.17	–	–	m³	39.17
maximum depth not exceeding 1.00 m	–	2.96	41.72	–	–	m³	41.72
maximum depth not exceeding 2.00 m	–	3.52	49.60	–	–	m³	49.60
To bench sloping ground to receive filling							
maximum depth not exceeding 0.25 m	–	1.30	18.32	–	–	m³	18.32
maximum depth not exceeding 1.00 m	–	1.48	20.86	–	–	m³	20.86
maximum depth not exceeding 2.00 m	–	1.67	23.54	–	–	m³	23.54
Extra over any types of excavating irrespective of depth							
excavating below ground water level	–	0.32	4.51	–	–	m³	4.51
next existing services	–	0.93	13.10	–	–	m³	13.10
around existing services crossing excavation	–	1.85	26.07	–	–	m³	26.07
Extra over any types of excavating irrespective of depth for breaking out existing materials							
rock	–	4.63	65.25	17.65	–	m³	82.90
concrete	–	4.16	58.62	14.71	–	m³	73.33
reinforced concrete	–	5.55	78.21	20.61	–	m³	98.82
brickwork, blockwork or stonework	–	2.78	39.17	8.83	–	m³	48.00
Extra over any types of excavating irrespective of depth for breaking out existing hard pavings, 60 mm thick							
precast concrete paving slabs	–	0.28	3.94	–	–	m²	3.94
Extra over any types of excavating irrespective of depth for breaking out existing hard pavings, 75 mm thick							
coated macadam or asphalt	–	0.37	5.21	1.18	–	m²	6.39
Extra over any types of excavating irrespective of depth for breaking out existing hard pavings, 150 mm thick							
concrete	–	0.65	9.16	2.07	–	m²	11.23
reinforced concrete	–	0.83	11.70	2.95	–	m²	14.65
coated macadam or asphalt and hardcore	–	0.46	6.48	1.47	–	m²	7.95
Working space allowance to excavations							
reduce levels, basements and the like	–	2.13	30.01	1.81	–	m²	31.82
pits	–	2.22	31.28	1.81	–	m²	33.09
trenches	–	1.94	27.34	1.81	–	m²	29.15
pile caps and ground beams between piles	–	2.31	32.55	1.81	–	m²	34.36
Extra over excavation for working space for backfilling with special materials							
hardcore	–	0.74	10.42	1.81	21.48	m²	33.71
sand	–	0.74	10.42	1.81	30.95	m²	43.18
40 mm–20 mm gravel	–	0.74	10.42	1.81	37.38	m²	49.61
plain in situ concrete ready mixed designated concrete; C7.5–40 mm aggregate	–	1.02	16.85	1.81	59.99	m²	78.65

05 EXCAVATE AND FILLING

Item	PC £	Labour hours	Labour £	Plant £	Material £	Unit	Total rate £
5.03 SUPPORT TO EXCAVATIONS							
Earthwork support (average risk prices)							
Maximum depth not exceeding 1.00 m							
distance between opposing faces not exceeding 2.00 m	–	0.10	1.34	–	–	m²	**1.34**
distance between opposing faces 2.00–4.00 m	–	0.10	1.47	–	–	m²	**1.47**
distance between opposing faces exceeding 4.00 m	–	0.11	1.61	–	–	m²	**1.61**
Maximum depth not exceeding 2.00 m							
distance between opposing faces not exceeding 2.00 m	–	0.11	1.61	–	–	m²	**1.61**
distance between opposing faces 2.00–4.00 m	–	0.12	1.74	–	–	m²	**1.74**
distance between opposing faces exceeding 4.00 m	–	0.13	1.87	–	–	m²	**1.87**
Maximum depth not exceeding 4.00 m							
distance between opposing faces not exceeding 2.00 m	–	0.15	2.14	–	–	m²	**2.14**
distance between opposing faces 2.00–4.00 m	–	0.15	2.14	–	–	m²	**2.14**
distance between opposing faces exceeding 4.00 m	–	0.17	2.41	–	–	m²	**2.41**
Maximum depth not exceeding 6.00 m							
distance between opposing faces not exceeding 2.00 m	–	0.17	2.41	–	–	m²	**2.41**
distance between opposing faces 2.00–4.00 m	–	0.18	2.54	–	–	m²	**2.54**
distance between opposing faces exceeding 4.00 m	–	0.21	2.95	–	–	m²	**2.95**
Maximum depth not exceeding 8.00 m							
distance between opposing faces not exceeding 2.00 m	–	0.22	3.08	–	–	m²	**3.08**
distance between opposing faces 2.00–4.00 m	–	0.27	3.75	–	–	m²	**3.75**
distance between opposing faces exceeding 4.00 m	–	0.31	4.42	–	–	m²	**4.42**
Earthwork support (open boarded)							
Maximum depth not exceeding 1.00 m							
distance between opposing faces not exceeding 2.00 m	–	0.27	3.75	–	–	m²	**3.75**
distance between opposing faces 2.00–4.00 m	–	0.29	4.15	–	–	m²	**4.15**
distance between opposing faces exceeding 4.00 m	–	0.33	4.69	–	–	m²	**4.69**
Maximum depth not exceeding 2.00 m							
distance between opposing faces not exceeding 2.00 m	–	0.33	4.69	–	–	m²	**4.69**
distance between opposing faces 2.00–4.00 m	–	0.37	5.22	–	–	m²	**5.22**
distance between opposing faces exceeding 4.00 m	–	0.42	5.89	–	–	m²	**5.89**

05 EXCAVATE AND FILLING

Item	PC £	Labour hours	Labour £	Plant £	Material £	Unit	Total rate £
5.03 SUPPORT TO EXCAVATIONS – cont							
Earthwork support (open boarded) – cont							
Maximum depth not exceeding 4.00 m							
distance between opposing faces not exceeding 2.00 m	–	0.42	5.89	–	–	m²	**5.89**
distance between opposing faces 2.00–4.00 m	–	0.47	6.70	–	–	m²	**6.70**
distance between opposing faces exceeding 4.00 m	–	0.53	7.50	–	–	m²	**7.50**
Maximum depth not exceeding 6.00 m							
distance between opposing faces not exceeding 2.00 m	–	0.53	7.50	–	–	m²	**7.50**
distance between opposing faces 2.00–4.00 m	–	0.58	8.17	–	–	m²	**8.17**
distance between opposing faces exceeding 4.00 m	–	0.67	9.37	–	–	m²	**9.37**
Maximum depth not exceeding 8.00 m							
distance between opposing faces not exceeding 2.00 m	–	0.70	9.91	–	–	m²	**9.91**
distance between opposing faces 2.00–4.00 m	–	0.79	11.11	–	–	m²	**11.11**
distance between opposing faces exceeding 4.00 m	–	0.92	12.99	–	–	m²	**12.99**
Earthwork support (closeboarded)							
Maximum depth not exceeding 1.00 m							
distance between opposing faces not exceeding 2.00 m	–	0.70	9.91	–	–	m²	**9.91**
distance between opposing faces 2.00–4.00 m	–	0.77	10.85	–	–	m²	**10.85**
distance between opposing faces exceeding 4.00 m	–	0.85	12.05	–	–	m²	**12.05**
Maximum depth not exceeding 2.00 m							
distance between opposing faces not exceeding 2.00 m	–	0.88	12.45	–	–	m²	**12.45**
distance between opposing faces 2.00–4.00 m	–	0.97	13.66	–	–	m²	**13.66**
distance between opposing faces exceeding 4.00 m	–	1.05	14.86	–	–	m²	**14.86**
Maximum depth not exceeding 4.00 m							
distance between opposing faces not exceeding 2.00 m	–	1.10	15.53	–	–	m²	**15.53**
distance between opposing faces 2.00–4.00 m	–	1.24	17.41	–	–	m²	**17.41**
distance between opposing faces exceeding 4.00 m	–	1.36	19.15	–	–	m²	**19.15**
Maximum depth not exceeding 6.00 m							
distance between opposing faces not exceeding 2.00 m	–	1.37	19.28	–	–	m²	**19.28**
distance between opposing faces 2.00–4.00 m	–	1.49	21.02	–	–	m²	**21.02**
distance between opposing faces exceeding 4.00 m	–	1.67	23.57	–	–	m²	**23.57**

05 EXCAVATE AND FILLING

Item	PC £	Labour hours	Labour £	Plant £	Material £	Unit	Total rate £
Maximum depth not exceeding 8.00 m							
distance between opposing faces not exceeding 2.00 m	–	1.67	23.57	–	–	m²	**23.57**
distance between opposing faces 2.00–4.00 m	–	1.84	25.97	–	–	m²	**25.97**
distance between opposing faces exceeding 4.00 m	–	2.11	29.72	–	–	m²	**29.72**
Extra over earthwork support for							
curved	–	0.02	0.27	–	–	m²	**0.27**
below ground water level	–	0.27	3.75	–	–	m²	**3.75**
unstable ground	–	0.44	6.16	–	–	m²	**6.16**
next to roadways	–	0.35	4.95	–	–	m²	**4.95**
left in	–	0.57	8.03	–	–	m²	**8.03**
Earthwork support (average risk prices – inside existing buildings)							
Maximum depth not exceeding 1.00 m							
distance between opposing faces not exceeding 2.00 m	–	0.18	2.53	–	–	m²	**2.53**
distance between opposing faces 2.00–4.00 m	–	0.19	2.68	–	–	m²	**2.68**
distance between opposing faces exceeding 4.00 m	–	0.22	3.10	–	–	m²	**3.10**
Maximum depth not exceeding 2.00 m							
distance between opposing faces not exceeding 2.00 m	–	0.22	3.10	–	–	m²	**3.10**
distance between opposing faces 2.00–4.00 m	–	0.24	3.38	–	–	m²	**3.38**
distance between opposing faces exceeding 4.00 m	–	0.32	4.51	–	–	m²	**4.51**
Maximum depth not exceeding 4.00 m							
distance between opposing faces not exceeding 2.00 m	–	0.28	3.94	–	–	m²	**3.94**
distance between opposing faces 2.00–4.00 m	–	0.31	4.37	–	–	m²	**4.37**
distance between opposing faces exceeding 4.00 m	–	0.34	4.79	–	–	m²	**4.79**
Maximum depth not exceeding 6.00 m							
distance between opposing faces not exceeding 2.00 m	–	0.34	4.79	–	–	m²	**4.79**
distance between opposing faces 2.00–4.00 m	–	0.38	5.36	–	–	m²	**5.36**
distance between opposing faces exceeding 4.00 m	–	0.43	6.06	–	–	m²	**6.06**
5.04 DISPOSAL							
Disposal of excavated material off site							
Basic Landfill tax rates (exclusive of site tipping charges)							
inactive waste	–	–	–	–	2.90	tonne	**2.90**
all other taxable wastes	–	–	–	–	91.35	tonne	**91.35**

05 EXCAVATE AND FILLING

Item	PC £	Labour hours	Labour £	Plant £	Material £	Unit	Total rate £
5.04 DISPOSAL – cont							
Disposal of excavated material off site – cont							
Excavated material deposited off site							
inactive waste off site; to tip not exceeding 13 km (using lorries); including Landfill Tax	–	0.10	1.37	14.51	16.59	m³	**32.47**
active non-hazardous waste off site; to tip not exceeding 13 km (using lorries); including Landfill Tax	–	0.06	1.01	14.51	239.43	m³	**254.95**
5.05 RETAINING EXCAVATED MATERIAL ON SITE							
Retain excavated material on site							
Excavated material deposited on site using dumpers to transport material							
on site; spreading; average 25 m distance	–	0.20	2.82	0.47	–	m³	**3.29**
on site; depositing in spoil heaps; average 50 m distance	–	–	–	1.14	–	m³	**1.14**
on site; spreading; average 50 m distance	–	0.20	2.82	0.88	–	m³	**3.70**
on site; depositing in spoil heaps; average 100 m distance	–	–	–	2.02	–	m³	**2.02**
on site; spreading; average 100 m distance	–	0.20	2.82	1.35	–	m³	**4.17**
on site; depositing in spoil heaps; average 200 m distance	–	–	–	2.55	–	m³	**2.55**
on site; spreading; average 200 m distance	–	0.20	2.82	1.81	–	m³	**4.63**
5.06 FILLING OBTAINED FROM EXCAVATED MATERIAL							
Filling to make up levels							
By machine							
average thickness over 50 mm not exceeding 500 mm	–	0.22	3.05	2.17	–	m³	**5.22**
average thickness exceeding 500 mm	–	0.18	2.53	1.57	–	m³	**4.10**
By hand							
average thickness over 50 mm not exceeding 500 mm	–	1.25	17.61	4.26	–	m³	**21.87**
average thickness exceeding 500 mm	–	1.02	14.38	3.47	–	m³	**17.85**
Surface treatments							
Surface packing to filling							
To vertical or battered faces	–	0.17	2.40	0.18	–	m²	**2.58**
Compacting							
bottoms of excavations	–	0.04	0.57	0.04	–	m²	**0.61**
Trimming							
sloping surfaces; by hand	–	0.17	2.40	–	–	m²	**2.40**

05 EXCAVATE AND FILLING

Item	PC £	Labour hours	Labour £	Plant £	Material £	Unit	Total rate £
5.07 IMPORTED FILLING							
Basic material prices delivered to site							
Supply only in full loads							
DOT type 1	–	–	–	–	19.11	tonne	**19.11**
DOT type 2	–	–	–	–	19.11	tonne	**19.11**
10 mm single size aggregate	–	–	–	–	21.78	tonne	**21.78**
20 mm single size aggregate	–	–	–	–	21.78	tonne	**21.78**
20 mm all in aggregate	–	–	–	–	24.88	tonne	**24.88**
40 mm all in aggregate	–	–	–	–	24.88	tonne	**24.88**
40 mm scalpings	–	–	–	–	21.78	tonne	**21.78**
hardcore	–	–	–	–	17.76	tonne	**17.76**
soft/building sand	–	–	–	–	22.66	tonne	**22.66**
recycled type 1	–	–	–	–	20.24	tonne	**20.24**
E-blend (50% type 1 and 50% recycled type 1)	–	–	–	–	16.52	tonne	**16.52**
Filling to make up levels; by machine							
Average thickness over 50 mm not exceeding 500 mm							
obtained off site; imported topsoil	18.90	0.22	3.05	1.81	19.47	m^3	**24.33**
obtained off site; hardcore	35.58	0.25	3.55	2.09	36.65	m^3	**42.29**
obtained off site; granular fill type one	43.86	0.25	3.55	3.64	45.18	m^3	**52.37**
obtained off site; granular fill type two	43.86	0.25	3.55	3.64	45.18	m^3	**52.37**
obtained off site; sand	52.03	0.25	3.55	3.64	53.59	m^3	**60.78**
Average thickness over 500 mm							
obtained off site; imported topsoil	18.90	0.18	2.53	2.04	19.47	m^3	**24.04**
obtained off site; hardcore	35.21	0.22	3.05	0.40	36.27	m^3	**39.72**
obtained off site; granular fill type one	43.86	0.22	3.05	2.84	45.18	m^3	**51.07**
obtained off site; granular fill type two	43.86	0.22	3.05	2.84	45.18	m^3	**51.07**
obtained off site; sand	52.03	0.22	3.05	2.84	53.59	m^3	**59.48**
Filling to make up levels; by hand							
Average thickness over 50 mm not exceeding 500 mm							
obtained off site; imported topsoil	18.90	1.25	17.61	4.25	19.47	m^3	**41.33**
obtained off site; hardcore	35.58	1.39	19.59	4.73	36.65	m^3	**60.97**
obtained off site; granular fill type one	43.86	1.54	21.70	5.23	45.18	m^3	**72.11**
obtained off site; granular fill type two	43.86	1.54	21.70	5.23	45.18	m^3	**72.11**
obtained off site; sand	52.03	1.54	21.70	5.23	53.59	m^3	**80.52**
Average thickness over 500 mm							
obtained off site; imported topsoil	18.90	1.02	14.38	3.47	19.47	m^3	**37.32**
obtained off site; hardcore	35.58	1.34	18.88	4.57	36.65	m^3	**60.10**
obtained off site; granular fill type one	43.86	1.43	20.15	4.89	45.18	m^3	**70.22**
obtained off site; granular fill type two	43.86	1.43	20.15	4.89	45.18	m^3	**70.22**
obtained off site; sand	52.03	1.43	20.15	4.89	53.59	m^3	**78.63**
Surface treatments							
Surface packing to filling							
To vertical or battered faces	–	0.17	2.40	0.18	–	m^2	**2.58**

05 EXCAVATE AND FILLING

Item	PC £	Labour hours	Labour £	Plant £	Material £	Unit	Total rate £
5.07 IMPORTED FILLING – cont							
Surface treatments – cont							
Compacting							
bottoms of excavations	–	0.04	0.57	0.04	–	m²	**0.61**
filling; blinding with sand 50 mm thick	–	0.04	0.57	0.04	2.18	m²	**2.79**
Trimming							
sloping surfaces; by hand	–	0.17	2.40	–	–	m²	**2.40**
sloping surfaces; in rock	–	0.93	13.00	4.11	–	m²	**17.11**
5.08 GEOTEXTILE FABRIC							
Erosion mats							
Filter membrane; one layer; laid on earth to receive granular material; over 500 mm wide							
Terram 500 filter membrane or other equal; one layer; laid on earth. NB mainly agricultural/landscape use	0.25	0.04	0.57	–	0.28	m²	**0.85**
Terram 700 filter membrane or other equal and approved; one layer; laid on earth	0.47	0.04	0.57	–	0.53	m²	**1.10**
Terram 1000; filter membrane or other equal and approved; one layer; laid on earth	0.39	0.04	0.57	–	0.43	m²	**1.00**
Terram 2000; filter membrane or other equal and approved; one layer; laid on earth	1.02	0.04	0.57	–	1.21	m²	**1.78**
Terram High Visibility; orange top surface to warn of potential hazardous material underneath	0.61	0.04	0.57	–	0.72	m²	**1.29**
5.09 MEMBRANES							
Sheeting to prevent moisture loss/gas barrier							
Building paper; lapped joints							
subsoil grade 410; horizontal on foundations	–	0.02	0.33	–	0.84	m²	**1.17**
standard grade 420; horizontal on slabs	–	0.04	0.66	–	1.25	m²	**1.91**
Polythene sheeting; lapped joints; horizontal on slabs							
250 microns; 0.25 mm thick	–	0.04	0.66	–	0.78	m²	**1.44**
Visqueen sheeting or other equal and approved; lapped joints; horizontal on slabs							
250 microns; 0.25 mm thick	–	0.04	0.66	–	0.65	m²	**1.31**
300 microns; 0.30 mm thick	–	0.05	0.82	–	0.48	m²	**1.30**
gas-resistant damp-proof membrane taped joints	–	0.05	0.82	–	5.91	m²	**6.73**
sealing at perimeter of slab	–	0.10	1.65	–	1.55	m	**3.20**
sealing around penetrations; drain pipes etc.	–	1.00	16.52	–	4.62	nr	**21.14**

07 PILING

Item	PC £	Labour hours	Labour £	Plant £	Material £	Unit	Total rate £
7.01 INTERLOCKING SHEET PILES							
Arcelor 600 mm wide 'U' shaped steel sheeting piling; or other equal approved; pitched and driven							
Provision of all plant for sheet pile installation; including bringing to and removing from site; maintenance, erection and dismantling; assuming one rig for 1500 m² of piling							
leader rig with vibratory hammer	–	–	–	–	–	item	4440.00
conventional rig	–	–	–	–	–	item	5800.00
silent vibrationless rig	–	–	–	–	–	item	6470.00
Supply only of standard sheet pile sections							
PU12	–	–	–	–	–	m²	115.00
PU18–1	–	–	–	–	–	m²	125.00
PU22–1	–	–	–	–	–	m²	143.00
PU25	–	–	–	–	–	m²	163.00
PU32	–	–	–	–	–	m²	200.00
Pitching and driving of sheet piles; using the following plant							
leader rig with vibratory hammer	–	–	–	–	–	m²	25.00
conventional rig	–	–	–	–	–	m²	34.00
silent vibrationless rig	–	–	–	–	–	m²	53.00
Provision of all plant for sheet pile extraction; including bringing to and removing from site; maintenance, erection and dismantling; assuming one rig for 1500 m² of piling							
leader rig with vibratory hammer	–	–	–	–	–	item	4550.00
conventional rig	–	–	–	–	–	item	5800.00
silent vibrationless rig	–	–	–	–	–	item	5800.00
Extraction of sheet piles; using the following plant							
leader rig with vibratory hammer	–	–	–	–	–	m²	18.00
conventional rig	–	–	–	–	–	m²	21.60
silent vibrationless rig	–	–	–	–	–	m²	31.20
Credit on extracted piles; recovered in re-usable lengths; for sheet pile sections							
PU12	–	–	–	–	–	m²	81.60
PU18–1	–	–	–	–	–	m²	90.00
PU22–1	–	–	–	–	–	m²	102.00
PU25	–	–	–	–	–	m²	115.00
PU32	–	–	–	–	–	m²	141.00
7.02 BORED PILES							
NOTE: The following approximate prices, for the quantities of piling quoted, are for work on clear open sites with reasonable access. They are based on normal concrete mix 35.00 N/mm²; reinforced and include up to 0.15 m of projecting reinforcement at top of pile. The prices do not allow for removal of spoil.							

07 PILING

Item	PC £	Labour hours	Labour £	Plant £	Material £	Unit	Total rate £
7.02 BORED PILES – cont							
Minipile cast-in-place concrete piles							
300 mm nominal dia.							
Establish equipment on site							
Provision of all plant (1 nr rig); including bringing to and removing from site; maintenance, erection and dismantling at each pile position for 50 nr piles	–	–	–	–	–	item	5100.00
Bored piles							
300 mm dia. piles; nominally reinforced; 15 m long	–	–	–	–	–	nr	225.00
set up at each pile position	–	–	–	–	–	nr	23.61
add for additional piles length	–	–	–	–	–	m	15.00
deduct for reduction in pile length	–	–	–	–	–	m	5.36
Delays							
rig standing time	–	–	–	–	–	hour	205.00
Pile tests							
integrity tests (minimum 20 per visit)	–	–	–	–	–	nr	16.00
300 mm nominal dia.; small area with difficult access and limited headroom							
Establish equipment on site							
Provision of all plant (1 nr rig); including bringing to and removing from site; maintenance, erection and dismantling at each pile position for 50 nr piles	–	–	–	–	–	item	2830.00
Bored piles							
300 mm dia. piles; nominally reinforced; 15 m long	–	–	–	–	–	nr	225.00
set up at each pile position	–	–	–	–	–	nr	23.61
add for additional piles length	–	–	–	–	–	m	48.30
deduct for reduction in pile length	–	–	–	–	–	m	19.32
Delays							
rig standing time	–	–	–	–	–	hour	203.89
Pile tests							
integrity tests (minimum 20 per visit)	–	–	–	–	–	nr	16.00
450 mm nominal dia.							
Establish equipment on site							
Provision of all plant (2 nr rigs); including bringing to and removing from site; maintenance, erection and dismantling at each pile position for 100 nr piles	–	–	–	–	–	item	23280.00
Bored piles							
450 mm dia. piles; nominally reinforced; 20 m long	–	–	–	–	–	nr	3300.00
add for additional piles length	–	–	–	–	–	m	160.00
deduct for reduction in pile length	–	–	–	–	–	m	21.46
Blind bored piles							
450 mm dia.	–	–	–	–	–	m	140.00

07 PILING

Item	PC £	Labour hours	Labour £	Plant £	Material £	Unit	Total rate £
Delays							
rig standing time	–	–	–	–	–	hour	204.00
Extra over piling							
breaking through obstructions	–	–	–	–	–	hour	204.00
Pile tests							
working to 600 kN; using tension piles as reaction; first pile	–	–	–	–	–	nr	11200.00
working to 600 kN; using tension piles as reaction; subsequent piles	–	–	–	–	–	nr	11200.00
integrity tests (minimum 20 per visit)	–	–	–	–	–	nr	16.00
Rotary CFA bored cast-in-place concrete piles							
450 mm dia. piles							
Provision of plant							
provision of all plant (1 nr rig); including bringing to and removing from site; maintenance, erection and dismantling at each pile position for 100 nr piles 20 m long	–	–	–	–	–	item	16275.00
Bored piles							
450 mm dia. piles; reinforced; 20 m long	–	–	–	–	–	nr	910.40
add for additional piles length	–	–	–	–	–	m	44.95
deduct for reduction in pile length	–	–	–	–	–	m	17.99
Delays							
rig standing time	–	–	–	–	–	hour	462.00
Pile tests							
working to 600 kN; using tension piles as reaction; first pile	–	–	–	–	–	nr	6400.00
working to 600 kN; using tension piles as reaction; subsequent piles	–	–	–	–	–	nr	6400.00
integrity tests (minimum 20 per visit)	–	–	–	–	–	nr	16.00
600 mm dia. piles							
Provision of plant							
provision of all plant (1 nr rig); including bringing to and removing from site; maintenance, erection and dismantling at each pile position for 100 nr piles 20 m long	–	–	–	–	–	item	16300.00
Bored piles							
600 mm dia. piles; reinforced; 20 m long	–	–	–	–	–	nr	1430.00
add for additional piles length	–	–	–	–	–	m	73.06
deduct for reduction in pile length	–	–	–	–	–	m	31.48
Delays							
rig standing time	–	–	–	–	–	hour	438.34
Pile tests							
working to 1000 kN; using tension piles as reaction; first pile	–	–	–	–	–	nr	7900.00
working to 1000 kN; using tension piles as reaction; subsequent piles	–	–	–	–	–	nr	7900.00
integrity test 600 mm dia. piles; min 20 per visit	–	–	–	–	–	nr	16.00

07 PILING

Item	PC £	Labour hours	Labour £	Plant £	Material £	Unit	Total rate £
7.02 BORED PILES – cont							
750 mm dia. piles							
Provision of plant							
provision of all plant (1 nr rig); including bringing to and removing from site; maintenance, erection and dismantling at each pile position for 100 nr piles 20 m long	–	–	–	–	–	item	19000.00
Bored piles							
750 mm dia. piles; reinforced; 20 m long	–	–	–	–	–	nr	2250.00
add for additional piles length	–	–	–	–	–	m	112.39
deduct for reduction in pile length	–	–	–	–	–	m	44.95
Delays							
rig standing time	–	–	–	–	–	hour	528.25
Pile tests							
working to 1500 kN; using tension piles as reaction; first pile	–	–	–	–	–	nr	9660.00
working to 1500 kN; using tension piles as reaction; subsequent piles	–	–	–	–	–	nr	9660.00
integrity test 600 mm dia. piles; min 20 per visit	–	–	–	–	–	nr	16.00
900 mm dia. piles							
Provision of plant							
provision of all plant (1 nr rig); including bringing to and removing from site; maintenance, erection and dismantling at each pile position for 100 nr piles 20 m long	–	–	–	–	–	item	19000.00
Bored piles							
900 mm dia. piles; reinforced; 20 m long	–	–	–	–	–	nr	3225.00
add for additional piles length	–	–	–	–	–	m	162.97
deduct for reduction in pile length	–	–	–	–	–	m	67.44
Delays							
rig standing time	–	–	–	–	–	hour	528.25
Pile tests							
working to 2000 kN; using tension piles as reaction; first pile	–	–	–	–	–	nr	12500.00
working to 2000 kN; using tension piles as reaction; subsequent piles	–	–	–	–	–	nr	12500.00
integrity test 900 mm dia. piles; min 20 per visit	–	–	–	–	–	nr	16.00
Pile tops							
Cut off and prepare reinforcement to receive reinforced concrete pile cap; not including disposal							
450 mm dia. piles	–	1.00	24.39	–	–	m	24.39
600 mm dia. piles	–	1.20	29.26	–	–	m	29.26
750 mm dia. piles	–	1.50	36.58	–	–	m	36.58
900 mm dia. piles	–	1.80	43.90	–	–	m	43.90

08 UNDERPINNING

Item	PC £	Labour hours	Labour £	Plant £	Material £	Unit	Total rate £
8.01 UNDERPINNING							
Excavating; by machine							
Preliminary trenches							
maximum depth not exceeding 1.00 m	–	0.23	3.24	6.64	–	m³	9.88
maximum depth not exceeding 2.00 m	–	0.28	3.94	8.00	–	m³	11.94
maximum depth not exceeding 4.00 m	–	0.32	4.51	9.36	–	m³	13.87
extra over preliminary trench excavating for breaking out existing hard pavings, 150 mm thick concrete	–	0.65	9.16	2.07	–	m²	11.23
Excavating; by hand							
Preliminary trenches							
maximum depth not exceeding 1.00 m	–	2.68	37.77	–	–	m³	37.77
maximum depth not exceeding 2.00 m	–	3.05	42.98	–	–	m³	42.98
maximum depth not exceeding 4.00 m	–	3.93	55.38	–	–	m³	55.38
extra over preliminary trench excavating for breaking out existing hard pavings, 150 mm thick concrete	–	2.00	28.18	2.84	–	m²	31.02
Underpinning pits; commencing from 1.00 m below existing ground level							
maximum depth not exceeding 0.25 m	–	4.07	57.35	–	–	m³	57.35
maximum depth not exceeding 1.00 m	–	4.44	62.57	–	–	m³	62.57
maximum depth not exceeding 2.00 m	–	5.32	74.97	–	–	m³	74.97
Underpinning pits; commencing from 2.00 m below existing ground level							
maximum depth not exceeding 0.25 m	–	5.00	70.46	–	–	m³	70.46
maximum depth not exceeding 1.00 m	–	5.37	75.67	–	–	m³	75.67
maximum depth not exceeding 2.00 m	–	6.24	87.93	–	–	m³	87.93
Underpinning pits; commencing from 4.00 m below existing ground level							
maximum depth not exceeding 0.25 m	–	5.92	83.42	–	–	m³	83.42
maximum depth not exceeding 1.00 m	–	6.29	88.64	–	–	m³	88.64
maximum depth not exceeding 2.00 m	–	7.17	101.04	–	–	m³	101.04
Extra over any types of excavating irrespective of depth							
excavating below ground water level	–	0.32	4.51	4.30	–	m³	8.81
Cutting away existing projecting foundations							
Concrete							
maximum width 150 mm; maximum depth 150 mm	–	0.15	2.11	0.20	–	m	2.31
maximum width 150 mm; maximum depth 225 mm	–	0.22	3.10	0.30	–	m	3.40
maximum width 150 mm; maximum depth 300 mm	–	0.30	4.22	0.40	–	m	4.62
maximum width 300 mm; maximum depth 300 mm	–	0.58	8.18	0.78	–	m	8.96
Masonry							
maximum width one brick thick; maximum depth one course high	–	0.04	0.57	0.07	–	m	0.64
maximum width one brick thick; maximum depth two courses high	–	0.13	1.83	0.18	–	m	2.01
maximum width one brick thick; maximum depth three courses high	–	0.25	3.52	0.34	–	m	3.86
maximum width one brick thick; maximum depth four courses high	–	0.42	5.92	0.56	–	m	6.48

Prices for Measured Works

08 UNDERPINNING

Item	PC £	Labour hours	Labour £	Plant £	Material £	Unit	Total rate £
8.01 UNDERPINNING – cont							
Preparing the underside of existing work to receive the pinning up of the new work							
Width of existing work							
380 mm wide	–	0.56	7.89	–	–	m	**7.89**
600 mm wide	–	0.74	10.42	–	–	m	**10.42**
900 mm wide	–	0.93	13.10	–	–	m	**13.10**
1200 mm wide	–	1.11	15.65	–	–	m	**15.65**
Disposal							
Excavated material							
off site; to tip not exceeding 13 km (using lorries); including Landfill Tax based on inactive waste	–	0.75	10.57	14.62	20.83	m³	**46.02**
Filling to excavations; by hand							
Average thickness exceeding 0.25 m							
arising from the excavations	–	0.93	13.10	–	–	m³	**13.10**
Surface treatments							
Compacting							
bottoms of excavations	–	0.04	0.57	0.04	–	m²	**0.61**
Wedging and pinning							
To underside of existing construction with slates in cement mortar (1:3)							
width of wall – half brick thick	–	1.02	22.37	–	10.96	m	**33.33**
width of wall – one brick thick	–	1.20	26.32	–	21.91	m	**48.23**
width of wall – one and a half brick thick	–	1.39	30.49	–	32.87	m	**63.36**
8.02 CONCRETE							
Plain in situ ready mixed designated concrete; C10–40 mm aggregate; poured against faces of excavation							
Underpinning							
thickness not exceeding 150 mm	–	3.42	56.51	–	111.14	m³	**167.65**
thickness 150–450 mm	–	2.87	47.42	–	111.14	m³	**158.56**
thickness exceeding 450 mm	–	2.50	41.31	–	111.14	m³	**152.45**
Plain in situ ready mixed designated concrete; C20–20 mm aggregate; poured against faces of excavation							
Underpinning							
thickness not exceeding 150 mm	–	3.42	56.51	–	115.20	m³	**171.71**
thickness 150–450 mm	–	2.87	47.42	–	115.20	m³	**162.62**
thickness exceeding 450 mm	–	2.50	41.31	–	115.20	m³	**156.51**
extra for working around reinforcement	–	0.28	4.62	–	–	m³	**4.62**

08 UNDERPINNING

Item	PC £	Labour hours	Labour £	Plant £	Material £	Unit	Total rate £
Sawn formwork; sides of foundations in underpinning							
Plain vertical							
height exceeding 1.00 m	–	1.48	28.94	–	–	m²	**28.94**
height not exceeding 250 mm	–	0.51	9.97	–	–	m²	**9.97**
height 250–500 mm	–	0.79	15.45	–	–	m²	**15.45**
height 500 mm–1.00 m	–	1.20	23.47	–	–	m²	**23.47**
8.03 FORMWORK							
Earthwork support to preliminary trenches (open boarded – in 3.00 m lengths)							
Maximum depth not exceeding 1.00 m							
distance between opposing faces not exceeding 2.00 m	–	0.37	5.21	–	–	m²	**5.21**
Maximum depth not exceeding 2.00 m							
distance between opposing faces not exceeding 2.00 m	–	0.46	6.48	–	–	m²	**6.48**
Maximum depth not exceeding 4.00 m							
distance between opposing faces not exceeding 2.00 m	–	0.59	8.31	–	–	m²	**8.31**
Earthwork support to underpinning pits (open boarded – in 3.00 m lengths)							
Maximum depth not exceeding 1.00 m							
distance between opposing faces not exceeding 2.00 m	–	0.41	5.78	–	–	m²	**5.78**
Maximum depth not exceeding 2.00 m							
distance between opposing faces not exceeding 2.00 m	–	0.51	7.19	–	–	m²	**7.19**
Maximum depth not exceeding 4.00 m							
distance between opposing faces not exceeding 2.00 m	–	0.65	9.16	–	–	m²	**9.16**
Earthwork support to preliminary trenches (closed boarded – in 3.00 m lengths)							
Maximum depth not exceeding 1.00 m							
1.00 m deep	–	0.93	13.10	–	–	m²	**13.10**
Maximum depth not exceeding 2.00 m							
distance between opposing faces not exceeding 2.00 m	–	1.16	16.35	–	–	m²	**16.35**
Maximum depth not exceeding 4.00 m							
distance between opposing faces not exceeding 2.00 m	–	1.43	20.15	–	–	m²	**20.15**

08 UNDERPINNING

Item	PC £	Labour hours	Labour £	Plant £	Material £	Unit	Total rate £
8.03 FORMWORK – cont							
Earthwork support to underpinning pits (closed boarded – in 3.00 m lengths)							
Maximum depth not exceeding 1.00 m							
distance between opposing faces not exceeding 2.00 m	–	1.02	14.38	–	–	m²	**14.38**
Maximum depth not exceeding 2.00 m							
distance between opposing faces not exceeding 2.00 m	–	1.28	18.04	–	–	m²	**18.04**
Maximum depth not exceeding 4.00 m							
distance between opposing faces not exceeding 2.00 m	–	1.57	22.12	–	–	m²	**22.12**
Extra over earthwork support for							
Left in	–	0.69	9.72	–	–	m²	**9.72**
8.04 REINFORCEMENT							
Reinforcement bar; hot rolled steel bars							
20 mm dia. nominal size							
straight	594.00	22.00	451.91	–	805.09	tonne	**1257.00**
bent	649.00	28.00	575.15	–	863.16	tonne	**1438.31**
16 mm dia. nominal size							
straight	594.00	28.00	575.15	–	827.34	tonne	**1402.49**
bent	649.00	36.00	739.49	–	885.41	tonne	**1624.90**
12 mm dia. nominal size							
straight	594.00	40.00	821.65	–	849.59	tonne	**1671.24**
bent	649.00	50.00	1027.06	–	907.66	tonne	**1934.72**
10 mm dia. nominal size							
straight	594.00	48.00	985.98	–	875.39	tonne	**1861.37**
bent	649.00	62.00	1273.56	–	933.45	tonne	**2207.01**
8 mm dia. nominal size							
straight	594.00	58.00	1191.39	–	882.34	tonne	**2073.73**
bent	649.00	74.00	1520.05	–	955.70	tonne	**2475.75**
8.05 BRICKWORK OR BLOCKWORK							
Common bricks; in cement mortar (1:3)							
Walls in underpinning							
one brick thick (PC £ per 1000)	380.00	2.22	48.69	–	58.52	m²	**107.21**
one and a half brick thick	–	3.05	66.89	–	87.01	m²	**153.90**
two brick thick	–	3.79	83.12	–	121.86	m²	**204.98**
Class A engineering bricks; in cement mortar (1:3)							
Walls in underpinning							
one brick thick (PC £ per 1000)	513.70	2.22	48.69	–	76.29	m²	**124.98**
one and a half brick thick	–	3.05	66.89	–	113.66	m²	**180.55**
two brick thick	–	3.79	83.12	–	157.38	m²	**240.50**

08 UNDERPINNING

Item	PC £	Labour hours	Labour £	Plant £	Material £	Unit	Total rate £
Class B engineering bricks; in cement mortar (1:3)							
Walls in underpinning							
one brick thick (PC £ per 1000)	438.10	2.22	48.69	–	66.24	m²	**114.93**
one and a half brick thick	–	3.05	66.89	–	98.59	m²	**165.48**
two brick thick	–	3.79	83.12	–	137.29	m²	**220.41**
Add or deduct for variation of £10.00/1000 in PC of bricks							
one brick thick	–	–	–	–	1.24	m²	**1.24**
one and a half bricks thick	–	–	–	–	1.85	m²	**1.85**
two bricks thick	–	–	–	–	2.47	m²	**2.47**
8.06 TANKING							
Damp-proof course							
Zedex CPT (Co-Polymer Thermoplastic) damp-proof course or other equal; 200 mm laps; in gauged mortar (1:1:6)							
Horizontal							
width exceeding 225 mm	–	0.23	5.05	–	7.17	m²	**12.22**
width not exceeding 225 mm	–	0.46	10.08	–	7.17	m²	**17.25**
Hyload (pitch polymer) damp-proof course or similar; 150 mm laps; in cement mortar (1:3)							
Horizontal							
width exceeding 225 mm	–	0.23	5.05	–	5.44	m²	**10.49**
width not exceeding 225 mm	–	0.46	10.08	–	5.56	m²	**15.64**
Alumite aluminium cored bitumen gas retardant damp-proof course or other equal and approved; 200 mm laps; in gauged mortar (1:1;6)							
Horizontal							
width exceeding 225 mm	–	0.31	6.80	–	7.56	m²	**14.36**
width not exceeding 225 mm	–	0.60	13.16	–	7.56	m²	**20.72**
Two courses of slates in cement mortar (1:3)							
Horizontal							
width exceeding 225 mm	–	1.39	30.49	–	44.38	m²	**74.87**
width not exceeding 225 mm	–	2.31	50.67	–	45.36	m²	**96.03**

09 DIAPHRAGM WALLS AND EMBEDDED RETAINING WALLS

Item	PC £	Labour hours	Labour £	Plant £	Material £	Unit	Total rate £
9.01 EMBEDDED RETAINING WALLS							
Diaphragm walls; contiguous panel construction; panel lengths not exceeding 5 m							
Plant							
provision of all plant; including bringing to and removing from site; maintenance, erection and dismantling; assuming one rig for 1000 m² of walling	–	–	–	–	–	item	270000.00
Excavation for diaphragm wall; excavated material removed from site; Bentonite slurry supplied and disposed of							
600 mm thick walls	–	–	–	–	–	m³	760.73
1000 mm thick walls	–	–	–	–	–	m³	543.38
Concrete							
ready mixed reinforced in situ concrete; normal portland cement; C30 in walls	–	–	–	–	–	m³	163.01
Reinforcement bar; BS 4449 cold rolled deformed square high yield steel bars; straight or bent							
25 mm–40 mm dia.	–	–	–	–	–	tonne	1195.42
20 mm dia.	–	–	–	–	–	tonne	1195.42
16 mm dia.	–	–	–	–	–	tonne	1195.42
Formwork							
formwork 75 mm thick to form chases	–	–	–	–	–	m²	97.81
construct twin guide walls in reinforced concrete; together with reinforcement and formwork along the axis of the diaphragm wall	–	–	–	–	–	m	1032.41
Delays							
rig standing	–	–	–	–	–	hour	2250.00

10 CRIB WALLS, GABIONS AND REINFORCED EARTH

Item	PC £	Labour hours	Labour £	Plant £	Material £	Unit	Total rate £
10.01 GABION BASKET WALLS							
Gabion baskets							
Wire mesh gabion baskets; Maccaferri Ltd or other equal and approved; galvanized mesh							
80 mm × 100 mm; filling with broken stones 125 mm–200 mm size							
2.00 m × 1.00 m × 0.50 m	21.10	1.00	24.39	12.94	81.91	nr	**119.24**
2.00 m × 1.00 m × 0.50 m PVC coated	26.68	1.00	24.39	12.94	88.49	nr	**125.82**
2.00 m × 1.00 m × 1.00 m	29.57	2.00	48.77	25.87	150.47	nr	**225.11**
2.00 m × 1.00 m × 1.00 m PVC coated	37.42	2.00	48.77	25.87	158.77	nr	**233.41**
Reno mattress gabion baskets or other equal and approved; Maccaferri Ltd; filling with broken stones 125 mm–200 mm size							
6.00 m × 2.00 m × 0.17 m	94.37	2.00	48.77	19.41	221.29	nr	**289.47**
6.00 m × 2.00 m × 0.23 m	102.16	2.50	60.97	25.87	272.43	nr	**359.27**
6.00 m × 2.00 m × 0.30 m	119.83	3.00	73.16	29.11	341.19	nr	**443.46**

11 IN SITU CONCRETE WORKS

Item	PC £	Labour hours	Labour £	Plant £	Material £	Unit	Total rate £
11.01 MASS CONCRETE							
Cement material prices							
Cements							
Ordinary Portland	–	–	–	–	132.64	tonne	**132.64**
high alumina	–	–	–	–	599.52	tonne	**599.52**
Sulfacrete sulphate-resisting	–	–	–	–	162.35	tonne	**162.35**
Ferrocrete rapid hardening	–	–	–	–	308.78	tonne	**308.78**
Snowcrete white cement	–	–	–	–	211.16	tonne	**211.16**
Cement admixtures							
Febtone colorant – red, marigold, yellow, brown, black	–	–	–	–	4.57	kg	**4.57**
Febproof waterproof	–	–	–	–	1.04	litre	**1.04**
Febond PVA bonding agent	–	–	–	–	2.49	litre	**2.49**
Febspeed frostproofer and hardener	–	–	–	–	1.38	litre	**1.38**
Basic mixed concrete prices							
NOTE: The following prices are for concrete ready for placing excluding any allowance for waste, discount or overheads and profit. Prices are based upon delivery to site within a 10 mile (16 km) radius of concrete mixing plant, using full loads							
GEN 0 (7.5N)	–	–	–	–	93.01	m^3	**93.01**
GEN 1 (10N)	–	–	–	–	93.99	m^3	**93.99**
GEN 2 (15N)	–	–	–	–	95.94	m^3	**95.94**
GEN 3 (20N)	–	–	–	–	97.42	m^3	**97.42**
RC20/25	–	–	–	–	102.31	m^3	**102.31**
RC25/30	–	–	–	–	103.77	m^3	**103.77**
RC30/37	–	–	–	–	105.24	m^3	**105.24**
RC35/45	–	–	–	–	110.13	m^3	**110.13**
RC40/50	–	–	–	–	111.12	m^3	**111.12**
RC50/60	–	–	–	–	113.07	m^3	**113.07**
FND3	–	–	–	–	110.13	m^3	**110.13**
FND4	–	–	–	–	112.09	m^3	**112.09**
Caltite waterprrof system to RC 35 concrete	–	–	–	–	190.86	m^3	**190.86**
Site mixed concrete (on site batching plant)							
mix 7.50 N/mm^2; cement (1:8)	–	–	–	–	102.74	m^3	**102.74**
mix 10.00 N/mm^2; cement (1:8)	–	–	–	–	107.16	m^3	**107.16**
mix 20.00 N/mm^2; cement (1:2:4)	–	–	–	–	123.97	m^3	**123.97**
mix 25.00 N/mm^2; cement to (1:1:5:3)	–	–	–	–	127.12	m^3	**127.12**
Add for							
rapid-hardening cement	–	–	–	–	12.08	m^3	**12.08**
polypropylene fibre additive	–	–	–	–	6.72	m^3	**6.72**
air entrained concrete	–	–	–	–	5.90	m^3	**5.90**
water repellent additive	–	–	–	–	6.30	m^3	**6.30**
foamed concrete	–	–	–	–	–	%	**25.00**
distance per mile in excess of 10 miles (16 km)	–	–	–	–	0.71	m^3	**0.71**

11 IN SITU CONCRETE WORKS

Item	PC £	Labour hours	Labour £	Plant £	Material £	Unit	Total rate £
11.02 HORIZONTAL WORK							
SUPPLY AND LAY PRICES							
NOTE: The following concrete material prices include an allowance for shrinkage and waste. PC Sums are concrete supply only prices							
Plain in situ ready mixed designated concrete; C7.5–40 mm aggregate							
Foundations	90.30	1.02	16.93	–	97.65	m³	**114.58**
Isolated foundations	90.30	1.13	18.59	–	97.65	m³	**116.24**
Beds							
thickness not exceeding 300 mm	90.30	1.02	16.93	–	97.65	m³	**114.58**
thickness exceeding 300 mm	90.30	0.93	15.37	–	97.65	m³	**113.02**
Screeded beds; protection to compressible formwork exceeding 600 mm wide							
50 mm thick	90.30	0.10	1.65	–	4.88	m²	**6.53**
75 mm thick	90.30	0.15	2.48	–	7.33	m²	**9.81**
100 mm thick	90.30	0.20	3.31	–	9.76	m²	**13.07**
Plain in situ ready mixed designated concrete; C10–40 mm aggregate							
Foundations	91.25	1.02	16.93	–	98.68	m³	**115.61**
Isolated foundations	91.25	1.13	18.63	–	98.68	m³	**117.31**
Beds							
thickness not exceeding 300 mm	91.25	1.02	16.93	–	98.68	m³	**115.61**
thickness exceeding 300 mm	91.25	0.93	15.29	–	98.68	m³	**113.97**
Plain in situ ready mixed designated concrete; C10–40 mm aggregate; poured on or against earth or unblinded hardcore							
Foundations	91.25	1.02	16.93	–	100.66	m³	**117.59**
Isolated foundations	91.25	1.13	18.63	–	100.66	m³	**119.29**
Beds							
thickness not exceeding 300 mm	91.25	1.02	16.93	–	100.66	m³	**117.59**
thickness exceeding 300 mm	91.25	0.93	15.29	–	100.66	m³	**115.95**
Plain in situ ready mixed designated concrete; C20–20 mm aggregate							
Foundations	94.58	0.98	16.20	–	102.29	m³	**118.49**
Isolated foundations	94.58	1.13	18.63	–	102.29	m³	**120.92**
Beds							
thickness not exceeding 300 mm	94.58	1.02	16.85	–	102.29	m³	**119.14**
thickness exceeding 300 mm	94.58	0.93	15.37	–	102.29	m³	**117.66**
Plain in situ ready mixed; 20N; poured on or against earth or unblinded hardcore							
Foundations	94.58	1.02	16.93	–	104.34	m³	**121.27**
Isolated foundations	94.58	1.13	18.63	–	104.34	m³	**122.97**

11 IN SITU CONCRETE WORKS

Item	PC £	Labour hours	Labour £	Plant £	Material £	Unit	Total rate £
11.02 HORIZONTAL WORK – cont							
Plain in situ ready mixed – cont							
Beds							
thickness not exceeding 300 mm	94.58	1.02	16.93	–	104.34	m³	**121.27**
thickness exceeding 300 mm	94.58	0.93	15.37	–	104.34	m³	**119.71**
Reinforced in situ ready mixed designated concrete; 25N							
Foundations	100.75	1.02	16.93	–	114.41	m³	**131.34**
Ground beams	100.75	2.59	42.80	–	108.96	m³	**151.76**
Isolated foundations	100.75	1.13	18.59	–	114.41	m³	**133.00**
Beds							
thickness not exceeding 300 mm	100.75	1.02	16.85	–	108.96	m³	**125.81**
thickness exceeding 300 mm	100.75	0.93	15.37	–	108.96	m³	**124.33**
Slabs							
thickness not exceeding 300 mm	100.75	1.25	20.65	–	108.96	m³	**129.61**
thickness exceeding 300 mm	100.75	1.15	19.00	–	108.96	m³	**127.96**
Coffered and troughed slabs							
thickness not exceeding 300 mm	100.75	2.96	48.90	–	108.96	m³	**157.86**
thickness exceeding 300 mm	100.75	2.59	42.80	–	108.96	m³	**151.76**
Extra over for sloping work							
not exceeding 15°	–	0.23	3.80	–	–	m³	**3.80**
exceeding 15°	–	0.46	7.60	–	–	m³	**7.60**
Beams							
isolated	100.75	3.70	61.14	–	108.96	m³	**170.10**
isolated deep	100.75	4.07	67.25	–	108.96	m³	**176.21**
attached deep	100.75	3.70	61.14	–	108.96	m³	**170.10**
Beam casings							
isolated	100.75	4.07	67.25	–	108.96	m³	**176.21**
isolated deep	100.75	4.44	73.37	–	108.96	m³	**182.33**
attached deep	100.75	4.07	67.25	–	108.96	m³	**176.21**
Staircases	100.75	5.25	86.75	–	108.96	m³	**195.71**
Upstands	100.75	3.30	54.53	–	108.96	m³	**163.49**
Reinforced in situ ready mixed designated concrete; 30N							
Foundations	102.17	1.02	16.93	–	110.50	m³	**127.43**
Ground beams	102.17	2.59	42.80	–	110.50	m³	**153.30**
Isolated foundations	102.17	1.13	18.67	–	110.50	m³	**129.17**
Beds							
thickness not exceeding 300 mm	102.17	1.02	16.85	–	110.50	m³	**127.35**
thickness exceeding 300 mm	102.17	0.95	15.70	–	110.50	m³	**126.20**
Slabs							
thickness not exceeding 300 mm	102.17	1.25	20.65	–	110.50	m³	**131.15**
thickness exceeding 300 mm	102.17	1.15	19.00	–	110.50	m³	**129.50**
Coffered and troughed slabs							
thickness not exceeding 300 mm	102.17	2.96	48.90	–	110.50	m³	**159.40**
thickness exceeding 300 mm	102.17	2.59	42.80	–	110.50	m³	**153.30**

11 IN SITU CONCRETE WORKS

Item	PC £	Labour hours	Labour £	Plant £	Material £	Unit	Total rate £
Extra over for sloping							
not exceeding 15°	–	0.23	3.80	–	–	m³	**3.80**
exceeding 15°	–	0.46	7.60	–	–	m³	**7.60**
Beams							
isolated	102.17	3.70	61.14	–	110.50	m³	**171.64**
isolated deep	102.17	4.07	67.25	–	110.50	m³	**177.75**
attached deep	102.17	3.70	61.14	–	110.50	m³	**171.64**
Beam casings							
isolated	102.17	4.07	67.25	–	110.50	m³	**177.75**
isolated deep	102.17	4.44	73.37	–	110.50	m³	**183.87**
attached deep	102.17	4.07	67.25	–	110.50	m³	**177.75**
Staircases	102.17	5.55	91.70	–	110.50	m³	**202.20**
Upstands	102.17	3.56	58.82	–	110.50	m³	**169.32**
Reinforced in situ ready mixed designated concrete; 35N							
Foundations	106.92	1.02	16.93	–	115.64	m³	**132.57**
Ground beams	106.92	2.59	42.80	–	115.64	m³	**158.44**
Isolated foundations	106.92	1.13	18.67	–	115.64	m³	**134.31**
Beds							
thickness not exceeding 300 mm	106.92	1.02	16.85	–	115.64	m³	**132.49**
thickness exceeding 300 mm	106.92	0.95	15.70	–	115.64	m³	**131.34**
Slabs							
thickness not exceeding 300 mm	106.92	1.25	20.65	–	115.64	m³	**136.29**
thickness exceeding 300 mm	106.92	1.15	19.00	–	115.64	m³	**134.64**
Coffered and troughed slabs							
thickness not exceeding 300 mm	106.92	2.96	48.90	–	115.64	m³	**164.54**
thickness exceeding 300 mm	106.92	2.59	42.80	–	115.64	m³	**158.44**
Extra over for sloping							
not exceeding 15°	–	0.23	3.80	–	–	m³	**3.80**
exceeding 15°	–	0.46	7.60	–	–	m³	**7.60**
Beams							
isolated	106.92	3.70	61.14	–	115.64	m³	**176.78**
isolated deep	106.92	4.07	67.25	–	115.64	m³	**182.89**
attached deep	106.92	3.70	61.14	–	115.64	m³	**176.78**
Beam casings							
isolated	106.92	4.07	67.25	–	115.64	m³	**182.89**
isolated deep	106.92	4.44	73.37	–	115.64	m³	**189.01**
attached deep	106.92	4.07	67.25	–	115.64	m³	**182.89**
Staircases	106.92	5.55	91.70	–	115.64	m³	**207.34**
Upstands	106.92	3.56	58.82	–	115.64	m³	**174.46**
Reinforced in situ ready mixed designated concrete; 40N							
Foundations	107.88	1.02	16.93	–	116.67	m³	**133.60**
Isolated foundations	107.88	1.13	18.63	–	116.67	m³	**135.30**
Ground beams	107.88	2.59	42.80	–	116.67	m³	**159.47**
Beds							
thickness not exceeding 300 mm	107.88	0.95	15.70	–	116.67	m³	**132.37**
thickness exceeding 300 mm	107.88	0.93	15.37	–	116.67	m³	**132.04**

11 IN SITU CONCRETE WORKS

Item	PC £	Labour hours	Labour £	Plant £	Material £	Unit	Total rate £
11.02 HORIZONTAL WORK – cont							
Reinforced in situ ready mixed designated concrete – cont							
Slabs							
thickness not exceeding 300 mm	107.88	1.20	19.83	–	116.67	m³	**136.50**
thickness exceeding 300 mm	107.88	1.15	19.00	–	116.67	m³	**135.67**
Coffered and troughed slabs							
thickness not exceeding 300 mm	107.88	2.96	48.90	–	116.67	m³	**165.57**
thickness exceeding 300 mm	107.88	2.59	42.80	–	116.67	m³	**159.47**
Extra over for sloping							
not exceeding 15°	–	0.23	3.80	–	–	m³	**3.80**
exceeding 15°	–	0.46	7.60	–	–	m³	**7.60**
Beams							
isolated	107.88	3.70	61.14	–	116.67	m³	**177.81**
isolated deep	107.88	4.07	67.25	–	116.67	m³	**183.92**
attached deep	107.88	3.70	61.14	–	116.67	m³	**177.81**
Beam casings							
isolated	107.88	4.07	67.25	–	116.67	m³	**183.92**
isolated deep	107.88	4.44	73.37	–	116.67	m³	**190.04**
attached deep	107.88	4.07	67.25	–	116.67	m³	**183.92**
Staircases	107.88	5.55	91.70	–	116.67	m³	**208.37**
Upstands	107.88	3.56	58.82	–	116.67	m³	**175.49**
Proprietary voided Bubbledeck, Cobiax or other equal and approved slab; concrete mix RC35; to achieve design loadings of 5.0 kN/m² live and 3.0 kN/m² dead; with trowelled finish							
Beds							
360 mm overall thickness	–	–	–	–	–	m²	**140.56**
additional concrete up to 600 mm wide at edges where formers omitted at junctions with walls etc.	–	–	–	–	–	m	**57.51**
11.03 VERTICAL WORK							
SUPPLY AND LAY PRICES							
NOTE: The following concrete material prices include an allowance for shrinkage and waste. PC Sums are concrete supply only prices							
Plain in situ ready mixed designated concrete; C7.5–40 mm aggregate							
Filling hollow walls							
thickness not exceeding 150 mm	90.30	3.15	52.05	–	97.65	m³	**149.70**
Column casings							
stub columns beneath suspended ground slabs	90.30	4.50	74.36	–	97.65	m³	**172.01**

11 IN SITU CONCRETE WORKS

Item	PC £	Labour hours	Labour £	Plant £	Material £	Unit	Total rate £
Plain in situ ready mixed designated concrete; C10–40 mm aggregate							
Filling hollow walls							
thickness not exceeding 300 mm	91.25	3.15	52.05	–	98.68	m³	**150.73**
Plain in situ ready mixed designated concrete; C20–20 mm aggregate							
Filling hollow walls							
thickness not exceeding 300 mm	94.58	3.00	49.57	–	102.29	m³	**151.86**
Reinforced in situ ready mixed designated concrete; 25N							
Walls							
thickness not exceeding 300 mm	100.75	3.15	52.05	–	108.96	m³	**161.01**
thickness exceeding 300 mm	100.75	2.40	39.74	–	108.96	m³	**148.70**
Columns	100.75	4.20	69.40	–	108.96	m³	**178.36**
Column casings	100.75	4.90	80.97	–	108.96	m³	**189.93**
Staircases	100.75	5.25	86.75	–	108.96	m³	**195.71**
Upstands	100.75	3.30	54.53	–	108.96	m³	**163.49**
Reinforced in situ ready mixed designated concrete; 30N							
Walls							
thickness not exceeding 300 mm	102.17	2.67	44.20	–	110.50	m³	**154.70**
thickness exceeding 300 mm	102.17	2.41	39.77	–	110.50	m³	**150.27**
Columns	102.17	4.44	73.37	–	110.50	m³	**183.87**
Column casings	102.17	4.90	80.97	–	110.50	m³	**191.47**
Staircases	102.17	5.55	91.70	–	110.50	m³	**202.20**
Upstands	102.17	3.56	58.82	–	110.50	m³	**169.32**
Reinforced in situ ready mixed designated concrete; 35N							
Walls							
thickness not exceeding 300 mm	106.92	2.67	44.20	–	115.64	m³	**159.84**
thickness exceeding 300 mm	106.92	2.41	39.77	–	115.64	m³	**155.41**
Columns	106.92	4.44	73.37	–	115.64	m³	**189.01**
Column casings	106.92	4.90	80.97	–	115.64	m³	**196.61**
Staircases	106.92	5.55	91.70	–	115.64	m³	**207.34**
Upstands	106.92	3.56	58.82	–	115.64	m³	**174.46**
Reinforced in situ ready mixed designated concrete; 40N							
Walls							
thickness not exceeding 300 mm	107.88	2.73	45.11	–	116.67	m³	**161.78**
thickness exceeding 300 mm	107.88	2.41	39.82	–	116.67	m³	**156.49**
Columns	107.88	4.44	73.37	–	116.67	m³	**190.04**
Column casings	107.88	4.90	80.97	–	116.67	m³	**197.64**
Staircases	107.88	5.55	91.70	–	116.67	m³	**208.37**
Upstands	107.88	3.56	58.82	–	116.67	m³	**175.49**

11 IN SITU CONCRETE WORKS

Item	PC £	Labour hours	Labour £	Plant £	Material £	Unit	Total rate £
11.04 SURFACE FINISHES							
Worked finishes							
Tamping by mechanical means	–	0.02	0.33	0.11	–	m²	**0.44**
Power floating	–	0.16	2.65	0.35	–	m²	**3.00**
Trowelling	–	0.31	5.12	–	–	m²	**5.12**
Hacking							
by mechanical means	–	0.31	5.12	0.41	–	m²	**5.53**
by hand	–	0.65	10.74	–	–	m²	**10.74**
Lightly shot blasting surface of concrete	–	0.37	6.12	–	–	m²	**6.12**
Blasting surface of concrete to produce textured finish	–	0.65	10.74	0.87	–	m²	**11.61**
Sand blasting (blast and vac method)	–	–	–	–	–	m²	**43.82**
Wood float finish	–	0.12	1.99	–	–	m²	**1.99**
Tamped finish							
level or to falls	–	0.06	0.99	–	–	m²	**0.99**
to falls	–	0.09	1.48	–	–	m²	**1.48**
Spade finish	–	0.14	2.32	–	–	m²	**2.32**
Diamond drilling/cutting							
Cutting holes in block wall							
up to 52 mm dia.; up to 150 mm thick	–	–	–	–	–	nr	**17.57**
up to 52 mm dia.; 200 mm thick	–	–	–	–	–	nr	**21.89**
up to 52 mm dia.; 250 mm thick	–	–	–	–	–	nr	**26.22**
up to 52 mm dia.; 300 mm thick	–	–	–	–	–	nr	**30.54**
79 mm–107 mm dia.; up to 150 mm thick	–	–	–	–	–	nr	**27.56**
79 mm–107 mm dia.; 200 mm thick	–	–	–	–	–	nr	**32.87**
79 mm–107 mm dia.; 250 mm thick	–	–	–	–	–	nr	**38.18**
79 mm–107 mm dia.; 300 mm thick	–	–	–	–	–	nr	**43.50**
251 mm–300 mm dia.; ne 150 mm thick	–	–	–	–	–	nr	**67.99**
251 mm–300 mm dia.; 200 mm thick	–	–	–	–	–	nr	**86.13**
251 mm–300 mm dia.; 250 mm thick	–	–	–	–	–	nr	**104.26**
251 mm–300 mm dia.; 300 mm thick	–	–	–	–	–	nr	**122.37**
Cutting holes in brick wall							
up to 52 mm dia.; up to 150 mm thick	–	–	–	–	–	nr	**20.06**
up to 52 mm dia.; 200 mm thick	–	–	–	–	–	nr	**25.01**
up to 52 mm dia.; 250 mm thick	–	–	–	–	–	nr	**29.97**
up to 52 mm dia.; 300 mm thick	–	–	–	–	–	nr	**34.93**
79 mm–107 mm dia.; up to 150 mm thick	–	–	–	–	–	nr	**31.47**
79 mm–107 mm dia.; 200 mm thick	–	–	–	–	–	nr	**37.55**
79 mm–107 mm dia.; 250 mm thick	–	–	–	–	–	nr	**43.62**
79 mm–107 mm dia.; 300 mm thick	–	–	–	–	–	nr	**49.70**
251 mm–300 mm dia.; ne 150 mm thick	–	–	–	–	–	nr	**77.71**
251 mm–300 mm dia.; 200 mm thick	–	–	–	–	–	nr	**98.43**
251 mm–300 mm dia.; 250 mm thick	–	–	–	–	–	nr	**119.13**
251 mm–300 mm dia.; 300 mm thick	–	–	–	–	–	nr	**139.87**
Cutting holes in concrete wall							
up to 52 mm dia.; up to 150 mm thick	–	–	–	–	–	nr	**25.08**
up to 52 mm dia.; 200 mm thick	–	–	–	–	–	nr	**31.27**
up to 52 mm dia.; 250 mm thick	–	–	–	–	–	nr	**37.45**

11 IN SITU CONCRETE WORKS

Item	PC £	Labour hours	Labour £	Plant £	Material £	Unit	Total rate £
up to 52 mm dia.; 300 mm thick	–	–	–	–	–	nr	43.63
79 mm–107 mm dia.; up to150 mm thick	–	–	–	–	–	nr	39.35
79 mm–107 mm dia.; 200 mm thick	–	–	–	–	–	nr	46.94
79 mm–107 mm dia.; 250 mm thick	–	–	–	–	–	nr	54.53
79 mm–107 mm dia.; 300 mm thick	–	–	–	–	–	nr	62.14
251 mm–300 mm dia.; ne 150 mm thick	–	–	–	–	–	nr	97.13
251 mm–300 mm dia.; 200 mm thick	–	–	–	–	–	nr	123.02
251 mm–300 mm dia.; 250 mm thick	–	–	–	–	–	nr	148.93
251 mm–300 mm dia.; 300 mm thick	–	–	–	–	–	nr	174.80
Cutting holes in timber wall							
up to 52 mm dia.; up to 150 mm thick	–	–	–	–	–	nr	23.81
up to 52 mm dia.; 200 mm thick	–	–	–	–	–	nr	29.69
up to 52 mm dia.; 250 mm thick	–	–	–	–	–	nr	35.59
up to 52 mm dia.; 300 mm thick	–	–	–	–	–	nr	41.46
79 mm–107 mm dia.; up to 150 mm thick	–	–	–	–	–	nr	37.39
79 mm–107 mm dia.; 200 mm thick	–	–	–	–	–	nr	44.60
79 mm–107 mm dia.; 250 mm thick	–	–	–	–	–	nr	51.83
79 mm–107 mm dia.; 300 mm thick	–	–	–	–	–	nr	59.02
251 mm–300 mm dia.; ne 150 mm thick	–	–	–	–	–	nr	92.28
251 mm–300 mm dia.; 200 mm thick	–	–	–	–	–	nr	116.88
251 mm–300 mm dia.; 250 mm thick	–	–	–	–	–	nr	141.49
251 mm–300 mm dia.; 300 mm thick	–	–	–	–	–	nr	166.09
Chasing for concealed conduits							
up to 35 mm deep, up to 50 mm wide in block wall	–	–	–	–	–	m	8.16
up to 35 mm deep, up to 50 mm wide in brick wall	–	–	–	–	–	m	9.40
up to 35 mm deep, up to 50 mm wide in concrete wall	–	–	–	–	–	m	31.35
up to 35 mm deep, up to 50 mm wide in concrete floor	–	–	–	–	–	m	37.62
up to 35 mm deep, 75 mm–100 mm wide in block wall	–	–	–	–	–	m	12.74
up to 35 mm deep, 75 mm–100 mm wide in brick wall	–	–	–	–	–	m	14.69
up to 35 mm deep, 75 mm–100 mm wide in concrete wall	–	–	–	–	–	m	48.99
up to 35 mm deep, 75 mm–100 mm wide in concrete floor	–	–	–	–	–	m	58.78
recess 100 mm × 100 mm × 35 mm deep in block wall	–	–	–	–	–	nr	4.37
recess 100 mm × 100 mm × 35 mm deep in brick wall	–	–	–	–	–	nr	5.07
recess 100 mm × 100 mm × 35 mm deep in concrete wall	–	–	–	–	–	nr	20.26
recess 100 mm × 100 mm × 35 mm deep in concrete floor	–	–	–	–	–	nr	16.87
Floor sawing							
diamond floor sawing; 25 mm depth	–	–	–	–	–	m	6.27
Ring sawing							
up to 25 mm deep in block wall	–	–	–	–	–	m	8.16
up to 25 mm deep in brick wall	–	–	–	–	–	m	9.40
up to 25 mm deep in concrete floor	–	–	–	–	–	m	12.54
up to 25 mm deep in concrete wall	–	–	–	–	–	m	15.05

11 IN SITU CONCRETE WORKS

Item	PC £	Labour hours	Labour £	Plant £	Material £	Unit	Total rate £
11.05 FORMWORK							
Sides of foundations; 18 mm thick external quality plywood; basic finish							
Plain vertical							
height not exceeding 250 mm	–	0.38	7.40	–	4.61	m	**12.01**
height not exceeding 250 mm; left in	–	0.38	7.40	–	7.10	m	**14.50**
height 250–500 mm	–	0.71	13.91	–	10.59	m	**24.50**
height 250–500 mm; left in	–	0.62	12.14	–	17.98	m	**30.12**
height 500 mm–1.00 m	–	1.00	19.54	–	14.30	m	**33.84**
height 500 mm–1.00 m; left in	–	0.95	18.65	–	25.30	m	**43.95**
height exceeding 1.00 m	–	1.33	26.05	–	13.81	m²	**39.86**
height exceeding 1.00 m; left in	–	1.17	22.89	–	24.33	m²	**47.22**
Sides of foundations; polystyrene sheet formwork; Cordek Claymaster or equal							
50 mm thick; plain vertical							
height not exceeding 250 mm; left in	–	0.09	1.76	–	1.49	m	**3.25**
height 250–500 mm; left in	–	0.14	2.81	–	2.99	m	**5.80**
height 500 mm–1.00 m; left in	–	0.22	4.22	–	5.97	m	**10.19**
height exceeding 1.00 m; left in	–	0.27	5.28	–	5.70	m²	**10.98**
75 mm thick; plain vertical							
height not exceeding 250 mm; left in	–	0.09	1.76	–	2.35	m	**4.11**
height 250–500 mm; left in	–	0.14	2.81	–	4.70	m	**7.51**
height 500 mm–1.00 m; left in	–	0.22	4.22	–	9.85	m	**14.07**
height exceeding 1.00 m; left in	–	0.27	5.28	–	9.39	m²	**14.67**
100 mm thick; plain vertical							
height not exceeding 250 mm; left in	–	0.09	1.76	–	3.28	m	**5.04**
height 250–500 mm; left in	–	0.14	2.81	–	6.55	m	**9.36**
height 500 mm–1.00 m; left in	–	0.22	4.22	–	13.10	m	**17.32**
height exceeding 1.00 m; left in	–	0.27	5.28	–	12.50	m²	**17.78**
150 mm thick; plain vertical							
height not exceeding 250 mm; left in	–	0.09	1.76	–	4.77	m	**6.53**
height 250–500 mm; left in	–	0.14	2.81	–	9.54	m	**12.35**
height 500 mm–1.00 m; left in	–	0.22	4.22	–	19.08	m	**23.30**
height exceeding 1.00 m; left in	–	0.28	5.38	–	18.21	m²	**23.59**
175 mm thick; plain vertical							
height not exceeding 250 mm; left in	–	0.10	1.96	–	5.74	m	**7.70**
height 250–500 mm; left in	–	0.15	2.94	–	11.47	m	**14.41**
height 500 mm–1.00 m; left in	–	0.22	4.31	–	22.94	m	**27.25**
height exceeding 1.00 m; left in	–	0.30	5.87	–	21.90	m²	**27.77**
200 mm thick; plain vertical							
height not exceeding 250 mm; left in	–	0.10	1.96	–	6.55	m	**8.51**
height 250–500 mm; left in	–	0.15	2.94	–	13.10	m	**16.04**
height 500 mm–1.00 m; left in	–	0.22	4.31	–	26.20	m	**30.51**
height exceeding 1.00 m; left in	–	0.30	5.87	–	25.01	m²	**30.88**
250 mm thick; plain vertical							
height not exceeding 250 mm; left in	–	0.15	2.94	–	8.24	m	**11.18**
height 250–500 mm; left in	–	0.17	3.42	–	16.48	m	**19.90**
height 500 mm–1.00 m; left in	–	0.25	4.89	–	32.96	m	**37.85**
height exceeding 1.00 m; left in	–	0.35	6.85	–	31.46	m²	**38.31**

11 IN SITU CONCRETE WORKS

Item	PC £	Labour hours	Labour £	Plant £	Material £	Unit	Total rate £
Combined heave pressure relief insulation and compressible board substructure formwork; Cordeck Cellcore CP or other equal and approved; butt joints; securely fixed in place							
Plain horizontal							
200 mm thick; beneath slabs; left in	–	0.54	10.56	–	25.70	m²	**36.26**
250 mm thick; beneath slabs; left in	–	0.58	11.44	–	28.59	m²	**40.03**
300 mm thick; beneath slabs; left in	–	0.63	12.32	–	30.81	m²	**43.13**
Dufaylite Clayboard void former; butt joints							
Residential (was KN30); compressive strength of 30 kN/m²; board thickness; typically for residential							
60 mm	–	0.05	0.70	–	22.33	m²	**23.03**
90 mm	–	0.05	0.70	–	25.31	m²	**26.01**
110 mm	–	0.05	0.70	–	26.72	m²	**27.42**
160 mm	–	0.05	0.70	–	30.17	m²	**30.87**
Commercial (was KN90); compressive strength of 90 kN/m²; board thickness; typically for commercial							
60 mm	–	0.06	0.84	–	22.69	m²	**23.53**
90 mm	–	0.06	0.84	–	24.85	m²	**25.69**
110 mm	–	0.06	0.84	–	26.92	m²	**27.76**
160 mm	–	0.06	0.84	–	28.48	m²	**29.32**
600 mm voidpack pipe							
36 mm dia.	–	0.05	0.70	–	8.52	nr	**9.22**
Sides of ground beams and edges of beds; basic finish; 18 mm thick external quality plywood; basic finish; four uses							
Plain vertical							
height not exceeding 250 mm	–	0.41	8.10	–	4.56	m	**12.66**
height 250–500 mm	–	0.75	14.61	–	10.05	m	**24.66**
height 500 mm–1.00 m	–	1.04	20.41	–	13.76	m	**34.17**
height exceeding 1.00 m	–	1.38	26.93	–	13.76	m²	**40.69**
Edges of suspended slabs; basic finish; 18 mm thick external quality plywood; basic finish; four uses							
Plain vertical							
height not exceeding 250 mm	–	0.62	12.14	–	4.67	m	**16.81**
height 250–500 mm	–	0.92	17.95	–	8.33	m	**26.28**
height 500 mm–1.00 m	–	1.46	28.51	–	13.86	m	**42.37**
Sides of upstands; basic finish; 18 mm thick external quality plywood; basic finish; four uses							
Plain vertical							
height not exceeding 250 mm	–	0.52	10.21	–	4.77	m	**14.98**
height 250–500 mm	–	0.84	16.37	–	10.26	m	**26.63**
height 500 mm–1.00 m	–	1.46	28.51	–	16.39	m	**44.90**
height exceeding 1.00 m	–	1.67	32.56	–	16.39	m²	**48.95**

11 IN SITU CONCRETE WORKS

Item	PC £	Labour hours	Labour £	Plant £	Material £	Unit	Total rate £
11.05 FORMWORK – cont							
Steps in top surfaces; basic finish; 18 mm thick external quality plywood; basic finish; four uses							
Plain vertical							
height not exceeding 250 mm	–	0.41	8.10	–	4.82	m	**12.92**
height 250–500 mm	–	0.67	13.03	–	10.31	m	**23.34**
Steps in soffits; basic finish; 18 mm thick external quality plywood; basic finish; four uses							
Plain vertical							
height not exceeding 250 mm	–	0.46	8.98	–	4.10	m	**13.08**
height 250–500 mm	–	0.73	14.26	–	7.67	m	**21.93**
Machine bases and plinths; basic finish; 18 mm thick external quality plywood; basic finish; four uses							
Plain vertical							
height not exceeding 250 mm	–	0.41	8.10	–	4.56	m	**12.66**
height 250–500 mm	–	0.71	13.91	–	10.05	m	**23.96**
height 500 mm–1.00 m	–	1.04	20.41	–	13.76	m	**34.17**
height exceeding 1.00 m	–	1.33	26.05	–	13.76	m²	**39.81**
Soffits of slabs; basic finish; 18 mm thick external quality plywood; basic finish; four uses							
Slab thickness not exceeding 200 mm							
horizontal; height to soffit not exceeding 1.50 m	–	1.50	29.40	–	11.98	m²	**41.38**
horizontal; height to soffit 1.50–2.40 m	–	1.46	28.51	–	12.08	m²	**40.59**
horizontal; height to soffit 2.40–2.70 m	–	1.38	26.93	–	10.85	m²	**37.78**
horizontal; height to soffit 2.70–3.00 m	–	1.33	26.05	–	9.61	m²	**35.66**
horizontal; height to soffit 3.00–4.50 m	–	1.41	27.63	–	12.08	m²	**39.71**
horizontal; height to soffit 4.50–6.00 m	–	1.50	29.40	–	12.60	m²	**42.00**
Slab thickness 200–300 mm							
horizontal; height to soffit 1.50–3.00 m	–	1.50	29.40	–	15.72	m²	**45.12**
Slab thickness 300–400 mm							
horizontal; height to soffit 1.50–3.00 m	–	1.54	30.10	–	16.93	m²	**47.03**
Slab thickness 400–500 mm							
horizontal; height to soffit 1.50–3.00 m	–	1.62	31.68	–	19.18	m²	**50.86**
Slab thickness 500–600 mm							
horizontal; height to soffit 1.50–3.00 m	–	1.75	34.14	–	19.18	m²	**53.32**
Extra over soffits of slabs for							
sloping not exceeding 15°	–	0.17	3.35	–	–	m²	**3.35**
sloping exceeding 15°	–	0.33	6.51	–	–	m²	**6.51**
Soffits of landings; basic finish; 18 mm thick external quality plywood; basic finish; four uses							
Slab thickness not exceeding 200 mm							
horizontal; height to soffit 1.50–3.00 m	–	1.50	29.40	–	13.88	m²	**43.28**

11 IN SITU CONCRETE WORKS

Item	PC £	Labour hours	Labour £	Plant £	Material £	Unit	Total rate £
Slab thickness 200–300 mm							
horizontal; height to soffit 1.50–3.00 m	–	1.58	30.98	–	17.12	m²	**48.10**
Slab thickness 300–400 mm							
horizontal; height to soffit 1.50–3.00 m	–	1.62	31.68	–	18.74	m²	**50.42**
Slab thickness 400–500 mm							
horizontal; height to soffit 1.50–3.00 m	–	1.71	33.44	–	20.35	m²	**53.79**
Slab thickness 500–600 mm							
horizontal; height to soffit 1.50–3.00 m	–	1.84	35.91	–	20.35	m²	**56.26**
Extra over soffits of landings for							
sloping not exceeding 15°	–	0.17	3.35	–	–	m²	**3.35**
sloping exceeding 15°	–	0.33	6.51	–	–	m²	**6.51**
Top formwork; basic finish; 18 mm thick external							
quality plywood; basic finish; four uses							
sloping exceeding 15°	–	1.25	24.46	–	10.61	m²	**35.07**
Walls; basic finish; 18 mm thick external quality							
plywood; basic finish; four uses							
vertical	–	1.50	29.40	–	13.30	m²	**42.70**
vertical; height exceeding 3.00 m above floor level	–	1.84	35.91	–	16.49	m²	**52.40**
vertical; interrupted	–	1.75	34.14	–	16.49	m²	**50.63**
vertical; to one side only	–	2.92	57.03	–	19.69	m²	**76.72**
battered	–	2.33	45.59	–	16.95	m²	**62.54**
Beams; basic finish; 18 mm thick external quality							
plywood; basic finish; four uses							
Attached to slabs							
regular shaped; square or rectangular; height to							
soffit 1.50–3.00 m	–	1.84	35.91	–	15.92	m²	**51.83**
regular shaped; square or rectangular; height to							
soffit 3.00–4.50 m	–	1.92	37.49	–	16.23	m²	**53.72**
regular shaped; square or rectangular; height to							
soffit 4.50–6.00 m	–	2.00	39.08	–	16.49	m²	**55.57**
Attached to walls							
regular shaped; square or rectangular; height to							
soffit 1.50–3.00 m	–	1.92	37.49	–	15.92	m²	**53.41**
Isolated							
regular shaped; square or rectangular; height to							
soffit 1.50–3.00 m	–	2.00	39.08	–	15.92	m²	**55.00**
regular shaped; square or rectangular; height to							
soffit 3.00–4.50 m	–	2.08	40.66	–	16.23	m²	**56.89**
regular shaped; square or rectangular; height to							
soffit 4.50–6.00 m	–	2.17	42.42	–	16.49	m²	**58.91**
Extra over beams for							
regular shaped; sloping not exceeding 15°	–	0.25	4.92	–	1.42	m²	**6.34**
regular shaped; sloping exceeding 15°	–	0.50	9.86	–	2.84	m²	**12.70**
Beam casings; basic finish; 18 mm thick external							
quality plywood; basic finish; four uses							
Attached to slabs							
regular shaped; square or rectangular; height to							
soffit 1.50–3.00 m	–	1.92	37.49	–	15.92	m²	**53.41**
regular shaped; square or rectangular; height to							
soffit 3.00–4.50 m	–	2.00	39.08	–	16.23	m²	**55.31**

11 IN SITU CONCRETE WORKS

Item	PC £	Labour hours	Labour £	Plant £	Material £	Unit	Total rate £
11.05 FORMWORK – cont							
Beam casings – cont							
Attached to walls							
regular shaped; square or rectangular; height to soffit 1.50–3.00 m	–	2.00	39.08	–	15.92	m²	**55.00**
Isolated							
regular shaped; square or rectangular; height to soffit 1.50–3.00 m	–	2.08	40.66	–	15.92	m²	**56.58**
regular shaped; square or rectangular; height to soffit 3.00–4.50 m	–	2.17	42.42	–	16.23	m²	**58.65**
Extra over beam casings for							
regular shaped; sloping not exceeding 15°	–	0.25	4.92	–	1.42	m²	**6.34**
regular shaped; sloping exceeding 15°	–	0.50	9.86	–	2.84	m²	**12.70**
Columns; basic finish; 18 mm thick external quality plywood; basic finish; four uses							
Attached to walls							
regular shaped; square or rectangular; height to soffit 1.50–3.00 m	–	1.84	35.91	–	13.81	m²	**49.72**
Isolated							
regular shaped; square or rectangular; height to soffit 1.50–3.00 m	–	1.92	37.49	–	13.30	m²	**50.79**
regular shaped; circular; not exceeding 300 mm dia.; height to soffit 1.50–3.00 m	–	3.33	65.13	–	20.58	m²	**85.71**
regular shaped; circular; 300–600 mm dia.; height to soffit 1.50–3.00 m	–	3.12	61.08	–	18.15	m²	**79.23**
regular shaped; circular; 600–900 mm dia.; height to soffit 1.50–3.00 m	–	2.92	57.03	–	18.15	m²	**75.18**
Column casings; basic finish; 18 mm thick external quality plywood; basic finish; four uses							
Attached to walls							
regular shaped; square or rectangular; height to soffit 1.50–3.00 m	–	1.92	37.49	–	13.81	m²	**51.30**
Isolated							
regular shaped; square or rectangular; height to soffit 1.50–3.00 m	–	2.00	39.08	–	13.81	m²	**52.89**
Special shapes and finishes							
Recesses or rebates							
12 × 12 mm	–	0.05	1.06	–	0.38	m	**1.44**
25 × 25 mm	–	0.05	1.06	–	0.79	m	**1.85**
25 × 50 mm	–	0.05	1.06	–	1.01	m	**2.07**
50 × 50 mm	–	0.05	1.06	–	0.98	m	**2.04**
Nibs							
50 × 50 mm	–	0.46	8.98	–	1.48	m	**10.46**
100 × 100 mm	–	0.65	12.67	–	2.22	m	**14.89**
100 × 200 mm	–	0.86	16.90	–	14.60	m	**31.50**

11 IN SITU CONCRETE WORKS

Item	PC £	Labour hours	Labour £	Plant £	Material £	Unit	Total rate £
Extra over a basic finish for fine formed finishes							
slabs	–	0.27	5.28	–	–	m²	**5.28**
walls	–	0.27	5.28	–	–	m²	**5.28**
beams	–	0.27	5.28	–	–	m²	**5.28**
columns	–	0.27	5.28	–	–	m²	**5.28**
Add to prices for basic formwork for							
curved radius 6.00 m	–	–	–	–	–	%	**50.00**
curved radius 2.00 m	–	–	–	–	–	%	**100.00**
coating with retardant agent	–	0.01	0.18	–	0.29	m²	**0.47**
Wall kickers; basic finish							
height 150 mm	–	0.41	8.10	–	3.96	m	**12.06**
height 225 mm	–	0.54	10.56	–	5.07	m	**15.63**
Suspended wall kickers; basic finish							
Height 150 mm	–	0.52	10.21	–	3.19	m	**13.40**
Wall ends, soffits and steps in walls; basic finish							
Plain							
width exceeding 1.00 m	–	1.58	30.98	–	16.23	m²	**47.21**
width not exceeding 250 mm	–	0.50	9.86	–	3.84	m	**13.70**
width 250–500 mm	–	0.79	15.49	–	8.63	m	**24.12**
width 500 mm–1.00 m	–	1.25	24.46	–	16.23	m	**40.69**
Openings in walls							
Plain							
width exceeding 1.00 m	–	1.75	34.14	–	16.23	m²	**50.37**
width not exceeding 250 mm	–	0.54	10.56	–	3.84	m	**14.40**
width 250–500 mm	–	0.92	17.95	–	8.63	m	**26.58**
width 500 mm–1.00 m	–	1.41	27.63	–	16.23	m	**43.86**
Stairflights							
Width 1.00 m; 150 mm waist; 150 mm undercut risers							
string, width 300 mm	–	4.17	81.49	–	34.58	m	**116.07**
Width 2.00 m; 200 mm waist; 150 mm undercut risers							
string, width 350 mm	–	7.50	146.62	–	104.92	m	**251.54**
Mortices							
Girth not exceeding 500 mm							
depth not exceeding 250 mm; circular	–	0.13	2.46	–	1.58	nr	**4.04**
Holes							
Girth not exceeding 500 mm							
depth not exceeding 250 mm; circular	–	0.17	3.35	–	1.85	nr	**5.20**
depth 250–500 mm; circular	–	0.25	4.92	–	4.65	nr	**9.57**
Girth 500 mm–1.00 m							
depth not exceeding 250 mm; circular	–	0.21	4.05	–	3.34	nr	**7.39**
depth 250–500 mm; circular	–	0.32	6.16	–	8.83	nr	**14.99**
Girth 1.00–2.00 m							
depth not exceeding 250 mm; circular	–	0.38	7.40	–	8.83	nr	**16.23**
depth 250–500 mm; circular	–	0.56	10.92	–	17.69	nr	**28.61**

11 IN SITU CONCRETE WORKS

Item	PC £	Labour hours	Labour £	Plant £	Material £	Unit	Total rate £
11.05 FORMWORK – cont							
Holes – cont							
Girth 2.00–3.00 m							
depth not exceeding 250 mm; circular	–	0.50	9.86	–	17.21	nr	**27.07**
depth 250–500 mm; circular	–	0.75	14.61	–	104.52	nr	**119.13**
11.06 REINFORCEMENT							
Bars; hot rolled deformed high steel bars; grade 500C							
40 mm dia. nominal size							
straight	–	7.00	143.79	–	739.22	tonne	**883.01**
bent	–	9.00	184.87	–	797.29	tonne	**982.16**
32 mm dia. nominal size							
straight	–	8.00	164.33	–	745.45	tonne	**909.78**
bent	–	10.00	205.41	–	803.52	tonne	**1008.93**
25 mm dia. nominal size							
straight	–	9.00	184.87	–	752.39	tonne	**937.26**
bent	–	12.00	246.50	–	810.47	tonne	**1056.97**
20 mm dia. nominal size							
straight	–	11.00	225.95	–	761.47	tonne	**987.42**
bent	–	14.00	287.58	–	819.54	tonne	**1107.12**
16 mm dia. nominal size							
straight	–	14.00	287.58	–	771.25	tonne	**1058.83**
bent	–	18.00	369.74	–	829.33	tonne	**1199.07**
12 mm dia. nominal size							
straight	–	20.00	410.83	–	787.27	tonne	**1198.10**
bent	–	25.00	513.53	–	845.34	tonne	**1358.87**
10 mm dia. nominal size							
straight	–	24.00	492.99	–	806.84	tonne	**1299.83**
bent	–	31.00	636.78	–	864.90	tonne	**1501.68**
8 mm dia. nominal size							
straight	–	29.00	595.70	–	822.86	tonne	**1418.56**
links	–	37.00	760.03	–	905.84	tonne	**1665.87**
bent	–	37.00	760.03	–	880.92	tonne	**1640.95**
Bars; stainless steel							
40 mm dia. nominal size							
straight	–	7.00	143.79	–	4027.91	tonne	**4171.70**
bent	–	9.00	184.87	–	4186.27	tonne	**4371.14**
32 mm dia. nominal size							
straight	–	17.00	349.20	–	4027.91	tonne	**4377.11**
bent	–	10.00	205.41	–	4186.27	tonne	**4391.68**
25 mm dia. nominal size							
straight	–	9.00	184.87	–	4019.28	tonne	**4204.15**
bent	–	12.00	246.50	–	4177.64	tonne	**4424.14**

11 IN SITU CONCRETE WORKS

Item	PC £	Labour hours	Labour £	Plant £	Material £	Unit	Total rate £
20 mm dia. nominal size							
straight	–	11.00	225.95	–	4031.46	tonne	**4257.41**
bent	–	14.00	287.58	–	4189.82	tonne	**4477.40**
16 mm dia. nominal size							
straight	–	14.00	287.58	–	4053.71	tonne	**4341.29**
bent	–	18.00	369.74	–	4212.07	tonne	**4581.81**
12 mm dia. nominal size							
straight	–	20.00	410.83	–	4075.96	tonne	**4486.79**
bent	–	25.00	513.53	–	4234.32	tonne	**4747.85**
10 mm dia. nominal size							
straight	–	24.00	492.99	–	4101.76	tonne	**4594.75**
bent	–	31.00	636.78	–	4260.12	tonne	**4896.90**
8 mm dia. nominal size							
straight	–	29.00	595.70	–	4124.01	tonne	**4719.71**
bent	–	37.00	760.03	–	4282.37	tonne	**5042.40**
Bars; stainless steel (low nickel alloys or lean duplexes)							
40 mm dia. nominal size							
straight	–	7.00	143.79	–	3204.42	tonne	**3348.21**
bent	–	9.00	184.87	–	3373.34	tonne	**3558.21**
32 mm dia. nominal size							
straight	–	8.00	164.33	–	3206.53	tonne	**3370.86**
bent	–	10.00	205.41	–	3373.34	tonne	**3578.75**
25 mm dia. nominal size							
straight	–	9.00	184.87	–	3195.79	tonne	**3380.66**
bent	–	12.00	246.50	–	3364.71	tonne	**3611.21**
20 mm dia. nominal size							
straight	–	11.00	225.95	–	3207.98	tonne	**3433.93**
bent	–	14.00	287.58	–	3376.90	tonne	**3664.48**
16 mm dia. nominal size							
straight	–	14.00	287.58	–	3230.22	tonne	**3517.80**
bent	–	18.00	369.74	–	3399.14	tonne	**3768.88**
12 mm dia. nominal size							
straight	–	20.00	410.83	–	3252.47	tonne	**3663.30**
bent	–	25.00	513.53	–	3421.39	tonne	**3934.92**
10 mm dia. nominal size							
straight	–	24.00	492.99	–	3278.27	tonne	**3771.26**
bent	–	31.00	636.78	–	3447.19	tonne	**4083.97**
8 mm dia. nominal size							
straight	–	29.00	595.70	–	3307.69	tonne	**3903.39**
bent	–	37.00	760.03	–	3469.44	tonne	**4229.47**
Fabric; in standard 4.8 m × 2.4 m sheets							
Ref D49 (0.77 kg/m^2)							
100 mm minimum laps; bent	–	0.24	4.93	–	2.62	m^2	**7.55**
Ref D98 (1.54 kg/m^2)							
400 mm minimum laps	–	0.12	2.46	–	1.56	m^2	**4.02**
strips in one width; 600 mm width	–	0.15	3.08	–	1.56	m^2	**4.64**
strips in one width; 900 mm width	–	0.14	2.87	–	1.56	m^2	**4.43**
strips in one width; 1200 mm width	–	0.13	2.67	–	1.56	m^2	**4.23**

11 IN SITU CONCRETE WORKS

Item	PC £	Labour hours	Labour £	Plant £	Material £	Unit	Total rate £
11.06 REINFORCEMENT – cont							
Fabric – cont							
Ref A142 (2.22 kg/m^2)							
400 mm minimum laps	–	0.12	2.46	–	1.97	m^2	**4.43**
strips in one width; 600 mm width	–	0.15	3.08	–	1.97	m^2	**5.05**
strips in one width; 900 mm width	–	0.14	2.87	–	1.97	m^2	**4.84**
strips in one width; 1200 mm width	–	0.13	2.67	–	1.97	m^2	**4.64**
Ref A193 (3.02 kg/m^2)							
400 mm minimum laps	–	0.12	2.46	–	2.67	m^2	**5.13**
strips in one width; 600 mm width	–	0.15	3.08	–	2.67	m^2	**5.75**
strips in one width; 900 mm width	–	0.14	2.87	–	2.67	m^2	**5.54**
strips in one width; 1200 mm width	–	0.13	2.67	–	2.67	m^2	**5.34**
Ref A252 (3.95 kg/m^2)							
400 mm minimum laps	–	0.13	2.67	–	3.45	m^2	**6.12**
strips in one width; 600 mm width	–	0.16	3.29	–	3.45	m^2	**6.74**
strips in one width; 900 mm width	–	0.15	3.08	–	3.45	m^2	**6.53**
strips in one width; 1200 mm width	–	0.14	2.87	–	3.45	m^2	**6.32**
Ref A393 (6.16 kg/m^2)							
400 mm minimum laps	–	0.15	3.08	–	5.38	m^2	**8.46**
strips in one width; 600 mm width	–	0.18	3.70	–	5.38	m^2	**9.08**
strips in one width; 900 mm width	–	0.17	3.49	–	5.38	m^2	**8.87**
strips in one width; 1200 mm width	–	0.16	3.29	–	5.38	m^2	**8.67**
Ref B196 (3.05 kg/m^2)							
400 mm minimum laps	–	0.12	2.46	–	3.24	m^2	**5.70**
strips in one width; 600 mm width	–	0.15	3.08	–	3.24	m^2	**6.32**
strips in one width; 900 mm width	–	0.14	2.87	–	3.24	m^2	**6.11**
strips in one width; 1200 mm width	–	0.13	2.67	–	3.24	m^2	**5.91**
Ref B283 (3.73 kg/m^2)							
400 mm minimum laps	–	0.12	2.46	–	3.39	m^2	**5.85**
strips in one width; 600 mm width	–	0.15	3.08	–	3.39	m^2	**6.47**
strips in one width; 900 mm width	–	0.14	2.87	–	3.39	m^2	**6.26**
strips in one width; 1200 mm width	–	0.13	2.67	–	3.39	m^2	**6.06**
Ref B385 (4.53 kg/m^2)							
400 mm minimum laps	–	0.13	2.67	–	4.13	m^2	**6.80**
strips in one width; 600 mm width	–	0.16	3.29	–	4.13	m^2	**7.42**
strips in one width; 900 mm width	–	0.15	3.08	–	4.13	m^2	**7.21**
strips in one width; 1200 mm width	–	0.14	2.87	–	4.13	m^2	**7.00**
Ref B503 (5.93 kg/m^2)							
400 mm minimum laps	–	0.15	3.08	–	5.33	m^2	**8.41**
strips in one width; 600 mm width	–	0.18	3.70	–	5.33	m^2	**9.03**
strips in one width; 900 mm width	–	0.17	3.49	–	5.33	m^2	**8.82**
strips in one width; 1200 mm width	–	0.16	3.29	–	5.33	m^2	**8.62**
Ref B785 (8.14 kg/m^2)							
400 mm minimum laps	–	0.17	3.49	–	7.30	m^2	**10.79**
strips in one width; 600 mm width	–	0.20	4.11	–	7.30	m^2	**11.41**
strips in one width; 900 mm width	–	0.19	3.90	–	7.30	m^2	**11.20**
strips in one width; 1200 mm width	–	0.18	3.70	–	7.30	m^2	**11.00**

11 IN SITU CONCRETE WORKS

Item	PC £	Labour hours	Labour £	Plant £	Material £	Unit	Total rate £
Ref B1131 (10.90 kg/m²)							
400 mm minimum laps	–	0.18	3.70	–	9.78	m²	**13.48**
strips in one width; 600 mm width	–	0.24	4.93	–	9.78	m²	**14.71**
strips in one width; 900 mm width	–	0.22	4.52	–	9.78	m²	**14.30**
strips in one width; 1200 mm width	–	0.20	4.11	–	9.78	m²	**13.89**
11.07 DESIGNED JOINTS							
Formed; Fosroc impregnated fibreboard joint filler or other equal							
Width not exceeding 150 mm							
12.50 mm thick	–	0.14	2.74	–	2.31	m	**5.05**
20 mm thick	–	0.19	3.72	–	2.90	m	**6.62**
25 mm thick	–	0.23	4.50	–	3.45	m	**7.95**
Width 150–300 mm							
12.50 mm thick	–	0.23	4.50	–	4.28	m	**8.78**
20 mm thick	–	0.23	4.50	–	5.70	m	**10.20**
25 mm thick	–	0.23	4.50	–	6.44	m	**10.94**
Width 300–450 mm							
12.50 mm thick	–	0.28	5.48	–	6.16	m	**11.64**
20 mm thick	–	0.28	5.48	–	8.10	m	**13.58**
25 mm thick	–	0.28	5.48	–	9.13	m	**14.61**
Formed; waterproof bonded cork joint filler board							
Width not exceeding 150 mm							
10 mm thick	–	0.14	2.74	–	4.42	m	**7.16**
13 mm thick	–	0.14	2.74	–	4.48	m	**7.22**
19 mm thick	–	0.14	2.74	–	5.84	m	**8.58**
25 mm thick	–	0.14	2.74	–	6.23	m	**8.97**
Width 150–300 mm							
10 mm thick	–	0.19	3.72	–	7.96	m	**11.68**
13 mm thick	–	0.19	3.72	–	8.10	m	**11.82**
19 mm thick	–	0.19	3.72	–	10.82	m	**14.54**
25 mm thick	–	0.19	3.72	–	11.60	m	**15.32**
Width 300–450 mm							
10 mm thick	–	0.23	4.50	–	12.20	m	**16.70**
13 mm thick	–	0.23	4.50	–	12.39	m	**16.89**
19 mm thick	–	0.23	4.50	–	16.48	m	**20.98**
25 mm thick	–	0.23	4.50	–	17.64	m	**22.14**
Sealants; Fosroc Nitoseal MB77 hot poured rubberized bituminous compound or other equal							
Width 10 mm							
25 mm depth	–	0.17	3.33	–	2.44	m	**5.77**
Width 12.50 mm							
25 mm depth	–	0.18	3.52	–	3.01	m	**6.53**

11 IN SITU CONCRETE WORKS

Item	PC £	Labour hours	Labour £	Plant £	Material £	Unit	Total rate £
11.07 DESIGNED JOINTS – cont							
Sealants – cont							
Width 20 mm							
25 mm depth	–	0.19	3.72	–	4.89	m	**8.61**
Width 25 mm							
25 mm depth	–	0.20	3.91	–	6.00	m	**9.91**
Sealants; Fosroc Thioflex 600 gun grade							
polysulphide or other equal							
Width 10 mm							
25 mm depth	–	0.05	0.98	–	3.39	m	**4.37**
Width 12.50 mm							
25 mm depth	–	0.06	1.17	–	4.24	m	**5.41**
Width 20 mm							
25 mm depth	–	0.07	1.37	–	6.79	m	**8.16**
Width 25 mm							
25 mm depth	–	0.08	1.57	–	8.49	m	**10.06**
Sealants; Grace Servicised Paraseal							
polysulphide compound or other equal; priming							
with Grace Servicised Primer P							
Width 10 mm							
25 mm depth	–	0.19	3.14	–	4.78	m	**7.92**
Width 13 mm							
25 mm depth	–	0.19	3.14	–	6.12	m	**9.26**
Width 19 mm							
25 mm depth	–	0.23	3.80	–	8.81	m	**12.61**
Width 25 mm							
25 mm depth	–	0.23	3.80	–	11.49	m	**15.29**
Waterstops; Grace Servicised or other equal							
Hydrophilic strip water stop; lapped joints; cast into							
concrete							
50 × 20 mm Adcor 500S	8.48	0.30	4.95	–	11.38	m	**16.33**
Servitite Internal 10 mm thick PVC water stop; flat							
dumbbell type; heat welded joints; cast into concrete							
Servitite 150; 150 mm wide	–	0.23	4.73	–	16.21	m	**20.94**
flat angle	–	0.28	5.75	–	36.58	nr	**42.33**
vertical angle	–	0.28	5.75	–	36.39	nr	**42.14**
flat three way intersection	–	0.37	7.60	–	53.22	nr	**60.82**
vertical three way intersection	–	0.37	7.60	–	60.06	nr	**67.66**
four way intersection	–	0.46	9.45	–	66.20	nr	**75.65**
Servitite 230; 230 mm wide	–	0.23	4.73	–	22.90	m	**27.63**
flat angle	–	0.28	5.75	–	44.79	nr	**50.54**
vertical angle	–	0.28	5.75	–	52.75	nr	**58.50**
flat three way intersection	–	0.37	7.60	–	65.54	nr	**73.14**
vertical three way intersection	–	0.37	7.60	–	112.23	nr	**119.83**
four way intersection	–	0.46	9.45	–	82.60	nr	**92.05**
Servitite AT200; 200 mm wide	–	0.23	4.73	–	25.15	m	**29.88**
flat angle	–	0.28	5.75	–	44.15	nr	**49.90**
vertical angle	–	0.28	5.75	–	47.68	nr	**53.43**

11 IN SITU CONCRETE WORKS

Item	PC £	Labour hours	Labour £	Plant £	Material £	Unit	Total rate £
flat three way intersection	–	0.37	7.60	–	76.11	nr	**83.71**
vertical three way intersection	–	0.37	7.60	–	60.34	nr	**67.94**
four way intersection	–	0.46	9.45	–	90.64	nr	**100.09**
Servitite K305; 305 mm wide	–	0.28	5.75	–	36.59	m	**42.34**
flat angle	–	0.32	6.57	–	76.35	nr	**82.92**
vertical angle	–	0.32	6.57	–	82.13	nr	**88.70**
flat three way intersection	–	0.42	8.63	–	109.09	nr	**117.72**
vertical three way intersection	–	0.42	8.63	–	127.18	nr	**135.81**
four way intersection	–	0.51	10.48	–	147.63	nr	**158.11**
Serviseal External PVC water stop; PVC water stop; centre bulb type; heat welded joints; cast into concrete							
Serviseal 195; 195 mm wide	–	0.23	4.73	–	11.11	m	**15.84**
flat angle	–	0.28	5.75	–	23.18	nr	**28.93**
vertical angle	–	0.28	5.75	–	37.90	nr	**43.65**
flat three way intersection	–	0.37	7.60	–	38.82	nr	**46.42**
four way intersection	–	0.46	9.45	–	57.21	nr	**66.66**
Serviseal 240; 240 mm wide	–	0.23	4.73	–	13.63	m	**18.36**
flat angle	–	0.28	5.75	–	26.73	nr	**32.48**
vertical angle	–	0.28	5.75	–	40.46	nr	**46.21**
flat three way intersection	–	0.37	7.60	–	44.19	nr	**51.79**
four way intersection	–	0.46	9.45	–	64.34	nr	**73.79**
Serviseal AT240; 240 mm wide	–	0.23	4.73	–	29.35	m	**34.08**
flat angle	–	0.28	5.75	–	43.50	nr	**49.25**
vertical angle	–	0.28	5.75	–	41.67	nr	**47.42**
flat three way intersection	–	0.37	7.60	–	66.60	nr	**74.20**
four way intersection	–	0.46	9.45	–	97.06	nr	**106.51**
Serviseal K320; 320 mm wide	–	0.28	5.75	–	17.66	m	**23.41**
flat angle	–	0.32	6.57	–	56.29	nr	**62.86**
vertical angle	–	0.32	6.57	–	30.61	nr	**37.18**
flat three way intersection	–	0.42	8.63	–	82.81	nr	**91.44**
four way intersection	–	0.51	10.48	–	104.21	nr	**114.69**
11.08 ACCESSORIES CAST IN							
Foundation bolt boxes							
Temporary plywood; for group of 4 nr bolts							
75 × 75 × 150 mm	–	0.42	8.21	–	1.63	nr	**9.84**
75 × 75 × 250 mm	–	0.42	8.21	–	1.91	nr	**10.12**
Expanded metal; Expamet Building Products Ltd or other equal and approved							
75 mm dia. × 150 mm long	–	0.28	5.48	–	75.34	nr	**80.82**
75 mm dia. × 300 mm long	–	0.28	5.48	–	84.65	nr	**90.13**
100 mm dia. × 450 mm long	–	0.28	5.48	–	2.79	nr	**8.27**
Foundation bolts and nuts							
Black hexagon							
10 mm dia. × 100 mm long	–	0.23	4.50	–	0.59	nr	**5.09**
12 mm dia. × 120 mm long	–	0.23	4.50	–	0.90	nr	**5.40**
16 mm dia. × 160 mm long	–	0.28	5.48	–	2.47	nr	**7.95**
20 mm dia. × 180 mm long	–	0.28	5.48	–	2.88	nr	**8.36**

11 IN SITU CONCRETE WORKS

Item	PC £	Labour hours	Labour £	Plant £	Material £	Unit	Total rate £
11.08 ACCESSORIES CAST IN – cont							
Masonry slots							
Stainless steel; dovetail slots; 1.20 mm thick; 18G							
1000 mm long	–	0.25	4.89	–	6.72	m	**11.61**
100 mm long	–	0.07	1.37	–	0.31	nr	**1.68**
Stainless steel; metal insert slots; Halfen Ltd Ribslot or other equal; 2.50 mm thick; end caps and foam filling							
41 × 41 mm; ref P3270	–	0.37	7.24	–	9.28	m	**16.52**
41 × 41 × 100 mm; ref P3250	–	0.09	1.76	–	1.40	nr	**3.16**
41 × 41 × 150 mm; ref P3251	–	0.09	1.76	–	1.03	nr	**2.79**
Cramps							
Stainless steel; once bent; one end shot fired into concrete; other end fanged and built into brickwork joint							
200 mm girth	–	0.14	3.21	–	1.28	nr	**4.49**
Column guards							
Corner protection; White nylon coated steel; plugging; screwing to concrete; 1.5 mm thick							
75 × 75 × 1200 mm	–	0.74	14.47	–	23.87	nr	**38.34**
Galvanized steel; 3 mm thick							
75 × 75 × 1000 mm	–	0.56	10.95	–	16.93	nr	**27.88**
Galvanized steel; 4.5 mm thick							
75 × 75 × 1000 mm	–	0.56	10.95	–	22.77	nr	**33.72**
Stainless steel; HKW or other equal; Halfen Ltd; 5 mm thick							
50 × 50 × 1200 mm	–	0.93	18.19	–	78.32	nr	**96.51**
50 × 50 × 2000 mm	–	1.11	21.71	–	129.40	nr	**151.11**
Channels							
Stainless steel; Halfen Ltd or other equal							
ref 38/17/HTA channel	–	0.32	6.26	–	45.24	m	**51.50**
ref 41/22/HZA; 80 mm long; including T headed bolts and plate washers	–	0.09	1.76	–	29.25	nr	**31.01**
Channel ties							
Stainless steel; Halfen Ltd or other equal							
ref HTS – B12; 150 mm projection; including insulation retainer	–	0.03	0.66	–	0.57	nr	**1.23**
ref HTS – B12; 200 mm projection; including insulation retainer	–	0.03	0.66	–	0.67	nr	**1.33**

11 IN SITU CONCRETE WORKS

Item	PC £	Labour hours	Labour £	Plant £	Material £	Unit	Total rate £
11.09 IN SITU CONCRETE SUNDRIES							
Grouting with cement mortar (1:1)							
Stanchion bases							
10 mm thick	–	0.93	15.37	–	0.17	nr	**15.54**
25 mm thick	–	1.16	19.17	–	0.41	nr	**19.58**
Grouting with epoxy resin							
Stanchion bases							
10 mm thick	–	1.16	19.17	–	5.64	nr	**24.81**
25 mm thick	–	1.39	22.97	–	28.23	nr	**51.20**
Grouting with Conbextra GP cementitious grout							
Stanchion bases							
10 mm thick	–	1.16	19.17	–	1.40	nr	**20.57**
25 mm thick	–	1.39	22.97	–	3.58	nr	**26.55**
Grouting with Conbextra HF flowable cementitious grout							
Stanchion bases							
10 mm thick	–	1.16	19.17	–	1.48	nr	**20.65**
25 mm thick	–	1.39	22.97	–	3.78	nr	**26.75**
Sundry in situ concrete work							
Filling; plain in situ concrete; mixed on site							
mortices	–	0.09	1.48	0.21	0.64	nr	**2.33**
holes	–	0.23	3.80	0.53	133.48	m³	**137.81**
chases not exceeding 300 mm wide or thick	–	0.14	2.32	0.32	1.34	m	**3.98**
chases exceeding 300 mm wide or thick	–	0.19	3.14	0.43	133.48	m³	**137.05**
extra for reinforcement content over 5%	–	0.51	8.43	–	–	m³	**8.43**

Prices for Measured Works

12 PRECAST/COMPOSITE CONCRETE

Item	PC £	Labour hours	Labour £	Plant £	Material £	Unit	Total rate £
12.01 PRECAST/COMPOSITE CONCRETE WORK							
Prestressed precast concrete structural suspended floors; Bison Hollowcore or other equal; supplied and fixed on hard level bearings, to areas of 500 m² per site visit; top surface screeding and ceiling finishes by others							
Floors to dwellings, offices, car parks, shop retail floors, hospitals, school teaching rooms, staff rooms and the like; superimposed load of 5.00 kN/m²							
floor spans up to 3.00 m; 1200 mm × 150 mm	–	–	–	–	–	m²	60.77
floor spans 3.00 m–6.00 m; 1200 mm × 150 mm	–	–	–	–	–	m²	62.24
floor spans 6.00 m–7.50 m; 1200 mm × 200 mm	–	–	–	–	–	m²	66.52
floor spans 7.50 m–9.50 m; 1200 mm × 250 mm	–	–	–	–	–	m²	70.80
floor spans 9.50 m–12.00 m; 1200 mm × 300 mm	–	–	–	–	–	m²	73.16
floor spans 12.00 m–12.50 m; 1200 mm × 350 mm	–	–	–	–	–	m²	75.51
floor spans 12.50 m–14.00 m; 1200 mm × 400 mm	–	–	–	–	–	m²	83.78
floor spans 14.00 m–15.00 m; 1200 mm × 450 mm	–	–	–	–	–	m²	84.96
Floors to shop stockrooms, light warehousing, schools, churches or similar places of assembly, light factory accommodation, laboratories and the like; superimposed load of 8.50 kN/m²							
floor spans up to 3.00 m; 1200 mm × 150 mm	–	–	–	–	–	m²	59.01
floor spans 3.00 m–6.00 m; 1200 mm × 200 mm	–	–	–	–	–	m²	61.37
floor spans 6.00 m–7.50 m; 1200 mm × 250 mm	–	–	–	–	–	m²	70.80
Floors to heavy warehousing, factories, stores and the like; superimposed load of 12.50 kN/m²							
floor spans up to 3.00 m; 1200 mm × 150 mm	–	–	–	–	–	m²	62.55
floor spans 3.00 m–6.00 m; 1200 mm × 250 mm	–	–	–	–	–	m²	70.80
Prestressed precast concrete staircase, supplied and fixed in conjunction with Bison Hollowcore flooring system or similar; comprising 2 nr 1100 mm wide flights with 7 nr 275 mm treads, 8 nr 185 mm risers and 150 mm waist; 1 nr 2200 mm × 1400 mm × 150 mm half landing and 1 nr top landing							
3.00 m storey height	–	–	–	–	–	nr	2899.28
Prestressed precast concrete beam and block floor; cement and sand grout brushed between beams and blocks; 440 × 215 × 100 mm concrete block infill							
Beam and block flooring at ground level							
beam and block flooring at ground level 150 mm thick beams; up to 3.30 m span	–	0.45	10.97	–	38.94	m²	49.91
beam and block flooring at ground level; 225 mm thick beams; up to 4.20 m span	–	0.45	10.97	–	75.05	m²	86.02
beam and poly block flooring at ground level 150 mm thick beams; up to 3.30 m span (U-value 0.20 W/m² K)	24.81	0.45	10.97	–	25.55	m²	36.52

12 PRECAST/COMPOSITE CONCRETE

Item	PC £	Labour hours	Labour £	Plant £	Material £	Unit	Total rate £
beam and poly block flooring at ground level 150 mm thick beams; up to 3.30 m span (U-value 0.15 W/m² K)	28.11	0.50	12.20	–	28.95	m²	**41.15**
beam and poly block flooring at ground level 150 mm thick beams; up to 3.30 m span (U-value 0.12 W/m² K)	32.52	0.50	12.20	–	33.50	m²	**45.70**
Composite floor comprising reinforced in situ ready-mixed concrete 30.00 N/mm²; on and including steel deck permanent shutting; complete with A142 anti-crack mesh; NOTE temporary props may be required, but have not been included in the following rates							
0.9 mm thick re-entrant type deck; 60 mm deep							
150 mm thick suspended slab							
1.50 m–3.00 m high to soffit	–	0.91	16.04	–	56.51	m²	**72.55**
3.00 m–4.50 m high to soffit	–	0.94	16.55	–	56.51	m²	**73.06**
4.50 m–6.00 m high to soffit	–	0.98	17.31	–	56.51	m²	**73.82**
200 mm thick suspended slab							
1.50 m–3.00 m high to soffit	–	0.95	16.71	–	62.15	m²	**78.86**
3.00 m–4.50 m high to soffit	–	0.97	17.21	–	62.15	m²	**79.36**
4.50 m–6.00 m high to soffit	–	1.00	17.64	–	62.15	m²	**79.79**
1.2 mm thick re-entrant type deck							
150 mm thick suspended slab							
1.50 m–3.00 m high to soffit	–	0.91	16.04	–	60.62	m²	**76.66**
3.00 m–4.50 m high to soffit	–	0.94	16.55	–	60.62	m²	**77.17**
4.50 m–6.00 m high to soffit	–	0.98	17.31	–	60.62	m²	**77.93**
200 mm thick suspended slab							
1.50 m–3.00 m high to soffit	–	0.95	16.71	–	66.26	m²	**82.97**
3.00 m–4.50 m high to soffit	–	0.97	17.21	–	66.26	m²	**83.47**
4.50 m–6.00 m high to soffit	–	1.00	17.64	–	66.26	m²	**83.90**
0.9 mm thick trapezoidal deck 60 mm deep							
150 mm thick suspended slab							
1.50 m–3.00 m high to soffit	–	0.91	16.04	–	53.64	m²	**69.68**
3.00 m–4.50 m high to soffit	–	0.94	16.55	–	53.64	m²	**70.19**
4.50 m–6.00 m high to soffit	–	0.98	17.31	–	53.64	m²	**70.95**
200 mm thick suspended slab							
1.50 m–3.00 m high to soffit	–	0.95	16.71	–	59.30	m²	**76.01**
3.00 m–4.50 m high to soffit	–	0.97	17.21	–	59.30	m²	**76.51**
4.50 m–6.00 m high to soffit	–	1.00	17.64	–	59.30	m²	**76.94**
0.9 mm thick trapezoidal deck 80 mm deep							
150 mm thick suspended slab							
1.50 m–3.00 m high to soffit	–	0.91	16.04	–	55.01	m²	**71.05**
3.00 m–4.50 m high to soffit	–	0.94	16.55	–	55.01	m²	**71.56**
4.50 m–6.00 m high to soffit	–	0.98	17.31	–	55.01	m²	**72.32**

12 PRECAST/COMPOSITE CONCRETE

Item	PC £	Labour hours	Labour £	Plant £	Material £	Unit	Total rate £
12.01 PRECAST/COMPOSITE CONCRETE WORK – cont							
0.9 mm thick trapezoidal deck 80 mm deep – cont							
200 mm thick suspended slab							
1.50 m–3.00 m high to soffit	–	0.95	16.71	–	60.67	m²	**77.38**
3.00 m–4.50 m high to soffit	–	0.97	17.21	–	60.67	m²	**77.88**
4.50 m–6.00 m high to soffit	–	1.00	17.64	–	60.67	m²	**78.31**
1.2 mm thick trapezoidal deck 80 mm deep							
150 mm thick suspended slab							
1.50 m–3.00 m high to soffit	–	0.91	16.04	–	58.75	m²	**74.79**
3.00 m–4.50 m high to soffit	–	0.94	16.55	–	58.75	m²	**75.30**
4.50 m–6.00 m high to soffit	–	0.98	17.31	–	58.75	m²	**76.06**
200 mm thick suspended slab							
1.50 m–3.00 m high to soffit	–	0.95	16.71	–	64.40	m²	**81.11**
3.00 m–4.50 m high to soffit	–	0.97	17.21	–	64.40	m²	**81.61**
4.50 m–6.00 m high to soffit	–	1.00	17.64	–	64.40	m²	**82.04**
Soffits of coffered or troughed slabs; basic finish							
Cordek Correx trough mould or other equal; 300 mm deep; ribs of mould at 600 mm centres and cross ribs at centres of bay; slab thickness 300–400 mm							
horizontal; height to soffit 1.50–3.00 m	–	2.08	40.66	–	22.16	m²	**62.82**
horizontal; height to soffit 3.00–4.50 m	–	2.17	42.42	–	22.41	m²	**64.83**
horizontal; height to soffit 4.50–6.00 m	–	2.25	44.00	–	22.57	m²	**66.57**

13 PRECAST CONCRETE

Item	PC £	Labour hours	Labour £	Plant £	Material £	Unit	Total rate £
13.01 PRECAST CONCRETE GOODS							
Contractor designed precast concrete staircases and landings; including all associated steel supports and fixing in position							
Straight staircases; 280 mm treads; 170 mm undercut risers							
1200 mm wide; 2750 mm rise	–	–	–	–	–	nr	**1916.87**
1200 mm wide; 3750 mm rise	–	–	–	–	–	nr	**2555.83**
Dogleg staircases							
1200 mm wide; one full width half landing; 2750 mm rise	–	–	–	–	–	nr	**2939.19**
1200 mm wide; one full width half landing; 3750 mm rise	–	–	–	–	–	nr	**3833.75**
1200 mm wide 200 mm thick concrete landing support walls	–	–	–	–	–	nr	**894.53**
1800 mm wide; one full width half landing; 2750 mm rise	–	–	–	–	–	nr	**4153.21**
1800 mm wide; one full width half landing; 3750 mm rise	–	–	–	–	–	nr	**5431.13**
1800 mm wide 200 mm thick concrete landing support walls	–	–	–	–	–	nr	**1380.14**
Precast concrete sill, lintels, copings							
Lintels; plate; prestressed bedded							
100 mm × 70 mm × 600 mm long	5.36	0.37	8.12	–	5.57	nr	**13.69**
100 mm × 70 mm × 900 mm long	8.00	0.37	8.12	–	8.28	nr	**16.40**
100 mm × 70 mm × 1100 mm long	9.80	0.37	8.12	–	10.15	nr	**18.27**
100 mm × 70 mm × 1200 mm long	10.67	0.37	8.12	–	11.04	nr	**19.16**
100 mm × 70 mm × 1500 mm long	13.35	0.46	10.08	–	13.83	nr	**23.91**
100 mm × 70 mm × 1800 mm long	16.01	0.46	10.08	–	16.56	nr	**26.64**
100 mm × 70 mm × 2100 mm long	18.68	0.56	12.28	–	19.35	nr	**31.63**
140 mm × 70 mm × 1200 mm long	15.70	0.46	10.08	–	16.24	nr	**26.32**
140 mm × 70 mm × 1500 mm long	19.62	0.56	12.28	–	20.31	nr	**32.59**
Lintels; rectangular; reinforced with mild steel bars; bedded							
100 mm × 145 mm × 900 mm long	3.95	0.56	12.28	–	4.12	nr	**16.40**
100 mm × 145 mm × 1050 mm long	4.62	0.56	12.28	–	4.81	nr	**17.09**
100 mm × 145 mm × 1200 mm long	5.28	0.56	12.28	–	5.49	nr	**17.77**
225 mm × 145 mm × 1200 mm long	20.48	0.74	16.23	–	21.21	nr	**37.44**
225 mm × 225 mm × 1800 mm long	30.64	1.39	30.49	–	31.67	nr	**62.16**
Lintels; boot; reinforced with mild steel bars; bedded							
250 mm × 225 mm × 1200 mm long	22.69	1.11	24.34	–	23.47	nr	**47.81**
275 mm × 225 mm × 1800 mm long	37.44	1.67	36.63	–	38.67	nr	**75.30**
Padstones							
100 mm × 200 mm × 440 mm	19.90	0.28	6.14	–	20.55	nr	**26.69**
150 mm × 215 mm × 440 mm	19.89	0.37	8.12	–	20.56	nr	**28.68**
150 mm × 225 mm × 225 mm	15.62	0.56	12.28	–	16.40	nr	**28.68**

13 PRECAST CONCRETE

Item	PC £	Labour hours	Labour £	Plant £	Material £	Unit	Total rate £
13.01 PRECAST CONCRETE GOODS – cont							
Precast concrete sill, lintels, copings – cont							
Copings; once weathered; once throated; bedded and pointed							
152 mm × 76 mm	5.96	0.65	14.26	–	6.29	m	**20.55**
178 mm × 64 mm	6.59	0.65	14.26	–	6.94	m	**21.20**
305 mm × 76 mm	11.11	0.74	16.23	–	11.80	m	**28.03**
extra for fair ends	–	–	–	–	4.39	nr	**4.39**
extra for angles	–	–	–	–	4.97	nr	**4.97**
Copings; twice weathered; twice throated; bedded and pointed							
152 mm × 76 mm	5.96	0.65	14.26	–	6.29	m	**20.55**
178 mm × 64 mm	6.54	0.65	14.26	–	6.89	m	**21.15**
305 mm × 76 mm	11.11	0.74	16.23	–	11.80	m	**28.03**
extra for fair ends	–	–	–	–	4.39	nr	**4.39**
extra for angles	–	–	–	–	4.97	nr	**4.97**
Sills; splayed top edge, stooled ends; bedded and pointed							
200 mm × 90 mm	27.59	0.75	16.45	–	28.57	m	**45.02**
200 mm × 90 mm; slip sill	30.50	0.75	16.45	–	31.57	m	**48.02**

14 MASONRY

Item	PC £	Labour hours	Labour £	Plant £	Material £	Unit	Total rate £
14.01 BRICK WALLING							
Basic mortar prices							
Mortar materials only							
cement	–	–	–	–	132.64	tonne	**132.64**
sand	–	–	–	–	23.23	tonne	**23.23**
lime	–	–	–	–	185.40	tonne	**185.40**
white cement	–	–	–	–	257.50	tonne	**257.50**
Cemplas Super mortar plasticizer	–	–	–	–	7.73	litre	**7.73**
Coloured mortar materials; (excluding cement)							
light	–	–	–	–	82.40	tonne	**82.40**
medium	–	–	–	–	82.40	tonne	**82.40**
dark	–	–	–	–	87.55	tonne	**87.55**
extra dark	–	–	–	–	87.55	tonne	**87.55**
SUPPLY ONLY BRICK PRICES							
Alternative prices below are for supply only bricks							
NOTE: These costs exclude labour, mortar wastage, overheads and profit recovery							
Forterra Brick							
Brown Rustic	–	–	–	–	464.74	1000	**464.74**
Rustic	–	–	–	–	713.48	1000	**713.48**
Brecken Grey	–	–	–	–	560.11	1000	**560.11**
Old English Brindled Red	–	–	–	–	481.83	1000	**481.83**
Chiltern	–	–	–	–	636.64	1000	**636.64**
Claydon Red Multi	–	–	–	–	561.35	1000	**561.35**
Cotswold	–	–	–	–	625.62	1000	**625.62**
Dapple Light	–	–	–	–	691.54	1000	**691.54**
Georgian	–	–	–	–	571.34	1000	**571.34**
Golden Buff	–	–	–	–	721.31	1000	**721.31**
Heather	–	–	–	–	636.64	1000	**636.64**
Hereward Light	–	–	–	–	581.13	1000	**581.13**
Honey Buff	–	–	–	–	550.33	1000	**550.33**
Ironstone	–	–	–	–	565.98	1000	**565.98**
Milton Buff	–	–	–	–	586.48	1000	**586.48**
Regency	–	–	–	–	594.72	1000	**594.72**
Sandfaced	–	–	–	–	614.70	1000	**614.70**
Saxon Gold	–	–	–	–	619.13	1000	**619.13**
Sunset Red	–	–	–	–	574.74	1000	**574.74**
Tudor	–	–	–	–	628.82	1000	**628.82**
Windsor	–	–	–	–	559.80	1000	**559.80**
Selected Regrades	–	–	–	–	295.61	1000	**295.61**
Sherbourne Red Pavers	–	–	–	–	27.67	m²	**27.67**
Coxmoor Rose Multi Pavers	–	–	–	–	27.67	m²	**27.67**
Ibstock Brick							
Aldridge Brown Blend	–	–	–	–	517.57	1000	**517.57**
Aldridge Leicester Anglican Red Rustic	–	–	–	–	517.57	1000	**517.57**
Chailey Stock	–	–	–	–	441.56	1000	**441.56**
Dorking Multi	–	–	–	–	413.75	1000	**413.75**
Funton Second Hard Stock	–	–	–	–	661.26	1000	**661.26**
Leicester Red Stock	–	–	–	–	397.06	1000	**397.06**
Roughdales Red Multi Rustic	–	–	–	–	517.57	1000	**517.57**

14 MASONRY

Item	PC £	Labour hours	Labour £	Plant £	Material £	Unit	Total rate £
14.01 BRICK WALLING – cont							
SUPPLY ONLY BRICK PRICES – cont							
Ibstock Brick – cont							
Stourbridge Himley Mixed Russet	–	–	–	–	784.65	1000	**784.65**
Stourbridge Kenilworth Multi	–	–	–	–	501.30	1000	**501.30**
Strattford Red Rustic	–	–	–	–	517.57	1000	**517.57**
Swanage Handmade Restoration	–	–	–	–	690.82	1000	**690.82**
Tonbridge Handmade Multi	–	–	–	–	1101.38	1000	**1101.38**
SUPPLY AND LAY PRICES							
Common bricks							
Walls; in gauged mortar (1:1:6)							
prime cost for bricks £ per 1000	380.00	–	–	–	380.00	1000	**380.00**
half brick thick	22.80	0.84	18.35	–	29.29	m²	**47.64**
half brick thick; building against other work;							
concrete	22.80	0.92	20.14	–	32.38	m²	**52.52**
half brick thick; building overhand	22.80	1.04	22.90	–	29.29	m²	**52.19**
half brick thick; curved; 6.00 m radii	22.80	1.08	23.69	–	29.29	m²	**52.98**
half brick thick; curved; 1.50 m radii	26.60	1.41	30.99	–	33.40	m²	**64.39**
one brick thick	45.60	1.41	30.99	–	58.59	m²	**89.58**
one brick thick; curved; 6.00 m radii	49.40	1.84	40.26	–	62.70	m²	**102.96**
one brick thick; curved; 1.50 m radii	49.40	2.29	50.13	–	64.24	m²	**114.37**
one and a half brick thick	68.40	1.92	42.04	–	87.88	m²	**129.92**
one and a half brick thick; battering	68.40	2.21	48.36	–	87.88	m²	**136.24**
two brick thick	91.20	2.33	51.12	–	117.17	m²	**168.29**
two brick thick; battering	91.20	2.75	60.20	–	117.17	m²	**177.37**
337 mm average thick; tapering, one side	68.40	2.41	52.90	–	87.88	m²	**140.78**
450 mm average thick; tapering, one side	91.20	3.12	68.50	–	117.17	m²	**185.67**
337 mm average thick; tapering, both sides	68.40	2.79	61.19	–	87.88	m²	**149.07**
450 mm average thick; tapering, both sides	91.20	3.50	76.78	–	118.72	m²	**195.50**
facework one side, half brick thick	22.80	0.92	20.14	–	29.29	m²	**49.43**
facework one side, one brick thick	45.60	1.50	32.96	–	58.59	m²	**91.55**
facework one side, one and a half brick thick	68.40	2.00	43.82	–	87.88	m²	**131.70**
facework one side, two brick thick	91.20	2.41	52.90	–	117.17	m²	**170.07**
facework both sides, half brick thick	22.80	1.00	21.91	–	29.29	m²	**51.20**
facework both sides, one brick thick	45.60	1.58	34.74	–	58.59	m²	**93.33**
facework both sides, one and a half brick thick	68.40	2.08	45.60	–	87.88	m²	**133.48**
facework both sides, two brick thick	91.20	2.50	54.87	–	117.17	m²	**172.04**
Isolated piers							
one brick thick	45.60	2.36	51.76	–	58.59	m²	**110.35**
two brick thick	91.20	3.70	81.14	–	118.72	m²	**199.86**
three brick thick	136.80	4.67	102.42	–	178.85	m²	**281.27**
Isolated casings							
half brick thick	22.80	1.20	26.32	–	29.29	m²	**55.61**
one brick thick	45.60	2.04	44.74	–	58.59	m²	**103.33**
Chimney stacks							
one brick thick	45.60	2.36	51.76	–	58.59	m²	**110.35**
two brick thick	91.20	3.70	81.14	–	118.72	m²	**199.86**
three brick thick	136.80	4.67	102.42	–	178.85	m²	**281.27**

14 MASONRY

Item	PC £	Labour hours	Labour £	Plant £	Material £	Unit	Total rate £
Projections							
225 mm width; 112 mm depth; vertical	5.07	0.28	6.14	–	6.03	m	**12.17**
225 mm width; 225 mm depth; vertical	10.13	0.56	12.28	–	12.04	m	**24.32**
337 mm width; 225 mm depth; vertical	15.20	0.83	18.20	–	18.07	m	**36.27**
440 mm width; 225 mm depth; vertical	20.27	0.93	20.39	–	24.08	m	**44.47**
Closing cavities							
width of cavity 50 mm, closing with common brickwork half brick thick; vertical	–	0.28	6.14	–	1.49	m	**7.63**
width of cavity 50 mm, closing with common brickwork half brick thick; horizontal	–	0.28	6.14	–	4.44	m	**10.58**
width of cavity 50 mm, closing with common brickwork half brick thick; including damp-proof course; vertical	–	0.37	8.12	–	2.54	m	**10.66**
width of cavity 50 mm, closing with common brickwork half brick thick; including damp-proof course; horizontal	–	0.32	7.01	–	5.48	m	**12.49**
width of cavity 75 mm, closing with common brickwork half brick thick; vertical	–	0.28	6.14	–	2.17	m	**8.31**
width of cavity 75 mm, closing with common brickwork half brick thick; horizontal	–	0.28	6.14	–	6.49	m	**12.63**
width of cavity 75 mm, closing with common brickwork half brick thick; including damp-proof course; vertical	–	0.37	8.12	–	3.22	m	**11.34**
width of cavity 75 mm, closing with common brickwork half brick thick; including damp-proof course; horizontal	–	0.32	7.01	–	7.54	m	**14.55**
Bonding to existing							
half brick thick	–	0.28	6.14	–	1.64	m	**7.78**
one brick thick	–	0.42	9.21	–	3.27	m	**12.48**
one and a half brick thick	–	0.65	14.26	–	4.90	m	**19.16**
two brick thick	–	0.88	19.30	–	6.54	m	**25.84**
Arches							
height on face 102 mm, width of exposed soffit 102 mm, shape of arch – segmental, one ring	2.53	1.57	30.35	–	18.09	m	**48.44**
height on face 102 mm, width of exposed soffit 215 mm, shape of arch – segmental, one ring	5.07	2.04	40.65	–	21.20	m	**61.85**
height on face 102 mm, width of exposed soffit 102 mm, shape of arch – semi-circular, one ring	2.53	1.99	39.56	–	18.09	m	**57.65**
height on face 102 mm, width of exposed soffit 215 mm, shape of arch – semi-circular, one ring	5.07	2.50	50.75	–	21.20	m	**71.95**
height on face 215 mm, width of exposed soffit 102 mm, shape of arch – segmental, two ring	5.07	1.99	39.56	–	21.21	m	**60.77**
height on face 215 mm, width of exposed soffit 215 mm, shape of arch – segmental, two ring	10.13	2.45	49.65	–	27.43	m	**77.08**
height on face 215 mm, width of exposed soffit 102 mm, shape of arch – semi-circular, two ring	5.07	2.68	54.69	–	21.21	m	**75.90**
height on face 215 mm, width of exposed soffit 215 mm, shape of arch – semi-circular, two ring	10.13	3.05	62.81	–	27.43	m	**90.24**

14 MASONRY

Item	PC £	Labour hours	Labour £	Plant £	Material £	Unit	Total rate £
14.01 BRICK WALLING – cont							
Common bricks – cont							
ADD or DEDUCT to walls for variation of £10.00/							
1000 in prime cost of common bricks							
half brick thick	–	–	–	–	0.62	m²	**0.62**
one brick thick	–	–	–	–	1.24	m²	**1.24**
one and a half brick thick	–	–	–	–	1.85	m²	**1.85**
two brick thick	–	–	–	–	2.47	m²	**2.47**
Class B engineering bricks							
Walls; in cement mortar (1:3)							
prime cost £ per 1000	438.10	–	–	–	438.10	1000	**438.10**
half brick thick	26.29	0.92	20.14	–	33.06	m²	**53.20**
one brick thick	52.57	1.50	32.96	–	66.13	m²	**99.09**
one brick thick; building against other work	52.57	1.79	39.27	–	70.76	m²	**110.03**
one brick thick; curved; 6.00 m radii	52.57	2.00	43.82	–	66.13	m²	**109.95**
one and a half brick thick	78.86	2.00	43.82	–	99.19	m²	**143.01**
one and a half brick thick; building against other work	78.86	2.41	52.90	–	99.19	m²	**152.09**
two brick thick	105.14	2.50	54.87	–	132.25	m²	**187.12**
337 mm thick; tapering, one side	78.86	2.58	56.65	–	99.19	m²	**155.84**
450 mm thick; tapering, one side	105.14	3.33	73.03	–	132.25	m²	**205.28**
337 mm thick; tapering, both sides	78.86	3.00	65.72	–	99.19	m²	**164.91**
450 mm thick; tapering, both sides	105.14	3.79	83.10	–	133.80	m²	**216.90**
facework one side, half brick thick	26.29	1.00	21.91	–	33.06	m²	**54.97**
facework one side, one brick thick	52.57	1.58	34.74	–	66.13	m²	**100.87**
facework one side, one and a half brick thick	78.86	2.08	45.60	–	99.19	m²	**144.79**
facework one side, two brick thick	105.14	2.58	56.65	–	132.25	m²	**188.90**
facework both sides, half brick thick	26.29	1.08	23.69	–	33.06	m²	**56.75**
facework both sides, one brick thick	52.57	1.67	36.51	–	66.13	m²	**102.64**
facework both sides, one and a half brick thick	78.86	2.17	47.57	–	99.19	m²	**146.76**
facework both sides, two brick thick	105.14	2.66	58.42	–	132.25	m²	**190.67**
Isolated piers							
one brick thick	52.57	2.59	56.80	–	66.13	m²	**122.93**
two brick thick	105.14	4.07	89.26	–	133.80	m²	**223.06**
three brick thick	157.72	5.00	109.65	–	201.47	m²	**311.12**
Isolated casings							
half brick thick	26.29	1.30	28.51	–	33.06	m²	**61.57**
one brick thick	52.57	2.22	48.69	–	66.13	m²	**114.82**
Projections							
225 mm width; 112 mm depth; vertical	5.84	0.32	7.01	–	6.86	m	**13.87**
225 mm width; 225 mm depth; vertical	11.68	0.60	13.16	–	13.72	m	**26.88**
337 mm width; 225 mm depth; vertical	17.52	0.88	19.30	–	20.58	m	**39.88**
440 mm width; 225 mm depth; vertical	23.37	1.02	22.37	–	27.43	m	**49.80**
Bonding to existing							
half brick thick	–	0.32	7.01	–	1.81	m	**8.82**
one brick thick	–	0.46	10.08	–	3.63	m	**13.71**
one and a half brick thick	–	0.65	14.26	–	5.43	m	**19.69**
two brick thick	–	0.97	21.27	–	7.24	m	**28.51**

14 MASONRY

Item	PC £	Labour hours	Labour £	Plant £	Material £	Unit	Total rate £
ADD or DEDUCT to walls for variation of £10.00/ 1000 in prime cost of bricks							
half brick thick	–	–	–	–	0.62	m²	**0.62**
one brick thick	–	–	–	–	1.24	m²	**1.24**
one and a half brick thick	–	–	–	–	1.85	m²	**1.85**
two brick thick	–	–	–	–	2.47	m²	**2.47**
Class A engineering bricks							
Walls; in cement mortar (1:3)							
bricks prime cost £ per 1000	513.70	–	–	–	513.70	1000	**513.70**
half brick thick	30.82	0.92	20.14	–	37.97	m²	**58.11**
one brick thick	61.64	1.50	32.96	–	75.94	m²	**108.90**
one brick thick; building against other work	61.64	1.79	39.27	–	80.58	m²	**119.85**
one brick thick; curved; 6.00 m radii	61.64	2.00	43.82	–	75.94	m²	**119.76**
one and a half brick thick	92.47	2.00	43.82	–	113.91	m²	**157.73**
one and a half brick thick; building against other work	92.47	2.41	52.90	–	113.91	m²	**166.81**
two brick thick	123.29	2.50	54.87	–	151.87	m²	**206.74**
337 mm thick; tapering, one side	92.47	2.58	56.65	–	113.91	m²	**170.56**
450 mm thick; tapering, one side	123.29	3.33	73.03	–	151.87	m²	**224.90**
337 mm thick; tapering, both sides	92.47	3.00	65.72	–	113.91	m²	**179.63**
450 mm thick; tapering, both sides	123.29	3.79	83.10	–	153.42	m²	**236.52**
facework one side, half brick thick	30.82	1.00	21.91	–	37.97	m²	**59.88**
facework one side, one brick thick	61.64	1.58	34.74	–	75.94	m²	**110.68**
facework one side, one and a half brick thick	92.47	2.08	45.60	–	113.91	m²	**159.51**
facework one side, two brick thick	123.29	2.58	56.65	–	151.87	m²	**208.52**
facework both sides, half brick thick	30.82	1.08	23.69	–	37.97	m²	**61.66**
facework both sides, one brick thick	61.64	1.67	36.51	–	75.94	m²	**112.45**
facework both sides, one and a half brick thick	92.47	2.17	47.57	–	113.91	m²	**161.48**
facework both sides, two brick thick	123.29	2.66	58.42	–	151.87	m²	**210.29**
Isolated piers							
one brick thick	61.64	2.59	56.80	–	75.94	m²	**132.74**
two brick thick	123.29	4.07	89.26	–	153.42	m²	**242.68**
three brick thick	184.93	5.00	109.65	–	230.91	m²	**340.56**
Isolated casings							
half brick thick	30.82	1.30	28.51	–	37.97	m²	**66.48**
one brick thick	61.64	2.22	48.69	–	75.94	m²	**124.63**
Projections							
225 mm width; 112 mm depth; vertical	6.85	0.32	7.01	–	7.95	m	**14.96**
225 mm width; 225 mm depth; vertical	13.70	0.60	13.16	–	15.89	m	**29.05**
337 mm width; 225 mm depth; vertical	20.55	0.88	19.30	–	23.84	m	**43.14**
440 mm width; 225 mm depth; vertical	27.40	1.02	22.37	–	31.80	m	**54.17**
Bonding to existing							
half brick thick	–	0.32	7.01	–	2.08	m	**9.09**
one brick thick	–	0.46	10.08	–	4.17	m	**14.25**
one and a half brick thick	–	0.65	14.26	–	6.25	m	**20.51**
two brick thick	–	0.97	21.27	–	8.33	m	**29.60**
ADD or DEDUCT to walls for variation of £10.00/ 1000 in prime cost of bricks							
half brick thick	–	–	–	–	0.62	m²	**0.62**
one brick thick	–	–	–	–	1.24	m²	**1.24**
one and a half brick thick	–	–	–	–	1.85	m²	**1.85**

14 MASONRY

Item	PC £	Labour hours	Labour £	Plant £	Material £	Unit	Total rate £
14.01 BRICK WALLING – cont							
Class A engineering bricks – cont							
ADD or DEDUCT to walls for variation of £10.00/ 1000 in prime cost of bricks – cont							
two brick thick	–	–	–	–	2.47	m²	**2.47**
Facing bricks; machine-made facings; in gauged mortar (1:1:6)							
Walls; in gauged mortar (1:1:6)							
bricks prime cost £ per 1000	500.00	–	–	–	500.00	1000	**500.00**
facework one side, half brick thick; stretcher bond	30.00	1.08	23.69	–	37.08	m²	**60.77**
facework one side, half brick thick, flemish bond with snapped headers	40.00	1.25	27.44	–	47.90	m²	**75.34**
facework one side, half brick thick, stretcher bond; building against other work; concrete	30.00	1.17	25.66	–	40.17	m²	**65.83**
facework one side, half brick thick; flemish bond with snapped headers; building against other work; concrete	40.00	1.33	29.21	–	50.99	m²	**80.20**
facework one side, half brick thick, stretcher bond; building overhand	30.00	1.33	29.21	–	37.08	m²	**66.29**
facework one side, half brick thick; flemish bond with snapped headers; building overhand	40.00	1.50	32.96	–	47.90	m²	**80.86**
facework one side, half brick thick; stretcher bond; curved; 6.00 m radii	30.00	1.58	34.74	–	37.08	m²	**71.82**
facework one side, half brick thick; flemish bond with snapped headers; curved; 6.00 m radii	40.00	1.79	39.27	–	47.90	m²	**87.17**
facework one side, half brick thick; stretcher bond; curved; 1.50 m radii	35.00	2.00	43.82	–	42.49	m²	**86.31**
facework one side, half brick thick; flemish bond with snapped headers; curved; 1.50 m radii	40.00	2.33	51.12	–	47.90	m²	**99.02**
facework both sides, one brick thick; two stretcher skins tied together	60.00	1.87	41.06	–	76.84	m²	**117.90**
facework both sides, one brick thick; flemish bond	60.00	1.92	42.04	–	74.16	m²	**116.20**
facework both sides, one brick thick; two stretcher skins tied together; curved; 6.00 m radii	60.00	2.58	56.65	–	76.84	m²	**133.49**
facework both sides, one brick thick; flemish bond; curved; 6.00 m radii	60.00	2.66	58.42	–	74.16	m²	**132.58**
facework both sides, one brick thick; two stretcher skins tied together; curved; 1.50 m radii	60.00	3.20	70.27	–	78.38	m²	**148.65**
facework both sides, one brick thick; flemish bond; curved; 1.50 m radii	60.00	3.33	73.03	–	75.71	m²	**148.74**
Isolated piers							
facework both sides, one brick thick; two stretcher skins tied together	60.00	2.45	53.74	–	78.37	m²	**132.11**
facework both sides, one brick thick; flemish bond	60.00	2.50	54.83	–	78.37	m²	**133.20**
Isolated casings							
facework one side, half brick thick; stretcher bond	30.00	1.85	40.57	–	37.08	m²	**77.65**
facework one side, half brick thick; flemish bond with snapped headers	40.00	2.04	44.74	–	47.90	m²	**92.64**

14 MASONRY

Item	PC £	Labour hours	Labour £	Plant £	Material £	Unit	Total rate £
Projections							
225 mm width; 112 mm depth; stretcher bond; vertical	6.67	0.28	6.14	–	7.76	m	**13.90**
225 mm width; 112 mm depth; flemish bond with snapped headers; vertical	6.67	0.37	8.12	–	7.76	m	**15.88**
225 mm width; 225 mm depth; flemish bond; vertical	13.33	0.60	13.16	–	22.37	m	**35.53**
328 mm width; 112 mm depth; stretcher bond; vertical	10.00	0.56	12.28	–	11.63	m	**23.91**
328 mm width; 112 mm depth; flemish bond with snapped headers; vertical	10.00	0.65	14.26	–	11.63	m	**25.89**
328 mm width; 225 mm depth; flemish bond; vertical	20.00	1.11	24.34	–	23.18	m	**47.52**
440 mm width; 112 mm depth; stretcher bond; vertical	13.33	0.83	18.20	–	15.50	m	**33.70**
440 mm width; 112 mm depth; flemish bond with snapped headers; vertical	13.33	0.88	19.30	–	15.50	m	**34.80**
440 mm width; 225 mm depth; flemish bond; vertical	26.67	1.62	35.52	–	31.00	m	**66.52**
Arches							
height on face 215 mm, width of exposed soffit 102 mm, shape of arch – flat	6.67	0.93	18.58	–	9.90	m	**28.48**
height on face 215 mm, width of exposed soffit 215 mm, shape of arch – flat	13.33	1.39	28.68	–	17.80	m	**46.48**
height on face 215 mm, width of exposed soffit 102 mm, shape of arch – segmental, one ring	6.67	1.76	34.06	–	22.80	m	**56.86**
height on face 215 mm, width of exposed soffit 215 mm, shape of arch segmental, one ring	13.33	2.13	42.18	–	30.00	m	**72.18**
height on face 215 mm, width of exposed soffit 102 mm, shape of arch – semi-circular, one ring	6.67	2.68	54.24	–	22.80	m	**77.04**
height on face 215 mm, width of exposed soffit 215 mm, shape of arch – semi-circular, one ring	13.33	3.61	74.63	–	30.00	m	**104.63**
height on face 215 mm, width of exposed soffit 102 mm, shape of arch – segmental, two ring	6.67	2.17	43.05	–	22.80	m	**65.85**
height on face 215 mm, width of exposed soffit 215 mm, shape of arch – segmental; two ring	13.33	2.82	57.31	–	30.00	m	**87.31**
height on face 215 mm, width of exposed soffit 102 mm, shape of arch – semi-circular, two ring	6.67	3.61	74.63	–	22.80	m	**97.43**
height on face 215 mm, width of exposed soffit 215 mm, shape of arch – semi-circular, two ring	13.33	5.00	105.12	–	30.00	m	**135.12**
Arches; cut voussoirs; prime cost £ per 1000	3450.00	–	–	–	3553.50	1000	**3553.50**
height on face 215 mm, width of exposed soffit 102 mm, shape of arch – segmental, one ring	46.00	1.80	34.94	–	65.34	m	**100.28**
height on face 215 mm, width of exposed soffit 215 mm, shape of arch – segmental, one ring	92.00	2.27	45.25	–	115.08	m	**160.33**
height on face 215 mm, width of exposed soffit 102 mm, shape of arch – semi-circular, one ring	46.00	2.04	40.20	–	65.34	m	**105.54**
height on face 215 mm, width of exposed soffit 215 mm, shape of arch – semi-circular, one ring	92.00	2.59	52.26	–	115.08	m	**167.34**

14 MASONRY

Item	PC £	Labour hours	Labour £	Plant £	Material £	Unit	Total rate £
14.01 BRICK WALLING – cont							
Facing bricks – cont							
Arches – cont							
height on face 320 mm, width of exposed soffit 102 mm, shape of arch – segmental, one and a half ring	92.00	2.41	48.32	–	115.40	m	**163.72**
height on face 320 mm, width of exposed soffit 215 mm, shape of arch – segmental, one and a half ring	184.00	3.15	64.55	–	232.50	m	**297.05**
Arches; bullnosed specials; prime cost £ per 1000	2000.00	–	–	–	2060.00	1000	**2060.00**
height on face 215 mm, width of exposed soffit 102 mm, shape of arch – flat	26.67	0.97	19.46	–	31.53	m	**50.99**
height on face 215 mm, width of exposed soffit 215 mm, shape of arch – flat	54.00	1.43	29.55	–	61.78	m	**91.33**
Bullseye windows; 600 mm dia.; facing bricks							
height on face 215 mm, width of exposed soffit 102 mm, two rings	14.00	4.63	97.01	–	22.16	nr	**119.17**
height on face 215 mm, width of exposed soffit 215 mm, two rings	28.00	6.48	137.58	–	38.88	nr	**176.46**
Bullseye windows; 600 mm; cut voussoirs; prime cost £ per 1000	3450.00	–	–	–	3553.50	1000	**3553.50**
height on face 215 mm, width of exposed soffit 102 mm, one ring	120.75	3.89	80.77	–	137.61	nr	**218.38**
height on face 215 mm, width of exposed soffit 215 mm, one ring	120.75	5.37	113.24	–	269.78	nr	**383.02**
Bullseye windows; 1200 mm dia.; facing bricks							
height on face 215 mm, width of exposed soffit 102 mm, two rings	28.00	7.22	153.81	–	53.14	nr	**206.95**
height on face 215 mm, width of exposed soffit 215 mm, two rings	56.00	10.36	222.68	–	88.04	nr	**310.72**
Bullseye windows; 1200 mm dia.; cut voussoirs	3450.00	–	–	–	3553.50	1000	**3553.50**
height on face 215 mm, width of exposed soffit 102 mm, one ring	207.00	6.11	129.46	–	248.27	nr	**377.73**
height on face 215 mm, width of exposed soffit 215 mm, one ring	414.00	8.70	186.27	–	475.22	nr	**661.49**
ADD or DEDUCT for variation of £10.00 per 1000 in PC of facing bricks in 102 mm high arches with 215 mm soffit	–	–	–	–	0.28	m	**0.28**
Facework sills							
150 mm × 102 mm; headers on edge; pointing top and one side; set weathering; horizontal	6.67	0.51	11.19	–	7.58	m	**18.77**
150 mm × 102 mm; cant headers on edge; pointing top and one side; set weathering; horizontal; prime cost £ per 1000	26.67	0.56	12.28	–	28.70	m	**40.98**
150 mm × 102 mm; bullnosed specials; headers on flat; pointing top and one side; horizontal; prime cost £ per 1000	26.67	0.46	10.08	–	28.70	m	**38.78**

14 MASONRY

Item	PC £	Labour hours	Labour £	Plant £	Material £	Unit	Total rate £
Facework copings							
215 mm × 102 mm; headers on edge; pointing top and both sides; horizontal	6.67	0.42	9.21	–	7.98	m	**17.19**
260 mm × 102 mm; headers on edge; pointing top and both sides; horizontal	10.00	0.65	14.26	–	11.79	m	**26.05**
215 mm × 102 mm; double bullnose specials; headers on edge; pointing top and both sides; horizontal; prime cost £ per 1000	2000.00	0.46	10.08	–	28.24	m	**38.32**
260 mm × 102 mm; single bullnose specials; headers on edge; pointing top and both sides; horizontal; prime cost £ per 1000	2000.00	0.65	14.26	–	55.91	m	**70.17**
ADD or DEDUCT for variation of £10.00 per 1000 in prime cost of facing bricks in copings 215 mm wide, 102 mm high	–	–	–	–	0.14	m	**0.14**
Extra over facing bricks for; facework ornamental bands and the like, plain bands							
flush; horizontal; 225 mm width; entirely of stretchers	–	0.19	4.17	–	0.72	m	**4.89**
Extra over facing brick for; facework quoins							
flush; mean girth 320 mm	–	0.28	6.14	–	0.72	m	**6.86**
Bonding to existing							
facework one side, half brick thick; stretcher bond	–	0.46	10.08	–	2.04	m	**12.12**
facework one side, half brick thick; flemish bond with snapped headers	–	0.46	10.08	–	2.04	m	**12.12**
facework both sides, one brick thick; two stretcher skins tied together	–	0.65	14.26	–	4.07	m	**18.33**
facework both sides, one brick thick; flemish bond	–	0.65	14.26	–	4.07	m	**18.33**
ADD or DEDUCT for variation of £10.00 per 1000 in prime cost of facing bricks; in walls built entirely of facings; in stretcher or flemish bond							
half brick thick	–	–	–	–	0.62	m²	**0.62**
one brick thick	–	–	–	–	1.24	m²	**1.24**
Facing bricks; handmade							
Walls in gauged mortar (1:1:6)							
prime cost for bricks £ per 1000	700.00	–	–	–	700.00	1000	**700.00**
facework one side, half brick thick; stretcher bond	42.00	1.08	23.69	–	50.06	m²	**73.75**
facework one side, half brick thick; flemish bond with snapped headers	56.00	1.25	27.44	–	65.20	m²	**92.64**
facework one side; half brick thick; stretcher bond; building against other work; concrete	42.00	1.17	25.66	–	53.15	m²	**78.81**
facework one side, half brick thick; flemish bond with snapped headers; building against other work; concrete	56.00	1.33	29.21	–	65.41	m²	**94.62**
facework one side, half brick thick; stretcher bond; building overhand	42.00	1.33	29.21	–	50.06	m²	**79.27**
facework one side, half brick thick; flemish bond with snapped headers; building overhand	56.00	1.50	32.96	–	65.20	m²	**98.16**

14 MASONRY

Item	PC £	Labour hours	Labour £	Plant £	Material £	Unit	Total rate £
14.01 BRICK WALLING – cont							
Facing bricks – cont							
Walls in gauged mortar (1:1:6) – cont							
facework one side, half brick thick; stretcher bond; curved; 6.00 m radii	42.00	1.58	34.74	–	50.06	m²	**84.80**
facework one side, half brick thick; flemish bond with snapped headers; curved; 6.00 m radii	56.00	1.79	39.27	–	65.20	m²	**104.47**
facework one side, half brick thick; stretcher bond; curved 1.50 m radii	42.00	2.00	43.82	–	50.06	m²	**93.88**
facework one side, half brick thick; flemish bond with snapped headers; curved; 1.50 m radii	56.00	2.33	51.12	–	65.20	m²	**116.32**
facework both sides, one brick thick; two stretcher skins tied together	84.00	1.87	41.06	–	102.79	m²	**143.85**
facework both sides, one brick thick; flemish bond	84.00	1.92	42.04	–	100.12	m²	**142.16**
facework both sides; one brick thick; two stretcher skins tied together; curved; 6.00 m radii	84.00	2.58	56.65	–	102.79	m²	**159.44**
facework both sides, one brick thick; flemish bond; curved; 6.00 m radii	84.00	2.66	58.42	–	100.12	m²	**158.54**
facework both sides, one brick thick; two stretcher skins tied together; curved; 1.50 m radii	84.00	3.20	70.27	–	104.34	m²	**174.61**
facework both sides, one brick thick; flemish bond; curved; 1.50 m radii	84.00	3.33	73.03	–	101.66	m²	**174.69**
Isolated piers							
facework both sides, one brick thick; two stretcher skins tied together	84.00	2.45	53.74	–	104.33	m²	**158.07**
facework both sides, one brick thick; flemish bond	84.00	2.50	54.83	–	104.33	m²	**159.16**
Isolated casings							
facework one side, half brick thick; stretcher bond	42.00	1.85	40.57	–	50.06	m²	**90.63**
facework one side, half brick thick; flemish bond with snapped headers	42.00	2.04	44.74	–	50.06	m²	**94.80**
Projections							
225 mm width; 112 mm depth; stretcher bond; vertical	9.33	0.28	6.14	–	10.64	m	**16.78**
225 mm width; 112 mm depth; flemish bond with snapped headers; vertical	9.33	0.37	8.12	–	10.64	m	**18.76**
225 mm width; 225 mm depth; flemish bond; vertical	18.67	0.60	13.16	–	21.27	m	**34.43**
328 mm width; 112 mm depth; stretcher bond; vertical	14.00	0.56	12.28	–	15.98	m	**28.26**
328 mm width; 112 mm depth; flemish bond with snapped headers; vertical	14.00	0.65	14.26	–	15.98	m	**30.24**
328 mm width; 225 mm depth; flemish bond; vertical	28.00	1.11	24.34	–	31.83	m	**56.17**
440 mm width; 112 mm depth; stretcher bond; vertical	18.67	0.83	18.20	–	21.27	m	**39.47**
440 mm width; 112 mm depth; flemish bond with snapped headers; vertical	18.67	0.88	19.30	–	21.27	m	**40.57**
440 mm width; 225 mm depth; flemish bond; vertical	37.33	1.62	35.52	–	42.54	m	**78.06**

14 MASONRY

Item	PC £	Labour hours	Labour £	Plant £	Material £	Unit	Total rate £
Arches							
height on face 215 mm, width of exposed soffit 102 mm, shape of arch – flat	9.33	0.93	18.58	–	12.78	m	**31.36**
height on face 215 mm, width of exposed soffit 215 mm, shape of arch – flat	18.90	1.39	28.68	–	23.82	m	**52.50**
height on face 215 mm, width of exposed soffit 102 mm, shape of arch – segmental, one ring	9.33	1.76	34.06	–	25.69	m	**59.75**
height on face 215 mm, width of exposed soffit 215 mm, shape of arch – segmental, one ring	18.67	2.13	42.18	–	35.77	m	**77.95**
height on face 215 mm, width of exposed soffit 102 mm, shape of arch – semi-circular, one ring	9.33	2.68	54.24	–	25.69	m	**79.93**
height on face 215 mm, width of exposed soffit 215 mm, shape of arch – semi-circular, one ring	18.67	3.61	74.63	–	35.77	m	**110.40**
height on face 215 mm, width of exposed soffit 102 mm, shape of arch – segmental, two ring	9.33	2.17	43.05	–	25.69	m	**68.74**
height on face 215 mm, width of exposed soffit 215 mm, shape of arch – segmental, two ring	18.67	2.82	57.31	–	35.77	m	**93.08**
height on face 215 mm, width of exposed soffit 102 mm, shape of arch – semi-circular, two ring	9.33	3.61	74.63	–	25.69	m	**100.32**
height on face 215 mm, width of exposed soffit 215 mm, shape of arch – semi-circular, two ring	18.67	5.00	105.12	–	35.77	m	**140.89**
Arches; cut voussoirs; prime cost £ per 1000	3450.00	–	–	–	3553.50	1000	**3553.50**
height on face 215 mm, width of exposed soffit 102 mm, shape of arch – segmental, one ring	46.00	1.80	34.94	–	65.34	m	**100.28**
height on face 215 mm, width of exposed soffit 215 mm, shape of arch – segmental, one ring	92.00	2.27	45.25	–	115.08	m	**160.33**
height on face 215 mm, width of exposed soffit 102 mm, shape of arch – semi-circular, one ring	46.00	2.04	40.20	–	65.34	m	**105.54**
height on face 215 mm, width of exposed soffit 215 mm, shape of arch – semi-circular, one ring	92.00	2.59	52.26	–	115.08	m	**167.34**
height on face 320 mm, width of exposed soffit 102 mm, shape of arch – segmental, one and a half ring	92.00	2.41	48.32	–	115.40	m	**163.72**
height on face 320 mm, width of exposed soffit 215 mm, shape of arch – segmental, one and a half ring	184.00	3.15	64.55	–	232.50	m	**297.05**
Arches; bullnosed specials; prime cost £ per 1000	2750.00	–	–	–	2832.50	1000	**2832.50**
height on face 215 mm, width of exposed soffit 102 mm, shape of arch – flat	36.67	0.97	19.46	–	42.34	m	**61.80**
height on face 215 mm, width of exposed soffit 215 mm, shape of arch – flat	74.25	1.43	29.55	–	83.69	m	**113.24**
Bullseye windows; 600 mm dia.							
height on face 215 mm, width of exposed soffit 102 mm, two ring	19.60	4.63	97.01	–	28.21	nr	**125.22**
height on face 215 mm, width of exposed soffit 215 mm, two ring	39.20	6.48	137.58	–	78.81	nr	**216.39**
Bullseye windows; 600 mm dia.; cut voussoirs; prime cost £ per 1000	4000.00	–	–	–	4120.00	1000	**4120.00**
height on face 215 mm, width of exposed soffit 102 mm, one ring	140.00	3.89	80.77	–	158.42	nr	**239.19**
height on face 215 mm, width of exposed soffit 215 mm, one ring	280.00	5.37	113.24	–	311.42	nr	**424.66**

Prices for Measured Works

14 MASONRY

Item	PC £	Labour hours	Labour £	Plant £	Material £	Unit	Total rate £
14.01 BRICK WALLING – cont							
Facing bricks – cont							
Bullseye windows; 1200 mm dia.							
height on face 215 mm, width of exposed soffit 102 mm, two ring	39.20	7.22	153.81	–	65.25	nr	**219.06**
height on face 215 mm, width of exposed soffit 215 mm, two ring	160.72	10.36	222.68	–	112.27	nr	**334.95**
Bullseye windows; 1200 mm dia.; cut voussoirs; prime cost £ per 1000	4000.00	–	–	–	4120.00	1000	**4120.00**
height on face 215 mm, width of exposed soffit 102 mm, one ring	240.00	6.11	129.46	–	283.96	nr	**413.42**
height on face 215 mm, width of exposed soffit 215 mm, one ring	480.00	8.70	186.27	–	546.60	nr	**732.87**
ADD or DEDUCT for variation of £10.00 per 1000 in prime cost of facing bricks in 102 mm high arches with 215 mm soffit	–	–	–	–	0.28	m	**0.28**
Facework sills							
150 mm × 102 mm; headers on edge; pointing top and one side; set weathering; horizontal; prime cost £ per m	9.33	0.51	11.19	–	10.63	m	**21.82**
150 mm × 102 mm; cant headers on edge; pointing top and one side; set weathering; prime cost £ per m	36.67	0.56	12.28	–	40.19	m	**52.47**
150 mm × 102 mm; bullnosed specials; headers on edge; pointing top and one side; horizontal; prime cost £ per m	36.67	0.46	10.08	–	40.19	m	**50.27**
Facework copings							
215 mm × 102 mm; headers on edge; pointing top and both sides; horizontal	9.33	0.42	9.21	–	10.87	m	**20.08**
260 mm × 102 mm; headers on edge; pointing top and both sides; horizontal	14.00	0.65	14.26	–	16.11	m	**30.37**
215 mm × 102 mm; double bullnose specials; headers on edge; pointing top and both sides; horizontal; prime cost £ per m	36.67	0.46	10.08	–	40.43	m	**50.51**
260 mm × 102 mm; single bullnose specials; headers on edge; pointing top and both sides; horizontal; prime cost £ per m	73.33	0.65	14.26	–	80.28	m	**94.54**
ADD or DEDUCT for variation of £10.00 per 1000 in prime cost of facing bricks in copings 215 mm wide, 102 mm high	–	–	–	–	0.14	m	**0.14**
Extra over facing bricks for; facework ornamental bands and the like, plain bands							
flush; horizontal; 225 mm width; entirely of stretchers; prime cost £ per m	–	0.19	4.17	–	0.72	m	**4.89**
Extra over facing bricks for; facework quoins							
flush mean girth 320 mm; prime cost £ per m	–	0.28	6.14	–	0.69	m	**6.83**

14 MASONRY

Item	PC £	Labour hours	Labour £	Plant £	Material £	Unit	Total rate £
Bonding ends to existing							
facework one side, half brick thick; stretcher bond	–	0.46	10.08	–	2.04	m	**12.12**
facework one side, half brick thick; flemish bond with snapped headers	–	0.46	10.08	–	2.04	m	**12.12**
facework both sides, one brick thick; two stretcher skins tied together	–	0.65	14.26	–	4.07	m	**18.33**
facework both sides, one brick thick; flemish bond	–	0.65	14.26	–	4.07	m	**18.33**
ADD or DEDUCT for variation of £10.00/1000 in prime cost of facing bricks; in walls built entirely of facings; in stretcher or flemish bond							
half brick thick	–	–	–	–	0.62	m²	**0.62**
one brick thick	–	–	–	–	1.24	m²	**1.24**
Facework steps							
Facework steps; machine cut specials in ce,emy mortar (1:3)							
prime cost £ per 1000 with class A engineering bricks	2000.00	–	–	–	2000.00	1000	**2000.00**
215 mm × 102 mm Class A engineering bricks; all headers-on-edge; edges set with bullnosed specials; pointing top and one side; set weathering; horizontal; specials	26.67	0.51	11.19	–	36.79	m	**47.98**
returned ends pointed	–	0.14	3.07	–	0.55	nr	**3.62**
430 mm × 102 mm; all headers-on-edge; edges set with bullnosed specials; pointing top and one side; set weathering; horizontal; engineering bricks	26.67	0.74	16.23	–	32.39	m	**48.62**
returned ends pointed	–	0.19	4.17	–	0.55	nr	**4.72**
Facing bricks; slips							
50 mm thick; in gauged mortar (1:1:6) built up against concrete including flushing up at back (ties not included)							
prime cost brick £ per 1000	1350.00	–	–	–	1350.00	1000	**1350.00**
walls facework one side, half brick thick; stretcher bond	81.00	1.85	40.57	–	259.10	m²	**299.67**
edges of suspended slabs; 200 mm wide	–	0.56	12.28	–	18.45	m	**30.73**
columns; 400 mm wide	–	1.11	24.34	–	36.89	m	**61.23**
Facing tile bricks; lbstock Tilebrick or other equal and approved; in gauged mortar (1:1:6)							
Walls							
standard tile brick; facework one side; half brick thick; stretcher bond	–	0.87	19.08	–	94.88	m²	**113.96**
three quarter tile brick; facework one side; half brick thick; stretcher bond	–	0.87	19.08	–	110.68	m²	**129.76**
Extra over facing tile bricks for							
fair ends; 79 mm long	–	0.28	6.14	–	41.34	m	**47.48**
fair ends; 163 mm long	–	0.28	6.14	–	41.34	m	**47.48**
90° × ½ external return	–	0.28	6.14	–	86.86	m	**93.00**

14 MASONRY

Item	PC £	Labour hours	Labour £	Plant £	Material £	Unit	Total rate £
14.01 BRICK WALLING – cont							
Facing tile bricks – cont							
Extra over facing tile bricks for – cont							
90° internal return	–	0.28	6.14	–	102.43	m	**108.57**
30° or 45° or 60° external return	–	0.28	6.14	–	86.86	m	**93.00**
30° or 45° or 60° internal return	–	0.28	6.14	–	86.86	m	**93.00**
angled verge	–	0.28	6.14	–	48.48	m	**54.62**
Air bricks							
Air bricks; red terracotta; building into prepared openings							
215 mm × 65 mm	–	0.07	1.53	–	3.94	nr	**5.47**
215 mm × 140 mm	–	0.07	1.53	–	5.85	nr	**7.38**
215 mm × 215 mm	–	0.07	1.53	–	14.58	nr	**16.11**
14.02 BLOCK WALLING							
SUPPLY ONLY BLOCK PRICES							
Alternative prices below are for supply only blocks.							
NOTE: These costs exclude labour, mortar wastage, overheads and profit recovery							
Celcon Standard blocks; 440 mm × 215 mm							
100 mm Standard; 3.6N	–	–	–	–	13.81	m²	**13.81**
140 mm Standard; 3.6N	–	–	–	–	19.37	m²	**19.37**
150 mm Standard; 3.6N	–	–	–	–	20.70	m²	**20.70**
215 mm Standard; 3.6N	–	–	–	–	29.67	m²	**29.67**
100 mm Standard; 7N	–	–	–	–	20.70	m²	**20.70**
140 mm Standard; 7N	–	–	–	–	34.74	m²	**34.74**
150 mm Standard; 7N	–	–	–	–	22.74	m²	**22.74**
215 mm Standard; 7N	–	–	–	–	44.64	m²	**44.64**
215 mm × 65 mm × 100 mm coursing block	–	–	–	–	2.48	m	**2.48**
Celcon Solar blocks; 440 mm × 215 mm							
100 mm Solar; 3.6N	–	–	–	–	14.74	m²	**14.74**
140 mm Solar; 3.6N	–	–	–	–	20.60	m²	**20.60**
150 mm Solar; 3.6N	–	–	–	–	22.05	m²	**22.05**
215 mm Solar; 3.6N	–	–	–	–	31.52	m²	**31.52**
265 mm Solar; 3.6N	–	–	–	–	38.95	m²	**38.95**
Celcon Foundation Blocks							
300 mm × 440 mm × 215 mm; 3.6N	–	–	–	–	8.03	m	**8.03**
Durox Supabloc; Aerated concrete; 620 mm × 215 mm (7 nr per m²)							
100 mm Supabloc; 3.6N	–	–	–	–	15.28	m²	**15.28**
140 mm Supabloc; 3.6N	–	–	–	–	21.40	m²	**21.40**
200 mm Supabloc; 3.6N	–	–	–	–	30.57	m²	**30.57**
215 mm Supabloc; 3.6N	–	–	–	–	32.86	m²	**32.86**
100 mm Supabloc4; 4.2N	–	–	–	–	18.15	m²	**18.15**
140 mm Supabloc4; 4.2N	–	–	–	–	25.40	m²	**25.40**
200 mm Supabloc4; 4.2N	–	–	–	–	36.29	m²	**36.29**

14 MASONRY

Item	PC £	Labour hours	Labour £	Plant £	Material £	Unit	Total rate £
215 mm Supabloc4; 4.2N	–	–	–	–	39.01	m²	**39.01**
100 mm Supabloc7; 7.3N	–	–	–	–	19.57	m²	**19.57**
140 mm Supabloc7; 7.3N	–	–	–	–	27.40	m²	**27.40**
200 mm Supabloc7; 7.3N	–	–	–	–	39.15	m²	**39.15**
215 mm Supabloc7; 7.3N	–	–	–	–	42.08	m²	**42.08**
Forterra Fenlite Background dense block							
100 mm solid; 3.6N	–	–	–	–	12.44	m²	**12.44**
100 mm solid; 7.3N	–	–	–	–	18.16	m²	**18.16**
Forterra Evalast Background Standard Dense Aggregate block							
100 mm solid	–	–	–	–	9.65	m²	**9.65**
140 mm hollow	–	–	–	–	13.51	m²	**13.51**
215 mm hollow	–	–	–	–	20.75	m²	**20.75**
Lignacite Lignacrete standard blocks; 440 mm × 215 mm; 7.3N							
100 mm	–	–	–	–	13.82	m²	**13.82**
140 mm	–	–	–	–	19.37	m²	**19.37**
150 mm	–	–	–	–	20.74	m²	**20.74**
190 mm	–	–	–	–	26.29	m²	**26.29**
215 mm	–	–	–	–	29.73	m²	**29.73**
Tarmac Hemelite; 440 mm × 215 mm							
100 mm solid; 3.50 N/mm²	–	–	–	–	15.76	m²	**15.76**
100 mm solid; 7.00 N/mm²	–	–	–	–	15.55	m²	**15.55**
140 mm solid; 7.00 N/mm²	–	–	–	–	16.87	m²	**16.87**
190 mm solid; 7.00 N/mm²	–	–	–	–	29.55	m²	**29.55**
215 mm solid; 7.00 N/mm²	–	–	–	–	33.44	m²	**33.44**
Tarmac Toplite standard blocks; 440 mm × 215 mm							
100 mm	–	–	–	–	15.89	m²	**15.89**
125 mm	–	–	–	–	19.89	m²	**19.89**
140 mm	–	–	–	–	22.26	m²	**22.26**
150 mm	–	–	–	–	23.89	m²	**23.89**
215 mm	–	–	–	–	34.15	m²	**34.15**
Tarmac Toplite GTi (thermal) blocks; 440 mm × 215 mm							
100 mm	–	–	–	–	18.81	m²	**18.81**
125 mm	–	–	–	–	23.56	m²	**23.56**
140 mm	–	–	–	–	26.37	m²	**26.37**
150 mm	–	–	–	–	28.21	m²	**28.21**
215 mm	–	–	–	–	40.42	m²	**40.42**
Plasmor concrete blocks; 440 mm × 215 mm							
100 mm Aglite, standard blocks 10.4N	–	–	–	–	14.47	m²	**14.47**
140 mm Aglite, standard blocks 10.4N	–	–	–	–	20.30	m²	**20.30**
190 mm Aglite, standard blocks 10.4N	–	–	–	–	33.17	m²	**33.17**
215 Aglite, standard blocks 10.4N	–	–	–	–	38.01	m²	**38.01**
100 mm Stranlite, standard blocks 10.4N	–	–	–	–	14.87	m²	**14.87**
140 mm Stranlite, standard blocks 10.4N	–	–	–	–	20.95	m²	**20.95**
190 mm Stranlite, standard blocks 10.4N	–	–	–	–	34.86	m²	**34.86**
215 mm Stranlite, standard blocks 10.4N	–	–	–	–	40.13	m²	**40.13**
100 mm Stranlite, paint grade blocks 7.3N	–	–	–	–	15.04	m²	**15.04**
140 mm Stranlite, paint grade blocks 7.3N	–	–	–	–	21.11	m²	**21.11**
190 mm Stranlite, paint grade blocks 7.3N	–	–	–	–	34.62	m²	**34.62**

Prices for Measured Works

14 MASONRY

Item	PC £	Labour hours	Labour £	Plant £	Material £	Unit	Total rate £
14.02 BLOCK WALLING – cont							
SUPPLY ONLY BLOCK PRICES – cont							
Plasmor concrete blocks – cont							
215 mm Stranlite, paint grade blocks 7.3N	–	–	–	–	39.95	m²	**39.95**
100 mm Stranlite, paint grade blocks 10.4N	–	–	–	–	15.77	m²	**15.77**
140 mm Stranlite, paint grade blocks 10.4N	–	–	–	–	22.09	m²	**22.09**
190 mm Stranlite, paint grade blocks 10.4N	–	–	–	–	35.98	m²	**35.98**
215 mm Stranlite, paint grade blocks 10.4N	–	–	–	–	42.87	m²	**42.87**
SUPPLY AND LAY PRICES							
Lightweight aerated concrete blocks; Thermalite Turbo blocks or other equal; in gauged mortar (1:2:9)							
Walls							
100 mm thick	14.74	0.41	9.08	–	17.36	m²	**26.44**
115 mm thick	16.96	0.41	9.08	–	19.95	m²	**29.03**
125 mm thick	18.43	0.41	9.08	–	21.69	m²	**30.77**
130 mm thick	19.17	0.41	9.08	–	22.55	m²	**31.63**
140 mm thick	20.64	0.46	10.06	–	24.29	m²	**34.35**
150 mm thick	22.12	0.46	10.06	–	26.03	m²	**36.09**
190 mm thick	28.01	0.50	11.05	–	32.96	m²	**44.01**
200 mm thick	29.49	0.50	11.05	–	34.70	m²	**45.75**
215 mm thick	31.70	0.50	11.05	–	37.31	m²	**48.36**
265 mm thick	39.07	0.55	12.06	–	45.92	m²	**57.98**
300 mm thick	44.22	0.55	12.06	–	52.06	m²	**64.12**
Isolated piers or chimney stacks							
190 mm thick	–	0.83	18.20	–	32.96	m²	**51.16**
215 mm thick	–	0.83	18.20	–	37.31	m²	**55.51**
Isolated casings							
100 mm thick	–	0.51	11.19	–	17.36	m²	**28.55**
115 mm thick	–	0.51	11.19	–	19.95	m²	**31.14**
125 mm thick	–	0.51	11.19	–	21.69	m²	**32.88**
140 mm thick	–	0.56	12.28	–	24.29	m²	**36.57**
Extra over for fair face; flush pointing							
walls; one side	–	0.04	0.88	–	–	m²	**0.88**
walls; both sides	–	0.09	1.98	–	–	m²	**1.98**
Closing cavities							
width of cavity 50 mm, closing with lightweight blockwork 100 mm thick; vertical	–	0.23	5.05	–	1.00	m	**6.05**
width of cavity 50 mm, closing with lightweight blockwork 100 mm thick; including damp-proof course; vertical	–	0.28	6.14	–	2.05	m	**8.19**
width of cavity 75 mm, closing with lightweight blockwork 100 mm thick; vertical	–	0.23	5.05	–	1.42	m	**6.47**
width of cavity 75 mm, closing with lightweight blockwork 100 mm thick; including damp-proof course; vertical	–	0.28	6.14	–	2.47	m	**8.61**

14 MASONRY

Item	PC £	Labour hours	Labour £	Plant £	Material £	Unit	Total rate £
Bonding ends to common brickwork							
100 mm thick	–	0.14	3.07	–	2.02	m	**5.09**
115 mm thick	–	0.14	3.07	–	2.32	m	**5.39**
125 mm thick	–	0.23	5.05	–	2.53	m	**7.58**
130 mm thick	–	0.23	5.05	–	2.64	m	**7.69**
140 mm thick	–	0.23	5.05	–	2.84	m	**7.89**
150 mm thick	–	0.23	5.05	–	3.03	m	**8.08**
190 mm thick	–	0.28	6.14	–	3.83	m	**9.97**
200 mm thick	–	0.28	6.14	–	4.05	m	**10.19**
215 mm thick	–	0.32	7.01	–	4.36	m	**11.37**
Lightweight aerated concrete blocks; Thermalite Aircrete Shield blocks or other equal; in thin joint mortar							
Walls							
90 mm thick	16.15	0.46	10.08	–	19.62	m²	**29.70**
100 mm thick	17.95	0.46	10.08	–	21.82	m²	**31.90**
115 mm thick	20.64	0.46	10.08	–	24.80	m²	**34.88**
125 mm thick	22.43	0.48	10.53	–	27.15	m²	**37.68**
140 mm thick	25.12	0.51	11.19	–	30.54	m²	**41.73**
150 mm thick	26.91	0.51	11.19	–	32.72	m²	**43.91**
190 mm thick	34.10	0.56	12.28	–	41.44	m²	**53.72**
200 mm thick	35.89	0.60	13.16	–	43.62	m²	**56.78**
Isolated piers or chimney stacks							
190 mm thick	–	0.60	13.16	–	41.44	m²	**54.60**
Isolated casings							
90 mm thick	–	0.35	7.67	–	19.83	m²	**27.50**
100 mm thick	–	0.35	7.67	–	21.82	m²	**29.49**
140 mm thick	–	0.38	8.33	–	30.54	m²	**38.87**
Lightweight aerated concrete blocks; Thermalite Aircrete Shield blocks or other equal; in gauged mortar (1:2:9)							
Walls							
90 mm thick	17.36	0.41	9.08	–	18.81	m²	**27.89**
100 mm thick	19.29	0.41	9.08	–	20.91	m²	**29.99**
115 mm thick	20.64	0.41	8.99	–	24.29	m²	**33.28**
125 mm thick	22.43	0.41	8.99	–	26.27	m²	**35.26**
140 mm thick	27.01	0.46	10.06	–	29.25	m²	**39.31**
150 mm thick	28.93	0.46	10.06	–	31.34	m²	**41.40**
190 mm thick	36.65	0.50	11.05	–	39.70	m²	**50.75**
200 mm thick	38.58	0.50	11.05	–	41.80	m²	**52.85**
Isolated piers or chimney stacks							
190 mm thick	–	0.83	18.20	–	39.70	m²	**57.90**
Isolated casings							
90 mm thick	–	0.51	11.19	–	18.81	m²	**30.00**
100 mm thick	–	0.51	11.19	–	20.91	m²	**32.10**
140 mm thick	–	0.56	12.28	–	29.25	m²	**41.53**

14 MASONRY

Item	PC £	Labour hours	Labour £	Plant £	Material £	Unit	Total rate £
14.02 BLOCK WALLING – cont							
Lightweight aerated concrete blocks – cont							
Extra over for fair face; flush pointing							
walls; one side	–	0.04	0.88	–	–	m²	**0.88**
walls; both sides	–	0.09	1.98	–	–	m²	**1.98**
Closing cavities							
width of cavity 50 mm, closing with lightweight blockwork 100 mm thick; vertical	–	0.23	5.05	–	1.18	m	**6.23**
width of cavity 50 mm, closing with lightweight blockwork 100 mm thick; including damp-proof course; vertical	–	0.28	6.14	–	2.22	m	**8.36**
width of cavity 75 mm, closing with lightweight blockwork 100 mm thick; vertical	–	0.23	5.05	–	1.69	m	**6.74**
width of cavity 75 mm, closing with lightweight blockwork 100 mm thick; including damp-proof course; vertical	–	0.28	6.14	–	2.74	m	**8.88**
Bonding ends to common brickwork							
90 mm thick	–	0.09	1.98	–	2.17	m	**4.15**
100 mm thick	–	0.14	3.07	–	2.41	m	**5.48**
140 mm thick	–	0.23	5.05	–	3.40	m	**8.45**
150 mm thick	–	0.23	5.05	–	3.62	m	**8.67**
190 mm thick	–	0.28	6.14	–	4.58	m	**10.72**
200 mm thick	–	0.28	6.14	–	4.83	m	**10.97**
Smooth face paint grade concrete block; in gauged mortar; flush pointing one side							
Walls							
100 mm thick	13.71	0.50	11.05	–	16.21	m²	**27.26**
140 mm thick	19.18	0.58	12.83	–	22.67	m²	**35.50**
150 mm thick	20.54	0.58	12.83	–	24.30	m²	**37.13**
190 mm thick	26.02	0.67	14.61	–	30.76	m²	**45.37**
200 mm thick	27.39	0.67	14.61	–	32.38	m²	**46.99**
215 mm thick	29.45	0.67	14.61	–	34.81	m²	**49.42**
100 mm thick; Paintgrade Smooth	16.36	0.55	12.08	–	19.16	m²	**31.24**
215 mm thick; Paintgrade Smooth	35.19	0.73	15.92	–	41.18	m²	**57.10**
Isolated piers or chimney stacks							
190 mm thick	–	0.93	20.39	–	30.76	m²	**51.15**
200 mm thick	–	0.93	20.39	–	32.38	m²	**52.77**
215 mm thick	–	0.93	20.39	–	34.81	m²	**55.20**
Isolated casings							
100 mm thick	–	0.69	15.13	–	16.21	m²	**31.34**
140 mm thick	–	0.74	16.23	–	22.67	m²	**38.90**
Extra over for fair face flush pointing							
walls; one side	–	0.04	0.88	–	–	m²	**0.88**
walls; both sides	–	0.09	1.98	–	–	m²	**1.98**
Bonding ends to common brickwork							
100 mm thick	–	0.23	5.05	–	1.93	m	**6.98**
140 mm thick	–	0.23	5.05	–	2.72	m	**7.77**
150 mm thick	–	0.28	6.14	–	2.89	m	**9.03**

14 MASONRY

Item	PC £	Labour hours	Labour £	Plant £	Material £	Unit	Total rate £
190 mm thick	–	0.32	7.01	–	3.66	m	**10.67**
200 mm thick	–	0.32	7.01	–	3.86	m	**10.87**
215 mm thick	–	0.32	7.01	–	4.17	m	**11.18**
Smooth face aerated concrete blocks; Party Wall block; in gauged mortar; flush pointing one side							
Walls							
100 mm thick	10.08	0.56	12.28	–	11.94	m²	**24.22**
215 mm thick	21.67	0.74	16.23	–	25.65	m²	**41.88**
Isolated piers or chimney stacks							
215 mm thick	–	0.93	20.39	–	25.65	m²	**46.04**
Isolated casings							
100 mm thick	–	0.69	15.13	–	11.94	m²	**27.07**
Extra over for fair face flush pointing							
walls; both sides	–	0.04	0.88	–	–	m²	**0.88**
Bonding ends to common brickwork							
100 mm thick	–	0.23	5.05	–	1.35	m	**6.40**
215 mm thick	–	0.32	7.01	–	2.95	m	**9.96**
High strength concrete blocks (7N); Thermalite High Strength 7 blocks or other equal and approved; in cement mortar (1:3)							
Walls							
100 mm thick	15.12	0.41	9.08	–	17.78	m²	**26.86**
140 mm thick	21.17	0.46	10.06	–	24.87	m²	**34.93**
150 mm thick	22.67	0.46	10.06	–	26.65	m²	**36.71**
190 mm thick	28.71	0.50	11.05	–	33.74	m²	**44.79**
200 mm thick	30.23	0.50	11.05	–	35.52	m²	**46.57**
215 mm thick	32.49	0.50	11.05	–	38.19	m²	**49.24**
Isolated piers or chimney stacks							
190 mm thick	–	0.83	18.20	–	33.74	m²	**51.94**
200 mm thick	–	0.83	18.20	–	35.52	m²	**53.72**
215 mm thick	–	0.83	18.20	–	38.19	m²	**56.39**
Isolated casings							
100 mm thick	–	0.51	11.19	–	17.78	m²	**28.97**
140 mm thick	–	0.56	12.28	–	24.87	m²	**37.15**
150 mm thick	–	0.56	12.28	–	26.65	m²	**38.93**
190 mm thick	–	0.69	15.13	–	33.74	m²	**48.87**
200 mm thick	–	0.69	15.13	–	35.52	m²	**50.65**
215 mm thick	–	0.69	15.13	–	38.19	m²	**53.32**
Extra over for flush pointing							
walls; one side	–	0.04	0.88	–	–	m²	**0.88**
walls; both sides	–	0.09	1.98	–	–	m²	**1.98**
Bonding ends to common brickwork							
100 mm thick	–	0.23	5.05	–	2.10	m	**7.15**
140 mm thick	–	0.23	5.05	–	2.98	m	**8.03**
150 mm thick	–	0.28	6.14	–	3.16	m	**9.30**
190 mm thick	–	0.32	7.01	–	4.00	m	**11.01**
200 mm thick	–	0.32	7.01	–	4.22	m	**11.23**
215 mm thick	–	0.32	7.01	–	4.55	m	**11.56**

14 MASONRY

Item	PC £	Labour hours	Labour £	Plant £	Material £	Unit	Total rate £
14.02 BLOCK WALLING – cont							
Dense smooth faced concrete blocks; Lignacite standard and paint grade; in gauged mortar (1:2:9); flush pointing one side							
Walls							
100 mm thick; 3.6N	18.27	0.56	12.24	–	19.86	m²	**32.10**
140 mm thick; 3.6N	25.59	0.65	14.21	–	27.79	m²	**42.00**
100 mm thick; 7N	14.86	0.56	12.24	–	16.34	m²	**28.58**
140 mm thick; 7N	20.82	0.65	14.21	–	22.89	m²	**37.10**
Isolated piers or chimney stacks							
100 mm thick; 7N	–	1.14	25.00	–	17.25	m²	**42.25**
140 mm thick; 7N	–	1.26	27.63	–	23.66	m²	**51.29**
Isolated casings							
100 mm thick; 7N	–	0.78	17.11	–	16.34	m²	**33.45**
140 mm thick; 7N	–	0.90	19.73	–	22.89	m²	**42.62**
Dense aggregate concrete blocks; in cement mortar (1:2:9)							
Walls or partitions or skins of hollow walls							
100 mm thick; solid	17.95	0.62	13.62	–	20.91	m²	**34.53**
140 mm thick; solid	25.12	0.75	16.39	–	29.25	m²	**45.64**
140 mm thick; hollow	25.12	0.67	14.61	–	29.25	m²	**43.86**
190 mm thick; hollow	34.10	0.84	18.35	–	39.70	m²	**58.05**
215 mm thick; hollow	35.89	0.92	20.14	–	41.95	m²	**62.09**
Isolated piers or chimney stacks							
140 mm thick; hollow	–	1.02	22.37	–	29.25	m²	**51.62**
190 mm thick; hollow	–	1.34	29.39	–	39.70	m²	**69.09**
215 mm thick; hollow	–	1.53	33.56	–	41.95	m²	**75.51**
Isolated casings							
100 mm thick; solid	–	0.74	16.23	–	20.91	m²	**37.14**
140 mm thick; solid	–	0.93	20.39	–	29.25	m²	**49.64**
Extra over for fair face; flush pointing							
walls; one side	–	0.09	1.98	–	–	m²	**1.98**
walls; both sides	–	0.14	3.07	–	–	m²	**3.07**
Bonding ends to common brickwork							
100 mm thick solid	–	0.23	5.05	–	2.46	m	**7.51**
140 mm thick solid	–	0.28	6.14	–	3.47	m	**9.61**
140 mm thick hollow	–	0.28	6.14	–	3.47	m	**9.61**
190 mm thick hollow	–	0.32	7.01	–	4.68	m	**11.69**
215 mm thick hollow	–	0.37	8.12	–	4.99	m	**13.11**
Dense aggregate paint grade concrete blocks; (7 N) Forticrete or similar; in cement mortar (1:3)							
Walls							
75 mm thick; solid	11.65	0.50	11.05	–	13.67	m²	**24.72**
100 mm thick; hollow	15.53	0.62	13.62	–	18.23	m²	**31.85**
100 mm thick; solid	15.62	0.62	13.62	–	18.32	m²	**31.94**
140 mm thick; hollow	21.74	0.67	14.61	–	25.50	m²	**40.11**
140 mm thick; solid	21.61	0.75	16.39	–	25.36	m²	**41.75**
190 mm thick; hollow	29.50	0.84	18.35	–	34.61	m²	**52.96**
190 mm thick; solid	29.50	0.92	20.14	–	34.61	m²	**54.75**

14 MASONRY

Item	PC £	Labour hours	Labour £	Plant £	Material £	Unit	**Total rate £**
215 mm thick; hollow	33.38	0.92	20.14	–	39.16	m²	**59.30**
215 mm thick; solid	33.38	0.94	20.61	–	39.16	m²	**59.77**
Dwarf support wall							
140 mm thick; solid	–	1.16	25.44	–	25.36	m²	**50.80**
190 mm thick; solid	–	1.34	29.39	–	34.61	m²	**64.00**
215 mm thick; solid	–	1.53	33.56	–	39.16	m²	**72.72**
Isolated piers or chimney stacks							
140 mm thick; hollow	–	1.02	22.37	–	25.50	m²	**47.87**
190 mm thick; hollow	–	1.34	29.39	–	34.61	m²	**64.00**
215 mm thick; hollow	–	1.53	33.56	–	39.16	m²	**72.72**
Isolated casings							
75 mm thick; solid	–	0.69	15.13	–	13.67	m²	**28.80**
100 mm thick; solid	–	0.74	16.23	–	18.32	m²	**34.55**
140 mm thick; solid	–	0.93	20.39	–	25.36	m²	**45.75**
Extra over for fair face; flush pointing							
walls; one side	–	0.09	1.98	–	–	m²	**1.98**
walls; both sides	–	0.14	3.07	–	–	m²	**3.07**
Bonding ends to common brickwork							
75 mm thick solid	–	0.14	3.07	–	1.62	m	**4.69**
100 mm thick solid	–	0.23	5.05	–	2.16	m	**7.21**
140 mm thick solid	–	0.28	6.14	–	3.03	m	**9.17**
190 mm thick solid	–	0.32	7.01	–	4.10	m	**11.11**
215 mm thick solid	–	0.37	8.12	–	4.67	m	**12.79**
Architectural finish block walls							
100 mm thick Fairface finish	–	1.50	32.90	–	32.49	m²	**65.39**
100 mm thick Shot-blasted finish	–	1.50	32.90	–	43.31	m²	**76.21**
100 mm thick Splitface finish	–	1.50	32.90	–	42.23	m²	**75.13**
100 mm thick Splitface finish	–	2.00	43.86	–	77.91	m²	**121.77**
High strength concrete blocks (10N); Thermalite High Strength 10 blocks; in cement mortar (1:3)							
Walls							
100 mm thick (NB other thicknesses as a special order item)	–	0.50	11.05	–	19.64	m²	**30.69**
Thermalite Trenchblock 3.6N; with tongued and grooved joints; in cement mortar (1:4)							
Walls							
255 mm thick	–	0.60	13.16	–	29.90	m²	**43.06**
275 mm thick	–	0.65	14.26	–	32.24	m²	**46.50**
305 mm thick	–	0.70	15.35	–	35.11	m²	**50.46**
355 mm thick	–	0.75	16.45	–	41.24	m²	**57.69**
Concrete blocks; Thermalite Trenchblock, 7N; with tongued and grooved joints; in cement mortar (1:4)							
Walls							
255 mm thick	33.06	0.70	15.35	–	38.77	m²	**54.12**
275 mm thick	35.68	0.75	16.45	–	41.82	m²	**58.27**
305 mm thick	38.92	0.80	17.54	–	45.57	m²	**63.11**
355 mm thick	46.03	0.85	18.64	–	53.59	m²	**72.23**

14 MASONRY

Item	PC £	Labour hours	Labour £	Plant £	Material £	Unit	Total rate £
14.02 BLOCK WALLING – cont							
Architectural blockwork							
Lignacite Ltd; 440 × 215 mm face size; solid blocks 17.5N; laid stretcher bond with class 3 Snowstorm mortar with white Portland cement and recessed joints							
Snowstorm Weathered							
100 mm thick weathered one face	–	1.00	21.93	–	49.56	m²	**71.49**
100 mm thick; weathered one face and one end	–	0.10	2.19	–	24.72	m	**26.91**
217 mm × 100 mm thick; cut half block; weathered one face	–	1.05	23.03	–	68.45	m²	**91.48**
quoins; weathered two external faces; 440 mm × 100 mm × 215 mm with 215 mm external return	–	0.11	2.41	–	100.01	m	**102.42**
100 mm × 440 mm × 65 mm Roman Bricks weathered one face	–	1.00	21.93	–	55.66	m²	**77.59**
Snowstorm Split							
100 mm thick; split one face	–	1.00	21.93	–	51.78	m²	**73.71**
100 mm thick; split one face and weathered one end	–	0.10	2.19	–	25.71	m	**27.90**
217 mm × 100 mm thick; cut half block; split one face	–	1.05	23.03	–	81.45	m²	**104.48**
quoins; split two external faces; 440 mm × 215 mm × 100 mm thick with 215 mm external return	–	0.11	2.41	–	114.79	m	**117.20**
100 mm × 440 mm × 65 mm Roman Bricks; split one face	–	1.00	21.93	–	60.30	m²	**82.23**
Snowstorm Polished							
Note: polished dimensions reduced by 3 mm							
100 mm thick; polished one face	–	1.10	24.12	–	85.42	m²	**109.54**
100 mm thick; polished one face one end	–	0.11	2.41	–	68.40	m	**70.81**
217 mm × 100 mm thick; cut half block; polished one face	–	1.18	25.77	–	103.39	m²	**129.16**
quoins; polished two external faces; 440 mm × 215 mm × 100 mm thick with 215 mm external return	–	0.12	2.64	–	166.24	m	**168.88**
100 mm × 440 mm × 65 mm Roman Bricks; polished one face	–	1.10	24.12	–	104.15	m²	**128.27**
Snowstorm Planished							
Note: planished dimensions reduced by 3 mm							
100 mm thick; planished one face	–	1.10	24.12	–	85.42	m²	**109.54**
100 mm thick; planished one face one end	–	0.11	2.41	–	68.40	m	**70.81**
100 mm thick; cut half block; planished one face	–	1.18	25.77	–	103.39	m²	**129.16**
quoins; planished two external faces; 440 mm × 215 mm × 100 mm thick with 215 mm external return	–	0.12	2.64	–	166.24	m	**168.88**
100 mm × 440 mm × 65 mm Roman Bricks; planished one face	–	1.10	24.12	–	84.84	m²	**108.96**

14 MASONRY

Item	PC £	Labour hours	Labour £	Plant £	Material £	Unit	Total rate £
Snowstorm and Crushed recycled Glass (Polished or Planished). Standard colours							
Note: planished dimensions reduced by 3 mm							
100 mm thick; planished one face	–	1.10	24.12	–	100.00	m²	**124.12**
100 mm thick; planished one face one end	–	0.11	2.41	–	74.02	m	**76.43**
100 mm thick; cut half block; planished one face	–	1.18	25.77	–	118.51	m²	**144.28**
quoins; planished two external faces;							
440 mm × 215 mm × 100 mm thick with 215 mm							
external return	–	0.12	2.64	–	172.75	m	**175.39**
100 mm × 440 mm × 65 mm Roman Bricks;							
planished one face	–	1.10	24.12	–	117.77	m²	**141.89**
Midnight Polished (Note 7.3N strength) 10 mm Black							
Granite facing bonded to 90 mm thick backing block							
100 mm thick; polished one face	–	1.15	25.22	–	143.22	m²	**168.44**
100 mm thick; polished one face one end	–	0.12	2.64	–	221.68	m	**224.32**
100 mm thick; cut half block; polished one face	–	1.19	25.99	–	203.29	m²	**229.28**
quoins; polished two external faces;							
440 mm × 215 mm × 110 mm thick with 215 mm							
external return	–	0.14	3.07	–	250.40	m	**253.47**
Dense aggregate coloured concrete blocks; Forticrete Bathstone or equal; in coloured gauged mortar (1:1:6); flush pointing one side							
Walls							
100 mm thick hollow	31.72	0.74	16.23	–	35.34	m²	**51.57**
100 mm thick solid	31.72	0.74	16.23	–	35.34	m²	**51.57**
140 mm thick hollow	45.94	0.83	18.20	–	51.12	m²	**69.32**
140 mm thick solid	45.94	0.93	20.39	–	51.12	m²	**71.51**
215 mm thick hollow	52.52	1.16	25.44	–	59.01	m²	**84.45**
Isolated piers or chimney stacks							
140 mm thick solid	–	1.25	27.42	–	51.12	m²	**78.54**
215 mm thick hollow	–	1.57	34.43	–	59.01	m²	**93.44**
Extra over blocks for							
100 mm thick half lintel blocks; ref D14	–	0.23	5.05	–	26.16	m	**31.21**
140 mm thick half lintel blocks; ref H14	–	0.28	6.14	–	46.80	m	**52.94**
140 mm thick quoin blocks; ref H16	–	0.32	7.01	–	39.96	m	**46.97**
140 mm thick cavity closer blocks; ref H17	–	0.32	7.01	–	42.85	m	**49.86**
140 mm thick sill blocks; ref H21	–	0.28	6.14	–	31.48	m	**37.62**
Glazed finish blocks; Forticrete Astra-Glaze or equal; in gauged mortar (1:1:6); joints raked out; gun applied latex grout to joints							
Walls or partitions or skins of hollow walls							
100 mm thick; glazed one side	–	0.93	20.39	–	130.81	m²	**151.20**
extra for glazed square end return	–	0.37	8.12	–	38.57	m	**46.69**
100 mm thick; glazed both sides	–	1.11	24.34	–	165.83	m²	**190.17**
100 mm thick lintel 200 mm high; glazed one side	–	0.83	15.93	–	36.06	m	**51.99**

14 MASONRY

Item	PC £	Labour hours	Labour £	Plant £	Material £	Unit	Total rate £
14.02 BLOCK WALLING – cont							
Fireborn terracotta blocks or other equal and approved; Ibstock Brick Ltd; in coloured gauged mortar (1:1:6); flush pointing one side							
Walls or partitions or skins of hollow walls							
102.50 mm thick; stretcher bond	–	0.33	7.24	–	51.96	m²	**59.20**
102.50 mm thick; stack bond	–	0.35	7.67	–	51.88	m²	**59.55**
14.03 GLASS BLOCK WALLING							
NOTE: The following specialist prices for glass block walling; assume standard blocks in panels of at least 50 m²; work in straight walls at ground level; and all necessary ancillary fixing; strengthening; easy access; pointing and expansion materials etc.							
Hollow glass block walling; Pittsburgh Corning sealed Thinline or other equal and approved; in cement mortar joints; reinforced with 6 mm dia. stainless steel rods; pointed both sides with mastic or other equal and approved							
Walls; facework both sides							
190 mm × 190 mm × 80 mm; clear blocks	69.00	6.00	131.58	–	113.04	m²	**244.62**
190 mm × 190 mm × 80 mm; craft blocks, standard colours	105.00	7.00	153.52	–	153.83	m²	**307.35**
Fire-rated walls							
190 mm × 190 mm × 100 mm glass blocks; 30 minute fire-rated	113.60	8.00	175.45	–	163.57	m²	**339.02**
190 mm × 190 mm × 160 mm glass blocks; 60 minute fire-rated	287.20	9.00	197.38	–	360.26	m²	**557.64**
14.04 NATURAL STONE							
Cotswold Guiting limestone or other equal and approved; laid dry							
Uncoursed random rubble walling							
275 mm thick	–	2.07	47.02	–	13.33	m²	**60.35**
350 mm thick	–	2.46	55.59	–	16.95	m²	**72.54**
425 mm thick	–	2.81	63.18	–	20.59	m²	**83.77**
500 mm thick	–	3.15	70.52	–	24.23	m²	**94.75**
Cotswold Guiting limestone or other equal; bedded; jointed and pointed in cement: lime mortar (1:2:9)							
Uncoursed random rubble walling; faced and pointed; both sides							
275 mm thick	–	1.98	44.83	–	22.67	m²	**67.50**
350 mm thick	–	2.18	48.76	–	28.85	m²	**77.61**
425 mm thick	–	2.39	52.93	–	35.04	m²	**87.97**
500 mm thick	–	2.59	56.87	–	41.22	m²	**98.09**

14 MASONRY

Item	PC £	Labour hours	Labour £	Plant £	Material £	Unit	Total rate £
Coursed random rubble walling; rough dressed; faced and pointed one side							
114 mm thick	–	1.48	31.81	–	59.03	m²	90.84
150 mm thick	–	1.76	41.66	–	71.29	m²	112.95
Fair returns on walling							
114 mm wide	–	0.02	0.44	–	–	m	0.44
150 mm wide	–	0.03	0.66	–	–	m	0.66
275 mm wide	–	0.06	1.32	–	–	m	1.32
350 mm wide	–	0.08	1.75	–	–	m	1.75
425 mm wide	–	0.10	2.19	–	–	m	2.19
500 mm wide	–	0.12	2.64	–	–	m	2.64
Fair raking cutting or circular cutting							
114 mm wide	–	0.20	4.44	–	8.27	m	12.71
150 mm wide	–	0.25	5.55	–	9.93	m	15.48
Level uncoursed rubble walling for damp-proof courses and the like							
275 mm wide	–	0.19	4.64	–	2.94	m	7.58
350 mm wide	–	0.20	4.88	–	3.63	m	8.51
425 mm wide	–	0.21	5.12	–	4.46	m	9.58
500 mm wide	–	0.22	5.37	–	5.29	m	10.66
Copings formed of rough stones; faced and pointed all round							
275 mm × 200 mm (average) high	–	0.56	12.95	–	9.75	m	22.70
350 mm × 250 mm (average) high	–	0.75	17.19	–	13.39	m	30.58
425 mm × 300 mm (average) high	–	0.97	22.08	–	17.96	m	40.04
500 mm × 300 mm (average) high	–	1.23	27.79	–	24.00	m	51.79
Sundries – stone walling							
Coating backs of stones with brush applied cold bitumen solution; two coats							
limestone facework	–	0.19	3.14	–	2.71	m²	5.85
Cutting grooves in limestone masonry for							
water bars or the like	–	–	–	–	–	m	13.24
Mortices in limestone masonry for							
metal dowel	–	–	–	–	–	nr	2.66
metal cramp	–	–	–	–	–	nr	5.29
14.05 CAST STONE							
Reconstructed walling; Bradstone masonry blocks; standard colours; or other equal; laid to pattern or course recommended; bedded; jointed and pointed in approved coloured cement: lime mortar (1:2:9)							
Walls; facing and pointing one side							
Fyfestone Enviromasonry Rustic	–	1.00	21.93	–	27.23	m²	49.16
Fyfestone Enviromasonry Fairfaced	–	1.00	21.93	–	21.05	m²	42.98
Fyfestone Enviromasonry Split	–	1.10	24.12	–	32.83	m²	56.95
Fyfestone Enviromasonry Textured	–	1.10	24.12	–	35.17	m²	59.29
Fyfestone Enviromasonry Polished	–	2.00	43.86	–	84.62	m²	128.48
masonry blocks; random uncoursed	–	1.04	22.80	–	46.40	m²	69.20

14 MASONRY

Item	PC £	Labour hours	Labour £	Plant £	Material £	Unit	Total rate £
14.05 CAST STONE – cont							
Reconstructed walling – cont							
Walls – cont							
extra for							
returned ends	–	0.37	8.12	–	36.11	m	**44.23**
plain L shaped quoins	–	0.12	2.64	–	45.25	m	**47.89**
traditional walling; coursed squared	–	1.30	28.51	–	46.40	m²	**74.91**
squared coursed rubble	–	1.25	27.42	–	48.56	m²	**75.98**
squared random rubble	–	1.30	28.51	–	48.26	m²	**76.77**
squared and pitched rock faced walling; coursed	–	1.34	29.39	–	48.26	m²	**77.65**
ashlar; 440 × 215 × 100 mm thick	–	1.10	24.12	–	49.25	m²	**73.37**
rough hewn rockfaced walling; random	–	1.39	30.49	–	47.95	m²	**78.44**
extra for							
returned ends	–	0.15	3.29	–	–	m	**3.29**
Isolated piers or chimney stacks; facing and pointing one side							
Enviromasonry Rustic	–	1.40	30.70	–	27.23	m²	**57.93**
masonry blocks; random uncoursed	–	1.43	31.36	–	46.40	m²	**77.76**
traditional walling; coursed squared	–	1.80	39.48	–	46.40	m²	**85.88**
squared coursed rubble	–	1.76	38.59	–	48.56	m²	**87.15**
squared random rubble	–	1.80	39.48	–	48.26	m²	**87.74**
squared and pitched rock faced walling; coursed	–	1.90	41.67	–	48.26	m²	**89.93**
ashlar; 440 × 215 × 100 mm thick	–	1.54	33.77	–	49.25	m²	**83.02**
rough hewn rockfaced walling; random	–	1.94	42.55	–	47.95	m²	**90.50**
Isolated casings; facing and pointing one side							
Enviromasonry Rustic	–	1.20	26.32	–	27.23	m²	**53.55**
masonry blocks; random uncoursed	–	1.25	27.42	–	46.40	m²	**73.82**
traditional walling; coursed squared	–	1.57	34.43	–	46.40	m²	**80.83**
squared coursed rubble	–	1.53	33.56	–	48.56	m²	**82.12**
squared random rubble	–	1.57	34.43	–	48.26	m²	**82.69**
squared and pitched rock faced walling; coursed	–	1.62	35.52	–	48.26	m²	**83.78**
ashlar; 440 × 215 × 100 mm thick	–	1.32	28.95	–	49.25	m²	**78.20**
rough hewn rockfaced walling; random	–	1.67	36.63	–	47.95	m²	**84.58**
Fair returns 100 mm wide							
Enviromasonry Rustic	–	0.10	2.19	–	–	m²	**2.19**
masonry blocks; random uncoursed	–	0.11	2.41	–	–	m²	**2.41**
traditional walling; coursed squared	–	0.14	3.07	–	–	m²	**3.07**
squared coursed rubble	–	0.13	2.85	–	–	m²	**2.85**
squared random rubble	–	0.14	3.07	–	–	m²	**3.07**
squared and pitched rock faced walling; coursed	–	0.14	3.07	–	–	m²	**3.07**
ashlar; 440 × 215 × 100 mm thick	–	0.14	3.07	–	–	m²	**3.07**
rough hewn rockfaced walling; random	–	0.15	3.29	–	–	m²	**3.29**
Fair raking cutting or circular cutting							
100 mm wide	–	0.17	3.73	–	–	m	**3.73**
Quoin							
ashlar; 440 × 215 × 215 × 100 mm thick	–	0.75	16.45	–	93.30	m	**109.75**

14 MASONRY

Item	PC £	Labour hours	Labour £	Plant £	Material £	Unit	Total rate £
Reconstructed limestone dressings; Bradstone Architectural dressings in weathered Cotswold or North Cerney shades or other equal and approved; bedded, jointed and pointed in approved coloured cement: lime mortar (1:2:9)							
Copings; twice weathered and throated							
305 mm × 76 mm; type A	–	0.37	8.12	–	31.21	m	**39.33**
extra for							
fair end	–	–	–	–	15.16	nr	**15.16**
returned mitred fair end	–	–	–	–	15.16	nr	**15.16**
Copings; once weathered and throated							
305 mm × 76 mm	–	0.37	8.12	–	31.21	m	**39.33**
356 mm × 76 mm	–	0.37	8.12	–	28.13	m	**36.25**
extra for							
fair end	–	–	–	–	15.16	nr	**15.16**
returned mitred fair end	–	–	–	–	15.16	nr	**15.16**
Pier caps; four times weathered and throated							
305 mm × 305 mm	–	0.23	5.05	–	18.19	nr	**23.24**
381 mm × 381 mm	–	0.23	5.05	–	26.58	nr	**31.63**
457 mm × 457 mm	–	0.28	6.14	–	34.54	nr	**40.68**
533 mm × 533 mm	–	0.28	6.14	–	51.17	nr	**57.31**
Splayed corbels							
479 mm × 100 mm × 215 mm	–	0.14	3.07	–	30.86	nr	**33.93**
665 mm × 100 mm × 215 mm	–	0.19	4.17	–	41.20	nr	**45.37**
100 mm × 140 mm lintels; rectangular; reinforced with mild steel bars							
all lengths to 2.07 m	–	0.26	5.71	–	48.93	m	**54.64**
100 mm × 215 mm lintels; rectangular; reinforced with mild steel bars							
all lengths to 2.85 m	–	0.30	6.58	–	50.88	m	**57.46**
Sills to suit standard windows; stooled 100 mm at ends							
150 mm x 140 mm; not exceeding 1.97 m long	–	0.28	6.14	–	63.93	m	**70.07**
197 mm x 140 mm; not exceeding 1.97 m long	–	0.28	6.14	–	71.66	m	**77.80**
Window surround; traditional with label moulding; for single light; sill 146 mm × 133 mm; jambs 146 mm × 146 mm; head 146 mm × 105 mm; including all dowels and anchors							
overall size 508 mm × 1479 mm	–	0.83	18.20	–	223.25	nr	**241.45**
Window surround; traditional with label moulding; three light; for windows 508 mm × 1219 mm; sill 146 mm × 133 mm; jambs 146 mm × 146 mm; head 146 mm × 103 mm; mullions 146 mm × 108 mm; including all dowels and anchors							
overall size 1975 mm × 1479 mm	–	2.17	47.59	–	526.33	nr	**573.92**
Door surround; moulded continuous jambs and head with label moulding; including all dowels and anchors							
door 839 mm × 1981 mm in 102 mm × 64 mm frame	–	1.53	33.56	–	476.12	nr	**509.68**

14 MASONRY

Item	PC £	Labour hours	Labour £	Plant £	Material £	Unit	Total rate £
14.05 CAST STONE – cont							
Portland Whitbed limestone bedded and jointed in cement–lime–mortar (1:2:9); slurrying with weak lime and stone dust mortar; flush pointing and cleaning on completion (cramps etc. not included)							
Facework; one face plain and rubbed; bedded against backing							
50 mm thick stones	–	–	–	–	–	m²	378.95
63 mm thick stones	–	–	–	–	–	m²	431.64
75 mm thick stones	–	–	–	–	–	m²	488.16
100 mm thick stones	–	–	–	–	–	m²	535.87
Fair returns on facework							
50 mm wide	–	–	–	–	–	m	5.29
63 mm wide	–	–	–	–	–	m	6.61
75 mm wide	–	–	–	–	–	m	9.26
100 mm wide	–	–	–	–	–	m	11.91
Fair raking cutting on facework							
50 mm thick	–	–	–	–	–	m	23.82
63 mm thick	–	–	–	–	–	m	26.46
75 mm thick	–	–	–	–	–	m	31.75
100 mm thick	–	–	–	–	–	m	34.41
Copings; once weathered; and throated; rubbed; set horizontal or raking							
250 mm × 50 mm	–	–	–	–	–	m	185.23
extra for							
external angle	–	–	–	–	–	nr	33.08
internal angle	–	–	–	–	–	nr	33.08
300 mm × 50 mm	–	–	–	–	–	m	195.83
extra for							
external angle	–	–	–	–	–	nr	33.08
internal angle	–	–	–	–	–	nr	39.70
350 mm × 75 mm	–	–	–	–	–	m	218.33
extra for							
external angle	–	–	–	–	–	nr	33.08
internal angle	–	–	–	–	–	nr	42.34
400 mm × 100 mm	–	–	–	–	–	m	261.98
extra for							
external angle	–	–	–	–	–	nr	39.70
internal angle	–	–	–	–	–	nr	55.58
450 mm × 100 mm	–	–	–	–	–	m	317.56
extra for							
external angle	–	–	–	–	–	nr	50.28
internal angle	–	–	–	–	–	nr	68.80
500 mm × 125 mm	–	–	–	–	–	m	482.95
extra for							
external angle	–	–	–	–	–	nr	68.80
internal angle	–	–	–	–	–	nr	86.00

14 MASONRY

Item	PC £	Labour hours	Labour £	Plant £	Material £	Unit	Total rate £
Band courses; plain; rubbed; horizontal							
225 mm × 112 mm	–	–	–	–	–	m	145.54
300 mm × 112 mm	–	–	–	–	–	m	194.50
extra for							
stopped ends	–	–	–	–	–	nr	7.94
external angles	–	–	–	–	–	nr	7.94
Band courses; moulded 100 mm girth on face; rubbed; horizontal							
125 mm × 75 mm	–	–	–	–	–	m	165.40
extra for							
stopped ends	–	–	–	–	–	nr	26.46
external angles	–	–	–	–	–	nr	33.08
internal angles	–	–	–	–	–	nr	66.16
150 mm × 75 mm	–	–	–	–	–	m	191.87
extra for							
stopped ends	–	–	–	–	–	nr	26.46
external angles	–	–	–	–	–	nr	46.32
internal angles	–	–	–	–	–	nr	79.40
200 mm × 100 mm	–	–	–	–	–	m	218.33
extra for							
stopped ends	–	–	–	–	–	nr	26.46
external angles	–	–	–	–	–	nr	66.16
internal angles	–	–	–	–	–	nr	119.08
250 mm × 150 mm	–	–	–	–	–	m	317.56
extra for							
stopped ends	–	–	–	–	–	nr	26.46
external angles	–	–	–	–	–	nr	66.16
internal angles	–	–	–	–	–	nr	132.32
300 mm × 250 mm	–	–	–	–	–	m	516.03
extra for							
stopped ends	–	–	–	–	–	nr	26.46
external angles	–	–	–	–	–	nr	92.62
internal angles	–	–	–	–	–	nr	158.78
Coping apex block; two sunk faces; rubbed							
650 mm × 450 mm × 225 mm	–	–	–	–	–	nr	648.35
Coping kneeler block; three sunk faces; rubbed							
350 mm × 350 mm × 375 mm	–	–	–	–	–	nr	516.03
450 mm × 450 mm × 375 mm	–	–	–	–	–	nr	595.42
Corbel; turned and moulded; rubbed							
225 mm × 225 mm × 375 mm	–	–	–	–	–	nr	423.41
Slab surrounds to openings; one face splayed; rubbed							
75 mm × 100 mm	–	–	–	–	–	m	92.62
75 mm × 200 mm	–	–	–	–	–	m	125.70
100 mm × 100 mm	–	–	–	–	–	m	112.47
125 mm × 100 mm	–	–	–	–	–	m	125.70
125 mm × 150 mm	–	–	–	–	–	m	152.16
175 mm × 175 mm	–	–	–	–	–	m	185.23
225 mm × 175 mm	–	–	–	–	–	m	218.33
300 mm × 175 mm	–	–	–	–	–	m	264.63
300 mm × 225 mm	–	–	–	–	–	m	344.02

14 MASONRY

Item	PC £	Labour hours	Labour £	Plant £	Material £	Unit	Total rate £
14.05 CAST STONE – cont							
Portland Whitbed limestone bedded and jointed in cement–lime–mortar (1:2:9) – cont							
Slab surrounds to openings; one face sunk splayed; rubbed							
75 mm × 100 mm	–	–	–	–	–	m	119.08
75 mm × 200 mm	–	–	–	–	–	m	152.16
100 mm × 100 mm	–	–	–	–	–	m	138.94
125 mm × 100 mm	–	–	–	–	–	m	152.16
125 mm × 150 mm	–	–	–	–	–	m	178.62
175 mm × 175 mm	–	–	–	–	–	m	211.70
225 mm × 175 mm	–	–	–	–	–	m	244.79
300 mm × 175 mm	–	–	–	–	–	m	291.10
300 mm × 225 mm	–	–	–	–	–	m	370.49
extra for							
throating	–	–	–	–	–	m	13.24
rebates and grooves	–	–	–	–	–	m	29.11
stooling	–	–	–	–	–	m	50.28
Eurobrick insulated brick cladding systems or other equal and approved; extruded polystyrene foam insulation; brick slips bonded to insulation panels with Eurobrick gun applied adhesive or other equal and approved; pointing with formulated mortar grout							
25 mm insulation to walls							
over 300 mm wide; fixing with proprietary screws and plates to timber	–	1.39	31.01	–	61.26	m²	92.27
50 mm insulation to walls							
over 300 mm wide; fixing with proprietary screws and plates; to timber	–	1.39	31.01	–	66.84	m²	97.85
14.06 FLUES AND FLUE LININGS							
Flues							
Scheidel Rite-Vent ICS Plus flue system; suitable for domestic multi-fuel appliances; stainless steel; twin wall; insulated; for use internally or externally							
80 mm pipes; including one locking band (fixing brackets measured separately)	–	0.90	17.89	–	103.54	m	121.43
extra for							
appliance connecter	–	0.80	15.90	–	16.02	nr	31.92
30° bend	–	1.80	35.77	–	77.95	nr	113.72
45° bend	–	1.80	35.77	–	74.06	nr	109.83
135° tee; fully welded	–	2.70	53.66	–	153.29	nr	206.95
inspection length	–	0.90	17.89	–	11.07	nr	28.96
drain plug and support	–	1.00	19.88	–	71.80	nr	91.68
damper	–	0.90	17.89	–	58.38	nr	76.27

14 MASONRY

Item	PC £	Labour hours	Labour £	Plant £	Material £	Unit	Total rate £
angled flashing including storm collar	–	1.25	24.84	–	77.34	nr	**102.18**
stub terminal	–	1.00	19.88	–	25.61	nr	**45.49**
tapered terminal	–	1.00	19.88	–	53.35	nr	**73.23**
floor support (2 piece)	–	1.50	29.81	–	42.14	nr	**71.95**
firestop floor support (2 piece)	–	1.50	29.81	–	23.62	nr	**53.43**
wall support (stainless steel)	–	1.00	19.88	–	89.49	nr	**109.37**
wall sleeve	–	1.20	23.85	–	35.51	nr	**59.36**
100 mm pipes; including one locking band (fixing brackets measured separately)	–	1.00	19.88	–	110.28	m	**130.16**
extra for							
appliance connecter	–	0.90	17.89	–	17.70	nr	**35.59**
30° bend	–	2.00	39.75	–	81.53	nr	**121.28**
45° bend	–	2.00	39.75	–	77.41	nr	**117.16**
135° tee; fully welded	–	3.00	59.63	–	148.83	nr	**208.46**
inspection length	–	1.00	19.88	–	253.82	nr	**273.70**
drain plug and support	–	1.10	21.87	–	74.35	nr	**96.22**
damper	–	1.00	19.88	–	61.47	nr	**81.35**
angled flashing including storm collar	–	1.40	27.83	–	86.19	nr	**114.02**
stub terminal	–	1.10	21.87	–	26.02	nr	**47.89**
tapered terminal	–	1.10	21.87	–	56.95	nr	**78.82**
floor support (2 piece)	–	1.65	32.80	–	47.92	nr	**80.72**
firestop floor support (2 piece)	–	1.65	32.80	–	26.03	nr	**58.83**
wall support (stainless steel)	–	1.10	21.87	–	94.55	nr	**116.42**
wall sleeve	–	1.35	26.83	–	40.31	nr	**67.14**
150 mm pipes; including one locking band (fixing brackets measured separately)	–	1.10	21.87	–	128.75	m	**150.62**
extra for							
appliance connecter	–	1.00	19.88	–	22.80	nr	**42.68**
30° bend	–	2.20	43.72	–	97.86	nr	**141.58**
45° bend	–	2.20	43.72	–	93.05	nr	**136.77**
135° tee; fully welded	–	3.30	65.59	–	171.89	nr	**237.48**
inspection length	–	1.10	21.87	–	266.18	nr	**288.05**
drain plug and support	–	1.20	23.85	–	94.95	nr	**118.80**
damper	–	1.10	21.87	–	80.34	nr	**102.21**
angled flashing including storm collar	–	1.55	30.81	–	87.46	nr	**118.27**
stub terminal	–	1.20	23.85	–	28.00	nr	**51.85**
tapered terminal	–	1.20	23.85	–	65.45	nr	**89.30**
floor support (2 piece)	–	1.80	35.77	–	47.92	nr	**83.69**
firestop floor support (2 piece)	–	1.80	35.77	–	26.03	nr	**61.80**
wall support (stainless steel)	–	1.20	23.85	–	104.72	nr	**128.57**
wall sleeve	–	1.50	29.81	–	40.31	nr	**70.12**
14.07 FORMING CAVITIES							
Forming cavities							
In hollow walls							
width of cavity 50 mm; polypropylene ties; three wall ties per m²	–	0.05	1.09	–	0.37	m²	**1.46**
width of cavity 50 mm; galvanized steel twisted wall ties; three wall ties per m²	–	0.05	1.09	–	0.79	m²	**1.88**
width of cavity 50 mm; stainless steel butterfly wall ties; three wall ties per m²	–	0.05	1.09	–	0.58	m²	**1.67**

14 MASONRY

Item	PC £	Labour hours	Labour £	Plant £	Material £	Unit	Total rate £
14.07 FORMING CAVITIES – cont							
Forming cavities – cont							
In hollow walls – cont							
width of cavity 50 mm; stainless steel twisted wall ties; three wall ties per m²	–	0.05	1.09	–	0.72	m²	**1.81**
width of cavity 75 mm; polypropylene ties; three wall ties per m²	–	0.05	1.09	–	0.37	m²	**1.46**
width of cavity 75 mm; galvanized steel twisted wall ties; three wall ties per m²	–	0.05	1.09	–	0.84	m²	**1.93**
width of cavity 75 mm; stainless steel butterfly wall ties; three wall ties per m²	–	0.05	1.09	–	0.81	m²	**1.90**
width of cavity 75 mm; stainless steel twisted wall ties; three wall ties per m²	–	0.05	1.09	–	0.80	m²	**1.89**
Pointing in							
Pointing with mastic							
wood frames or sills	–	0.09	1.57	–	1.16	m	**2.73**
Pointing with polysulphide sealant							
wood frames or sills	–	0.09	1.57	–	2.95	m	**4.52**
Wedging and pinning							
To underside of existing construction with slates in cement: mortar (1:3)							
width of wall – one brick thick	–	0.74	16.23	–	4.24	m	**20.47**
width of wall – one and a half brick thick	–	0.93	20.39	–	8.49	m	**28.88**
width of wall – two brick thick	–	1.11	24.34	–	12.72	m	**37.06**
Slate and tile sills							
Sills; two courses of machine-made plain roofing tiles							
set weathering; bedded and pointed	–	0.56	12.28	–	5.74	m	**18.02**
Sundries							
Weep holes							
weep vents to perpends; plastic	–	0.02	0.44	–	0.35	nr	**0.79**
Chimney pots; red terracotta; plain roll top or cannon-head; setting and flaunching in cement mortar (1:3)							
185 mm dia. × 450 mm long	46.74	1.67	36.63	–	51.23	nr	**87.86**
185 mm dia. × 600 mm long	71.60	1.85	40.57	–	76.84	nr	**117.41**
185 mm dia. × 900 mm long	135.41	1.85	40.57	–	142.56	nr	**183.13**
Halfen Channels and Brick Support (NB Supply only rates)							
cavity brick tie; Halfen HTS-C12; 225 mm stainless steel A2	–	–	–	–	0.42	nr	**0.42**
cavity brick tie; Halfen HTS-C12; 225 mm pre-galvanized	–	–	–	–	0.17	nr	**0.17**

14 MASONRY

Item	PC £	Labour hours	Labour £	Plant £	Material £	Unit	Total rate £
frame cramp, with slot; Halfen HTS-FS12; 150 mm stainless steel A2	–	–	–	–	0.32	nr	**0.32**
frame cramp, with slot; Halfen HTS-FS12; 150 mm pre-galvanized	–	–	–	–	0.14	nr	**0.14**
head restraint; Halfen CHR-V telescopic concealed tube and HSC-9 tie, in A2/304 stainless steel	–	–	–	–	2.62	nr	**2.62**
brick tie channel; Halfen FRS 25/14 2700; supplied in 2700 mm lengths with fixing holes at 110 mm centres. grade 304 stainless steel	–	–	–	–	2.18	nr	**2.18**
channel tie; Halfen FRS-03; 150 mm long safety end to suit 25/14 FRS channel, stainless steel	–	–	–	–	0.42	nr	**0.42**
ribslot brick tie channel; Halfen; 3000 mm long, stainless steel A2, complete with polystyrene filler (vf) and nail holes, to suit HTS ties	–	–	–	–	12.73	nr	**12.73**
channel tie; Halfen HTS-B12; 200 mm long, grade 304 stainless steel safety end to suit 28/15, 25/17HH and ribslot channels	–	–	–	–	0.45	nr	**0.45**
ribslot brick tie channel; Halfen; 3000 mm long, pre-galvanized mild steel, complete with polystyrene filler (vf) and nail holes, to suit HTS ties	–	–	–	–	6.81	nr	**6.81**
channel tie; Halfen HTS-B12; 150 mm long, safety end pre-galvanized to suit 28/15, 25/17 HH and ribslot channels	–	–	–	–	0.17	nr	**0.17**
brick support angle swystem; Halfen HK4-U-100 cavity/6kN inc fixings stainless steel, A4	–	–	–	–	60.47	m	**60.47**
stone support anchor, grout-in; Halfen UMA-10-1; 150 mm stainless steel	–	–	–	–	1.56	nr	**1.56**
stone restraint anchor, grout-in; Halfen UHA-5-1; 180 mm stainless steel A4	–	–	–	–	0.85	nr	**0.85**
windpost, brickwork; Halfen BW1-1406 × 2500 mm complete with fixings & ties, stainless steel, A4	–	–	–	–	230.72	nr	**230.72**
windpost, cavity; Halfen CW2-3544 × 2500 mm stainless steel complete with fixings & ties	–	–	–	–	150.38	nr	**150.38**
cast-in CE marked channel, cold rolled; Halfen HTA-K-38/17; 200 mm anchor centres, stainless steel A2-70, complete with combi filler (kf) and nail holes supplied in 3000 mm lengths	–	–	–	–	21.58	m	**21.58**
'T' bolt and nut to suit; Halfen HS-38/17 channel; M12 × 50 mm, stainless steel A2-70	–	–	–	–	2.03	nr	**2.03**
cast-in CE marked channel, cold rolled; Halfen HTA-K-38/17; 200 mm anchor centres, hot-dip galvanized (fv), complete with combi filler (kf) and nail holes supplies in 3000 mm lengths	–	–	–	–	12.79	m	**12.79**
'T' bolt and nut to suit Halfen HS-38/17 channel; M12 × 50 mm, hot-dip galvanized (fv) 4.6 mild steel	–	–	–	–	0.79	nr	**0.79**

14 MASONRY

Item	PC £	Labour hours	Labour £	Plant £	Material £	Unit	Total rate £
14.07 FORMING CAVITIES – cont							
Sundries – cont							
Halfen Channels and Brick Support – cont							
cast-in CE marked channel, hot rolled; Halfen HTA-K–40/22; hot-dip galvanized (fv), complete with combi filler (kf) and nail holes supplied in 3000 mm lengths	–	–	–	–	43.16	m	**43.16**
'T' bolt and nut to suit Halfen HS–40/22 channel; M16 × 50 mm, hot-dip galvanized (fv) 4.6 mild steel	–	–	–	–	1.48	nr	**1.48**
cast-in channel, hot rolled; Halfen HTA–52/34; hot-dip galvanized (fv), complete with polystyrene filler (vf) and nail holes supplied in 6070 mm lengths	–	–	–	–	37.29	m	**37.29**
'T' bolt and nut, to suit Halfen HS – 52/34 channel; M16 × 60 mm hot-dip galvanized (fv) 4.6 mild steel	–	–	–	–	2.81	nr	**2.81**
cast-in CE marked channel, hot rolled; Halfen HTA–52/34; stainless steel A4, complete with polystyrene filler (vf) and nail holes supplied in 6070 mm lengths	–	–	–	–	115.05	m	**115.05**
'T' bolt and nut, to suit Halfen HS–52/34 channel; M16 × 60 mm stainless steel A4 (fv) 4.6 mild steel	–	–	–	–	7.46	nr	**7.46**
reinforcement, female coupler; Halfen HBM-F16/ M20; 810 mm long	–	–	–	–	3.30	nr	**3.30**
reinforcement, male coupler; Halfen HBM-M16/ M20; 770 mm long	–	–	–	–	4.43	nr	**4.43**
reinforcement continuity system; Halfen Kwikastrip M17Sb12C150; 1200 mm	–	–	–	–	36.59	nr	**36.59**
shear dowel set; Halfen CRET–124 stainless steel A4	–	–	–	–	56.44	nr	**56.44**
shear dowel set; Halfen CRET–10; 20 × 300 stainless steel A4	–	–	–	–	8.24	nr	**8.24**
shear sleeve; Halfen CRET-J–20; 160 stainless steel A4	–	–	–	–	6.80	nr	**6.80**
concrete balcony insulated connection Unit Halfen HIT HP; 80 mm unit for a 200 mm slab for a 1500 mm cantilever balcony (Passivhaus certified)	–	–	–	–	87.55	nr	**87.55**
concrete balcony insulated connection Unit Halfen HIT HP; 80 mm Unit for a 200 mm slab for a 2000 mm cantilever balcony (Passivhaus certified)	–	–	–	–	128.75	nr	**128.75**
steel balcony connection SBC-HBM; M20x4 BZP Cast-in Socket Anchor Assembly, 190 × 110 mm	–	–	–	–	39.14	nr	**39.14**
steel balcony connection; Halfen; SBC-TSS–10; 330 × 200 mm thermal separator pad, 4no. 22 mm holes at 190 × 110 mm	–	–	–	–	92.70	nr	**92.70**
precast lifting spread anchor CE marked; Halfen TPA-FS; 4.0 tonne × 180 mm carbon steel (wb)	–	–	–	–	2.11	nr	**2.11**
precast ring clutch CE marked; Halfen TPA R1; 5.0T (T)	–	–	–	–	117.42	nr	**117.42**

14 MASONRY

Item	PC £	Labour hours	Labour £	Plant £	Material £	Unit	Total rate £
External door and window cavity closures							
Thermabate or equivalent; inclusive of flange clips;							
jointing strips; wall fixing ties and adhesive tape							
closing cavities; width of cavity 50 mm–60 mm	–	0.14	3.07	–	3.92	m	**6.99**
closing cavities; width of cavity 75 mm–84 mm	–	0.14	3.07	–	4.26	m	**7.33**
closing cavities; width of cavity 90 mm–99 mm	–	0.14	3.07	–	4.67	m	**7.74**
closing cavities; width of cavity 100 mm–110 mm	–	0.14	3.07	–	4.18	m	**7.25**
Kooltherm or equivalent; inclusive of flange clips;							
jointing strips; wall fixing ties and adhesive tape; 1 hr							
fire-rating							
closing cavities; width of cavity 50 mm	–	0.14	3.07	–	4.66	m	**7.73**
closing cavities; width of cavity 75 mm	–	0.14	3.07	–	4.84	m	**7.91**
closing cavities; width of cavity 100 mm	–	0.14	3.07	–	5.18	m	**8.25**
closing cavities; width of cavity 125 mm	–	0.14	3.07	–	6.16	m	**9.23**
closing cavities; width of cavity 150 mm	–	0.14	3.07	–	6.45	m	**9.52**
Type H cavicloser or simialr; uPVC universal cavity							
closer, insulator and damp-proof course by Cavity							
Trays Ltd; built into cavity wall as work proceeds,							
complete with face closer and ties							
closing cavities; width of cavity 50 mm–100 mm	–	0.07	1.53	–	2.09	m	**3.62**
Type L durropolyethelene lintel stop ends or other							
equal and approved; Cavity Trays Ltd; fixing with							
butyl anchoring strip; building in as the work							
proceeds							
adjusted to lintel as required	–	0.04	0.88	–	0.74	nr	**1.62**
Type W polypropylene weeps/vents or other equal							
and approved; Cavity Trays Ltd; built into cavity wall							
as work proceeds							
100/115 mm × 65 mm × 10 mm including lock fit							
wedges	–	0.04	0.88	–	0.56	nr	**1.44**
extra; extension duct							
200/225 mm × 65 mm × 10 mm	–	0.07	1.53	–	0.97	nr	**2.50**
Type X polypropylene abutment cavity tray or other							
equal and approved; Cavity Trays Ltd; built into							
facing brickwork as the work proceeds; complete							
with Code 4 flashing; intermediate/catchment tray							
with short leads (requiring soakers); to suit roof of							
17–20° pitch	–	0.05	1.09	–	6.91	nr	**8.00**
21–25° pitch	–	0.05	1.09	–	6.44	nr	**7.53**
26–45° pitch	–	0.05	1.09	–	6.15	nr	**7.24**
Type X polypropylene abutment cavity tray or other							
equal and approved; Cavity Trays Ltd; built into							
facing brickwork as the work proceeds; complete							
with Code 4 flashing; intermediate/catchment tray							
with long leads (suitable only for corrugated roof							
tiles); to suit roof of							
17–20° pitch	–	0.05	1.09	–	9.33	nr	**10.42**
21–25° pitch	–	0.05	1.09	–	8.60	nr	**9.69**
26–45° pitch	–	0.05	1.09	–	7.92	nr	**9.01**

14 MASONRY

Item	PC £	Labour hours	Labour £	Plant £	Material £	Unit	Total rate £
14.07 FORMING CAVITIES – cont							
External door and window cavity closures – cont							
Type X polypropylene abutment cavity tray or other equal and approved; Cavity Trays Ltd; built into facing brickwork as the work proceeds; complete with Code 4 flashing; ridge tray with short/long leads; to suit roof of							
17–20° pitch	–	0.05	1.09	–	18.87	nr	**19.96**
21–25° pitch	–	0.05	1.09	–	14.58	nr	**15.67**
26–45° pitch	–	0.05	1.09	–	12.98	nr	**14.07**
Expamet stainless steel wall starters or other equal and approved; plugged and screwed							
to suit walls 60 mm–75 mm thick	–	0.23	3.21	0.13	2.51	m	**5.85**
to suit walls 100 mm–115 mm thick	–	0.23	3.21	0.13	25.32	m	**28.66**
to suit walls 125 mm–180 mm thick	–	0.37	5.17	0.20	3.41	m	**8.78**
to suit walls 190 mm–260 mm thick	–	0.46	6.43	0.25	4.23	m	**10.91**
Galvanized steel lintels; Catnic or other equal; built into brickwork or blockwork							
70/100 range open back lintel for cavity wall; 70 mm–85 mm cavity wall							
1200 mm long	–	0.32	7.01	–	103.71	nr	**110.72**
1500 mm long	–	0.37	8.12	–	129.52	nr	**137.64**
1800 mm long	–	0.42	9.21	–	162.11	nr	**171.32**
2100 mm long	–	0.46	10.08	–	185.39	nr	**195.47**
2400 mm long	–	0.56	12.28	–	219.75	nr	**232.03**
2700 mm long	–	0.65	14.26	–	253.61	nr	**267.87**
3000 mm long	–	0.74	16.23	–	360.84	nr	**377.07**
90/100 range open back lintel for cavity wall; 90 mm–105 mm cavity wall							
1200 mm long	–	0.32	7.01	–	107.76	nr	**114.77**
1500 mm long	–	0.37	8.12	–	134.70	nr	**142.82**
1800 mm long	–	0.42	9.21	–	162.11	nr	**171.32**
2100 mm long	–	0.46	10.08	–	192.80	nr	**202.88**
2400 mm long	–	0.56	12.28	–	232.44	nr	**244.72**
2700 mm long	–	0.65	14.26	–	278.32	nr	**292.58**
3000 mm long	–	0.74	16.23	–	362.04	nr	**378.27**
3300 mm long	–	0.83	18.20	–	410.35	nr	**428.55**
CN92 single lintel; for 75 mm internal walls							
1050 mm long	–	0.28	6.14	–	14.05	nr	**20.19**
1200 mm long	–	0.32	7.01	–	15.89	nr	**22.90**
CN102 single lintel; for 100 mm internal walls							
1050 mm long	–	0.28	6.14	–	17.77	nr	**23.91**
1200 mm long	–	0.32	7.01	–	19.60	nr	**26.61**
CN100 single lintel; for 75 mm internal walls							
1050 mm long	–	0.28	6.14	–	43.29	nr	**49.43**
1200 mm long	–	0.32	7.01	–	53.77	nr	**60.78**

14 MASONRY

Item	PC £	Labour hours	Labour £	Plant £	Material £	Unit	Total rate £
BHD140 single lintel; for 140 mm internal walls							
1050 mm long	–	0.28	6.14	–	94.41	nr	**100.55**
1200 mm long	–	0.32	7.01	–	107.89	nr	**114.90**
1500 mm long	–	0.32	7.01	–	134.85	nr	**141.86**
Precast concrete sill, lintels, copings							
Lintels; plate; prestressed bedded							
100 mm × 70 mm × 600 mm long	5.36	0.37	8.12	–	5.57	nr	**13.69**
100 mm × 70 mm × 900 mm long	8.00	0.37	8.12	–	8.28	nr	**16.40**
100 mm × 70 mm × 1100 mm long	9.80	0.37	8.12	–	10.15	nr	**18.27**
100 mm × 70 mm × 1200 mm long	10.67	0.37	8.12	–	11.04	nr	**19.16**
100 mm × 70 mm × 1500 mm long	13.35	0.46	10.08	–	13.83	nr	**23.91**
100 mm × 70 mm × 1800 mm long	16.01	0.46	10.08	–	16.56	nr	**26.64**
100 mm × 70 mm × 2100 mm long	18.68	0.56	12.28	–	19.35	nr	**31.63**
140 mm × 70 mm × 1200 mm long	15.70	0.46	10.08	–	16.24	nr	**26.32**
140 mm × 70 mm × 1500 mm long	19.62	0.56	12.28	–	20.31	nr	**32.59**
Lintels; rectangular; reinforced with mild steel bars; bedded							
100 mm × 145 mm × 900 mm long	3.95	0.56	12.28	–	4.12	nr	**16.40**
100 mm × 145 mm × 1050 mm long	4.62	0.56	12.28	–	4.81	nr	**17.09**
100 mm × 145 mm × 1200 mm long	5.28	0.56	12.28	–	5.49	nr	**17.77**
225 mm × 145 mm × 1200 mm long	20.48	0.74	16.23	–	21.21	nr	**37.44**
225 mm × 225 mm × 1800 mm long	30.64	1.39	30.49	–	31.67	nr	**62.16**
Lintels; boot; reinforced with mild steel bars; bedded							
250 mm × 225 mm × 1200 mm long	22.69	1.11	24.34	–	23.47	nr	**47.81**
275 mm × 225 mm × 1800 mm long	37.44	1.67	36.63	–	38.67	nr	**75.30**
14.08 DAMP-PROOF COURSES							
Damp-proof courses							
Polythene damp-proof course or other equal and approved; 200 mm laps; in gauged mortar (1:1:6)							
100 mm width exceeding horizontal in brick or block course	–	0.01	0.22	–	0.14	m	**0.36**
150 mm width exceeding horizontal in brick or block course	–	0.01	0.22	–	0.21	m	**0.43**
300 mm width exceeding horizontal in brick or block course	–	0.01	0.22	–	0.41	m	**0.63**
width exceeding 225 mm; horizontal	–	0.23	5.05	–	0.59	m²	**5.64**
width exceeding 225 mm; forming cavity gutters in hollow walls; horizontal	–	0.37	8.12	–	0.59	m²	**8.71**
width not exceeding 225 mm; horizontal	–	0.46	10.08	–	0.59	m²	**10.67**
width not exceeding 225 mm; vertical	–	0.69	15.13	–	0.59	m²	**15.72**
Icopal Polymeric damp-proof course; Xtra-Load Pro-Build or other equal and approved; 200 mm laps; in gauged morter (1:1:6)							
width exceeding 225 mm; horizontal	–	0.23	5.05	–	16.79	m²	**21.84**
width exceeding 225 mm; forming cavity gutters in hollow walls; horizontal	–	0.37	8.12	–	16.79	m²	**24.91**
width not exceeding 225 mm; horizontal	–	0.46	10.08	–	16.79	m²	**26.87**
width not exceeding 225 mm; vertical	–	0.69	15.13	–	16.79	m²	**31.92**

14 MASONRY

Item	PC £	Labour hours	Labour £	Plant £	Material £	Unit	Total rate £
14.08 DAMP-PROOF COURSES – cont							
Damp-proof courses – cont							
Zedex CPT (Co-Polymer Thermoplastic) damp-proof course or other equal and approved; 200 mm laps; in gauged mortar (1:1:6)							
width exceeding 225 mm; horizontal	–	0.23	5.05	–	7.17	m²	**12.22**
width exceeding 225 mm wide; forming cavity gutters in hollow walls; horizontal	–	0.37	8.12	–	7.17	m²	**15.29**
width not exceeding 225 mm; horizontal	–	0.46	10.08	–	7.17	m²	**17.25**
width not exceeding 225 mm; vertical	–	0.69	15.13	–	7.17	m²	**22.30**
Hyload (pitch polymer) damp-proof course or other equal and approved; 150 mm laps; in gauged mortar (1:1:6)							
width exceeding 225 mm; horizontal	–	0.23	5.05	–	5.06	m²	**10.11**
width exceeding 225 mm; forming cavity gutters in hollow walls; horizontal	–	0.37	8.12	–	5.06	m²	**13.18**
width not exceeding 225 mm; horizontal	–	0.46	10.08	–	5.06	m²	**15.14**
width not exceeding 225 mm; vertical	–	0.69	15.13	–	5.06	m²	**20.19**
Nubit bitumen and polyester-based damp-proof course or other equal and approved; 200 mm laps; in gauged mortar (1:1:6)							
width exceeding 225 mm; horizontal	–	0.23	5.05	–	10.09	m²	**15.14**
width exceeding 225 mm wide; forming cavity gutters in hollow walls; horizontal	–	0.37	8.12	–	10.09	m²	**18.21**
width not exceeding 225 mm; horizontal	–	0.46	10.08	–	10.09	m²	**20.17**
width not exceeding 225 mm; vertical	–	0.69	15.13	–	10.09	m²	**25.22**
Permabit bitumen polymer damp-proof course or other equal and approved; 150 mm laps; in gauged mortar (1:1:6)							
width exceeding 225 mm; horizontal	–	0.23	5.05	–	10.89	m²	**15.94**
width exceeding 225 mm; forming cavity gutters in hollow walls; horizontal	–	0.37	8.12	–	10.89	m²	**19.01**
width not exceeding 225 mm; horizontal	–	0.46	10.08	–	10.89	m²	**20.97**
width not exceeding 225 mm; vertical	–	0.69	15.13	–	10.89	m²	**26.02**
Alumite aluminium cored bitumen gas retardant damp-proof course or other equal and approved; 200 mm laps; in gauged mortar (1:1;6)							
width exceeding 225 mm; horizontal	–	0.31	6.80	–	7.56	m²	**14.36**
width exceeding 225 mm; forming cavity gutters in hollow walls; horizontal	–	0.49	10.74	–	7.56	m²	**18.30**
width not exceeding 225 mm; horizontal	–	0.60	13.16	–	7.56	m²	**20.72**
width not exceeding 225 mm; vertical	–	0.83	18.20	–	7.56	m²	**25.76**
Milled lead damp-proof course; BS 1178; 1.80 mm thick (code 4), 175 mm laps; in cement: lime mortar (1:2:9)							
width exceeding 225 mm; horizontal (PC £/kg)	–	1.85	40.57	–	39.94	m²	**80.51**
width not exceeding 225 mm; horizontal	–	2.78	60.97	–	39.94	m²	**100.91**

14 MASONRY

Item	PC £	Labour hours	Labour £	Plant £	Material £	Unit	Total rate £
Two courses slates in cement: mortar (1:3)							
width exceeding 225 mm; horizontal	–	1.39	30.49	–	20.36	m²	**50.85**
width exceeding 225 mm; vertical	–	2.08	45.62	–	20.36	m²	**65.98**
Synthaprufe damp-proof membrane or other equal and approved; three coats brushed on							
width not exceeding 150 mm; vertical	–	0.31	4.34	–	8.12	m²	**12.46**
width 150 mm–225 mm; vertical	–	0.30	4.19	–	8.12	m²	**12.31**
width 225 mm–300 mm; vertical	–	0.28	3.91	–	8.12	m²	**12.03**
width exceeding 300 mm wide; vertical	–	0.26	3.64	–	8.12	m²	**11.76**
14.09 JOINT REINFORCEMENT							
Joint reinforcement							
Brickforce galvanized steel joint reinforcement or other equal and approved							
width 60 mm; ref GBF40W60B25	–	0.05	1.09	–	1.09	m	**2.18**
width 100 mm; ref GBF40W100B25	–	0.07	1.53	–	1.28	m	**2.81**
width 175 mm; ref GBF40W175B25	–	0.10	2.19	–	1.36	m	**3.55**
Brickforce stainless steel joint reinforcement or other equal and approved							
width 60 mm; ref SBF35W60BSC	–	0.05	1.09	–	2.54	m	**3.63**
width 100 mm; ref SBF35W100BSC	–	0.07	1.53	–	2.59	m	**4.12**
width 175 mm; ref SBF35W175BSC	–	0.10	2.19	–	3.05	m	**5.24**
Bricktor galvanized or stainless steel joint reinforcement or similar							
width 160 mm, galvanized steel; ref GBT160CCR	–	0.12	2.64	–	4.99	m	**7.63**
width 160 mm stainless steel; ref SBT160CCR	–	0.13	2.85	–	6.60	m	**9.45**
width 180 mm galvanized steel; ref GBT180CCR	–	0.14	3.07	–	5.54	m	**8.61**
width 180 mm stainless steel; ref SBT180CCR	–	0.14	3.07	–	7.42	m	**10.49**
14.10 FILLETS							
Weather fillets							
Weather fillets in cement: mortar (1:3)							
50 mm face width	–	0.11	2.41	–	0.10	m	**2.51**
100 mm face width	–	0.19	4.17	–	0.39	m	**4.56**
Angle fillets							
Angle fillets in cement: mortar (1:3)							
50 mm face width	–	0.11	2.41	–	0.10	m	**2.51**
100 mm face width	–	0.19	4.17	–	0.39	m	**4.56**

14 MASONRY

Item	PC £	Labour hours	Labour £	Plant £	Material £	Unit	Total rate £
14.11 JOINTS							
Joints							
Hacking joints and faces of brickwork or blockwork							
to form key for plaster	–	0.24	3.36	–	–	m²	**3.36**
Raking out joint in brickwork or blockwork for							
turned-in edge of flashing							
horizontal	–	0.14	3.07	–	–	m	**3.07**
stepped	–	0.19	4.17	–	–	m	**4.17**
Raking out and enlarging joint in brickwork or							
blockwork for nib of asphalt							
horizontal	–	0.19	4.17	–	–	m	**4.17**
Cutting grooves in brickwork or blockwork							
for water bars and the like	–	0.23	3.21	0.94	–	m	**4.15**
for nib of asphalt; horizontal	–	0.23	3.21	0.94	–	m	**4.15**
Preparing to receive new walls							
top existing 215 mm wall	–	0.19	4.17	–	–	m	**4.17**
Cleaning and priming both faces; filling with							
preformed closed cell joint filler and pointing one							
side with polysulphide sealant; 12 mm deep							
expansion joints; 12 mm wide	–	0.23	4.59	–	5.39	m	**9.98**
expansion joints; 20 mm wide	–	0.28	5.46	–	8.08	m	**13.54**
expansion joints; 25 mm wide	–	0.32	6.11	–	9.85	m	**15.96**
Fire-resisting horizontal expansion joints; filling with							
joint filler; fixed with high temperature slip adhesive;							
between top of wall and soffit							
wall not exceeding 215 mm wide; 10 mm wide							
joint with 30 mm deep filler (one hour fire seal)	–	0.23	5.05	–	5.77	m	**10.82**
wall not exceeding 215 mm wide; 10 mm wide							
joint with 30 mm deep filler (two hour fire seal)	–	0.23	5.05	–	5.77	m	**10.82**
wall not exceeding 215 mm wide; 20 mm wide							
joint with 45 mm deep filler (two hour fire seal)	–	0.28	6.14	–	8.92	m	**15.06**
wall not exceeding 215 mm wide; 30 mm wide							
joint with 75 mm deep filler (three hour fire seal)	–	0.32	7.01	–	22.95	m	**29.96**
Fire-resisting vertical expansiojn joints; filling with							
joint filler; fixed with high temperature slip adhesive;							
with polysulphide sealant one side; between end of							
wall and concrete							
wall not exceeding 215 mm wide; 20 mm wide							
joint with 45 mm deep filler (two hour fire seal)	–	0.37	7.66	–	14.81	m	**22.47**

15 STRUCTURAL METALWORK

Item	PC £	Labour hours	Labour £	Plant £	Material £	Unit	Total rate £
15.01 FRAMED MEMBERS, FRAMING, FABRICATION							
BASIC STEEL PRICES – MATERIAL SUPPLY ONLY							
TATA Steel have recently revised ther pricing structure and now publish pries for extras only. The extras cost needs to be added to their steel Base Price. The Base Price given here is a guide price only and subject to normal trading conditions with regards to discounts or plussages. The Base Price assumes a minimum order of 5 tonnes, S355JR steel and standard lengths.							
Universal beams and columns							
Base price for steel	–	–	–	–	669.50	tonne	**669.50**
Extra to be added to Base Price for: (kg/m)							
1016 × 305 mm (222, 249, 272, 314, 349, 393, 438, 487) – Price on application							
914 × 419 mm (343, 388)	–	–	–	–	206.00	tonne	**206.00**
914 × 305 mm (201, 224, 253, 289)	–	–	–	–	175.10	tonne	**175.10**
838 × 292 mm (226)	–	–	–	–	175.10	tonne	**175.10**
838 × 292 mm (176, 194)	–	–	–	–	164.80	tonne	**164.80**
762 × 267 mm (197)	–	–	–	–	164.80	tonne	**164.80**
762 × 267 mm (134, 147, 173)	–	–	–	–	144.20	tonne	**144.20**
686 × 254 mm (152, 170)	–	–	–	–	164.80	tonne	**164.80**
686 × 254 mm (125, 140)	–	–	–	–	144.20	tonne	**144.20**
610 × 305 mm (238)	–	–	–	–	175.10	tonne	**175.10**
610 × 305 mm (149, 179)	–	–	–	–	144.20	tonne	**144.20**
610 × 229 mm (125, 140)	–	–	–	–	10.30	tonne	**10.30**
610 × 229 mm (101, 113)	–	–	–	–	10.30	tonne	**10.30**
610 × 178 mm (82, 92, 100) – Price on application							
533 × 312 mm (150, 182, 219, 272) – Price on application							
533 × 210 mm (138)	–	–	–	–	185.40	tonne	**185.40**
533 × 210 mm (101, 109, 122)	–	–	–	–	10.30	tonne	**10.30**
533 × 210 mm (82, 92)	–	–	–	–	10.30	tonne	**10.30**
533 × 165 mm (66, 74, 85) – Price on application							
457 × 191 mm (161)	–	–	–	–	206.00	tonne	**206.00**
457 × 191 mm (133)	–	–	–	–	185.40	tonne	**185.40**
457 × 191 mm (106)	–	–	–	–	51.50	tonne	**51.50**
457 × 191 mm (74, 82, 89, 98)	–	–	–	–	10.30	tonne	**10.30**
457 × 191 mm (67)	–	–	–	–	10.30	tonne	**10.30**
457 × 152 mm (74, 82)	–	–	–	–	51.50	tonne	**51.50**
457 × 152 mm (67)	–	–	–	–	51.50	tonne	**51.50**
457 × 152 mm (52, 60)	–	–	–	–	10.30	tonne	**10.30**
406 × 178 mm (85)	–	–	–	–	51.50	tonne	**51.50**
406 × 178 mm (74)	–	–	–	–	10.30	tonne	**10.30**
406 × 178 mm (54, 60, 67)	–	–	–	–	10.30	tonne	**10.30**

15 STRUCTURAL METALWORK

Item	PC £	Labour hours	Labour £	Plant £	Material £	Unit	Total rate £
15.01 FRAMED MEMBERS, FRAMING, FABRICATION – cont							
Universal beams and columns – cont							
Extra to be added to Base Price for – cont							
406 × 140 mm (53)	–	–	–	–	51.50	tonne	**51.50**
406 × 140 mm (46)	–	–	–	–	10.30	tonne	**10.30**
406 × 140 mm (39)	–	–	–	–	10.30	tonne	**10.30**
356 × 171 mm (45, 51, 57, 67)	–	–	–	–	10.30	tonne	**10.30**
305 × 165 mm (40, 46, 54)	–	–	–	–	10.30	tonne	**10.30**
305 × 127 mm (42, 48)	–	–	–	–	10.30	tonne	**10.30**
305 × 127 mm (37)	–	–	–	–	10.30	tonne	**10.30**
305 × 102 mm (28, 33)	–	–	–	–	10.30	tonne	**10.30**
305 × 102 mm (25)	–	–	–	–	10.30	tonne	**10.30**
254 × 102 mm (28)	–	–	–	–	20.60	tonne	**20.60**
254 × 102 mm (22, 25)	–	–	–	–	10.30	tonne	**10.30**
254 × 146 (43)	–	–	–	–	10.30	tonne	**10.30**
254 × 146 (31, 37)	–	–	–	–	10.30	tonne	**10.30**
203 × 133 mm (25, 30)	–	–	–	–	10.30	tonne	**10.30**
203 × 102 mm (23)	–	–	–	–	10.30	tonne	**10.30**
178 × 102 mm (19)	–	–	–	–	20.60	tonne	**20.60**
152 × 89 mm (16)	–	–	–	–	20.60	tonne	**20.60**
127 × 76 mm (13)	–	–	–	–	20.60	tonne	**20.60**
Asymmetric Slimflor Beams (ASB) Price on Application							
Universal columns (kg/m). NB: These are to be added to the base steel price. Base price for steel	–	–	–	–	669.50	tonne	**669.50**
356 × 406 mm (634)	–	–	–	–	309.00	tonne	**309.00**
356 × 406 mm (340, 393, 467, 551)	–	–	–	–	206.00	tonne	**206.00**
356 × 406 mm (235, 287)	–	–	–	–	175.10	tonne	**175.10**
356 × 368 mm (177, 202)	–	–	–	–	175.10	tonne	**175.10**
356 × 368 mm (153)	–	–	–	–	164.80	tonne	**164.80**
356 × 368 mm (129)	–	–	–	–	144.20	tonne	**144.20**
305 × 305 mm (240, 283)	–	–	–	–	20.60	tonne	**20.60**
305 × 305 mm (137, 158, 198)	–	–	–	–	10.30	tonne	**10.30**
305 × 305 mm (97, 118)	–	–	–	–	10.30	tonne	**10.30**
254 × 254 mm (132, 167)	–	–	–	–	10.30	tonne	**10.30**
254 × 254 mm (73, 89, 107)	–	–	–	–	10.30	tonne	**10.30**
203 × 203 mm (127)	–	–	–	–	185.40	tonne	**185.40**
203 × 203 mm (100, 113)	–	–	–	–	51.50	tonne	**51.50**
203 × 203 mm (71, 86)	–	–	–	–	10.30	tonne	**10.30**
203 × 203 mm (46, 52, 60)	–	–	–	–	10.30	tonne	**10.30**
152 × 152 mm (44, 51)	–	–	–	–	51.50	tonne	**51.50**
152 × 152 mm (23, 30, 37)	–	–	–	–	10.30	tonne	**10.30**
Channels (kg/m). NB: These are to be added to the base steel price. Base price for steel	–	–	–	–	669.50	tonne	**669.50**
430 × 100 mm (64.4)	–	–	–	–	303.85	tonne	**303.85**
380 × 100 mm (54.0)	–	–	–	–	283.25	tonne	**283.25**
300 × 100 mm (45.5)	–	–	–	–	118.45	tonne	**118.45**
300 × 90 mm (41.4)	–	–	–	–	128.75	tonne	**128.75**

15 STRUCTURAL METALWORK

Item	PC £	Labour hours	Labour £	Plant £	Material £	Unit	Total rate £
260 × 90 mm (34.8)	–	–	–	–	118.45	tonne	**118.45**
260 × 75 mm (27.6)	–	–	–	–	72.10	tonne	**72.10**
230 × 90 mm (32.2)	–	–	–	–	128.75	tonne	**128.75**
230 × 75 mm (25.7)	–	–	–	–	72.10	tonne	**72.10**
200 × 90 mm (29.7)	–	–	–	–	118.45	tonne	**118.45**
200 × 75 mm (23.4)	–	–	–	–	41.20	tonne	**41.20**
180 × 90 mm (26.1)	–	–	–	–	118.45	tonne	**118.45**
180 × 75 mm (20.3)	–	–	–	–	41.20	tonne	**41.20**
150 × 90 mm (23.9)	–	–	–	–	118.45	tonne	**118.45**
150 × 75 mm (17.9)	–	–	–	–	20.60	tonne	**20.60**
125 × 65 mm (14.8)	–	–	–	–	20.60	tonne	**20.60**
Equal angles (mm). NB: These are to be added to the base steel price. Base price for steel	–	–	–	–	669.50	tonne	**669.50**
200 × 200 mm (16,18,20,24)	–	–	–	–	66.95	tonne	**66.95**
150 × 150 mm (10,12,15,18)	–	–	–	–	66.95	tonne	**66.95**
120 × 120 mm (8, 10, 12, 15)	–	–	–	–	10.30	tonne	**10.30**
Unequal angles (mm). NB: These are to be added to the base steel price. Base price for steel	–	–	–	–	669.50	tonne	**669.50**
200 × 150 mm (12,15, 18)	–	–	–	–	118.45	tonne	**118.45**
200 × 100 mm (10, 12, 15)	–	–	–	–	66.95	tonne	**66.95**
Specification extras							
for Advance275JR steel	–	–	–	–	30.90	tonne	**30.90**
for Advance275JO steel	–	–	–	–	41.20	tonne	**41.20**
for Advance275J2 steel	–	–	–	–	82.40	tonne	**82.40**
for Advance355JO steel	–	–	–	–	10.30	tonne	**10.30**
for Advance355J2 steel	–	–	–	–	41.20	tonne	**41.20**
for Advance355 K2 steel	–	–	–	–	82.40	tonne	**82.40**
Hollow sections							
NOTE: The following prices are for basic quantities of 10 tonnes and over in one size, thickness, length, steelgrade and surface finish and include delivery (for delivery to outer London)							
circular hollow sections	–	–	–	–	1030.00	tonne	**1030.00**
rectangular hollow section	–	–	–	–	1030.00	tonne	**1030.00**
square hollow sections	–	–	–	–	1030.00	tonne	**1030.00**
SUPPLY AND FIX PRICES							
Framing, fabrication; weldable steel; BS EN 10025: 2004 Grade S275; hot rolled structural steel sections; welded fabrication							
Columns							
weight not exceeding 40 kg/m	–	–	–	–	–	tonne	**1674.25**
weight not exceeding 40 kg/m; cellular (Fabsec)	–	–	–	–	–	tonne	**2081.25**
weight not exceeding 40 kg/m; curved	–	–	–	–	–	tonne	**2381.88**
weight not exceeding 40 kg/m; square hollow section	–	–	–	–	–	tonne	**2243.13**
weight not exceeding 40 kg/m; circular hollow section	–	–	–	–	–	tonne	**2243.13**
weight 40–100 kg/m	–	–	–	–	–	tonne	**1577.13**
weight 40–100 kg/m; cellular (Fabsec)	–	–	–	–	–	tonne	**1822.25**

15 STRUCTURAL METALWORK

Item	PC £	Labour hours	Labour £	Plant £	Material £	Unit	Total rate £
15.01 FRAMED MEMBERS, FRAMING, FABRICATION – cont							
Framing, fabrication – cont							
Columns – cont							
weight 40–100 kg/m; curved	–	–	–	–	–	tonne	**1854.63**
weight 40–100 kg/m; square hollow section	–	–	–	–	–	tonne	**1826.88**
weight 40–100 kg/m; circular hollow section	–	–	–	–	–	tonne	**1826.88**
weight exceeding 100 kg/m	–	–	–	–	–	tonne	**1526.25**
weight exceeding 100 kg/m; cellular (Fabsec)	–	–	–	–	–	tonne	**1739.00**
weight exceeding 100 kg/m; curved	–	–	–	–	–	tonne	**1729.75**
weight exceeding 100 kg/m; square hollow section	–	–	–	–	–	tonne	**2881.38**
weight exceeding 100 kg/m; circular hollow section	–	–	–	–	–	tonne	**2881.38**
Beams							
weight not exceeding 40 kg/m	–	–	–	–	–	tonne	**1965.63**
weight not exceeding 40 kg/m; cellular (Fabsec)	–	–	–	–	–	tonne	**2335.63**
weight not exceeding 40 kg/m; curved	–	–	–	–	–	tonne	**2446.63**
weight not exceeding 40 kg/m; square hollow section	–	–	–	–	–	tonne	**2483.63**
weight not exceeding 40 kg/m; circular hollow section	–	–	–	–	–	tonne	**2483.63**
weight 40–100 kg/m	–	–	–	–	–	tonne	**1577.13**
weight 40–100 kg/m; cellular (Fabsec)	–	–	–	–	–	tonne	**2109.00**
weight 40–100 kg/m; curved	–	–	–	–	–	tonne	**1854.63**
weight 40–100 kg/m; square hollow section	–	–	–	–	–	tonne	**2321.75**
weight 40–100 kg/m; circular hollow section	–	–	–	–	–	tonne	**2321.75**
weight exceeding 100 kg/m	–	–	–	–	–	tonne	**1526.25**
weight exceeding 100 kg/m; cellular (Fabsec)	–	–	–	–	–	tonne	**1748.25**
weight exceeding 100 kg/m; curved	–	–	–	–	–	tonne	**1799.13**
weight exceeding 100 kg/m; square hollow section	–	–	–	–	–	tonne	**2881.38**
weight exceeding 100 kg/m; circular hollow section	–	–	–	–	–	tonne	**2881.38**
Bracings							
weight not exceeding 40 kg/m	–	–	–	–	–	tonne	**1965.63**
weight not exceeding 40 kg/m; square hollow section	–	–	–	–	–	tonne	**2483.63**
weight not exceeding 40 kg/m; circular hollow section	–	–	–	–	–	tonne	**2483.63**
weight 40–100 kg/m	–	–	–	–	–	tonne	**1577.13**
weight 40–100 kg/m; square hollow section	–	–	–	–	–	tonne	**2321.75**
weight 40–100 kg/m; circular hollow section	–	–	–	–	–	tonne	**2321.75**
weight exceeding 100 kg/m	–	–	–	–	–	tonne	**1526.25**
weight exceeding 100 kg/m; square hollow section	–	–	–	–	–	tonne	**2881.38**
weight exceeding 100 kg/m; circular hollow section	–	–	–	–	–	tonne	**2881.38**
Purlins and cladding rails							
weight not exceeding 40 kg/m	–	–	–	–	–	tonne	**1965.63**
weight not exceeding 40 kg/m; square hollow section	–	–	–	–	–	tonne	**2483.63**
weight not exceeding 40 kg/m; circular hollow section	–	–	–	–	–	tonne	**2483.63**

15 STRUCTURAL METALWORK

Item	PC £	Labour hours	Labour £	Plant £	Material £	Unit	Total rate £
weight 40–100 kg/m	–	–	–	–	–	tonne	1577.13
weight 40–100 kg/m; square hollow section	–	–	–	–	–	tonne	2321.75
weight 40–100 kg/m; circular hollow section	–	–	–	–	–	tonne	2321.75
weight exceeding 100 kg/m	–	–	–	–	–	tonne	1526.25
weight exceeding 100 kg/m; square hollow section	–	–	–	–	–	tonne	2881.38
weight exceeding 100 kg/m; circular hollow section	–	–	–	–	–	tonne	2881.38
Grillages							
weight not exceeding 40 kg/m	–	–	–	–	–	tonne	1965.63
weight 40–100 kg/m	–	–	–	–	–	tonne	1965.63
weight exceeding 100 kg/m	–	–	–	–	–	tonne	1577.13
Trestles, towers and built up columns							
straight	–	–	–	–	–	tonne	2483.63
Trusses and built up girders							
straight	–	–	–	–	–	tonne	2483.63
curved	–	–	–	–	–	tonne	264.55
Fittings							
general steel fittings; nuts, bolts, plates etc.	–	–	–	–	–	tonne	2284.75
Add to the aforementioned prices for							
grade 355 steelwork	–	–	–	–	–	%	3.12
Framing, erection							
Trial erection	–	–	–	–	–	tonne	208.13
Permanent erection on site	–	–	–	–	–	tonne	249.75
Metsec Lightweight Steel Framing System (SFS); or other equal and approved; as inner leaf to external wall; studs typically at 600 mm centres; including provision for all openings, abutments, junctions and head details etc.							
Inner leaf; with supports and perimeter sections; 12 mm plasterboard internally; 10 mm cement fibre substrate externally; (insulation and external cladding measured separately)							
100 mm thick steel walling	–	–	–	–	–	m²	74.94
150 mm thick steel walling	–	–	–	–	–	m²	80.84
200 mm thick steel walling	–	–	–	–	–	m²	86.76
Inner leaf; with 16 mm Pyroc sheething board							
100 mm thick steel walling	–	–	–	–	–	m²	92.68
150 mm thick steel walling	–	–	–	–	–	m²	98.58
200 mm thick steel walling	–	–	–	–	–	m²	104.51
16 mm Pyroc sheething board fixed to slab perimeter not exceeding 300 mm	–	–	–	–	–	m	9.20
Inner leaf; with 16 mm Pyroc sheething board and Thermawall TW50 insulation supported by halfen channels type 28/15 fixed to studs at 450 mm centres.							
100 mm thick steel walling with 50 mm insulation	–	–	–	–	–	m²	106.99
150 mm thick steel walling with 75 mm insulation	–	–	–	–	–	m²	118.83
200 mm thick steel walling with 100 mm insulatiom	–	–	–	–	–	m²	130.00

Prices for Measured Works

15 STRUCTURAL METALWORK

Item	PC £	Labour hours	Labour £	Plant £	Material £	Unit	Total rate £
15.01 FRAMED MEMBERS, FRAMING, FABRICATION – cont							
Metsec Lightweight Steel Framing System (SFS) – cont							
16 mm Pyroc sheething board and 40 mm Thermawall TW55 insulation fixed to slab perimeter							
not exceeding 300 mm	–	–	–	–	–	m	**10.51**
Storage costs							
Costs for storing fabricated steelwork							
storage off site	–	–	–	–	18.54	t/week	**18.54**
storage on extending trailers	–	–	–	–	30.90	t/week	**30.90**
15.02 ISOLATED STRUCTURAL METAL MEMBERS							
Isolated structural member; weldable steel; BS EN 10025: 2004 Grade S275; hot rolled structural steel sections							
Plain member; beams							
weight not exceeding 40 kg/m	–	–	–	–	–	tonne	**1462.00**
weight 40–100 kg/m	–	–	–	–	–	tonne	**1462.00**
weight exceeding 100 kg/m	–	–	–	–	–	tonne	**1462.00**
Metsec open web steel lattice beams or other equal and approved; in single members; raised 3.50 m above ground; ends built in							
Beams; one coat zinc phosphate primer at works							
220 mm deep; to span 6.00 m (11.50 kg/m); ref B22	–	0.19	4.64	–	30.97	m	**35.61**
270 mm deep; to span 7.00 m (11.50 kg/m); ref B27	–	0.19	4.64	–	30.97	m	**35.61**
300 mm deep; to span 8.00 m (12.50 kg/m); ref B30	–	0.23	5.61	–	33.60	m	**39.21**
350 mm deep; to span 9.00 m (14.00 kg/m); ref B35	–	0.23	5.61	–	37.54	m	**43.15**
350 mm deep; to span 10.00 m (20.00 kg/m); ref D35	–	0.28	6.83	–	53.30	m	**60.13**
450 mm deep; to span 11.00 m (21.00 kg/m); ref D45	–	0.32	7.81	–	55.93	m	**63.74**
450 mm deep; to span 12.00 m (32.50 kg/m); ref G45	–	0.46	11.22	–	86.13	m	**97.35**
Beams; galvanized							
220 mm deep; to span 6.00 m (11.50 kg/m); ref B22	–	0.19	4.64	–	35.10	m	**39.74**
270 mm deep; to span 7.00 m (11.50 kg/m); ref B27	–	0.19	4.64	–	35.10	m	**39.74**
300 mm deep; to span 8.00 m (12.50 kg/m); ref B30	–	0.23	5.61	–	38.07	m	**43.68**

15 STRUCTURAL METALWORK

Item	PC £	Labour hours	Labour £	Plant £	Material £	Unit	Total rate £
350 mm deep; to span 9.00 m (14.00 kg/m); ref B35	–	0.23	5.61	–	42.57	m	**48.18**
350 mm deep; to span 10.00 m (20.00 kg/m); ref D35	–	0.28	6.83	–	60.45	m	**67.28**
450 mm deep; to span 11.00 m (21.00 kg/m); ref D45	–	0.32	7.81	–	63.45	m	**71.26**
450 mm deep; to span 12.00 m (32.50 kg/m); ref G45	–	0.46	11.22	–	97.76	m	**108.98**
15.03 COLD ROLLED STEEL PURLINS AND CLADDING RAILS							
Cold formed galvanized steel; Kingspan Multibeam or other equal and approved							
Cold rolled purlins and cladding rails							
175 × 65 × 1.40 mm gauge purlins or rails; fixed to steelwork	–	0.04	0.82	–	7.89	m	**8.71**
175 × 65 × 1.60 mm gauge purlins or rails; fixed to steelwork	–	0.04	0.82	–	8.39	m	**9.21**
175 × 65 × 2.00 mm gauge purlins or rails; fixed to steelwork	–	0.04	0.82	–	10.15	m	**10.97**
205 × 65 × 1.40 mm gauge purlins or rails; fixed to steelwork	–	0.04	0.82	–	8.70	m	**9.52**
205 × 65 × 1.60 mm gauge purlins or rails; fixed to steelwork	–	0.04	0.82	–	9.50	m	**10.32**
205 × 65 × 2.00 mm gauge purlins or rails; fixed to steelwork	–	0.04	0.82	–	10.80	m	**11.62**
Cleats							
weld-on for 175 mm purlin or rail	–	0.10	2.05	–	2.29	nr	**4.34**
bolt-on for 175 mm purlin or rail; including fixing bolts	–	0.02	0.41	–	4.91	m	**5.32**
weld-on for 205 mm purlin or rail	–	0.10	2.05	–	2.60	nr	**4.65**
bolt-on for 205 mm purlin or rail; including fixing bolts	–	0.02	0.41	–	5.28	m	**5.69**
Tubular ties							
1800 mm long; bolted diagonally across purlins or cladding rails	–	0.02	0.41	–	5.78	m	**6.19**
15.04 PROFILED METAL DECKING							
Soffits of slabs; galvanized steel permanent re-entrant type shuttering; including safety net							
Slab thickness not exceeding 200 mm							
0.9 mm decking; height to soffit 1.50–3.00 m	31.72	0.27	5.76	–	36.14	m²	**41.90**
0.9 mm decking; height to soffit 3.00–4.50 m	31.72	0.30	6.40	–	36.14	m²	**42.54**
1.2 mm decking; height to soffit 3.00–4.50 m	35.62	0.27	5.76	–	40.16	m²	**45.92**
Edge trim and restraints to decking							
Edge trim 1.2 mm × 300 mm girth	–	0.17	3.52	–	14.13	m	**17.65**
Edge trim 1.2 mm × 350 mm girth	–	0.17	3.52	–	14.13	m	**17.65**
Edge trim 1.2 mm × 400 mm girth	–	0.17	3.52	–	14.13	m	**17.65**

15 STRUCTURAL METALWORK

Item	PC £	Labour hours	Labour £	Plant £	Material £	Unit	Total rate £
15.04 PROFILED METAL DECKING – cont							
Soffits of slabs – cont							
Bearings to decking; connection to steel work with thru-deck welded shear studs							
1995 × 95 mm high studs at 100 mm centres	–	–	–	–	15.80	m	**15.80**
1995 × 95 mm high studs at 200 mm centres	–	–	–	–	7.90	m	**7.90**
1995 × 95 mm high studs at 300 mm centres	–	–	–	–	5.26	m	**5.26**
19120 × 120 mm high studs at 100 mm centres	–	–	–	–	15.80	m	**15.80**
19120 × 120 mm high studs at 200 mm centres	–	–	–	–	7.90	m	**7.90**
19120 × 120 mm high studs at 300 mm centres	–	–	–	–	5.26	m	**5.26**
Soffits of slabs; galvanized steel permanent trapazoidal type shuttering; including safety net							
Slab thickness not exceeding 200 mm							
0.9 mm decking; height to soffit 1.50–3.00 m	29.01	0.27	5.76	–	33.35	m²	**39.11**
0.9 mm decking; height to soffit 3.00–4.50 m	29.01	0.30	6.40	–	33.35	m²	**39.75**
1.2 mm decking; height to soffit 3.00–4.50 m	33.85	0.27	5.76	–	38.34	m²	**44.10**
15.05 SURFACE TREATMENTS							
Surface preparation							
At works							
blast cleaning	–	–	–	–	–	m²	**2.96**
Surface treatment							
At works							
galvanizing	–	–	–	–	–	tonne	**297.44**
galvanizing	–	–	–	–	–	m²	**20.02**
shotblasting and priming to SA 2.5	–	–	–	–	–	m²	**8.16**
touch up primer and one coat of two pack epoxy zinc phosphate primer	–	–	–	–	–	m²	**5.10**
intumescent paint fire protection (30 minutes); spray applied	–	–	–	–	–	m²	**22.05**
intumescent paint fire protection (60 minutes); spray applied	–	–	–	–	–	m²	**22.05**
intumescent paint fire protection (90 minutes); spray applied	–	–	–	–	–	m²	**44.10**
intumescent paint fire protection (120 minutes); spray applied	–	–	–	–	–	m²	**66.15**
extra over for; separate decorative sealer top coat	–	–	–	–	–	m²	**4.13**
On site							
intumescent paint fire protection (30 minutes); spray applied	–	–	–	–	–	m²	**11.68**
intumescent paint fire protection (30 minutes) to circular columns etc.; spray applied	–	–	–	–	–	m²	**19.57**
intumescent paint fire protection (60 minutes) to UBs etc.; spray applied	–	–	–	–	–	m²	**14.80**
intumescent paint fire protection (60 minutes) to circular columns etc.; spray applied	–	–	–	–	–	m²	**24.82**
extra for separate decorative sealer top coat	–	–	–	–	–	m²	**3.88**

16 CARPENTRY

Item	PC £	Labour hours	Labour £	Plant £	Material £	Unit	Total rate £
16.01 PRIMARY AND STRUCTURAL TIMBERS							
SUPPLY ONLY TIMBER PRICES							
Hardwood; Joinery quality; 25 mm thicknes (£/m^3)							
Black Walnut	–	–	–	–	2060.00	m^3	**2060.00**
American White Ash	–	–	–	–	947.60	m^3	**947.60**
American White Oak	–	–	–	–	2111.50	m^3	**2111.50**
Pretreatment of timber by vacuum/pressure impregnation, excluding transport costs and any subsequent seasoning:							
interior work; minimum salt retention 4.00 kg/m^3	–	–	–	–	84.52	m^3	**84.52**
exterior work; minimum salt retention 5.30 kg/m^3	–	–	–	–	97.29	m^3	**97.29**
SUPPLY AND FIX PRICES							
Sawn softwood; untreated							
Floor members							
38 mm × 100 mm	–	0.11	2.52	–	2.34	m	**4.86**
38 mm × 150 mm	–	0.13	2.99	–	3.22	m	**6.21**
47 mm × 75 mm	–	0.11	2.52	–	1.49	m	**4.01**
47 mm × 100 mm	–	0.13	2.99	–	1.92	m	**4.91**
47 mm × 125 mm	–	0.13	2.99	–	2.45	m	**5.44**
47 mm × 150 mm	–	0.14	3.21	–	2.81	m	**6.02**
47 mm × 175 mm	–	0.14	3.21	–	3.32	m	**6.53**
47 mm × 200 mm	–	0.15	3.45	–	3.59	m	**7.04**
47 mm × 225 mm	–	0.15	3.45	–	4.14	m	**7.59**
47 mm × 250 mm	–	0.16	3.68	–	4.66	m	**8.34**
75 mm × 125 mm	–	0.15	3.45	–	5.61	m	**9.06**
75 mm × 150 mm	–	0.15	3.45	–	6.08	m	**9.53**
75 mm × 175 mm	–	0.15	3.45	–	7.31	m	**10.76**
75 mm × 200 mm	–	0.16	3.68	–	8.20	m	**11.88**
75 mm × 225 mm	–	0.16	3.68	–	8.93	m	**12.61**
75 mm × 250 mm	–	0.17	3.90	–	13.58	m	**17.48**
100 mm × 150 mm	–	0.20	4.59	–	7.83	m	**12.42**
100 mm × 200 mm	–	0.21	4.83	–	10.42	m	**15.25**
100 mm × 250 mm	–	0.23	5.28	–	13.07	m	**18.35**
100 mm × 300 mm	–	0.25	5.75	–	16.89	m	**22.64**
Wall or partition members							
25 mm × 25 mm	–	0.06	1.38	–	0.92	m	**2.30**
25 mm × 38 mm	–	0.06	1.38	–	0.90	m	**2.28**
25 mm × 75 mm	–	0.08	1.84	–	1.28	m	**3.12**
38 mm × 38 mm	–	0.08	1.84	–	1.19	m	**3.03**
38 mm × 50 mm	–	0.08	1.84	–	1.52	m	**3.36**
38 mm × 75 mm	–	0.11	2.52	–	1.86	m	**4.38**
38 mm × 100 mm	–	0.14	3.21	–	2.34	m	**5.55**
47 mm × 50 mm	–	0.11	2.52	–	1.16	m	**3.68**
47 mm × 75 mm	–	0.14	3.21	–	1.55	m	**4.76**
47 mm × 100 mm	–	0.17	3.90	–	1.97	m	**5.87**
47 mm × 125 mm	–	0.18	4.14	–	2.50	m	**6.64**
75 mm × 75 mm	–	0.17	3.90	–	3.45	m	**7.35**
75 mm × 100 mm	–	0.19	4.37	–	4.69	m	**9.06**
100 mm × 100 mm	–	0.19	4.37	–	5.71	m	**10.08**

16 CARPENTRY

Item	PC £	Labour hours	Labour £	Plant £	Material £	Unit	Total rate £
16.01 PRIMARY AND STRUCTURAL TIMBERS – cont							
Sawn softwood – cont							
Joist strutting; herringbone							
47 mm × 50 mm; depth of joist 150 mm	–	0.46	10.57	–	2.84	m	**13.41**
47 mm × 50 mm; depth of joist 175 mm	–	0.46	10.57	–	2.89	m	**13.46**
47 mm × 50 mm; depth of joist 200 mm	–	0.46	10.57	–	2.95	m	**13.52**
47 mm × 50 mm; depth of joist 225 mm	–	0.46	10.57	–	3.00	m	**13.57**
47 mm × 50 mm; depth of joist 250 mm	–	0.46	10.57	–	3.05	m	**13.62**
Joist strutting; block							
47 mm × 150 mm; depth of joist 150 mm	–	0.28	6.44	–	3.47	m	**9.91**
47 mm × 175 mm; depth of joist 175 mm	–	0.28	6.44	–	3.98	m	**10.42**
47 mm × 200 mm; depth of joist 200 mm	–	0.28	6.44	–	4.25	m	**10.69**
47 mm × 225 mm; depth of joist 225 mm	–	0.28	6.44	–	4.80	m	**11.24**
47 mm × 250 mm; depth of joist 250 mm	–	0.28	6.44	–	5.31	m	**11.75**
Cleats							
225 mm × 100 mm × 75 mm	–	0.19	4.37	–	0.92	nr	**5.29**
Extra for stress grading to above timbers							
general structural (GS) grade	–	–	–	–	35.87	m³	**35.87**
special structural (SS) grade	–	–	–	–	71.74	m³	**71.74**
Extra for protecting and flameproofing timber with							
Celgard CF protection or other equal and approved							
small sections	–	–	–	–	145.23	m³	**145.23**
large sections	–	–	–	–	140.22	m³	**140.22**
Wrot surfaces							
plain; 50 mm wide	–	0.02	0.46	–	–	m	**0.46**
plain; 100 mm wide	–	0.03	0.69	–	–	m	**0.69**
plain; 150 mm wide	–	0.04	0.92	–	–	m	**0.92**
Sawn softwood; tanalized							
Floor members							
38 mm × 75 mm	–	0.11	2.52	–	2.05	m	**4.57**
38 mm × 100 mm	–	0.11	2.52	–	2.58	m	**5.10**
38 mm × 150 mm	–	0.13	2.99	–	3.57	m	**6.56**
47 mm × 75 mm	–	0.11	2.52	–	1.74	m	**4.26**
47 mm × 100 mm	–	0.13	2.99	–	2.24	m	**5.23**
47 mm × 125 mm	–	0.13	2.99	–	2.85	m	**5.84**
47 mm × 150 mm	–	0.14	3.21	–	3.29	m	**6.50**
47 mm × 175 mm	–	0.14	3.21	–	3.87	m	**7.08**
47 mm × 200 mm	–	0.15	3.45	–	4.23	m	**7.68**
47 mm × 225 mm	–	0.15	3.45	–	4.84	m	**8.29**
47 mm × 250 mm	–	0.16	3.68	–	5.44	m	**9.12**
75 mm × 125 mm	–	0.15	3.45	–	6.20	m	**9.65**
75 mm × 150 mm	–	0.15	3.45	–	6.79	m	**10.24**
75 mm × 175 mm	–	0.15	3.45	–	8.14	m	**11.59**
75 mm × 200 mm	–	0.16	3.68	–	9.14	m	**12.82**
75 mm × 225 mm	–	0.16	3.68	–	9.99	m	**13.67**
75 mm × 250 mm	–	0.17	3.90	–	14.75	m	**18.65**
100 mm × 150 mm	–	0.20	4.59	–	8.77	m	**13.36**
100 mm × 200 mm	–	0.21	4.83	–	11.68	m	**16.51**

16 CARPENTRY

Item	PC £	Labour hours	Labour £	Plant £	Material £	Unit	Total rate £
100 mm × 250 mm	–	0.23	5.28	–	14.64	m	**19.92**
100 mm × 300 mm	–	0.25	5.75	–	18.78	m	**24.53**
Wall or partition members							
25 mm × 25 mm	–	0.06	1.38	–	0.96	m	**2.34**
25 mm × 38 mm	–	0.06	1.38	–	0.96	m	**2.34**
25 mm × 75 mm	–	0.08	1.84	–	1.39	m	**3.23**
38 mm × 38 mm	–	0.08	1.84	–	1.29	m	**3.13**
38 mm × 50 mm	–	0.08	1.84	–	1.65	m	**3.49**
38 mm × 75 mm	–	0.11	2.52	–	2.05	m	**4.57**
38 mm × 100 mm	–	0.14	3.21	–	2.58	m	**5.79**
47 mm × 50 mm	–	0.11	2.52	–	1.33	m	**3.85**
47 mm × 75 mm	–	0.14	3.21	–	1.78	m	**4.99**
47 mm × 100 mm	–	0.17	3.90	–	2.28	m	**6.18**
47 mm × 125 mm	–	0.18	4.14	–	2.90	m	**7.04**
75 mm × 75 mm	–	0.17	3.90	–	3.80	m	**7.70**
75 mm × 100 mm	–	0.19	4.37	–	5.16	m	**9.53**
100 mm × 100 mm	–	0.19	4.37	–	6.33	m	**10.70**
Roof members; flat							
38 mm × 75 mm	–	0.13	2.99	–	2.05	m	**5.04**
38 mm × 100 mm	–	0.13	2.99	–	2.58	m	**5.57**
38 mm × 125 mm	–	0.13	2.99	–	3.07	m	**6.06**
38 mm × 150 mm	–	0.13	2.99	–	3.57	m	**6.56**
47 mm × 100 mm	–	0.13	2.99	–	2.24	m	**5.23**
47 mm × 125 mm	–	0.13	2.99	–	2.85	m	**5.84**
47 mm × 150 mm	–	0.14	3.21	–	3.29	m	**6.50**
47 mm × 175 mm	–	0.14	3.21	–	3.87	m	**7.08**
47 mm × 200 mm	–	0.15	3.45	–	4.23	m	**7.68**
47 mm × 225 mm	–	0.15	3.45	–	4.84	m	**8.29**
47 mm × 250 mm	–	0.16	3.68	–	5.44	m	**9.12**
75 mm × 150 mm	–	0.15	3.45	–	6.79	m	**10.24**
75 mm × 175 mm	–	0.15	3.45	–	8.14	m	**11.59**
75 mm × 200 mm	–	0.16	3.68	–	9.14	m	**12.82**
75 mm × 225 mm	–	0.16	3.68	–	9.99	m	**13.67**
75 mm × 250 mm	–	0.17	3.90	–	14.75	m	**18.65**
Roof members; pitched							
25 mm × 100 mm	–	0.11	2.52	–	2.02	m	**4.54**
25 mm × 125 mm	–	0.11	2.52	–	2.77	m	**5.29**
25 mm × 150 mm	–	0.14	3.21	–	3.31	m	**6.52**
25 mm × 175 mm	–	0.16	3.68	–	3.87	m	**7.55**
25 mm × 200 mm	–	0.17	3.90	–	4.43	m	**8.33**
38 mm × 100 mm	–	0.14	3.21	–	2.58	m	**5.79**
38 mm × 125 mm	–	0.14	3.21	–	3.07	m	**6.28**
38 mm × 150 mm	–	0.14	3.21	–	3.57	m	**6.78**
38 mm × 175 mm	–	0.16	3.68	–	4.24	m	**7.92**
38 mm × 200 mm	–	0.17	3.90	–	4.87	m	**8.77**
47 mm × 50 mm	–	0.11	2.52	–	1.28	m	**3.80**
47 mm × 75 mm	–	0.14	3.21	–	1.74	m	**4.95**
47 mm × 100 mm	–	0.17	3.90	–	2.24	m	**6.14**
47 mm × 125 mm	–	0.17	3.90	–	2.85	m	**6.75**
47 mm × 150 mm	–	0.19	4.37	–	3.29	m	**7.66**

16 CARPENTRY

Item	PC £	Labour hours	Labour £	Plant £	Material £	Unit	Total rate £
16.01 PRIMARY AND STRUCTURAL TIMBERS – cont							
Sawn softwood – cont							
Roof members – cont							
47 mm × 175 mm	–	0.19	4.37	–	3.87	m	**8.24**
47 mm × 200 mm	–	0.19	4.37	–	4.23	m	**8.60**
47 mm × 225 mm	–	0.19	4.37	–	4.84	m	**9.21**
75 mm × 100 mm	–	0.23	5.28	–	5.06	m	**10.34**
75 mm × 125 mm	–	0.23	5.28	–	6.20	m	**11.48**
75 mm × 150 mm	–	0.23	5.28	–	6.79	m	**12.07**
100 mm × 150 mm	–	0.28	6.44	–	8.82	m	**15.26**
100 mm × 175 mm	–	0.28	6.44	–	10.28	m	**16.72**
100 mm × 200 mm	–	0.28	6.44	–	11.68	m	**18.12**
100 mm × 225 mm	–	0.31	7.13	–	13.12	m	**20.25**
100 mm × 250 mm	–	0.31	7.13	–	14.64	m	**21.77**
Plates							
38 mm × 75 mm	–	0.11	2.52	–	2.11	m	**4.63**
38 mm × 100 mm	–	0.14	3.21	–	2.58	m	**5.79**
47 mm × 75 mm	–	0.14	3.21	–	1.74	m	**4.95**
47 mm × 100 mm	–	0.17	3.90	–	2.24	m	**6.14**
75 mm × 100 mm	–	0.19	4.37	–	5.06	m	**9.43**
75 mm × 125 mm	–	0.22	5.06	–	6.15	m	**11.21**
75 mm × 150 mm	–	0.25	5.75	–	6.74	m	**12.49**
Plates; fixing by bolting							
38 mm × 75 mm	–	0.20	4.59	–	2.05	m	**6.64**
38 mm × 100 mm	–	0.23	5.28	–	2.58	m	**7.86**
47 mm × 75 mm	–	0.23	5.28	–	1.74	m	**7.02**
47 mm × 100 mm	–	0.26	5.97	–	2.24	m	**8.21**
75 mm × 100 mm	–	0.29	6.66	–	5.06	m	**11.72**
75 mm × 125 mm	–	0.31	7.13	–	6.15	m	**13.28**
75 mm × 150 mm	–	0.34	7.82	–	6.74	m	**14.56**
Joist strutting; herringbone							
47 mm × 50 mm; depth of joist 150 mm	–	0.46	10.57	–	3.16	m	**13.73**
47 mm × 50 mm; depth of joist 175 mm	–	0.46	10.57	–	3.22	m	**13.79**
47 mm × 50 mm; depth of joist 200 mm	–	0.46	10.57	–	3.29	m	**13.86**
47 mm × 50 mm; depth of joist 225 mm	–	0.46	10.57	–	3.34	m	**13.91**
47 mm × 50 mm; depth of joist 250 mm	–	0.46	10.57	–	3.40	m	**13.97**
Joist strutting; block							
47 mm × 150 mm; depth of joist 150 mm	–	0.28	6.44	–	3.94	m	**10.38**
47 mm × 175 mm; depth of joist 175 mm	–	0.28	6.44	–	4.53	m	**10.97**
47 mm × 200 mm; depth of joist 200 mm	–	0.28	6.44	–	4.89	m	**11.33**
47 mm × 225 mm; depth of joist 225 mm	–	0.28	6.44	–	5.50	m	**11.94**
47 mm × 250 mm; depth of joist 250 mm	–	0.28	6.44	–	6.10	m	**12.54**
Cleats							
225 mm × 100 mm × 75 mm	–	0.19	4.37	–	1.02	nr	**5.39**
Extra for stress grading to above timbers							
general structural (GS) grade	–	–	–	–	35.87	m³	**35.87**
special structural (SS) grade	–	–	–	–	71.74	m³	**71.74**

16 CARPENTRY

Item	PC £	Labour hours	Labour £	Plant £	Material £	Unit	Total rate £
Extra for protecting and flameproofing timber with Celgard CF protection or other equal and approved							
small sections	–	–	–	–	145.23	m³	**145.23**
large sections	–	–	–	–	140.22	m³	**140.22**
Wrot surfaces							
plain; 50 mm wide	–	0.02	0.46	–	–	m	**0.46**
plain; 100 mm wide	–	0.03	0.69	–	–	m	**0.69**
plain; 150 mm wide	–	0.04	0.92	–	–	m	**0.92**
16.02 ENGINEERED OR PREFABRICATED ITEMS							
Trussed rafters, stress graded sawn softwood pressure impregnated; raised through two storeys and fixed in position. Roof trusses are always project specific and prices always need to be obtained from a manufacturer for any particular roof design							
W type truss (Fink); 22.5° pitch; 450 mm eaves overhang							
5.00 m span	–	1.48	34.02	–	59.93	nr	**93.95**
7.60 m span	–	1.62	37.23	–	79.39	nr	**116.62**
10.00 m span	–	1.85	42.52	–	144.28	nr	**186.80**
W type truss (Fink); 30° pitch; 450 mm eaves overhang							
5.00 m span	–	1.48	34.02	–	59.93	nr	**93.95**
7.60 m span	–	1.62	37.23	–	81.56	nr	**118.79**
10.00 m span	–	1.85	42.52	–	142.12	nr	**184.64**
W type truss (Fink); 45° pitch; 450 mm eaves overhang							
4.60 m span	–	1.48	34.02	–	131.30	nr	**165.32**
7.00 m span	–	1.62	37.23	–	254.60	nr	**291.83**
Mono type truss; 17.5° pitch; 450 mm eaves overhang							
3.30 m span	–	1.30	29.88	–	53.44	nr	**83.32**
5.60 m span	–	1.48	34.02	–	72.90	nr	**106.92**
7.00 m span	–	1.71	39.30	–	83.72	nr	**123.02**
Attic type truss; 45° pitch; 450 mm eaves overhang							
5.00 m span	–	2.91	66.89	–	66.41	nr	**133.30**
7.60 m span	–	3.05	70.10	–	98.86	nr	**168.96**
9.00 m span	–	3.24	74.47	–	287.04	nr	**361.51**
Glulam timber beams whitewood; pressure impregnated; adhesive; clean planed finish. **Note: hoisting into position not included**							
Laminated roof beams upto 8 m long							
approximate cubic rate for glulam beams material only	–	–	–	–	1030.00	m³	**1030.00**
66 mm × 225 mm	–	0.20	4.59	–	16.22	m	**20.81**
66 mm × 315 mm	–	0.25	5.75	–	22.71	m	**28.46**

16 CARPENTRY

Item	PC £	Labour hours	Labour £	Plant £	Material £	Unit	Total rate £
16.02 ENGINEERED OR PREFABRICATED ITEMS – cont							
Glulam timber beams whitewood – cont							
Laminated roof beams upto 8 m long – cont							
90 mm × 315 mm	–	0.40	9.20	–	28.84	m	**38.04**
90 mm × 405 mm	–	0.50	11.49	–	26.78	m	**38.27**
115 mm × 405 mm	–	0.50	11.49	–	48.41	m	**59.90**
115 mm × 495 mm	–	0.75	17.24	–	58.71	m	**75.95**
115 mm × 630 mm	–	0.90	20.68	–	74.16	m	**94.84**
140 mm × 405 mm	–	0.60	13.79	–	58.71	m	**72.50**
140 mm × 495 mm	–	0.75	17.24	–	71.07	m	**88.31**
140 mm × 630 mm	–	1.00	22.98	–	95.17	m	**118.15**
16.03 BOARDING TO FLOORS							
Chipboard boarding and flooring							
Boarding to floors; butt joints							
18 mm thick	4.52	0.28	6.44	–	5.51	m²	**11.95**
Boarding to floors; tongued and grooved joints							
18 mm thick	5.69	0.30	6.89	–	6.81	m²	**13.70**
22 mm thick	6.62	0.32	7.35	–	7.84	m²	**15.19**
Acoustic chipboard flooring							
Boarding to floors; tongued and grooved joints							
chipboard on blue bat bearers	–	–	–	–	–	m²	**25.95**
chipboard on New Era levelling system	–	–	–	–	–	m²	**35.05**
Laminated engineered board flooring; 180 or 240 mm face widths; with 6 mm wear surface down to tongue; pre-finished laquered, oiled or untreated							
Boarding to floors; micro-bevel or square edge							
Country laquered; on 10 mm Pro Foam	–	–	–	–	–	m²	**76.28**
Rustic laquered; on 10 mm Pro Foam	–	–	–	–	–	m²	**72.47**
Plywood flooring							
Boarding to floors; tongued and grooved joints							
18 mm thick	9.38	0.41	9.42	–	10.90	m²	**20.32**
22 mm thick	11.37	0.45	10.34	–	13.09	m²	**23.43**
Strip boarding; wrought softwood							
Boarding to floors; butt joints							
19 mm × 75 mm boards	–	0.56	12.88	–	17.44	m²	**30.32**
19 mm × 125 mm boards	–	0.51	11.72	–	13.37	m²	**25.09**
22 mm × 150 mm boards	–	0.46	10.57	–	14.83	m²	**25.40**
25 mm × 100 mm boards	–	0.51	11.72	–	16.21	m²	**27.93**
25 mm × 150 mm boards	–	0.46	10.57	–	16.48	m²	**27.05**

16 CARPENTRY

Item	PC £	Labour hours	Labour £	Plant £	Material £	Unit	Total rate £
Boarding to floors; tongued and grooved joints							
19 mm × 75 mm boards	–	0.65	14.94	–	18.77	m²	**33.71**
19 mm × 125 mm boards	–	0.60	13.79	–	15.04	m²	**28.83**
22 mm × 150 mm boards	–	0.56	12.88	–	15.27	m²	**28.15**
25 mm × 100 mm boards	–	0.60	13.79	–	19.25	m²	**33.04**
25 mm × 150 mm boards	–	0.56	12.88	–	18.04	m²	**30.92**
16.04 BOARDING TO CEILINGS							
Plywood (Eastern European); internal quality							
Lining to ceilings 4 mm thick							
over 300 mm wide	2.41	0.46	10.57	–	3.07	m²	**13.64**
not exceeding 300 mm wide	–	0.30	6.89	–	0.95	m	**7.84**
holes for pipes and the like	–	0.02	0.46	–	–	nr	**0.46**
Lining to ceilings 6 mm thick							
over 300 mm wide	3.48	0.49	11.26	–	4.25	m²	**15.51**
not exceeding 300 mm wide	–	0.32	7.35	–	1.31	m	**8.66**
holes for pipes and the like	–	0.02	0.46	–	–	nr	**0.46**
Lining to ceilings 12 mm thick							
over 300 mm wide	6.46	0.56	12.88	–	7.56	m²	**20.44**
not exceeding 300 mm wide	–	0.37	8.51	–	2.30	m	**10.81**
holes for pipes and the like	–	0.03	0.69	–	–	nr	**0.69**
Lining to ceilings 18 mm thick							
over 300 mm wide	9.44	0.60	13.79	–	10.86	m²	**24.65**
not exceeding 300 mm wide	–	0.40	9.20	–	3.29	m	**12.49**
holes for pipes and the like	–	0.03	0.69	–	–	nr	**0.69**
Plywood (Eastern European); external quality							
Lining to ceilings 4 mm thick							
over 300 mm wide	5.96	0.46	10.57	–	7.00	m²	**17.57**
not exceeding 300 mm wide	–	0.30	6.89	–	2.13	m	**9.02**
holes for pipes and the like	–	0.02	0.46	–	–	nr	**0.46**
Lining to ceilings 6.5 mm thick							
over 300 mm wide	6.67	0.49	11.26	–	7.79	m²	**19.05**
not exceeding 300 mm wide	–	0.32	7.35	–	2.37	m	**9.72**
holes for pipes and the like	–	0.02	0.46	–	–	nr	**0.46**
Lining to ceilings 9 mm thick							
over 300 mm wide	8.58	0.53	12.18	–	9.91	m²	**22.09**
not exceeding 300 mm wide	–	0.34	7.82	–	3.01	m	**10.83**
holes for pipes and the like	–	0.03	0.69	–	–	nr	**0.69**
Lining to ceilings 12 mm thick							
over 300 mm wide	10.72	0.56	12.88	–	12.28	m²	**25.16**
not exceeding 300 mm wide	–	0.37	8.51	–	3.71	m	**12.22**
holes for pipes and the like	–	0.03	0.69	–	–	nr	**0.69**
Extra over Linings fixed with screws	–	0.10	2.30	–	0.08	m²	**2.38**

16 CARPENTRY

Item	PC £	Labour hours	Labour £	Plant £	Material £	Unit	Total rate £
16.04 BOARDING TO CEILINGS – cont							
Masterboard or other equal; sanded finish							
Lining to ceilings 6 mm thick							
over 300 mm wide	13.19	0.41	9.42	–	14.91	m²	**24.33**
not exceeding 300 mm wide	–	0.25	5.75	–	4.48	m	**10.23**
holes for pipes and the like	–	0.02	0.46	–	–	nr	**0.46**
Lining to ceilings 9 mm thick							
over 300 mm wide	30.49	0.42	9.65	–	34.06	m²	**43.71**
not exceeding 300 mm wide	–	0.27	6.20	–	10.23	m	**16.43**
holes for pipes and the like	–	0.03	0.69	–	–	nr	**0.69**
Supalux or other equal; sanded finish							
Lining to ceilings 6 mm thick							
over 300 mm wide	25.45	0.41	9.42	–	28.49	m²	**37.91**
not exceeding 300 mm wide	–	0.25	5.75	–	8.56	m	**14.31**
holes for pipes and the like	–	0.03	0.69	–	–	nr	**0.69**
Lining to ceilings 9 mm thick							
over 300 mm wide	32.40	0.42	9.65	–	36.18	m²	**45.83**
not exceeding 300 mm wide	–	0.27	6.20	–	10.87	m	**17.07**
holes for pipes and the like	–	0.03	0.69	–	–	nr	**0.69**
Lining to ceilings 12 mm thick							
over 300 mm wide	44.00	0.49	11.26	–	49.03	m²	**60.29**
not exceeding 300 mm wide	–	0.30	6.89	–	14.72	m	**21.61**
holes for pipes and the like	–	0.04	0.92	–	–	nr	**0.92**
Extra over Linings fixed with screws	–	0.10	2.30	–	0.08	m²	**2.38**
Strip boarding; wrought softwood							
Boarding to internal ceilings							
12 mm × 100 mm boards	–	0.93	21.37	–	17.24	m²	**38.61**
16 mm × 100 mm boards	–	0.93	21.37	–	18.65	m²	**40.02**
19 mm × 100 mm boards	–	0.93	21.37	–	21.14	m²	**42.51**
19 mm × 125 mm boards	–	0.88	20.23	–	19.29	m²	**39.52**
19 mm × 125 mm boards; chevron pattern	–	1.30	29.88	–	19.29	m²	**49.17**
25 mm × 125 mm boards	–	0.88	20.23	–	16.59	m²	**36.82**
12 mm × 100 mm boards; knotty pine	–	0.93	21.37	–	10.61	m²	**31.98**
16.05 BOARDING TO ROOFS							
Plywood; external quality; 18 mm thick							
Boarding to roofs; butt joints							
flat to falls	12.55	0.37	8.51	–	14.41	m²	**22.92**
sloping	12.55	0.40	9.20	–	14.41	m²	**23.61**
vertical	12.55	0.53	12.18	–	14.41	m²	**26.59**
Plywood; external quality; 12 mm thick							
Boarding to roofs; butt joints							
flat to falls	8.47	0.37	8.51	–	9.89	m²	**18.40**
sloping	8.47	0.40	9.20	–	9.89	m²	**19.09**
vertical	8.47	0.53	12.18	–	9.89	m²	**22.07**

16 CARPENTRY

Item	PC £	Labour hours	Labour £	Plant £	Material £	Unit	Total rate £
Sawn softwood; untreated							
Boarding to roofs; 150 mm wide boards; butt joints							
19 mm thick; flat; over 600 mm wide	–	0.42	9.65	–	12.77	m²	**22.42**
19 mm thick; flat; not exceeding 600 mm wide	–	0.55	12.64	–	7.77	m	**20.41**
19 mm thick; sloping; over 600 mm wide	–	0.46	10.57	–	12.77	m²	**23.34**
19 mm thick; sloping; not exceeding 600 mm wide	–	0.62	14.26	–	7.77	m	**22.03**
19 mm thick; sloping; laid diagonally; over 600 mm wide	–	0.58	13.33	–	12.77	m²	**26.10**
19 mm thick; sloping; laid diagonally; not exceeding 600 mm wide	–	0.75	17.24	–	7.77	m	**25.01**
25 mm thick; flat; over 600 mm wide	–	0.42	9.65	–	20.82	m²	**30.47**
25 mm thick; flat; not exceeding 600 mm wide	–	0.56	12.88	–	12.59	m	**25.47**
25 mm thick; sloping; over 600 mm wide	–	0.46	10.57	–	20.82	m²	**31.39**
25 mm thick; sloping; not exceeding 600 mm wide	–	0.62	14.26	–	12.59	m	**26.85**
25 mm thick; sloping; laid diagonally; over 600 mm wide	–	0.58	13.33	–	20.82	m²	**34.15**
25 mm thick; sloping; laid diagonally; not exceeding 600 mm wide	–	0.74	17.01	–	12.59	m	**29.60**
Boarding to tops or cheeks of dormers; 150 mm wide boards; butt joints							
19 mm thick; laid diagonally; over 600 mm wide	–	0.74	17.01	–	12.77	m²	**29.78**
19 mm thick; laid diagonally; not exceeding 600 mm wide	–	0.92	21.15	–	7.77	m	**28.92**
19 mm thick; laid diagonally; area not exceeding 1.00 m² irrespective of width	–	0.93	21.37	–	12.11	nr	**33.48**
Sawn softwood; tanalized							
Boarding to roofs; 150 wide boards; butt joints							
19 mm thick; flat; over 600 mm wide	–	0.42	9.65	–	13.97	m²	**23.62**
19 mm thick; flat; not exceeding 600 mm wide	–	0.56	12.88	–	8.49	m	**21.37**
19 mm thick; sloping; over 600 mm wide	–	0.46	10.57	–	13.97	m²	**24.54**
19 mm thick; sloping; not exceeding 600 mm wide	–	0.62	14.26	–	8.49	m	**22.75**
19 mm thick; sloping; laid diagonally; over 600 mm wide	–	0.58	13.33	–	13.97	m²	**27.30**
19 mm thick; sloping; laid diagonally; not exceeding 600 mm wide	–	0.74	17.01	–	8.49	m	**25.50**
25 mm thick; flat; over 600 mm wide	–	0.42	9.65	–	22.38	m²	**32.03**
25 mm thick; flat; not exceeding 600 mm wide	–	0.56	12.88	–	13.31	m	**26.19**
25 mm thick; sloping; over 600 mm wide	–	0.46	10.57	–	22.38	m²	**32.95**
25 mm thick; sloping; not exceeding 600 mm wide	–	0.62	14.26	–	13.31	m	**27.57**
25 mm thick; sloping; laid diagonally; over 600 mm wide	–	0.58	13.33	–	22.38	m²	**35.71**
25 mm thick; sloping; laid diagonally; not exceeding 600 mm wide	–	0.74	17.01	–	13.31	m	**30.32**
Boarding to tops or cheeks of dormers; 150 mm wide boards; butt joints							
19 mm thick; laid diagonally; over 600 mm wide	–	0.74	17.01	–	13.97	m²	**30.98**
19 mm thick; laid diagonally; not exceeding 600 mm wide	–	0.92	21.15	–	8.49	m	**29.64**
19 mm thick; laid diagonally; area not exceeding 1.00 m² irrespective of width	–	0.93	21.37	–	13.31	nr	**34.68**

16 CARPENTRY

Item	PC £	Labour hours	Labour £	Plant £	Material £	Unit	Total rate £
16.05 BOARDING TO ROOFS – cont							
Wrought softwood							
Boarding to roofs; tongued and grooved joints							
19 mm thick; flat to falls	–	0.51	11.72	–	17.66	m²	**29.38**
19 mm thick; sloping	–	0.56	12.88	–	17.66	m²	**30.54**
19 mm thick; sloping; laid diagonally	–	0.72	16.55	–	17.66	m²	**34.21**
25 mm thick; flat to falls	–	0.51	11.72	–	17.79	m²	**29.51**
25 mm thick; sloping	–	0.56	12.88	–	17.79	m²	**30.67**
Boarding to tops or cheeks of dormers; tongued and grooved joints							
19 mm thick; laid diagonally	–	0.93	21.37	–	17.66	m²	**39.03**
Wrought softwood; tanalized							
Boarding to roofs; tongued and grooved joints							
19 mm thick; flat to falls	–	0.51	11.72	–	18.86	m²	**30.58**
19 mm thick; sloping	–	0.56	12.88	–	18.86	m²	**31.74**
19 mm thick; sloping; laid diagonally	–	0.72	16.55	–	18.86	m²	**35.41**
25 mm thick; flat to falls	–	0.51	11.72	–	19.35	m²	**31.07**
25 mm thick; sloping	–	0.56	12.88	–	19.35	m²	**32.23**
Boarding to tops or cheeks of dormers; tongued and grooved joints							
19 mm thick; laid diagonally	–	0.93	21.37	–	18.86	m²	**40.23**
Masterboard or other equal and approved; 6 mm thick							
Eaves, verge soffit boards, fascia boards and the like							
over 300 mm wide	13.19	0.65	14.94	–	15.63	m²	**30.57**
75 mm wide	–	0.19	4.37	–	1.19	m	**5.56**
150 mm wide	–	0.22	5.06	–	2.35	m	**7.41**
225 mm wide	–	0.26	5.97	–	3.49	m	**9.46**
300 mm wide	–	0.28	6.44	–	4.64	m	**11.08**
Plywood; external quality; 12 mm thick							
Eaves, verge soffit boards, fascia boards and the like							
over 300 mm wide	8.47	0.76	17.47	–	10.39	m²	**27.86**
75 mm wide	–	0.23	5.28	–	0.80	m	**6.08**
150 mm wide	–	0.27	6.20	–	1.56	m	**7.76**
225 mm wide	–	0.31	7.13	–	2.32	m	**9.45**
300 mm wide	–	0.34	7.82	–	3.07	m	**10.89**
Plywood; external quality; 15 mm thick							
Eaves, verge soffit boards, fascia boards and the like							
over 300 mm wide	10.72	0.76	17.47	–	12.89	m²	**30.36**
75 mm wide	–	0.23	5.28	–	0.99	m	**6.27**
150 mm wide	–	0.27	6.20	–	1.94	m	**8.14**
225 mm wide	–	0.31	7.13	–	2.87	m	**10.00**
300 mm wide	–	0.34	7.82	–	3.81	m	**11.63**

16 CARPENTRY

Item	PC £	Labour hours	Labour £	Plant £	Material £	Unit	Total rate £
Plywood; external quality; 18 mm thick							
Eaves, verge soffit boards, fascia boards and the like							
over 300 mm wide	12.55	0.76	17.47	–	14.91	m²	**32.38**
75 mm wide	–	0.23	5.28	–	1.14	m	**6.42**
150 mm wide	–	0.27	6.20	–	2.24	m	**8.44**
225 mm wide	–	0.31	7.13	–	3.33	m	**10.46**
300 mm wide	–	0.34	7.82	–	4.43	m	**12.25**
Plywood; marine quality; 18 mm thick							
Gutter boards; butt joints							
over 300 mm wide	12.37	0.86	19.77	–	14.71	m²	**34.48**
150 mm wide	–	0.31	7.13	–	2.20	m	**9.33**
225 mm wide	–	0.34	7.82	–	3.34	m	**11.16**
300 mm wide	–	0.38	8.73	–	4.41	m	**13.14**
Eaves, verge soffit boards, fascias boards and the like							
over 300 mm wide	12.37	0.76	17.47	–	14.71	m²	**32.18**
75 mm wide	–	0.23	5.28	–	1.13	m	**6.41**
150 mm wide	–	0.27	6.20	–	2.20	m	**8.40**
225 mm wide	–	0.31	7.13	–	3.29	m	**10.42**
300 mm wide	–	0.34	7.82	–	4.36	m	**12.18**
Plywood; marine quality; 25 mm thick							
Gutter boards; butt joints							
over 300 mm wide	17.18	0.93	21.37	–	20.04	m²	**41.41**
150 mm wide	–	0.32	7.35	–	3.01	m	**10.36**
225 mm wide	–	0.37	8.51	–	4.53	m	**13.04**
300 mm wide	–	0.42	9.65	–	6.02	m	**15.67**
Eaves, verge soffit boards, fascia baords and the like							
over 300 mm wide	17.18	0.81	18.61	–	20.04	m²	**38.65**
75 mm wide	–	0.24	5.52	–	1.52	m	**7.04**
150 mm wide	–	0.29	6.66	–	3.01	m	**9.67**
225 mm wide	–	0.29	6.66	–	4.48	m	**11.14**
300 mm wide	–	0.37	8.51	–	5.96	m	**14.47**
Sawn softwood; untreated							
Gutter boards; butt joints							
19 mm thick; 150 mm wide; sloping	–	1.16	26.66	–	13.53	m²	**40.19**
19 mm thick; 75 mm wide	–	0.32	7.35	–	1.05	m	**8.40**
19 mm thick; 150 mm wide	–	0.37	8.51	–	1.97	m	**10.48**
19 mm thick; 225 mm wide	–	0.42	9.65	–	3.58	m	**13.23**
25 mm thick; sloping	–	1.16	26.66	–	21.58	m²	**48.24**
25 mm thick; 75 mm wide	–	0.32	7.35	–	1.33	m	**8.68**
25 mm thick; 150 mm wide	–	0.37	8.51	–	3.17	m	**11.68**
25 mm thick; 225 mm wide	–	0.42	9.65	–	5.00	m	**14.65**
Cesspools with 25 mm thick sides and bottom							
225 mm × 225 mm × 150 mm	–	1.11	25.51	–	4.19	nr	**29.70**
300 mm × 300 mm × 150 mm	–	1.30	29.88	–	5.50	nr	**35.38**

16 CARPENTRY

Item	PC £	Labour hours	Labour £	Plant £	Material £	Unit	Total rate £
16.05 BOARDING TO ROOFS – cont							
Sawn softwood – cont							
Individual supports; firrings							
50 mm wide × 36 mm average depth	–	0.14	3.21	–	2.81	m	**6.02**
50 mm wide × 50 mm average depth	–	0.14	3.21	–	4.27	m	**7.48**
50 mm wide × 75 mm average depth	–	0.14	3.21	–	5.56	m	**8.77**
Individual supports; bearers							
25 mm × 50 mm	–	0.09	2.07	–	1.17	m	**3.24**
38 mm × 50 mm	–	0.09	2.07	–	1.63	m	**3.70**
50 mm × 50 mm	–	0.09	2.07	–	1.22	m	**3.29**
50 mm × 75 mm	–	0.09	2.07	–	1.60	m	**3.67**
Individual supports; angle fillets							
38 mm × 38 mm	–	0.09	2.07	–	1.12	m	**3.19**
50 mm × 50 mm	–	0.09	2.07	–	1.43	m	**3.50**
75 mm × 75 mm	–	0.11	2.52	–	2.95	m	**5.47**
Individual supports; tilting fillets							
19 mm × 38 mm	–	0.09	2.07	–	0.68	m	**2.75**
25 mm × 50 mm	–	0.09	2.07	–	1.10	m	**3.17**
38 mm × 75 mm	–	0.09	2.07	–	1.69	m	**3.76**
50 mm × 75 mm	–	0.09	2.07	–	2.16	m	**4.23**
75 mm × 100 mm	–	0.14	3.21	–	4.02	m	**7.23**
Individual supports; grounds or battens							
13 mm × 19 mm	–	0.04	0.92	–	0.54	m	**1.46**
13 mm × 32 mm	–	0.04	0.92	–	0.54	m	**1.46**
25 mm × 50 mm	–	0.04	0.92	–	1.07	m	**1.99**
Individual supports; grounds or battens; plugged and screwed							
13 mm × 19 mm	–	0.14	3.21	–	0.53	m	**3.74**
13 mm × 32 mm	–	0.14	3.21	–	0.53	m	**3.74**
25 mm × 50 mm	–	0.14	3.21	–	1.05	m	**4.26**
Framed supports; open-spaced grounds or battens; at 300 mm centres one way							
25 mm × 50 mm	–	0.14	3.21	–	3.54	m²	**6.75**
25 mm × 50 mm; plugged and screwed	–	0.42	9.65	–	3.51	m²	**13.16**
Framed supports; at 300 mm centres one way and 600 mm centres the other way							
25 mm × 50 mm	–	0.69	15.86	–	5.31	m²	**21.17**
38 mm × 50 mm	–	0.69	15.86	–	7.58	m²	**23.44**
50 mm × 50 mm	–	0.69	15.86	–	5.54	m²	**21.40**
50 mm × 75 mm	–	0.69	15.86	–	7.44	m²	**23.30**
75 mm × 75 mm	–	0.69	15.86	–	16.95	m²	**32.81**
Framed supports; at 300 mm centres one way and 600 mm centres the other way; plugged and screwed							
25 mm × 50 mm	–	1.16	26.66	–	5.46	m²	**32.12**
38 mm × 50 mm	–	1.16	26.66	–	7.73	m²	**34.39**
50 mm × 50 mm	–	1.16	26.66	–	5.69	m²	**32.35**
50 mm × 75 mm	–	1.16	26.66	–	7.57	m²	**34.23**
75 mm × 75 mm	–	1.16	26.66	–	17.09	m²	**43.75**

16 CARPENTRY

Item	PC £	Labour hours	Labour £	Plant £	Material £	Unit	Total rate £
Framed supports; at 500 mm centres both ways							
25 mm × 50 mm; to bath panels	–	0.83	19.08	–	6.92	m²	**26.00**
Framed supports; as bracketing and cradling around steelwork							
25 mm × 50 mm	–	1.30	29.88	–	7.51	m²	**37.39**
50 mm × 50 mm	–	1.39	31.95	–	7.83	m²	**39.78**
50 mm × 75 mm	–	1.48	34.02	–	10.48	m²	**44.50**
Sawn softwood; tanalized							
Gutter boards; butt joints							
19 mm thick; 150 mm; sloping	–	1.16	26.66	–	14.73	m²	**41.39**
19 mm thick; 75 mm wide	–	0.32	7.35	–	1.13	m	**8.48**
19 mm thick; 150 mm wide	–	0.37	8.51	–	2.15	m	**10.66**
19 mm thick; 225 mm wide	–	0.42	9.65	–	3.85	m	**13.50**
25 mm thick; sloping	–	1.16	26.66	–	23.14	m²	**49.80**
25 mm thick; 75 mm wide	–	0.32	7.35	–	1.44	m	**8.79**
25 mm thick; 150 mm wide	–	0.37	8.51	–	3.41	m	**11.92**
25 mm thick; 225 mm wide	–	0.42	9.65	–	5.35	m	**15.00**
Cesspools with 25 mm thick sides and bottom							
225 mm × 225 mm × 150 mm	–	1.11	25.51	–	4.50	nr	**30.01**
300 mm × 300 mm × 150 mm	–	1.30	29.88	–	5.92	nr	**35.80**
Individual supports; firrings							
50 mm wide × 36 mm average depth	–	0.14	3.21	–	2.93	m	**6.14**
50 mm wide × 50 mm average depth	–	0.14	3.21	–	4.43	m	**7.64**
50 mm wide × 75 mm average depth	–	0.14	3.21	–	5.80	m	**9.01**
Individual supports; bearers							
25 mm × 50 mm	–	0.09	2.07	–	1.26	m	**3.33**
38 mm × 50 mm	–	0.09	2.07	–	1.74	m	**3.81**
50 mm × 50 mm	–	0.09	2.07	–	1.38	m	**3.45**
50 mm × 75 mm	–	0.09	2.07	–	1.83	m	**3.90**
Individual supports; angle fillets							
38 mm × 38 mm	–	0.09	2.07	–	1.16	m	**3.23**
50 mm × 50 mm	–	0.09	2.07	–	1.51	m	**3.58**
75 mm × 75 mm	–	0.11	2.52	–	3.12	m	**5.64**
Individual supports; tilting fillets							
19 mm × 38 mm	–	0.09	2.07	–	0.70	m	**2.77**
25 mm × 50 mm	–	0.09	2.07	–	1.13	m	**3.20**
38 mm × 75 mm	–	0.09	2.07	–	1.77	m	**3.84**
50 mm × 75 mm	–	0.09	2.07	–	2.29	m	**4.36**
75 mm × 100 mm	–	0.14	3.21	–	4.25	m	**7.46**
Individual supports; grounds or battens							
13 mm × 19 mm	–	0.04	0.92	–	0.56	m	**1.48**
13 mm × 32 mm	–	0.04	0.92	–	0.57	m	**1.49**
25 mm × 50 mm	–	0.04	0.92	–	1.15	m	**2.07**
Individual supports; grounds or battens; plugged and screwed							
13 mm × 19 mm	–	0.14	3.21	–	0.54	m	**3.75**
13 mm × 32 mm	–	0.14	3.21	–	0.55	m	**3.76**
25 mm × 50 mm	–	0.14	3.21	–	1.13	m	**4.34**

16 CARPENTRY

Item	PC £	Labour hours	Labour £	Plant £	Material £	Unit	Total rate £
16.05 BOARDING TO ROOFS – cont							
Sawn softwood – cont							
Framed supports; open-spaced grounds or battens; at 300 mm centres one way							
25 mm × 50 mm	–	0.14	3.21	–	3.80	m²	**7.01**
25 mm × 50 mm; plugged and screwed	–	0.42	9.65	–	3.78	m²	**13.43**
Framed supports; at 300 mm centres one way and 600 mm centres the other way							
25 mm × 50 mm	–	0.69	15.86	–	5.71	m²	**21.57**
38 mm × 50 mm	–	0.69	15.86	–	8.18	m²	**24.04**
50 mm × 50 mm	–	0.69	15.86	–	6.32	m²	**22.18**
50 mm × 75 mm	–	0.69	15.86	–	8.61	m²	**24.47**
75 mm × 75 mm	–	0.69	15.86	–	18.72	m²	**34.58**
Framed supports; at 300 mm centres one way and 600 mm centres the other way; plugged and screwed							
25 mm × 50 mm	–	1.16	26.66	–	5.85	m²	**32.51**
38 mm × 50 mm	–	1.16	26.66	–	8.32	m²	**34.98**
50 mm × 50 mm	–	1.16	26.66	–	6.46	m²	**33.12**
50 mm × 75 mm	–	1.16	26.66	–	8.76	m²	**35.42**
75 mm × 75 mm	–	1.16	26.66	–	18.85	m²	**45.51**
Framed supports; at 500 mm centres both ways							
25 mm × 50 mm; to bath panels	–	0.83	19.08	–	7.43	m²	**26.51**
Framed supports; as bracketing and cradling around steelwork							
25 mm × 50 mm	–	1.30	29.88	–	8.06	m²	**37.94**
50 mm × 50 mm	–	1.39	31.95	–	8.92	m²	**40.87**
50 mm × 75 mm	–	1.48	34.02	–	12.13	m²	**46.15**
Wrought softwood							
Gutter boards; tongued and grooved joints							
19 mm thick; 150 mm; sloping	–	1.39	31.95	–	18.43	m²	**50.38**
19 mm thick; 75 mm wide	–	0.37	8.51	–	1.34	m	**9.85**
19 mm thick; 150 mm wide	–	0.42	9.65	–	2.70	m	**12.35**
19 mm thick; 225 mm wide	–	0.46	10.57	–	3.93	m	**14.50**
25 mm thick; sloping	–	1.39	31.95	–	19.06	m²	**51.01**
25 mm thick; 75 mm wide	–	0.37	8.51	–	1.48	m	**9.99**
25 mm thick; 150 mm wide	–	0.42	9.65	–	2.67	m	**12.32**
25 mm thick; 225 mm wide	–	0.46	10.57	–	4.00	m	**14.57**
Eaves, verge soffit boards, fascia boards and the like							
19 mm thick; over 300 mm wide	–	1.15	26.43	–	18.46	m²	**44.89**
19 mm thick; 150 mm wide; once grooved	–	0.19	4.37	–	3.24	m	**7.61**
25 mm thick; 150 mm wide; once grooved	–	0.19	4.37	–	3.94	m	**8.31**
25 mm thick; 175 mm wide; once grooved	–	0.19	4.37	–	3.88	m	**8.25**
32 mm thick; 225 mm wide; once grooved	–	0.23	5.28	–	6.12	m	**11.40**

16 CARPENTRY

Item	PC £	Labour hours	Labour £	Plant £	Material £	Unit	Total rate £
Wrought softwood; tanalized							
Gutter boards; tongued and grooved joints							
19 mm thick; 150 mm; sloping	–	1.39	31.95	–	19.62	m²	**51.57**
19 mm thick; 75 mm wide	–	0.37	8.51	–	1.43	m	**9.94**
19 mm thick; 150 mm wide	–	0.42	9.65	–	2.88	m	**12.53**
19 mm thick; 225 mm wide	–	0.46	10.57	–	4.20	m	**14.77**
25 mm thick; sloping	–	1.39	31.95	–	20.62	m²	**52.57**
25 mm thick; 75 mm wide	–	0.37	8.51	–	1.60	m	**10.11**
25 mm thick; 150 mm wide	–	0.42	9.65	–	2.90	m	**12.55**
25 mm thick; 225 mm wide	–	0.46	10.57	–	4.35	m	**14.92**
Eaves, verge soffit boards, fascia boards and the like							
19 mm thick; over 300 mm wide	–	1.15	26.43	–	19.65	m²	**46.08**
19 mm thick; 150 mm wide; once grooved	–	0.19	4.37	–	3.42	m	**7.79**
25 mm thick; 150 mm wide; once grooved	–	0.19	4.37	–	4.18	m	**8.55**
25 mm thick; 175 mm wide; once grooved	–	0.20	4.59	–	4.15	m	**8.74**
32 mm thick; 225 mm wide; once grooved	–	0.23	5.28	–	6.56	m	**11.84**
16.06 CASINGS							
Chipboard (plain)							
Two-sided 15 mm thick pipe casing; to softwood framing (not included)							
300 mm girth	–	0.56	12.88	–	1.27	m	**14.15**
600 mm girth	–	0.65	14.94	–	2.24	m	**17.18**
Three-sided 15 mm thick pipe casing; to softwood framing (not included)							
450 mm girth	–	1.16	26.66	–	1.91	m	**28.57**
900 mm girth	–	1.39	31.95	–	3.40	m	**35.35**
extra for 400 mm × 400 mm removable access panel; brass cups and screws; additional framing	–	0.93	21.37	–	1.42	nr	**22.79**
Plywood (Eastern European); internal quality							
Two-sided 6 mm thick pipe casings; to softwood framing (not included)							
300 mm girth	–	0.74	17.01	–	1.46	m	**18.47**
600 mm girth	–	0.93	21.37	–	2.62	m	**23.99**
Three-sided 6 mm thick pipe casing; to softwood framing (not included)							
450 mm girth	–	1.06	24.36	–	2.19	m	**26.55**
900 mm girth	–	1.25	28.73	–	3.98	m	**32.71**
Plywood (Eastern European); external quality							
Two-sided 6.5 mm thick pipe casings; to softwood framing (not included)							
300 mm girth	–	0.74	17.01	–	2.52	m	**19.53**
600 mm girth	–	0.93	21.37	–	4.74	m	**26.11**

16 CARPENTRY

Item	PC £	Labour hours	Labour £	Plant £	Material £	Unit	Total rate £
16.06 CASINGS – cont							
Plywood (Eastern European) – cont							
Three-sided 6.5 mm thick pipe casing; to softwood framing (not included)							
450 mm girth	–	1.06	24.36	–	3.78	m	**28.14**
900 mm girth	–	1.25	28.73	–	7.15	m	**35.88**
Two-sided 12 mm thick pipe casing; to softwood framing (not included)							
300 mm girth	–	0.69	15.86	–	3.86	m	**19.72**
600 mm girth	–	0.83	19.08	–	7.43	m	**26.51**
Three-sided 12 mm thick pipe casing; to softwood framing (not included)							
450 mm girth	–	0.93	21.37	–	5.80	m	**27.17**
900 mm girth	–	1.11	25.51	–	11.19	m	**36.70**
extra for 400 mm × 400 mm removable access panel; brass cups and screws; additional framing	–	1.00	22.98	–	1.42	nr	**24.40**
Preformed white melamine faced plywood casings; Pendock Profiles Ltd or other equal and approved; to softwood battens (not included)							
Skirting trunking profile; plain butt joints in the running length							
45 mm × 150 mm; ref TK150	–	0.11	2.52	–	32.87	m	**35.39**
extra for stop end	–	0.04	0.92	–	20.48	nr	**21.40**
extra for external corner	–	0.09	2.07	–	28.28	nr	**30.35**
extra for internal corner	–	0.09	2.07	–	17.18	nr	**19.25**
Casing profiles							
150 mm × 150 mm; ref MX150/150; 5 mm thick	–	0.11	2.52	–	27.41	m	**29.93**
extra for stop end	–	0.04	0.92	–	7.08	nr	**8.00**
extra for external corner	–	0.09	2.07	–	43.28	nr	**45.35**
extra for internal corner	–	0.09	2.07	–	17.18	nr	**19.25**
16.07 METAL FIXINGS, FIXINGS, FASTENINGS AND FITTINGS							
Straps; mild steel; galvanized							
Standard twisted vertical restraint; fixing to softwood and brick or blockwork							
27.5 mm × 2.5 mm × 400 mm girth	–	0.23	5.28	–	2.35	nr	**7.63**
27.5 mm × 2.5 mm × 600 mm girth	–	0.24	5.52	–	3.27	nr	**8.79**
27.5 mm × 2.5 mm × 800 mm girth	–	0.25	5.75	–	4.74	nr	**10.49**
27.5 mm × 2.5 mm × 1000 mm girth	–	0.28	6.44	–	6.16	nr	**12.60**
27.5 mm × 2.5 mm × 1200 mm girth	–	–	–	–	–	nr	**–**
Hangers; mild steel; galvanized							
Joist hangers 0.90 mm thick; The Expanded Metal Company Ltd Speedy or other equal and approved; for fixing to softwood; joist sizes							
50 mm wide; all sizes to 225 mm deep	3.17	0.11	2.52	–	3.56	nr	**6.08**
75 mm wide; all sizes to 225 mm deep	3.31	0.14	3.21	–	3.86	nr	**7.07**
100 mm wide; all sizes to 225 mm deep	3.56	0.17	3.90	–	4.26	nr	**8.16**

16 CARPENTRY

Item	PC £	Labour hours	Labour £	Plant £	Material £	Unit	Total rate £
Joist hangers 2.50 mm thick; for building in; joist sizes							
50 mm × 150 mm	2.62	0.09	2.05	–	3.14	nr	**5.19**
50 mm × 175 mm	2.01	0.09	2.05	–	2.47	nr	**4.52**
50 mm × 200 mm	3.18	0.11	2.50	–	3.68	nr	**6.18**
50 mm × 225 mm	3.06	0.11	2.50	–	3.71	nr	**6.21**
75 mm × 150 mm	2.22	0.09	2.05	–	2.70	nr	**4.75**
75 mm × 175 mm	2.94	0.09	2.05	–	3.48	nr	**5.53**
75 mm × 200 mm	3.12	0.11	2.50	–	3.77	nr	**6.27**
75 mm × 225 mm	3.38	0.11	2.50	–	4.06	nr	**6.56**
75 mm × 250 mm	6.00	0.13	2.97	–	6.99	nr	**9.96**
100 mm × 200 mm	9.60	0.11	2.50	–	10.78	nr	**13.28**
Metal connectors; mild steel; galvanized							
Round toothed plate; for 10 mm or 12 mm dia. bolts							
38 mm dia.; single sided	–	0.01	0.23	–	0.88	nr	**1.11**
38 mm dia.; double sided	–	0.01	0.23	–	0.97	nr	**1.20**
50 mm dia.; single sided	–	0.01	0.23	–	0.94	nr	**1.17**
50 mm dia.; double sided	–	0.01	0.23	–	1.05	nr	**1.28**
63 mm dia.; single sided	–	0.01	0.23	–	1.37	nr	**1.60**
63 mm dia.; double sided	–	0.01	0.23	–	1.51	nr	**1.74**
75 mm dia.; single sided	–	0.01	0.23	–	2.02	nr	**2.25**
75 mm dia.; double sided	–	0.01	0.23	–	2.10	nr	**2.33**
framing anchor	–	0.14	3.21	–	1.65	nr	**4.86**
Bolts; mild steel; galvanized							
Fixing only bolts; 50 mm–200 mm long							
6 mm dia.	–	0.03	0.69	–	–	nr	**0.69**
8 mm dia.	–	0.03	0.69	–	–	nr	**0.69**
10 mm dia.	–	0.04	0.92	–	–	nr	**0.92**
12 mm dia.	–	0.04	0.92	–	–	nr	**0.92**
16 mm dia.	–	0.05	1.15	–	–	nr	**1.15**
20 mm dia.	–	0.05	1.15	–	–	nr	**1.15**
Bolts							
Expanding bolts; Rawlbolt projecting type or other equal and approved; Rawl Fixings; plated; one nut; one washer							
6 mm dia.; ref M6 10P	–	0.09	2.07	–	0.66	nr	**2.73**
6 mm dia.; ref M6 25P	–	0.09	2.07	–	0.77	nr	**2.84**
6 mm dia.; ref M6 60P	–	0.09	2.07	–	0.77	Unit	**2.84**
8 mm dia.; ref M8 25P	–	0.09	2.07	–	0.76	nr	**2.83**
8 mm dia.; ref M8 60P	–	0.09	2.07	–	0.77	nr	**2.84**
10 mm dia.; ref M10 15P	–	0.09	2.07	–	1.02	nr	**3.09**
10 mm dia.; ref M10 30P	–	0.09	2.07	–	1.07	nr	**3.14**
10 mm dia.; ref M10 60P	–	0.09	2.07	–	1.04	nr	**3.11**
12 mm dia.; ref M12 15P	–	0.09	2.07	–	1.88	nr	**3.95**
12 mm dia.; ref M12 30P	–	0.10	2.30	–	0.16	nr	**2.46**
12 mm dia.; ref M12 75P	–	0.09	2.07	–	0.22	nr	**2.29**
16 mm dia.; ref M16 35P	–	0.09	2.07	–	4.01	nr	**6.08**
16 mm dia.; ref M16 75P	–	0.09	2.07	–	4.14	nr	**6.21**

16 CARPENTRY

Item	PC £	Labour hours	Labour £	Plant £	Material £	Unit	Total rate £
16.07 METAL FIXINGS, FIXINGS, FASTENINGS AND FITTINGS – cont							
Bolts – cont							
Expanding bolts; Rawlbolt loose bolt type or other equal; Rawl Fixings; plated; one bolt; one washer							
6 mm dia.; ref M6 10L	–	0.09	2.07	–	0.71	nr	**2.78**
6 mm dia.; ref M6 25L	–	0.09	2.07	–	0.71	nr	**2.78**
6 mm dia.; ref M6 40L	–	0.09	2.07	–	0.83	nr	**2.90**
8 mm dia.; ref M8 25L	–	0.09	2.07	–	0.92	nr	**2.99**
8 mm dia.; ref M8 40L	–	0.09	2.07	–	0.96	nr	**3.03**
10 mm dia.; ref M10 10L	–	0.09	2.07	–	0.97	nr	**3.04**
10 mm dia.; ref M10 25L	–	0.09	2.07	–	1.08	nr	**3.15**
10 mm dia.; ref M10 50L	–	0.09	2.07	–	1.08	nr	**3.15**
10 mm dia.; ref M10 75L	–	0.09	2.07	–	1.18	nr	**3.25**
12 mm dia.; ref M12 10L	–	0.09	2.07	–	1.87	nr	**3.94**
12 mm dia.; ref M12 25L	–	0.09	2.07	–	1.95	nr	**4.02**
12 mm dia.; ref M12 40L	–	0.09	2.07	–	1.93	nr	**4.00**
12 mm dia.; ref M12 60L	–	0.09	2.07	–	2.44	nr	**4.51**
16 mm dia.; ref M16 30L	–	0.09	2.07	–	4.10	nr	**6.17**
16 mm dia.; ref M16 60L	–	0.09	2.07	–	4.79	nr	**6.86**
Truss clips							
Truss clips; fixing to softwood; joist size							
38 mm wide	1.24	0.14	3.21	–	1.88	nr	**5.09**
50 mm wide	1.18	0.14	3.21	–	1.81	nr	**5.02**
Sole plate angles; mild steel galvanized							
Sole plate angle; fixing to softwood and concrete							
112 mm × 40 mm × 76 mm	1.42	0.19	4.37	–	3.19	nr	**7.56**
Chemical anchors							
R-CAS Spin-in epoxy acrylate capsules and standard studs or other equal and approved; Rawl Fixings; with nuts and washers; drilling masonry							
capsule ref 60–408; stud ref 60–448	–	0.25	5.75	–	1.03	nr	**6.78**
capsule ref 60–410; stud ref 60–454	–	0.28	6.44	–	1.19	nr	**7.63**
capsule ref 60–412; stud ref 60–460	–	0.31	7.13	–	1.62	nr	**8.75**
capsule ref 60–416; stud ref 60–472	–	0.34	7.82	–	2.01	nr	**9.83**
capsule ref 60–420; stud ref 60–478	–	0.36	8.27	–	2.61	nr	**10.88**
capsule ref 60–424; stud ref 60–484	–	0.40	9.20	–	8.32	nr	**17.52**
R-CAS Spin-in epoxy acrylate capsules and stainless steel studs or other equal and approved; Rawl Fixings; with nuts and washers; drilling masonry							
capsule ref 60–408; stud ref 60–905	–	0.25	5.75	–	2.08	nr	**7.83**
capsule ref 60–410; stud ref 60–910	–	0.28	6.44	–	3.66	nr	**10.10**
capsule ref 60–412; stud ref 60–915	–	0.31	7.13	–	4.66	nr	**11.79**
capsule ref 60–416; stud ref 60–920	–	0.34	7.82	–	9.93	nr	**17.75**
capsule ref 60–420; stud ref 60–925	–	0.36	8.27	–	19.29	nr	**27.56**
capsule ref 60–424; stud ref 60–930	–	0.40	9.20	–	34.35	nr	**43.55**

16 CARPENTRY

Item	PC £	Labour hours	Labour £	Plant £	Material £	Unit	Total rate £
R-CAS Spin-in epoxy acrylate capsules and standard internal threaded sockets or other equal and approved; Rawl Fixings; drilling masonry							
capsule ref 60–408; socket ref 60–650	–	0.25	5.75	–	1.42	nr	**7.17**
capsule ref 60–410; socket ref 60–656	–	0.28	6.44	–	1.47	nr	**7.91**
capsule ref 60–412; socket ref 60–662	–	0.31	7.13	–	1.57	nr	**8.70**
capsule ref 60–416; socket ref 60–668	–	0.34	7.82	–	2.22	nr	**10.04**
capsule ref 60–420; socket ref 60–674	–	0.36	8.27	–	1.63	nr	**9.90**
capsule ref 60–424; socket ref 60–676	–	0.40	9.20	–	5.12	nr	**14.32**
R-CAS Spin-in epoxy acrylate capsules and stainless steel internal threaded sockets or other equal and approved; Rawl Fixings; drilling masonry							
capsule ref 60–408; socket ref 60–943	–	0.25	5.75	–	2.91	nr	**8.66**
capsule ref 60–410; socket ref 60–945	–	0.28	6.44	–	2.97	nr	**9.41**
capsule ref 60–412; socket ref 60–947	–	0.31	7.13	–	3.70	nr	**10.83**
capsule ref 60–416; socket ref 60–949	–	0.34	7.82	–	5.40	nr	**13.22**
capsule ref 60–420; socket ref 60–951	–	0.36	8.27	–	5.72	nr	**13.99**
capsule ref 60–424; socket ref 60–955	–	0.40	9.20	–	13.74	nr	**22.94**
R-CAS Spin-in epoxy acrylate capsules, perforated sleeves and standard studs or other equal and approved; Rawl Fixings; in low density material; with nuts and washers; drilling masonry							
capsule ref 60–408; sleeve ref 60–538; stud ref 60–448	–	0.25	5.75	–	2.56	nr	**8.31**
capsule ref 60–410; sleeve ref 60–544; stud ref 60–454	–	0.28	6.44	–	2.91	nr	**9.35**
capsule ref 60–412; sleeve ref 60–550; stud ref 60–460	–	0.31	7.13	–	3.40	nr	**10.53**
capsule ref 60–416; sleeve ref 60–562; stud ref 60–472	–	0.34	7.82	–	3.80	nr	**11.62**
R-CAS Spin-in epoxy acrylate capsules, perforated sleeves and stainless steel studs or other equal and approved; Rawl Fixings; in low density material; with nuts and washers; drilling masonry							
capsule ref 60 408; sleeve ref 60–538; stud ref 60–905	–	0.25	5.75	–	3.62	nr	**9.37**
capsule ref 60–410; sleeve ref 60–544; stud ref 60–910	–	0.28	6.44	–	5.38	nr	**11.82**
capsule ref 60–412; sleeve ref 60–550; stud ref 60–915	–	0.31	7.13	–	6.44	nr	**13.57**
capsule ref 60–416; sleeve ref 60–562; stud ref 60–920	–	0.34	7.82	–	11.72	nr	**19.54**

16 CARPENTRY

Item	PC £	Labour hours	Labour £	Plant £	Material £	Unit	Total rate £
16.07 METAL FIXINGS, FIXINGS, FASTENINGS AND FITTINGS – cont							
Chemical anchors – cont							
R-CAS Spin-in epoxy acrylate capsules, perforated sleeves and standard internal threaded sockets or other equal and approved; The Rawlplug Company; in low density material; with nuts and washers; drilling masonry							
capsule ref 60–408; sleeve ref 60–538; socket ref 60–650	–	0.25	5.75	–	2.96	nr	**8.71**
capsule ref 60–410; sleeve ref 60–544; socket ref 60–656	–	0.28	6.44	–	3.19	nr	**9.63**
capsule ref 60–412; sleeve ref 60–550; socket ref 60–662	–	0.31	7.13	–	3.35	nr	**10.48**
R-CAS Spin-in epoxy acrylate capsules, perforated sleeves and stainless steel internal threaded sockets or other equal and approved; The Rawlplug Company; in low density material; drilling masonry							
capsule ref 60–416; sleeve ref 60–562; socket ref 60–668	–	0.34	7.82	–	4.03	nr	**11.85**
capsule ref 60–408; sleeve ref 60–538; socket ref 60–943	–	0.25	5.75	–	4.45	nr	**10.20**
capsule ref 60–410; sleeve ref 60–544; socket ref 60–945	–	0.28	6.44	–	4.69	nr	**11.13**
capsule ref 60–412; sleeve ref 60–550; socket ref 60–947	–	0.31	7.13	–	5.47	nr	**12.60**
capsule ref 60–416; sleeve ref 60–562; socket ref 60–949	–	0.34	7.82	–	7.19	nr	**15.01**

17 SHEET ROOF COVERINGS

Item	PC £	Labour hours	Labour £	Plant £	Material £	Unit	Total rate £
17.01 BUILT UP FELT ROOF COVERINGS							
NOTE: The following items of felt roofing, unless otherwise described, include for conventional lapping, laying and bonding between layers and to base; and laying flat or to falls, crossfalls or to slopes not exceeding 10° – but exclude any insulation etc.							
Reinforced bitumen membranes							
Three layer coverings							
type S1P1 bitumen glass fibre based felt	–	–	–	–	–	m²	**17.16**
cover with and bed in hot bitumen 13 mm thick stone chippings	–	–	–	–	–	m²	**4.98**
two base layers type S2P3 bitumen polyester-based felt; top layer type S4P4 polyester-based mineral surfaced felt; 10 mm stone chipping covering; bitumen bonded	–	–	–	–	–	m²	**29.22**
cover with and bed in hot bitumen 300 mm × 300 mm × 8 mm g.r.p. tiles	–	–	–	–	–	m²	**53.15**
Skirtings; three layer; top layer mineral surfaced; dressed over tilting fillet; turned into groove							
not exceeding 200 mm girth	–	–	–	–	–	m	**12.78**
200 mm–400 mm girth	–	–	–	–	–	m	**15.79**
Coverings to kerbs; three layer							
400 mm–600 mm girth	–	–	–	–	–	m	**20.44**
Linings to gutters; three layer							
400 mm–600 mm girth	–	–	–	–	–	m	**24.82**
Collars around pipes and the like; three layer mineral surface; 150 mm high							
not exceeding 55 mm nominal size	–	–	–	–	–	nr	**13.57**
55 mm–110 mm nominal size	–	–	–	–	–	nr	**13.57**
Outlets and dishing to gullies							
300 mm dia.	–	–	–	–	–	nr	**14.71**
Polyester-based roofing systems							
Andersons high performance polyester-based roofing system or other equal							
two layer coverings; first layer HT 125 underlay; second layer HT 350; fully bonded to wood; fibre or cork base	–	–	–	–	–	m²	**24.22**
top layer mineral surfaced	–	–	–	–	–	m²	**2.06**
13 mm thick stone chippings	–	–	–	–	–	m²	**4.98**
third layer of type 3B as underlay for concrete or screeded base	–	–	–	–	–	m²	**6.33**
working into outlet pipes and the like	–	–	–	–	–	nr	**14.70**
Skirtings; two layer; top layer mineral surfaced; dressed over tilting fillet; turned into groove							
not exceeding 200 mm girth	–	–	–	–	–	m	**124.77**
200 mm–400 mm girth	–	–	–	–	–	m	**16.15**

Prices for Measured Works

17 SHEET ROOF COVERINGS

Item	PC £	Labour hours	Labour £	Plant £	Material £	Unit	Total rate £
17.01 BUILT UP FELT ROOF COVERINGS – cont							
Polyester-based roofing systems – cont							
Coverings to kerbs; two layer							
400 mm–600 mm girth	–	–	–	–	–	m	**20.93**
Linings to gutters; three layer							
400 mm–600 mm girth	–	–	–	–	–	m	**22.49**
Collars around pipes and the like; two layer; 150 mm high							
not exceeding 55 mm nominal size	–	–	–	–	–	nr	**14.70**
55 mm–110 mm nominal size	–	–	–	–	–	nr	**14.70**
Ruberoid Challenger SBS high performance roofing or other equal							
two layer coverings; first and second layers Ruberglas 120 GP; fully bonded to wood, fibre or cork base	–	–	–	–	–	m²	**15.94**
top layer mineral surfaced	–	–	–	–	–	m²	**5.64**
13 mm thick stone chippings	–	–	–	–	–	m²	**4.98**
third layer of Rubervent 3G as underlay for concrete or screeded base	–	–	–	–	–	m²	**6.31**
working into outlet pipes and the like	–	–	–	–	–	nr	**14.58**
Skirtings; two layer; top layer mineral surfaced; dressed over tilting fillet; turned into groove							
not exceeding 200 mm girth	–	–	–	–	–	m	**12.17**
200 mm–400 mm girth	–	–	–	–	–	m	**15.92**
Coverings to kerbs; two layer							
400 mm–600 mm girth	–	–	–	–	–	m	**20.64**
Linings to gutters; three layer							
400 mm–600 mm girth	–	–	–	–	–	m	**22.09**
Collars around pipes and the like; two layer, 150 mm high							
not exceeding 55 mm nominal size	–	–	–	–	–	nr	**14.58**
55 mm–110 mm nominal size	–	–	–	–	–	nr	**14.58**
Ruberfort HP 350 high performance roofing or other equal							
two layer coverings; first layer Ruberfort HP 180; second layer Ruberfort HP 350; fully bonded; to wood; fibre or cork base	–	–	–	–	–	m²	**18.78**
top layer mineral surfaced	–	–	–	–	–	m²	**7.77**
13 mm thick stone chippings	–	–	–	–	–	m²	**4.98**
third layer of Rubervent 3G; as underlay for concrete or screeded base	–	–	–	–	–	m²	**6.31**
working into outlet pipes and the like	–	–	–	–	–	nr	**14.76**
Skirtings; two layer; top layer mineral surface; dressed over tilting fillet; turned into groove							
not exceeding 200 mm girth	–	–	–	–	–	m	**12.41**
200 mm–400 mm girth	–	–	–	–	–	m	**16.23**
Coverings to kerbs; two layer							
400 mm–600 mm girth	–	–	–	–	–	m	**21.05**

17 SHEET ROOF COVERINGS

Item	PC £	Labour hours	Labour £	Plant £	Material £	Unit	Total rate £
Linings to gutters; three layer							
400 mm–600 mm girth	–	–	–	–	–	m	**27.20**
Collars around pipes and the like; two layer; 150 mm high							
not exceeding 55 mm nominal size	–	–	–	–	–	nr	**14.76**
55 mm–110 mm nominal size	–	–	–	–	–	nr	**14.76**
Ruberoid Superflex Firebloc high performance roofing or other equal (15 year guarantee specification)							
two layer coverings; first layer Superflex 180; second layer Superflex 250; fully bonded to wood; fibre or cork base	–	–	–	–	–	m²	**23.23**
top layer mineral surfaced	–	–	–	–	–	m²	**5.52**
13 mm thick stone chippings	–	–	–	–	–	m²	**4.98**
third layer of Rubervent 3G as underlay for concrete or screeded base	–	–	–	–	–	m²	**6.31**
working into outlet pipes and the like	–	–	–	–	–	nr	**16.76**
Skirtings; two layer; top layer mineral surfaced; dressed over tilting fillet; turned into groove							
not exceeding 200 mm girth	–	–	–	–	–	m	**14.51**
200 mm–400 mm girth	–	–	–	–	–	m	**19.14**
Coverings to kerbs; two layer							
400 mm–600 mm girth	–	–	–	–	–	m	**25.56**
Linings to gutters; three layer							
400 mm–600 mm girth	–	–	–	–	–	m	**27.67**
Collars around pipes and the like; two layer; 150 mm high							
not exceeding 55 mm nominal size	–	–	–	–	–	nr	**16.76**
55 mm–110 mm nominal size	–	–	–	–	–	nr	**16.76**
Ruberoid Ultra Prevent high performance roofing or other equal							
two layer coverings; first layer Ultra prevENt underlay; second layer Ultra prevENt mineral surface cap sheet.	–	–	–	–	–	m²	**42.75**
extra over for							
third layer of Rubervent 3G as underlay for concrete or screeded base	–	–	–	–	–	m²	**6.31**
working into outlet pipes and the like	–	–	–	–	–	nr	**20.27**
Skirtings; two layer; dressed over tilting fillet; turned into groove							
not exceeding 200 mm girth	–	–	–	–	–	m	**18.16**
200 mm–400 mm girth	–	–	–	–	–	m	**24.19**
Coverings to kerbs; two layer							
400 mm–600 mm girth	–	–	–	–	–	m	**33.39**
Linings to gutters; three layer							
400 mm–600 mm girth	–	–	–	–	–	m	**34.90**
Collars around pipes and the like; two layer; 150 mm high							
not exceeding 55 mm nominal size	–	–	–	–	–	nr	**20.24**
55 mm–110 mm nominal size	–	–	–	–	–	nr	**20.24**

17 SHEET ROOF COVERINGS

Item	PC £	Labour hours	Labour £	Plant £	Material £	Unit	Total rate £
17.01 BUILT UP FELT ROOF COVERINGS – cont							
Accessories							
Eaves trim; extruded aluminium alloy; working felt into trim							
Rubertrim; type FL/G; 65 mm face	–	–	–	–	–	m	**16.20**
extra for external angle	–	–	–	–	–	nr	**16.31**
Roof screed ventilator – aluminium alloy							
Extr-aqua-vent or other equal and approved – set on screed over and including dished sinking and collar	–	–	–	–	–	nr	**51.25**
Insulation board underlays							
Vapour barrier							
reinforced; metal lined	–	–	–	–	–	m²	**15.50**
Rockwool; Duorock flat insulation board							
140 mm thick (0.25 U-value)	–	–	–	–	–	m²	**42.46**
Kingspan Thermaroof TR21 zero OPD urethene insulation board							
50 mm thick	–	–	–	–	–	m²	**26.09**
90 mm thick	–	–	–	–	–	m²	**44.96**
100 mm thick (0.25 U-value)	–	–	–	–	–	m²	**49.94**
Wood fibre boards; impregnated; density 220–350 kg/m³							
12.70 mm thick	–	–	–	–	–	m²	**7.04**
Tapered insulation board underlays							
Tapered insulation £/m² prices can vary dramatically depending upon the factors which determine the scheme layout; these primarily being gutter/outlet locations and the length of fall involved. The following guide assumes a U-value of 0.18 W/m² K as a benchmark.							
As the required insulation value will vary from project to project the required U-value should be determined by calculating the buildings energy consumption at the design stage. Due to tapered insulation scheme prices varying by project, the following prices are indicative. Please contact a specialist for a project specific quotation. U-value must be calculated in accordance with BSENISO 6946:2007 Annex C							
Tapered PIR (Polyisocyanurate) boards; bedded in hot bitumen							
effective thickness achieving 0.18 W/m² K	30.30	–	–	–	–	m²	**66.64**
minimum thickness achieving 0.18 W/m² K	36.05	–	–	–	–	m²	**73.99**
Tapered PIR (Polyisocyanurate) board; mechanically fastened							
effective thickness achieving 0.18 W/m² K	30.30	–	–	–	–	m²	**69.56**

17 SHEET ROOF COVERINGS

Item	PC £	Labour hours	Labour £	Plant £	Material £	Unit	Total rate £
Tapered Aspire; Hybrid EPS/PIR boards; bedded in hot bitumen							
effective thickness achieving 0.18 W/m²K	28.21	–	–	–	–	m²	**66.64**
minimum thickness achieving 0.18 W/m²K	33.96	–	–	–	–	m²	**73.99**
Tapered PIR (Polyisocyanurate) board; mechanically fastened							
effective thickness achieving 0.18 W/m²K	28.21	–	–	–	–	m²	**69.56**
minimum thickness achieving 0.18 W/m²K	33.96	–	–	–	–	m²	**76.91**
Tapered Rockwool boards; bedded in hot bitumen							
effective thickness achieving 0.18 W/m²K	54.13	–	–	–	–	m²	**98.04**
minimum thickness achieving 0.18 W/m²K	66.88	–	–	–	–	m²	**111.24**
Tapered Rockwool boards; mechanically fastened							
effective thickness achieving 0.18 W/m²K	54.13	–	–	–	–	m²	**100.97**
minimum thickness achieving 0.18 W/m²K	66.88	–	–	–	–	m²	**109.54**
Tapered EPS (Expanded polystyrene) boards; bedded in hot bitumen							
effective thickness achieving 0.18 W/m²K	24.77	–	–	–	–	m²	**59.38**
minimum thickness achieving 0.18 W/m²K	30.30	–	–	–	–	m²	**66.64**
Tapered EPS (Expanded polystyrene) boards; mechanicaly fastened							
effective thickness achieving 0.18 W/m²K	24.77	–	–	–	–	m²	**62.21**
minimum thickness achieving 0.18 W/m²K	30.30	–	–	–	–	m²	**69.55**
Insulation board overlays							
Dow Roofmate SL extruded polystyrene foam boards or other equal and approved; Thermal conductivity – 0.028 W/mK							
50 mm thick	–	–	–	–	–	m²	**17.27**
140 mm thick	–	–	–	–	–	m²	**30.15**
160 mm thick	–	–	–	–	–	m²	**32.66**
Dow Roofmate LG extruded polystyrene foam boards or other equal and approved; Thermal conductivity – 0.028 W/mK							
80 mm thick	–	–	–	–	–	m²	**61.51**
100 mm thick	–	–	–	–	–	m²	**65.92**
120 mm thick	–	–	–	–	–	m²	**70.40**
17.02 FIBRE BITUMEN THERMOPLASTIC SHEET COVERINGS/FLASHINGS							
Glass fibre reinforced bitumen strip slates; Ruberglas 105 or other equal and approved; 1000 mm × 336 mm mineral finish; to external quality plywood boarding (boarding not included)							
Roof coverings	12.87	0.23	5.10	–	14.77	m²	**19.87**
Wall coverings	12.87	0.37	8.20	–	14.77	m²	**22.97**
Extra over coverings for							
double course at eaves; felt soaker	–	0.19	4.21	–	9.87	m	**14.08**
verges; felt soaker	–	0.14	3.10	–	8.20	m	**11.30**
valley slate; cut to shape; felt soaker and cutting both sides	–	0.42	9.30	–	12.88	m	**22.18**

17 SHEET ROOF COVERINGS

Item	PC £	Labour hours	Labour £	Plant £	Material £	Unit	Total rate £
17.02 FIBRE BITUMEN THERMOPLASTIC SHEET COVERINGS/FLASHINGS – cont							
Glass fibre reinforced bitumen strip slates – cont							
Extra over coverings for – cont							
ridge slate; cut to shape	–	0.28	6.20	–	8.20	m	**14.40**
hip slate; cut to shape; felt soaker and cutting							
both sides	–	0.42	9.30	–	12.79	m	**22.09**
holes for pipes and the like	–	0.48	10.63	–	–	nr	**10.63**
Bostik Findley Flashband Plus sealing strips and flashings or other equal and approved; special grey finish							
Flashings; wedging at top if required; pressure bonded; to walls							
100 mm girth	–	0.23	6.10	–	1.61	m	**7.71**
150 mm girth	–	0.31	8.21	–	2.37	m	**10.58**
225 mm girth	–	0.37	9.81	–	3.18	m	**12.99**
300 mm girth	–	0.42	11.12	–	4.76	m	**15.88**
450 mm girth	–	0.45	11.92	–	9.52	m	**21.44**
600 mm girth	–	0.47	12.59	–	14.26	m	**26.85**
17.03 SINGLE LAYER PLASTIC ROOF COVERINGS							
Kingspan KS1000TD composite single ply roof panels for roof pitches greater than 0.7° (after deflection)							
1.5 mm single ply external covering, internal coating bright white polyester (steel)							
71 mm thick panel; U-value 0.25	–	–	–	–	–	m²	**51.33**
100 mm thick panel; U-value 0.18	–	–	–	–	–	m²	**56.78**
120 mm thick panel; U-value 0.15	–	–	–	–	–	m²	**60.06**
Trocal S PVC roofing or other equal and approved							
Coverings	–	–	–	–	–	m²	**19.32**
Skirtings; dressed over metal upstands							
not exceeding 200 mm girth	–	–	–	–	–	m	**15.02**
200 mm–400 mm girth	–	–	–	–	–	m	**18.44**
Coverings to kerbs							
400 mm–600 mm girth	–	–	–	–	–	m	**33.78**
Collars around pipes and the like; 150 mm high							
not exceeding 55 mm nominal size	–	–	–	–	–	nr	**10.33**
55 mm–110 mm nominal size	–	–	–	–	–	nr	**10.33**
Trocal metal upstands or other equal and approved							

17 SHEET ROOF COVERINGS

Item	PC £	Labour hours	Labour £	Plant £	Material £	Unit	Total rate £
Sarnafil polymeric waterproofing membrane; cold roof							
Roof coverings							
pitch not exceeding 5°; to metal decking or the like	–	–	–	–	–	m²	**35.95**
pitch not exceeding 5°; to concrete base or the like; prime concrete with spririt priming solution	–	–	–	–	–	m²	**35.95**
Sarnafil polymeric waterproofing membrane; 1.2 mm thick fleece backed membrane; cold roof							
Roof coverings							
pitch not exceeding 5°; to metal decking or the like	–	–	–	–	–	m²	**35.95**
pitch not exceeding 5°; to concrete base or the like; prime concrete with spririt priming solution	–	–	–	–	–	m²	**35.95**
Sarnafil polymeric waterproofing membrane; 120 mm thick Sarnaform G CFC & HCFC free insulation board; vapour control layer; prime concrete with spirit priming solution							
Mechanically fastened system							
Roof coverings							
pitch not exceeding 5°; to metal decking or the like	–	–	–	–	–	m²	**54.57**
pitch not exceeding 5°; to concrete base or the like	–	–	–	–	–	m²	**63.74**
Coverings to kerbs; parapet flashing; Sarnatrim 50 mm deep on face 100 mm fixing arm; standard Sarnafil detail 1.1							
not exceeding 200 mm girth	–	–	–	–	–	m	**30.60**
200 mm–400 mm girth	–	–	–	–	–	m	**34.87**
400 mm–600 mm girth	–	–	–	–	–	m	**39.11**
Eaves detail; Sarnatrim drip edge to gutter; standard Sarnafil detail 1.3							
not exceeding 200 mm girth	–	–	–	–	–	m	**29.43**
Skirtings/Upstands; skirting to brickwork with galvanized steel counter flashing to top edge; standard Sarnafil detail 2.3							
not exceeding 200 mm girth	–	–	–	–	–	m	**35.79**
200 mm–400 mm girth	–	–	–	–	–	m	**43.04**
400 mm–600 mm girth	–	–	–	–	–	m	**50.38**
Skirtings/Upstands; skirting to brickwork with Sarnametal Raglet to chase; standard Sarnafil detail 2.8							
not exceeding 200 mm girth	–	–	–	–	–	m	**35.79**
200 mm–400 mm girth	–	–	–	–	–	m	**43.04**
400 mm–600 mm girth	–	–	–	–	–	m	**50.38**

17 SHEET ROOF COVERINGS

Item	PC £	Labour hours	Labour £	Plant £	Material £	Unit	Total rate £
17.03 SINGLE LAYER PLASTIC ROOF COVERINGS – cont							
Mechanically fastened system – cont							
Collars around pipe standards, and the like							
50 mm dia. × 150 mm high	–	–	–	–	–	nr	46.38
100 mm dia. × 150 mm high	–	–	–	–	–	nr	46.38
Outlets and dishing to gullies							
fix Sarnadrain PVC rainwater outlet; 110 mm dia.;							
weld membrane to same; fit plastic leafguard	–	–	–	–	–	nr	109.64
Fully adhered system							
Roof coverings							
pitch not exceeding 5°; to metal decking or the like	–	–	–	–	–	m²	60.89
pitch not exceeding 5°; to concrete base or the like	–	–	–	–	–	m²	67.92
Coverings to kerbs; parapet flashing; Sarnatrim 50 mm deep on face 100 mm fixing arm; standard Sarnafil detail 1.1							
not exceeding 200 mm girth	–	–	–	–	–	m	29.14
200 mm–400 mm girth	–	–	–	–	–	m	33.52
400 mm–600 mm girth	–	–	–	–	–	m	37.91
Eaves detail; Sarnametal drip edge to gutter; standard Sarnafil detail 1.3							
not exceeding 200 mm girth	–	–	–	–	–	m	27.91
Skirtings/Upstands; skirting to brickwork with galvanized steel counter flashing to top edge; standard Sarnafil detail 2.3							
not exceeding 200 mm girth	–	–	–	–	–	m	34.47
200 mm–400 mm girth	–	–	–	–	–	m	41.99
400 mm–600 mm girth	–	–	–	–	–	m	47.01
Skirtings/Upstands; skirting to brickwork with Sarnametal Raglet to chase; standard Sarnafil detail 2.8							
not exceeding 200 mm girth	–	–	–	–	–	m	35.73
200 mm–400 mm girth	–	–	–	–	–	m	49.57
400 mm–600 mm girth	–	–	–	–	–	m	49.57
Collars around pipe standards, and the like							
50 mm dia. × 150 mm high	–	–	–	–	–	nr	47.97
100 mm dia. × 150 mm high	–	–	–	–	–	nr	47.97
Outlets and dishing to gullies							
fix Sarnadrain PVC rainwater outlet; 110 mm dia.;							
weld membrane to same; fit plastic leafguard	–	–	–	–	–	nr	113.41
Options							
extra over for 1.2 mm fleece backed membrane	–	–	–	–	–	m²	6.18

17 SHEET ROOF COVERINGS

Item	PC £	Labour hours	Labour £	Plant £	Material £	Unit	Total rate £
Landscape roofing							
SarnaVert extensive biodiverse roof; sedum blanket; 100 mm growing medium; aquafrain; 1.5 mm thick membrane; 120 mm thick insulation board; vapour control layer							
Pitch not exceeding 5°; to metal decking or the like	–	–	–	–	–	m²	**146.38**
Pitch not exceeding 5°; to concrete base or the like; prime concrete with spririt priming solution	–	–	–	–	–	m²	**152.32**
Kerb and eaves; standard Sarnafil details							
Sarnafil kerb; 150 mm above roof level	–	–	–	–	–	m	**36.25**
Sarnafil eaves detail with gravel stop ne 200 mm girth	–	–	–	–	–	m	**60.07**
Collars around pipes and the like							
50 mm dia. × 150 mm high	–	–	–	–	–	nr	**46.60**
100 mm dia. × 150 mm high	–	–	–	–	–	nr	**46.60**
Sarnafil rainwater outlet	–	–	–	–	–	nr	**187.47**
17.04 GLASS REINFORCED PLASTIC PANEL							
Glass fibre translucent sheeting grade AB class 3							
Roof cladding; sloping not exceeding 50°; fixing to timber purlins with drive screws; to suit							
Profile 3 or other equal and approved	15.00	0.18	3.99	–	21.22	m²	**25.21**
Profile 6 or other equal and approved	15.00	0.23	5.10	–	21.22	m²	**26.32**
Roof cladding; sloping not exceeding 50°; fixing to timber purlins with hook bolts; to suit							
Profile 3 or other equal and approved	15.00	0.23	5.10	–	22.17	m²	**27.27**
Profile 6 or other equal and approved	15.00	0.28	6.20	–	22.17	m²	**28.37**
Longrib 1000 or other equal and approved	15.00	0.28	6.20	–	22.17	m²	**28.37**
17.05 LEAD SHEET COVERINGS/FLASHINGS							
Milled Lead; BS EN 12588; on and including Geotec underlay							
Roof and dormer coverings							
1.80 mm thick (code 4) roof coverings							
flat (in wood roll construction) (Prime Cost £ per kg)	3.47	1.80	47.69	–	48.40	m²	**96.09**
pitched (in wood roll construction)	–	2.00	52.98	–	48.64	m²	**101.62**
pitched (in welded seam construction)	–	1.80	47.69	–	48.40	m²	**96.09**
vertical (in welded seam construction)	–	2.00	52.98	–	46.20	m²	**99.18**
1.80 mm thick (code 4) dormer coverings							
flat (in wood roll construction) (Prime Cost £ per kg)	3.47	1.35	35.76	–	47.85	m²	**83.61**
pitched (in wood roll construction)	–	1.50	39.74	–	48.03	m²	**87.77**
pitched (in welded seam construction)	–	1.35	35.76	–	47.85	m²	**83.61**
vertical (in welded seam construction)	–	3.00	79.47	–	46.20	m²	**125.67**

17 SHEET ROOF COVERINGS

Item	PC £	Labour hours	Labour £	Plant £	Material £	Unit	Total rate £
17.05 LEAD SHEET COVERINGS/FLASHINGS – cont							
Roof and dormer coverings – cont							
2.24 mm thick (code 5) roof coverings							
flat (in wood roll construction) (Prime Cost £ per kg)	3.47	1.89	50.07	–	58.34	m²	**108.41**
pitched (in wood roll construction)	–	2.10	55.63	–	58.60	m²	**114.23**
pitched (in welded seam construction)	–	1.89	50.07	–	58.34	m²	**108.41**
vertical (in welded seam construction)	–	2.10	55.63	–	56.03	m²	**111.66**
2.24 mm thick (code 5) dormer coverings							
flat (in wood roll construction) (Prime Cost £ per kg)	3.47	1.42	37.56	–	57.77	m²	**95.33**
pitched (in wood roll construction)	–	1.58	41.75	–	57.96	m²	**99.71**
pitched (in welded seam construction)	–	1.42	37.56	–	57.77	m²	**95.33**
vertical (in welded seam construction)	–	3.15	83.45	–	56.03	m²	**139.48**
2.65 mm thick (code 6) roof coverings							
flat (in wood roll construction) (Prime Cost £ per kg)	3.47	1.98	52.46	–	67.62	m²	**120.08**
pitched (in wood roll construction)	–	2.20	58.28	–	67.89	m²	**126.17**
pitched (in welded seam construction)	–	1.98	52.46	–	67.62	m²	**120.08**
vertical (in welded seam construction)	–	2.20	58.28	–	65.20	m²	**123.48**
2.65 mm thick (code 6) dormer coverings							
flat (in wood roll construction) (Prime Cost £ per kg)	3.47	1.49	39.37	–	67.01	m²	**106.38**
pitched (in wood roll construction)	–	1.65	43.71	–	67.21	m²	**110.92**
pitched (in welded seam construction)	–	1.49	39.37	–	67.01	m²	**106.38**
vertical (in welded seam construction)	–	3.30	87.43	–	65.20	m²	**152.63**
3.15 mm thick (code 7) roof coverings							
flat (in wood roll construction) (Prime Cost £ per kg)	3.47	2.12	56.05	–	78.96	m²	**135.01**
pitched (in wood roll construction)	–	2.35	62.25	–	79.24	m²	**141.49**
pitched (in welded seam construction)	–	2.12	56.05	–	78.96	m²	**135.01**
vertical (in welded seam construction)	–	2.35	62.25	–	76.37	m²	**138.62**
3.15 mm thick (code 7) dormer coverings							
flat (in wood roll construction) (Prime Cost £ per kg)	3.47	1.59	42.01	–	78.31	m²	**120.32**
pitched (in wood roll construction)	–	1.76	46.68	–	78.53	m²	**125.21**
pitched (in welded seam construction)	–	1.59	42.01	–	78.31	m²	**120.32**
vertical (in welded seam construction)	–	3.53	93.41	–	76.37	m²	**169.78**
3.55 mm thick (code 8) roof coverings							
flat (in wood roll construction) (Prime Cost £ per kg)	3.47	2.30	60.82	–	88.13	m²	**148.95**
pitched (in wood roll construction)	–	2.55	67.56	–	88.44	m²	**156.00**
pitched (in welded seam construction)	–	2.30	60.82	–	88.13	m²	**148.95**
vertical (in welded seam construction)	–	2.55	67.56	–	85.32	m²	**152.88**
3.55 mm thick (code 8) dormer coverings							
flat (in wood roll construction) (Prime Cost £ per kg)	3.47	1.72	45.62	–	87.42	m²	**133.04**
pitched (in wood roll construction)	–	1.91	50.66	–	87.66	m²	**138.32**
pitched (in welded seam construction)	–	1.72	45.62	–	87.42	m²	**133.04**
vertical (in welded seam construction)	–	3.83	101.35	–	85.32	m²	**186.67**

17 SHEET ROOF COVERINGS

Item	PC £	Labour hours	Labour £	Plant £	Material £	Unit	Total rate £
Sundries							
patination oil to finished work surfaces	–	0.05	1.33	–	0.24	m²	**1.57**
chalk slurry to underside of panels	–	0.67	17.64	–	2.14	m²	**19.78**
provision of 45 × 45 mm wood rolls at 600 mm centres (per m)	–	0.20	5.29	–	1.01	m	**6.30**
dressing over glazing bars and glass	–	0.50	13.25	–	0.66	m	**13.91**
soldered nail head	–	0.02	0.42	–	0.05	nr	**0.47**
1.32 mm thick (code 3) lead flashings, etc.							
Soakers							
200 × 200 mm	–	0.03	0.79	–	1.08	nr	**1.87**
300 × 300 mm	–	0.03	0.79	–	2.47	nr	**3.26**
1.80 mm thick (code 4) lead flashings, etc.							
Flashings; wedging into grooves							
150 mm girth	–	0.50	13.25	–	6.26	m	**19.51**
200 mm girth	–	0.50	13.25	–	8.34	m	**21.59**
240 mm girth	–	0.50	13.25	–	10.01	m	**23.26**
300 mm girth	–	0.50	13.25	–	12.51	m	**25.76**
Stepped flashings; wedging into grooves							
180 mm girth	–	1.00	26.49	–	7.51	m	**34.00**
270 mm girth	–	1.00	26.49	–	11.26	m	**37.75**
Linings to sloping gutters							
390 mm girth	–	0.80	21.20	–	16.27	m	**37.47**
450 mm girth	–	0.90	23.84	–	18.77	m	**42.61**
600 mm girth	–	1.10	29.14	–	25.03	m	**54.17**
Cappings to hips or ridges							
450 mm girth	–	1.00	26.49	–	18.77	m	**45.26**
600 mm girth	–	1.20	31.79	–	25.03	m	**56.82**
Saddle flashings; at intersections of hips and ridges; dressing and bossing							
450 × 450 mm	–	1.00	26.49	–	10.89	nr	**37.38**
600 × 200 mm	–	1.00	26.49	–	17.46	nr	**43.95**
Slates; with 150 mm high collar							
450 × 450 mm; to suit 50 mm dia. pipe	–	1.50	39.74	–	13.10	nr	**52.84**
450 × 450 mm; to suit 100 mm dia. pipe	–	1.50	39.74	–	14.08	nr	**53.82**
450 × 450 mm; to suit 150 mm dia. pipe	–	1.50	39.74	–	15.06	nr	**54.80**
2.24 mm thick (code 5) lead flashings, etc.							
Flashings; wedging into grooves							
150 mm girth	–	0.50	13.25	–	7.69	m	**20.94**
200 mm girth	–	0.50	13.25	–	10.25	m	**23.50**
240 mm girth	–	0.50	13.25	–	12.30	m	**25.55**
300 mm girth	–	0.50	13.25	–	15.38	m	**28.63**
Stepped flashings; wedging into grooves							
180 mm girth	–	1.00	26.49	–	9.23	m	**35.72**
270 mm girth	–	1.00	26.49	–	13.84	m	**40.33**
Linings to sloping gutters							
390 mm girth	–	0.80	21.20	–	19.99	m	**41.19**
450 mm girth	–	0.90	23.84	–	23.07	m	**46.91**
600 mm girth	–	1.10	29.14	–	30.75	m	**59.89**

17 SHEET ROOF COVERINGS

Item	PC £	Labour hours	Labour £	Plant £	Material £	Unit	Total rate £
17.05 LEAD SHEET COVERINGS/FLASHINGS – cont							
2.24 mm thick (code 5) lead flashings, etc. – cont							
Cappings to hips or ridges							
450 mm girth	–	1.00	26.49	–	23.07	m	**49.56**
600 mm girth	–	1.20	31.79	–	30.75	m	**62.54**
Saddle flashings; at intersections of hips and ridges; dressing and bossing							
450 × 450 mm	–	1.00	26.49	–	11.60	nr	**38.09**
600 × 200 mm	–	1.00	26.49	–	19.67	nr	**46.16**
Slates; with 150 mm high collar							
450 × 450 mm; to suit 50 mm dia. pipe	–	1.50	39.74	–	13.42	nr	**53.16**
450 × 450 mm; to suit 100 mm dia. pipe	–	1.50	39.74	–	14.62	nr	**54.36**
450 × 450 mm; to suit 150 mm dia. pipe	–	1.50	39.74	–	15.84	nr	**55.58**
17.06 ALUMINIUM SHEET COVERINGS/ FLASHINGS							
Aluminium roofing; commercial grade; on and including Geotec underlay							
The following rates are based upon nett 'deck' or 'wall' areas							
Roof, dormer and wall coverings							
0.7 mm thick roof coverings; mill finish							
flat (in wood roll construction) (Prime Cost rate £ per kg)	5.25	2.00	52.98	–	21.74	m²	**74.72**
eaves detail ED1	–	0.40	10.60	–	2.70	m	**13.30**
abutment upstands at perimeters	–	0.66	17.49	–	1.08	m	**18.57**
pitched over 3° (in standing seam construction)	–	1.50	39.74	–	18.18	m²	**57.92**
vertical (in angled or flat seam construction)	–	1.60	42.38	–	18.18	m²	**60.56**
0.7 mm thick dormer coverings; mill finish							
flat (in wood roll construction)	–	3.00	79.47	–	21.42	m²	**100.89**
eaves detail ED1	–	0.40	10.60	–	2.70	m	**13.30**
pitched over 3° (in standing seam construction)	–	2.50	66.23	–	18.18	m²	**84.41**
vertical (in angled or flat seam construction)	–	2.70	71.52	–	18.18	m²	**89.70**
0.7 mm thick roof coverings; Pvf2 finish							
flat (in wood roll construction) (Prime Cost rate £ per kg)	6.67	2.00	52.98	–	25.54	m²	**78.52**
eaves detail ED1	–	0.40	10.60	–	3.37	m	**13.97**
abutment upstands at perimeters	–	0.66	17.49	–	1.35	m	**18.84**
pitched over 3° (in standing seam construction)	–	1.50	39.74	–	22.18	m²	**61.92**
vertical (in angled or flat seam construction)	–	1.60	42.38	–	22.18	m²	**64.56**
0.7 mm thick dormer coverings; Pvf2 finish							
flat (in wood roll construction)	–	3.00	79.47	–	25.54	m²	**105.01**
eaves detail ED1	–	0.40	10.60	–	3.37	m	**13.97**
pitched over 3° (in standing seam construction)	–	2.50	66.23	–	22.18	m²	**88.41**
vertical (in angled or flat seam construction)	–	2.70	71.52	–	22.18	m²	**93.70**

17 SHEET ROOF COVERINGS

Item	PC £	Labour hours	Labour £	Plant £	Material £	Unit	Total rate £
0.7 mm thick aluminium flashings, etc.							
Flashings; wedging into grooves; mill finish							
150 mm girth (PC per kg)	5.25	0.50	13.25	–	1.62	m	**14.87**
240 mm girth	–	0.50	13.25	–	2.60	m	**15.85**
300 mm girth	–	0.50	13.25	–	3.24	m	**16.49**
Stepped flashings; wedging into grooves; mill finish							
180 mm girth	–	1.00	26.49	–	1.95	m	**28.44**
270 mm girth	–	1.00	26.49	–	2.92	m	**29.41**
Flashings; wedging into grooves; Pvf2 finish							
150 mm girth (PC per kg)	6.67	0.50	13.25	–	2.02	m	**15.27**
240 mm girth	–	0.50	13.25	–	3.23	m	**16.48**
300 mm girth	–	0.50	13.25	–	4.04	m	**17.29**
Stepped flashings; wedging into grooves; Pvf2 finish							
180 mm girth	–	1.00	26.49	–	2.42	m	**28.91**
270 mm girth	–	1.00	26.49	–	3.63	m	**30.12**
Sundries							
provision of square batten roll at 500 mm centres (per m)	–	0.20	5.29	–	1.34	m	**6.63**
Standing seam aluminium roofing							
Kalzip; 65 mm seam, 400 cover width, Ref BS AW 3004 standard natural aluminium, stucco embossed finish, 0.9 mm thick; ST clips fixed with stainless steel fasteners; 37 Plus 180 mm Glassfibre Insulation compressed to 165 mm (0.25 U-value); vapour control layer, clear reinforced polyethelyne 530 MNs/g all laps sealed; Liner Sheets, profiled steel, 1000 mm cover width, bright white polyester paint finish Ref TR35/200S, 0.7 mm thick, fixed with stainless steel fasteners.							
roof coverings (twin skin construction); pitch not less than 1.5°; fixed to cold rolled purlins (not included	–	–	–	–	–	m²	**66.51**
Eaves details							
40 × 20 mm extruded aluminium drip angle fixed to Kalzip sheet using aluminium blind sealed rivets; black solid rubber eaves filler blocks; ST clips fixed with stainless steel fasteners	–	–	–	–	–	m	**21.56**
0.90 mm thick stucco embossed natural aluminium external eaves closure; 375 mm girth twice bent	–	–	–	–	–	m	**9.04**
0.70 mm thick bright white polyester liner sheet closure internal flashing; 200 mm girth once bent with stainless steel fasteners, black solid rubber profiled liner small flute filler sealed top and bottom with sealant tape	–	–	–	–	–	m	**10.42**

17 SHEET ROOF COVERINGS

Item	PC £	Labour hours	Labour £	Plant £	Material £	Unit	Total rate £
17.06 ALUMINIUM SHEET COVERINGS/ FLASHINGS – cont							
Standing seam aluminium roofing – cont							
Verge details							
extruded aluminium gable end channel fixed to kalzip seam using aluminium blind rivets; extruded aluminium gable end clips fixed to ST clips with stainless steel fasteners; extruded aluminium gable tolerence clip hooked over gable end channel	–	–	–	–	–	m	**17.58**
0.90 mm thick stucco embossed natural aluminium external verge closure 600 mm girth four times bent, fixed to extruded aluminium gable tolerence clip and vertical cladding with stainless steel fasteners, black profiled filler blocks to vertical clddding	–	–	–	–	–	m	**17.66**
0.70 mm thick bright white polyester liner sheet closure internal flashing 200 mm girth once bent fixed with stainless steel fasteners, black solid rubber profiled filler blocks sealed top and bottom with sealant tape	–	–	–	–	–	m	**10.60**
Duo-Ridge details							
2 Nr Extruded aluminium zed sections fixed to kalzip seams using aluminium blind sealed rivets; 2 Nr natural aluminium stucco embossed U Type ridge closures fixed to kalzip seams using aluminium blind sealed rivets; 2 Nr black solid rubber ridge filler blocks, 2 Nr ST clips fixed with stainless steel fasteners; fix seam of kaizip sheet to ST clips using aluminium blind sealed rivets (for fixed point); turn up kalzip 400 sheets both sides	–	–	–	–	–	m	**34.04**
0.90 mm thick stucco embossed natural aluminium external ridge closure; 600 mm girth three times bent, fixed to U Type Ridge closure with stainless steel fasteners	–	–	–	–	–	m	**13.52**
0.70 mm thick bright white polyester liner sheet closure flashing 600 mm girth once bent fixed with stainless steel fasteners, black solid rubber profiled filler blocks sealed top and bottom with sealant tape	–	–	–	–	–	m	**11.80**
Accessories							
smooth curving Kalzip sheets	–	–	–	–	–	m^2	**7.83**
crimp curving liner (below 52.5 m convex radius)	–	–	–	–	–	sheet	**13.20**
polyster coating Kalzip sheets	–	–	–	–	–	m^2	**4.07**
PvDF coating Kalzip sheets	–	–	–	–	–	m^2	**4.71**
vapour control layer, foil encapsulated polythene 4300 MNs/g	–	–	–	–	–	m^2	**0.65**
200 mm thick thermal insulation quilt	–	–	–	–	–	m^2	**0.21**
30 mm thick semi-rigid acoustic insulation slab	–	–	–	–	–	m^2	**4.88**

17 SHEET ROOF COVERINGS

Item	PC £	Labour hours	Labour £	Plant £	Material £	Unit	Total rate £
1.0 mm flashings etc.; fixing/wedging into grooves							
flashing; 500 mm girth	–	–	–	–	–	m	**14.42**
flashing; 750 mm girth	–	–	–	–	–	m	**19.04**
flashing; 1000 mm girth	–	–	–	–	–	m	**26.79**
1.2 mm flashings etc.; fixing/wedging into grooves							
flashing; 500 mm girth	–	–	–	–	–	m	**15.86**
flashing; 750 mm girth	–	–	–	–	–	m	**20.95**
flashing; 1000 mm girth	–	–	–	–	–	m	**26.48**
1.4 mm flashings etc.; fixing/wedging into grooves							
flashing; 500 mm girth	–	–	–	–	–	m	**18.24**
flashing; 750 mm girth	–	–	–	–	–	m	**24.08**
flashing; 1000 mm girth	–	–	–	–	–	m	**36.19**
Aluminium Alumasc Skyline coping system; polyester powder coated							
Coping; fixing straps plugged and screwed to brickwork							
362 mm wide; for parapet wall 241–300 mm wide	–	0.50	13.25	–	38.13	m	**51.38**
90° angle	–	0.25	6.62	–	76.28	nr	**82.90**
90° tee junction	–	0.35	9.27	–	79.15	nr	**88.42**
stop end	–	0.15	3.98	–	53.29	nr	**57.27**
stop end upstand	–	0.20	5.29	–	54.49	nr	**59.78**
17.07 COPPER STRIP SHEET COVERINGS/ FLASHINGS							
Copper roofing; BS EN 504; on and including Geotec underlay							
The following rates are based upon nett deck or wall areas							
Roof and dormer coverings							
0.6 mm thick roof coverings; mill finish							
flat (in wood roll construction) (Prime Cost £ per kg)	6.24	2.20	58.28	–	60.35	m²	**118.63**
eaves detail ED1	–	0.40	10.60	–	7.34	m	**17.94**
abutment upstands at perimeters	–	0.66	17.49	–	3.67	m	**21.16**
pitched over 3° (in standing seam construction)	–	1.70	45.03	–	49.35	m²	**94.38**
vertical (in angled or flat seam construction)	–	1.80	47.69	–	49.35	m²	**97.04**
0.6 mm thick dormer coverings; mill finish							
flat (in wood roll construction)	6.24	3.20	84.77	–	60.35	m²	**145.12**
eaves detail ED1	–	0.40	10.60	–	7.34	m	**17.94**
pitched over 3° (in standing seam construction)	–	2.50	66.23	–	49.35	m²	**115.58**
vertical (in angled or flat seam construction)	–	2.70	71.52	–	49.35	m²	**120.87**
0.6 mm thick roof coverings; oxide finish							
flat (in wood roll construction) (Prime Cost £ per kg)	7.65	2.20	58.28	–	73.21	m²	**131.49**
eaves detail ED1	–	0.40	10.60	–	9.01	m	**19.61**
abutment upstands at perimeters	–	0.66	17.49	–	4.50	m	**21.99**
pitched over 3° (in standing seam construction)	–	1.70	45.03	–	60.11	m²	**105.14**
vertical (in angled or flat seam construction)	–	1.60	42.38	–	60.11	m²	**102.49**

17 SHEET ROOF COVERINGS

Item	PC £	Labour hours	Labour £	Plant £	Material £	Unit	Total rate £
17.07 COPPER STRIP SHEET COVERINGS/ FLASHINGS – cont							
Roof and dormer coverings – cont							
0.6 mm thick dormer coverings; oxide finish							
flat (in wood roll construction)	7.65	3.00	79.47	–	73.21	m²	**152.68**
eaves detail ED1	–	0.40	10.60	–	9.01	m	**19.61**
pitched over 3° (in standing seam construction)	–	2.50	66.23	–	60.11	m²	**126.34**
vertical (in angled or flat seam construction)	–	2.70	71.52	–	60.11	m²	**131.63**
0.6 mm thick roof coverings; KME pre-patinated finish							
flat (in wood roll construction)	64.00	2.20	58.28	–	108.14	m²	**166.42**
eaves detail ED1	–	0.40	10.60	–	13.70	m	**24.30**
abutment upstands at perimeters	–	0.66	17.49	–	6.85	m	**24.34**
pitched over 3° (in standing seam construction)	–	1.70	45.03	–	87.59	m²	**132.62**
vertical (in angled or flat seam construction)	–	1.80	47.69	–	87.59	m²	**135.28**
0.6 mm thick dormer coverings; KME pre-patinated finish							
flat (in wood roll construction)	64.00	3.00	79.47	–	108.14	m²	**187.61**
eaves detail ED1	–	0.40	10.60	–	13.70	m	**24.30**
pitched over 3° (in standing seam construction)	–	2.50	66.23	–	87.59	m²	**153.82**
vertical (in angled or flat seam construction)	–	2.70	71.52	–	87.59	m²	**159.11**
0.7 mm thick roof coverings; mill finish							
flat (in wood roll construction) (Prime Cost £ per kg)	6.24	2.00	52.98	–	68.36	m²	**121.34**
eaves detail ED1	–	0.40	10.60	–	8.40	m	**19.00**
abutment upstands at perimeters	–	0.66	17.49	–	4.20	m	**21.69**
pitched over 3° (in standing seam construction)	–	1.50	39.74	–	55.75	m²	**95.49**
vertical (in angled or flat seam construction)	–	1.60	42.38	–	55.75	m²	**98.13**
0.7 mm thick dormer coverings; mill finish							
flat (in wood roll construction)	6.24	3.00	79.47	–	68.36	m²	**147.83**
eaves detail ED1	–	0.40	10.60	–	8.40	m	**19.00**
pitched over 3° (in standing seam construction)	–	2.50	66.23	–	55.72	m²	**121.95**
vertical (in angled or flat seam construction)	–	2.70	71.52	–	55.68	m²	**127.20**
0.7 mm thick roof coverings; oxide finish							
flat (in wood roll construction) (Prime Cost £ per kg)	7.65	2.00	52.98	–	92.86	m²	**145.84**
eaves detail ED1	–	0.40	10.60	–	10.32	m	**20.92**
abutment upstands at perimeters	–	0.66	17.49	–	5.16	m	**22.65**
pitched over 3° (in standing seam construction)	–	1.50	39.74	–	75.67	m²	**115.41**
vertical (in angled or flat seam construction)	–	1.60	42.38	–	67.48	m²	**109.86**
0.7 mm thick dormer coverings; oxide finish							
flat (in wood roll construction)	7.65	3.00	79.47	–	83.04	m²	**162.51**
eaves detail ED1	–	0.40	10.60	–	10.32	m	**20.92**
pitched over 3° (in standing seam construction)	–	2.50	66.23	–	67.48	m²	**133.71**
vertical (in angled or flat seam construction)	–	2.70	71.52	–	67.48	m²	**139.00**
0.7 mm thick roof coverings; KME pre-patinated finish							
flat (in wood roll construction)	73.65	2.20	58.28	–	124.10	m²	**182.38**
eaves detail ED1	–	0.40	10.60	–	15.76	m	**26.36**
abutment upstands at perimeters	–	0.66	17.49	–	7.88	m	**25.37**

17 SHEET ROOF COVERINGS

Item	PC £	Labour hours	Labour £	Plant £	Material £	Unit	Total rate £
pitched over 3° (in standing seam construction)	–	1.70	45.03	–	100.47	m²	145.50
vertical (in angled or flat seam construction)	–	1.80	47.69	–	100.47	m²	148.16
0.7 mm thick dormer coverings; KME pre-patinated finish							
flat (in wood roll construction)	73.65	3.00	79.47	–	124.10	m²	203.57
eaves detail ED1	–	0.40	10.60	–	15.76	m	26.36
pitched over 3° (in standing seam construction)	–	2.50	66.23	–	100.47	m²	166.70
vertical (in angled or flat seam construction)	–	2.70	71.52	–	100.47	m²	171.99
0.6 mm thick copper flashings, etc.							
Flashings; wedging into grooves; mill finish							
150 mm girth (Prime Cost £ per kg)	6.55	0.50	13.25	–	4.40	m	17.65
240 mm girth	–	0.50	13.25	–	8.80	m	22.05
300 mm girth	–	0.50	13.25	–	11.00	m	24.25
Stepped flashings; wedging into grooves; mill finish							
180 mm girth	–	1.00	26.49	–	6.60	m	33.09
270 mm girth	–	1.00	26.49	–	9.90	m	36.39
Flashings; wedging into grooves; oxide finish							
150 mm girth (Prime Cost £ per kg)	8.03	0.50	13.25	–	6.75	m	20.00
240 mm girth	–	0.50	13.25	–	10.80	m	24.05
300 mm girth	–	0.50	13.25	–	13.50	m	26.75
Stepped flashings; wedging into grooves; oxide finish							
180 mm girth	–	1.00	26.49	–	8.10	m	34.59
270 mm girth	–	1.00	26.49	–	12.15	m	38.64
Flashings; wedging into grooves; KME pre-patinated finish							
150 mm girth (Prime Cost £ per kg)	67.20	0.50	13.25	–	10.27	m	23.52
240 mm girth	–	0.50	13.25	–	16.44	m	29.69
300 mm girth	–	0.50	13.25	–	20.55	m	33.80
Stepped flashings; wedging into grooves; KME pre-patinated finish							
180 mm girth	–	1.00	26.49	–	12.33	m	38.82
270 mm girth	–	1.00	26.49	–	18.49	m	44.98
0.7 mm thick copper flashings, etc.							
Flashings; wedging into grooves; mill finish							
150 mm girth (Prime Cost £ per kg)	6.24	0.50	13.25	–	6.30	m	19.55
240 mm girth	–	0.50	13.25	–	10.08	m	23.33
300 mm girth	–	0.50	13.25	–	12.60	m	25.85
Stepped flashings; wedging into grooves; mill finish							
180 mm girth	–	1.00	26.49	–	7.56	m	34.05
270 mm girth	–	1.00	26.49	–	11.34	m	37.83
Flashings; wedging into grooves; oxide finish							
150 mm girth (Prime Cost £ per kg)	7.65	0.50	13.25	–	7.73	m	20.98
240 mm girth	–	0.50	13.25	–	12.37	m	25.62
300 mm girth	–	0.50	13.25	–	15.47	m	28.72
Stepped flashings; wedging into grooves; oxide finish							
180 mm girth	–	1.00	26.49	–	9.28	m	35.77
270 mm girth	–	1.00	26.49	–	13.92	m	40.41

17 SHEET ROOF COVERINGS

Item	PC £	Labour hours	Labour £	Plant £	Material £	Unit	Total rate £
17.07 COPPER STRIP SHEET COVERINGS/ FLASHINGS – cont							
0.7 mm thick copper flashings, etc. – cont							
Flashings; wedging into grooves; KME pre-patinated finish							
150 mm girth (Prime Cost £ per kg)	73.65	0.50	13.25	–	11.82	m	**25.07**
240 mm girth	–	0.50	13.25	–	18.91	m	**32.16**
300 mm girth	–	0.50	13.25	–	23.64	m	**36.89**
Stepped flashings; wedging into grooves; KME pre-patinated finish							
180 mm girth	–	1.00	26.49	–	14.18	m	**40.67**
270 mm girth	–	1.00	26.49	–	21.27	m	**47.76**
Sundries							
provision of square batten roll at 500 mm centres (per m)	–	0.20	5.29	–	1.34	m	**6.63**
17.08 ZINC STRIP SHEET COVERINGS/ FLASHINGS							
Zinc roofing; CP 143–5; on and including Delta Trella underlay							
The following rates are based upon nett deck or wall areas							
Natural Bright Rheinzink							
Roof, dormer and wall coverings							
0.7 mm thick roof coverings							
flat (in wood roll construction) (Prime Cost £ per kg)	3.29	2.00	52.98	–	31.72	m²	**84.70**
eaves detail ED1	–	0.40	10.60	–	5.35	m	**15.95**
abutment upstands at perimeters	–	0.66	17.49	–	2.68	m	**20.17**
pitched over 3° (in standing seam construction)	–	1.50	39.74	–	26.35	m²	**66.09**
0.7 mm thick dormer coverings							
flat (in wood roll construction) (Prime Cost £ per kg)	3.29	3.00	79.47	–	31.72	m²	**111.19**
eaves detail ED1	–	0.40	10.60	–	5.35	m	**15.95**
pitched over 3° (in standing seam construction)	–	2.50	66.23	–	26.35	m²	**92.58**
0.8 mm thick wall coverings							
vertical (in angled or flat seam construction)	–	1.60	42.38	–	29.51	m²	**71.89**
0.8 mm thick dormer coverings							
vertical (in angled or flat seam construction)	–	2.70	71.52	–	29.61	m²	**101.13**
0.8 mm thick zinc flashings, etc.; Natural Bright Rheinzink							
Flashings; wedging into grooves							
150 mm girth	–	0.50	13.25	–	3.07	m	**16.32**
240 mm girth	–	0.50	13.25	–	4.92	m	**18.17**
300 mm girth	–	0.50	13.25	–	6.15	m	**19.40**

17 SHEET ROOF COVERINGS

Item	PC £	Labour hours	Labour £	Plant £	Material £	Unit	Total rate £
Stepped flashings; wedging into grooves							
180 mm girth	–	1.00	26.49	–	3.69	m	**30.18**
270 mm girth	–	1.00	26.49	–	5.53	m	**32.02**
Integral box gutter							
900 mm girth; 2 × bent; 2 × welted	–	2.00	52.98	–	25.41	m	**78.39**
Valley gutter							
600 mm girth; 2 × bent; 2 × welted	–	1.50	39.74	–	15.24	m	**54.98**
Hips and ridges							
450 mm girth; 2 × bent; 2 × welted	–	2.00	52.98	–	9.22	m	**62.20**
Natural Bright Rheinzink PRO							
Roof, dormer and wall coverings							
0.7 mm thick roof coverings							
flat (in wood roll construction) (Prime Cost £ per kg)	4.48	2.00	52.98	–	43.22	m²	**96.20**
eaves detail ED1	–	0.40	10.60	–	7.29	m	**17.89**
abutment upstands at perimeters	–	0.66	17.49	–	3.65	m	**21.14**
pitched over 3° (in standing seam construction)	–	1.50	39.74	–	35.90	m²	**75.64**
0.7 mm thick dormer coverings							
flat (in wood roll construction) (Prime Cost £ per kg)	4.48	3.00	79.47	–	43.22	m²	**122.69**
eaves detail ED1	–	0.40	10.60	–	7.29	m	**17.89**
pitched over 3° (in standing seam construction)	–	2.50	66.23	–	35.89	m²	**102.12**
0.8 mm thick wall coverings							
vertical (in angled or flat seam construction)	–	1.60	42.38	–	40.21	m²	**82.59**
0.8 mm thick dormer coverings							
vertical (in angled or flat seam construction)	–	2.70	71.52	–	40.21	m²	**111.73**
0.8 mm thick zinc flashings, etc.; Natural Bright Rheinzink PRO							
Flashings; wedging into grooves							
150 mm girth	–	0.50	13.25	–	4.19	m	**17.44**
240 mm girth	–	0.50	13.25	–	6.70	m	**19.95**
300 mm girth	–	0.50	13.25	–	8.38	m	**21.63**
Stepped flashings; wedging into grooves							
180 mm girth	–	1.00	26.49	–	5.02	m	**31.51**
270 mm girth	–	1.00	26.49	–	7.54	m	**34.03**
Integral box gutter							
900 mm girth; 2 × bent; 2 × welted	–	2.00	52.98	–	34.63	m	**87.61**
Valley gutter							
600 mm girth; 2 × bent; 2 × welted	–	1.50	39.74	–	20.77	m	**60.51**
Hips and ridges							
450 mm girth; 2 × bent; 2 × welted	–	2.00	52.98	–	12.56	m	**65.54**

17 SHEET ROOF COVERINGS

Item	PC £	Labour hours	Labour £	Plant £	Material £	Unit	Total rate £
17.08 ZINC STRIP SHEET COVERINGS/ FLASHINGS – cont							
Pre-weathered Rheinzink							
Roof, dormer and wall coverings							
0.7 mm thick roof coverings; pre-weathered Rheinzink							
flat (in wood roll construction) (Prime Cost £ per kg)	3.80	2.00	52.98	–	36.70	m²	**89.68**
eaves detail ED1	–	0.40	10.60	–	6.19	m	**16.79**
abutment upstands at perimeters	–	0.66	17.49	–	3.10	m	**20.59**
pitched over 3° (in standing seam construction)	–	1.50	39.74	–	30.49	m²	**70.23**
0.7 mm thick dormer coverings; pre-weathered Rheinzink							
flat (in wood roll construction) (Prime Cost £ per kg)	3.80	3.00	79.47	–	36.70	m²	**116.17**
eaves detail ED1	–	0.40	10.60	–	6.19	m	**16.79**
pitched over 3° (in standing seam construction)	–	2.50	66.23	–	30.49	m²	**96.72**
0.8 mm thick wall coverings; pre-weathered Rheinzink							
vertical (in angled or flat seam construction)	–	1.60	42.38	–	34.15	m²	**76.53**
0.8 mm thick dormer coverings; pre-weathered Rheinzink							
vertical (in angled or flat seam construction)	–	2.70	71.52	–	34.15	m²	**105.67**
0.8 mm thick zinc flashings, etc.; pre-weathered Rheinzink							
Flashings; wedging into grooves							
150 mm girth	–	0.50	13.25	–	3.56	m	**16.81**
240 mm girth	–	0.50	13.25	–	5.69	m	**18.94**
300 mm girth	–	0.50	13.25	–	7.11	m	**20.36**
Stepped flashings; wedging into grooves							
180 mm girth	–	1.00	26.49	–	4.27	m	**30.76**
270 mm girth	–	1.00	26.49	–	6.40	m	**32.89**
Integral box gutter							
900 mm girth; 2 × bent; 2 × welted	–	2.00	52.98	–	29.41	m	**82.39**
Valley gutter							
600 mm girth; 2 × bent; 2 × welted	–	1.50	39.74	–	17.64	m	**57.38**
Hips and ridges							
450 mm girth; 2 × bent; 2 × welted	–	2.00	52.98	–	10.66	m	**63.64**
Pre-weathered Rheinzink PRO							
Roof, dormer and wall coverings							
0.7 mm thick roof coverings							
flat (in wood roll construction) (Prime Cost £ per kg)	4.98	2.00	52.98	–	48.10	m²	**101.08**
eaves detail ED1	–	0.40	10.60	–	8.12	m	**18.72**
abutment upstands at perimeters	–	0.66	17.49	–	4.06	m	**21.55**
pitched over 3° (in standing seam construction)	–	1.50	39.74	–	39.96	m²	**79.70**

17 SHEET ROOF COVERINGS

Item	PC £	Labour hours	Labour £	Plant £	Material £	Unit	Total rate £
0.7 mm thick dormer coverings							
flat (in wood roll construction) (Prime Cost £ per kg)	4.98	3.00	79.47	–	48.10	m²	**127.57**
eaves detail ED1	–	0.40	10.60	–	8.12	m	**18.72**
pitched over 3° (in standing seam construction)	–	2.50	66.23	–	39.96	m²	**106.19**
0.8 mm thick wall coverings							
vertical (in angled or flat seam construction)	–	1.60	42.38	–	44.76	m²	**87.14**
0.8 mm thick dormer coverings							
vertical (in angled or flat seam construction)	–	2.70	71.52	–	44.76	m²	**116.28**
0.8 mm thick zinc flashings, etc.; Pre-weathered Rheinzink PRO							
Flashings; wedging into grooves							
150 mm girth	–	0.50	13.25	–	4.66	m	**17.91**
240 mm girth	–	0.50	13.25	–	7.46	m	**20.71**
300 mm girth	–	0.50	13.25	–	9.32	m	**22.57**
Stepped flashings; wedging into grooves							
180 mm girth	–	1.00	26.49	–	5.59	m	**32.08**
270 mm girth	–	1.00	26.49	–	8.39	m	**34.88**
Integral box gutter							
900 mm girth; 2 × bent; 2 × welted	–	–	–	–	–	m	**38.54**
Valley gutter							
600 mm girth; 2 × bent; 2 × welted	–	1.50	39.74	–	23.12	m	**62.86**
Hips and ridges							
450 mm girth; 2 × bent; 2 × welted	–	2.00	52.98	–	13.98	m	**66.96**
VM Natural Bright							
Roof, dormer and wall coverings							
0.7 mm thick roof coverings							
flat (in wood roll construction) (Prime Cost £ per kg)	3.29	2.00	52.98	–	31.72	m²	**84.70**
eaves detail ED1	–	0.40	10.60	–	5.35	m	**15.95**
abutment upstands at perimeters	–	0.66	17.49	–	2.68	m	**20.17**
pitched over 3° (in standing seam construction)	–	1.50	39.74	–	26.35	m²	**66.09**
0.7 mm thick dormer coverings							
flat (in wood roll construction) (Prime Cost £ per kg)	3.29	3.00	79.47	–	31.72	m²	**111.19**
eaves detail ED1	–	0.40	10.60	–	5.35	m	**15.95**
pitched over 3° (in standing seam construction)	–	2.50	66.23	–	26.35	m²	**92.58**
0.8 mm thick wall coverings							
vertical (in angled or flat seam construction)	–	1.60	42.38	–	29.51	m²	**71.89**
0.8 mm thick dormer coverings							
vertical (in angled or flat seam construction)	–	2.70	71.52	–	29.51	m²	**101.03**
0.8 mm thick zinc flashings, etc.; VM Natural Bright							
Flashings; wedging into grooves							
150 mm girth	–	0.50	13.25	–	3.07	m	**16.32**
240 mm girth	–	0.50	13.25	–	4.92	m	**18.17**
300 mm girth	–	0.50	13.25	–	6.15	m	**19.40**

17 SHEET ROOF COVERINGS

Item	PC £	Labour hours	Labour £	Plant £	Material £	Unit	Total rate £
17.08 ZINC STRIP SHEET COVERINGS/ FLASHINGS – cont							
0.8 mm thick zinc flashings, etc. – cont							
Stepped flashings; wedging into grooves							
180 mm girth	–	1.00	26.49	–	3.69	m	**30.18**
270 mm girth	–	1.00	26.49	–	5.53	m	**32.02**
Integral box gutter							
900 mm girth; 2 × bent; 2 × welted	–	2.00	52.98	–	25.41	m	**78.39**
Valley gutter							
600 mm girth; 2 × bent; 2 × welted	–	1.50	39.74	–	15.24	m	**54.98**
Hips and ridges							
450 mm girth; 2 × bent; 2 × welted	–	2.00	52.98	–	8.37	m	**61.35**
VM Natural Bright PLUS							
Roof, dormer and wall coverings							
0.7 mm thick roof coverings							
flat (in wood roll construction) (Prime Cost £ per kg)	4.48	2.00	52.98	–	43.22	m²	**96.20**
eaves detail ED1	–	0.40	10.60	–	7.29	m	**17.89**
abutment upstands at perimeters	–	0.66	17.49	–	3.65	m	**21.14**
pitched over 3° (in standing seam construction)	–	1.50	39.74	–	35.90	m²	**75.64**
0.7 mm thick dormer coverings							
flat (in wood roll construction) (Prime Cost £ per kg)	4.48	3.00	79.47	–	43.22	m²	**122.69**
eaves detail ED1	–	0.40	10.60	–	7.29	m	**17.89**
pitched over 3° (in standing seam construction)	–	2.50	66.23	–	35.90	m²	**102.13**
0.8 mm thick wall coverings							
vertical (in angled or flat seam construction)	–	1.60	42.38	–	40.21	m²	**82.59**
0.8 mm thick dormer coverings							
vertical (in angled or flat seam construction)	–	2.70	71.52	–	40.21	m²	**111.73**
0.8 mm thick zinc flashings, etc.; VM Natural Bright PLUS							
Flashings; wedging into grooves							
150 mm girth	–	0.50	13.25	–	4.19	m	**17.44**
240 mm girth	–	0.50	13.25	–	6.70	m	**19.95**
300 mm girth	–	0.50	13.25	–	8.38	m	**21.63**
Stepped flashings; wedging into grooves							
180 mm girth	–	1.00	26.49	–	5.02	m	**31.51**
270 mm girth	–	1.00	26.49	–	7.54	m	**34.03**
Integral box gutter							
900 mm girth; 2 × bent; 2 × welted	–	2.00	52.98	–	34.63	m	**87.61**
Valley gutter							
600 mm girth; 2 × bent; 2 × welted	–	1.50	39.74	–	20.77	m	**60.51**
Hips and ridges							
450 mm girth; 2 × bent; 2 × welted	–	2.00	52.98	–	12.56	m	**65.54**

17 SHEET ROOF COVERINGS

Item	PC £	Labour hours	Labour £	Plant £	Material £	Unit	Total rate £
VM Quartz (pre-weathered)							
Roof, dormer and wall coverings							
0.7 mm thick roof coverings							
flat (in wood roll construction) (Prime Cost £ per kg)	4.08	2.00	52.98	–	39.48	m²	**92.46**
eaves detail ED1	–	0.40	10.60	–	6.66	m	**17.26**
abutment upstands at perimeters	–	0.66	17.49	–	3.33	m	**20.82**
pitched over 3° (in standing seam construction)	–	1.50	39.74	–	32.80	m²	**72.54**
0.7 mm thick dormer coverings							
flat (in wood roll construction) (Prime Cost £ per kg)	4.08	3.00	79.47	–	39.48	m²	**118.95**
eaves detail ED1	–	0.40	10.60	–	6.66	m	**17.26**
pitched over 3° (in standing seam construction)	–	2.50	66.23	–	32.80	m²	**99.03**
0.8 mm thick wall coverings							
vertical (in angled or flat seam construction)	–	1.60	42.38	–	36.73	m²	**79.11**
0.8 mm thick dormer coverings							
vertical (in angled or flat seam construction)	–	2.70	71.52	–	36.73	m²	**108.25**
0.8 mm thick zinc flashings, etc.; VM Quartz (pre-weathered)							
Flashings; wedging into grooves							
150 mm girth	–	0.50	13.25	–	3.83	m	**17.08**
240 mm girth	–	0.50	13.25	–	6.12	m	**19.37**
300 mm girth	–	0.50	13.25	–	7.65	m	**20.90**
Stepped flashings; wedging into grooves							
180 mm girth	–	1.00	26.49	–	4.59	m	**31.08**
270 mm girth	–	1.00	26.49	–	6.89	m	**33.38**
Integral box gutter							
900 mm girth; 2 × bent; 2 × welted	–	2.00	52.98	–	31.63	m	**84.61**
Valley gutter							
600 mm girth; 2 × bent; 2 × welted	–	1.50	39.74	–	18.97	m	**58.71**
Hips and ridges							
450 mm girth; 2 × bent; 2 × welted	–	2.00	52.98	–	11.47	m	**64.45**
VM Quartz (pre-weathered) PLUS							
Roof, dormer and wall coverings							
0.7 mm thick roof coverings							
flat (in wood roll construction) (Prime Cost £ per kg)	5.27	2.00	52.98	–	50.98	m²	**103.96**
eaves detail ED1	–	0.40	10.60	–	8.60	m	**19.20**
abutment upstands at perimeters	–	0.66	17.49	–	4.30	m	**21.79**
pitched over 3° (in standing seam construction)	–	1.50	39.74	–	42.35	m²	**82.09**
0.7 mm thick dormer coverings							
flat (in wood roll construction) (Prime Cost £ per kg)	5.27	3.00	79.47	–	50.98	m²	**130.45**
eaves detail ED1	–	0.40	10.60	–	8.60	m	**19.20**
pitched over 3° (in standing seam construction)	–	2.50	66.23	–	42.35	m²	**108.58**

17 SHEET ROOF COVERINGS

Item	PC £	Labour hours	Labour £	Plant £	Material £	Unit	Total rate £
17.08 ZINC STRIP SHEET COVERINGS/ FLASHINGS – cont							
Roof, dormer and wall coverings – cont							
0.8 mm thick wall coverings							
vertical (in angled or flat seam construction)	–	1.60	42.38	–	47.43	m²	**89.81**
0.8 mm thick dormer coverings							
vertical (in angled or flat seam construction)	–	2.70	71.52	–	47.43	m²	**118.95**
0.8 mm thick zinc flashings, etc.; VM Quartz (pre-weathered) PLUS							
Flashings; wedging into grooves							
150 mm girth	–	0.50	13.25	–	4.94	m	**18.19**
240 mm girth	–	0.50	13.25	–	7.91	m	**21.16**
300 mm girth	–	0.50	13.25	–	9.88	m	**23.13**
Stepped flashings; wedging into grooves							
180 mm girth	–	1.00	26.49	–	5.93	m	**32.42**
270 mm girth	–	1.00	26.49	–	8.89	m	**35.38**
Integral box gutter							
900 mm girth; 2 × bent; 2 × welted	–	2.00	52.98	–	40.85	m	**93.83**
Valley gutter							
600 mm girth; 2 × bent; 2 × welted	–	1.50	39.74	–	24.50	m	**64.24**
Hips and ridges							
450 mm girth; 2 × bent; 2 × welted	–	2.00	52.98	–	14.81	m	**67.79**
Sundries and accessories							
Klober breather membrane/underlay	–	0.20	5.29	–	3.87	m²	**9.16**
Delta Trela Chestwig underlay	–	0.20	5.29	–	6.61	m²	**11.90**
Delta Trela Football Studs underlay	–	0.20	5.29	–	1.22	m²	**6.51**
Trapezoidal batten roll at 500 mm centres (per m)	–	0.20	5.29	–	1.22	m	**6.51**
Zinflash; 0.6 mm thick lead look flashing (no patination oil required)							
Flashings; wedging into grooves							
150 mm girth	–	0.50	13.25	–	5.46	m	**18.71**
250 mm girth	–	0.50	13.25	–	9.09	m	**22.34**
300 mm girth	–	0.50	13.25	–	10.91	m	**24.16**
380 mm girth	–	0.50	13.25	–	13.83	m	**27.08**
450 mm girth	–	0.50	13.25	–	16.38	m	**29.63**
Stepped flashings; wedging into grooves							
150 mm girth	–	1.00	26.49	–	5.46	m	**31.95**
250 mm girth	–	1.00	26.49	–	9.09	m	**35.58**
300 mm girth	–	1.00	26.49	–	10.91	m	**37.40**
380 mm girth	–	1.00	26.49	–	13.83	m	**40.32**
450 mm girth	–	1.00	26.49	–	16.38	m	**42.87**

17 SHEET ROOF COVERINGS

Item	PC £	Labour hours	Labour £	Plant £	Material £	Unit	Total rate £
17.09 STAINLESS STEEL SHEET COVERINGS/ FLASHINGS							
Terne-coated stainless steel roofing; Associated Lead Mills Ltd; or other equal and approved: on and including Metmatt underlay							
The following rates are based upon nett deck or wall areas							
Roof, dormer and wall coverings in Uginox grade 316; marine							
0.4 mm thick roof coverings							
flat (in wood roll construction) (Prime Cost £ per kg)	7.65	2.00	52.98	–	42.24	m²	**95.22**
eaves detail ED1	–	0.40	10.60	–	4.99	m	**15.59**
abutment upstands at perimeters	–	0.66	17.49	–	2.49	m	**19.98**
pitched over 3° (in standing seam construction)	–	1.50	39.74	–	34.76	m²	**74.50**
0.5 mm thick dormer coverings							
flat (in wood roll construction) (Prime Cost £ per kg)	7.14	3.00	79.47	–	48.96	m²	**128.43**
eaves detail ED1	–	0.40	10.60	–	4.64	m	**15.24**
pitched over 3° (in standing seam construction)	–	2.50	66.23	–	40.14	m²	**106.37**
0.5 mm thick wall coverings							
vertical (in angled or flat seam construction)	7.14	1.60	42.38	–	40.14	m²	**82.52**
vertical (with Coulisseau joint construction)	–	2.50	66.23	–	41.44	m²	**107.67**
0.5 mm thick Uginox grade 316 flashings, etc.							
Flashings; wedging into grooves							
150 mm girth (Prime Cost £ per kg)	7.14	0.50	13.25	–	4.42	m	**17.67**
240 mm girth	–	0.50	13.25	–	7.07	m	**20.32**
300 mm girth	–	0.50	13.25	–	8.83	m	**22.08**
Stepped flashings; wedging into grooves							
180 mm girth	–	1.00	26.49	–	5.30	m	**31.79**
270 mm girth	–	1.00	26.49	–	7.95	m	**34.44**
Fan apron							
250 mm girth	–	0.50	13.25	–	7.36	m	**20.61**
Integral box gutter							
900 mm girth; 2 × bent; 2 × welted	–	2.00	52.98	–	30.33	m	**83.31**
Valley gutter							
600 mm girth; 2 × bent; 2 × welted	–	1.50	39.74	–	21.55	m	**61.29**
Hips and ridges							
450 mm girth; 2 × bent; 2 × welted	–	2.00	52.98	–	13.25	m	**66.23**
Roof, dormer and wall coverings in Ugitop grade 304							
0.4 mm thick roof coverings							
flat (in wood roll construction) (Prime Cost £ per kg)	5.78	2.00	52.98	–	36.23	m²	**89.21**
eaves detail ED1	–	0.40	10.60	–	3.76	m	**14.36**
abutment upstands at perimeters	–	0.66	17.49	–	1.88	m	**19.37**
pitched over 3° (in standing seam construction)	–	1.50	39.74	–	27.44	m²	**67.18**

17 SHEET ROOF COVERINGS

Item	PC £	Labour hours	Labour £	Plant £	Material £	Unit	Total rate £
17.09 STAINLESS STEEL SHEET COVERINGS/ FLASHINGS – cont							
Roof, dormer and wall coverings in Ugitop grade 304 – cont							
0.5 mm thick dormer coverings							
flat (in wood roll construction) (Prime Cost £ per kg)	5.50	3.00	79.47	–	38.84	m²	**118.31**
eaves detail ED1	–	0.40	10.60	–	3.58	m	**14.18**
pitched over 3° (in standing seam construction)	–	2.50	66.23	–	32.04	m²	**98.27**
0.5 mm thick wall coverings							
vertical (in angled or flat seam construction)	–	1.60	42.38	–	32.04	m²	**74.42**
vertical (with Coulisseau joint construction)	–	2.50	66.23	–	33.04	m²	**99.27**
0.5 mm thick Ugitop grade 304 flashings, etc.							
Flashings; wedging into grooves							
150 mm girth (Prime Cost £ per kg)	5.50	0.50	13.25	–	3.29	m	**16.54**
240 mm girth	–	0.50	13.25	–	5.72	m	**18.97**
300 mm girth	–	0.50	13.25	–	7.15	m	**20.40**
Stepped flashings; wedging into grooves							
180 mm girth	–	1.00	26.49	–	4.29	m	**30.78**
270 mm girth	–	1.00	26.49	–	6.44	m	**32.93**
Fan apron							
250 mm girth	–	0.50	13.25	–	5.96	m	**19.21**
Integral box gutter							
900 mm girth; 2 × bent; 2 × welted	–	2.00	52.98	–	24.55	m	**77.53**
Valley gutter							
600 mm girth; 2 × bent; 2 × welted	–	1.50	39.74	–	17.45	m	**57.19**
Hips and ridges							
450 mm girth; 2 × bent; 2 × welted	–	2.00	52.98	–	10.73	m	**63.71**
Roof, dormer and wall coverings in Ugitop grade 316							
0.4 mm thick roof coverings							
flat (in wood roll construction) (Prime Cost £ per kg)	7.14	2.00	52.98	–	43.55	m²	**96.53**
eaves detail ED1	–	0.40	10.60	–	4.64	m	**15.24**
abutment upstands at perimeters	–	0.66	17.49	–	2.32	m	**19.81**
pitched over 3° (in standing seam construction)	–	1.50	39.74	–	32.71	m²	**72.45**
0.5 mm thick dormer coverings							
flat (in wood roll construction) (Prime Cost £ per kg)	6.80	3.00	79.47	–	46.94	m²	**126.41**
eaves detail ED1	–	0.40	10.60	–	4.43	m	**15.03**
pitched over 3° (in standing seam construction)	–	2.50	66.23	–	38.52	m²	**104.75**
0.5 mm thick wall coverings							
vertical (in angled or flat seam construction)	–	1.60	42.38	–	38.52	m²	**80.90**
vertical (with Coulisseau joint construction)	–	2.50	66.23	–	33.04	m²	**99.27**

17 SHEET ROOF COVERINGS

Item	PC £	Labour hours	Labour £	Plant £	Material £	Unit	Total rate £
0.5 mm thick Ugitop grade 316 flashings, etc.							
Flashings; wedging into grooves							
150 mm girth	–	0.50	13.25	–	4.42	m	**17.67**
240 mm girth	–	0.50	13.25	–	7.07	m	**20.32**
300 mm girth	–	0.50	13.25	–	8.83	m	**22.08**
Stepped flashings; wedging into grooves							
180 mm girth	–	1.00	26.49	–	5.30	m	**31.79**
270 mm girth	–	1.00	26.49	–	7.95	m	**34.44**
Fan apron							
250 mm girth	–	0.50	13.25	–	7.36	m	**20.61**
Integral box gutter							
900 mm girth; 2 × bent; 2 × welted	–	2.00	52.98	–	30.33	m	**83.31**
Valley gutter							
600 mm girth; 2 × bent; 2 × welted	–	1.50	39.74	–	21.55	m	**61.29**
Hips and ridges							
450 mm girth; 2 × bent; 2 × welted	–	2.00	52.98	–	13.25	m	**66.23**
Sundries							
provision of square batten roll at 500 mm centres (per m)	–	0.20	5.29	–	1.34	m	**6.63**

18 TILE AND SLATE ROOF AND WALL COVERINGS

Item	PC £	Labour hours	Labour £	Plant £	Material £	Unit	Total rate £
18.01 ALTERNATIVE SUPPLY ONLY TILE PRICES							
Clay tiles; plain, interlocking and pantiles							
Dreadnought							
Handformed Classic Staffordshire Blue	–	–	–	–	639.63	1000	**639.63**
Handformed Classic Purple Brown	–	–	–	–	639.63	1000	**639.63**
Handformed Classic Bronze	–	–	–	–	899.19	1000	**899.19**
Handformed Classic Deep Red	–	–	–	–	871.38	1000	**871.38**
Red smooth/sandfaced	–	–	–	–	556.20	1000	**556.20**
Country brown smooth/sandfaced	–	–	–	–	602.55	1000	**602.55**
Brown Antique smooth/sandfaced	–	–	–	–	546.93	1000	**546.93**
Blue/Dark Heather	–	–	–	–	639.63	1000	**639.63**
Sandtoft pantiles							
Bridgewater Double Roman	–	–	–	–	5432.22	1000	**5432.22**
Gaelic	–	–	–	–	2771.73	1000	**2771.73**
Arcadia	–	–	–	–	1520.28	1000	**1520.28**
William Blyth pantiles							
Celtic (French)	–	–	–	–	1133.00	1000	**1133.00**
Concrete tiles; plain and interlocking							
Marley Eternit roof tiles							
Anglia	–	–	–	–	667.44	1000	**667.44**
Ashmore	–	–	–	–	804.33	1000	**804.33**
Duo Modern	–	–	–	–	1006.72	1000	**1006.72**
Pewter Mendip	–	–	–	–	1148.97	1000	**1148.97**
Malvern	–	–	–	–	1072.33	1000	**1072.33**
Plain	–	–	–	–	382.95	1000	**382.95**
Redland roof tiles							
Redland 49	–	–	–	–	699.68	1000	**699.68**
Mini Stoneworld	–	–	–	–	988.80	1000	**988.80**
Grovebury	–	–	–	–	3726.54	1000	**3726.54**
18.02 PLAIN TILING							
NOTE: The following items of tile roofing unless otherwise described, include for conventional fixing assuming normal exposure with appropriate nails and/or rivets or clips to pressure impregnated softwood battens fixed with galvanized nails; prices also include for all bedding and pointing at verges, beneath ridge tiles, etc.							
Clay interlocking plain tiles; Sandtoft 20/20 natural red faced or other equal; 75 mm lap; on 25 mm × 38 mm battens and type 1F reinforced underlay							
Tiles 370 mm × 223 mm (Prime Cost £ per 1000)	747.00	–	–	–	747.00	1000	**747.00**
roof coverings	11.21	0.50	11.08	–	16.96	m²	**28.04**
extra over coverings for;							
fixing every tile	–	0.02	0.44	–	1.19	m²	**1.63**
double course at eaves	–	0.28	6.20	–	13.73	m	**19.93**
verges; extra single undercloak course of plain tiles	–	0.28	6.20	–	6.98	m	**13.18**

18 TILE AND SLATE ROOF AND WALL COVERINGS

Item	PC £	Labour hours	Labour £	Plant £	Material £	Unit	Total rate £
open valleys; cutting both sides	–	0.17	3.77	–	3.23	m	**7.00**
dry ridge tiles	–	0.56	12.41	–	17.50	m	**29.91**
dry hips; cutting both sides	–	0.69	15.29	–	13.70	m	**28.99**
holes for pipes and the like	–	0.19	4.21	–	–	nr	**4.21**
Clay pantiles; Sandtoft Old English; red sand faced or other equal; 75 mm lap; on 25 mm × 38 mm battens and type 1F reinforced underlay							
Pantiles 342 mm × 241 mm (Prime Cost £ per 1000)	1278.00	–	–	–	1278.00	1000	**1278.00**
roof coverings	20.45	0.50	11.08	–	27.56	m²	**38.64**
extra over coverings for;							
fixing every tile	–	0.02	0.44	–	3.00	m²	**3.44**
double course at eaves	–	0.31	6.87	–	6.53	m	**13.40**
verges; extra single undercloak course of plain tiles	–	0.28	6.20	–	16.39	m	**22.59**
open valleys; cutting both sides	–	0.17	3.77	–	5.53	m	**9.30**
ridge tiles; tile slips	–	0.56	12.41	–	50.40	m	**62.81**
hips; cutting both sides	–	0.69	15.29	–	55.92	m	**71.21**
holes for pipes and the like	–	0.19	4.21	–	–	nr	**4.21**
Clay pantiles; William Blyth's Lincoln natural or other equal; 75 mm lap; on 19 mm × 38 mm battens and type 1F reinforced underlay							
Pantiles 343 mm × 280 mm (Prime Cost £ per 1000)	1320.00	–	–	–	1320.00	1000	**1320.00**
roof coverings	20.46	0.50	11.08	–	27.57	m²	**38.65**
extra over coverings for;							
fixing every tile	–	0.02	0.44	–	3.00	m²	**3.44**
other colours	–	–	–	–	1.76	m²	**1.76**
double course at eaves	–	0.31	6.87	–	6.71	m	**13.58**
verges; extra single undercloak course of plain tiles	–	0.28	6.20	–	15.08	m	**21.28**
open valleys; cutting both sides	–	0.17	3.77	–	5.71	m	**9.48**
ridge tiles; tile slips	–	0.56	12.41	–	31.28	m	**43.69**
hips; cutting both sides	–	0.69	15.29	–	37.00	m	**52.29**
holes for pipes and the like	–	0.19	4.21	–	–	nr	**4.21**
Clay plain tiles; Hinton, Perry and Davenhill Dreadnought smooth red machine-made or other equal; on 19 mm × 38 mm battens and type 1F reinforced underlay							
Tiles 265 mm × 165 mm (Prime Cost £ per 1000)	540.00	–	–	–	540.00	1000	**540.00**
roof coverings; to 64 mm lap	32.40	0.50	11.08	–	46.20	m²	**57.28**
wall coverings; to 38 mm lap	28.62	0.75	16.61	–	40.09	m²	**56.70**
extra over coverings for;							
ornamental tiles	–	–	–	–	44.16	m²	**44.16**
double course at eaves	–	0.23	5.10	–	8.92	m	**14.02**
verges	–	0.28	6.20	–	13.70	m	**19.90**

18 TILE AND SLATE ROOF AND WALL COVERINGS

Item	PC £	Labour hours	Labour £	Plant £	Material £	Unit	Total rate £
18.02 PLAIN TILING – cont							
Clay plain tiles – cont							
swept valleys; cutting both sides	–	0.60	13.30	–	9.34	m	**22.64**
bonnet hips; cutting both sides	–	0.74	16.40	–	118.19	m	**134.59**
external vertical angle tiles; supplementary nail fixings	–	0.37	8.20	–	87.57	m	**95.77**
half round ridge tiles	–	0.56	12.41	–	17.32	m	**29.73**
holes for pipes and the like	–	0.19	4.21	–	–	nr	**4.21**
Concrete plain tiles; BS EN 490 group A; on 25 mm × 38 mm battens and type 1F reinforced underlay							
Tiles 267 mm × 165 mm (Prime Cost £ per 1000)	360.00	–	–	–	360.00	1000	**360.00**
roof coverings; to 64 mm lap	21.60	0.25	5.54	–	34.52	m²	**40.06**
wall coverings; to 38 mm lap	19.08	0.30	6.64	–	29.77	m²	**36.41**
extra over coverings for;							
ornamental tiles	–	–	–	–	36.79	m²	**36.79**
double course at eaves	–	0.23	5.10	–	4.25	m	**9.35**
verges	–	0.31	6.87	–	1.50	m	**8.37**
swept valleys; cutting both sides	–	0.60	13.30	–	44.04	m	**57.34**
bonnet hips; cutting both sides	–	0.74	16.40	–	44.19	m	**60.59**
external vertical angle tiles; supplementary nail fixings	–	0.37	8.20	–	34.08	m	**42.28**
half round ridge tiles	–	0.46	10.19	–	9.95	m	**20.14**
third round hip tiles; cutting both sides	–	0.46	10.19	–	12.29	m	**22.48**
holes for pipes and the like	–	0.19	4.21	–	–	nr	**4.21**
18.03 INTERLOCKING TILING							
NOTE: The following items of tile roofing unless otherwise described, include for conventional fixing assuming normal exposure with appropriate nails and/or rivets or clips to pressure impregnated softwood battens fixed with galvanized nails; prices also include for all bedding and pointing at verges, beneath ridge tiles, etc.							
Concrete interlocking tiles; Marley Eternit Anglia granule finish tiles or other equal; 75 mm lap; on 25 mm × 38 mm battens and type 1F reinforced underlay							
Tiles 387 mm × 230 mm (Prime Cost £ per 1000)	565.20	–	–	–	565.20	1000	**565.20**
roof coverings	8.87	0.42	9.30	–	14.31	m²	**23.61**
extra over coverings for;							
fixing every tile	–	0.02	0.44	–	0.49	m²	**0.93**
eaves; eaves filler	–	0.04	0.89	–	12.43	m	**13.32**
verges; 150 mm wide asbestos free strip undercloak	–	0.21	4.66	–	2.19	m	**6.85**

18 TILE AND SLATE ROOF AND WALL COVERINGS

Item	PC £	Labour hours	Labour £	Plant £	Material £	Unit	Total rate £
valley trough tiles; cutting both sides	–	0.51	11.30	–	28.51	m	39.81
segmental ridge tiles; tile slips	–	0.51	11.30	–	15.05	m	26.35
segmental hip tiles; tile slips; cutting both sides	–	0.65	14.40	–	16.88	m	31.28
dry ridge tiles; segmental including batten							
sections; unions and filler pieces	–	0.28	6.20	–	21.20	m	27.40
segmental mono-ridge tiles	–	0.51	11.30	–	23.41	m	34.71
gas ridge terminal	–	0.46	10.19	–	76.86	nr	87.05
holes for pipes and the like	–	0.19	4.21	–	–	nr	4.21
Concrete interlocking tiles; Marley Eternit							
Ludlow Major granule finish tiles or other equal;							
75 mm lap; on 25 mm × 38 mm battens and type							
1F reinforced underlay							
Tiles 420 mm × 330 mm (Prime Cost £ per 1000)	968.90	–	–	–	968.90	1000	968.90
roof coverings	9.97	0.32	7.09	–	14.15	m²	21.24
extra over coverings for;							
fixing every tile	–	0.02	0.44	–	0.49	m²	0.93
eaves; eaves filler	–	0.04	0.89	–	0.43	m	1.32
verges; 150 mm wide asbestos free strip							
undercloak	–	0.21	4.66	–	2.19	m	6.85
dry verge system; extruded white PVC	–	0.14	3.10	–	13.23	m	16.33
segmental ridge cap to dry verge	–	0.02	0.44	–	4.37	m	4.81
valley trough tiles; cutting both sides	–	0.51	11.30	–	29.39	m	40.69
segmental ridge tiles	–	0.46	10.19	–	9.81	m	20.00
segmental hip tiles; cutting both sides	–	0.60	13.30	–	12.95	m	26.25
dry ridge tiles; segmental including batten							
sections; unions and filler pieces	–	0.28	6.20	–	21.24	m	27.44
segmental mono-ridge tiles	–	0.46	10.19	–	20.36	m	30.55
gas ridge terminal	–	0.46	10.19	–	76.86	nr	87.05
holes for pipes and the like	–	0.19	4.21	–	–	nr	4.21
Concrete interlocking tiles; Marley Eternit							
Mendip granule finish double pantiles or other							
equal; 75 mm lap; on 22 mm × 38 mm battens and							
type 1F reinforced underlay							
Tiles 420 mm × 330 mm (Prime Cost £ per 1000)	908.30	–	–	–	908.30	1000	908.30
roof coverings (PC £ per 1000)	9.25	0.32	7.09	–	13.41	m²	20.50
extra over coverings for;							
fixing every tile	–	0.02	0.44	–	0.49	m²	0.93
eaves; eaves filler	–	0.02	0.44	–	12.13	m	12.57
verges; 150 mm wide asbestos free strip							
undercloak	–	0.21	4.66	–	2.19	m	6.85
dry verge system; extruded white PVC	–	0.14	3.10	–	13.23	m	16.33
segmental ridge cap to dry verge	–	0.02	0.44	–	4.37	m	4.81
valley trough tiles; cutting both sides	–	0.51	11.30	–	29.25	m	40.55
segmental ridge tiles	–	0.51	11.30	–	15.05	m	26.35
segmental hip tiles; cutting both sides	–	0.65	14.40	–	17.99	m	32.39
dry ridge tiles; segmental including batten							
sections; unions and filler pieces	–	0.28	6.20	–	21.24	m	27.44
segmental mono-ridge tiles	–	0.46	10.19	–	22.98	m	33.17
gas ridge terminal	–	0.46	10.19	–	76.86	nr	87.05
holes for pipes and the like	–	0.19	4.21	–	–	nr	4.21

18 TILE AND SLATE ROOF AND WALL COVERINGS

Item	PC £	Labour hours	Labour £	Plant £	Material £	Unit	Total rate £
18.03 INTERLOCKING TILING – cont							
Concrete interlocking tiles; Marley Eternit Modern smooth finish tiles or other equal; 75 mm lap; on 25 mm × 38 mm battens and type 1F reinforced underlay							
Tiles 420 mm × 220 mm (Prime Cost £ per 1000)	807.40	–	–	–	807.40	1000	**807.40**
roof coverings	8.39	0.32	7.09	–	12.86	m²	**19.95**
extra over coverings for;							
fixing every tile	–	0.02	0.44	–	0.49	m²	**0.93**
verges; 150 wide asbestos free strip undercloak	–	0.21	4.66	–	2.19	m	**6.85**
dry verge system; extruded white PVC	–	0.19	4.21	–	13.23	m	**17.44**
Modern ridge cap to dry verge	–	0.02	0.44	–	4.37	m	**4.81**
valley trough tiles; cutting both sides	–	0.51	11.30	–	29.04	m	**40.34**
Modern ridge tiles	–	0.46	10.19	–	12.24	m	**22.43**
Modern hip tiles; cutting both sides	–	0.60	13.30	–	14.85	m	**28.15**
dry ridge tiles; Modern; including batten sections; unions and filler pieces	–	0.28	6.20	–	23.66	m	**29.86**
Modern mono-ridge tiles	–	0.46	10.19	–	20.36	m	**30.55**
gas ridge terminal	–	0.46	10.19	–	76.86	nr	**87.05**
holes for pipes and the like	–	0.19	4.21	–	–	nr	**4.21**
Concrete interlocking tiles; Marley Eternit Ecologic Ludlow Major granule finish tiles or other equal; 75 mm lap; on 25 mm × 38 mm battens and type 1F reinforced underlay							
Tiles 420 mm × 330 mm (Prime Cost £ per 1000)	958.80	–	–	–	958.80	1000	**958.80**
roof coverings	9.40	0.32	7.09	–	14.05	m²	**21.14**
extra over coverings for:							
fixing every tile	–	0.02	0.44	–	0.49	m²	**0.93**
eaves; eaves filler	–	0.04	0.89	–	0.43	m	**1.32**
verges; 150 mm wide asbestos free strip undercloak	–	0.21	4.66	–	2.19	m	**6.85**
dry verge system; extruded white PVC	–	0.14	3.10	–	13.23	m	**16.33**
segmental ridge cap to dry verge	–	0.02	0.44	–	4.37	m	**4.81**
valley trough tiles; cutting both sides	–	0.51	11.30	–	29.37	m	**40.67**
segmental ridge tiles	–	0.46	10.19	–	9.81	m	**20.00**
segmental hip tiles; cutting both sides	–	0.60	13.30	–	12.92	m	**26.22**
dry ridge tiles; segmental including batten sections; unions and filler pieces	–	0.28	6.20	–	21.24	m	**27.44**
segmental mono-ridge tiles	–	0.46	10.19	–	20.36	m	**30.55**
gas ridge terminal	–	0.46	10.19	–	76.86	nr	**87.05**
holes for pipes and the like	–	0.19	4.21	–	–	nr	**4.21**
Concrete interlocking tiles; Marley Eternit Wessex smooth finish tiles or other equal; 75 mm lap; on 25 mm × 38 mm battens and type 1F reinforced underlay							
Tiles 413 mm × 330 mm (Prime Cost £ per 1000)	1413.00	–	–	–	1413.00	1000	**1413.00**
roof coverings	13.99	0.32	7.09	–	19.34	m²	**26.43**

18 TILE AND SLATE ROOF AND WALL COVERINGS

Item	PC £	Labour hours	Labour £	Plant £	Material £	Unit	Total rate £
extra over coverings for;							
fixing every tile	–	0.02	0.44	–	0.49	m²	**0.93**
verges; 150 mm wide asbestos free strip							
undercloak	–	0.21	4.66	–	2.19	m	**6.85**
dry verge system; extruded white PVC	–	0.19	4.21	–	13.23	m	**17.44**
Modern ridge cap to dry verge	–	0.02	0.44	–	4.37	m	**4.81**
valley trough tiles; cutting both sides	–	0.51	11.30	–	30.34	m	**41.64**
Modern ridge tiles	–	0.46	10.19	–	12.24	m	**22.43**
Modern hip tiles; cutting both sides	–	0.60	13.30	–	16.82	m	**30.12**
dry ridge tiles; Modern; including batten sections;							
unions and filler pieces	–	0.28	6.20	–	23.63	m	**29.83**
Modern mono-ridge tiles	–	0.46	10.19	–	20.36	m	**30.55**
gas ridge terminal	–	0.46	10.19	–	76.86	nr	**87.05**
holes for pipes and the like	–	0.19	4.21	–	–	nr	**4.21**
Concrete interlocking tiles; Fenland Pantile smooth finish pantiles or other equal; 75 mm lap; on 25 mm × 38 mm battens and type 1F reinforced underlay							
Tiles 381 mm × 229 mm (Prime Cost £ per 1000)	684.00	–	–	–	684.00	1000	**684.00**
roof coverings	12.78	0.42	9.30	–	17.93	m²	**27.23**
extra over coverings for;							
fixing every tile	–	0.04	0.89	–	0.21	m²	**1.10**
eaves; eaves filler	–	0.04	0.89	–	1.51	m	**2.40**
verges; extra single undercloak course of plain							
tiles	–	0.28	6.20	–	8.99	m	**15.19**
valley trough tiles; cutting both sides	–	0.56	12.41	–	41.38	m	**53.79**
universal ridge tiles	–	0.46	10.19	–	18.07	m	**28.26**
universal hip tiles; cutting both sides	–	0.60	13.30	–	21.76	m	**35.06**
universal gas flue ridge tile	–	0.46	10.19	–	92.72	nr	**102.91**
universal ridge vent tile with 110 mm dia. adaptor	–	0.50	11.08	–	119.43	nr	**130.51**
holes for pipes and the like	–	0.19	4.21	–	–	nr	**4.21**
Concrete interlocking tiles; Redland Renown granule finish tiles or other equal and approved; 418 mm × 330 mm; to 75 mm lap; on 25 mm × 38 mm battens and type 1F reinforced underlay							
Tiles 418 mm × 330 mm (Prime Cost £ per 1000)	856.90	–	–	–	856.90	1000	**856.90**
roof coverings	8.31	0.32	7.09	–	13.30	m²	**20.39**
extra over coverings for;							
fixing every tile	–	0.02	0.44	–	0.25	m²	**0.69**
verges; extra single undercloak course of plain							
tiles	–	0.23	5.10	–	4.96	m	**10.06**
cloaked verge system	–	0.14	3.10	–	11.38	m	**14.48**
valley trough tiles; cutting both sides	–	0.51	11.30	–	40.46	m	**51.76**
universal ridge tiles	–	0.46	10.19	–	18.07	m	**28.26**
universal hip tiles; cutting both sides	–	0.60	13.30	–	20.85	m	**34.15**
dry ridge system; universal ridge tiles	–	0.23	5.10	–	47.29	m	**52.39**

18 TILE AND SLATE ROOF AND WALL COVERINGS

Item	PC £	Labour hours	Labour £	Plant £	Material £	Unit	Total rate £
18.03 INTERLOCKING TILING – cont							
Concrete interlocking tiles – cont							
universal half round mono-pitch ridge tiles	–	0.51	11.30	–	38.72	m	**50.02**
universal gas flue ridge tile	–	0.46	10.19	–	92.72	nr	**102.91**
universal ridge vent tile with 110 mm dia. adaptor	–	0.46	10.19	–	119.43	nr	**129.62**
holes for pipes and the like	–	0.19	4.21	–	–	nr	**4.21**
Concrete interlocking tiles; Redland Regent granule finish bold roll tiles or other equal; 75 mm lap; on 25 mm × 38 mm battens and type 1F reinforced underlay							
Tiles 418 mm × 332 mm (Prime Cost £ per 1000)	909.10	–	–	–	909.10	1000	**909.10**
roof coverings	8.82	0.32	7.09	–	13.84	m²	**20.93**
extra over coverings for;							
fixing every tile	–	0.03	0.67	–	0.81	m²	**1.48**
eaves; eaves filler	–	0.04	0.89	–	1.23	m	**2.12**
verges; extra single undercloak course of plain tiles	–	0.23	5.10	–	4.04	m	**9.14**
cloaked verge system	–	0.14	3.10	–	11.22	m	**14.32**
valley trough tiles; cutting both sides	–	0.51	11.30	–	40.63	m	**51.93**
universal ridge tiles	–	0.46	10.19	–	18.07	m	**28.26**
universal hip tiles; cutting both sides	–	0.60	13.30	–	21.02	m	**34.32**
dry ridge system; universal ridge tiles	–	0.23	5.10	–	59.32	m	**64.42**
universal half round mono-pitch ridge tiles	–	0.51	11.30	–	38.72	m	**50.02**
universal gas flue ridge tile	–	0.46	10.19	–	92.72	nr	**102.91**
universal ridge vent tile with 110 mm dia. adaptor	–	0.46	10.19	–	119.43	nr	**129.62**
holes for pipes and the like	–	0.19	4.21	–	–	nr	**4.21**
Concrete interlocking slates; Redland Mini Stonewold concrete slates or other equal; 75 mm lap; on 25 mm × 38 mm battens and type 1F reinforced underlay							
Slates 418 mm × 334 mm (Prime Cost £ per 1000)	960.00	–	–	–	960.00	1000	**960.00**
roof coverings	10.27	0.32	7.09	–	15.53	m²	**22.62**
extra over coverings for;							
fixing every tile	–	0.02	0.44	–	2.96	m²	**3.40**
eaves; eaves filler	–	0.02	0.44	–	4.08	m	**4.52**
verges; extra single undercloak course of plain tiles	–	0.28	6.20	–	4.96	m	**11.16**
ambi-dry verge system	–	0.19	4.21	–	17.37	m	**21.58**
ambi-dry verge eave/ridge end piece	–	0.02	0.44	–	6.23	m	**6.67**
valley trough tiles; cutting both sides	–	0.51	11.30	–	40.64	m	**51.94**
universal angle ridge tiles	–	0.46	10.19	–	14.95	m	**25.14**
universal hip tiles; cutting both sides	–	0.60	13.30	–	31.49	m	**44.79**
dry ridge system; universal angle ridge tiles	–	0.23	5.10	–	23.19	m	**28.29**
universal mono-pitch angle ridge tiles	–	0.51	11.30	–	29.26	m	**40.56**
universal gas flue angle ridge tile	–	0.46	10.19	–	103.26	nr	**113.45**
universal angle ridge vent tile with 110 mm dia. adaptor	–	0.46	10.19	–	131.80	nr	**141.99**
holes for pipes and the like	–	0.19	4.21	–	–	nr	**4.21**

18 TILE AND SLATE ROOF AND WALL COVERINGS

Item	PC £	Labour hours	Labour £	Plant £	Material £	Unit	Total rate £
Concrete interlocking slates; Redland Richmond 10 Slates laid broken bonded, normally from right to left; smooth finish tiles or other equal; 75 mm lap; on 25 mm × 38 mm battens and type 1F reinforced underlay							
Slates 418 mm × 330 mm (Prime Cost £ per 1000)	1520.00	–	–	–	1520.00	1000	**1520.00**
roof coverings	16.28	0.32	7.09	–	22.08	m²	**29.17**
extra over coverings for;							
fixing every tile	–	0.02	0.44	–	0.92	m²	**1.36**
eaves; eaves filler	–	0.02	0.44	–	3.81	m	**4.25**
verges; extra single undercloak course of plain tiles	–	0.23	5.10	–	5.11	m	**10.21**
ambi-dry verge system	–	0.19	4.21	–	17.37	m	**21.58**
ambi-dry verge eave/ridge end piece	–	0.02	0.44	–	6.23	m	**6.67**
universal valley trough tiles; cutting both sides	–	0.56	12.41	–	48.36	m	**60.77**
universal hip tiles; cutting both sides	–	0.60	13.30	–	19.88	m	**33.18**
universal angle ridge tiles	–	0.46	10.19	–	14.95	m	**25.14**
dry ridge system; universal angle ridge tiles	–	0.23	5.10	–	23.19	m	**28.29**
universal mono-pitch angle ridge tiles	–	0.51	11.30	–	29.26	m	**40.56**
gas ridge terminal	–	0.46	10.19	–	101.91	nr	**112.10**
ridge vent with 110 mm dia. flexible adaptor	–	0.46	10.19	–	130.45	nr	**140.64**
holes for pipes and the like	–	0.19	4.21	–	–	nr	**4.21**
Concrete interlocking slates; Redland Stonewold II smooth finish tiles or other equal; 75 mm lap; on 25 mm × 38 mm battens and type 1F reinforced underlay							
Slates 430 mm × 380 mm (Prime Cost £ per 1000)	5099.80	–	–	–	5099.80	1000	**5099.80**
roof coverings	49.47	0.32	7.09	–	58.86	m²	**65.95**
extra over coverings for;							
fixing every tile	–	0.02	0.44	–	2.96	m²	**3.40**
eaves; eaves filler	–	0.02	0.44	–	4.07	m	**4.51**
verges; extra single undercloak course of plain tiles	–	0.28	6.20	–	5.28	m	**11.48**
ambi-dry verge system	–	0.19	4.21	–	17.37	m	**21.58**
ambi-dry verge eave/ridge end piece	–	0.02	0.44	–	6.23	m	**6.67**
valley trough tiles; cutting both sides	–	0.51	11.30	–	54.23	m	**65.53**
universal angle ridge tiles	–	0.46	10.19	–	14.95	m	**25.14**
universal hip tiles; cutting both sides	–	0.60	13.30	–	31.49	m	**44.79**
dry ridge system; universal angle ridge tiles	–	0.23	5.10	–	23.19	m	**28.29**
universal mono-pitch angle ridge tiles	–	0.51	11.30	–	29.26	m	**40.56**
universal gas flue angle ridge tile	–	0.46	10.19	–	103.26	nr	**113.45**
universal angle ridge vent tile with 110 mm dia. adaptor	–	0.46	10.19	–	131.80	nr	**141.99**
holes for pipes and the like	–	0.19	4.21	–	–	nr	**4.21**
Concrete interlocking slates; Redland Cambrian Slates made from 60% recycled Welsh slate or other equal; 50 mm lap; on 25 mm × 38 mm battens and type 1F reinforced underlay							
Slates 300 mm × 336 mm (Prime Cost £ per 1000)	2070.00	–	–	–	2070.00	1000	**2070.00**
roof coverings	32.91	0.32	7.09	–	43.45	m²	**50.54**

18 TILE AND SLATE ROOF AND WALL COVERINGS

Item	PC £	Labour hours	Labour £	Plant £	Material £	Unit	Total rate £
18.03 INTERLOCKING TILING – cont							
Concrete interlocking slates – cont							
extra over coverings for;							
fixing every tile	–	0.02	0.44	–	1.30	m²	**1.74**
eaves; eaves filler	–	0.02	0.44	–	4.52	m	**4.96**
verges; extra single undercloak course of plain tiles	–	0.23	5.10	–	13.14	m	**18.24**
ambi-dry verge system	–	0.19	4.21	–	17.37	m	**21.58**
ambi-dry verge eave/ridge end piece	–	0.02	0.44	–	6.23	m	**6.67**
universal valley trough tiles; cutting both sides	–	0.56	12.41	–	48.36	m	**60.77**
universal hip tiles; cutting both sides	–	0.60	13.30	–	21.34	m	**34.64**
universal angle ridge tiles	–	0.46	10.19	–	14.95	m	**25.14**
dry ridge system; universal angle ridge tiles	–	0.23	5.10	–	23.19	m	**28.29**
universal mono-pitch angle ridge tiles	–	0.51	11.30	–	29.26	m	**40.56**
gas ridge terminal	–	0.46	10.19	–	101.91	nr	**112.10**
ridge vent with 110 mm dia. flexible adaptor	–	0.46	10.19	–	130.45	nr	**140.64**
holes for pipes and the like	–	0.19	4.21	–	–	nr	**4.21**
Sundries and accessories							
Hip irons							
galvanized mild steel; fixing with screws	–	0.09	2.00	–	3.41	nr	**5.41**
Rytons Clip strip or other equal and approved; continuous soffit ventilator							
51 mm wide; plastic; code CS351	–	0.28	6.20	–	1.01	m	**7.21**
Rytons over fascia ventilator or other equal and approved; continuous eaves ventilator							
40 mm wide; plastic; code OFV890	–	0.09	2.00	–	1.41	m	**3.41**
Rytons roof ventilator or other equal and approved; to suit rafters at 600 mm centres							
250 mm deep × 43 mm high; plastic; code TV600	–	0.09	2.00	–	1.62	m	**3.62**
Rytons push and lock ventilators or other equal and approved; circular							
83 mm dia.; plastic; code PL235	–	0.04	0.92	–	0.36	nr	**1.28**
Fixing only							
lead soakers (supply cost not included)	–	0.07	1.85	–	–	nr	**1.85**
Pressure impregnated softwood counter battens; 25 mm × 50 mm							
450 mm centres	–	0.06	1.33	–	2.71	m²	**4.04**
600 mm centres	–	0.04	0.89	–	2.05	m²	**2.94**
Underlays							
Underlay; BS EN 13707 type 1B; bitumen felt weighing 14 kg/10 m²; 75 mm laps							
To sloping or vertical surfaces	0.69	0.02	0.44	–	1.12	m²	**1.56**
Underlay; BS EN 13707 type 1F; reinforced bitumen felt weighing 22.50 kg/10 m²; 75 mm laps							
To sloping or vertical surfaces	0.87	0.02	0.44	–	1.31	m²	**1.75**

18 TILE AND SLATE ROOF AND WALL COVERINGS

Item	PC £	Labour hours	Labour £	Plant £	Material £	Unit	Total rate £
Underlay; Visqueen HP vapour barrier or other equal; multilayer reinforce LDPE membrane with an aluminium core							
To sloping or vertical surfaces	0.52	0.02	0.44	–	1.02	m²	**1.46**
Underlay; Visqueen HP vapour barrier or other equal; multilayer reinforce LDPE membrane with an aluminium core							
To sloping or vertical surfaces	0.52	0.02	0.44	–	1.02	m²	**1.46**
Underlay; reinforced vapour permeable membrane; 75 mm laps							
To sloping or vertical surfaces	0.71	0.02	0.44	–	1.14	m²	**1.58**
Underlay; Klober Permo air breathable reinforced vapour permeable membrane; 75 mm laps							
To sloping or vertical surfaces	3.22	0.03	0.56	–	3.73	m²	**4.29**
UV Breather membranes; Web Dynamics Ltd							
Web UV 10 standard grade breather membrane draped between open rafters; 100 mm vertical laps, 150 mm horizontal laps, secured with galvanized nails	0.79	0.03	0.67	–	1.01	m²	**1.68**
Web UV 15 professional grade breather membrane draped between open rafters or on sarking board; 100 mm vertical laps, 150 mm horizontal laps secured with galvanized nails	0.79	0.03	0.67	–	1.01	m²	**1.68**
Web UV 25 heavy dutyl grade breather membrane draped between open rafters or on sarking board; 100 mm vertical laps, 150 mm horizontal laps secured with galvanized nails	0.82	0.04	0.77	–	1.04	m²	**1.81**
18.04 FIBRE CEMENT SLATING							
Fibre cement artificial slates; Eternit Garsdale/ E2000T or other equal to 75 mm lap; on 19 mm × 50 mm battens and type 1F reinforced underlay							
Coverings; 600 mm × 600 mm slates							
roof coverings	–	0.60	13.30	–	20.25	m²	**33.55**
wall coverings	–	0.74	16.40	–	21.83	m²	**38.23**
Coverings; 600 mm × 300 mm slates							
roof coverings	–	0.46	10.19	–	20.39	m²	**30.58**
wall coverings	–	0.60	13.30	–	20.39	m²	**33.69**
Extra over slate coverings for							
double course at eaves	–	0.23	5.10	–	4.87	m	**9.97**
verges; extra single undercloak course	–	0.31	6.87	–	0.95	m	**7.82**
open valleys; cutting both sides	–	0.19	4.21	–	3.81	m	**8.02**
stop end	–	0.09	2.00	–	11.34	nr	**13.34**
roll top ridge tiles	–	0.56	12.41	–	33.23	m	**45.64**
stop end	–	0.09	2.00	–	18.36	nr	**20.36**
mono-pitch ridge tiles	–	0.46	10.19	–	41.09	m	**51.28**
stop end	–	0.09	2.00	–	48.65	nr	**50.65**
duo-pitch ridge tiles	–	0.46	10.19	–	39.36	m	**49.55**
stop end	–	0.09	2.00	–	31.69	nr	**33.69**
half round hip tiles; cutting both sides	–	0.19	4.21	–	74.40	m	**78.61**

18 TILE AND SLATE ROOF AND WALL COVERINGS

Item	PC £	Labour hours	Labour £	Plant £	Material £	Unit	Total rate £
18.04 FIBRE CEMENT SLATING – cont							
Fibre cement artificial slates – cont							
Extra over slate coverings for – cont							
holes for pipes and the like	–	0.19	4.21	–	–	nr	**4.21**
18.05 NATURAL SLATING							
NOTE: The following items of slate roofing unless otherwise described, include for conventional fixing assuming normal exposure with appropriate nails and/or rivets or clips to pressure impregnated softwood battens fixed with galvanized nails; prices also include for all bedding and pointing at verges; beneath verge tiles etc.							
Natural slates; First quality Spanish blue grey; uniform size; to 75 mm lap; on 25 mm × 50 mm battens and type 1F reinforced underlay							
Coverings; 400 mm × 250 mm slates (Prime Cost £ per 1000)	950.00	–	–	–	978.50	1000	**978.50**
roof coverings	23.27	1.20	26.58	–	34.35	m²	**60.93**
wall coverings	23.27	1.06	23.48	–	34.35	m²	**57.83**
Coverings; 500 mm × 250 mm slates (Prime Cost £ per 1000)	1170.00	–	–	–	1205.10	1000	**1205.10**
roof coverings	21.88	1.20	26.58	–	30.88	m²	**57.46**
wall coverings	21.88	0.88	19.50	–	30.88	m²	**50.38**
Coverings; 600 mm × 300 mm slates (Prime Cost £ per 1000)	2100.00	–	–	–	2163.00	1000	**2163.00**
roof coverings	26.46	1.20	26.58	–	34.57	m²	**61.15**
wall coverings	26.46	0.69	15.29	–	34.57	m²	**49.86**
Extra over coverings for							
double course at eaves	–	0.28	6.20	–	8.84	m	**15.04**
verges; extra single undercloak course	–	0.39	8.64	–	4.65	m	**13.29**
open valleys; cutting both sides	–	0.20	4.43	–	18.17	m	**22.60**
blue/black glass reinforced concrete 152 mm half round ridge tiles	–	0.46	10.19	–	14.14	m	**24.33**
blue/black glass reinforced concrete 125 mm × 125 mm plain angle ridge tiles	–	0.46	10.19	–	14.14	m	**24.33**
mitred hips; cutting both sides	–	0.20	4.43	–	18.17	m	**22.60**
blue/black glass reinforced concrete 152 mm half round hip tiles; cutting both sides	–	0.65	14.40	–	32.31	m	**46.71**
blue/black glass reinforced concrete 125 mm × 125 mm plain angle hip tiles; cutting both sides	–	0.65	14.40	–	32.30	m	**46.70**
holes for pipes and the like	–	0.19	4.21	–	–	nr	**4.21**

18 TILE AND SLATE ROOF AND WALL COVERINGS

Item	PC £	Labour hours	Labour £	Plant £	Material £	Unit	Total rate £
Natural slates; Welsh blue grey; 7 mm nominal thickness; uniform size; to 75 mm lap; on 25 mm × 50 mm battens and type 1F reinforced underlay							
Coverings; 400 mm × 250 mm slates (Prime Cost £ per 1000)	1575.00	–	–	–	1622.25	1000	**1622.25**
roof coverings	38.59	0.70	15.51	–	50.91	m²	**66.42**
wall coverings	38.59	1.00	22.16	–	50.91	m²	**73.07**
Coverings; 500 mm × 250 mm slate (Prime Cost £ per 1000)	4464.00	–	–	–	4597.92	1000	**4597.92**
roof coverings	83.48	0.60	13.30	–	97.50	m²	**110.80**
wall coverings	83.48	0.80	17.73	–	97.50	m²	**115.23**
Coverings; 500 mm × 300 mm slates (Prime Cost £ per 1000)	5148.00	–	–	–	5302.44	1000	**5302.44**
roof coverings	80.31	0.60	13.30	–	94.07	m²	**107.37**
wall coverings	80.31	0.75	16.61	–	94.07	m²	**110.68**
Coverings; 600 mm × 300 mm slates (Prime Cost £ per 1000)	9387.00	–	–	–	9668.61	1000	**9668.61**
roof coverings	118.28	0.50	11.08	–	133.87	m²	**144.95**
wall coverings	118.28	0.65	14.40	–	133.87	m²	**148.27**
Extra over coverings for							
double course at eaves	–	0.25	5.54	–	34.84	m	**40.38**
verges; extra single undercloak course	–	0.35	7.76	–	20.40	m	**28.16**
open valleys; cutting both sides	–	0.20	4.43	–	81.22	m	**85.65**
blue/black glazed ware 152 mm half round ridge tiles	–	0.46	10.19	–	17.89	m	**28.08**
blue/black glazed ware 125 mm × 125 mm plain angle ridge tiles	–	0.46	10.19	–	20.66	m	**30.85**
mitred hips; cutting both sides	–	0.20	4.43	–	81.22	m	**85.65**
blue/black glazed ware 152 mm half round hip tiles; cutting both sides	–	0.65	14.40	–	99.11	m	**113.51**
blue/black glazed ware 125 mm × 125 mm plain angle hip tiles; cutting both sides	–	0.65	14.40	–	101.88	m	**116.28**
holes for pipes and the like	–	0.19	4.21	–	–	nr	**4.21**
Natural slates; Westmoreland green; random lengths; 457 mm–229 mm proportionate widths to 75 mm lap; in diminishing courses; on 25 mm × 50 mm battens and type 1F underlay							
Coverings (Prime Cost £ per tonne; approximately 17.5 m²)	2850.00	–	–	–	2935.50	tonne	**2935.50**
roof coverings	162.45	1.00	22.16	–	183.76	m²	**205.92**
wall coverings	162.45	1.30	28.80	–	183.76	m²	**212.56**
Extra over coverings for							
double course at eaves	–	0.60	13.30	–	32.30	m	**45.60**
verges; extra single undercloak course slates 152 mm wide	–	0.67	14.84	–	28.07	m	**42.91**
holes for pipes and the like	–	0.25	5.54	–	–	nr	**5.54**

18 TILE AND SLATE ROOF AND WALL COVERINGS

Item	PC £	Labour hours	Labour £	Plant £	Material £	Unit	Total rate £
18.06 NATURAL OR ARTIFICIAL STONE SLATING							
Reconstructed stone slates; Hardrow Slates or similar; standard colours; 75 mm lap; on 25 mm × 50 mm battens and type 1F reinforced underlay							
Coverings; 457 mm × 305 mm slates (Prime Cost £ per 1000)	2241.00	–	–	–	2308.23	1000	**2308.23**
roof coverings	38.55	0.74	16.40	–	49.96	m²	**66.36**
wall coverings	38.55	0.93	20.60	–	49.96	m²	**70.56**
Coverings; 457 mm × 457 mm slates (Prime Cost £ per 1000)	2241.00	–	–	–	2308.23	1000	**2308.23**
roof coverings	25.77	0.60	13.30	–	35.83	m²	**49.13**
wall coverings	25.77	0.79	17.50	–	35.83	m²	**53.33**
Extra over 457 mm × 305 mm coverings for							
double course at eaves	–	0.28	6.20	–	9.35	m	**15.55**
verges; pointed	–	0.39	8.64	–	–3.02	m	**5.62**
open valleys; cutting both sides	–	0.20	4.43	–	24.24	m	**28.67**
ridge tiles	–	0.46	10.19	–	49.98	m	**60.17**
hip tiles; cutting both sides	–	0.65	14.40	–	43.31	m	**57.71**
holes for pipes and the like	–	0.19	4.21	–	–	nr	**4.21**
Reconstructed stone slates; Bradstone Cotswold style or similar; random lengths 550 mm– 300 mm; proportional widths; to 80 mm lap; in diminishing courses; on 25 mm × 50 mm battens and type 1F reinforced underlay							
Roof coverings							
All-in rate inclusive of eaves and verges, (Prime Cost £ per m²)	34.78	0.97	21.49	–	45.88	m²	**67.37**
Extra over coverings for							
open valleys/mitred hips; cutting both sides	–	0.42	9.30	–	16.92	m²	**26.22**
ridge tiles	–	0.61	13.51	–	22.87	m	**36.38**
hip tiles; cutting both sides	–	0.97	21.49	–	38.30	m	**59.79**
holes for pipes and the like	–	0.28	6.20	–	–	nr	**6.20**
Reconstructed stone slates; Bradstone Moordale style or similar; random lengths 550 mm– 450 mm; proportional widths; to 80 mm lap; in diminishing course; on 25 mm × 50 mm battens and type 1F reinforced underlay							
Roof coverings							
All-in rate inclusive of eaves and verges, (Prime Cost £ per m²)	32.72	0.97	21.49	–	43.65	m²	**65.14**
Extra over coverings for							
open valleys/mitred hips; cutting both sides	–	0.42	9.30	–	15.92	m²	**25.22**
ridge tiles	–	0.61	13.51	–	22.87	m	**36.38**
holes for pipes and the like	–	0.28	6.20	–	–	nr	**6.20**

18 TILE AND SLATE ROOF AND WALL COVERINGS

Item	PC £	Labour hours	Labour £	Plant £	Material £	Unit	Total rate £
18.07 TIMBER SHINGLES							
Red Cedar sawn shingles preservative treated; uniform length 400 mm; to 125 mm gauge; on 25 mm × 38 mm battens and type 1F reinforced underlay							
Shingles; uniform length 400 mm (Prime Cost £ per bundle)	53.44	–	–	–	55.04	bundle	**55.04**
roof coverings; 125 mm gauge (2.28 m² per bundle)	23.46	0.97	21.49	–	35.27	m²	**56.76**
wall coverings; 190 mm gauge (3.47 m²/bundle)	15.39	0.74	16.40	–	23.63	m²	**40.03**
Extra over for							
double course at eaves	–	0.19	4.21	–	3.28	m	**7.49**
open valleys; cutting both sides	–	0.19	4.21	–	6.21	m	**10.42**
preformed ridge capping	–	0.28	6.20	–	4.58	m	**10.78**
preformed hip capping; cutting both sides	–	0.46	10.19	–	10.78	m	**20.97**
double starter course to cappings	–	0.09	2.00	–	4.36	m	**6.36**
holes for pipes and the like	–	0.14	3.10	–	–	nr	**3.10**

19 WATERPROOFING

Item	PC £	Labour hours	Labour £	Plant £	Material £	Unit	Total rate £
19.01 MASTIC ASPHALT ROOFING							
Mastic asphalt to BS 6925 Type R988							
20 mm thick two coat coverings; felt isolating membrane; to concrete (or timber) base; flat or to falls or slopes not exceeding 10° from horizontal							
over 300 mm wide	–	–	–	–	–	m²	25.31
225 mm–300 mm wide	–	–	–	–	–	m²	39.08
150 mm–225 mm wide	–	–	–	–	–	m²	45.65
not exceeding 150 mm wide	–	–	–	–	–	m²	58.76
Add to the above for covering with							
10 mm thick limestone chippings in hot bitumen	–	–	–	–	–	m²	4.13
coverings with solar reflective paint	–	–	–	–	–	m²	4.64
300 mm × 300 mm × 8 mm g.r.p. tiles in hot bitumen	–	–	–	–	–	m²	69.98
Cutting to line; jointing to old asphalt	–	–	–	–	–	m	8.01
13 mm thick two coat skirtings to brickwork base							
not exceeding 150 mm girth	–	–	–	–	–	m	17.25
150 mm–225 mm girth	–	–	–	–	–	m	19.80
225 mm–300 mm girth	–	–	–	–	–	m	24.23
13 mm thick three coat skirtings; expanded metal lathing reinforcement nailed to timber base							
not exceeding 150 mm girth	–	–	–	–	–	m	28.99
150 mm–225 mm girth	–	–	–	–	–	m	34.53
225 mm–300 mm girth	–	–	–	–	–	m	40.42
13 mm thick two coat fascias to concrete base							
not exceeding 150 mm girth	–	–	–	–	–	m	17.25
150 mm–225 mm girth	–	–	–	–	–	m	19.80
20 mm thick two coat linings to channels to concrete base							
not exceeding 150 mm girth	–	–	–	–	–	m	37.92
150 mm–225 mm girth	–	–	–	–	–	m	43.12
225 mm–300 mm girth	–	–	–	–	–	m	44.37
20 mm thick two coat lining to cesspools							
250 mm × 150 mm × 150 mm deep	–	–	–	–	–	nr	37.16
Collars around pipes, standards and like members	–	–	–	–	–	nr	26.58
Accessories							
Eaves trim; extruded aluminium alloy; working asphalt into trim							
Alutrim; type A roof edging or other equal and approved	–	–	–	–	–	m	15.84
extra; angle	–	–	–	–	–	nr	8.87
Roof screed ventilator – aluminium alloy							
Extr-aqua-vent or other equal and approved; set on screed over and including dished sinking; working collar around ventilator	–	–	–	–	–	nr	30.70

19 WATERPROOFING

Item	PC £	Labour hours	Labour £	Plant £	Material £	Unit	Total rate £
Bituminous lightweight insulating roof screeds							
Bit-Ag or similar roof screed or other equal and approved; to falls or cross-falls; bitumen felt vapour barrier; over 300 mm wide							
75 mm (average) thick	–	–	–	–	–	m²	**55.83**
100 mm (average) thick	–	–	–	–	–	m²	**70.75**
19.02 APPLIED LIQUID APPLIED TANKING							
Tanking and damp-proofing							
Synthaprufe or other equal and approved; blinding with sand; horizontal on slabs							
two coats	–	0.19	3.14	–	3.81	m²	**6.95**
three coats	–	0.26	4.30	–	5.61	m²	**9.91**
One coat Vandex Super 0.75 kg/m² slurry or other equal and approved; one consolidating coat of Vandex BB75 1 kg/m² slurry or other equal and approved; horizontal on beds							
over 225 mm wide	–	0.32	5.28	–	8.65	m²	**13.93**
Intergritank; Methacrylate resin based structural waterproffing membrane; in two separate colour coded coats; minimumm 2 mm overalll dry film finish; on a primed substrate. Typical project size between 250 m² and 10,000 m², over 250 mm wide							
100 m²–499 m²	–	–	–	–	–	m²	**58.94**
500 m²–1999 m²	–	–	–	–	–	m²	**48.65**
over 2000 m²	–	–	–	–	–	m²	**39.29**
19.03 MASTIC ASPHALT TANKING AND DAMP-PROOF MEMBRANES							
Mastic asphalt to BS 6925 Type T 1097							
13 mm thick one coat coverings to concrete base; flat; subsequently covered							
over 300 mm wide	–	–	–	–	–	m²	**19.01**
225 mm–300 mm wide	–	–	–	–	–	m²	**54.60**
150 mm–225 mm wide	–	–	–	–	–	m²	**59.89**
not exceeding 150 mm wide	–	–	–	–	–	m²	**74.81**
20 mm thick two coat coverings to concrete base; flat; subsequently covered							
over 300 mm wide	–	–	–	–	–	m²	**23.92**
225 mm–300 mm wide	–	–	–	–	–	m²	**49.29**
150 mm–225 mm wide	–	–	–	–	–	m²	**68.97**
not exceeding 150 mm wide	–	–	–	–	–	m²	**80.59**
30 mm thick three coat coverings to concrete base; flat; subsequently covered							
over 300 mm wide	–	–	–	–	–	m²	**38.40**
225 mm–300 mm wide	–	–	–	–	–	m²	**79.14**
150 mm–225 mm wide	–	–	–	–	–	m²	**85.87**
not exceeding 150 mm wide	–	–	–	–	–	m²	**104.62**

19 WATERPROOFING

Item	PC £	Labour hours	Labour £	Plant £	Material £	Unit	Total rate £
19.03 MASTIC ASPHALT TANKING AND DAMP-PROOF MEMBRANES – cont							
Mastic asphalt to BS 6925 Type T 1097 – cont							
13 mm thick two coat coverings to brickwork base; vertical; subsequently covered							
over 300 mm wide	–	–	–	–	–	m²	52.83
225 mm–300 mm wide	–	–	–	–	–	m²	75.94
150 mm–225 mm wide	–	–	–	–	–	m²	82.01
not exceeding 150 mm wide	–	–	–	–	–	m²	107.18
20 mm thick three coat coverings to brickwork base; vertical; subsequently covered							
over 300 mm wide	–	–	–	–	–	m²	85.47
225 mm–300 mm wide	–	–	–	–	–	m²	102.32
150 mm–225 mm wide	–	–	–	–	–	m²	112.28
not exceeding 150 mm wide	–	–	–	–	–	m²	145.56
Turning into groove 20 mm deep	–	–	–	–	–	m	1.03
Internal angle fillets; subsequently covered	–	–	–	–	–	m	5.95
19.04 FLEXIBLE SHEET TANKING AND DAMP-PROOF MEMBRANES							
Sheet tanking							
Preprufe pre-applied self-adhesive waterproofing membranes for use below concrete slabs or behind concrete walls. HPE film with a pressure sensitive adhesive and weather resistant protective coating							
Preprufe 300R heavy duty grade for use below slabs and on rafts							
over 300 mm wide; horizontal	–	0.10	1.65	–	19.19	m²	20.84
not exceeding 300 mm wide; horizontal	–	0.11	1.81	–	8.04	m	9.85
Preprufe 300R Plus heavy duty grade for use below slabs and on rafts designed to accept placing of heavy reinforcement							
over 300 mm wide; horizontal	–	0.11	1.81	–	20.35	m²	22.16
not exceeding 300 mm wide; horizontal	–	0.12	2.00	–	8.14	m	10.14
Preprufe 160R thinner grade for use blindside, zero property line applications against soil retention							
over 300 mm wide; horizontal	–	0.10	1.65	–	16.86	m²	18.51
not exceeding 300 mm wide; horizontal	–	0.11	1.81	–	7.07	m	8.88
Preprufe 160R Plus thinner grade for use blindside, zero property line applications against soil retention. Vertical use ony							
over 300 mm wide; horizontal	–	0.11	1.81	–	17.89	m²	19.70
not exceeding 300 mm wide; horizontal	–	0.12	1.99	–	7.16	m	9.15
Preprufe 800PA reinforced cross laminated HDPE film for use below ground car parks, basements, underground reservoirs and tanks							
over 300 mm wide; horizontal	–	0.15	2.48	–	10.85	m²	13.33
not exceeding 300 mm wide; horizontal	–	0.17	2.73	–	4.54	m	7.27

19 WATERPROOFING

Item	PC £	Labour hours	Labour £	Plant £	Material £	Unit	Total rate £
Visqueen self-adhesive damp-proof membrane							
over 300 mm wide; horizontal	–	–	–	–	–	m²	**8.06**
not exceeding 300 mm wide; horizontal	–	–	–	–	–	m	**3.11**
Tanking primer for self-adhesive dpm							
over 300 mm wide; horizontal	–	–	–	–	–	m²	**5.58**
not exceeding 300 mm wide; horizontal	–	–	–	–	–	m	**2.54**
Bituthene sheeting or other equal and approved; lapped joints; horizontal on slabs							
8000 grade	–	0.10	1.65	–	24.58	m²	**26.23**
5000HD heavy duty grade	–	0.12	1.99	–	23.77	m²	**25.76**
Bituthene sheeting or other equal and approved; lapped joints; dressed up vertical face of concrete							
8000 grade	–	0.17	2.81	–	24.58	m²	**27.39**
RIW Structureseal tanking and damp-proof membrane; or other equal and approved							
over 300 mm wide; horizontal	–	–	–	–	–	m²	**8.53**
Structureseal Fillet							
40 mm × 40 mm	–	–	–	–	–	m	**6.45**
Ruberoid Plasfrufe 2000SA self-adhesive damp-proof membrane							
over 300 mm wide; horizontal	–	–	–	–	–	m²	**15.33**
not exceeding 300 mm wide; horizontal	–	–	–	–	–	m	**5.87**
Extra for 50 mm thick sand blinding	–	–	–	–	–	m²	**2.70**
Servipak protection board or other equal; butt jointed; taped joints; to horizontal surfaces;							
3 mm thick	–	0.14	2.32	–	12.59	m²	**14.91**
6 mm thick	–	0.14	2.32	–	14.38	m²	**16.70**
12 mm thick	–	0.19	3.14	–	24.62	m²	**27.76**
Servipak protection board or other equal; butt jointed; taped joints; to vertical surfaces							
3 mm thick	–	0.19	3.14	–	12.59	m²	**15.73**
6 mm thick	–	0.19	3.14	–	14.38	m²	**17.52**
12 mm thick	–	0.23	3.80	–	24.62	m²	**28.42**
Expandite Famflex hot bitumen bonded waterproof tanking or other equal and approved; 150 mm laps							
horizontal; over 300 mm wide	–	0.37	6.12	–	12.80	m²	**18.92**
vertical; over 300 mm wide	–	0.60	9.92	–	12.80	m²	**22.72**

19.05 SPECIALIST WATERPROOF RENDERING

Sika waterproof rendering or other equal; steel trowelled

Item	PC £	Labour hours	Labour £	Plant £	Material £	Unit	Total rate £
20 mm work to walls; three coat; to concrete base							
over 300 mm wide	–	–	–	–	–	m²	**37.84**
not exceeding 300 mm wide	–	–	–	–	–	m²	**57.32**
25 mm work to walls; three coat; to concrete base							
over 300 mm wide	–	–	–	–	–	m²	**44.71**
not exceeding 300 mm wide	–	–	–	–	–	m²	**68.79**

19 WATERPROOFING

Item	PC £	Labour hours	Labour £	Plant £	Material £	Unit	Total rate £
19.05 SPECIALIST WATERPROOF RENDERING – cont							
Sika waterproof rendering or other equal – cont							
40 mm work to walls; four coat; to concrete base							
over 300 mm wide	–	–	–	–	–	m²	**65.93**
not exceeding 300 mm wide	–	–	–	–	–	m²	**103.19**
Sto External render only system; comprising glassfibre mesh reinforcement embedded in 10 mm Sto Levell Cote with Sto Armat Classic Basecoat Render and Stolit K 1.5 Decorative Topcoat Render (white)							
15 mm thick work to walls; two coats; to brickwork or blockwork base							
over 300 mm wide	–	–	–	–	–	m²	**58.51**
extra over for;							
bellcast bead	–	–	–	–	–	m	**5.57**
external angle with PVC mesh angle bead	–	–	–	–	–	m	**5.13**
internal angle with Sto Armor angle	–	–	–	–	–	m	**5.13**
render stop bead	–	–	–	–	–	m	**5.13**
K-Rend render or similar through-colour render system							
18 mm thick work to walls; two coats; to brickwork or blockwork base; first coat 8 mm standard base coat; second coat 10 mm K-rend silicone WP/FT							
over 300 mm wide	–	–	–	–	–	m²	**71.45**

20 PROPRIETARY LININGS AND PARTITIONS

Item	PC £	Labour hours	Labour £	Plant £	Material £	Unit	Total rate £
20.01 PARTITIONS AND WALLS							
Gyproc metal stud proprietary partitions or other equal; metal Gypframe C studs at 600 mm centres; floor and ceiling channels							
73 mm partition; 48 mm studs and channels; one layer of 12.5 mm Gyproc Wallboard each side; joints filled with joint filler and joint tape to receive direct decoration							
average height 2.00 m	–	0.63	15.24	–	18.18	m²	**33.42**
average height 3.00 m	–	0.58	14.22	–	16.70	m²	**30.92**
Rate per m (SMM7 measurement rule)							
height 2.10 m–2.40 m	–	1.50	36.58	–	43.62	m	**80.20**
height 2.40 m–2.70 m	–	1.69	41.21	–	47.05	m	**88.26**
height 2.70 m–3.00 m	–	1.75	42.67	–	50.10	m	**92.77**
78 mm partition; 48 mm studs and channels; one layer of 15 mm Gyproc Wallboard each side; joints filled with joint filler and joint tape to receive direct decoration							
average height 2.00 m	–	0.65	15.85	–	19.71	m²	**35.56**
average height 3.00 m	–	0.61	14.87	–	18.24	m²	**33.11**
Rate per m (SMM7 measurement rule)							
height 2.10 m–2.40 m	–	1.57	38.41	–	47.32	m	**85.73**
height 2.40 m–2.70 m	–	1.77	43.16	–	51.21	m	**94.37**
height 2.70 m–3.00 m	–	1.99	48.52	–	61.02	m	**109.54**
95 mm partition; 70 mm studs and channels; one layer of 12.5 mm Gyproc Wallboard each side; joints filled with joint filler and joint tape to receive direct decoration							
average height 2.00 m	–	0.63	15.24	–	18.81	m²	**34.05**
average height 3.00 m	–	0.58	14.22	–	17.44	m²	**31.66**
Rate per m (SMM7 measurement rule)							
height 2.10 m–2.40 m	–	1.50	36.58	–	45.13	m	**81.71**
height 2.40 m–2.70 m	–	1.69	41.21	–	48.77	m	**89.98**
height 2.70 m–3.00 m	–	1.75	42.67	–	52.31	m	**94.98**
height 3.00 m–3.30 m	–	2.00	48.77	–	55.85	m	**104.62**
100 mm partition; 70 mm studs and channels; one layer of 15 mm Gyproc Wallboard each side; joints filled with joint filler and joint tape to receive direct decoration							
average height 2.00 m	–	0.65	15.85	–	20.32	m²	**36.17**
average height 3.00 m	–	0.61	14.87	–	21.00	m²	**35.87**
Rate per m (SMM7 measurement rule)							
height 2.10 m–2.40 m	–	1.57	38.41	–	48.83	m	**87.24**
height 2.40 m–2.70 m	–	1.77	43.16	–	52.93	m	**96.09**
height 2.70 m–3.00 m	–	1.99	48.52	–	63.01	m	**111.53**
height 3.00 m–3.30 m	–	2.10	51.21	–	66.67	m	**117.88**

20 PROPRIETARY LININGS AND PARTITIONS

Item	PC £	Labour hours	Labour £	Plant £	Material £	Unit	Total rate £
20.01 PARTITIONS AND WALLS – cont							
120 mm partition; 70 mm studs and channels; two layers of 12.5 mm Gyproc Wallboard each side; joints filled with joint filler and joint tape to receive direct decoration							
average height 2.00 m	–	0.73	17.80	–	26.09	m²	**43.89**
average height 3.00 m	–	0.68	16.67	–	22.10	m²	**38.77**
Rate per m (SMM7 measurement rule)							
height 2.10 m–2.40 m	–	1.75	42.67	–	62.62	m	**105.29**
height 2.40 m–2.70 m	–	1.99	48.52	–	68.44	m	**116.96**
height 2.70 m–3.00 m	–	2.05	49.99	–	74.17	m	**124.16**
height 3.00 m–3.30 m	–	2.50	60.97	–	79.90	m	**140.87**
130 mm partition; 70 mm studs and channels; two layers of 15 mm Gyproc Wallboard each side; joints filled with joint filler and joint tape to receive direct decoration							
average height 2.00 m	–	0.80	19.51	–	29.17	m²	**48.68**
average height 3.00 m	–	0.75	18.29	–	27.90	m²	**46.19**
Rate per m (SMM7 measurement rule)							
height 2.10 m–2.40 m	–	1.82	44.38	–	70.02	m	**114.40**
height 2.40 m–2.70 m	–	2.07	50.48	–	53.21	m	**103.69**
height 2.70 m–3.00 m	–	2.29	55.85	–	89.49	m	**145.34**
height 3.00 m–3.30 m	–	2.40	58.52	–	95.81	m	**154.33**
Gypwall metal stud proprietary partitions or other equal; metal GypWall Acoustic C studs at 600 mm centres; deep flange floor and ceiling channels							
95 mm partition; 70 mm studs and channels; one layer of 12.5 mm Gyproc Soundbloc each side; joints filled with joint filler and joint tape to receive direct decoration							
average height 2.00 m	–	0.63	15.24	–	28.28	m²	**43.52**
average height 3.00 m	–	–	–	–	–	m²	**-**
Rate per m (SMM7 measurement rule)							
height 2.10 m–2.40 m	–	1.50	36.58	–	67.88	m	**104.46**
height 2.40 m–2.70 m	–	1.69	41.21	–	73.75	m	**114.96**
height 2.70 m–3.00 m	–	1.75	42.67	–	80.89	m	**123.56**
100 mm partition; 70 mm studs and channels; one layer of 15 mm Gyproc Soundbloc each side; joints filled with joint filler and joint tape to receive direct decoration							
average height 2.00 m	–	0.63	15.24	–	30.36	m²	**45.60**
average height 3.00 m	–	0.58	14.22	–	29.05	m²	**43.27**
Rate per m (SMM7 measurement rule)							
height 2.10 m–2.40 m	–	1.57	38.41	–	72.87	m	**111.28**
height 2.40 m–2.70 m	–	1.77	43.16	–	79.38	m	**122.54**
height 2.70 m–3.00 m	–	1.99	48.52	–	85.41	m	**133.93**

20 PROPRIETARY LININGS AND PARTITIONS

Item	PC £	Labour hours	Labour £	Plant £	Material £	Unit	Total rate £
120 mm partition; 70 mm studs and channels; two layers of 12.5 mm Gyproc Soundbloc each side; joints filled with joint filler and joint tape to receive direct decoration							
average height 2.00 m	–	0.80	19.51	–	38.98	m²	**58.49**
average height 3.00 m	–	0.75	18.29	–	37.66	m²	**55.95**
Rate per m (SMM7 measurement rule)							
height 2.10 m–2.40 m	–	2.00	48.77	–	93.94	m	**142.71**
height 2.40 m–2.70 m	–	2.20	53.65	–	102.91	m	**156.56**
height 2.70 m–3.00 m	–	2.30	56.08	–	113.29	m	**169.37**
130 mm partition; 70 mm studs and channels; two layers of 15 mm Gyproc Soundbloc each side; joints filled with joint filler and joint tape to receive direct decoration							
average height 2.00 m	–	0.80	19.51	–	43.15	m²	**62.66**
average height 3.00 m	–	0.75	18.29	–	41.83	m²	**60.12**
Rate per m (SMM7 measurement rule)							
height 2.10 m–2.40 m	–	2.00	48.77	–	103.94	m	**152.71**
height 2.40 m–2.70 m	–	2.20	53.65	–	114.17	m	**167.82**
height 2.70 m–3.00 m	–	2.30	56.08	–	125.79	m	**181.87**
Gyproc metal stud proprietary partitions or other equal; metal Gypframe C studs at 600 mm centres; floor and ceiling channels							
95 mm partition; 70 mm studs and channels; one layer of 12.5 mm Gyproc Fireline each side; joints filled with joint filler and joint tape to receive direct decoration							
average height 2.00 m	–	0.80	19.51	–	36.16	m²	**55.67**
average height 3.00 m	–	0.75	18.29	–	34.84	m²	**53.13**
Rate per m (SMM7 measurement rule)							
height 2.10 m–2.40 m	–	1.50	36.58	–	64.50	m	**101.08**
height 2.40 m–2.70 m	–	1.70	41.46	–	69.96	m	**111.42**
height 2.70 m–3.00 m	–	1.75	42.67	–	74.94	m	**117.61**
100 mm partition; 70 mm studs and channels; one layer of 15 mm Gyproc Fireline each side; joints filled with joint filler and joint tape to receive direct decoration							
average height 2.00 m	–	0.80	19.51	–	39.93	m²	**59.44**
average height 3.00 m	–	0.75	18.29	–	38.61	m²	**56.90**
Rate per m (SMM7 measurement rule)							
height 2.10 m–2.40 m	–	1.50	36.58	–	69.02	m	**105.60**
height 2.40 m–2.70 m	–	1.70	41.46	–	75.04	m	**116.50**
height 2.70 m–3.00 m	–	1.75	42.67	–	80.59	m	**123.26**
120 mm partition; 70 mm studs and channels; two layers of 12.5 mm Gyproc Fireline each side; joints filled with joint filler and joint tape to receive direct decoration							
average height 2.00 m	–	1.10	26.82	–	36.27	m²	**63.09**
average height 3.00 m	–	1.15	28.05	–	33.65	m²	**61.70**

20 PROPRIETARY LININGS AND PARTITIONS

Item	PC £	Labour hours	Labour £	Plant £	Material £	Unit	Total rate £
20.01 PARTITIONS AND WALLS – cont							
Gyproc metal stud proprietary partitions or other equal – cont							
Rate per m (SMM7 measurement rule)							
height 2.10 m–2.40 m	–	2.00	48.77	–	87.05	m	**135.82**
height 2.40 m–2.70 m	–	2.20	53.65	–	95.33	m	**148.98**
height 2.70 m–3.00 m	–	2.30	56.08	–	100.96	m	**157.04**
130 mm partition; 70 mm studs and channels; two layers of 15 mm Gyproc Fireline each side; joints filled with joint filler and joint tape to receive direct decoration							
average height 2.00 m	–	1.10	26.82	–	40.04	m²	**66.86**
average height 3.00 m	–	1.15	28.05	–	37.41	m²	**65.46**
Rate per m (SMM7 measurement rule)							
height 2.10 m–2.40 m	–	2.00	48.77	–	96.08	m	**144.85**
height 2.40 m–2.70 m	–	2.20	53.65	–	105.48	m	**159.13**
height 2.70 m–3.00 m	–	2.30	56.08	–	114.07	m	**170.15**
Gypwall Rapid/db Plus metal stud housing partitioning system; or other equal; floor and ceiling channels plugged and screwed to concrete or nailed to timber							
75 mm partition; 43 mm studs at 600 mm centres; one layer of 15 mm SoundBloc Rapid each side; joints filled with joint filler and joint tape to receive direct decoration							
average height 2.00 m	–	0.63	14.37	–	19.15	m²	**33.52**
average height 3.00 m	–	0.83	19.15	–	24.31	m²	**43.46**
Rate per m (SMM7 measurement rule)							
height 2.10 m–2.40 m	–	1.50	34.47	–	45.97	m	**80.44**
height 2.10 m–2.40 m	–	1.75	40.22	–	50.81	m	**91.03**
height 2.40 m–2.70 m	–	2.00	45.97	–	56.06	m	**102.03**
Angles, corners and similar							
T-junctions	–	0.17	4.02	–	5.14	m	**9.16**
splayed corners	–	0.13	2.87	–	3.45	m	**6.32**
corners	–	0.10	2.30	–	1.76	m	**4.06**
fair ends	–	0.11	2.52	–	0.93	m	**3.45**
Cavity insulation to stud partitions pinned to board							
25 mm Isover APR							
average height 2.00 m	–	0.06	1.43	–	1.19	m	**2.62**
average height 3.00	–	0.07	1.61	–	1.19	m	**2.80**
Rate per m (SMM7 measurement rule)							
height 2.10 m–2.40 m	–	0.15	3.45	–	2.86	m	**6.31**
height 2.40 m–2.70 m	–	0.16	3.68	–	3.22	m	**6.90**
height 3.00 m–3.30 m	–	0.17	3.90	–	3.94	m	**7.84**

20 PROPRIETARY LININGS AND PARTITIONS

Item	PC £	Labour hours	Labour £	Plant £	Material £	Unit	Total rate £
50 mm Isover APR							
average height 2.00 m	–	0.06	1.43	–	1.96	m	3.39
average height 3.00 m	–	0.07	1.61	–	1.96	m	3.57
Rate per m (SMM7 measurement rule)							
height 2.10 m–2.40 m	–	0.15	3.45	–	4.71	m	8.16
height 2.40 m–2.70 m	–	0.16	3.68	–	5.29	m	8.97
height 3.00 m–3.30 m	–	0.17	3.90	–	6.47	m	10.37
Terrazzo faced partitions; polished on two faces							
Precast reinforced terrazzo faced WC partitions							
38 mm thick; over 300 mm wide	–	–	–	–	–	m²	329.86
50 mm thick; over 300 mm wide	–	–	–	–	–	m²	338.51
Wall post; once rebated							
64 mm × 102 mm	–	–	–	–	–	m	151.41
64 mm × 152 mm	–	–	–	–	–	m	164.39
Centre post; twice rebated							
64 mm × 102 mm	–	–	–	–	–	m	158.98
64 mm × 152 mm	–	–	–	–	–	m	175.20
Lintel; once rebated							
64 mm × 102 mm	–	–	–	–	–	m	164.39
Pair of brass topped plates or sockets cast into posts for fixings (not included)	–	–	–	–	–	nr	37.85
Brass indicator bolt lugs cast into posts for fixings (not included)	–	–	–	–	–	nr	16.22
Demountable Partitions							
Fire-resistant concertina partition							
Insulated panel and two-hour fire wall system for warehouses etc., comprising white polyester coated galvanized steel frame and 0.55 mm galvanized steel panels either side of rockwool infill							
100 mm thick wall: 31 Rw dB acoustic rating	–	–	–	–	–	m²	51.12
150 mm thick wall: 31 Rw dB acoustic rating	–	–	–	–	–	m²	54.95
intumescent mastic sealant; bedding frames at perimeter of metal fire walls	–	–	–	–	–	m	4.48
Room divider moveable wall							
Laminated both sides top hung movable acoustic panel wall with concealed uPVC vertical edge profiles, 1106 m × 3000 mm panels and two point panel support system							
105 mm thick wall: 47 Rw dB acoustic rating	–	–	–	–	–	m²	472.82
105 mm thick wall: 50 Rw dB acoustic rating	–	–	–	–	–	m²	511.16
105 mm thick wall: 53 Rw dB acoustic rating	–	–	–	–	–	m²	549.50

Prices for Measured Works

20 PROPRIETARY LININGS AND PARTITIONS

Item	PC £	Labour hours	Labour £	Plant £	Material £	Unit	Total rate £
20.02 LININGS TO WALLS							
SUPPLY ONLY SHEET LINING – ALTERNATIVE MATERIAL PRICES							
Fibreboard; 19 mm Decorative faced							
Ash	–	–	–	–	11.38	m²	**11.38**
Beech	–	–	–	–	10.88	m²	**10.88**
Oak	–	–	–	–	11.21	m²	**11.21**
Edgings; self-adhesive							
22 mm Ash	–	–	–	–	0.32	m	**0.32**
22 mm Beech	–	–	–	–	0.32	m	**0.32**
22 mm Oak	–	–	–	–	0.32	m	**0.32**
Chipboard Standard Grade							
12 mm	–	–	–	–	2.14	m²	**2.14**
18 mm	–	–	–	–	3.03	m²	**3.03**
22 mm	–	–	–	–	3.71	m²	**3.71**
25 mm	–	–	–	–	4.24	m²	**4.24**
Chipboard; melamine faced							
15 mm	–	–	–	–	3.15	m²	**3.15**
18 mm	–	–	–	–	3.44	m²	**3.44**
Medium density fibreboard; external quality							
6 mm	–	–	–	–	4.47	m²	**4.47**
9 mm	–	–	–	–	5.93	m²	**5.93**
19 mm	–	–	–	–	9.62	m²	**9.62**
25 mm	–	–	–	–	13.42	m²	**13.42**
Wallboard plank							
9.5 mm	–	–	–	–	3.27	m²	**3.27**
12.5 mm	–	–	–	–	3.27	m²	**3.27**
15 mm	–	–	–	–	3.97	m²	**3.97**
Moisture-resistant board							
12.5 mm	–	–	–	–	5.41	m²	**5.41**
15 mm	–	–	–	–	6.50	m²	**6.50**
Fireline board							
12.5 mm	–	–	–	–	4.22	m²	**4.22**
15 mm	–	–	–	–	5.08	m²	**5.08**
Wall linings; Gyproc GypLyner IWL walling system or other equal; comprising 48 mm wide metal I stud frame; 50 mm wide metal C stud floor and head channels; plugged and screwed to concrete							
62.5 mm lining; outer skin of 12.50 mm thick tapered edge wallboard one side; joints filled with joint filler and joint tape to receive direct decoration							
average height 2.00 m	–	0.53	12.18	–	9.24	m²	**21.42**
average height 3.00 m	–	0.53	12.18	–	9.08	m²	**21.26**
average height 4.00 m	–	0.53	12.18	–	9.05	m²	**21.23**

20 PROPRIETARY LININGS AND PARTITIONS

Item	PC £	Labour hours	Labour £	Plant £	Material £	Unit	Total rate £
Rate per m (SMM7 measurement rule)							
height 2.10 m–2.40 m	–	1.30	29.88	–	21.17	m	**51.05**
height 2.40 m–2.70 m	–	1.45	33.33	–	24.95	m	**58.28**
height 2.70 m–3.00 m	–	1.60	36.77	–	27.25	m	**64.02**
height 3.00 m–3.30 m	–	1.75	40.22	–	29.63	m	**69.85**
height 3.30 m–3.60 m	–	1.90	43.67	–	32.05	m	**75.72**
height 3.60 m–3.90 m	–	2.05	47.11	–	34.47	m	**81.58**
height 3.90 m–4.20 m	–	2.20	50.56	–	36.90	m	**87.46**
65 mm lining; outer skin of 15.00 mm thick tapered edge wallboard one side; joints filled with joint filler and joint tape to receive direct decoration							
average height 2.00 m	–	0.53	12.18	–	10.29	m²	**22.47**
average height 3.00 m	–	0.53	12.18	–	9.98	m²	**22.16**
average height 4.00 m	–	0.53	12.18	–	9.60	m²	**21.78**
Rate per m (SMM7 measurement rule)							
height 2.10 m–2.40 m	–	1.30	29.88	–	22.94	m	**52.82**
height 2.40 m–2.70 m	–	1.45	33.33	–	26.93	m	**60.26**
height 2.70 m–3.00 m	–	1.60	36.77	–	29.47	m	**66.24**
height 3.00 m–3.30 m	–	1.75	40.22	–	32.05	m	**72.27**
height 3.30 m–3.60 m	–	1.90	43.67	–	34.70	m	**78.37**
height 3.60 m–3.90 m	–	2.05	47.11	–	37.35	m	**84.46**
height 3.90 m–4.20 m	–	2.20	50.56	–	39.99	m	**90.55**
62.5 mm lining; outer skin of 12.50 mm thick tapered edge Fireline board one side; joints filled with joint filler and joint tape to receive direct decoration							
average height 2.00 m	–	0.53	12.18	–	10.43	m²	**22.61**
average height 3.00 m	–	0.53	12.18	–	9.96	m²	**22.14**
average height 4.00 m	–	0.53	12.18	–	9.75	m²	**21.93**
Rate per m (SMM7 measurement rule)							
height 2.10 m–2.40 m	–	1.30	30.01	–	23.59	m	**53.60**
height 2.40 m–2.70 m	–	1.45	33.46	–	27.67	m	**61.13**
height 2.70 m–3.00 m	–	1.60	36.92	–	30.27	m	**67.19**
height 3.00 m–3.30 m	–	1.75	40.37	–	32.95	m	**73.32**
height 3.30 m–3.60 m	–	1.90	43.81	–	35.67	m	**79.48**
height 3.60 m–3.90 m	–	2.05	47.26	–	38.40	m	**85.66**
height 3.90 m–4.20 m	–	2.20	50.71	–	41.13	m	**91.84**
65 mm partition; outer skin of 15.00 mm thick tapered edge Fireline board one side; joints filled with joint filler and joint tape to receive direct decoration							
average height 2.00 m	–	0.53	12.18	–	11.33	m²	**23.51**
average height 3.00 m	–	0.53	12.18	–	10.86	m²	**23.04**
average height 4.00 m	–	0.53	12.18	–	10.65	m²	**22.83**
Rate per m (SMM7 measurement rule)							
height 2.10 m–2.40 m	–	1.30	30.01	–	25.74	m	**55.75**
height 2.40 m–2.70 m	–	1.45	33.46	–	30.09	m	**63.55**
height 2.70 m–3.00 m	–	1.60	36.92	–	32.97	m	**69.89**
height 3.00 m–3.30 m	–	1.75	40.37	–	35.91	m	**76.28**
height 3.30 m–3.60 m	–	1.90	43.81	–	38.90	m	**82.71**
height 3.60 m–3.90 m	–	2.05	47.26	–	41.90	m	**89.16**
height 3.90 m–4.20 m	–	2.20	50.71	–	44.90	m	**95.61**

20 PROPRIETARY LININGS AND PARTITIONS

Item	PC £	Labour hours	Labour £	Plant £	Material £	Unit	Total rate £
20.02 LININGS TO WALLS – cont							
Wall linings – cont							
62.5 mm lining; outer skin of 12.50 mm thick tapered edge wallboard one side; filling cavity with 50 mm Isover insulation; wallboard joints filled with joint filler and joint tape to receive direct decoration							
average height 2.00 m	–	0.65	14.94	–	11.18	m²	**26.12**
average height 3.00 m	–	0.65	14.94	–	10.70	m²	**25.64**
average height 4.00 m	–	0.65	14.94	–	10.57	m²	**25.51**
Rate per m (SMM7 measurement rule)							
height 2.10 m–2.40 m	–	1.42	32.77	–	25.45	m	**58.22**
height 2.40 m–2.70 m	–	1.60	36.92	–	29.77	m	**66.69**
height 2.70 m–3.00 m	–	1.65	37.92	–	32.61	m	**70.53**
height 3.00 m–3.30 m	–	1.92	44.16	–	35.51	m	**79.67**
height 3.30 m–3.60 m	–	2.08	47.95	–	38.47	m	**86.42**
height 3.60 m–3.90 m	–	2.25	51.74	–	41.43	m	**93.17**
height 3.90 m–4.20 m	–	2.41	55.53	–	44.39	m	**99.92**
65 mm lining; outer skin of 15.00 mm thick tapered edge wallboard one side; filling cavity with 50 mm Isover slabs; wallboard joints filled with joint filler and joint tape to receive direct decoration							
average height 2.00 m	–	0.65	14.94	–	11.91	m²	**26.85**
average height 3.00 m	–	0.65	14.94	–	11.43	m²	**26.37**
average height 4.00 m	–	0.65	14.94	–	11.31	m²	**26.25**
Rate per m (SMM7 measurement rule)							
height 2.10 m–2.40 m	–	1.42	32.77	–	27.22	m	**59.99**
height 2.40 m–2.70 m	–	1.60	36.92	–	31.74	m	**68.66**
height 2.70 m–3.00 m	–	1.65	37.92	–	34.81	m	**72.73**
height 3.00 m–3.30 m	–	1.92	44.16	–	37.93	m	**82.09**
height 3.30 m–3.60 m	–	2.08	47.95	–	41.12	m	**89.07**
height 3.60 m–3.90 m	–	2.25	51.74	–	44.30	m	**96.04**
height 3.90 m–4.20 m	–	2.41	55.53	–	47.48	m	**103.01**
Gypsum plasterboard; fixing on dabs or with nails; joints filled with joint filler and joint tape to receive direct decoration; to softwood base (measured elsewhere)							
Plain grade tapered edge wallboard							
9.50 mm board to walls							
average height 2.00 m	–	0.16	3.68	–	4.75	m²	**8.43**
average height 3.00 m	–	0.16	3.56	–	4.75	m²	**8.31**
average height 4.00 m	–	0.15	3.45	–	4.75	m²	**8.20**
Rate per m (SMM7 measurement rule)							
wall height 2.40 m–2.70 m	–	0.43	10.02	–	12.83	m	**22.85**
wall height 2.70 m–3.00 m	–	0.46	10.71	–	14.26	m	**24.97**
wall height 3.00 m–3.30 m	–	0.50	11.63	–	15.69	m	**27.32**
wall height 3.30 m–3.60 m	–	0.59	13.70	–	17.19	m	**30.89**

20 PROPRIETARY LININGS AND PARTITIONS

Item	PC £	Labour hours	Labour £	Plant £	Material £	Unit	Total rate £
9.50 mm board to reveals and soffits of openings and recesses							
not exceeding 300 mm wide	–	0.14	3.29	–	2.35	m	**5.64**
300 mm–600 mm wide	–	0.22	5.13	–	3.54	m	**8.67**
9.50 mm board to faces of columns – 4 nr faces							
not exceeding 300 mm total girth	–	0.20	4.59	–	3.57	m	**8.16**
300 mm–600 mm total girth	–	0.20	4.59	–	4.81	m	**9.40**
600 mm–900 mm total girth	–	0.20	4.59	–	6.06	m	**10.65**
900 mm–1200 mm total girth	–	0.20	4.59	–	7.30	m	**11.89**
12.50 mm board to walls							
average height 2.00 m	–	0.16	3.68	–	6.65	m²	**10.33**
average height 3.00 m	–	0.16	3.56	–	6.65	m²	**10.21**
average height 4.00 m	–	0.15	3.45	–	6.65	m²	**10.10**
Rate per m (SMM7 measurement rule)							
wall height 2.40 m–2.70 m	–	0.45	10.34	–	17.98	m	**28.32**
wall height 2.70 m–3.00 m	–	0.50	11.49	–	19.98	m	**31.47**
wall height 3.00 m–3.30 m	–	0.55	12.64	–	22.09	m	**34.73**
wall height 3.30 m–3.60 m	–	0.60	13.79	–	24.06	m	**37.85**
12.50 mm board to reveals and soffits of openings and recesses							
not exceeding 300 mm wide	–	0.19	4.51	–	2.93	m	**7.44**
300 mm–600 mm wide	–	0.37	8.79	–	4.69	m	**13.48**
12.50 mm board to faces of columns – 4 nr faces							
not exceeding 300 mm total girth	–	0.30	6.89	–	4.09	m	**10.98**
300 mm–600 mm total girth	–	0.30	6.89	–	5.86	m	**12.75**
600 mm–900 mm total girth	–	0.30	6.89	–	7.62	m	**14.51**
900 mm–1200 mm total girth	–	0.30	6.89	–	9.39	m	**16.28**
Tapered edge wallboard TEN							
12.50 mm board to walls							
average height 2.00 m	–	0.16	3.68	–	4.91	m²	**8.59**
average height 3.00 m	–	0.16	3.56	–	4.91	m²	**8.47**
average height 4.00 m	–	0.15	3.45	–	4.91	m²	**8.36**
Rate per m (SMM7 measurement rule)							
wall height 2.40 m–2.70 m	–	0.47	10.94	–	13.29	m	**24.23**
wall height 2.70 m–3.00 m	–	0.56	13.08	–	14.76	m	**27.84**
wall height 3.00 m–3.30 m	–	0.65	15.22	–	16.24	m	**31.46**
wall height 3.30 m–3.60 m	–	0.78	18.28	–	17.79	m	**36.07**
12.50 mm board to reveals and soffits of openings and recesses							
not exceeding 300 mm wide	–	0.19	4.51	–	2.41	m	**6.92**
300 mm–600 mm wide	–	0.37	8.79	–	3.65	m	**12.44**
12.50 mm board to faces of columns – 4 nr faces							
not exceeding 300 mm total girth	–	0.25	6.10	–	3.69	m	**9.79**
300 mm–600 mm total girth	–	0.20	4.59	–	4.92	m	**9.51**
600 mm–900 mm total girth	–	0.20	4.59	–	6.17	m	**10.76**
900 mm–1200 mm total girth	–	0.20	4.59	–	7.44	m	**12.03**

20 PROPRIETARY LININGS AND PARTITIONS

Item	PC £	Labour hours	Labour £	Plant £	Material £	Unit	Total rate £
20.02 LININGS TO WALLS – cont							
Tapered edge plank							
19 mm plank to walls							
average height 2.00 m	–	0.16	3.68	–	8.18	m²	**11.86**
average height 3.00 m	–	0.16	3.56	–	8.18	m²	**11.74**
average height 4.00 m	–	0.15	3.45	–	8.18	m²	**11.63**
Rate per m (SMM7 measurement rule)							
wall height 2.40 m–2.70 m	–	1.02	24.29	–	22.10	m	**46.39**
wall height 2.70 m–3.00 m	–	1.20	28.56	–	24.56	m	**53.12**
wall height 3.00 m–3.30 m	–	1.30	31.00	–	27.02	m	**58.02**
wall height 3.30 m–3.60 m	–	1.53	36.43	–	29.55	m	**65.98**
19 mm plank to reveals and soffits of openings and recesses							
not exceeding 300 mm wide	–	0.20	4.74	–	3.39	m	**8.13**
300 mm–600 mm wide	–	0.42	9.93	–	5.60	m	**15.53**
19 mm plank to faces of columns – 4 nr faces							
not exceeding 300 mm total girth	–	0.80	19.51	–	4.66	m	**24.17**
300 mm–600 mm total girth	–	0.80	18.39	–	6.89	m	**25.28**
600 mm–900 mm total girth	–	0.80	18.39	–	9.14	m	**27.53**
900 mm–1200 mm total girth	–	0.80	18.39	–	11.36	m	**29.75**
ThermaLine Plus Board (0.19 W/mK)							
27 mm board to walls							
average height 2.00 m	–	0.16	3.68	–	15.02	m²	**18.70**
average height 3.00 m	–	0.16	3.56	–	15.02	m²	**18.58**
average height 4.00 m	–	0.15	3.45	–	15.02	m²	**18.47**
Rate per m (SMM7 measurement rule)							
wall height 2.40 m–2.70 m	–	1.06	24.36	–	40.56	m	**64.92**
wall height 2.70 m–3.00 m	–	1.23	28.27	–	45.06	m	**73.33**
wall height 3.00 m–3.30 m	–	1.34	30.80	–	49.57	m	**80.37**
wall height 3.30 m–3.60 m	–	1.62	37.23	–	54.16	m	**91.39**
27 mm board to reveals and soffits of openings and recesses							
not exceeding 300 mm wide	–	0.21	4.83	–	5.44	m	**10.27**
300 mm–600 mm wide	–	0.43	9.89	–	9.70	m	**19.59**
27 mm board to faces of columns – 4 nr faces							
not exceeding 300 mm total girth	–	0.52	11.95	–	10.88	m	**22.83**
300 mm–600 mm total girth	–	0.65	14.94	–	10.88	m	**25.82**
600 mm–900 mm total girth	–	0.75	17.24	–	19.42	m	**36.66**
900 mm–1200 mm total girth	–	1.00	22.98	–	19.43	m	**42.41**
48 mm board to walls							
average height 2.00 m	–	0.16	3.68	–	20.48	m²	**24.16**
average height 3.00 m	–	0.16	3.56	–	20.48	m²	**24.04**
average height 4.00 m	–	0.15	3.45	–	20.48	m²	**23.93**
Rate per m (SMM7 measurement rule)							
wall height 2.40 m–2.70 m	–	1.06	24.36	–	55.32	m	**79.68**
wall height 2.70 m–3.00 m	–	1.30	29.88	–	61.47	m	**91.35**
wall height 3.00 m–3.30 m	–	1.43	32.87	–	67.62	m	**100.49**
wall height 3.30 m–3.60 m	–	1.71	39.30	–	73.84	m	**113.14**

20 PROPRIETARY LININGS AND PARTITIONS

Item	PC £	Labour hours	Labour £	Plant £	Material £	Unit	Total rate £
48 mm board to reveals and soffits of openings and recesses							
not exceeding 300 mm wide	–	0.23	5.28	–	7.08	m	**12.36**
300 mm–600 mm wide	–	0.46	10.57	–	12.99	m	**23.56**
48 mm board to faces of columns – 4 nr faces							
not exceeding 300 mm total girth	–	0.52	11.95	–	14.15	m	**26.10**
300 mm–600 mm total girth	–	0.65	14.94	–	14.15	m	**29.09**
600 mm–900 mm total girth	–	0.75	17.24	–	25.97	m	**43.21**
900 mm–1200 mm total girth	–	1.00	22.98	–	25.98	m	**48.96**
ThermaLine Super boards							
50 mm board to walls							
average height 2.00 m	–	0.40	9.20	–	27.44	m²	**36.64**
average height 3.00 m	–	0.43	9.89	–	27.44	m²	**37.33**
average height 4.00 m	–	0.47	10.92	–	27.44	m²	**38.36**
Rate per m (SMM7 measurement rule)							
wall height 2.40 m–2.70 m	–	1.06	24.36	–	74.10	m	**98.46**
wall height 2.70 m–3.00 m	–	1.30	29.88	–	82.33	m	**112.21**
wall height 3.00 m–3.30 m	–	1.43	32.87	–	90.57	m	**123.44**
wall height 3.30 m–3.60 m	–	1.71	39.30	–	98.88	m	**138.18**
50 mm board to reveals and soffits of openings and recesses							
not exceeding 300 mm wide	–	0.52	11.95	–	9.17	m	**21.12**
300 mm–600 mm wide	–	0.46	10.57	–	17.16	m	**27.73**
50 mm board to faces of columns – 4 nr faces							
not exceeding 300 mm total girth	–	0.52	11.95	–	10.34	m	**22.29**
300–600 mm total girth	–	0.65	14.94	–	18.33	m	**33.27**
600 mm–900 mm total girth	–	0.75	17.24	–	26.34	m	**43.58**
900 mm–1200 mm total girth	–	1.00	22.98	–	34.34	m	**57.32**
60 mm board to walls							
average height 2.00 m	–	0.40	9.20	–	32.26	m²	**41.46**
average height 3.00 m	–	0.43	9.89	–	32.26	m²	**42.15**
average height 4.00 m	–	0.47	10.92	–	32.26	m²	**43.18**
Rate per m (SMM7 measurement rule)							
wall height 2.40 m–2.70 m	–	1.06	24.36	–	87.09	m	**111.45**
wall height 2.70 m–3.00 m	–	1.30	29.88	–	96.77	m	**126.65**
wall height 3.00 m–3.30 m	–	1.43	32.87	–	106.45	m	**139.32**
wall height 3.30 m–3.60 m	–	1.71	39.30	–	116.20	m	**155.50**
70 mm board to walls							
average height 2.00 m	–	0.40	9.20	–	38.14	m²	**47.34**
average height 3.00 m	–	0.43	9.89	–	38.14	m²	**48.03**
average height 4.00 m	–	0.47	10.92	–	38.14	m²	**49.06**
Rate per m (SMM7 measurement rule)							
wall height 2.40 m–2.70 m	–	1.06	24.36	–	102.98	m	**127.34**
wall height 2.70 m–3.00 m	–	1.30	29.88	–	114.41	m	**144.29**
wall height 3.00 m–3.30 m	–	1.43	32.87	–	125.87	m	**158.74**
wall height 3.30 m–3.60 m	–	1.71	39.30	–	137.38	m	**176.68**
70 mm board to reveals and soffits of openings and recesses							
not exceeding 300 mm wide	–	0.23	5.28	–	12.37	m	**17.65**
300 mm–600 mm wide	–	0.46	10.57	–	23.58	m	**34.15**

20 PROPRIETARY LININGS AND PARTITIONS

Item	PC £	Labour hours	Labour £	Plant £	Material £	Unit	Total rate £
20.02 LININGS TO WALLS – cont							
ThermaLine Super boards – cont							
70 mm board to faces of columns – 4 nr faces							
not exceeding 300 mm total girth	–	0.52	11.95	–	13.55	m	**25.50**
300–600 mm total girth	–	0.65	14.94	–	24.75	m	**39.69**
600 mm–900 mm total girth	–	0.75	17.24	–	35.97	m	**53.21**
900 mm–1200 mm total girth	–	1.00	22.98	–	47.17	m	**70.15**
80 mm board to walls							
average height 2.00 m	–	0.40	9.20	–	44.10	m²	**53.30**
average height 3.00 m	–	0.43	9.89	–	44.10	m²	**53.99**
average height 4.00 m	–	0.47	10.92	–	44.10	m²	**55.02**
Rate per m (SMM7 measurement rule)							
wall height 2.40 m–2.70 m	–	1.06	24.36	–	119.10	m	**143.46**
wall height 2.70 m–3.00 m	–	1.30	29.88	–	132.32	m	**162.20**
wall height 3.00 m–3.30 m	–	1.43	32.87	–	145.56	m	**178.43**
wall height 3.30 m–3.60 m	–	1.71	39.30	–	158.88	m	**198.18**
80 mm board to reveals and soffits of openings and recesses							
not exceeding 300 mm wide	–	0.23	5.28	–	14.16	m	**19.44**
300 mm–600 mm wide	–	0.46	10.57	–	27.16	m	**37.73**
80 mm board to faces of columns – 4 nr faces							
not exceeding 300 mm total girth	–	0.52	11.95	–	15.35	m	**27.30**
300–600 mm total girth	–	0.65	14.94	–	28.34	m	**43.28**
600 mm–900 mm total girth	–	0.75	17.24	–	41.33	m	**58.57**
900 mm–1200 mm total girth	–	1.00	22.98	–	54.33	m	**77.31**
90 mm board to walls							
average height 2.00 m	–	0.40	9.20	–	46.64	m²	**55.84**
average height 3.00 m	–	0.43	9.89	–	46.64	m²	**56.53**
average height 4.00 m	–	0.47	10.92	–	46.64	m²	**57.56**
Rate per m (SMM7 measurement rule)							
wall height 2.40 m–2.70 m	–	1.06	24.36	–	125.93	m	**150.29**
wall height 2.70 m–3.00 m	–	1.30	29.88	–	139.92	m	**169.80**
wall height 3.00 m–3.30 m	–	1.43	32.87	–	153.91	m	**186.78**
wall height 3.30 m–3.60 m	–	1.71	39.30	–	167.98	m	**207.28**
90 mm board to reveals and soffits of openings and recesses							
not exceeding 300 mm wide	–	0.23	5.28	–	14.92	m	**20.20**
300 mm–600 mm wide	–	0.46	10.57	–	28.68	m	**39.25**
90 mm board to faces of columns – 4 nr faces							
not exceeding 300 mm total girth	–	0.52	11.95	–	16.10	m	**28.05**
300–600 mm total girth	–	0.65	14.94	–	29.85	m	**44.79**
600 mm–900 mm total girth	–	0.75	17.24	–	43.61	m	**60.85**
900 mm–1200 mm total girth	–	1.00	22.98	–	57.37	m	**80.35**

20 PROPRIETARY LININGS AND PARTITIONS

Item	PC £	Labour hours	Labour £	Plant £	Material £	Unit	Total rate £
Kingspan Kooltherm K18 insulated plasterboard; fixing with nails; joints filled with joint filler and joint tape to receive direct decoration; to softwood base (measured elsewhere)							
12.5 mm plasterboard bonded to CFC/HCFC free rigid phenolic insulation; overall thickness (0.21 W/m.K)							
37.5 mm thick panel	–	0.23	5.17	–	21.49	m²	**26.66**
42.5 mm thick panel	–	0.23	5.17	–	13.89	m²	**19.06**
52.5 mm thick panel	–	0.25	5.75	–	17.39	m²	**23.14**
62.5 mm thick panel	–	0.25	5.75	–	15.44	m²	**21.19**
72.5 mm thick panel	–	0.25	5.75	–	18.52	m²	**24.27**
Plasterboard jointing system; filling joint with jointing compounds							
To walls and ceilings							
to suit 9.50 mm or 12.50 mm thick boards	–	0.09	2.07	–	2.27	m	**4.34**
Angle trim; plasterboard edge support system							
To walls and ceilings							
to suit 9.50 mm or 12.50 mm thick boards	–	0.09	2.07	–	2.13	m	**4.20**
Gyproc SoundBloc tapered edge plasterboard with higher density core; fixing on dabs or with nails; joints filled with joint filler and joint tape to receive direct decoration; to softwood base							
12.50 mm board to walls							
average height 2.00 m	–	0.36	8.27	–	5.90	m²	**14.17**
average height 3.00 m	–	0.38	8.73	–	5.90	m²	**14.63**
average height 4.00 m	–	0.40	9.20	–	5.98	m²	**15.18**
Rate per m (SMM7 measurement rule)							
wall height 2.40 m–2.70 m	–	0.97	22.30	–	15.92	m	**38.22**
wall height 2.70 m–3.00 m	–	1.11	25.51	–	17.70	m	**43.21**
wall height 3.00 m–3.30 m	–	1.32	30.34	–	19.47	m	**49.81**
wall height 3.30 m–3.60 m	–	1.43	32.87	–	21.31	m	**54.18**
12.50 mm board to ceilings							
over 300 mm wide	–	0.41	9.42	–	5.90	m²	**15.32**
15.00 mm board to walls							
average height 2.00 m	–	0.36	8.27	–	6.89	m²	**15.16**
average height 3.00 m	–	0.38	8.73	–	6.89	m²	**15.62**
average height 4.00 m	–	0.40	9.20	–	6.98	m²	**16.18**
Rate per m (SMM7 measurement rule)							
wall height 2.40 m–2.70 m	–	1.00	22.98	–	18.63	m	**41.61**
wall height 2.70 m–3.00 m	–	1.14	26.20	–	20.69	m	**46.89**
wall height 3.00 m–3.30 m	–	1.27	29.19	–	22.77	m	**51.96**
wall height 3.30 m–3.60 m	–	1.46	33.56	–	24.92	m	**58.48**
15.00 mm board to reveals and soffits of openings and recesses							
not exceeding 300 mm wide	–	0.20	4.59	–	3.00	m	**7.59**
300 mm–600 mm wide	–	0.38	8.73	–	4.83	m	**13.56**

20 PROPRIETARY LININGS AND PARTITIONS

Item	PC £	Labour hours	Labour £	Plant £	Material £	Unit	Total rate £
20.02 LININGS TO WALLS – cont							
Gyproc SoundBloc tapered edge plasterboard with higher density core – cont							
15.00 mm board to ceilings							
over 300 mm wide	–	0.43	9.89	–	6.90	m²	**16.79**
Two layers of gypsum plasterboard; plain grade square and tapered edge wallboard; fixing on dabs or with nails; joints filled with joint filler and joint tape; top layer to receive direct decoration; to softwood base							
19 mm two layer board to walls							
average height 2.00 m	–	0.48	11.03	–	8.20	m²	**19.23**
average height 3.00 m	–	0.51	11.72	–	8.20	m²	**19.92**
average height 4.00 m	–	0.54	12.41	–	8.29	m²	**20.70**
Rate per m (SMM7 measurement rule)							
wall height 2.40 m–2.70 m	–	1.30	29.88	–	22.13	m	**52.01**
wall height 2.70 m–3.00 m	–	1.48	34.02	–	24.60	m	**58.62**
wall height 3.00 m–3.30 m	–	1.67	38.39	–	27.06	m	**65.45**
wall height 3.30 m–3.60 m	–	1.94	44.59	–	29.59	m	**74.18**
19 mm two layer board to reveals and soffits of openings and recesses							
not exceeding 300 mm wide	–	0.28	6.44	–	3.39	m	**9.83**
300 mm–600 mm wide	–	0.56	12.88	–	5.61	m	**18.49**
19 mm two layer board to faces of columns – 4 nr							
not exceeding 300 mm total girth	–	0.50	11.49	–	5.73	m	**17.22**
300–600 mm total girth	–	0.69	15.86	–	6.80	m	**22.66**
600 mm–900 mm total girth	–	0.75	17.24	–	9.01	m	**26.25**
900 mm–1200 mm total girth	–	1.00	22.98	–	10.74	m	**33.72**
25 mm two layer board to walls							
average height 2.00 m	–	0.48	11.03	–	10.10	m²	**21.13**
average height 3.00 m	–	0.51	11.72	–	10.10	m²	**21.82**
average height 4.00 m	–	0.54	12.41	–	10.20	m²	**22.61**
Rate per m (SMM7 measurement rule)							
wall height 2.40 m–2.70 m	–	1.39	31.95	–	27.30	m	**59.25**
wall height 2.70 m–3.00 m	–	1.57	36.08	–	30.32	m	**66.40**
wall height 3.00 m–3.30 m	–	1.76	40.45	–	33.36	m	**73.81**
wall height 3.30 m–3.60 m	–	2.04	46.89	–	36.47	m	**83.36**
25 mm two layer board to reveals and soffits of openings and recesses							
not exceeding 300 mm wide	–	0.28	6.44	–	3.97	m	**10.41**
300 mm–600 mm wide	–	0.56	12.88	–	6.76	m	**19.64**
25 mm two layer board to faces of columns – 4 nr							
not exceeding 300 mm total girth	–	0.35	8.04	–	5.13	m	**13.17**
300–600 mm total girth	–	0.69	15.86	–	7.95	m	**23.81**
600 mm–900 mm total girth	–	1.00	22.98	–	10.73	m	**33.71**
900 mm–1200 mm total girth	–	1.25	28.78	–	13.49	m	**42.27**

20 PROPRIETARY LININGS AND PARTITIONS

Item	PC £	Labour hours	Labour £	Plant £	Material £	Unit	Total rate £
Gyproc Dri-Wall dry lining system or other equal or approved; plain grade tapered edge wallboard; fixed to walls with adhesive; joints filled with joint filler and joint tape; to receive direct decoration							
9.50 mm board to walls							
average height 2.00 m	–	0.41	9.45	–	6.21	m²	**15.66**
average height 3.00 m	–	0.43	9.95	–	6.21	m²	**16.16**
average height 4.00 m	–	0.51	11.63	–	6.21	m²	**17.84**
Rate per m (SMM7 measurement rule)							
wall height 2.40 m–2.70 m	–	1.11	25.51	–	16.77	m	**42.28**
wall height 2.70 m–3.00 m	–	1.28	29.42	–	18.61	m	**48.03**
wall height 3.00 m–3.30 m	–	1.43	32.87	–	20.46	m	**53.33**
wall height 3.30 m–3.60 m	–	1.67	38.39	–	22.37	m	**60.76**
9.50 mm board to reveals and soffits of openings and recesses							
not exceeding 300 mm wide	–	0.23	5.43	–	2.76	m	**8.19**
300 mm–600 mm wide	–	0.46	10.86	–	4.38	m	**15.24**
9.50 mm board to faces of columns – 4 nr faces							
300 mm total girth	–	0.45	10.34	–	2.82	m	**13.16**
300–600 mm total girth	–	0.58	13.33	–	5.53	m	**18.86**
600 mm–900 mm total girth	–	0.75	17.24	–	6.77	m	**24.01**
900 mm–1200 mm total girth	–	0.95	21.84	–	7.96	m	**29.80**
Angle; with joint tape bedded and covered with Jointex or other equal							
internal	–	0.05	1.22	–	0.58	m	**1.80**
external	–	0.11	2.68	–	0.58	m	**3.26**
Gyproc Dri-Wall M/F dry lining system or other equal; mild steel channel fixed to walls with adhesive; tapered edge wallboard screwed to channel; joints filled with joint filler and joint tape							
9.50 mm board to walls							
average height 2.00 m	–	0.55	12.64	–	9.30	m²	**21.94**
average height 3.00 m	–	0.57	13.21	–	9.29	m²	**22.50**
average height 4.00 m	–	0.62	14.17	–	9.31	m²	**23.48**
Rate per m (SMM7 measurement rule)							
wall height 2.40 m–2.70 m	–	1.48	34.02	–	25.10	m	**59.12**
wall height 2.70 m–3.00 m	–	1.69	39.82	–	27.87	m	**67.69**
wall height 3.00 m–3.30 m	–	1.90	43.67	–	30.67	m	**74.34**
wall height 3.30 m–3.60 m	–	2.22	51.03	–	33.54	m	**84.57**
9.50 mm board to reveals and soffits of openings and recesses							
not exceeding 300 mm wide	–	0.23	5.43	–	2.45	m	**7.88**
300 mm–600 mm wide	–	0.46	10.86	–	3.74	m	**14.60**
12.50 mm board to walls							
average height 2.00 m	–	0.55	12.64	–	11.21	m²	**23.85**
average height 3.00 m	–	0.57	13.21	–	11.20	m²	**24.41**
average height 4.00 m	–	0.62	14.17	–	11.22	m²	**25.39**

20 PROPRIETARY LININGS AND PARTITIONS

Item	PC £	Labour hours	Labour £	Plant £	Material £	Unit	Total rate £
20.02 LININGS TO WALLS – cont							
Gyproc Dri-Wall M/F dry lining system or other equal – cont							
Rate per m (SMM7 measurement rule)							
wall height 2.40 m–2.70 m	–	1.48	34.02	–	30.24	m	**64.26**
wall height 2.70 m–3.00 m	–	1.69	39.82	–	33.59	m	**73.41**
wall height 3.00 m–3.30 m	–	1.90	43.67	–	36.95	m	**80.62**
wall height 3.30 m–3.60 m	–	2.22	51.03	–	40.39	m	**91.42**
12.50 mm board to reveals and soffits of openings and recesses							
not exceeding 300 mm wide	–	0.23	5.43	–	3.02	m	**8.45**
300 mm–600 mm wide	–	0.46	10.86	–	4.88	m	**15.74**
Megadeco wallboard; fixing on dabs or with screws; joints filled with joint filler and joint tape to receive direct decoration; to softwood							
12.50 mm board to walls							
average height 2.00 m	–	0.36	8.27	–	5.87	m²	**14.14**
average height 3.00 m	–	0.38	8.70	–	5.87	m²	**14.57**
average height 4.00 m	–	0.40	9.20	–	5.87	m²	**15.07**
Rate per m (SMM7 measurement rule)							
wall height 2.40 m–2.70 m	–	0.97	22.30	–	18.25	m	**40.55**
wall height 2.70 m–3.00 m	–	1.11	26.49	–	20.28	m	**46.77**
wall height 3.00 m–3.30 m	–	1.25	28.73	–	22.31	m	**51.04**
wall height 3.30 m–3.60 m	–	1.43	32.87	–	24.34	m	**57.21**
12.50 mm board to ceilings							
over 300 mm wide	–	0.41	9.70	–	6.76	m²	**16.46**
15 mm board to walls							
average height 2.00 m	–	0.38	8.68	–	6.76	m²	**15.44**
average height 3.00 m	–	0.40	9.14	–	6.76	m²	**15.90**
average height 4.00 m	–	0.42	9.65	–	6.76	m²	**16.41**
Rate per m (SMM7 measurement rule)							
wall height 2.40 m–2.70 m	–	1.02	23.41	–	18.25	m	**41.66**
wall height 2.70 m–3.00 m	–	1.17	27.82	–	20.28	m	**48.10**
wall height 3.00 m–3.30 m	–	1.31	30.17	–	22.31	m	**52.48**
wall height 3.30 m–3.60 m	–	1.50	34.52	–	24.34	m	**58.86**
15 mm board to ceilings							
over 300 mm wide	–	0.43	10.19	–	6.76	m²	**16.95**
Gypsum cladding; Glasroc Firecase S board or other equal; fixed with adhesive; joints pointed in adhesive							
25 mm thick column linings, faces = 4; 2 hour fire protection rating							
not exceeding 300 mm girth	–	0.30	6.89	–	23.05	m	**29.94**
300 mm–600 mm girth	–	0.30	6.89	–	31.29	m	**38.18**
600 mm–900 mm girth	–	0.45	10.34	–	39.54	m	**49.88**
900 mm–1200 mm girth	–	0.55	12.64	–	47.79	m	**60.43**
1200 mm–1500 mm girth	–	0.60	13.79	–	56.04	m	**69.83**

20 PROPRIETARY LININGS AND PARTITIONS

Item	PC £	Labour hours	Labour £	Plant £	Material £	Unit	Total rate £
30 mm thick beam linings, faces = 3; 2 hour fire protection rating							
not exceeding 300 mm girth	–	0.60	13.79	–	20.97	m	**34.76**
300 mm–600 mm girth	–	0.60	13.79	–	30.29	m	**44.08**
600 mm–900 mm girth	–	0.75	17.24	–	39.60	m	**56.84**
900 mm–1200 mm girth	–	0.90	20.68	–	48.93	m	**69.61**
1200 mm–1500 mm girth	–	1.20	27.58	–	58.25	m	**85.83**
Vermiculite gypsum cladding; Vermiculux board or other equal approved; fixed with adhesive; joints pointed in adhesive							
25 mm thick column linings, faces = 4; 2 hour fire protection rating							
not exceeding 300 mm girth	–	0.20	4.59	–	12.11	m	**16.70**
300 mm–600 mm girth	–	0.30	6.89	–	24.24	m	**31.13**
600 mm–900 mm girth	–	0.45	10.34	–	35.97	m	**46.31**
900 mm–1200 mm girth	–	0.60	13.79	–	47.96	m	**61.75**
1200 mm–1500 mm girth	–	0.70	16.09	–	59.82	m	**75.91**
30 mm thick beam linings, faces = 3; 2 hour fire protection rating							
not exceeding 300 mm girth	–	0.50	11.49	–	15.24	m	**26.73**
300 mm–600 mm girth	–	0.60	13.79	–	30.50	m	**44.29**
600 mm–900 mm girth	–	0.90	20.68	–	48.79	m	**69.47**
900 mm–1200 mm girth	–	1.00	22.98	–	60.38	m	**83.36**
1200 mm–1500 mm girth	–	1.20	27.58	–	75.38	m	**102.96**
55 mm thick column linings, faces = 4; 4 hour fire protection rating							
not exceeding 300 mm girth	–	0.60	13.79	–	32.31	m	**46.10**
300 mm–600 mm girth	–	0.75	17.24	–	64.62	m	**81.86**
600 mm–900 mm girth	–	0.80	18.39	–	96.55	m	**114.94**
900 mm–1200 mm girth	–	1.00	22.98	–	128.74	m	**151.72**
1200 mm–1500 mm girth	–	1.02	23.44	–	160.79	m	**184.23**
60 mm thick beam linings, faces = 3; 4 hour fire protection rating							
not exceeding 300 mm girth	–	0.60	13.79	–	34.91	m	**48.70**
300 mm–600 mm girth	–	0.75	17.24	–	69.81	m	**87.05**
600 mm–900 mm girth	–	0.80	18.39	–	104.35	m	**122.74**
900 mm–1200 mm girth	–	1.00	22.98	–	139.13	m	**162.11**
1200 mm–1500 mm girth	–	1.02	23.44	–	173.78	m	**197.22**
Add to the above rates for working at height							
for work 3.50 m–5.00 m high	–	–	–	–	–	%	**2.50**
for work 5.00 m–6.50 m high	–	–	–	–	–	%	**5.00**
for work 6.50 m–8.00 m high	–	–	–	–	–	%	**12.50**
for work over 8.00 m high	–	–	–	–	–	%	**17.50**
Cutting and fitting around steel joints, angles, trunking, ducting, ventilators, pipes, tubes, etc.							
not exceeding 0.30 m girth	–	0.28	6.44	–	–	nr	**6.44**
0.30 m–1 m girth	–	0.37	8.51	–	–	nr	**8.51**
1 m–2 m girth	–	0.51	11.72	–	–	nr	**11.72**
over 2 m girth	–	0.42	9.65	–	–	m	**9.65**

20 PROPRIETARY LININGS AND PARTITIONS

Item	PC £	Labour hours	Labour £	Plant £	Material £	Unit	Total rate £
20.02 LININGS TO WALLS – cont							
Glass reinforced gypsum Glasroc Multi-board or other equal and approved; fixing with nails; joints filled with joint filler and joint tape; finishing with Jointex or other equal and approved to receive decoration; to softwood base							
10 mm board to walls							
average wall height 2.00 m	–	0.37	8.43	–	32.69	m²	**41.12**
average wall height 3.00 m	–	0.36	8.35	–	32.69	m²	**41.04**
average wall height 4.00 m	–	0.36	8.27	–	32.69	m²	**40.96**
Rate per m (SMM7 measurement rule)							
wall height 2.40 m–2.70 m	–	0.93	21.37	–	88.25	m	**109.62**
wall height 2.70 m–3.00 m	–	1.06	24.36	–	98.06	m	**122.42**
wall height 3.00 m–3.30 m	–	1.07	24.51	–	107.87	m	**132.38**
wall height 3.30 m–3.60 m	–	1.39	31.95	–	117.75	m	**149.70**
12.50 mm board to walls							
average wall height 2.00 m	–	0.39	8.85	–	42.76	m²	**51.61**
average wall height 3.00 m	–	0.38	8.78	–	42.76	m²	**51.54**
average wall height 4.00 m	–	0.38	8.68	–	42.76	m²	**51.44**
Rate per m (SMM7 measurement rule)							
wall height 2.40 m–2.70 m	–	0.97	22.30	–	115.46	m	**137.76**
wall height 2.70 m–3.00 m	–	1.11	25.51	–	128.29	m	**153.80**
wall height 3.00 m–3.30 m	–	1.25	28.73	–	141.12	m	**169.85**
wall height 3.30 m–3.60 m	–	1.14	26.28	–	154.03	m	**180.31**
Blockboard (birch faced)							
Lining to walls 18 mm thick							
over 600 wide	7.21	0.46	10.57	–	7.83	m²	**18.40**
not exceeding 600 wide	–	0.60	13.79	–	4.76	m	**18.55**
holes for pipes and the like	–	0.04	0.92	–	–	nr	**0.92**
Chipboard (plain)							
Lining to walls 12 mm thick							
over 600 mm wide	2.65	0.35	8.04	–	3.14	m²	**11.18**
not exceeding 600 mm wide	–	0.40	9.20	–	1.79	m	**10.99**
holes for pipes and the like	–	0.02	0.46	–	–	nr	**0.46**
Lining to walls 15 mm thick							
over 600 mm wide	3.12	0.37	8.51	–	3.63	m²	**12.14**
not exceeding 600 mm wide	–	0.44	10.11	–	2.24	m	**12.35**
holes for pipes and the like	–	0.03	0.69	–	–	nr	**0.69**
Lining to walls 18 mm thick							
over 600 mm wide	3.74	0.39	8.96	–	4.37	m²	**13.33**
not exceeding 600 mm wide	–	0.50	11.49	–	2.62	m	**14.11**
holes for pipes and the like	–	0.04	0.92	–	–	nr	**0.92**

20 PROPRIETARY LININGS AND PARTITIONS

Item	PC £	Labour hours	Labour £	Plant £	Material £	Unit	Total rate £
Fire-retardant chipboard/MDF; Class 1 spread of flame							
Lining to walls 12 mm thick							
over 600 mm wide	–	0.35	8.04	–	7.94	m²	**15.98**
not exceeding 600 mm wide	–	0.40	9.20	–	4.83	m	**14.03**
holes for pipes and the like	–	0.02	0.46	–	–	nr	**0.46**
Lining to walls 18 mm thick							
over 600 mm wide	–	0.39	8.96	–	11.78	m²	**20.74**
not exceeding 600 mm wide	–	0.50	11.49	–	7.13	m	**18.62**
holes for pipes and the like	–	0.04	0.92	–	–	nr	**0.92**
Lining to walls 25 mm thick							
over 600 mm wide	–	0.41	9.42	–	16.54	m²	**25.96**
not exceeding 600 mm wide	–	0.56	12.88	–	9.98	m	**22.86**
holes for pipes and the like	–	0.05	1.15	–	–	nr	**1.15**
Chipboard melamine faced; white matt finish; laminated masking strips							
Lining to walls 15 mm thick							
over 600 mm wide	4.02	0.97	22.30	–	4.87	m²	**27.17**
not exceeding 600 mm wide	–	1.26	28.96	–	3.01	m	**31.97**
holes for pipes and the like	–	0.06	1.38	–	–	nr	**1.38**
Insulation board to BS EN 622							
Lining to walls 12 mm thick							
over 600 mm wide	2.14	0.22	5.06	–	2.61	m²	**7.67**
not exceeding 600 mm wide	–	0.26	5.97	–	1.63	m	**7.60**
holes for pipes and the like	–	0.01	0.23	–	–	nr	**0.23**
Masterboard or other equal; sanded finish							
Lining to walls 6 mm thick							
over 600 mm wide	14.18	0.31	7.13	–	14.91	m²	**22.04**
not exceeding 600 mm wide	–	0.38	8.73	–	8.97	m	**17.70**
Lining to walls 9 mm thick							
over 600 mm wide	32.78	0.33	7.58	–	34.06	m²	**41.64**
not exceeding 600 mm wide	–	0.38	8.73	–	20.46	m	**29.19**
Supalux or other equal; sanded finish							
Lining to walls 6 mm thick							
over 600 mm wide	27.36	0.31	7.13	–	28.49	m²	**35.62**
not exceeding 600 mm wide	–	0.38	8.73	–	17.11	m	**25.84**
Lining to walls 9 mm thick							
over 600 mm wide	34.83	0.33	7.58	–	36.18	m²	**43.76**
not exceeding 600 mm wide	–	0.38	8.73	–	21.73	m	**30.46**
Lining to walls 12 mm thick							
over 600 mm wide	47.30	0.37	8.51	–	49.03	m²	**57.54**
not exceeding 600 mm wide	–	0.44	10.11	–	29.44	m	**39.55**

20 PROPRIETARY LININGS AND PARTITIONS

Item	PC £	Labour hours	Labour £	Plant £	Material £	Unit	Total rate £
20.02 LININGS TO WALLS – cont							
Monolux 40 or other equal; 6 mm × 50 mm							
Supalux cover fillets or other equal and							
approved one side							
Lining to walls 19 mm thick							
over 600 mm wide	95.47	0.65	14.94	–	102.68	m²	**117.62**
not exceeding 600 mm wide	–	0.92	21.15	–	64.29	m	**85.44**
Lining to walls 25 mm thick							
over 600 mm wide	114.47	0.69	15.86	–	122.25	m²	**138.11**
not exceeding 600 mm wide	68.68	0.98	22.53	–	76.03	m	**98.56**
Cement particle consisting of a mixture of wood							
fibre, cement and additive, Class 0, Cemboard or							
similar rigid high performance board as a							
sheathing							
Lining to walls 8 mm thick							
over 600 mm wide	6.79	0.65	14.94	–	11.33	m²	**26.27**
not exceeding 600 mm wide	–	0.92	21.15	–	9.48	m	**30.63**
Lining to walls 10 mm thick							
over 600 mm wide	7.82	0.65	14.94	–	12.40	m²	**27.34**
not exceeding 600 mm wide	–	0.92	21.15	–	10.12	m	**31.27**
Lining to walls 12 mm thick							
over 600 mm wide	9.27	0.65	14.94	–	13.88	m²	**28.82**
not exceeding 600 mm wide	–	0.92	21.15	–	11.01	m	**32.16**
Lining to walls 16 mm thick							
over 600 mm wide	12.89	0.65	14.94	–	17.61	m²	**32.55**
not exceeding 600 mm wide	–	0.92	21.15	–	13.26	m	**34.41**
Lining to walls 18 mm thick							
over 600 mm wide	14.35	0.65	14.94	–	19.13	m²	**34.07**
not exceeding 600 mm wide	–	0.92	21.15	–	14.16	m	**35.31**
Lining to walls 24 mm thick							
over 600 mm wide	20.32	0.65	14.94	–	25.28	m²	**40.22**
not exceeding 600 mm wide	–	0.92	21.15	–	17.85	m	**39.00**
Fibre cement board consisting of fibre cement							
sheet; smooth finish for paint; A1							
non-combustible; high performance board							
Lining to walls 6 mm thick							
over 600 mm wide	6.82	0.65	14.94	–	11.36	m²	**26.30**
not exceeding 600 mm wide	–	0.92	21.15	–	9.50	m	**30.65**
Lining to walls 9 mm thick							
over 600 mm wide	10.64	0.65	14.94	–	15.30	m²	**30.24**
not exceeding 600 mm wide	–	0.92	21.15	–	11.87	m	**33.02**
Lining to walls 12 mm thick							
over 600 mm wide	14.11	0.65	14.94	–	18.87	m²	**33.81**
not exceeding 600 mm wide	–	0.92	21.15	–	14.01	m	**35.16**

20 PROPRIETARY LININGS AND PARTITIONS

Item	PC £	Labour hours	Labour £	Plant £	Material £	Unit	Total rate £
Plywood (Eastern European); internal quality							
Lining to walls 4 mm thick							
over 600 mm wide	2.59	0.34	7.82	–	3.07	m²	**10.89**
not exceeding 600 mm wide	–	0.44	10.11	–	1.91	m	**12.02**
Lining to walls 6 mm thick							
over 600 mm wide	3.74	0.37	8.51	–	4.25	m²	**12.76**
not exceeding 600 mm wide	–	0.48	11.03	–	2.62	m	**13.65**
Lining to walls 12 mm thick							
over 600 mm wide	6.95	0.43	9.89	–	7.56	m²	**17.45**
not exceeding 600 mm wide	–	0.56	12.88	–	4.59	m	**17.47**
Lining to walls 18 mm thick							
over 600 mm wide	10.15	0.46	10.57	–	10.86	m²	**21.43**
not exceeding 600 mm wide	–	0.60	13.79	–	6.57	m	**20.36**
Plywood (Eastern European); external quality							
Lining to walls 4 mm thick							
over 600 mm wide	6.41	0.34	7.82	–	7.00	m²	**14.82**
not exceeding 600 mm wide	–	0.44	10.11	–	4.26	m	**14.37**
Lining to walls 6.5 mm thick							
over 600 mm wide	7.17	0.37	8.51	–	7.79	m²	**16.30**
not exceeding 600 mm wide	–	0.48	11.03	–	4.74	m	**15.77**
Lining to walls 9 mm thick							
over 600 mm wide	9.23	0.40	9.20	–	9.91	m²	**19.11**
not exceeding 600 mm wide	–	0.52	11.95	–	6.00	m	**17.95**
Lining to walls 12 mm thick							
over 600 mm wide	11.52	0.43	9.89	–	12.28	m²	**22.17**
not exceeding 600 mm wide	–	0.56	12.88	–	7.43	m	**20.31**
holes for pipes and the like	–	0.03	0.69	–	–	nr	**0.69**
Extra over Linings fixed with screws	–	0.10	2.30	–	0.08	m²	**2.38**
Internal quality American Cherry veneered plywood; 6 mm thick							
Lining to walls							
over 600 mm wide	8.40	0.41	9.42	–	8.96	m²	**18.38**
not exceeding 600 mm wide	–	0.54	12.41	–	5.50	m	**17.91**
Glazed hardboard to BS EN 622; on and including 38 mm × 38 mm sawn softwood framing							
3.20 mm thick panel							
to side of bath	–	1.67	38.39	–	9.45	nr	**47.84**
to end of bath	–	0.65	14.94	–	3.02	nr	**17.96**

20 PROPRIETARY LININGS AND PARTITIONS

Item	PC £	Labour hours	Labour £	Plant £	Material £	Unit	Total rate £
20.02 LININGS TO WALLS – cont							
Rigid Sheet Acoustic Panel Linings							
Perforated steel acoustic wall panels; Eckel type HD EFP or other equal and approved; polyurethene enamel finish; fibrous glass acoustic insulation Walls							
average height 3.00 m; fixed to timber or masonry	–	–	–	–	–	m²	**215.10**
Acoustic Panel linings; Troldekt Ultrafine 1200 mm × 600 mm × 25 mm on and including 50 mm × 70 mm timber framework at 600 mm centres both ways and supported from the roof with 50 mm × 50 mm hangers 400 mm–600 mm long Ceilings							
Allow for ceiling tiles to be clipped to prevent wind uplift when external doors are opened	–	–	–	–	–	m²	**118.71**
Perimeter edge detail with 50 mm × 50 mm support batten	–	–	–	–	–	m	**7.52**
20.03 LININGS TO CEILINGS							
Plaster; one coat Thistle board finish or similar; steel trowelled 3 mm work to ceilings; one coat on and including gypsum plasterboard; fixing with nails; 3 mm joints filled with plaster and jute scrim cloth; to softwood base; plain grade baseboard or lath with rounded edges 9.50 mm thick boards to ceilings							
over 600 mm wide	–	0.89	14.63	–	5.01	m²	**19.64**
over 600 mm wide; 3.50 m–5.00 m high	–	1.03	17.07	–	5.01	m²	**22.08**
not exceeding 600 mm wide	–	0.43	7.48	–	1.50	m	**8.98**
9.50 mm thick boards to ceilings; in staircase areas or plant rooms							
over 600 mm wide	–	0.98	16.19	–	5.01	m²	**21.20**
not exceeding 600 mm wide	–	0.47	8.18	–	1.50	m	**9.68**
9.50 mm thick boards to isolated beams							
over 600 mm wide	–	1.05	17.41	–	5.01	m²	**22.42**
not exceeding 600 mm wide	–	0.50	8.70	–	1.50	m	**10.20**
12.50 mm thick boards to ceilings							
over 600 mm wide	–	0.95	15.68	–	5.01	m²	**20.69**
over 600 mm wide; 3.50 m–5.00 m high	–	1.06	17.58	–	5.01	m²	**22.59**
not exceeding 600 mm wide	–	0.45	7.83	–	1.50	m	**9.33**
12.50 mm thick boards to ceilings; in staircase areas or plant rooms							
over 600 mm wide	–	1.06	17.58	–	5.01	m²	**22.59**
not exceeding 600 mm wide	–	0.51	8.87	–	1.50	m	**10.37**
12.50 mm thick boards to isolated beams							
over 600 mm wide	–	1.15	19.15	–	5.01	m²	**24.16**
not exceeding 600 mm wide	–	0.56	9.74	–	1.50	m	**11.24**

21 CLADDING AND COVERING

Item	PC £	Labour hours	Labour £	Plant £	Material £	Unit	Total rate £
21.01 CURTAIN WALLING AND GLAZED ROOFING							
Stick curtain walling system; proprietary solution from system supplier; e.g. Schuco or equivalent Polyester powder coated solid colour matt finish or natural anodized finish mullions spaced 1.5 m apart and spanning typical storey height of 3.8 m. Floor to ceiling glass sealed units with 8.8 mm low-e coated laminated inner pane, air filled cavity and 8 mm clear monolithic heat strengthened outer pane, retained by external pressure plates and caps. Rates to include glass fronted solid spandrel panels, all brackets, membranes, fire stopping between floors, trade contractor preliminaries, including external access equipment							
Flat system	–	–	–	–	–	m²	512.45
extra over for							
neutral selective high performance coating on surface #2 in lieu of low-e coating on surface #3 (G-values circa 0.25–0.30); for assisting in solar control	–	–	–	–	–	m²	42.14
inner laminated glass to be heat strengthened laminated to mitigate thermal fracture risk	–	–	–	–	–	m²	42.14
outer glass to be heat strengthened laminated in lieu of monolithic heat strengthened	–	–	–	–	–	m²	42.14
ceramic fritting glass on surface #2 for visual and/or performance requirements (minimal G-value improvement)	–	–	–	–	–	m²	56.94
flush glass finish without external face caps, achieved by concealed toggle fixings locating within perimeter channels within sealed units including silicone sealing between glass panes	–	–	–	–	–	m²	56.94
typical coping detail, including pressed aluminium profiles, membranes, seals, etc.	–	–	–	–	–	m	307.47
typical sill detail, including pressed aluminium profiles, membranes, seals, etc.	–	–	–	–	–	m	250.53
intermediate transoms (per transom)	–	–	–	–	–	m	54.66

21 CLADDING AND COVERING

Item	PC £	Labour hours	Labour £	Plant £	Material £	Unit	Total rate £
21.01 CURTAIN WALLING AND GLAZED ROOFING – cont							
Unitized curtain walling system; proprietary solution from system supplier e.g. Schuco or equivalent alternatively pre-designed system solution from specialist facade contractor Polyester powder coated solid colour matt finish or natural anodized curtain walling Element widths of 1.5 m spanning typical storey height of 3.8 m. Floor to ceiling glass sealed units with 8.8 mm low-e coated laminated inner pane, air filled cavity and 8 mm monolithic heat strengthed outer pane, retained by external beading system. Rates include 1.1 m solid spandrel panels, all brackets, membranes, fire stopping between floors, trade contractor preliminaries, including external access equipment							
Flat system	–	–	–	–	–	m²	1059.07
extra over for							
neutral selective high performance coating on surface #2 in lieu of low-e coating on surface #3 (G-values circa 0.25–0.30); for assisting in solar control	–	–	–	–	–	m²	42.14
inner laminated glass to be heat strengthened laminated to mitigate thermal fracture risk	–	–	–	–	–	m²	42.14
flush glass finish without external face caps, achieved by carrier frames with glass sealed units factory silicone bonded; often referred to as SSG (Structural Silicone Glazing)	–	–	–	–	–	m²	68.33
typical coping detail, including pressed aluminium profiles, membranes, seals, etc.	–	–	–	–	–	m	307.47
typical sill detail, including pressed aluminium profiles, membranes, seals, etc.	–	–	–	–	–	m	250.53
Other curtain walling systems/costs Unitized curtain walling system; bespoke project specific solution via specialist facade contractor mostly based in mainland Europe							
generally as described for system supplier solution but profiles catered to specific performance and visual requirements, thus additional design development. Note: These rates are subject to currency fluctuations between £ and €.	–	–	–	–	–	m²	1059.07
flush glass finish finish (SEG) assuming direct bonding in he factory but with base chassis design to accommodate carrier frame for glass replacement, avoiding site bending	–	–	–	–	–	m²	34.17
project specific unitized curtain walling generally requires project specific performance testing. This rate is for a single wall type	–	–	–	–	–	nr	102489.75

21 CLADDING AND COVERING

Item	PC £	Labour hours	Labour £	Plant £	Material £	Unit	Total rate £
visual mock-ups are often required for project specific unitized curtain walling solutions and in cases for proprietary unitized and stick curtain walling projects. This rate is for a single wall type all curtain walling should be site hose tested. The	–	–	–	–	–	nr	34163.25
rate depends upon the quantum of joints to be tested, generally 5%. Assume 5 days @ £1000 extra over for	–	–	–	–	–	nr	6832.65
neutral selective high performance coating on surface #2 in lieu of low-e coating on surface #3 (G-values circa 0.25–0.30); for assisting in solar control	–	–	–	–	–	m²	42.14
inner laminated glass to be heat strengthened laminated to mitigate thermal fracture risk	–	–	–	–	–	m²	42.14
outer glass to be heat strengthened laminated in lieu of monolithic heat strengthened	–	–	–	–	–	m²	42.14
alternative solutions for achieving Part L and comfort criteria, include increasing the amount of solidity	–	–	–	–	–	m²	102.48
Patent glazing; aluminium alloy bars 2.55 m long at 622 mm centres; fixed to supports							
Roof cladding							
single glazed with 6.4 mm laminated glass	–	–	–	–	–	m²	167.42
single glazed with 7 mm thick Georgian wired cast glass	–	–	–	–	–	m²	173.20
thermally broken and double glazed with low-e clear toughened and laminated double glazed units; aluminium finished RAL matt colour	–	–	–	–	–	m²	456.08
Opening roof vents							
600 mm × 900 mm top hung opening roof vent; manually operated	–	–	–	–	–	nr	513.82
600 mm × 900 mm top hung opening roof vent; electrically operated	–	–	–	–	–	nr	658.15
Skylight							
Self-supporting hipped or gable ended lantern/ skylight thermally broken and double glazed with low-e clear toughened and laminated double glazed units; aluminium finished RAL matt colour	–	–	–	–	–	m²	912.17
Associated code 4 lead flashings							
top flashing; 210 mm girth	–	–	–	–	–	m	71.58
bottom flashing; 240 mm girth	–	–	–	–	–	m	81.98
end flashing; 300 mm girth	–	–	–	–	–	m	90.06
Wall cladding							
single glazed with 6.4 mm laminated glass	–	–	–	–	–	m²	166.26
single glazed with 7 mm thick Georgian wired cast glass	–	–	–	–	–	m²	184.75
thermally broken and double glazed with low-e clear toughened and laminated double glazed units; aluminium finished RAL matt colour	–	–	–	–	–	m²	475.72

21 CLADDING AND COVERING

Item	PC £	Labour hours	Labour £	Plant £	Material £	Unit	Total rate £
21.01 CURTAIN WALLING AND GLAZED ROOFING – cont							
Patent glazing – cont							
Extra for aluminium alloy perimeter members							
38 mm × 38 mm × 3 mm angle jamb	–	–	–	–	–	m	**25.41**
pressed sill member	–	–	–	–	–	m	**50.80**
pressed channel head and PVC case	–	–	–	–	–	m	**50.80**
21.02 RIGID SHEET CLADDING							
Resoplan sheet or other equal and approved; Eternit UK Ltd; flexible neoprene gasket joints; fixing with stainless steel screws and coloured caps							
6 mm thick cladding to walls							
over 300 mm wide	–	1.94	44.59	–	67.21	m²	**111.80**
not exceeding 300 mm wide	–	0.65	14.94	–	24.60	m	**39.54**
Eternit 2000 Glasal sheet or other equal and approved; Eternit UK Ltd; flexible neoprene gasket joints; fixing with stainless steel screws and coloured caps							
7.50 mm thick cladding to walls							
over 300 mm wide	–	1.94	44.59	–	59.20	m²	**103.79**
not exceeding 300 mm wide	–	0.65	14.94	–	22.20	m	**37.14**
external angle trim	–	0.09	2.07	–	11.45	m	**13.52**
7.50 mm thick cladding to eaves, verge soffit boards, fascia boards or the like							
100 mm wide	–	0.46	10.57	–	11.34	m	**21.91**
200 mm wide	–	0.56	12.88	–	16.77	m	**29.65**
300 mm wide	–	0.65	14.94	–	22.20	m	**37.14**
21.03 WEATHERBOARDING							
Prodema ProdEX high density resin-bonded cellulose fibre weatherboarding panels; including secondary supports and fixing							
Walls							
8 mm panels face fixed on to timber battens	–	–	–	–	–	m²	**194.37**
8 mm panels face fixed on to aluminium rails	–	–	–	–	–	m²	**220.63**
8 mm panels adhesive fixed on to timber battens or aluminium rails	–	–	–	–	–	m²	**236.40**
10 mm panels secret fixed on to helping hand aluminium system	–	–	–	–	–	m²	**288.93**

21 CLADDING AND COVERING

Item	PC £	Labour hours	Labour £	Plant £	Material £	Unit	Total rate £
21.04 PROFILED SHEET CLADDING							
Galvanized steel strip troughed sheets; Corus Products or other equal							
Roof cladding or decking; sloping not exceeding 50°; fixing to steel purlins with plastic headed self-tapping screws							
0.7 mm thick; 46 profile	–	–	–	–	–	m²	**13.04**
0.7 mm thick; 60 profile	–	–	–	–	–	m²	**14.29**
0.7 mm thick; 100 profile	–	–	–	–	–	m²	**15.53**
Galvanized steel strip troughed sheets; PMF Strip Mill Products or other equal							
Roof cladding; sloping not exceeding 50°; fixing to steel purlins with plastic headed self-tapping screws							
0.7 mm thick type HPS200 13.5/3 corrugated	–	–	–	–	–	m²	**14.71**
0.7 mm thick type HPS200 R32/1000	–	–	–	–	–	m²	**13.36**
0.7 mm thick type Arcline 40; plasticol finished	–	–	–	–	–	m²	**19.63**
Extra over last for aluminium roof cladding or decking	–	–	–	–	–	m²	**8.07**
Accessories for roof cladding							
HPS200 Drip flashing; 250 mm girth	–	–	–	–	–	m	**4.71**
HPS200 Ridge flashing; 375 mm girth	–	–	–	–	–	m	**5.97**
HPS200 Gable flashing; 500 mm girth	–	–	–	–	–	m	**7.59**
HPS200 Internal angle; 625 mm girth	–	–	–	–	–	m	**8.75**
GRP Transluscent rooflights; factory assembled							
Rooflight; vertical fixing to steel purlins (measured elsewhere)							
double skin; class 3 over 1	–	–	–	–	–	m²	**49.15**
triple skin; class 3 over 1	–	–	–	–	–	m²	**54.60**
Lightweight galvanized steel roof tiles; Decra Roof Systems; or other equal; coated finish							
Zinc coated tile panels, dry fixed							
Classic profile; clay or concrete tile appearance	–	0.23	5.10	–	20.16	m²	**25.26**
Stratos profile; slate or concrete tiles appearance	–	0.23	5.10	–	25.04	m²	**30.14**
Accessories for roof cladding							
pitched D ridge	–	0.09	2.00	–	11.94	m	**13.94**
barge cover (handed)	–	0.09	2.00	–	11.76	m	**13.76**
in line air vent	–	0.09	2.00	–	68.78	nr	**70.78**
in line soil vent	–	0.09	2.00	–	68.78	nr	**70.78**
gas flue terminal	–	0.19	4.21	–	139.05	nr	**143.26**
Kingspan KS100RW composite roof panels for roof pitches greater than 4° (after deflection)							
External coating XL Forte (steel), internal coating bright white polyester (steel)							
80 mm thick panel; U-value 0.25 W/m² K	–	–	–	–	–	m²	**42.59**
115 mm thick panel; U-value 0.18 W/m² K	–	–	–	–	–	m²	**44.73**
150 mm thick panel; U-value 0.14 W/m² K	–	–	–	–	–	m²	**60.06**
KS1000PC polycarbonate rooflight	–	–	–	–	–	m²	**13.11**

21 CLADDING AND COVERING

Item	PC £	Labour hours	Labour £	Plant £	Material £	Unit	Total rate £
21.04 PROFILED SHEET CLADDING – cont							
Kingspan KS100RW composite roof panels for roof pitches greater than 4° (after deflection) – **cont**							
Associated flashings							
eaves – 200 mm girth	–	–	–	–	–	m	**3.28**
ridge – 620 mm girth	–	–	–	–	–	m	**19.65**
hip – 620 mm girth	–	–	–	–	–	m	**19.65**
Kingspan KS1000 KZ composite Standing Seam roof panels for roof pitches greater than 1.5° (after deflection)							
External Coating XL Forte (Steel) Internal Coating Bright White Polyester (steel)							
90 mm thick panel; U-value 0.25 W/m² K	–	–	–	–	–	m²	**68.80**
110 mm thick panel; U-value 0.20 W/m² K	–	–	–	–	–	m²	**74.26**
125 mm thick panel; U-value 0.18 W/m² K	–	–	–	–	–	m²	**77.54**
Associated flashings							
eaves – 200 mm girth	–	–	–	–	–	m	**3.28**
ridge – 620 mm girth	–	–	–	–	–	m	**19.65**
hip – 620 mm girth	–	–	–	–	–	m	**19.65**
Fibre cement corrugated sheets; Eternit 2000 or other equal							
Roof cladding; sloping not exceeding 50°; fixing to steel purlins with hook bolts							
Profile 3; natural grey	–	0.23	5.10	–	20.90	m²	**26.00**
Profile 3; coloured	–	0.23	5.10	–	23.76	m²	**28.86**
Profile 6; natural grey	–	0.28	6.20	–	15.71	m²	**21.91**
Profile 6; coloured	–	0.28	6.20	–	17.49	m²	**23.69**
Profile 6; natural grey; insulated 100 mm glass fibre infill; white finish steel lining panel	–	0.46	10.19	–	31.15	m²	**41.34**
Profile 6; coloured; insulated 100 mm glass fibre infill; white finish steel lining panel	–	0.46	10.19	–	32.47	m²	**42.66**
Accessories; to Profile 3 cladding; natural grey							
eaves filler	–	0.09	2.00	–	12.32	m	**14.32**
external corner piece	–	0.11	2.44	–	9.11	m	**11.55**
apron flashing	–	0.11	2.44	–	12.32	m	**14.76**
plain wing or close fitting two piece adjustable capping to ridge	–	0.16	3.54	–	11.57	m	**15.11**
ventilating two piece adjustable capping to ridge	–	0.16	3.54	–	17.78	m	**21.32**
Accessories; to Profile 6 cladding; natural grey							
eaves filler	–	0.09	2.00	–	7.43	m	**9.43**
external corner piece	–	0.11	2.44	–	8.46	m	**10.90**
apron flashing	–	0.11	2.44	–	8.25	m	**10.69**
underglazing flashing	–	0.11	2.44	–	10.89	m	**13.33**
plain cranked crown to ridge	–	0.16	3.54	–	22.11	m	**25.65**
plain wing or close fitting two piece adjustable capping to ridge	–	0.16	3.54	–	14.85	m	**18.39**
ventilating two piece adjustable capping to ridge	–	0.16	3.54	–	18.99	m	**22.53**

21 CLADDING AND COVERING

Item	PC £	Labour hours	Labour £	Plant £	Material £	Unit	Total rate £
21.05 WALL CLADDING – METAL							
Extended, hard skinned, foamed PVC-UE profiled sections; Swish Celuka or other equal and approved; Class 1 fire-rated to BS 476; Part 7; in white finish							
Wall cladding; vertical; fixing to timber							
100 mm shiplap profiles; Code 001	–	0.35	8.04	–	67.78	m²	**75.82**
150 mm shiplap profiles; Code 002	–	0.32	7.35	–	60.06	m²	**67.41**
125 mm feather-edged profiles; Code C208	–	0.34	7.82	–	65.46	m²	**73.28**
vertical angles	–	0.19	4.37	–	6.96	m	**11.33**
raking cutting	–	0.14	3.21	–	–	m	**3.21**
Kingspan KS1000RW Composite Wall Panels							
External coating XL Forte (steel) internal coating bright white polyester (steel)							
60 mm thick; U-value 0.35 W/m² K	–	–	–	–	–	m²	**49.49**
80 mm thick; U-value 0.26 W/m² K	–	–	–	–	–	m²	**52.08**
100 mm thick; U-value 0.20 W/m² K	–	–	–	–	–	m²	**55.33**
Kingspan KS1000MR Composite Wall Panels							
External coating Spectum (steel) internal coating bright white polyester (steel)							
60 mm thick; U-value 0.35 W/m² K	–	–	–	–	–	m²	**88.54**
80 mm thick; U-value 0.26 W/m² K	–	–	–	–	–	m²	**95.06**
100 mm thick; U-value 0.20 W/m² K	–	–	–	–	–	m²	**101.57**
Kingspan KS900MR Composite Wall Panels							
External coating Spectum (steel) internal coating bright white polyester (steel)							
60 mm thick; U-value 0.35 W/m² K	–	–	–	–	–	m²	**101.57**
80 mm thick; U-value 0.26 W/m² K	–	–	–	–	–	m²	**109.38**
100 mm thick; U-value 0.20 W/m² K	–	–	–	–	–	m²	**117.18**
Kingspan KS600MR Composite Wall Panels							
External coating Sprectum (steel) internal coating bright white polyester (steel)							
60 mm thick; U-value 0.35 W/m² K	–	–	–	–	–	m²	**136.72**
80 mm thick; U-value 0.26 W/m² K	–	–	–	–	–	m²	**148.45**
100 mm thick; U-value 0.20 W/m² K	–	–	–	–	–	m²	**160.16**
Kingspan KS1000 Optimo FLAT Composite Wall Panels							
External coating Spectum (steel) internal coating bright white polyester (steel)							
60 mm thick; U-value 0.35 W/m² K	–	–	–	–	–	m²	**148.45**
80 mm thick; U-value 0.26 W/m² K	–	–	–	–	–	m²	**156.25**
100 mm thick; U-value 0.20 W/m² K	–	–	–	–	–	m²	**164.07**

21 CLADDING AND COVERING

Item	PC £	Labour hours	Labour £	Plant £	Material £	Unit	Total rate £
21.05 WALL CLADDING – METAL – cont							
Kingspan KS900 Optimo FLAT Composite Wall Panels							
External coating Spectrum (steel) internal coating bright white polyester (steel)							
60 mm thick; U-value 0.35 W/m² K	–	–	–	–	–	m²	169.27
80 mm thick; U-value 0.26 W/m² K	–	–	–	–	–	m²	177.09
100 mm thick; U-value 0.20 W/m² K	–	–	–	–	–	m²	184.88
Kingspan KS600 Optimo FLAT Composite Wall Panels							
External coating Spectrum (steel) internal coating bright white polyester (steel)							
60 mm thick; U-value 0.35 W/m² K	–	–	–	–	–	m²	221.36
80 mm thick; U-value 0.26 W/m² K	–	–	–	–	–	m²	234.38
100 mm thick; U-value 0.20 W/m² K	–	–	–	–	–	m²	247.41
Associated flashings							
0.7 mm pre-coated steel							
cill; 300 mm girth	–	–	–	–	–	m	23.43
composite top hat	–	–	–	–	–	m	27.34
Miscellaneous items							
panel bearers at 1500 mm centres	–	–	–	–	–	nr	10.42
preformed corners to horizontally laid panels	–	–	–	–	–	nr	117.18
21.06 WALL CLADDING – RAINSCREEN							
Timber							
Western Red Cedar tongued and grooved wall cladding on and including treated softwood battens on breather mambrane, 10 mm Eternit Blueclad board and 50 mm insulation board; the whole fixed to Metsec frame system; including sealing all joints etc.							
26 mm thick cladding to walls; boards laid horizontally	–	–	–	–	–	m²	115.00
Aluminium							
Reynobond rainscreen cladding; aluminium composite material cassettes with thermoplastic cores, back ventilated, including insulation, vapour control membrane and aluminium support system							
4 mm thick cladding; fixed to walls	–	–	–	–	–	m²	204.48
Clay							
Terracotta clay rainscreen cladding; including insulation, vapour control membrane and aluminium support system							
400 × 200 × 30 mm tile cladding; fixed to walls	–	–	–	–	–	m²	370.60

22 GENERAL JOINERY

Item	PC £	Labour hours	Labour £	Plant £	Material £	Unit	Total rate £
22.01 SOFTWOOD							
Wrought softwood							
Skirtings, picture rails, dado rails and the like; splayed or moulded							
19 mm × 44 mm; splayed	–	0.09	2.07	–	3.11	m	**5.18**
19 mm × 44 mm; moulded	–	0.09	2.07	–	3.31	m	**5.38**
19 mm × 69 mm; splayed	–	0.09	2.07	–	3.35	m	**5.42**
19 mm × 69 mm; moulded	–	0.09	2.07	–	3.35	m	**5.42**
19 mm × 94 mm; splayed	–	0.09	2.07	–	3.74	m	**5.81**
19 mm × 94 mm; moulded	–	0.09	2.07	–	3.74	m	**5.81**
19 mm × 144 mm; moulded	–	0.11	2.52	–	4.39	m	**6.91**
19 mm × 169 mm; moulded	–	0.11	2.52	–	4.65	m	**7.17**
25 mm × 50 mm; moulded	–	0.09	2.07	–	3.24	m	**5.31**
25 mm × 69 mm; splayed	–	0.09	2.07	–	3.58	m	**5.65**
25 mm × 94 mm; splayed	–	0.09	2.07	–	4.00	m	**6.07**
25 mm × 144 mm; splayed	–	0.11	2.52	–	4.82	m	**7.34**
25 mm × 144 mm; moulded	–	0.11	2.52	–	4.82	m	**7.34**
25 mm × 169 mm; moulded	–	0.11	2.52	–	5.36	m	**7.88**
25 mm × 219 mm; moulded	–	0.13	2.99	–	6.94	m	**9.93**
returned ends	–	0.14	3.21	–	–	nr	**3.21**
mitres	–	0.09	2.07	–	–	nr	**2.07**
Architraves, cover fillets and the like; half round; splayed or moulded							
13 mm × 25 mm; half round	–	0.11	2.52	–	2.93	m	**5.45**
13 mm × 50 mm; moulded	–	0.11	2.52	–	3.11	m	**5.63**
16 mm × 32 mm; half round	–	0.11	2.52	–	3.24	m	**5.76**
16 mm × 38 mm; moulded	–	0.11	2.52	–	3.24	m	**5.76**
16 mm × 50 mm; moulded	–	0.11	2.52	–	3.24	m	**5.76**
19 mm × 50 mm; splayed	–	0.11	2.52	–	3.24	m	**5.76**
19 mm × 63 mm; splayed	–	0.11	2.52	–	3.35	m	**5.87**
19 mm × 69 mm; splayed	–	0.11	2.52	–	3.55	m	**6.07**
25 mm × 44 mm; splayed	–	0.11	2.52	–	3.19	m	**5.71**
25 mm × 50 mm; moulded	–	0.11	2.52	–	3.36	m	**5.88**
25 mm × 63 mm; splayed	–	0.11	2.52	–	3.50	m	**6.02**
25 mm × 69 mm; splayed	–	0.11	2.52	–	3.98	m	**6.50**
32 mm × 88 mm; moulded	–	0.11	2.52	–	4.00	m	**6.52**
38 mm × 38 mm; moulded	–	0.11	2.52	–	3.54	m	**6.06**
50 mm × 50 mm; moulded	–	0.11	2.52	–	4.11	m	**6.63**
returned ends	–	0.14	3.21	–	–	nr	**3.21**
mitres	–	0.09	2.07	–	–	nr	**2.07**
Stops; screwed on							
16 mm × 38 mm	–	0.09	2.07	–	1.41	m	**3.48**
16 mm × 50 mm	–	0.09	2.07	–	1.52	m	**3.59**
19 mm × 38 mm	–	0.09	2.07	–	1.41	m	**3.48**
25 mm × 38 mm	–	0.09	2.07	–	1.53	m	**3.60**
25 mm × 50 mm	–	0.09	2.07	–	1.59	m	**3.66**
Glazing beads and the like							
13 mm × 16 mm	–	0.04	0.92	–	1.72	m	**2.64**
13 mm × 19 mm	–	0.04	0.92	–	1.72	m	**2.64**
13 mm × 25 mm	–	0.04	0.92	–	1.76	m	**2.68**
13 mm × 25 mm; screwed	–	0.04	0.92	–	2.85	m	**3.77**
13 mm × 25 mm; fixing with brass cups and screws	–	0.04	0.92	–	3.69	m	**4.61**

22 GENERAL JOINERY

Item	PC £	Labour hours	Labour £	Plant £	Material £	Unit	Total rate £
22.01 SOFTWOOD – cont							
Glazing beads and the like – cont							
16 mm × 25 mm; screwed	–	0.04	0.92	–	2.85	m	**3.77**
16 mm quadrant	–	0.04	0.92	–	2.63	m	**3.55**
19 mm quadrant or scotia	–	0.04	0.92	–	2.63	m	**3.55**
19 mm × 36 mm; screwed	–	0.04	0.92	–	2.90	m	**3.82**
25 mm × 38 mm; screwed	–	0.04	0.92	–	3.05	m	**3.97**
25 mm quadrant or scotia	–	0.04	0.92	–	2.77	m	**3.69**
38 mm scotia	–	0.04	0.92	–	3.34	m	**4.26**
50 mm scotia	–	0.04	0.92	–	3.91	m	**4.83**
Isolated shelves, worktops, seats and the like							
19 mm × 150 mm	–	0.15	3.45	–	3.65	m	**7.10**
19 mm × 200 mm	–	0.20	4.59	–	5.03	m	**9.62**
25 mm × 150 mm	–	0.15	3.45	–	4.15	m	**7.60**
25 mm × 200 mm	–	0.20	4.59	–	5.91	m	**10.50**
32 mm × 150 mm	–	0.15	3.45	–	4.85	m	**8.30**
32 mm × 200 mm	–	0.20	4.59	–	6.62	m	**11.21**
Isolated shelves, worktops, seats and the like; cross-tongued joints							
19 mm × 300 mm	–	0.26	5.97	–	14.98	m	**20.95**
19 mm × 450 mm	–	0.31	7.13	–	22.58	m	**29.71**
19 mm × 600 mm	–	0.37	8.51	–	29.21	m	**37.72**
25 mm × 300 mm	–	0.26	5.97	–	16.06	m	**22.03**
25 mm × 450 mm	–	0.31	7.13	–	24.34	m	**31.47**
25 mm × 600 mm	–	0.37	8.51	–	31.66	m	**40.17**
32 mm × 300 mm	–	0.26	5.97	–	17.02	m	**22.99**
32 mm × 450 mm	–	0.31	7.13	–	25.87	m	**33.00**
32 mm × 600 mm	–	0.37	8.51	–	33.75	m	**42.26**
Isolated shelves, worktops, seats and the like; slatted with 50 wide slats at 75 mm centres							
19 mm thick	–	0.60	13.79	–	36.84	m	**50.63**
25 mm thick	–	0.60	13.79	–	37.67	m	**51.46**
32 mm thick	–	0.60	13.79	–	38.38	m	**52.17**
Window boards, nosings, bed moulds and the like; rebated and rounded							
19 mm × 75 mm	–	0.17	3.90	–	4.77	m	**8.67**
19 mm × 150 mm	–	0.19	4.37	–	5.86	m	**10.23**
19 mm × 225 mm; in one width	–	0.24	5.52	–	7.22	m	**12.74**
19 mm × 300 mm; cross-tongued joints	–	0.28	6.44	–	16.46	m	**22.90**
25 mm × 75 mm	–	0.17	3.90	–	5.03	m	**8.93**
25 mm × 150 mm	–	0.19	4.37	–	6.41	m	**10.78**
25 mm × 225 mm; in one width	–	0.24	5.52	–	8.06	m	**13.58**
25 mm × 300 mm; cross-tongued joints	–	0.28	6.44	–	17.79	m	**24.23**
32 mm × 75 mm	–	0.17	3.90	–	5.25	m	**9.15**
32 mm × 150 mm	–	0.19	4.37	–	6.90	m	**11.27**
32 mm × 225 mm; in one width	–	0.24	5.52	–	8.80	m	**14.32**
32 mm × 300 mm; cross-tongued joints	–	0.28	6.44	–	18.91	m	**25.35**
38 mm × 75 mm	–	0.17	3.90	–	5.75	m	**9.65**
38 mm × 150 mm	–	0.19	4.37	–	7.93	m	**12.30**
38 mm × 225 mm; in one width	–	0.24	5.52	–	10.26	m	**15.78**

22 GENERAL JOINERY

Item	PC £	Labour hours	Labour £	Plant £	Material £	Unit	Total rate £
38 mm × 300 mm; cross-tongued joints	–	0.28	6.44	–	21.16	m	**27.60**
returned and fitted ends	–	0.14	3.21	–	–	nr	**3.21**
Handrails; mopstick							
50 mm dia.	–	0.23	5.28	–	10.44	m	**15.72**
Handrails; rounded							
44 mm × 50 mm	–	0.23	5.28	–	10.12	m	**15.40**
50 mm × 75 mm	–	0.25	5.75	–	11.05	m	**16.80**
63 mm × 87 mm	–	0.28	6.44	–	12.27	m	**18.71**
75 mm × 100 mm	–	0.32	7.35	–	15.03	m	**22.38**
Handrails; moulded							
44 mm × 50 mm	–	0.23	5.28	–	10.12	m	**15.40**
50 mm × 75 mm	–	0.25	5.75	–	11.05	m	**16.80**
63 mm × 87 mm	–	0.28	6.44	–	12.27	m	**18.71**
75 mm × 100 mm	–	0.32	7.35	–	15.03	m	**22.38**
Sundries on softwood							
Extra over fixing with nails for							
gluing and pinning	–	0.02	0.40	–	0.04	m	**0.44**
masonry nails	–	0.02	0.41	–	0.13	m	**0.54**
steel screws	–	0.02	0.38	–	0.10	m	**0.48**
self-tapping screws	–	0.02	0.40	–	0.10	m	**0.50**
steel screws; gluing	–	0.03	0.67	–	0.10	m	**0.77**
steel screws; sinking; filling heads	–	0.04	0.85	–	0.10	m	**0.95**
steel screws; sinking; pellating over	–	0.08	1.85	–	0.10	m	**1.95**
brass cups and screws	–	0.10	2.29	–	0.26	m	**2.55**
Extra over for							
countersinking	–	0.01	0.34	–	–	m	**0.34**
pellating	–	0.07	1.61	–	–	m	**1.61**
Head or nut in softwood							
let in flush	–	0.04	0.85	–	–	nr	**0.85**
Head or nut; in hardwood							
let in flush	–	0.06	1.27	–	–	nr	**1.27**
let in over; pellated	–	0.13	2.97	–	–	nr	**2.97**
22.02 HARDWOOD							
Selected Sapele							
Skirtings, picture rails, dado rails and the like;							
splayed or moulded							
19 mm × 44 mm; splayed	4.37	0.13	2.99	–	4.71	m	**7.70**
19 mm × 44 mm; moulded	4.37	0.13	2.99	–	4.71	m	**7.70**
19 mm × 69 mm; splayed	5.08	0.13	2.99	–	5.44	m	**8.43**
19 mm × 69 mm; moulded	5.08	0.13	2.99	–	5.44	m	**8.43**
19 mm × 94 mm; splayed	5.93	0.13	2.99	–	6.31	m	**9.30**
19 mm × 94 mm; moulded	5.93	0.13	2.99	–	6.31	m	**9.30**
19 mm × 144 mm; moulded	7.88	0.15	3.45	–	8.32	m	**11.77**
19 mm × 169 mm; moulded	8.72	0.15	3.45	–	9.19	m	**12.64**
25 mm × 44 mm; moulded	4.91	0.13	2.99	–	5.26	m	**8.25**
25 mm × 69 mm; splayed	5.85	0.13	2.99	–	6.23	m	**9.22**
25 mm × 94 mm; splayed	7.17	0.13	2.99	–	7.59	m	**10.58**

22 GENERAL JOINERY

Item	PC £	Labour hours	Labour £	Plant £	Material £	Unit	Total rate £
22.02 HARDWOOD – cont							
Selected Sapele – cont							
Skirtings, picture rails, dado rails and the like – cont							
25 mm × 144 mm; splayed	9.40	0.15	3.45	–	9.89	m	**13.34**
25 mm × 144 mm; moulded	9.40	0.15	3.45	–	9.89	m	**13.34**
25 mm × 169 mm; moulded	10.55	0.15	3.45	–	11.07	m	**14.52**
25 mm × 219 mm; moulded	12.05	0.17	3.90	–	12.61	m	**16.51**
returned ends	–	0.20	4.59	–	–	nr	**4.59**
mitres	–	0.14	3.21	–	–	nr	**3.21**
Architraves, cover fillets and the like; half round; splayed or moulded							
13 mm × 25 mm; half round	2.65	0.15	3.45	–	2.94	m	**6.39**
13 mm × 50 mm; moulded	4.24	0.15	3.45	–	4.57	m	**8.02**
16 mm × 32 mm; half round	2.69	0.15	3.45	–	2.98	m	**6.43**
16 mm × 38 mm; moulded	4.09	0.15	3.45	–	4.42	m	**7.87**
16 mm × 50 mm; moulded	4.37	0.15	3.45	–	4.71	m	**8.16**
19 mm × 50 mm; splayed	4.37	0.15	3.45	–	4.71	m	**8.16**
19 mm × 63 mm; splayed	4.74	0.15	3.45	–	5.08	m	**8.53**
19 mm × 69 mm; splayed	5.08	0.15	3.45	–	5.44	m	**8.89**
25 mm × 44 mm; splayed	4.73	0.15	3.45	–	5.08	m	**8.53**
25 mm × 50 mm; moulded	4.91	0.15	3.45	–	5.26	m	**8.71**
25 mm × 63 mm; splayed	5.49	0.15	3.45	–	5.85	m	**9.30**
25 mm × 69 mm; splayed	5.85	0.15	3.45	–	6.23	m	**9.68**
32 mm × 88 mm; moulded	7.13	0.15	3.45	–	7.55	m	**11.00**
38 mm × 38 mm; moulded	5.59	0.15	3.45	–	5.96	m	**9.41**
50 mm × 50 mm; moulded	7.32	0.15	3.45	–	7.74	m	**11.19**
returned ends	–	0.20	4.59	–	–	nr	**4.59**
mitres	–	0.14	3.21	–	–	nr	**3.21**
Stops; screwed on							
16 mm × 38 mm	1.80	0.14	3.21	–	1.85	m	**5.06**
16 mm × 50 mm	1.94	0.14	3.21	–	2.00	m	**5.21**
19 mm × 38 mm	1.80	0.14	3.21	–	1.85	m	**5.06**
25 mm × 38 mm	2.28	0.14	3.21	–	2.35	m	**5.56**
25 mm × 50 mm	2.65	0.14	3.21	–	2.73	m	**5.94**
Glazing beads and the like							
13 mm × 16 mm	2.35	0.06	1.38	–	2.42	m	**3.80**
13 mm × 19 mm	2.35	0.06	1.38	–	2.42	m	**3.80**
13 mm × 25 mm	2.54	0.06	1.38	–	2.62	m	**4.00**
13 mm × 25 mm; screwed	3.43	0.06	1.38	–	3.53	m	**4.91**
13 mm × 25 mm; fixing with brass cups and screws	4.23	0.06	1.38	–	4.36	m	**5.74**
16 mm × 25 mm; screwed	3.43	0.06	1.38	–	3.53	m	**4.91**
16 mm quadrant	3.31	0.06	1.38	–	3.41	m	**4.79**
19 mm quadrant or scotia	3.31	0.06	1.38	–	3.41	m	**4.79**
19 mm × 36 mm; screwed	4.23	0.06	1.38	–	4.36	m	**5.74**
25 mm × 38 mm; screwed	4.61	0.06	1.38	–	4.75	m	**6.13**
25 mm quadrant or scotia	3.77	0.06	1.38	–	3.88	m	**5.26**
38 mm scotia	5.59	0.06	1.38	–	5.76	m	**7.14**
50 mm scotia	7.32	0.06	1.38	–	7.54	m	**8.92**

22 GENERAL JOINERY

Item	PC £	Labour hours	Labour £	Plant £	Material £	Unit	Total rate £
Isolated shelves; worktops, seats and the like							
19 mm × 150 mm	8.02	0.20	4.59	–	8.26	m	**12.85**
19 mm × 200 mm	9.49	0.28	6.44	–	9.77	m	**16.21**
25 mm × 150 mm	9.40	0.20	4.59	–	9.68	m	**14.27**
25 mm × 200 mm	11.30	0.28	6.44	–	11.64	m	**18.08**
32 mm × 150 mm	10.61	0.20	4.59	–	10.93	m	**15.52**
32 mm × 200 mm	12.84	0.28	6.44	–	13.23	m	**19.67**
Isolated shelves, worktops, seats and the like; cross-tongued joints							
19 mm × 300 mm	21.71	0.35	8.04	–	22.36	m	**30.40**
19 mm × 450 mm	33.99	0.42	9.65	–	35.01	m	**44.66**
19 mm × 600 mm	45.14	0.51	11.72	–	46.49	m	**58.21**
25 mm × 300 mm	24.38	0.35	8.04	–	25.11	m	**33.15**
25 mm × 450 mm	38.30	0.42	9.65	–	39.45	m	**49.10**
25 mm × 600 mm	50.90	0.51	11.72	–	52.43	m	**64.15**
32 mm × 300 mm	26.66	0.35	8.04	–	27.46	m	**35.50**
32 mm × 450 mm	41.98	0.42	9.65	–	43.24	m	**52.89**
32 mm × 600 mm	55.82	0.51	11.72	–	57.49	m	**69.21**
Isolated shelves, worktops, seats and the like; slatted with 50 wide slats at 75 mm centres							
19 mm thick	62.56	0.80	18.39	–	65.20	m²	**83.59**
25 mm thick	66.97	0.80	18.39	–	69.74	m²	**88.13**
32 mm thick	70.75	0.80	18.39	–	73.63	m²	**92.02**
Window boards, nosings, bed moulds and the like; rebated and rounded							
19 mm × 75 mm	6.19	0.22	5.06	–	6.83	m	**11.89**
19 mm × 150 mm	8.98	0.25	5.75	–	9.70	m	**15.45**
19 mm × 225 mm; in one width	11.02	0.33	7.58	–	11.80	m	**19.38**
19 mm × 300 mm; cross-tongued joints	22.45	0.37	8.51	–	23.59	m	**32.10**
25 mm × 75 mm	6.82	0.22	5.06	–	7.48	m	**12.54**
25 mm × 150 mm	10.00	0.25	5.75	–	10.75	m	**16.50**
25 mm × 225 mm; in one width	13.13	0.33	7.58	–	13.98	m	**21.56**
25 mm × 300 mm; cross-tongued joints	26.09	0.37	8.51	–	27.33	m	**35.84**
32 mm × 75 mm	7.39	0.22	5.06	–	8.08	m	**13.14**
32 mm × 150 mm	11.12	0.25	5.75	–	11.92	m	**17.67**
32 mm × 225 mm; in one width	14.83	0.33	7.58	–	15.74	m	**23.32**
32 mm × 300 mm; cross-tongued joints	28.69	0.37	8.51	–	30.01	m	**38.52**
returned and fitted ends	–	0.21	4.83	–	–	nr	**4.83**
Handrails; rounded							
44 mm × 50 mm	13.62	0.31	7.13	–	14.03	m	**21.16**
50 mm × 75 mm	16.43	0.33	7.58	–	16.92	m	**24.50**
63 mm × 87 mm	19.28	0.37	8.51	–	19.86	m	**28.37**
75 mm × 100 mm	23.96	0.42	9.65	–	24.68	m	**34.33**
Handrails; moulded							
44 mm × 50 mm	15.17	0.31	7.13	–	15.63	m	**22.76**
50 mm × 75 mm	17.95	0.33	7.58	–	18.49	m	**26.07**
63 mm × 87 mm	20.82	0.37	8.51	–	21.44	m	**29.95**
75 mm × 100 mm	25.49	0.42	9.65	–	26.25	m	**35.90**

22 GENERAL JOINERY

Item	PC £	Labour hours	Labour £	Plant £	Material £	Unit	Total rate £
22.02 HARDWOOD – cont							
Sundries on hardwood							
Extra over fixing with nails for							
gluing and pinning	–	0.02	0.40	–	0.04	m	**0.44**
masonry nails	–	0.02	0.41	–	0.13	m	**0.54**
steel screws	–	0.02	0.38	–	0.10	m	**0.48**
self-tapping screws	–	0.02	0.40	–	0.10	m	**0.50**
steel screws; gluing	–	0.03	0.67	–	0.10	m	**0.77**
steel screws; sinking; filling heads	–	0.04	0.85	–	0.10	m	**0.95**
steel screws; sinking; pellating over	–	0.08	1.85	–	0.10	m	**1.95**
brass cups and screws	–	0.10	2.29	–	0.26	m	**2.55**
Extra over for							
countersinking	–	0.01	0.34	–	–	m	**0.34**
pellating	–	0.07	1.61	–	–	m	**1.61**
Head or nut in softwood							
let in flush	–	0.04	0.85	–	–	nr	**0.85**
Head or nut in hardwood							
let in flush	–	0.06	1.27	–	–	nr	**1.27**
let in over; pellated	–	0.13	2.97	–	–	nr	**2.97**
22.03 MEDIUM DENSITY FIBREBOARD							
Medium density fibreboard; Sapele veneered one side; 18 mm thick							
Window boards and the like; rebated; hardwood lipped on one edge							
18 mm × 200 mm	–	0.25	5.75	–	16.12	m	**21.87**
18 mm × 250 mm	–	0.28	6.44	–	16.95	m	**23.39**
18 mm × 300 mm	–	0.31	7.13	–	17.38	m	**24.51**
18 mm × 350 mm	–	0.33	7.58	–	18.63	m	**26.21**
returned and fitted ends	–	0.20	4.59	–	3.12	nr	**7.71**
Medium density fibreboard; American White Ash veneered one side; 18 mm thick							
Window boards and the like; rebated; hardwood lipped on one edge							
18 mm × 200 mm	–	0.25	5.75	–	16.76	m	**22.51**
18 mm × 250 mm	–	0.28	6.44	–	17.80	m	**24.24**
18 mm × 300 mm	–	0.31	7.13	–	18.31	m	**25.44**
18 mm × 350 mm	–	0.33	7.58	–	19.87	m	**27.45**
returned and fitted ends	–	0.20	4.59	–	3.12	nr	**7.71**
Medium density fibreboard							
Skirtings, picture rails, dado rails and the like; splayed or moulded							
18 mm × 50 mm; splayed	–	0.09	2.07	–	3.05	m	**5.12**
18 mm × 50 mm; moulded	–	0.09	2.07	–	3.05	m	**5.12**
18 mm × 75 mm; splayed	–	0.09	2.07	–	3.17	m	**5.24**

22 GENERAL JOINERY

Item	PC £	Labour hours	Labour £	Plant £	Material £	Unit	Total rate £
18 mm × 75 mm; moulded	–	0.09	2.07	–	3.17	m	**5.24**
18 mm × 100 mm; splayed	–	0.09	2.07	–	3.31	m	**5.38**
18 mm × 100 mm; moulded	–	0.09	2.07	–	3.31	m	**5.38**
18 mm × 150 mm; moulded	–	0.11	2.52	–	3.59	m	**6.11**
18 mm × 175 mm; moulded	–	0.11	2.52	–	3.72	m	**6.24**
22 mm × 100 mm; splayed	–	0.09	2.07	–	5.26	m	**7.33**
25 mm × 50 mm; moulded	–	0.09	2.07	–	3.19	m	**5.26**
25 mm × 75 mm; splayed	–	0.09	2.07	–	3.38	m	**5.45**
25 mm × 100 mm; splayed	–	0.09	2.07	–	3.58	m	**5.65**
25 mm × 150 mm; splayed	–	0.11	2.52	–	4.02	m	**6.54**
25 mm × 150 mm; moulded	–	0.11	2.52	–	4.02	m	**6.54**
25 mm × 175 mm; moulded	–	0.11	2.52	–	4.23	m	**6.75**
25 mm × 225 mm; moulded	–	0.13	2.99	–	4.53	m	**7.52**
returned ends	–	0.14	3.21	–	–	nr	**3.21**
mitres	–	0.09	2.07	–	–	nr	**2.07**
Architraves, cover fillets and the like; half round; splayed or moulded							
12 mm × 25 mm; half round	–	0.11	2.52	–	2.93	m	**5.45**
12 mm × 50 mm; moulded	–	0.11	2.52	–	3.01	m	**5.53**
15 mm × 32 mm; half round	–	0.11	2.52	–	2.94	m	**5.46**
15 mm × 38 mm; moulded	–	0.11	2.52	–	2.97	m	**5.49**
15 mm × 50 mm; moulded	–	0.11	2.52	–	3.01	m	**5.53**
18 mm × 50 mm; splayed	–	0.11	2.52	–	3.01	m	**5.53**
18 mm × 63 mm; splayed	–	0.11	2.52	–	3.13	m	**5.65**
18 mm × 75 mm; splayed	–	0.11	2.52	–	3.19	m	**5.71**
25 mm × 44 mm; splayed	–	0.11	2.52	–	3.19	m	**5.71**
25 mm × 50 mm; moulded	–	0.11	2.52	–	3.19	m	**5.71**
25 mm × 63 mm; splayed	–	0.11	2.52	–	3.31	m	**5.83**
25 mm × 75 mm; splayed	–	0.11	2.52	–	3.41	m	**5.93**
30 mm × 88 mm; moulded	–	0.11	2.52	–	4.39	m	**6.91**
38 mm × 38 mm; moulded	–	0.11	2.52	–	3.78	m	**6.30**
50 mm × 50 mm; moulded	–	0.11	2.52	–	3.97	m	**6.49**
returned ends	–	0.14	3.21	–	–	nr	**3.21**
mitres	–	0.09	2.07	–	–	nr	**2.07**
Stops; screwed on							
15 mm × 38 mm	–	0.09	2.07	–	1.62	m	**3.69**
15 mm × 50 mm	–	0.09	2.07	–	1.67	m	**3.74**
18 mm × 38 mm	–	0.09	2.07	–	1.65	m	**3.72**
25 mm × 38 mm	–	0.09	2.07	–	1.74	m	**3.81**
25 mm × 50 mm	–	0.09	2.07	–	1.82	m	**3.89**
Glazing beads and the like							
12 mm × 16 mm	–	0.04	0.92	–	1.85	m	**2.77**
12 mm × 19 mm	–	0.04	0.92	–	1.86	m	**2.78**
12 mm × 25 mm	–	0.04	0.92	–	1.88	m	**2.80**
12 mm × 25 mm; screwed	–	0.04	0.92	–	2.70	m	**3.62**
12 mm × 25 mm; fixing with brass cups and screws	–	0.04	0.92	–	3.11	m	**4.03**
15 mm × 25 mm; screwed	–	0.04	0.92	–	2.81	m	**3.73**
15 mm quadrant	–	0.04	0.92	–	2.68	m	**3.60**

22 GENERAL JOINERY

Item	PC £	Labour hours	Labour £	Plant £	Material £	Unit	Total rate £
22.03 MEDIUM DENSITY FIBREBOARD – cont							
Medium density fibreboard – cont							
Glazing beads and the like – cont							
18 mm quadrant or scotia	–	0.04	0.92	–	2.69	m	**3.61**
18 mm × 36 mm; screwed	–	0.04	0.92	–	2.85	m	**3.77**
25 mm × 38 mm; screwed	–	0.04	0.92	–	2.99	m	**3.91**
25 mm quadrant or scotia	–	0.04	0.92	–	2.80	m	**3.72**
38 mm scotia	–	0.04	0.92	–	2.67	m	**3.59**
50 mm scotia	–	0.04	0.92	–	3.09	m	**4.01**
Isolated shelves, worktops, seats and the like							
18 mm × 150 mm	–	0.15	3.45	–	3.50	m	**6.95**
18 mm × 200 mm	–	0.20	4.59	–	3.70	m	**8.29**
25 mm × 150 mm	–	0.15	3.45	–	4.03	m	**7.48**
25 mm × 200 mm	–	0.20	4.59	–	4.31	m	**8.90**
30 mm × 150 mm	–	0.15	3.45	–	5.65	m	**9.10**
30 mm × 200 mm	–	0.20	4.59	–	6.26	m	**10.85**
Isolated shelves, worktops, seats and the like; cross-tongued joints							
18 mm × 300 mm	–	0.26	5.97	–	11.44	m	**17.41**
18 mm × 450 mm	–	0.31	7.13	–	13.06	m	**20.19**
18 mm × 600 mm	–	0.37	8.51	–	21.85	m	**30.36**
25 mm × 300 mm	–	0.26	5.97	–	12.06	m	**18.03**
25 mm × 450 mm	–	0.31	7.13	–	14.75	m	**21.88**
25 mm × 600 mm	–	0.37	8.51	–	21.28	m	**29.79**
30 mm × 300 mm	–	0.26	5.97	–	13.92	m	**19.89**
30 mm × 450 mm	–	0.31	7.13	–	16.57	m	**23.70**
30 mm × 600 mm	–	0.37	8.51	–	24.10	m	**32.61**
Isolated shelves, worktops, seats and the like; slatted with 50 wide slats at 75 mm centres							
18 mm thick	–	0.60	13.79	–	36.17	m	**49.96**
25 mm thick	–	0.60	13.79	–	38.37	m	**52.16**
30 mm thick	–	0.60	13.79	–	40.35	m	**54.14**
Window boards, nosings, bed moulds and the like; rebated and rounded							
18 mm × 75 mm	–	0.17	3.90	–	3.58	m	**7.48**
18 mm × 150 mm	–	0.19	4.37	–	4.03	m	**8.40**
18 mm × 225 mm	–	0.24	5.52	–	4.37	m	**9.89**
18 mm × 300 mm	–	0.28	6.44	–	4.80	m	**11.24**
25 mm × 75 mm	–	0.17	3.90	–	3.74	m	**7.64**
25 mm × 150 mm	–	0.19	4.37	–	4.37	m	**8.74**
25 mm × 225 mm	–	0.24	5.52	–	4.87	m	**10.39**
25 mm × 300 mm	–	0.28	6.44	–	5.45	m	**11.89**
30 mm × 75 mm	–	0.17	3.90	–	5.04	m	**8.94**
30 mm × 150 mm	–	0.19	4.37	–	6.18	m	**10.55**
30 mm × 225 mm	–	0.24	5.52	–	7.07	m	**12.59**
30 mm × 300 mm	–	0.28	6.44	–	8.11	m	**14.55**
38 mm × 75 mm	–	0.17	3.90	–	5.69	m	**9.59**
38 mm × 150 mm	–	0.19	4.37	–	7.06	m	**11.43**
38 mm × 225 mm	–	0.24	5.52	–	8.11	m	**13.63**
38 mm × 300 mm	–	0.28	6.44	–	9.37	m	**15.81**
returned and fitted ends	–	–	–	–	1.10	nr	**1.10**

22 GENERAL JOINERY

Item	PC £	Labour hours	Labour £	Plant £	Material £	Unit	Total rate £
Pin-boards; medium board							
Sundeala A pin-board or other equal and approved; fixed with adhesive to backing (not included); over 300 mm wide							
6 mm thick	–	0.56	12.88	–	6.73	m²	**19.61**
9 mm thick Colourboard	–	0.56	12.88	–	11.53	m²	**24.41**
22.04 FRAMED PANEL CUBICLE PARTITIONS							
Toilet cubicle partitions and IPS systems; Amwells or other equal and approved; standard colours and ironmongery; assembling and screwing to floor and wall							
Axis MFC standard cubic system, for use in small offices, retail and community halls etc.; standard cubicle set; 800 mm × 1500 mm × 1980 mm high per cubicle, with polished aluminium framing; 19 mm melamine-faced chipboard divisions and doors							
one cubicle set; 2 nr panels; 1 nr door	–	7.00	185.44	–	375.40	nr	**560.84**
range of 3 cubicle sets; 4 nr panels; 3 nr doors	–	21.50	569.57	–	902.53	nr	**1472.10**
range of 6 cubicle sets; 7 nr panels; 6 nr doors	–	42.50	1125.89	–	1693.20	nr	**2819.09**
reduction of 1 nr panel for end unit adjoining side wall	–	–	–	–	–85.78	nr	**–85.78**
Splash SGL cubic system for heavy use environments like schools, swimming pools and prisons; standard cubicle set; 800 mm × 1500 mm × 1980 mm high per cubicle, with polished aluminium framing; 19 mm melamine-faced chipboard divisions and doors							
one cubicle set; 2 nr panels; 1 nr door	–	7.00	185.44	–	771.96	nr	**957.40**
range of 3 cubicle sets; 4 nr panels; 3 nr doors	–	21.50	569.57	–	1781.20	nr	**2350.77**
range of 6 cubicle sets; 7 nr panels; 6 nr door	–	42.50	1125.89	–	3295.05	nr	**4420.94**
reduction of 1 nr panel for end unit adjoining side wall	–	–	–	–	–237.27	nr	**–237.27**
Minima Stylish stainless steel framed cubicle system in MFC, HPL, SGL, Real Wood veneer or Glass; 800 mm × 1500 mm × 2100 mm high per cubicle, with satin polished stainless steel framing; 18 mm high pressure laminated (HPL) chipboard divisions and doors							
one cubicle set; 2 nr panels; 1 nr door	–	7.00	185.44	–	902.29	nr	**1087.73**
range of 3 cubicle sets; 4 nr panels; 3 nr doors	–	21.50	569.57	–	2011.12	nr	**2580.69**
range of 6 cubicle set; 7 nr panels; 6 nr doors	–	42.50	1125.89	–	3674.34	nr	**4800.23**
reduction of 1 nr panel for end unit adjoining side wall	–	–	–	–	–200.51	nr	**–200.51**

22 GENERAL JOINERY

Item	PC £	Labour hours	Labour £	Plant £	Material £	Unit	Total rate £
22.04 FRAMED PANEL CUBICLE PARTITIONS – cont							
Toilet cubicle partitions and IPS systems – cont							
Urban Flush fronted, floor to ceiling option cubicle system in either 13 mm or 20 mm panels, material options HPL, SGL or Real Wood Veneer: cubicle set; 800 mm × 1500 mm × 2400 mm high per cubicle, with sating finished stainless steel ironmongery; 20 mm divisions, doors and pilasters							
one cubicle set; 2 nr panels; 1 nr door	–	8.00	211.93	–	1373.49	nr	**1585.42**
range of 3 cubicle set; 4 nr panels; 3 nr doors	–	23.50	622.55	–	3378.60	nr	**4001.15**
range of 6 cubicle sets; 7 nr panels; 6 nr doors	–	46.80	1239.81	–	6386.25	nr	**7626.06**
reduction of 1 nr panel for end unit adjoining side wall	–	–	–	–	–266.23	nr	**–266.23**
Sylan High end specification flush fronted system floor to ceiling 44 mm doors, no visible fixings. Finishes in Real wood Veneer, HPL, high Gloss paint and lacquer; cubicle set; 800 mm × 1500 mm × 2400 mm high per cubicle, with satin finished stainless steel ironmongery; 30 mm high pressure laminated (HPL) chipboard divisions and 44 mm solid cored real wood veneered doors and pilasters							
one cubicle set; 2 nr panels; 1 nr door	–	10.00	264.92	–	2553.16	nr	**2818.08**
range of 3 cubicle set; 4 nr panels; 3 nr doors	–	30.00	794.75	–	6014.19	nr	**6808.94**
range of 6 cubicle sets; 7 nr panels; 6 nr doors	–	60.00	1589.50	–	11205.17	nr	**12794.67**
reduction of 1 nr panel for end unit adjoining side wall	–	–	–	–	–443.35	nr	**–443.35**
IPS Panel Systems; 12.50 mm melamine faced chipboard; fixed to existing timber subframe							
IPS back panel system to accomodate urinals; to conceal pipework; approx 2.70 m high × 2.10 m wide	–	–	–	–	–	nr	**1041.74**
IPS back panel system to accomodate wash hand basins; to conceal pipework; approx 2.70 m high × 5.70 m wide	–	–	–	–	–	nr	**2053.79**
IPS back panel system to accomodate wash hand basins; to conceal pipework; approx 2.70 m high × 2.10 m wide	–	–	–	–	–	nr	**1160.47**

22 GENERAL JOINERY

Item	PC £	Labour hours	Labour £	Plant £	Material £	Unit	Total rate £
22.05 ASSOCIATED METALWORK							
Metalwork; mild steel							
Angle section bearers; for building in							
90 mm × 90 mm × 6 mm	–	0.31	7.56	–	7.03	m	**14.59**
120 mm × 120 mm × 8 mm	–	0.32	7.81	–	10.50	m	**18.31**
200 mm × 150 mm × 12 mm	–	0.37	9.02	–	26.73	m	**35.75**
Metalwork; mild steel; galvanized							
Waterbars; groove in timber							
6 mm × 30 mm	–	0.46	10.57	–	5.51	m	**16.08**
6 mm × 40 mm	–	0.46	10.57	–	6.92	m	**17.49**
6 mm × 50 mm	–	0.46	10.57	–	5.10	m	**15.67**
Angle section bearers; for building in							
90 mm × 90 mm × 6 mm	–	0.31	7.56	–	16.08	m	**23.64**
120 mm × 120 mm × 8 mm	–	0.32	7.81	–	28.39	m	**36.20**
200 mm × 150 mm × 12 mm	–	0.37	9.02	–	63.24	m	**72.26**
Dowels; mortice in timber							
8 mm dia. × 100 mm long	–	0.04	0.92	–	0.76	nr	**1.68**
10 mm dia. × 50 mm long	–	0.04	0.92	–	1.19	nr	**2.11**
Cramps							
25 mm × 3 mm × 230 mm girth; one end bent, holed and screwed to softwood; other end fishtailed for building in	–	0.06	1.38	–	1.79	nr	**3.17**
Metalwork; stainless steel							
Angle section bearers; for building in							
90 mm × 90 mm × 6 mm	–	0.31	7.56	–	35.24	m	**42.80**
120 mm × 120 mm × 8 mm	–	0.32	7.81	–	62.19	m	**70.00**
200 mm × 150 mm × 12 mm	–	0.37	9.02	–	136.40	m	**145.42**

23 WINDOWS, SCREENS AND LIGHTS

Item	PC £	Labour hours	Labour £	Plant £	Material £	Unit	Total rate £
23.01 WINDOWS							
SUPPLY ONLY PRICES							
NOTE: The following supply only prices are for purpose-made components, to which fixings, sealants etc. labour and overheads and profit need to be added, before they may be used to arrive at a guide price for a complete window. The reader is then referred to the subsequent SUPPLY AND FIX pages for fixing costs based on the overall window size.							
Purpose-made window casements; treated wrought softwood							
Casements; rebated; moulded							
44 mm thick	–	–	–	–	54.36	m²	**54.36**
57 mm thick	–	–	–	–	57.07	m²	**57.07**
Casements; rebated; moulded; in medium panes							
44 mm thick	–	–	–	–	86.90	m²	**86.90**
57 mm thick	–	–	–	–	90.54	m²	**90.54**
Casements; rebated; moulded; with semi-circular head							
44 mm thick	–	–	–	–	113.61	m²	**113.61**
57 mm thick	–	–	–	–	117.17	m²	**117.17**
Casements; rebated; moulded; to bullseye window							
44 mm thick; 600 mm dia.	–	–	–	–	180.10	nr	**180.10**
44 mm thick; 900 mm dia.	–	–	–	–	214.56	nr	**214.56**
57 mm thick; 600 mm dia.	–	–	–	–	188.59	nr	**188.59**
57 mm thick; 900 mm dia.	–	–	–	–	25.95	nr	**25.95**
Fitting and hanging casements (in factory)							
square or rectangular	–	–	–	–	12.71	nr	**12.71**
semi-circular	–	–	–	–	20.65	nr	**20.65**
bullseye	–	–	–	–	25.95	nr	**25.95**
Purpose-made window casements; selected Sapele							
Casements; rebated; moulded							
44 mm thick	–	–	–	–	61.78	m²	**61.78**
57 mm thick	–	–	–	–	67.69	m²	**67.69**
Casements; rebated; moulded; in medium panes							
44 mm thick	–	–	–	–	100.06	m²	**100.06**
57 mm thick	–	–	–	–	107.95	m²	**107.95**
Casements; rebated; moulded with semi-circular head							
44 mm thick	–	–	–	–	126.03	m²	**126.03**
57 mm thick	–	–	–	–	133.74	m²	**133.74**
Casements; rebated; moulded; to bullseye window							
44 mm thick; 600 mm dia.	–	–	–	–	222.75	nr	**222.75**
44 mm thick; 900 mm dia.	–	–	–	–	268.05	nr	**268.05**
57 mm thick; 600 mm dia.	–	–	–	–	241.12	nr	**241.12**
57 mm thick; 900 mm dia.	–	–	–	–	291.47	nr	**291.47**

23 WINDOWS, SCREENS AND LIGHTS

Item	PC £	Labour hours	Labour £	Plant £	Material £	Unit	Total rate £
Fitting and hanging casements (in factory)							
square or rectangular	–	–	–	–	13.77	nr	**13.77**
semi-circular	–	–	–	–	22.75	nr	**22.75**
bullseye	–	–	–	–	29.11	nr	**29.11**
Purpose-made window frames; treated wrought softwood							
Frames; rounded; rebated check grooved							
44 mm × 69 mm	–	–	–	–	14.78	m	**14.78**
44 mm × 94 mm	–	–	–	–	15.49	m	**15.49**
44 mm × 119 mm	–	–	–	–	16.19	m	**16.19**
57 mm × 94 mm	–	–	–	–	16.23	m	**16.23**
69 mm × 144 mm	–	–	–	–	21.38	m	**21.38**
90 mm × 140 mm	–	–	–	–	30.36	m	**30.36**
Mullions and transoms; twice rounded, rebated and check grooved							
57 mm × 69 mm	–	–	–	–	17.21	m	**17.21**
57 mm × 94 mm	–	–	–	–	18.09	m	**18.09**
69 mm × 94 mm	–	–	–	–	20.44	m	**20.44**
69 mm × 144 mm	–	–	–	–	30.05	m	**30.05**
Sill; sunk weathered, rebated and grooved							
69 mm × 94 mm	–	–	–	–	35.89	m	**35.89**
69 mm × 144 mm	–	–	–	–	38.01	m	**38.01**
Purpose-made window frames; selected Sapele							
Frames; rounded; rebated check grooved							
44 mm × 69 mm	–	–	–	–	18.86	m	**18.86**
44 mm × 94 mm	–	–	–	–	20.29	m	**20.29**
44 mm × 119 mm	–	–	–	–	21.71	m	**21.71**
57 mm × 94 mm	–	–	–	–	23.75	m	**23.75**
69 mm × 144 mm	–	–	–	–	34.03	m	**34.03**
90 mm × 140 mm	–	–	–	–	48.33	m	**48.33**
Mullions and transoms; twice rounded, rebated and check grooved							
57 mm × 69 mm	–	–	–	–	21.44	m	**21.44**
57 mm × 94 mm	–	–	–	–	24.95	m	**24.95**
69 mm × 94 mm	–	–	–	–	29.90	m	**29.90**
69 mm × 144 mm	–	–	–	–	45.46	m	**45.46**
Sill; sunk weathered, rebated and grooved							
69 mm × 94 mm	–	–	–	–	42.37	m	**42.37**
69 mm × 144 mm	–	–	–	–	46.97	m	**46.97**

23 WINDOWS, SCREENS AND LIGHTS

Item	PC £	Labour hours	Labour £	Plant £	Material £	Unit	Total rate £
23.01 WINDOWS – cont							
SUPPLY AND FIX PRICES							
Standard windows; treated wrought softwood; Jeld-Wen or other equal							
Casement windows; factory glazed with low-e 24 mm double glazing (U-value = 1.6 W/m² K); with 140 mm wide softwood sills; opening casements and ventilators hung on rustproof hinges; fitted with aluminized lacquered finish casement stays and fasteners							
488 mm × 750 mm; ref LEWN07V	288.97	0.65	14.94	–	297.78	nr	**312.72**
488 mm × 900 mm; ref LEWN09V	293.34	0.74	17.01	–	302.28	nr	**319.29**
630 mm × 750 mm; ref LEW107C	263.90	0.74	17.01	–	271.96	nr	**288.97**
630 mm × 750 mm; ref LEW107V	303.57	0.74	17.01	–	312.82	nr	**329.83**
630 mm × 900 mm; ref LEW109V	308.69	0.83	19.08	–	318.09	nr	**337.17**
630 mm × 900 mm; ref LEW109CH	276.37	0.74	17.01	–	284.81	nr	**301.82**
630 mm × 1050 mm; ref LEW110C	289.33	0.93	21.37	–	298.20	nr	**319.57**
630 mm × 1050 mm; ref LEW110V	318.44	0.74	17.01	–	328.19	nr	**345.20**
915 mm × 900 mm; ref LEW2NO9W	380.86	1.02	23.44	–	392.43	nr	**415.87**
915 mm × 1050 mm; ref LEW2N1OW	393.47	1.06	24.36	–	405.47	nr	**429.83**
915 mm × 1200 mm; ref LEW2N12W	406.08	1.11	25.51	–	418.46	nr	**443.97**
915 mm × 1350 mm; ref LEW2N13W	418.69	1.25	28.73	–	431.44	nr	**460.17**
915 mm × 1500 mm; ref LEW2N15W	452.74	1.30	29.88	–	466.56	nr	**496.44**
1200 mm × 750 mm; ref LEW2O7C	390.37	1.06	24.36	–	402.28	nr	**426.64**
1200 mm × 750 mm; ref LEW2O7CV	492.07	1.06	24.36	–	507.02	nr	**531.38**
1200 mm × 900 mm; ref LEW2O9C	411.07	1.11	25.51	–	423.59	nr	**449.10**
1200 mm × 900 mm; ref LEW2O9W	424.63	1.11	25.51	–	437.55	nr	**463.06**
1200 mm × 900 mm; ref LEW2O9CV	508.13	1.11	25.51	–	523.57	nr	**549.08**
1200 mm × 1050 mm; ref LEW210C	432.25	1.25	28.73	–	445.46	nr	**474.19**
1200 mm × 1050 mm; ref LEW210W	440.08	1.25	28.73	–	453.52	nr	**482.25**
1200 mm × 1050 mm; ref LEW210T	529.32	1.25	28.73	–	545.45	nr	**574.18**
1200 mm × 1050 mm; ref LEW210CV	737.44	1.25	28.73	–	759.81	nr	**788.54**
1200 mm × 1200 mm; ref LEW212C	454.34	1.34	30.80	–	468.26	nr	**499.06**
1200 mm × 1200 mm; ref LEW212W	455.54	1.34	30.80	–	469.49	nr	**500.29**
1200 mm × 1200 mm; ref LEW212TX	583.73	1.34	30.80	–	601.54	nr	**632.34**
1200 mm × 1200 mm; ref LEW212CV	551.41	1.34	30.80	–	568.24	nr	**599.04**
1200 mm × 1350 mm; ref LEW213W	470.99	1.43	32.87	–	485.42	nr	**518.29**
1200 mm × 1350 mm; ref LEW213CV	580.87	1.43	32.87	–	598.58	nr	**631.45**
1200 mm × 1500 mm; ref LEW215W	515.32	1.57	36.08	–	531.12	nr	**567.20**
1770 mm × 750 mm; ref LEW307CC	576.37	1.30	29.88	–	593.95	nr	**623.83**
1770 mm × 900 mm; ref LEW309CC	607.97	1.57	36.08	–	626.51	nr	**662.59**
1770 mm × 1050 mm; ref LEW310C	535.01	1.62	37.23	–	551.35	nr	**588.58**
1770 mm × 1050 mm; ref LEW310T	632.43	1.57	36.08	–	651.69	nr	**687.77**
1770 mm × 1050 mm; ref LEW310CC	640.57	1.30	29.88	–	660.08	nr	**689.96**
1770 mm × 1050 mm; ref LEW310CW	644.73	1.30	29.88	–	664.36	nr	**694.24**
1770 mm × 1200 mm; ref LEW312C	562.79	1.67	38.39	–	580.01	nr	**618.40**
1770 mm × 1200 mm; ref LEW312T	658.81	1.67	38.39	–	678.90	nr	**717.29**
1770 mm × 1200 mm; ref LEW312CC	674.96	1.67	38.39	–	695.55	nr	**733.94**
1770 mm × 1200 mm; ref LEW312CW	665.89	1.67	38.39	–	686.21	nr	**724.60**

23 WINDOWS, SCREENS AND LIGHTS

Item	PC £	Labour hours	Labour £	Plant £	Material £	Unit	Total rate £
1770 mm × 1200 mm; ref LEW312CVC	771.83	1.67	38.39	–	795.32	nr	**833.71**
1770 mm × 1350 mm; ref LEW313CC	724.13	1.76	40.45	–	746.19	nr	**786.64**
1770 mm × 1350 mm; ref LEW313CW	597.97	1.76	40.45	–	616.24	nr	**656.69**
1770 mm × 1350 mm; ref LEW313CVC	821.00	1.76	40.45	–	845.97	nr	**886.42**
1770 mm × 1500 mm; ref LEW315T	761.30	1.85	42.52	–	784.48	nr	**827.00**
2340 mm × 1050 mm; ref LEW410CWC	854.54	1.80	41.38	–	880.52	nr	**921.90**
2340 mm × 1200 mm; ref LEW412CWC	894.63	1.90	43.67	–	921.86	nr	**965.53**
2340 mm × 1350 mm; ref LEW413CWC	949.49	2.04	46.89	–	978.41	nr	**1025.30**
Top-hung casement windows; factory glazed with low-e 24 mm double glazing (U-value = 1.6 W/m² K); with 140 mm wide softwood sills; opening casements and ventilators hung on rustproof hinges; fitted with aluminized lacquered finish casement stays							
630 mm × 750 mm; ref LEW107 A	271.93	0.74	17.01	–	280.23	nr	**297.24**
630 mm × 900 mm; ref LEW109 A	284.39	0.83	19.08	–	293.07	nr	**312.15**
630 mm × 1050 mm; ref LEW110 A	291.14	0.93	21.37	–	300.07	nr	**321.44**
915 mm × 750 mm; ref LEW2N07 A	339.12	0.97	22.30	–	349.44	nr	**371.74**
915 mm × 900 mm; ref LEW2N09 A	354.43	1.02	23.44	–	365.21	nr	**388.65**
915 mm × 1050 mm; ref LEW2N10 A	364.03	1.06	24.36	–	375.15	nr	**399.51**
915 mm × 1350 mm; ref LEW2N13 AS	379.34	1.25	28.73	–	390.96	nr	**419.69**
1200 mm × 750 mm; ref LEW207 A	377.89	1.06	24.36	–	389.42	nr	**413.78**
1200 mm × 900 mm; ref LEW209 A	396.05	1.11	25.51	–	408.13	nr	**433.64**
1200 mm × 1050 mm; ref LEW210 A	408.50	1.25	28.73	–	420.99	nr	**449.72**
1200 mm × 1200 mm; ref LEW212 A	426.66	1.34	30.80	–	439.71	nr	**470.51**
1200 mm × 1350 mm; ref LEW213 AS	483.31	1.43	32.87	–	498.10	nr	**530.97**
1200 mm × 1500 mm; ref LEW215 AS	508.65	1.57	36.08	–	524.20	nr	**560.28**
1770 mm × 1050 mm; ref LEW310 AE	632.43	1.57	36.08	–	651.69	nr	**687.77**
1770 mm × 1200 mm; ref LEW312 AE	658.81	1.67	38.39	–	678.86	nr	**717.25**
Standard Oak windows; Jeld-Wen or other equal; factory applied preservative stain base coat							
Side-hung casement windows; factory glazed with low-e 24 mm double glazing (U-Value 1.4 W/m² K); 45 mm × 140 mm hardwood sills; weather stripping; opening sashes on canopy hinges; fitted with fasteners; brown finish ironmongery							
630 mm × 750 mm; ref LEW107C	570.00	0.88	20.23	–	587.24	nr	**607.47**
630 mm × 900 mm; ref LEW109C	612.00	1.11	25.51	–	630.50	nr	**656.01**
630 mm × 1050 mm; ref LEW110C	648.00	1.20	27.58	–	667.58	nr	**695.16**
630 mm × 1195 mm; ref LEW112C	686.40	1.20	27.58	–	707.14	nr	**734.72**
1195 mm × 1195 mm; ref LEW212C	1008.00	1.39	31.95	–	1038.44	nr	**1070.39**
1195 mm × 1345 mm; ref LEW213C	1008.00	1.39	31.95	–	1038.44	nr	**1070.39**

23 WINDOWS, SCREENS AND LIGHTS

Item	PC £	Labour hours	Labour £	Plant £	Material £	Unit	Total rate £
23.01 WINDOWS – cont							
Standard Oak windows – cont							
Casement windows; factory glazed with low-e 24 mm double glazing (U-value = 1.4W/m^{2}K); 45 mm × 140 mm hardwood sills; weather stripping; opening sashes on canopy hinges; fitted with fasteners; brown finish ironmongery							
630 mm × 900 mm; ref LEW109 AH	618.00	0.88	20.23	–	636.68	nr	**656.91**
630 mm × 1050 mm; ref LEW110 AH	648.00	1.20	27.58	–	667.64	nr	**695.22**
915 mm × 900 mm; ref LEW2N09 AH	777.00	1.39	31.95	–	800.45	nr	**832.40**
915 mm × 1050 mm; ref LEW2N10 AH	810.00	1.48	34.02	–	834.50	nr	**868.52**
1200 mm × 895 mm; ref LEW213 ASH	888.00	1.67	38.39	–	914.84	nr	**953.23**
1200 mm × 1050 mm; ref LEW210 AH	924.00	1.57	36.08	–	951.92	nr	**988.00**
1770 mm × 1050 mm; ref LEW310 AEH	1200.00	1.80	41.38	–	1236.24	nr	**1277.62**
Purpose-made double-hung sash windows; treated wrought softwood							
Cased frames of 100 mm × 25 mm grooved inner linings; 114 mm × 25 mm grooved outer linings; 125 mm × 38 mm twice rebated head linings; 125 mm × 32 mm twice rebated grooved pulley stiles; 150 mm × 13 mm linings; 50 mm × 19 mm parting slips; 25 mm × 19 mm inside beads; 150 mm × 75 mm Oak twice sunk weathered throated sill; 50 mm thick rebated and moulded sashes; moulded horns							
over 1.25 m^2 each; both sashes in medium panes; including spiral spring balances	397.30	2.08	47.80	–	483.45	m^2	**531.25**
up to 1.25 m^2 each; both sashes in medium panes; including spiral spring balances with cased mullions	451.59	2.31	53.10	–	539.38	m^2	**592.48**
Purpose-made double-hung sash windows; selected Sapele							
Cased frames of 100 mm × 25 mm grooved inner linings; 114 mm × 25 mm grooved outer linings; 125 mm × 38 mm twice rebated head linings; 125 mm × 32 mm twice rebated grooved pulley stiles; 150 mm × 13 mm linings; 50 mm × 19 mm parting slips; 25 mm × 19 mm inside beads; 150 mm × 75 mm Oak twice sunk weathered throated sill; 50 mm thick rebated and moulded sashes; moulded horns							
over 1.25 m^2 each; both sashes in medium panes; including spiral sash balances	442.15	2.78	63.89	–	529.66	m^2	**593.55**
over 1.25 m^2 each; both sashes in medium panes; including spiral sash balances with cased mullions	474.44	3.08	70.79	–	562.91	m^2	**633.70**

23 WINDOWS, SCREENS AND LIGHTS

Item	PC £	Labour hours	Labour £	Plant £	Material £	Unit	Total rate £
Clements EB24 range of factory finished steel fixed light; casement and fanlight windows and doors; with a U-value of 2.0 W/m² K (part L compliant); to EN ISO 9001 2000; polyester powder coated; factory glazed with low-e double glazing; fixed in position; including lugs plugged and screwed to brickwork or blockwork							
Basic fixed light including easy-glaze snap-on beads							
508 mm × 292 mm	189.70	2.00	52.98	–	195.48	nr	**248.46**
508 mm × 457 mm	206.95	2.00	52.98	–	213.25	nr	**266.23**
508 mm × 628 mm	224.20	2.00	52.98	–	231.07	nr	**284.05**
508 mm × 923 mm	258.69	2.00	52.98	–	266.65	nr	**319.63**
508 mm × 1218 mm	293.18	2.50	66.23	–	302.22	nr	**368.45**
Basic 'Tilt and Turn' window; including easy-glaze snap-on beads							
508 mm × 292 mm	448.39	2.00	52.98	–	461.93	nr	**514.91**
508 mm × 457 mm	465.64	2.00	52.98	–	479.70	nr	**532.68**
508 mm × 628 mm	482.88	2.00	52.98	–	497.51	nr	**550.49**
508 mm × 923 mm; including fixed light	569.11	2.00	52.98	–	586.38	nr	**639.36**
508 mm × 1218 mm; including fixed light	603.60	2.50	66.23	–	621.96	nr	**688.19**
Basic casement; including easy-glaze snap-on beads							
508 mm × 628 mm	534.62	2.00	52.98	–	550.80	nr	**603.78**
508 mm × 923 mm	569.11	2.00	52.98	–	586.38	nr	**639.36**
508 mm × 1218 mm	603.60	2.50	66.23	–	621.96	nr	**688.19**
Double door							
1143 mm × 2057 mm	3380.18	3.50	92.72	–	3481.93	nr	**3574.65**
Extra for							
pressed steel sills; to suit above windows	51.74	0.50	11.49	–	53.34	m	**64.83**
G + bar	103.47	–	–	–	106.57	m	**106.57**
simulated leaded light	103.47	–	–	–	106.57	m	**106.57**
Thermally broken composite double glazed aluminium/timber windows; Velfac 200 or other approved; with a maximum glazing U value of 1.5 W/m² K; argon filled cavity; low-e glazing with laminated glass unless otherwise specified; including multipoint espagnolette locking mechanisms and other ironmongery							
Standard fixed casement windows							
900 mm × 900 mm single fixed pane; low-e glass 6/14/4	220.00	2.70	62.06	–	232.35	nr	**294.41**
900 mm × 2000 mm single fixed pane; low-e glass 6/14/4	465.00	5.94	136.53	–	487.60	nr	**624.13**
1200 mm × 1200 mm single fixed pane; low-e glass 6/14/4	300.00	4.75	109.17	–	316.31	nr	**425.48**
1200 mm × 2200 mm three fixed panes; low-e glass 6/14/4	600.00	7.90	181.57	–	628.32	nr	**809.89**
2200 mm × 2200 mm single fixed pane; low-e glass 6/12/6	750.00	12.00	275.80	–	787.28	nr	**1063.08**

23 WINDOWS, SCREENS AND LIGHTS

Item	PC £	Labour hours	Labour £	Plant £	Material £	Unit	Total rate £
23.01 WINDOWS – cont							
Thermally broken composite double glazed aluminium/timber windows – cont							
Outward opening standard sash casement windows							
900 mm × 900 mm top-hung sash; low-e glass 6/14/4	520.00	2.67	61.37	–	541.35	nr	**602.72**
900 mm × 2200 mm with small top-hung sash; fixed lower pane; low-e glass 6/14/4	520.00	5.90	135.61	–	540.78	nr	**676.39**
900 mm × 3000 mm with small top-hung sash; fixed lower pane; low-e glass 6/14/4	655.00	8.00	183.88	–	686.24	nr	**870.12**
1600 mm × 1600 mm with two side-hung sashes; low-e glass 4/16/4	610.00	2.60	59.76	–	637.55	nr	**697.31**
1600 mm × 1600 mm with two side-hung projecting sashes; low-e glass 6/14/4	675.00	2.60	59.76	–	699.96	nr	**759.72**
1800 mm × 900 mm with two side-hung projecting sashes; low-e glass 6/14 E	450.00	5.50	126.41	–	466.78	nr	**593.19**
1800 mm × 3000 mm with two side-hung projecting sashes; two top-hung sashes; low-e glass 6/14/4	1390.00	15.00	344.76	–	1446.53	nr	**1791.29**
2000 mm × 1600 mm with one side-hung sash next to a top-hung projecting sash over a fixed sash; low-e glass 6/16/4	745.00	9.50	218.35	–	774.98	nr	**993.33**
1200 mm × 2200 mm with fixed lower sash and top-hung projecting upper sash; lower low-e glass 4 toughened/16/6.4; upper low-e glass 6/14/4	590.00	7.90	181.57	–	618.82	nr	**800.39**
1200 mm × 2200 mm with fixed lower sash and fully reversible upper sash; lower 6 toughened/14/4; upper low-e glass 4/16/4	642.00	7.90	181.57	–	672.38	nr	**853.95**
1800 mm × 900 mm with two side-hung projecting sashes; low-e glass 6/14/4; 60 minute fire integrity	675.00	5.50	126.41	–	701.32	nr	**827.73**
Outward opening standard doors							
1800 mm × 2200 mm French casement patio door; low-e toughened glass 4/16/4; deadlock; handles; cylinder	3000.00	13.00	298.79	–	3104.73	nr	**3403.52**
Allternative cavity fill							
Extra for Krypton cavity fill (over Argon)	–	–	–	–	25.75	m²	**25.75**
uPVC windows; Profile 22 or other equal and approved; reinforced where appropriate with aluminium alloy; including standard ironmongery; sills and factory glazed with low-e 24 mm double glazing; fixed in position; including lugs plugged and screwed to brickwork or blockwork							
Casement/fixed light; including e.p.d.m. glazing gaskets and weather seals							
630 mm × 900 mm; ref P109C	60.51	2.50	66.23	–	62.52	nr	**128.75**
630 mm × 1200 mm; ref P112V	70.48	2.50	66.23	–	72.84	nr	**139.07**
1200 mm × 1200 mm; ref P212C	108.30	2.50	66.23	–	111.84	nr	**178.07**
1770 mm × 1200 mm; ref P312CC	129.19	3.00	79.47	–	133.35	nr	**212.82**

23 WINDOWS, SCREENS AND LIGHTS

Item	PC £	Labour hours	Labour £	Plant £	Material £	Unit	Total rate £
Casement/fixed light; including vents; e.p.d.m. glazing gaskets and weather seals							
630 mm × 900 mm; ref P109V	74.19	2.50	66.23	–	76.61	nr	**142.84**
630 mm × 1200 mm; ref P112C	77.94	2.50	66.23	–	80.52	nr	**146.75**
1200 mm × 1200 mm; ref P212W	115.98	2.50	66.23	–	119.75	nr	**185.98**
1200 mm × 1200 mm; ref P212CV	146.62	2.50	66.23	–	151.31	nr	**217.54**
1770 mm × 1200 mm; ref P312WW	138.73	3.00	79.47	–	143.18	nr	**222.65**
Secured by Design accreditation							
Casement/fixed light; including e.p.d.m. glazing gaskets and weather seals							
630 mm × 900 mm; ref P109C	63.09	2.50	66.23	–	65.18	nr	**131.41**
630 mm × 1200 mm; ref P112V	73.06	2.50	66.23	–	75.50	nr	**141.73**
1200 mm × 1200 mm; ref P212C	110.88	2.50	66.23	–	114.49	nr	**180.72**
1770 mm × 1200 mm; ref P312CC	131.78	3.00	79.47	–	136.02	nr	**215.49**
Casement/fixed light; including vents; e.p.d.m. glazing gaskets and weather seals							
630 mm × 900 mm; ref P109V	76.78	2.50	66.23	–	79.28	nr	**145.51**
630 mm × 1200 mm; ref P112C	80.52	2.50	66.23	–	83.17	nr	**149.40**
1200 mm × 1200 mm; ref P212W	118.57	2.50	66.23	–	122.42	nr	**188.65**
1200 mm × 1200 mm; ref P212CV	151.82	2.50	66.23	–	156.66	nr	**222.89**
1770 mm × 1200 mm; ref P312WW	141.31	3.00	79.47	–	145.84	nr	**225.31**
WER A rating							
Casement/fixed light; including e.p.d.m. glazing gaskets and weather seals							
630 mm × 900 mm; ref P109C	61.35	2.50	66.23	–	63.39	nr	**129.62**
630 mm × 1200 mm; ref P112V	71.64	2.50	66.23	–	74.04	nr	**140.27**
1200 mm × 1200 mm; ref P212C	110.98	2.50	66.23	–	114.60	nr	**180.83**
1770 mm × 1200 mm; ref P312CC	133.57	3.00	79.47	–	137.87	nr	**217.34**
Casement/fixed light; including vents; e.p.d.m. glazing gaskets and weather seals							
630 mm × 900 mm; ref P109V	75.87	2.50	66.23	–	78.34	nr	**144.57**
630 mm × 1200 mm; ref P112C	79.78	2.50	66.23	–	82.42	nr	**148.65**
1200 mm × 1200 mm; ref P212W	119.34	2.50	66.23	–	123.21	nr	**189.44**
1200 mm × 1200 mm; ref P212CV	149.64	2.50	66.23	–	154.42	nr	**220.65**
1770 mm × 1200 mm; ref P312WW	143.51	3.00	79.47	–	148.10	nr	**227.57**
WER C rating							
Casement/fixed light; including e.p.d.m. glazing gaskets and weather seals							
630 mm × 900 mm; ref P109C	60.51	2.50	66.23	–	62.52	nr	**128.75**
630 mm × 1200 mm; ref P112V	70.48	2.50	66.23	–	72.84	nr	**139.07**
1200 mm × 1200 mm; ref P212C	108.30	2.50	66.23	–	111.84	nr	**178.07**
1770 mm × 1200 mm; ref P312CC	129.19	3.00	79.47	–	133.35	nr	**212.82**
Casement/fixed light; including vents; e.p.d.m. glazing gaskets and weather seals							
630 mm × 900 mm; ref P109V	74.19	2.50	66.23	–	76.61	nr	**142.84**
630 mm × 1200 mm; ref P112C	77.94	2.50	66.23	–	80.52	nr	**146.75**
1200 mm × 1200 mm; ref P212W	115.98	2.50	66.23	–	119.75	nr	**185.98**
1200 mm × 1200 mm; ref P212CV	146.62	2.50	66.23	–	151.31	nr	**217.54**
1770 mm × 1200 mm; ref P312WW	138.73	3.00	79.47	–	143.18	nr	**222.65**

23 WINDOWS, SCREENS AND LIGHTS

Item	PC £	Labour hours	Labour £	Plant £	Material £	Unit	Total rate £
23.01 WINDOWS – cont							
Colour finish uPVC windows							
Casement/fixed light; including e.p.d.m. glazing							
gaskets and weather seals							
630 mm × 900 mm; ref P109C	85.27	2.50	66.23	–	88.02	nr	**154.25**
630 mm × 1200 mm; ref P112V	96.66	2.50	66.23	–	99.80	nr	**166.03**
1200 mm × 1200 mm; ref P212C	147.17	2.50	66.23	–	151.87	nr	**218.10**
1770 mm × 1200 mm; ref P312CC	176.48	3.00	79.47	–	182.06	nr	**261.53**
Casement/fixed light; including vents; e.p.d.m.							
glazing gaskets and weather seals							
630 mm × 900 mm; ref P109V	100.64	2.50	66.23	–	103.85	nr	**170.08**
630 mm × 1200 mm; ref P112C	105.39	2.50	66.23	–	108.80	nr	**175.03**
1200 mm × 1200 mm; ref P212W	156.28	2.50	66.23	–	161.26	nr	**227.49**
1200 mm × 1200 mm; ref P212CV	194.50	2.50	66.23	–	200.62	nr	**266.85**
1770 mm × 1200 mm; ref P312WW	185.59	2.50	66.23	–	191.45	nr	**257.68**
uPVC windows; Profile 22 or other equal and							
approved; reinforced where appropriate with							
aluminium alloy; in refurbishment work,							
including standard ironmongery; sills and							
factory glazed with low-e 24 mm double glazing;							
removing existing windows and fixing new in							
position; including lugs plugged and screwed to							
brickwork or blockwork							
Casement/fixed light; including e.p.d.m. glazing							
gaskets and weather seals							
630 mm × 900 mm; ref P109C	60.51	5.00	132.46	–	62.52	nr	**194.98**
630 mm × 1200 mm; ref P112V	77.94	5.00	132.46	–	80.52	nr	**212.98**
1200 mm × 1200 mm; ref P212C	108.30	6.00	158.95	–	111.84	nr	**270.79**
1770 mm × 1200 mm; ref P312CC	129.19	6.50	172.20	–	133.35	nr	**305.55**
Casement/fixed light; including vents; e.p.d.m.							
glazing gaskets and weather seals							
630 mm × 900 mm; ref P109V	74.19	5.00	132.46	–	76.61	nr	**209.07**
630 mm × 1200 mm; ref P112C	70.48	5.50	145.70	–	72.84	nr	**218.54**
1200 mm × 1200 mm; ref P212W	115.98	6.00	158.95	–	119.75	nr	**278.70**
1200 mm × 1200 mm; ref P212CV	146.62	6.00	158.95	–	151.31	nr	**310.26**
1770 mm × 1200 mm; ref P312WW	138.73	6.50	172.20	–	143.18	nr	**315.38**
1770 mm × 1200 mm; ref P312CV	146.62	6.50	172.20	–	151.31	nr	**323.51**
Aluminium windows; Schuco AWS 50 (or similar)							
proprietary system or equal and approved							
Polyester powder coated solid colour matt finish or							
natural anodized window system of glass sealed							
units with 6.4 mm low-e coated laminated inner							
pane, air filled cavity and 6 mm clear annealed outer							
pane. Rates to include all brackets, membranes,							
cills, silicone seals, trade contractor preliminaries,							
including external access equipment							
Ribbon construction windows 1.5 m high	–	–	–	–	–	m²	**568.70**
Punched hole windows fixing into prepared							
apertures by others	–	–	–	–	–	m²	**610.50**

23 WINDOWS, SCREENS AND LIGHTS

Item	PC £	Labour hours	Labour £	Plant £	Material £	Unit	Total rate £
Extra for							
1.25 m wide × 1.5 m high opening vents, assuming tilt and turn operation	–	–	–	–	–	m²	**181.50**
neutral selective high performance coating in lieu of low-e, for assisting in solar control	–	–	–	–	–	m²	**47.00**
outer glass pane to be toughened and heat soak tested or heat strengthened in lieu of annealed	–	–	–	–	–	m²	**33.57**
inner laminated glass to be toughened and heat soak tested laminated, or heat strengthened laminated	–	–	–	–	–	m²	**66.00**
23.02 ROOFLIGHTS							
Rooflights, skylights, roof windows and frames; pre-glazed; treated Nordic Red Pine and aluminium trimmed Velux windows GGU 3070 or other equal; type U flashings and soakers (for tiles and pantiles up to 45 mm deep), and sealed double glazing unit (trimming opening not included)							
Roof windows; U-value = 1.3 W/m²K; size:							
550 mm × 780 mm	291.60	1.85	42.52	–	367.24	nr	**409.76**
550 mm × 980 mm	308.40	2.08	47.80	–	385.78	nr	**433.58**
660 mm × 1180 mm	358.00	2.31	53.10	–	445.54	nr	**498.64**
780 mm × 980 mm	334.00	2.31	53.10	–	420.82	nr	**473.92**
780 mm × 1180 mm	378.00	2.78	63.89	–	469.93	nr	**533.82**
780 mm × 1400 mm	418.00	2.31	53.10	–	516.05	nr	**569.15**
940 mm × 1600 mm	516.00	2.78	63.89	–	630.61	nr	**694.50**
1140 mm × 1180 mm	502.00	2.78	63.89	–	612.48	nr	**676.37**
1340 mm × 1400 mm	585.00	2.78	63.89	–	710.95	nr	**774.84**
extra for electric powered windows plugged in to adjacent power supply	–	–	–	–	309.00	nr	**309.00**
Rooflights, skylights, roof windows and frames; uPVC; plugged and screwed to concrete; or screwed to timber							
Rooflight; Cox Suntube range or other equal and approved; double skin polycarbonate dome							
230 mm dia.; for flat roof using felt or membrane	285.03	2.50	57.46	–	294.18	nr	**351.64**
230 mm dia.; for up to 30° pitch roof with standard tiles	315.14	3.00	68.95	–	325.09	nr	**394.04**
230 mm dia.; for up to 30° pitch roof with bold roll tiles	292.55	3.00	68.95	–	301.87	nr	**370.82**
300 mm dia.; for flat roof using felt or membrane	462.49	2.50	57.46	–	476.96	nr	**534.42**
300 mm dia.; for up to 30° pitch roof with standard tiles	462.49	3.00	68.95	–	476.86	nr	**545.81**
300 mm dia.; for up to 30° pitch roof with bold roll tiles	495.82	3.00	68.95	–	511.24	nr	**580.19**

Prices for Measured Works

23 WINDOWS, SCREENS AND LIGHTS

Item	PC £	Labour hours	Labour £	Plant £	Material £	Unit	Total rate £
23.02 ROOFLIGHTS – cont							
Rooflights, skylights, roof windows and frames – cont							
Rooflight; Cox Galaxy range or other equal and approved; double skin polycarbonate dome only; fitting to existing kerb							
600 mm × 600 mm	132.29	1.50	34.47	–	136.59	nr	**171.06**
900 mm × 900 mm	244.14	1.75	40.22	–	251.91	nr	**292.13**
1200 mm × 1800 mm	716.31	2.00	45.97	–	738.35	nr	**784.32**
Rooflight; Cox Galaxy range or other equal and approved; triple skin polycarbonate dome only; fitting to existing kerb							
600 mm × 600 mm rooflight	204.35	1.50	34.47	–	210.81	nr	**245.28**
900 mm × 900 mm rooflight	366.76	1.75	40.22	–	378.20	nr	**418.42**
1200 mm × 1800 mm rooflight	1108.89	2.00	45.97	–	1142.70	nr	**1188.67**
Rooflight; Cox Trade range or other equal and approved; double skin polycarbonate fixed light dome on 150 mm PVC upstand							
600 mm × 600 mm	226.94	2.00	45.97	–	234.13	nr	**280.10**
900 mm × 900 mm	366.76	2.25	51.72	–	378.25	nr	**429.97**
1200 mm × 1800 mm	725.99	2.50	57.46	–	748.37	nr	**805.83**
Rooflight; Cox Trade range or other equal and approved; double skin polycarbonate manual opening light dome on 150 mm PVC upstand							
600 mm × 600 mm	354.93	2.50	57.46	–	366.40	nr	**423.86**
900 mm × 900 mm	514.11	2.75	63.20	–	530.45	nr	**593.65**
1200 mm × 1800 mm	966.91	3.10	71.25	–	996.96	nr	**1068.21**
Rooflight; Cox Trade range or other equal and approved; double skin polycarbonate electric opening light dome on 150 mm PVC upstand (not including adjacent power supply)							
600 mm × 600 mm	603.38	2.75	63.20	–	622.30	nr	**685.50**
900 mm × 900 mm	803.43	3.00	68.95	–	828.46	nr	**897.41**
1200 mm × 1800 mm	1347.66	3.50	80.44	–	1389.12	nr	**1469.56**
Rooflight; Cox Trade range or other equal and approved; triple skin polycarbonate fixed light dome on 150 mm PVC upstand							
600 mm × 600 mm	266.73	2.00	45.97	–	275.11	nr	**321.08**
900 mm × 900 mm	438.81	2.00	45.97	–	452.36	nr	**498.33**
1200 m × 1800 mm	889.48	2.50	57.46	–	916.76	nr	**974.22**
Rooflight; Cox Trade range or other equal and approved; triple skin polycarbonate manual opening light dome on 150 mm PVC upstand							
600 mm × 600 mm	394.72	2.50	57.46	–	407.38	nr	**464.84**
900 mm × 900 mm	586.16	2.50	57.46	–	604.56	nr	**662.02**
1200 m × 1800 mm	1130.40	3.10	71.25	–	1165.34	nr	**1236.59**

23 WINDOWS, SCREENS AND LIGHTS

Item	PC £	Labour hours	Labour £	Plant £	Material £	Unit	Total rate £
Rooflight; Cox Trade range or other equal and approved; triple skin polycarbonate electric opening light dome on 150 mm PVC upstand							
600 mm × 600 mm light	643.17	2.75	63.20	–	663.28	nr	**726.48**
900 mm × 900 mm light	875.48	2.75	63.20	–	902.57	nr	**965.77**
1200 mm × 1800 mm light	1511.14	3.50	80.44	–	1557.51	nr	**1637.95**
Rooflight; Cox 2000 range or other equal and approved double skin polycarbonate fixed light dome on 235 mm solid core PVC upstand							
600 mm × 600 mm	725.99	2.00	45.97	–	748.15	nr	**794.12**
900 mm × 900 mm	984.62	2.50	57.46	–	1014.54	nr	**1072.00**
1200 mm × 1800 mm	2626.23	3.00	68.95	–	2705.45	nr	**2774.40**
Rooflight; Cox 2000 range or other equal and approved double skin polycarbonate manual opening light dome on 235 mm solid core PVC upstand							
600 mm × 600 mm	997.04	2.50	57.46	–	1027.76	nr	**1085.22**
900 mm × 900 mm	1255.66	3.00	68.95	–	1294.14	nr	**1363.09**
1200 mm × 1800 mm	3146.80	3.60	82.74	–	3242.07	nr	**3324.81**
Rooflight; Cox 2000 range or other equal and approved double skin polycarbonate electric opening light dome on 235 mm solid core PVC upstand (not including adjacent power supply)							
600 mm × 600 mm	1312.16	2.75	63.20	–	1352.35	nr	**1415.55**
900 mm × 900 mm	1570.79	3.25	74.70	–	1618.73	nr	**1693.43**
1200 mm × 1800 mm	3542.60	4.00	91.94	–	3649.74	nr	**3741.68**
Rooflight; Cox 2000 range or other equal and approved triple skin polycarbonate fixed light dome on 235 mm solid core PVC upstand							
600 mm × 600 mm	977.67	2.00	45.97	–	1007.38	nr	**1053.35**
900 mm × 900 mm	1428.38	2.50	57.46	–	1471.61	nr	**1529.07**
1200 mm × 1800 mm	3508.32	3.00	68.95	–	3614.00	nr	**3682.95**
Rooflight; Cox 2000 range or other equal and approved triple skin polycarbonate manual opening light dome on 235 mm solid core PVC upstand							
600 mm × 600 mm	1248.71	2.50	57.46	–	1287.00	nr	**1344.46**
900 mm × 900 mm	1699.43	3.00	68.95	–	1751.23	nr	**1820.18**
1200 mm × 1800 mm	4028.89	3.60	82.74	–	4150.62	nr	**4233.36**
Rooflight; Cox 2000 range or other equal and approved triple skin polycarbonate electric opening light dome on 235 mm solid core PVC upstand (not incluing adjacent power supply)							
600 mm × 600 mm	1563.84	2.75	63.20	–	1611.57	nr	**1674.77**
900 mm × 900 mm	2014.55	3.25	74.70	–	2075.80	nr	**2150.50**
1200 mm × 1800 mm	4424.69	4.00	91.94	–	4558.30	nr	**4650.24**

23 WINDOWS, SCREENS AND LIGHTS

Item	PC £	Labour hours	Labour £	Plant £	Material £	Unit	Total rate £
23.03 LOUVRES							
Louvres, Brise Soleils and frames; polyester powder coated aluminium; fixing in position including brackets							
Brise soleil, to mitigate the effects of solar gain. This rate assumes a single natural anodized extruded aluminium fin, with brackets and orientated either horizontally or vertically. The quantity of fins per storey height should be calculated to achieve desired shading							
300 mm deep	–	–	–	–	–	m	**170.82**

24 DOORS, SHUTTERS AND HATCHES

Item	PC £	Labour hours	Labour £	Plant £	Material £	Unit	Total rate £
24.01 EXTERNAL DOORS AND FRAMES							
EXTERNAL DOOR FRAMES							
External softwood door frame composite standard joinery sets							
External door frame composite set; 56 mm × 78 mm wide (finished); for external doors							
762 mm × 1981 mm × 44 mm	68.21	0.75	17.24	–	70.45	nr	**87.69**
813 mm × 1981 mm × 44 mm	70.02	0.75	17.24	–	72.32	nr	**89.56**
838 mm × 1981 mm × 44 mm	69.56	0.75	17.24	–	71.84	nr	**89.08**
External door frame composite set; 56 mm × 78 mm wide (finished); with 45 mm × 140 mm (finished) hardwood sill; for external doors							
686 mm × 1981 mm × 44 mm	84.53	1.00	22.98	–	87.26	nr	**110.24**
762 mm × 1981 mm × 44 mm	87.51	1.00	22.98	–	90.33	nr	**113.31**
838 mm × 1981 mm × 44 mm	103.18	1.00	22.98	–	106.46	nr	**129.44**
826 mm × 2040 mm × 44 mm	90.48	1.00	22.98	–	93.39	nr	**116.37**
EXTERNAL DOORS							
Note: door ironmogery is not included unless stated							
Doors; standard matchboarded; wrought softwood							
Matchboarded, framed, ledged and braced doors; 44 mm thick overall; 19 mm thick tongued, grooved and V-jointed boarding; one side vertical boarding							
762 mm × 1981 mm	60.52	1.67	38.39	–	62.34	nr	**100.73**
838 mm × 1981 mm	66.13	1.67	38.39	–	68.11	nr	**106.50**
Flush door; external quality; skeleton or cellular core; plywood faced both sides; lipped all round							
762 mm × 1981 mm × 54 mm	77.38	1.62	37.23	–	79.70	nr	**116.93**
838 mm × 1981 mm × 54 mm	79.05	1.62	37.23	–	81.42	nr	**118.65**
Fire doors							
Flush door; half-hour fire-resisting; external quality; skeleton or cellular core; plywood faced both sides; lipped on all four edges							
762 mm × 1981 mm × 44 mm	163.04	1.71	39.30	–	167.93	nr	**207.23**
838 mm × 1981 mm × 44 mm	164.71	1.71	39.30	–	169.65	nr	**208.95**
Flush door; half-hour fire-resisting; external quality with 6 mm Georgian wired polished plate glass opening; skeleton or cellular core; plywood faced both sides; lipped on all four edges; including glazing beads							
762 mm × 1981 mm × 44 mm	163.04	1.71	39.30	–	167.93	nr	**207.23**
838 mm × 1981 mm × 44 mm	164.71	1.71	39.30	–	169.65	nr	**208.95**
726 mm × 2040 mm × 44 mm	163.04	1.71	39.30	–	167.93	nr	**207.23**
826 mm × 2040 mm × 44 mm	164.78	1.71	39.30	–	169.72	nr	**209.02**
926 mm × 2040 mm × 44 mm	173.87	1.71	39.30	–	179.09	nr	**218.39**

24 DOORS, SHUTTERS AND HATCHES

Item	PC £	Labour hours	Labour £	Plant £	Material £	Unit	Total rate £
24.01 EXTERNAL DOORS AND FRAMES – cont							
External softwood door frame composite standard joinery sets							
External door frame composite set; 56 mm × 78 mm wide (finished); for external doors							
762 mm × 1981 mm × 44 mm	68.21	0.75	17.24	–	70.45	nr	**87.69**
813 mm × 1981 mm × 44 mm	70.02	0.75	17.24	–	72.32	nr	**89.56**
838 mm × 1981 mm × 44 mm	69.56	0.75	17.24	–	71.84	nr	**89.08**
Doorsets; Anti-Vandal Security door and frame units; Bastion Security Ltd or other equal; to BS 5051; factory primed; fixing with frame anchors to masonry; cutting mortices; external							
46 mm thick insulated door with birch grade plywood; sheet steel bonded into door core; 2 mm thick polyester coated laminate finish; hardwood lippings all edges; 95 mm × 65 mm hardwood frame; polyester coated standard ironmongery; weather stripping all round; low projecting aluminium threshold; plugging; screwing							
for 980 mm × 2100 mm structural opening; single doorsets; panic bolt	–	–	–	–	–	nr	**2300.00**
for 1830 mm × 2100 mm structural opening; double doorsets; panic bolt	–	–	–	–	–	nr	**3000.00**
Doorsets; galvanized steel door and frame units; treated softwood frame, primed hardwood sill; fixing in position; plugged and screwed to brickwork or blockwork							
Door and frame							
838 mm × 1981 mm	–	2.78	63.89	–	520.93	nr	**584.82**
Doorsets; steel security door and frame; Hormann or other equal; including ironmongery, weather seals and all necessary fixing accessories							
Horman ref E55–1 doorset.							
To suit structural opening 1100 × 2105 mm; fire-rating 30 minutes; acoustic rating 38dB; including stainless steel ironmongery	–	–	–	–	–	nr	**2236.34**
To suit structural opening 2000 × 2105 mm; fire-rating 30 minutes; acoustic rating 38dB; including stainless steel ironmongery	–	–	–	–	–	nr	**3322.57**

24 DOORS, SHUTTERS AND HATCHES

Item	PC £	Labour hours	Labour £	Plant £	Material £	Unit	Total rate £
Doorsets; steel bullet-resistant door and frame units; Wormald Doors or other equal; Medite laquered panels; ironmongery							
Door and frame							
1000 mm × 2060 mm overall; fixed to masonry	–	–	–	–	–	nr	**4664.39**
uPVC doors; Profile 22 or other equal; reinforced where appropriate with aluminium alloy; including standard ironmongery; thresholds and factory glazed with low-e 24 mm double glazing; fixed in position; including lugs plugged and screwed to brickwork or blockwork							
Fixed light; including e.p.d.m. glazing gaskets and weather seals							
2100 mm × 900 mm uPVC door with midrail; half glazed	288.00	3.00	79.47	–	296.86	nr	**376.33**
2100 mm × 900 mm uPVC door with midrail; glazed	284.00	3.00	79.47	–	292.77	nr	**372.24**
Secured by Design accrediation							
Fixed light; including e.p.d.m. glazing gaskets and weather seals							
2100 mm × 900 mm uPVC door with midrail; half glazed	328.00	4.00	105.97	–	338.06	nr	**444.03**
2100 mm × 900 mm uPVC door with midrail; glazed	332.00	3.00	79.47	–	342.21	nr	**421.68**
WER A rating							
Fixed light; including e.p.d.m. glazing gaskets and weather seals							
2100 mm × 900 mm uPVC door with midrail; half glazed	296.00	4.00	105.97	–	305.10	nr	**411.07**
2100 mm × 900 mm uPVC door with midrail; glazed	289.00	3.00	79.47	–	297.92	nr	**377.39**
WER C rating							
Fixed light; including e.p.d.m. glazing gaskets and weather seals							
2100 mm × 900 mm uPVC door with midrail; half glazed	288.00	4.00	105.97	–	296.86	nr	**402.83**
2100 mm × 900 mm uPVC door with midrail; glazed	284.00	3.00	79.47	–	292.77	nr	**372.24**
Colour finish							
Fixed light; including e.p.d.m. glazing gaskets and weather seals							
2100 mm × 900 mm uPVC door with midrail; half glazed	441.00	4.00	105.97	–	454.45	nr	**560.42**
2100 mm × 900 mm uPVC door with midrail; glazed	352.00	3.00	79.47	–	362.81	nr	**442.28**

24 DOORS, SHUTTERS AND HATCHES

Item	PC £	Labour hours	Labour £	Plant £	Material £	Unit	Total rate £
24.01 EXTERNAL DOORS AND FRAMES – cont							
Sliding/folding; aluminium double glazed sliding patio doors; white acrylic finish; with and including 18 thick annealed double glazing; fixed in position; including lugs plugged and screwed to brickwork or blockwork							
Patio doors							
1800 mm × 2100 mm	1218.00	16.00	423.87	–	1255.12	nr	**1678.99**
2400 mm × 2100 mm	1239.00	16.00	423.87	–	1276.75	nr	**1700.62**
2700 mm × 2100 mm	1470.00	16.00	423.87	–	1514.68	nr	**1938.55**
24.02 REVOLVING DOORS							
Automatic revolving doors; structural glass							
ASSA ABLOY Automatic Glass Revolving Door System providing hands free operation and Hi-Tech functionality. Modern modular design comprising four door leaves, curved glazed outer walls, glass ceiling/ canopy, complete with in-floor mounted operator and imbedded Configurable Digital Control. System including normal installation to comply with BS7036–0-2014 / BS-EN16005, complete with the installer's Declaration of Conformity in accordance with Machinery Regulation							
ASSA ABLOY RD300 structural glass Automatic Revolving Door; 3 or 4 wing 1800 mm dia., 2200 mm internal height	–	–	–	–	–	nr	**38110.00**
ASSA ABLOY RD300 structural glass Automatic Revolving Door; 3 or 4 wing 2100 mm dia., 2200 mm internal height	–	–	–	–	–	nr	**38934.00**
ASSA ABLOY RD300 structural glass Automatic Revolving Door; 3 or 4 wing 2400 mm dia., 2200 mm internal height	–	–	–	–	–	nr	**39758.00**
ASSA ABLOY RD300 structural glass Automatic Revolving Door; 3 or 4 wing 2700 mm dia., 2200 mm internal height	–	–	–	–	–	nr	**40376.00**
ASSA ABLOY RD300 structural glass Automatic Revolving Door; 3 or 4 wing 3000 mm dia., 2200 mm internal height	–	–	–	–	–	nr	**41200.00**

24 DOORS, SHUTTERS AND HATCHES

Item	PC £	Labour hours	Labour £	Plant £	Material £	Unit	Total rate £
Automatic revolving doors; framed							
ASSA ABLOY Framed automatic revolving door System providing hands free operation and ESCAPE functionality. Modern modular design comprising four collapsible door leaves with electromechanical breakout, framed curved glazed outer walls, radially segmented ceiling and semi-circular dustcovers, complete with operator and imbedded Configurable Digital Control. System including normal installation to comply with BS7036–0-2014 / BS-EN16005, complete with the installer's Declaration of Conformity in accordance with Machinery Regulations							
RD3/RD4 Framed Automatic Revolving Door; 3 or 4 wing 1800 mm dia., 2200 mm internal height	–	–	–	–	–	nr	**25544.00**
RD3/RD4 Framed Automatic Revolving Door; 3 or 4 wing 2100 mm dia., 2200 mm internal height	–	–	–	–	–	nr	**26059.00**
RD3/RD4 Framed Automatic Revolving Door; 3 or 4 wing 2400 mm dia., 2200 mm internal height	–	–	–	–	–	nr	**26574.00**
RD3/RD4 Framed Automatic Revolving Door; 3 or 4 wing 2700 mm dia., 2200 mm internal height	–	–	–	–	–	nr	**27089.00**
RD3/RD4 Framed Automatic Revolving Door; 3 or 4 wing 3000 mm dia., 2200 mm internal height	–	–	–	–	–	nr	**27604.00**
RD3/RD4 Framed Automatic Revolving Door; 3 or 4 wing 3300 mm dia., 2200 mm internal height	–	–	–	–	–	nr	**30179.00**
RD3/RD4 Framed Automatic Revolving Door; 3 or 4 wing 3600 mm dia., 2200 mm internal height	–	–	–	–	–	nr	**30694.00**
Access control revolving doors; framed							
ASSA ABLOY Framed automatic revolving door System providing controlled single or bi-directional controlled access. Modern modular design comprising four door leaves, framed curved glazed outer walls, radially segmented ceiling and semi-circular dustcovers, complete with operator, anti-tailgate and imbedded Configurable Digital Control. System including normal installation to comply with BS7036–0-2014 / BS-EN16005, complete with the installer's Declaration of Conformity in accordance with Machinery Regulations							
RD4-A1 Framed Access Control Revolving Door; 4 wing 1800 mm dia., 2200 mm internal height	–	–	–	–	–	nr	**27089.00**
RD4-A2 Framed Access & Egress Control Revolving Door; 4 wing 1800 mm dia., 2200 mm internal height	–	–	–	–	–	nr	**27604.00**
RD4-A1 Framed Access Control Revolving Door; 4 wing 2100 mm dia., 2200 mm internal height	–	–	–	–	–	nr	**27604.00**

24 DOORS, SHUTTERS AND HATCHES

Item	PC £	Labour hours	Labour £	Plant £	Material £	Unit	Total rate £
24.02 REVOLVING DOORS – cont							
Access control revolving doors – cont							
ASSA ABLOY Framed automatic revolving door System providing controlled single or bi-directional controlled access. – cont							
RD4-A2 Framed Access & Egress Control Revolving Door; 4 wing 2100 mm dia., 2200 mm internal height	–	–	–	–	–	nr	28119.00
RD4-A1 Framed Access Control Revolving Door; 4 wing 2400 mm dia., 2200 mm internal height	–	–	–	–	–	nr	28119.00
RD4-A2 Framed Access & Egress Control Revolving Door; 4 wing 2400 mm dia., 2200 mm internal height	–	–	–	–	–	nr	28634.00
Manually operated revolving doors; framed							
ASSA ABLOY Intelligent Manually Operated Revolving Door System providing Comfort, Convenience and Safety automatically at the user's pace. Modern modular design comprising four door leaves, framed curved glazed outer walls, radially segmented ceiling and semi-circular dustcovers, complete with electromechanical operator and Configurable Digital Control. System including normal installation to comply with BS7036–0-2014 / BS-EN16005, complete with the installer's Declaration of Conformity in accordance with Machinery Regulations							
RD100 Framed Intelligent Manually Operated Revolving Door; 3 or 4 wing 1800 mm dia., 2600 mm internal height	–	–	–	–	–	nr	22454.00
RD100 Framed Intelligent anually Operated Automatic Revolving Door; 3 or 4 wing 2100 mm dia., 2600 mm internal height	–	–	–	–	–	nr	22969.00
RD100 Framed Intelligent anually Operated Automatic Revolving Door; 3 or 4 wing 2400 mm dia., 2600 mm internal height	–	–	–	–	–	nr	23484.00
RD100 Framed Intelligent anually Operated Automatic Revolving Door; 3 or 4 wing 2700 mm dia., 2600 mm internal height	–	–	–	–	–	nr	23999.00
RD100 Framed Intelligent anually Operated Automatic Revolving Door; 3 or 4 wing 3000 mm dia., 2600 mm internal height	–	–	–	–	–	nr	24514.00

24 DOORS, SHUTTERS AND HATCHES

Item	PC £	Labour hours	Labour £	Plant £	Material £	Unit	Total rate £
Automatic revolving doors; framed							
ASSA ABLOY Framed automatic revolving door system providing hands free operation and safe adaptive Manual Speed control. Modern modular design comprising four door leaves, curved glazed outer walls, radially segmented ceiling and semi-circular dustcovers, complete with operator and imbedded Configurable Digital Control. System including normal installation to comply with BS7036–0-2014 / BS-EN16005, complete with the installer's Declaration of Conformity in accordance with Machinery Regulations							
RD150 Framed Automatic Revolving Door; 3 or 4 wing 1800 mm dia., 2200 mm internal height	–	–	–	–	–	nr	24308.00
RD150 Framed Automatic Revolving Door; 3 or 4 wing 2100 mm dia., 2200 mm internal height	–	–	–	–	–	nr	24823.00
RD150 Framed Automatic Revolving Door; 3 or 4 wing 2400 mm dia., 2200 mm internal height	–	–	–	–	–	nr	25338.00
RD150 Framed Automatic Revolving Door; 3 or 4 wing 2700 mm dia., 2200 mm internal height	–	–	–	–	–	nr	25853.00
RD150 Framed Automatic Revolving Door; 3 or 4 wing 3000 mm dia., 2200 mm internal height	–	–	–	–	–	nr	26368.00
Automatic high capacity revolving doors; 2-wing							
ASSA ABLOY UniTurn providing safe hands free operation with capacity for multiple occupants and trolleys per compartment. Modern modular design with two centre passage manually pivoting break out door leaves arranged between two rotating showcases. Curved glazed outer walls and canopy enclose a whole rotating radially segmented ceiling, fixed semi-circular dustcovers and load-bearing peripheral drive with imbedded Configurable Digital Control. System including normal installation to comply with BS7036–0-2014 / BS-EN16005, complete with the installer's Declaration of Conformity in accordance with Machinery Regulations							
UniTurn High Capacity Automatic Revolving Door; 2 wing 3600 mm dia., 2200 mm internal height	–	–	–	–	–	nr	53560.00
UniTurn High Capacity Automatic Revolving Door; 2 wing 4200 mm dia., 2200 mm internal height	–	–	–	–	–	nr	55929.00
UniTurn High Capacity Automatic Revolving Door; 2 wing 4800 mm dia., 2200 mm internal height	–	–	–	–	–	nr	58298.00
UniTurn High Capacity Automatic Revolving Door; 2 wing 5400 mm dia., 2200 mm internal height	–	–	–	–	–	nr	60667.00

24 DOORS, SHUTTERS AND HATCHES

Item	PC £	Labour hours	Labour £	Plant £	Material £	Unit	Total rate £
24.02 REVOLVING DOORS – cont							
Automatic high capacity revolving doors; 3-wing, panic exit							
ASSA ABLOY RD3L System providing safe hands free operation, capacity for multiple occupants with trolleys. Modern modular design comprising three fixed door leaves arranged around the centre shaft and electromechanical emergency break out leaves folding around secondary pivot points. Curved glazed outer walls and canopy enclose a whole rotating radially segmented ceiling, fixed semi-circular dustcovers and load compensating peripheral drive with imbedded Configurable Digital Control. System including normal installation to comply with BS7036–0-2014 / BS-EN16005, complete with the installer's Declaration of Conformity in accordance with Machinery Regulations							
RD3L High Capacity Automatic Revolving Door; 3 wing 4200 mm dia., 2600 mm internal height. For up to five occupants per compartment	–	–	–	–	–	nr	**56547.00**
RD3L High Capacity Automatic Revolving Door; 3 wing 4800 mm dia., 2600 mm internal height. For up to seven occupants per compartment	–	–	–	–	–	nr	**58504.00**
RD3L High Capacity Automatic Revolving Door; 3 wing 5400 mm dia., 2600 mm internal height. For up to nine occupants per compartment	–	–	–	–	–	nr	**60358.00**
RD3L High Capacity Automatic Revolving Door; 3 wing 6200 mm dia., 2600 mm internal height. For up to twelve occupants per compartment	–	–	–	–	–	nr	**62212.00**

24 DOORS, SHUTTERS AND HATCHES

Item	PC £	Labour hours	Labour £	Plant £	Material £	Unit	Total rate £
24.03 SLIDING DOOR ENTRANCES							
Sliding door entrances (including doors and screens)							
Automatic Sliding Door EN1627: RC2 Anti-Burglar Security Entrance System. ASSA ABLOY SL510 P RC2 Anti-Burglar Sliding Door System complete with 100 mm operator providing automated operation of the sliding door leaves. RAL powder coated or anodized with insulated P4 A security glass. Providing full retraction of the doors to maximize the clear open width. Operator should optimize hold open time in relation to changing traffic flow. The door should change opening width in relation to changing traffic flow. The timeframe for change should be flexible and possible to program. The electromechanical lock must block all active door leaves separately. Third Party Verified system EPDs to be a demand. System including normal installation to comply with BS7036–0-2014 / BS-EN16005, complete with the installer's Declaration of Conformity in accordance with Machinery Regulations							
ASSA ABLOY SL510 P EN1627: RC2 1P; Single Door Leaf and 100 mm Operator with Pocket Safety Screen 1200 mm wide × 2300 mm high	–	–	–	–	–	nr	**6695.00**
ASSA ABLOY SL510 P EN1627: RC2 2PL/R; Single Door Leaf with One Sidescreen and 100 mm Operator with Pocket Safety Screen 2000 mm wide × 2300 mm high	–	–	–	–	–	nr	**7725.00**
ASSA ABLOY SL510 P EN1627: RC2 2P; Double Bi-Parting Door Leaves and 100 mm Operator with Pocket Safety Screens 2000 mm wide × 2300 mm high	–	–	–	–	–	nr	**8755.00**
ASSA ABLOY SL510 P EN1627: RC2 4P; Double Bi-Parting Door Leaves with Two Sidescreens and 100 mm Operator with Pocket Safety Screens 4000 mm wide × 2300 mm high	–	–	–	–	–	nr	**11845.00**

24 DOORS, SHUTTERS AND HATCHES

Item	PC £	Labour hours	Labour £	Plant £	Material £	Unit	Total rate £
24.03 SLIDING DOOR ENTRANCES – cont							
Sliding door entrances (including doors and screens) – cont							
Automatic Sliding Door EN1627: RC3 Anti-Burglar Security Entrance System. ASSA ABLOY SL510 P RC3 Anti-Burglar Sliding Door System complete with 100 mm operator providing automated operation of the sliding door leaves. RAL powder coated or anodized with insulated P5 A security glass. Providing full retraction of the doors to maximize the clear open width. Operator should optimize hold open time in relation to changing traffic flow. The door should change opening width in relation to changing traffic flow. The timeframe for change should be flexible and possible to program. The electromechanical lock must block all active door leaves separately. Third Party Verified system EPDs to be a demand. System including normal installation to comply with BS7036–0-2014 / BS-EN16005, complete with the installer's Declaration of Conformity in accordance with Machinery Regulations							
ASSA ABLOY SL510 P RC3 1P; Single Door Leaf and 100 mm Operator with Pocket Safety Screen 1200 mm wide × 2300 mm high	–	–	–	–	–	nr	–
ASSA ABLOY SL510 P RC3 2PL/R; Single Door Leaf with One Sidescreen and 100 mm Operator with Pocket Safety Screen 2000 mm wide × 2300 mm high	–	–	–	–	–	nr	8240.00
ASSA ABLOY SL510 P RC3 2P; Double Bi-Parting Door Leaves and 100 mm Operator with Pocket Safety Screens 2000 mm high × 2300 mm high	–	–	–	–	–	nr	9270.00
ASSA ABLOY SL510 P RC3 4P; Double Bi-Parting Door Leaves with Two Sidescreens and 100 mm Operator with Pocket Safety Screens 4000 mm wide × 2300 mm high	–	–	–	–	–	nr	12360.00

24 DOORS, SHUTTERS AND HATCHES

Item	PC £	Labour hours	Labour £	Plant £	Material £	Unit	Total rate £
Automatic Sliding Door Rugged Framed Entrance System. ASSA ABLOY SL510 F Sliding Door System complete with 100 mm operator providing automated operation of the sliding door leaves. RAL powder coated or anodized with insulated glass. Providing full retraction of the doors to maximize the clear open width. Operator should optimize hold open time in relation to changing traffic flow. The door should change opening width in relation to changing traffic flow. The timeframe for change should be flexible and possible to program. The electromechanical lock must block all active door leaves separately. Third Party Verified system EPDs to be a demand. System including normal installation to comply with BS7036–0-2014 / BS-EN16005, complete with the installer's Declaration of Conformity in accordance with Machinery Regulations							
ASSA ABLOY SL510 F 1P; Single Door Leaf and 100 mm Operator with Pocket Safety Screen 1200 mm wide × 2300 mm high	–	–	–	–	–	nr	3605.00
ASSA ABLOY SL510 F 2PL/R; Single Door Leaf with One Sidescreen and 100 mm Operator with Pocket Safety Screen 2000 mm wide × 2300 mm high	–	–	–	–	–	nr	4635.00
ASSA ABLOY SL510 F 2P; Double Bi-Parting Door Leaves and 100 mm Operator with Pocket Safety Screens 2000 mm wide × 2300 mm high	–	–	–	–	–	nr	4944.00
ASSA ABLOY SL510 F 4P; Double Bi-Parting Door Leaves with Two Sidescreens and 100 mm Operator with Pocket Safety Screens 4000 mm wide × 2300 mm high	–	–	–	–	–	nr	6180.00

24 DOORS, SHUTTERS AND HATCHES

Item	PC £	Labour hours	Labour £	Plant £	Material £	Unit	Total rate £
24.03 SLIDING DOOR ENTRANCES – cont							
Sliding door entrances (including doors and screens) – cont							
Automatic Sliding Door Narrow Framed Entrance System. ASSA ABLOY SL510 SE Sliding Door System complete with 100 mm operator providing automated operation of the sliding door leaves. RAL powder coated or anodized with insulated glass. Providing full retraction of the doors to maximize the clear open width. Operator should optimize hold open time in relation to changing traffic flow. The door should change opening width in relation to changing traffic flow. The timeframe for change should be flexible and possible to program. The electromechanical lock must block all active door leaves separately. Third Party Verified system EPDs to be a demand. System including normal installation to comply with BS7036–0-2014 / BS-EN16005, complete with the installer's Declaration of Conformity in accordance with Machinery Regulations							
ASSA ABLOY SL510 SE 1P; Single Door Leaf and 100 mm Operator with Pocket Safety Screen 1200 mm wide × 2300 mm high	–	–	–	–	–	nr	**3605.00**
ASSA ABLOY SL510 SE 2PL/R; Single Door Leaf with One Sidescreen and 100 mm Operator with Pocket Safety Screen 2000 mm wide × 2300 mm high	–	–	–	–	–	nr	**4635.00**
ASSA ABLOY SL510 SE 2P; Double Bi-Parting Door Leaves and 100 mm Operator with Pocket Safety Screens 2000 mm wide × 2300 mm high	–	–	–	–	–	nr	**4944.00**
ASSA ABLOY SL510 SE 4P; Double Bi-Parting Door Leaves with Two Sidescreens and 100 mm Operator with Pocket Safety Screens 4000 mm wide × 2300 mm high	–	–	–	–	–	nr	**6180.00**

24 DOORS, SHUTTERS AND HATCHES

Item	PC £	Labour hours	Labour £	Plant £	Material £	Unit	Total rate £
24.04 SWING DOORS							
Swing door operators (mounted onto doors supplied by others)							
Automatic Low Kinetic Energy Swing Door Operator. ASSA ABLOY SW100 Surface Mounted Door Operator for internal doors. The SW100 is an electromechanical operator and has been developed to comply with European standards and requirements for low-energy power operated doors. Mounted onto a suitable transom by others. Door and frame by others. Fire-rated upto 60 min. EPD verified by 3rd Party							
ASSA ABLOY SW100–1 Single Door Leaf Low Energy Door Operator; Low risk users environment	–	–	–	–	–	nr	**1802.50**
ASSA ABLOY SW100–1 Single Door Leaf Low Energy Door Operator; High risk users environment	–	–	–	–	–	nr	**1905.50**
ASSA ABLOY SW100–2 Double Door Leaf Low Energy Door Operator; Low risk users environment	–	–	–	–	–	nr	**2472.00**
ASSA ABLOY SW100–2 Double Door Leaf Low Energy Door Operator; High risk users environment	–	–	–	–	–	nr	**2884.00**
ASSA ABLOY SW300 70 mm Surface Mounted Door Operator for external and internal doors. The SW300 is an electromechanical operator and has been developed to comply with European standards and requirements for low and high energy power operated doors. Door can be used manually with power assist or by push pad(s). Includes intelligent control for wind and stack pressure management. Mounted onto a suitable transom by others. Door and frame by others. Fire-rated upto 60 min. EPD verified by 3rd Party							
ASSA ABLOY SW300–1 Single Door Leaf Door Operator; Low risk users environment	–	–	–	–	–	nr	**2266.00**
ASSA ABLOY SW300–1 Single Door Leaf Door Operator; High risk users environment	–	–	–	–	–	nr	**2472.00**
ASSA ABLOY SW300–2 Double Door Leaf Door Operator; Low risk users environment	–	–	–	–	–	nr	**3090.00**
ASSA ABLOY SW300–2 Double Door Leaf Door Operator; High risk users environment	–	–	–	–	–	nr	**3502.00**

Prices for Measured Works

24 DOORS, SHUTTERS AND HATCHES

Item	PC £	Labour hours	Labour £	Plant £	Material £	Unit	Total rate £
24.04 SWING DOORS – cont							
Swing door operators (mounted onto doors supplied by others) – cont							
ASSA ABLOY SW300 70 mm Surface Mounted Door Operator for external and internal doors. The SW300 is an electromechanical operator and has been developed to comply with European standards and requirements for low and high energy power operated doors. Allows the door to open manually with power assist in both directions with one direction automatically by push pad(s). Includes intelligent control for wind and stack pressure management. Mounted onto a suitable transom by others. Door and frame by others. Fire-rated upto 60 min. EPD verified by 3rd Party							
ASSA ABLOY SW300–1DA Single Door Leaf Door Operator; Low risk users environment	–	–	–	–	–	nr	2369.00
ASSA ABLOY SW300–1DA Single Door Leaf Door Operator; High risk users environment	–	–	–	–	–	nr	2575.00
ASSA ABLOY SW300–2DA Double Door Leaf Door Operator; Low risk users environment	–	–	–	–	–	nr	3296.00
ASSA ABLOY SW300–2DA Double Door Leaf Door Operator; High risk users environment	–	–	–	–	–	nr	3708.00
Swing door systems (including doors and screens)							
ASSA ABLOY SW200 I, anti-ligature, digitally controlled and intelligent complete swing door system. Includes built-in anti-finger trap door leaf mounted into a tight sealed frame. Complete with an overhead concealed swing door operator and ball-raced heavy duty bottom pivot. Door automatically opens inwards or outwards and can also be opened manually with a maximum of 22N opening force. Tested to 1 m cycles. Verified with 3rd party EPD documentation. System including normal installation to comply with BS7036–0/EN16005							
ASSA ABLOY SW200 I–1 Single Door Leaf Concealed In-Head Swing Door Entrance; High risk users environment	–	–	–	–	–	nr	4429.00
ASSA ABLOY SW200 I–1 Double Door Leaf Concealed In-Head Swing Door Entrance; High risk users environment	–	–	–	–	–	nr	5974.00

24 DOORS, SHUTTERS AND HATCHES

Item	PC £	Labour hours	Labour £	Plant £	Material £	Unit	Total rate £
ASSA ABLOY SW200 B Balance Door System, consisting of a liner frame, one door leaf and transom. The Balance Door System is very effective in windy conditions, combining the benefits of a swing and sliding door system. The door system should be fully functional in wind loads according to table 1013665. A perfect solution for narrow passages or in corridors. 3rd party EPD verification							
ASSA ABLOY SW200 B–1 Single Door Leaf Balance Door Entrance; High risk users environment	–	–	–	–	–	nr	5150.00
ASSA ABLOY SW200 B–2 Double Door Leaf Balance Door Entrance; High risk users environment	–	–	–	–	–	nr	8240.00
24.05 SHUTTER DOORS AND PANEL DOORS							
ASSA ABLOY Entrance Systems Ltd; Overhead Sectional Insulated Doors ASSA ABLOY; Thermal transmittance (EN12428) = 0.43 W/m² K (Installed Door 1.1 W/m² K); Water penetration (EN12425) – Class 3; Air permeability (EN12426) – Class 3; Wind-Load (EN12424) – Class 3 reinforced steel panel; spring break device; versatile track system; Single phase electrically operated; manual override, 'deadman' control, pre-coated panel in a standard colour; 1 row DAOP vision panels							
ASSA ABLOY 1042P, Standard Lift; 2500 mm × 3000 mm; 42 mm thick insulated sandwich panels	–	–	–	–	–	nr	2305.00
ASSA ABLOY OH1042P, Standard Lift; 3000 mm × 3000 mm; 42 mm thick insulated sandwich panels	–	–	–	–	–	nr	3125.00
ASSA ABLOY OH1042P, Standard Lift; 4000 mm × 5000 mm; 42 mm thick insulated sandwich panels	–	–	–	–	–	nr	4000.00
ASSA ABLOY OH1042P, Standard Lift; 6000 mm × 6000 mm; 42 mm thick insulated sandwich panels	–	–	–	–	–	nr	6660.00
ASSA ABLOY OH1042P, Standard Lift; 8000 mm × 6000 mm; 42 mm thick insulated sandwich panels	–	–	–	–	–	nr	9600.00

24 DOORS, SHUTTERS AND HATCHES

Item	PC £	Labour hours	Labour £	Plant £	Material £	Unit	Total rate £
24.05 SHUTTER DOORS AND PANEL DOORS – cont							
ASSA ABLOY Entrance Systems Ltd – cont							
Insulated overhead sectional speed doors; ASSA ABLOY; Thermal transmittance (EN12428) = 0.43 W/m^2K (Installed Door 1.1 W/m^2K); Water penetration (EN12425) – Class 3; Air permeability (EN12426) – Class 3; Wind-Load (EN12424) – Class 3 reinforced steel panel; spring break device; versatile track system; Single phase electrically operated; manual override, 'deadman' control, pre-coated panel in a standard colour; 1 row DAOP vision panels							
ASSA ABLOY OH1042S, Standard Lift; 2500 mm × 3000 mm; 42 mm thick insulated sandwich panels	–	–	–	–	–	nr	3800.00
ASSA ABLOY OH1042S, Standard Lift; 3000 mm × 3000 mm; 42 mm thick insulated sandwich panels	–	–	–	–	–	nr	4000.00
ASSA ABLOY OH1042S, Standard Lift; 4000 mm × 5000 mm; 42 mm thick insulated sandwich panels	–	–	–	–	–	nr	5000.00
ASSA ABLOY Entrance Systems Ltd; High Speed Doors							
ASSA ABLOY RR3000 ISO – a low noise high speed insulated secure door, fabricated using 50 mm thick thermally insulated laths. The anti-cold bridge laths are covered in a hard wearing steel skin. A doubled sealing system links the laths ensuring excellent wind and water tightness and a smooth operation of the door. Patented V-Drive Technology supports the door curtain movement, thereby ensuring longevity. The laths are wound onto winding discs without contact and travel within a specially designed channel guide. Using this technology, the door can be opened and closed very quickly, with a soft start and stop feature with minimal wear. Should accidental damage occur then each lath can be replaced individually, rapidly and easily thanks to their individual suspension from the rubber drive belt. Optionally double glazed window laths can be integrated within the door curtain. Door can complete up to 5 opening/closing cycles per minute. Opening speed up to 2.1 m/s closing at 0.7 m/s							
ASSA ABLOY RR3000 ISO, external 3000 mm × 3000 mm	–	–	–	–	–	nr	14800.00
ASSA ABLOY RR3000 ISO, external 5000 mm × 5000 mm	–	–	–	–	–	nr	20100.00
ASSA ABLOY RR3000 ISO, external 7000 mm × 6000 mm	–	–	–	–	–	nr	23000.00

24 DOORS, SHUTTERS AND HATCHES

Item	PC £	Labour hours	Labour £	Plant £	Material £	Unit	Total rate £
High speed internal fabric door; ASSA ABLOY; breakaway facility with auto repair system; electrically operated frequency inverter motor with opening speeds to 2.4 m/s; lifetime expectations 1,000,000 cycles; galvanized steel guide bodies; soft bottom edge for safety; high tensile fabric curtain; stop and return safety-edge, safety photo cells, vision strip, push button operation to both sides							
ASSA ABLOY HS9020G internal; one row windows 2000 mm × 2500 mm	–	–	–	–	–	nr	4695.00
ASSA ABLOY HS9010P internal; one row windows 3000 mm × 3000 mm	–	–	–	–	–	nr	7453.00
ASSA ABLOY HS9010P internal; one row windows 4000 mm × 4000 mm	–	–	–	–	–	nr	7950.00
High Speed external fabric doors; ASSA ABLOY; breakaway facility with auto repair system; air permeability (EN12426) Class 2; Wind load resistance Class 4; lifetime expectations 1,000,000 cycles; electrically operated frequency inverter motor, opening speeds up to 2.4 m/s; galvanized steel guide bodies; soft bottom edge for safety; high tensile fabric curtain; stop and return safety-edge, safety photo cells, push button operation to both sides, vision strip							
ASSA ABLOY HS8010P external; one row windows 3000 mm × 3000 mm	–	–	–	–	–	nr	7940.00
ASSA ABLOY HS8010P external; one row windows 4000 mm × 4500 mm	–	–	–	–	–	nr	9200.00
ASSA ABLOY HS8010P external; one row windows 5500 mm × 5500 mm	–	–	–	–	–	nr	9750.00

24 DOORS, SHUTTERS AND HATCHES

Item	PC £	Labour hours	Labour £	Plant £	Material £	Unit	Total rate £
24.05 SHUTTER DOORS AND PANEL DOORS – cont							
ASSA ABLOY Entrance Systems Ltd; High Speed Clean Room Doors							
Fraunhofer Institute certified high vertical opening door. Certified according to ISO 5 and GMP class C. Compliant to the European norms and regulations in cleanroom applications according to DIN EN ISO 14644–1 and GMP. V2 A stainless steel. The low particle emissions and sealing properties are supported by the unique clean room design. Hidden containment of all mechanical and electronic parts solution for controlled environmental areas. The standard smooth clear PVC curtain has vertical coloured fabric stripes for optimal transparency, person recognition and cleaning. Optional anti-static curtain available, some restrictions in performance, size or certification may apply. Standard alternative would be a silicon free curtain.Opening/closing speed: Up to 2.0/1.0 m/s. Leakage: max. 20 m³/h at a pressure difference of 25 Pa, at a door size of 3 m²							
ASSA ABLOY RR300 Clean internal; clear PVC curtain 2000 mm × 2500 mm	–	–	–	–	–	nr	8666.00
ASSA ABLOY RR300 Clean internal; clear PVC curtain 3000 mm × 3000 mm	–	–	–	–	–	nr	9177.00
DOCK LEVELLERS							
Swingdock leveller from ASSA ABLOY Entrance Systems; electrohydraulic operation with two lifting cylinders; integrated frame to suit various pit applications; Control for door and leveller from one communal control panel, including interlocking of leveller/door; hold-to-run operation; mains power switch; load-bearing capacity 60 kN in line with EN1398; colour RAL5010; toe guards, warning signs							
ASSA ABLOY DL6010S 500 mm swing lip; 2500 mm length × 2000 mm wide × 600 mm height	–	–	–	–	–	nr	3280.00
ASSA ABLOY DL6010S 500 mm swing lip; 4000 mm length × 2000 mm wide × 1100 mm height	–	–	–	–	–	nr	3945.00

24 DOORS, SHUTTERS AND HATCHES

Item	PC £	Labour hours	Labour £	Plant £	Material £	Unit	Total rate £
Teledock leveller from ASSA ABLOY Entrance Systems; electrohydraulic operation with two lifting cylinders; one lip ram cylinder; movable telescopic lip; integrated frame to suit various pit applications; Control for door and leveller from one communal control panel, including interlocking of leveller/door; hold-to-run operation; mains power switch; load-bearing capacity 60 kN in line with EN1398; colour RAL5010; toe guards, warning signs							
ASSA ABLOY DL6020T 500 mm telescopic lip; 2500 mm length × 2000 mm wide	–	–	–	–	–	nr	**3485.00**
ASSA ABLOY DL6020T 1000 mm telescopic lip; 4000 mm length × 2000 mm wide	–	–	–	–	–	nr	**4355.00**
DOCK SHELTERS							
Curtain mechanical shelter; ASSA ABLOY Entrance Systems; extruded aluminium frame; spring loaded-frame connected with parallel bracing arms; 2 side curtains, one top curtain; double-layered high quality polyester, coated on both sides, longitudinal strength: approx. 750 N, transverse strength: approx. 900 N; 4500 mm above outside yard level, 600 mm projection, 1000 mm top curtain, top curtain to have one slit at the contact point on each side, integrated rain-channels							
ASSA ABLOY DS6060 A; 3200 mm height × 3250 mm width	–	–	–	–	–	nr	**1230.00**
ASSA ABLOY DS6060 A; 600 mm projection, 1000 mm top curtain; 3400 mm height × 3250 width	–	–	–	–	–	nr	**1330.00**
ASSA ABLOY DS6060 A; 600 mm projection, 1000 mm top curtain; 3400 mm height × 3450 width	–	–	–	–	–	nr	**1432.00**
Inflatable mechanical shelter; ASSA ABLOY Entrance Systems; hot dipped galvanized surface treatment, polyester painted, corrosion protection class 3; top bag with polyester fabric panels, 1000 mm extension; side bags with polyester fabric panels, 650 mm extension, strong blower for straight movement of top and side bags; side section with steel guards; colour from standard range							
ASSA ABLOY DS6070B; 3755 mm height × 3600 width × 770 mm depth	–	–	–	–	–	nr	**4100.00**
ASSA ABLOY DS6070B; 4055 mm height × 3600 width × 770 mm depth	–	–	–	–	–	nr	**4200.00**
ASSA ABLOY DS6070B; 4555 mm height × 3600 width × 770 mm depth	–	–	–	–	–	nr	**4407.00**

24 DOORS, SHUTTERS AND HATCHES

Item	PC £	Labour hours	Labour £	Plant £	Material £	Unit	Total rate £
24.05 SHUTTER DOORS AND PANEL DOORS – cont							
LOAD HOUSES							
Stand-alone Load House; ASSA ABLOY Entrance Systems; a complete loading unit attached externally to building; modular steel skin/insulated cladding; two wall elements, one roof element; hot-dipped galvanized steel frame; installed at 90°/45° to building; NL 2000/3000 mm; NW3300/3600 mm; DH 950/1300 mm; 4° roof slope; drainpipe, gutter; standard colour range							
ASSA ABLOY LH6081L Load house, 6010SA Swingdock c/w Autodock platform; 2000 mm length × 3300 mm width × 3495 mm height	–	–	–	–	–	nr	6660.00
ASSA ABLOY LH6081L Load house, 6010SA Swingdock c/w Autodock platform; 3000 mm length × 3600 mm width × 3575 mm height	–	–	–	–	–	nr	8405.00
Rolling shutters and collapsible gates; steel counter shutters; push-up, self-coiling; polyester power coated; fixing by bolting							
Shutter doors							
3000 mm × 1000 mm	–	–	–	–	–	nr	1549.80
4000 mm × 1000 mm; in two panels	–	–	–	–	–	nr	2679.86
Rolling shutters with collapsible gates; galvanized steel; one hour fire-resisting; self-coiling; activated by fusible link; fixing with bolts							
Shutter doors							
1000 mm × 2750 mm	–	–	–	–	–	nr	1851.15
1500 mm × 2750 mm	–	–	–	–	–	nr	1953.39
2400 mm × 2750 mm	–	–	–	–	–	nr	2335.46

24 DOORS, SHUTTERS AND HATCHES

Item	PC £	Labour hours	Labour £	Plant £	Material £	Unit	Total rate £
Translucent GRP stacking door (78% translucency); U-value = 2.6 W/m² K; electrically operated; Envirodoor Ltd; fully enclosed aluminium track system with SAA finish; manual over-ride, lock interlock, stop and return safety cage, deadmans down button, anti-flip device, photoelectric cell and beam deflectors; fixing by bolting; standard panel finishes							
Stacking doors							
HT40 stacking door 2500 mm × 3000 mm; 40 mm thick × 500 mm high twin walled GRP translucent panels	–	–	–	–	–	nr	**5849.16**
HT40 stacking door 4500 mm × 6000 mm; 40 mm thick × 500 mm high twin walled translucent panels	–	–	–	–	–	nr	**10870.13**
HT60-N stacking door 6000 mm × 6000 mm: 60 mm thick × 500 mm high twinn walled translucent panels	–	–	–	–	–	nr	**16564.00**
HT60-H stacking door 7000 mm × 8000 mm: 60 mm thick × 1000 mm high twin walled translucent panels	–	–	–	–	–	nr	**23293.13**
HT80–400 stacking door 10000 mm × 10000 mm; 80 mm thick × 1000 mm high twin walled translucent panels	–	–	–	–	–	nr	**57559.90**
Doors; galvanized steel up and over type garage doors; Catnic Horizon 90 or other equal and approved; spring counter balanced; fixed to timber frame (not included)							
Garage door							
2135 mm × 1980 mm	440.00	3.70	86.69	–	462.17	nr	**548.86**
2135 mm × 2135 mm	481.60	3.70	86.69	–	505.85	nr	**592.54**
2400 mm × 2135 mm	609.60	3.70	86.69	–	640.28	nr	**726.97**
3965 mm × 2135 mm	983.20	5.55	130.04	–	1032.76	nr	**1162.80**
Grilles; Galaxy nylon rolling counter grille or other equal and approved; Bolton Brady Ltd; colour, off-white; self-coiling; fixing by bolting							
Grilles							
3000 mm × 1000 mm	–	–	–	–	–	nr	**1100.00**
4000 mm × 1000 mm	–	–	–	–	–	nr	**1800.00**
Sliding/folding partitions; Alco Beldan Ltd or equal and approved							
Sliding/folding partitions							
ref. NW100 Moveable Wall; 5000 mm (wide) × 2495 mm (high) comprising 4 nr 954 mm (wide) standard panels and 1 n. 954 mm (wide) telescopic panel; sealing; fixing	–	–	–	–	–	nr	**14700.00**

24 DOORS, SHUTTERS AND HATCHES

Item	PC £	Labour hours	Labour £	Plant £	Material £	Unit	Total rate £
24.06 INTERNAL DOORS AND FRAMES							
DOOR FRAMES AND LININGS							
Internal softwood door lining set; with loose stops							
32 mm × 115 mm wide (finished) set; with loose stops; for internal doors							
686 / 762 / 838 wide doors	13.89	0.70	16.09	–	14.49	nr	**30.58**
32 mm × 138 mm wide (finished) set; with loose stops; for internal doors							
686 / 762 / 838 wide doors	15.87	0.70	16.09	–	16.54	nr	**32.63**
Internal softwood fire door door lining set; with loose stops							
38 mm × 115 mm wide (finished) set; with loose stops; for internal doors							
686 / 762 / 838 wide doors	20.92	0.70	16.09	–	21.73	nr	**37.82**
38 mm × 138 mm wide (finished) set; with loose stops; for internal doors							
686 / 762 / 838 wide doors	24.82	0.70	16.09	–	25.76	nr	**41.85**
Door frames and door linings, sets; purpose-made; wrought softwood							
Jambs and heads; as linings							
32 mm × 63 mm	–	0.16	3.68	–	5.73	m	**9.41**
32 mm × 100 mm	–	0.16	3.68	–	6.45	m	**10.13**
32 mm × 140 mm	–	0.16	3.68	–	6.87	m	**10.55**
Jambs and heads; as frames; rebated, rounded and grooved							
44 mm × 75 mm	–	0.16	3.68	–	9.21	m	**12.89**
44 mm × 100 mm	–	0.16	3.68	–	9.91	m	**13.59**
44 mm × 115 mm	–	0.16	3.68	–	9.96	m	**13.64**
44 mm × 140 mm	–	0.19	4.37	–	10.44	m	**14.81**
57 mm × 100 mm	–	0.19	4.37	–	10.60	m	**14.97**
57 mm × 125 mm	–	0.19	4.37	–	11.20	m	**15.57**
69 mm × 88 mm	–	0.19	4.37	–	10.78	m	**15.15**
69 mm × 100 mm	–	0.19	4.37	–	11.56	m	**15.93**
69 mm × 125 mm	–	0.20	4.59	–	12.26	m	**16.85**
69 mm × 150 mm	–	0.20	4.59	–	12.98	m	**17.57**
94 mm × 100 mm	–	0.23	5.28	–	17.51	m	**22.79**
94 mm × 150 mm	–	0.23	5.28	–	20.41	m	**25.69**
Mullions and transoms; in linings							
32 mm × 63 mm	–	0.11	2.52	–	7.63	m	**10.15**
32 mm × 100 mm	–	0.11	2.52	–	8.36	m	**10.88**
32 mm × 140 mm	–	0.11	2.52	–	8.72	m	**11.24**
Mullions and transoms; in frames; twice rebated, rounded and grooved							
44 mm × 75 mm	–	0.11	2.52	–	11.57	m	**14.09**
44 mm × 100 mm	–	0.11	2.52	–	12.05	m	**14.57**
44 mm × 115 mm	–	0.11	2.52	–	12.05	m	**14.57**

24 DOORS, SHUTTERS AND HATCHES

Item	PC £	Labour hours	Labour £	Plant £	Material £	Unit	Total rate £
44 mm × 140 mm	–	0.13	2.99	–	12.52	m	**15.51**
57 mm × 100 mm	–	0.13	2.99	–	12.68	m	**15.67**
57 mm × 125 mm	–	0.13	2.99	–	13.28	m	**16.27**
69 mm × 88 mm	–	0.13	2.99	–	12.52	m	**15.51**
69 mm × 100 mm	–	0.13	2.99	–	13.26	m	**16.25**
Add 5% to the above material prices for selected softwood for staining							
Door frames and door linings, sets; purpose-made; medium density fireboard							
Jambs and heads; as linings							
18 mm × 126 mm	–	0.16	3.68	–	6.89	m	**10.57**
22 mm × 126 mm	–	0.16	3.68	–	7.14	m	**10.82**
25 mm × 126 mm	–	0.16	3.68	–	7.28	m	**10.96**
Door frames and door linings, sets; purpose-made; selected Sapele							
Jambs and heads; as linings							
32 mm × 63 mm	8.59	0.21	4.83	–	8.90	m	**13.73**
32 mm × 100 mm	10.61	0.21	4.83	–	10.98	m	**15.81**
32 mm × 140 mm	11.61	0.21	4.83	–	12.05	m	**16.88**
Jambs and heads; as frames; rebated, rounded and grooved							
44 mm × 75 mm	14.07	0.21	4.83	–	14.54	m	**19.37**
44 mm × 100 mm	16.03	0.21	4.83	–	16.56	m	**21.39**
44 mm × 115 mm	16.57	0.21	4.83	–	17.16	m	**21.99**
44 mm × 140 mm	17.37	0.25	5.75	–	17.98	m	**23.73**
57 mm × 100 mm	17.76	0.25	5.75	–	18.39	m	**24.14**
57 mm × 125 mm	19.43	0.25	5.75	–	20.12	m	**25.87**
69 mm × 88 mm	17.74	0.25	5.75	–	18.31	m	**24.06**
69 mm × 100 mm	19.73	0.25	5.75	–	20.41	m	**26.16**
69 mm × 125 mm	21.73	0.28	6.44	–	22.47	m	**28.91**
69 mm × 150 mm	23.71	0.28	6.44	–	24.51	m	**30.95**
94 mm × 100 mm	27.70	0.28	6.44	–	28.62	m	**35.06**
94 mm × 150 mm	33.97	0.28	6.44	–	35.09	m	**41.53**
Mullions and transoms; in linings							
32 mm × 63 mm	10.78	0.15	3.45	–	11.10	m	**14.55**
32 mm × 100 mm	12.79	0.15	3.45	–	13.17	m	**16.62**
32 mm × 140 mm	13.81	0.15	3.45	–	14.22	m	**17.67**
Mullions and transoms; in frames; twice rebated, rounded and grooved							
44 mm × 75 mm	17.33	0.15	3.45	–	17.85	m	**21.30**
44 mm × 100 mm	18.64	0.15	3.45	–	19.20	m	**22.65**
44 mm × 115 mm	19.18	0.15	3.45	–	19.76	m	**23.21**
44 mm × 140 mm	19.98	0.17	3.90	–	20.58	m	**24.48**
57 mm × 100 mm	20.38	0.17	3.90	–	20.99	m	**24.89**
57 mm × 125 mm	22.05	0.17	3.90	–	22.71	m	**26.61**
69 mm × 88 mm	19.98	0.17	3.90	–	20.58	m	**24.48**
69 mm × 100 mm	22.48	0.17	3.90	–	23.15	m	**27.05**

Prices for Measured Works

24 DOORS, SHUTTERS AND HATCHES

Item	PC £	Labour hours	Labour £	Plant £	Material £	Unit	Total rate £
24.06 INTERNAL DOORS AND FRAMES – cont							
Door frames and door linings, sets – cont							
Sills; once sunk weathered; once rebated, three times grooved							
63 mm × 175 mm	49.77	0.31	7.13	–	51.26	m	**58.39**
75 mm × 125 mm	47.94	0.31	7.13	–	49.38	m	**56.51**
75 mm × 150 mm	50.19	0.31	7.13	–	51.70	m	**58.83**
Door frames and door linings, sets; European Oak							
Sills; once sunk weathered; once rebated, three times grooved							
63 mm × 175 mm	81.81	0.31	7.13	–	84.26	m	**91.39**
75 mm × 125 mm	80.77	0.31	7.13	–	83.19	m	**90.32**
75 mm × 150 mm	88.30	0.31	7.13	–	90.95	m	**98.08**
Fire-resisting door frame; internal and external; fitted with 15 mm × 4 mm intumescent strips; 12 mm deep rebates; screwed to masonry/ concrete							
Softwood frames; no sill – open in or out. Door size:							
762 mm × 1981 mm × 44 mm; FD30; intemescent strip only	83.91	0.75	17.24	–	86.61	nr	**103.85**
838 mm × 1981 mm × 44 mm; FD30; intemescent strip only	86.20	0.75	17.24	–	88.98	nr	**106.22**
826 mm × 2040 mm × 44 mm; FD30; intemescent strip only	98.92	0.75	17.24	–	102.08	nr	**119.32**
762 mm × 1981 mm × 44 mm; FD30; intemescent strip/smoke seal	89.28	0.75	17.24	–	92.15	nr	**109.39**
838 mm × 1981 mm × 44 mm; FD30; intemescent strip/smoke seal	91.24	0.75	17.24	–	94.17	nr	**111.41**
826 mm × 2040 mm × 44 mm; FD30; intemescent strip/smoke seal	91.24	0.75	17.24	–	94.17	nr	**111.41**
Hardwood frames; no sill – open in or out. Door size:							
762 mm × 1981 mm × 44 mm; FD60; intemescent strip/smoke seal	169.24	0.75	17.24	–	174.51	nr	**191.75**
838 mm × 1981 mm × 44 mm; FD60; intemescent strip/smoke seal	173.04	0.75	17.24	–	178.43	nr	**195.67**
826 mm × 2040 mm × 44 mm; FD60; intemescent strip/smoke seal	175.67	0.75	17.24	–	181.13	nr	**198.37**

24 DOORS, SHUTTERS AND HATCHES

Item	PC £	Labour hours	Labour £	Plant £	Material £	Unit	Total rate £
INTERNAL DOORS							
Note: door frames, linings and ironmogery is not included							
Doors; standard flush; softwood composition							
Moulded panel doors; white based coated facings suitable for paint finish only; two, four or six panel options							
533 mm × 1981 mm × 35 mm	33.08	0.75	17.24	–	34.07	nr	**51.31**
610 mm × 1981 mm × 35 mm	33.08	0.75	17.24	–	34.07	nr	**51.31**
762 mm × 1981 mm × 35 mm	33.08	0.75	17.24	–	34.07	nr	**51.31**
838 mm × 1981 mm × 35 mm	35.72	0.75	17.24	–	36.79	nr	**54.03**
914 mm × 1981 mm × 35 mm	39.69	0.75	17.24	–	40.88	nr	**58.12**
526 mm × 2040 mm × 40 mm	33.08	0.75	17.24	–	34.07	nr	**51.31**
626 mm × 2040 mm × 40 mm	33.08	0.75	17.24	–	34.07	nr	**51.31**
726 mm × 2040 mm × 40 mm	33.08	0.75	17.24	–	34.07	nr	**51.31**
826 mm × 2040 mm × 40 mm	35.72	0.75	17.24	–	36.79	nr	**54.03**
926 mm × 2040 mm × 40 mm	39.69	0.75	17.24	–	40.88	nr	**58.12**
Flush door; internal quality; skeleton or cellular core; faced both sides; lipped on two long edges; Jeld-Wen paint grade veneer or other equal							
457 mm × 1981 mm × 35 mm	49.96	1.16	26.66	–	51.46	nr	**78.12**
533 mm × 1981 mm × 35 mm	50.90	1.16	26.66	–	52.43	nr	**79.09**
610 mm × 1981 mm × 35 mm	50.90	1.16	26.66	–	52.43	nr	**79.09**
686 mm × 1981 mm × 35 mm	50.90	1.16	26.66	–	52.43	nr	**79.09**
762 mm × 1981 mm × 35 mm	50.90	1.16	26.66	–	52.43	nr	**79.09**
838 mm × 1981 mm × 35 mm	52.92	1.16	26.66	–	54.51	nr	**81.17**
864 mm × 1981 mm × 35 mm	65.77	1.16	26.66	–	67.74	nr	**94.40**
426 mm × 2040 mm × 40 mm	54.56	1.16	26.66	–	56.20	nr	**82.86**
526 mm × 2040 mm × 40 mm	54.56	1.16	26.66	–	56.20	nr	**82.86**
626 mm × 2040 mm × 40 mm	54.56	1.16	26.66	–	56.20	nr	**82.86**
726 mm × 2040 mm × 40 mm	55.69	1.16	26.66	–	57.36	nr	**84.02**
826 mm × 2040 mm × 40 mm	63.44	1.16	26.66	–	65.34	nr	**92.00**
926 mm × 2040 mm × 40 mm	65.20	1.16	26.66	–	67.16	nr	**93.82**
Flush door; internal quality; skeleton or cellular core; chipboard veneered; faced both sides; lipped on two long edges; Jeld-Wen Sapele veneered or other equal and approved							
457 mm × 1981 mm × 35 mm	72.45	1.25	28.73	–	74.62	nr	**103.35**
533 mm × 1981 mm × 35 mm	72.45	1.25	28.73	–	74.62	nr	**103.35**
610 mm × 1981 mm × 35 mm	72.45	1.25	28.73	–	74.62	nr	**103.35**
686 mm × 1981 mm × 35 mm	77.36	1.25	28.73	–	79.68	nr	**108.41**
762 mm × 1981 mm × 35 mm	77.36	1.25	28.73	–	79.68	nr	**108.41**
838 mm × 1981 mm × 35 mm	82.28	1.25	28.73	–	84.75	nr	**113.48**
526 mm × 2040 mm × 40 mm	77.36	1.25	28.73	–	79.68	nr	**108.41**
626 mm × 2040 mm × 40 mm	82.28	1.25	28.73	–	84.75	nr	**113.48**
726 mm × 2040 mm × 40 mm	82.28	1.25	28.73	–	84.75	nr	**113.48**
826 mm × 2040 mm × 40 mm	87.19	1.25	28.73	–	89.81	nr	**118.54**
926 mm × 2040 mm × 40 mm	92.11	1.25	28.73	–	94.87	nr	**123.60**

24 DOORS, SHUTTERS AND HATCHES

Item	PC £	Labour hours	Labour £	Plant £	Material £	Unit	Total rate £
24.06 INTERNAL DOORS AND FRAMES – cont							
Doors; purpose-made panelled; wrought softwood							
Panelled doors; one open panel for glass; including glazing beads							
686 mm × 1981 mm × 44 mm	96.64	1.62	37.23	–	99.54	nr	**136.77**
762 mm × 1981 mm × 44 mm	97.44	1.62	37.23	–	100.36	nr	**137.59**
838 mm × 1981 mm × 44 mm	98.24	1.62	37.23	–	101.19	nr	**138.42**
Panelled doors; two open panel for glass; including glazing beads							
686 mm × 1981 mm × 44 mm	135.11	1.62	37.23	–	139.16	nr	**176.39**
762 mm × 1981 mm × 44 mm	136.28	1.62	37.23	–	140.37	nr	**177.60**
838 mm × 1981 mm × 44 mm	137.45	1.62	37.23	–	141.57	nr	**178.80**
Panelled doors; four 19 mm thick plywood panels; mouldings worked on solid both sides							
686 mm × 1981 mm × 44 mm	205.18	1.62	37.23	–	211.34	nr	**248.57**
762 mm × 1981 mm × 44 mm	207.83	1.62	37.23	–	214.06	nr	**251.29**
838 mm × 1981 mm × 44 mm	210.48	1.62	37.23	–	216.79	nr	**254.02**
Panelled doors; six 25 mm thick panels raised and fielded; mouldings worked on solid both sides							
686 mm × 1981 mm × 44 mm	377.52	1.94	44.59	–	388.85	nr	**433.44**
762 mm × 1981 mm × 44 mm	381.29	1.94	44.59	–	392.73	nr	**437.32**
838 mm × 1981 mm × 44 mm	385.06	1.94	44.59	–	396.61	nr	**441.20**
rebated edges beaded	–	–	–	–	2.34	m	**2.34**
rounded edges or heels	–	–	–	–	0.54	m	**0.54**
weatherboard fixed to bottom rail	–	0.23	5.28	–	8.30	m	**13.58**
stopped groove for weatherboard	–	–	–	–	2.66	m	**2.66**
Doors; purpose-made panelled; selected Sapele							
Panelled doors; one open panel for glass; including glazing beads							
686 mm × 1981 mm × 44 mm	130.83	2.31	53.10	–	134.75	nr	**187.85**
762 mm × 1981 mm × 44 mm	132.60	2.31	53.10	–	136.58	nr	**189.68**
838 mm × 1981 mm × 44 mm	134.38	2.31	53.10	–	138.41	nr	**191.51**
686 mm × 1981 mm × 57 mm	139.84	2.54	58.38	–	144.04	nr	**202.42**
762 mm × 1981 mm × 57 mm	141.94	2.54	58.38	–	146.20	nr	**204.58**
838 mm × 1981 mm × 57 mm	144.04	2.54	58.38	–	148.36	nr	**206.74**
Panelled doors; 250 mm wide cross tongued intermediate rail; two open panels for glass; mouldings worked on the solid one side; 19 mm × 13 mm beads one side; fixing with brass cups and screws							
686 mm × 1981 mm × 44 mm	200.01	2.31	53.10	–	206.01	nr	**259.11**
762 mm × 1981 mm × 44 mm	203.39	2.31	53.10	–	209.49	nr	**262.59**
838 mm × 1981 mm × 44 mm	213.28	2.31	53.10	–	219.68	nr	**272.78**
686 mm × 1981 mm × 57 mm	213.28	2.54	58.38	–	219.68	nr	**278.06**
762 mm × 1981 mm × 57 mm	217.27	2.54	58.38	–	223.79	nr	**282.17**
838 mm × 1981 mm × 57 mm	221.36	2.54	58.38	–	228.00	nr	**286.38**

24 DOORS, SHUTTERS AND HATCHES

Item	PC £	Labour hours	Labour £	Plant £	Material £	Unit	Total rate £
Panelled doors; four panels (19 mm thick for 44 mm doors, 25 mm thick for 57 mm doors); mouldings worked on solid both sides							
686 mm × 1981 mm × 44 mm	281.05	2.31	53.10	–	289.48	nr	**342.58**
762 mm × 1981 mm × 44 mm	302.96	2.31	53.10	–	312.05	nr	**365.15**
838 mm × 1981 mm × 44 mm	317.94	2.31	53.10	–	327.48	nr	**380.58**
686 mm × 1981 mm × 57 mm	286.73	2.54	58.38	–	295.33	nr	**353.71**
762 mm × 1981 mm × 57 mm	310.45	2.54	58.38	–	319.76	nr	**378.14**
838 mm × 1981 mm × 57 mm	292.42	2.54	58.38	–	301.19	nr	**359.57**
Panelled doors; 150 mm wide stiles in one width; 430 mm wide cross tongued bottom rail; six panels raised and fielded one side (19 mm thick for 44 mm doors, 25 mm thick for 57 mm doors); mouldings worked on solid both sides							
686 mm × 1981 mm × 44 mm	479.77	2.31	53.10	–	494.16	nr	**547.26**
762 mm × 1981 mm × 44 mm	529.45	2.31	53.10	–	545.33	nr	**598.43**
838 mm × 1981 mm × 44 mm	539.51	2.31	53.10	–	555.70	nr	**608.80**
686 mm × 1981 mm × 57 mm	510.89	2.54	58.38	–	526.22	nr	**584.60**
762 mm × 1981 mm × 57 mm	563.34	2.54	58.38	–	580.24	nr	**638.62**
838 mm × 1981 mm × 57 mm	576.16	2.54	58.38	–	593.44	nr	**651.82**
rebated edges beaded	–	–	–	–	2.98	m	**2.98**
rounded edges or heels	–	–	–	–	0.79	m	**0.79**
weatherboard fixed to bottom rail	–	0.31	7.13	–	11.23	m	**18.36**
stopped groove for weatherboard	–	–	–	–	2.76	m	**2.76**
Internal white foiled moisture-resistant MDF door lining composite standard joinery set							
22 mm × 77 mm wide (finished) set; with loose stops; for internal doors							
610 mm × 1981 mm × 35 mm	10.44	0.70	16.09	–	10.95	nr	**27.04**
686 mm × 1981 mm × 35 mm	10.39	0.70	16.09	–	10.90	nr	**26.99**
762 mm × 1981 mm × 35 mm	10.39	0.70	16.09	–	10.90	nr	**26.99**
838 mm × 1981 mm × 35 mm	11.44	0.70	16.09	–	11.98	nr	**28.07**
864 mm × 1981 mm × 35 mm	10.39	0.70	16.09	–	10.90	nr	**26.99**
22 mm × 150 mm wide (finished) set; with loose stops; for internal doors							
610 mm × 1981 mm × 35 mm	11.49	0.70	16.09	–	12.03	nr	**28.12**
686 mm × 1981 mm × 35 mm	11.44	0.70	16.09	–	11.98	nr	**28.07**
762 mm × 1981 mm × 35 mm	11.44	0.70	16.09	–	11.98	nr	**28.07**
838 mm × 1981 mm × 35 mm	11.44	0.70	16.09	–	11.98	nr	**28.07**
864 mm × 1981 mm × 35 mm	11.44	0.70	16.09	–	11.98	nr	**28.07**
FIRE DOORS							
Flush door; half-hour fire-resisting (FD30); chipboard veneered; faced both sides; lipped on two long edges; Jeld-Wen paint grade veneer or other equal and approved							
610 mm × 1981 mm × 44 mm	71.13	1.62	37.23	–	73.26	nr	**110.49**
686 mm × 1981 mm × 44 mm	71.13	1.62	37.23	–	73.26	nr	**110.49**
762 mm × 1981 mm × 44 mm	71.13	1.62	37.23	–	73.26	nr	**110.49**

24 DOORS, SHUTTERS AND HATCHES

Item	PC £	Labour hours	Labour £	Plant £	Material £	Unit	Total rate £
24.06 INTERNAL DOORS AND FRAMES – cont							
FIRE DOORS – cont							
Flush door – cont							
838 mm × 1981 mm × 44 mm	73.14	1.62	37.23	–	75.33	nr	**112.56**
526 mm × 2040 mm × 44 mm	74.78	1.62	37.23	–	77.02	nr	**114.25**
626 mm × 2040 mm × 44 mm	74.78	1.62	37.23	–	77.02	nr	**114.25**
726 mm × 2040 mm × 44 mm	75.92	1.62	37.23	–	78.20	nr	**115.43**
826 mm × 2040 mm × 44 mm	83.66	1.62	37.23	–	86.17	nr	**123.40**
Moulded panel doors; half-hour fire-resisting (FD30); white based coated facings suitable for paint finish only; two, four or six panel options; Premdor Fireshield or other equal and approved							
526 mm × 2040 mm × 44 mm	82.03	1.10	25.29	–	84.49	nr	**109.78**
626 mm × 2040 mm × 44 mm	82.03	1.10	25.29	–	84.49	nr	**109.78**
726 mm × 2040 mm × 44 mm	280.48	1.10	25.29	–	288.89	nr	**314.18**
826 mm × 2040 mm × 44 mm	86.66	1.10	25.29	–	89.26	nr	**114.55**
926 mm × 2040 mm × 44 mm	92.61	1.10	25.29	–	95.39	nr	**120.68**
Flush door; half-hour fire-resisting (FD30); faced both sides; lipped on two long edges; Jeld-Wen Sapele veneered or other equal and approved							
610 mm × 1981 mm × 44 mm	86.63	1.71	39.30	–	89.23	nr	**128.53**
686 mm × 1981 mm × 44 mm	91.54	1.71	39.30	–	94.29	nr	**133.59**
762 mm × 1981 mm × 44 mm	91.54	1.71	39.30	–	94.29	nr	**133.59**
838 mm × 1981 mm × 44 mm	96.45	1.71	39.30	–	99.34	nr	**138.64**
726 mm × 2040 mm × 44 mm	96.45	1.71	39.30	–	99.34	nr	**138.64**
826 mm × 2040 mm × 44 mm	101.37	1.71	39.30	–	104.41	nr	**143.71**
926 mm × 2040 mm × 44 mm	106.28	1.71	39.30	–	109.47	nr	**148.77**
Flush door; half-hour fire-resisting (FD30); chipboard for painting; hardwood lipping two long edges; Premdor or other equal							
526 mm × 2040 mm × 44 mm	51.60	1.62	37.23	–	53.15	nr	**90.38**
626 mm × 2040 mm × 44 mm	51.60	1.62	37.23	–	53.15	nr	**90.38**
726 mm × 2040 mm × 44 mm	51.60	1.62	37.23	–	53.15	nr	**90.38**
826 mm × 2040 mm × 44 mm	52.92	1.62	37.23	–	54.51	nr	**91.74**
926 mm × 2040 mm × 44 mm	59.53	1.62	37.23	–	61.32	nr	**98.55**
826 mm × 2040 mm × 44 mm; single side vision panel 150 mm × 700 mm; factory fitted clear fire-rated glass	154.13	1.71	39.30	–	158.75	nr	**198.05**
826 mm × 2040 mm × 44 mm; two side vision panels 150 mm × 700 mm; factory fitted clear fire-rated glass	212.34	1.71	39.30	–	218.71	nr	**258.01**
926 mm × 2040 mm × 44 mm; single side vision panel 150 mm × 700 mm; factory fitted clear fire-rated glass	160.74	1.71	39.30	–	165.56	nr	**204.86**
926 mm × 2040 mm × 44 mm; two side vision panels 150 mm × 700 mm; factory fitted clear fire-rated glass	218.96	1.71	39.30	–	225.53	nr	**264.83**

24 DOORS, SHUTTERS AND HATCHES

Item	PC £	Labour hours	Labour £	Plant £	Material £	Unit	Total rate £
Flush door; half-hour fire-resisting (FD30); White Oak veneer; hardwood lipping all edges; Premdor Fireshield or other equal and approved;							
526 mm × 2040 mm × 44 mm	87.32	1.62	37.23	–	89.94	nr	**127.17**
626 mm × 2040 mm × 44 mm	87.32	1.62	37.23	–	89.94	nr	**127.17**
726 mm × 2040 mm × 44 mm	87.32	1.62	37.23	–	89.94	nr	**127.17**
826 mm × 2040 mm × 44 mm	90.63	1.62	37.23	–	93.35	nr	**130.58**
926 mm × 2040 mm × 44 mm	97.90	1.62	37.23	–	100.84	nr	**138.07**
826 mm × 2040 mm × 44 mm; single side vision panel 508 mm × 1649 mm; factory fitted clear fire-rated glass	201.10	1.71	39.30	–	207.13	nr	**246.43**
826 mm × 2040 mm × 44 mm; two side vision panels 150 mm × 775 mm and 150 mm × 700 mm; factory fitted clear fire-rated glass	218.96	1.71	39.30	–	225.53	nr	**264.83**
926 mm × 2040 mm × 44 mm; single side vision panel 508 mm × 1649 mm; factory fitted clear fire-rated glass	208.37	1.71	39.30	–	214.62	nr	**253.92**
926 mm × 2040 mm × 44 mm; two side vision panels 150 mm × 775 mm and 150 mm × 700 mm; factory fitted clear fire-rated glass	226.23	1.71	39.30	–	233.02	nr	**272.32**
Flush door; one-hour fire-resisting (FD60); chipboard for painting; hardwood lipping two long edges; Premdor Firemaster or other equal and approved							
626 mm × 2040 mm × 54 mm	191.84	1.71	39.30	–	197.60	nr	**236.90**
726 mm × 2040 mm × 54 mm	191.84	1.71	39.30	–	197.60	nr	**236.90**
826 mm × 2040 mm × 54 mm	212.34	1.71	39.30	–	218.71	nr	**258.01**
926 mm × 2040 mm × 54 mm	218.96	1.71	39.30	–	225.53	nr	**264.83**
Moulded panel doors; half-hour fire-resisting (FD60); white based coated facings suitable for paint finish only; two, four or six panel options; Premdor Firemaster or other equal and approved							
626 mm × 2040 mm × 54 mm	257.99	1.71	39.30	–	265.73	nr	**305.03**
726 mm × 2040 mm × 54 mm	257.99	1.71	39.30	–	265.73	nr	**305.03**
826 mm × 2040 mm × 54 mm	263.94	1.71	39.30	–	271.86	nr	**311.16**
926 mm × 2040 mm × 54 mm	272.54	1.71	39.30	–	280.72	nr	**320.02**
Flush door; one-hour fire-resisting (FD60); White Oak veneer; hardwood lipping all edges; Premdor Firemaster or other equal and approved							
626 mm × 2040 mm × 54 mm	216.31	1.94	44.59	–	222.80	nr	**267.39**
726 mm × 2040 mm × 54 mm	216.31	1.94	44.59	–	222.80	nr	**267.39**
826 mm × 2040 mm × 54 mm	228.88	1.94	44.59	–	235.75	nr	**280.34**
926 mm × 2040 mm × 54 mm	236.16	1.94	44.59	–	243.24	nr	**287.83**
Flush door; one-hour fire-resisting (FD60); Steamed Beech veneer; hardwood lipping all edges; Premdor Firemaster or other equal and approved							
626 mm × 2040 mm × 54 mm	217.63	1.94	44.59	–	224.16	nr	**268.75**
726 mmx 2040 mm × 54 mm	217.63	1.94	44.59	–	224.16	nr	**268.75**
826 mm × 2040 mm × 54 mm	232.19	1.94	44.59	–	239.16	nr	**283.75**
926 mm × 2040 mm × 54 mm	239.46	1.94	44.59	–	246.64	nr	**291.23**

24 DOORS, SHUTTERS AND HATCHES

Item	PC £	Labour hours	Labour £	Plant £	Material £	Unit	Total rate £
24.06 INTERNAL DOORS AND FRAMES – cont							
HYGENIC DOORSETS							
Altro Whiterock flush doors complete with frame and stops, leaf and integral hinges. NOTE EXCLUDING IRONMONGERY							
946 mm × 2040 mm single leaf doorset; non fire-rated	1035.00	3.00	68.95	–	1066.25	nr	**1135.20**
946 mm × 2040 mm single leaf doorset; FD30	1100.00	3.00	68.95	–	1133.20	nr	**1202.15**
946 mm × 2040 mm single leaf doorset; FD60	1350.00	3.00	68.95	–	1390.70	nr	**1459.65**
1350 mm × 2040 mm leaf and & half doorset; non fire-rated	1490.00	4.00	91.94	–	1534.94	nr	**1626.88**
1350 mm × 2040 mm leaf & half doorset; FD30	1580.00	4.00	91.94	–	1627.64	nr	**1719.58**
1350 mm × 2040 mm leaf and half doorset; FD60	2000.00	4.00	91.94	–	2060.24	nr	**2152.18**
1800 mm × 2040 mm double leaf doorset; non fire-rated	1650.00	6.00	137.91	–	1699.79	nr	**1837.70**
1800 mm × 2040 mm double leaf doorset; FD30	1800.00	6.00	137.91	–	1854.29	nr	**1992.20**
1800 mm × 2040 mm double leaf doorset; FD60	2200.00	6.00	137.91	–	2266.29	nr	**2404.20**
Intumescent strips							
Factory installed intumescent smoke seals to fire doors; 3 edges, jambs and head; site fixed intumescent strips	–	–	–	–	27.41	door	**27.41**
fire and smoke intumescent strips							
15 mm × 4 mm – FD30 doors	–	0.15	3.45	–	0.76	m	**4.21**
fire and smoke intumescent strips							
20 mm × 4 mm – FD60 doors	–	0.15	3.45	–	1.07	m	**4.52**
Bedding and pointing frames							
Pointing wood frames or sills with mastic							
one side	–	0.09	1.57	–	1.16	m	**2.73**
both sides	–	0.19	3.31	–	2.32	m	**5.63**
Pointing wood frames or sills with polysulphide sealant							
one side	–	0.09	1.57	–	2.95	m	**4.52**
both sides	–	0.19	3.31	–	5.89	m	**9.20**
Bedding wood frames in cement mortar (1:3) and point							
one side	–	0.07	1.53	–	0.16	m	**1.69**
both sides	–	0.09	1.98	–	0.21	m	**2.19**
one side in mortar; other side in mastic	–	0.19	3.76	–	1.32	m	**5.08**
24.07 ASSOCIATED IRONMONGERY							
NOTE: Ironmongery is largely a matter of selection and specification and prices vary considerably; indicative prices for reasonable quantities of good quality ironmongery are given below.							

24 DOORS, SHUTTERS AND HATCHES

Item	PC £	Labour hours	Labour £	Plant £	Material £	Unit	Total rate £
Allgood Ltd or other equal; to softwood							
Bolts							
75 × 35 mm Modric anodized aluminium straight barrel bolt	9.17	0.30	6.89	–	9.45	nr	**16.34**
150 × 35 mm Modric anodized aluminium straight barrel bolt	10.47	0.30	6.89	–	10.78	nr	**17.67**
75 × 35 mm Modric anodized aluminium necked barrel bolt	10.27	0.30	6.89	–	10.58	nr	**17.47**
150 × 35 mm Modric anodized aluminium necked barrel bolt	13.13	0.30	6.89	–	13.52	nr	**20.41**
11 mm Easiclean socket for wood or stone	4.48	0.10	2.30	–	4.61	nr	**6.91**
Security hinge bolt chubb WS12	9.29	0.50	11.49	–	9.57	nr	**21.06**
203 × 19 × 11 mm lever action flush bolt set, with coil spring and intumescent pack for FD30 and FD60 fire doors	31.51	0.60	13.79	–	32.46	nr	**46.25**
609 × 19 mm lever action flush bolt set, with coil sprin and intumescent pack for FD30 and FD60 fire doors	91.85	0.60	13.79	–	94.61	nr	**108.40**
Stainless steel indicating bolt complete with outside indicator and emergency release	65.83	0.60	13.79	–	67.80	nr	**81.59**
Catches							
Magnetic catch	0.46	0.20	4.59	–	0.47	nr	**5.06**
Door closers and furniture							
13 mm stainless steel rebate component for 7104/08/78/79/86	27.31	0.60	13.79	–	28.13	nr	**41.92**
70 × 70 mm Modric anodized aluminium electrically powered hold open wall magnet (excluding power supply and connection)	128.47	0.40	9.20	–	132.32	nr	**141.52**
Modric anodized aluminium bathroom configuration with quadaxial assembly, turn, release and optional indicator	50.06	0.80	18.39	–	51.56	nr	**69.95**
Concealed jamb door closer check action	149.85	1.00	22.98	–	154.35	nr	**177.33**
75 × 57 × 170 mm Modric anodized aluminium door coordinator for pairs of rebated leaves, CE Marked to BS EN1158 3–5-3/5–1-1-0	33.16	0.80	18.39	–	34.15	nr	**52.54**
290 × 48 × 50 mm Modric anodized aluminium rectangular overhead door closer with adjustable power and adjustable backcheck intumescent protected arm heavy duty UL & certifire listed & CE Marked to BS EN1154 4–8-2/4–1-1-3 and kite-marked	96.27	1.00	22.98	–	99.16	nr	**122.14**
Stainless steel overhead door closer. Projecting armset, Power EN 2–5, CE marked, c/w Backcheck, Latch action and Speed control. Max door width 1100 mm, Max door weight 100 kg	96.00	1.00	22.98	–	98.88	nr	**121.86**

Prices for Measured Works

24 DOORS, SHUTTERS AND HATCHES

Item	PC £	Labour hours	Labour £	Plant £	Material £	Unit	Total rate £
24.07 ASSOCIATED IRONMONGERY – cont							
Door closers and furniture – cont							
288 × 45 × 32 mm fully concealed overhead door closer complete with track and arm for single action doors, adjustable power, latch action and backcheck. Certifire approved	160.84	0.80	18.39	–	165.67	nr	**184.06**
75 × 45 mm heavy duty floor pivot set with thrust roller bearing 200 kg load capacity. Complete with forged steel intumescent protected double action strap with 10 mm height adjustment and matching cover plate	245.11	2.30	52.86	–	252.46	nr	**305.32**
Cavalier floor spring unit with cover plate and loose box for concrete, adjustable power 25 mm offset strap & top centre, intumescent pack. Certfire listed	259.94	2.30	52.86	–	267.74	nr	**320.60**
Double action pivot set for door maximum width 1100 mm and maximum weight 80 kg	82.06	2.30	52.86	–	84.52	nr	**137.38**
Surface vertical rod push bar panic bolt, reversible, to suit doors 2500 × 1100 mm maximum, silver finish, CE marked to EN1125 class 3–7-5–1–1-3-2–2-A	137.13	1.50	34.47	–	141.24	nr	**175.71**
Rim push bar panic latch, reversible, to suit doors 1100 mm wide maximum, silver finish, CE marked to EN1125 class 3–7-5-1–1-3-2–2-A	97.77	1.30	29.88	–	100.70	nr	**130.58**
76 × 51 × 13 mm adjustable heavy roller catch, stainless steel forend and strike complete with satin nickel plate roller bolt	7.73	0.60	13.79	–	7.96	nr	**21.75**
External access device for use with XX10280/2 panic hardware to suit door thickness 45–55 mm, complete with SS3006N lever, SS755 rose, SS796 profile escutcheon and spindle. For use with MA7420 A51 or MA7420 A55 profile cylinders	32.48	1.30	29.88	–	33.45	nr	**63.33**
142 × 22 mm Ø Concealed jamb door closer light duty	17.39	0.80	18.39	–	17.91	nr	**36.30**
80 × 40 × 45 mm Emergency release door stop with holdback facility	82.06	1.00	22.98	–	84.52	nr	**107.50**
Modric anodized aluminium quadaxial lever MA3503 assembly tested to BS EN1906 4/7/-/1/1/4/0/U	30.42	0.80	18.39	–	31.33	pair	**49.72**
Modric anodized aluminium quadaxial lever MA3502 assembly Tested to BS EN1906 4/7/-/1/1/4/0/U	32.43	0.80	18.39	–	33.40	pair	**51.79**
Modric anodized aluminium quadaxial lever assembly Tested to BS EN1906 4/7/-/1/1/4/0/U with Biocote® antibacterial protection	58.48	0.80	18.39	–	60.23	pair	**78.62**
Modric stainless steel quadaxial lever assembly Tested to BS EN1906 4/7/-/1/1/4/0/U	55.68	0.80	18.39	–	57.35	pair	**75.74**
152 × 38 × 13 mm Modric anodized aluminium security door chain leather covered	43.38	0.40	9.20	–	44.68	nr	**53.88**

24 DOORS, SHUTTERS AND HATCHES

Item	PC £	Labour hours	Labour £	Plant £	Material £	Unit	Total rate £
50 Ø × 3 mm Modric anodized aluminium circular covered rose for profile cylinder	4.26	0.10	2.30	–	4.39	nr	**6.69**
50 Ø × 3 mm Modric anodized aluminium circular covered rose with indicator and emergency release	8.79	0.15	3.45	–	9.05	nr	**12.50**
50 Ø × 3 mm Modric anodized aluminium circular covered rose with heavy turn, 5–8 mm spindle	15.43	0.15	3.45	–	15.89	nr	**19.34**
Budget lock escutcheon – satin stainless steel 316	8.40	0.10	2.30	–	8.65	nr	**10.95**
50 Ø × 3 mm Stainless steel circular covered rose for profile cylinder	6.83	0.10	2.30	–	7.03	nr	**9.33**
50 Ø × 3 mm Stainless steel circular covered rose with indicator and emergency release	9.21	0.15	3.45	–	9.49	nr	**12.94**
50 Ø × 3 mm Stainless steel circular covered rose with heavy turn, 5–8 mm spindle	18.98	0.15	3.45	–	19.55	nr	**23.00**
330 × 76 × 1.6 mm Modric anodized aluminium push plate	3.36	0.15	3.45	–	3.46	nr	**6.91**
330 × 76 × 1.6 mm Stainless steel push plate	7.78	0.15	3.45	–	8.01	nr	**11.46**
800 × 150 × 1.5 mm Modric anodized aluminium kicking plate, drilled and countersunk with screws	8.28	0.25	5.75	–	8.53	nr	**14.28**
900 × 150 × 1.5 mm Modric anodized aluminium kicking plate, drilled and countersunk with screws	9.31	0.25	5.75	–	9.59	nr	**15.34**
1000 × 150 × 1.5 mm Modric anodized aluminium kicking plate, drilled & countersunk with screws	10.33	0.25	5.75	–	10.64	nr	**16.39**
800 × 150 × 1.5 mm Stainless steel kicking plate, drilled and countersunk with screws	14.65	0.25	5.75	–	15.09	nr	**20.84**
900 × 150 × 1.5 mm Stainless steel kicking plate, drilled and countersunk with screws	16.48	0.25	5.75	–	16.97	nr	**22.72**
1000 × 150 × 1.5 mm Stainless steel kicking plate, drilled & countersunk with screws	18.31	0.25	5.75	–	18.86	nr	**24.61**
610 × 70 × 19 mm Ø Modric anodized aluminium grab handle bolt through fixing for doors 10 to 55 mm thick	29.93	0.40	9.20	–	30.83	nr	**40.03**
400 × 19 mm Ø Stainless steel D line straight pull handle with M8 threaded holes, fixing centres 300 mm	46.31	0.33	7.58	–	52.77	nr	**60.35**
Hinges							
100 × 75 × 3 mm Stainless steel triple knuckle concealed twin bearings, button tipped butt hinges, jig drilled for metal doors/frames, complete with M6x12MT 'undercut' machine screws, stainless steel 316 CE marked to EN1935	18.39	0.25	5.75	–	18.94	pair	**24.69**
100 × 100 × 3 mm Stainless steel triple knuckle concealed twin Newton bearings, button tipped hinges, jig drilled, stainless steel grade 316 CE marked to EN1935	32.64	0.25	5.75	–	33.62	pair	**39.37**

24 DOORS, SHUTTERS AND HATCHES

Item	PC £	Labour hours	Labour £	Plant £	Material £	Unit	Total rate £
24.07 ASSOCIATED IRONMONGERY – cont							
Latches							
Modric anodized aluminium round cylinder for rim night latch, 2 keyed satin nickel plated	25.84	0.40	9.20	–	26.62	nr	**35.82**
93 × 75 mm Cylinder rim non-deadlocking night latch case only 60 mm backset	22.14	0.40	9.20	–	22.80	nr	**32.00**
71 series mortice latch, case only, low friction latchbolt, griptight follower, heavy spring for levers. Radius forend and sq strike. CE marked to BS EN12209 3/X/8/1/0G/-/B/02/0	16.74	0.80	18.39	–	17.24	nr	**35.63**
Modric anodized aluminium latch configuration with quadaxial assembly	30.23	0.80	18.39	–	31.14	nr	**49.53**
Modric anodized aluminium Nightlatch configuration with quadaxial assembly and single cylinder	56.07	0.80	18.39	–	57.75	nr	**76.14**
Locks							
44 mm case Bright zinc plated steel mortice budget lock with slotted strike plate 33 mm backset	28.41	0.80	18.39	–	29.26	nr	**47.65**
76 × 58 mm b/s Stainless steel cubicle mortice deadlock with 8 mm follower	13.46	0.80	18.39	–	13.86	nr	**32.25**
'A' length European profile double cylinder lock, 2 keyed satin nickel plated	26.50	0.80	18.39	–	27.30	nr	**45.69**
'A' length European profile cylinder and large turn, 2 keyed satin nickel plated	30.11	0.80	18.39	–	31.01	nr	**49.40**
'A' length European profile cylinder and large turn, 2 keyed under master key, satin nickel plated	27.71	0.80	18.39	–	28.54	nr	**46.93**
'A' length European profile single cylinder, 2 keyed satin nickel plated	20.49	0.80	18.39	–	21.10	nr	**39.49**
'A' length European profile single cylinder, 2 keyed under master key, satin nickel plated	20.49	0.80	18.39	–	21.10	nr	**39.49**
93 × 60 mm b/s 71 series profile cylinder mortice deadlock, case only. Single throw 22 mm deadbolt. Radius forend and square strike. CE marked to BS EN12209 3/X/8/1/0/G/4/B/A/0/0	16.74	0.80	18.39	–	17.24	nr	**35.63**
92 × 60 mm b/s 71 series bathroom lock, case only, low friction latchbolt, griptight follower, heavy spring for levers, twin 8 mm followers at 78 mm centres. Radius forend and square strike. CE marked to BS EN12209 3/X/8/0/0/G-/B/0/2/0	19.83	0.80	18.39	–	20.42	nr	**38.81**
93 × 60 mm b/s 71 series profile cylinder mortice lock, case only, low friction latchbolt, griptight follower. Heavy spring for levers, 22 mm throw deadbolt, cylinder withdraws bolt bolts. Radius forend and square strike. CE marked to BS EN12209 3/X/8/1/0G/4/B/A2/0	19.83	0.80	18.39	–	20.42	nr	**38.81**
92 × 60 mm b/s71 series profile cylinder emergency lock, case only. Low friction latchbolt, griptight follower, heavy spring for lever, single throw 22 mm deadbolt, lever can withdraw both bolts. Radius forend and strike	65.25	0.80	18.39	–	67.21	nr	**85.60**

24 DOORS, SHUTTERS AND HATCHES

Item	PC £	Labour hours	Labour £	Plant £	Material £	Unit	Total rate £
Modric anodized aluminium lock configuration with quadaxial assembly and cylinder with turn	86.38	0.80	18.39	–	88.97	nr	**107.36**
Sundries							
76 mm Ø Modric anodized aluminium circular sex symbol male	4.69	0.08	1.84	–	4.83	nr	**6.67**
76 mm Ø Modric anodized aluminium circular symbol fire door keep locked	4.69	0.08	1.84	–	4.83	nr	**6.67**
76 mm Ø Modric anodized aluminium circular symbol fire door keep shut	4.69	0.10	2.30	–	4.83	nr	**7.13**
38 × 47 mm Ø Modric anodized aluminium heavy circular floor door stop with cover	9.48	0.10	2.30	–	9.76	nr	**12.06**
38 × 47 mm Ø Stainless steel heavy circular floor door stop with cover	16.93	0.10	2.30	–	17.44	nr	**19.74**
63 × 19 mm Ø Modric anodized aluminium Circular heavy duty skirting buffer with thief-resistant insert	6.20	0.10	2.30	–	6.39	nr	**8.69**
102 × 25 mm Ø Stainless steel circular heavy duty skirting buffer with thief-resistant insert	10.74	0.10	2.30	–	11.06	nr	**13.36**
152 mm Cabin hook satin chrome on brass	22.46	0.15	3.45	–	23.13	nr	**26.58**
14 mm Ø × 145 × 94 mm Toilet roll holder, length 145 mm, colour white, satin stainless steel 316	62.52	0.15	3.45	–	66.08	nr	**69.53**
Towel rail with bushes, fixing centres 450 mm, satin stainless steel 316	84.97	0.25	5.75	–	92.59	nr	**98.34**
Toilet brush holder with toilet brush, with bushes, satin stainless steel 316	131.94	0.20	4.59	–	139.28	nr	**143.87**
Bathline 850 mm lift up support rail	199.85	0.75	17.24	–	205.85	set	**223.09**
Bathline 600 × 95 × 35 mm support rail with concealed fixing roses	102.44	0.50	11.49	–	105.51	set	**117.00**
Bathline 400 × 250 × 35 mm backrest rail with concealed fixing roses	102.44	0.50	11.49	–	105.51	set	**117.00**
Allgood Ltd or other equal; to hardwood							
Bolts							
75 × 35 mm Modric anodized aluminium straight barrel bolt	9.17	0.40	9.20	–	9.45	nr	**18.65**
150 × 35 mm Modric anodized aluminium straight barrel bolt	10.47	0.40	9.20	–	10.78	nr	**19.98**
75 × 35 mm Modric anodized aluminium necked barrel bolt	10.27	0.40	9.20	–	10.58	nr	**19.78**
150 × 35 mm Modric anodized aluminium necked barrel bolt	13.13	0.40	9.20	–	13.52	nr	**22.72**
11 mm Easiclean socket for wood or stone	4.48	0.15	3.45	–	4.61	nr	**8.06**
Security hinge bolt chubb WS12	9.29	0.65	14.94	–	9.57	nr	**24.51**
203 × 19 × 11 mm lever action flush bolt set with coil spring and intumescent pack for FD30 and FD60 fire doors	31.51	0.80	18.39	–	32.46	nr	**50.85**
609 × 19 mm lever action flush bolt set with coil spring and intumescent pack for FD30 and FD60 fire doors	91.85	0.80	18.39	–	94.61	nr	**113.00**
Stainless steel indicating bolt complete with outside indicator and emergency release	65.83	0.80	18.39	–	67.80	nr	**86.19**

24 DOORS, SHUTTERS AND HATCHES

Item	PC £	Labour hours	Labour £	Plant £	Material £	Unit	Total rate £
24.07 ASSOCIATED IRONMONGERY – cont							
Allgood Ltd or other equal – cont							
Catches							
Magnetic catch	0.46	0.25	5.75	–	0.47	nr	**6.22**
Door closers and furniture							
13 mm stainless steel rebate component for 7104/ 08/78/79/86	27.31	0.80	18.39	–	28.13	nr	**46.52**
70 × 70 mm Modric anodized aluminium electrically powered hold open wall magnet. CE marked to BS EN1155:1997 & A1:2002 3–5-6/ 3–1-1–3	128.47	0.55	12.64	–	132.32	nr	**144.96**
Modric anodized aluminium bathroom configuration with quadaxial assembly, turn, release and optional indicator	50.06	1.05	24.13	–	51.56	nr	**75.69**
495 × 17.5 × 16 mm concealed overhead door restraining stay with aluminium channel. Stainless steel plated arm and bracket with adjustable friction slide. Block and spring buffer to cussion door opening. Not for use with door closing devices	25.76	1.35	31.02	–	26.53	nr	**57.55**
Concealed jamb door closer check action	149.85	1.35	31.02	–	154.35	nr	**185.37**
75 × 57 × 170 mm Modric anodized aluminium door coordinator for pairs of rebated leaves, CE Marked to BS EN1158 3–5-3/5–1-1–0	33.16	1.05	24.13	–	34.15	nr	**58.28**
290 × 48 × 50 mm Modric anodized aluminium rectangular overhead door closer with adjustable power and adjustable backcheck intumescent protected arm heavy duty UL & certifire listed & CE Marked to BS EN1154 4–8-2/4–1-1–3 and kite-marked.	96.27	1.35	31.02	–	99.16	nr	**130.18**
Stainless steel overhead door closer. Projecting armset, Power EN 2–5, CE marked, c/w Backcheck, Latch action and Speed control. Max door width 1100 mm, Max door weight 100 kg	96.00	1.35	31.02	–	98.88	nr	**129.90**
288 × 45 × 32 mm fully concealed overhead door closer complete with track and arm for single action doors, adjustable power, latch action and backcheck. Certifire approved	160.84	1.05	24.13	–	165.67	nr	**189.80**
75 × 45 mm heavy duty floor pivot set with thrust roller bearing 200 kg load capacity. Complete with forged steel intumescent protected double action strap with 10 mm height adjustment and matching cover plate	245.11	3.05	70.10	–	252.46	nr	**322.56**
Cavalier floor spring unit with cover plate and loose box for concrete, adjustable power 25 mm offset strap & top centre with intumescent pack. Certfire listed	259.94	2.30	52.86	–	267.74	nr	**320.60**

24 DOORS, SHUTTERS AND HATCHES

Item	PC £	Labour hours	Labour £	Plant £	Material £	Unit	Total rate £
Double action pivot set for door maximum width 1100 mm and maximum weight 80 kg	82.06	3.05	70.10	–	84.52	nr	**154.62**
Surface vertical rod push bar panic bolt, reversible, to suit doors 2500x1100 mm maximum, silver finish, CE marked to EN1125 class 3–7-5-1–1-3-2–2-A	137.13	2.00	45.97	–	141.24	nr	**187.21**
Rim push bar panic latch, reversible, to suit doors 1100 mm wide maximum, silver finish, CE marked to EN1125 class 3–7-5-1–1-3-2–2-A	97.77	1.75	40.22	–	100.70	nr	**140.92**
76 × 51 × 13 mm adjustable heavy roller catch, stainless steel forend and strike complete with satin nickel roller bolt satin chrome	7.73	0.80	18.39	–	7.96	nr	**26.35**
External access device for use with XX10280/2 panic hardware to suit door thickness 45–55 mm, complete with SS3006N lever, SS755 rose, SS796 profile escutcheon and spindle. For use with MA7420 A51 or MA7420 A55 profile cylinders	32.48	1.75	40.22	–	33.45	nr	**73.67**
142 × 22 mm Ø Concealed jamb door closer light duty	17.39	1.05	24.13	–	17.91	nr	**42.04**
80 × 40 × 45 mm Emergency release door stop with holdback facility	82.06	1.35	31.02	–	84.52	nr	**115.54**
Modric anodized aluminium quadaxial lever MA3503 assembly tested to BS EN1906 4/7/-/1/1/4/0/U	30.42	1.05	24.13	–	31.33	pair	**55.46**
Modric anodized aluminium quadaxial lever MA3502 assembly Tested to BS EN1906 4/7/-/1/1/4/0/U	32.43	1.05	24.13	–	33.40	pair	**57.53**
Modric anodized aluminium quadaxial lever assembly Tested to BS EN1906 4/7/-/1/1/4/0/U with Biocote® antibacterial protection	58.48	1.05	24.13	–	60.23	pair	**84.36**
Modric stainless steel quadaxial lever assembly Tested to BS EN1906 4/7/-/1/1/4/0/U	55.68	1.05	24.13	–	57.35	pair	**81.48**
152 × 38 × 13 mm Modric anodized aluminium security door chain leather covered	43.38	0.55	12.64	–	44.68	nr	**57.32**
50 Ø × 3 mm Modric anodized aluminium circular covered rose for profile cylinder	4.26	0.15	3.45	–	4.39	nr	**7.84**
50 Ø × 3 mm Modric anodized aluminium circular covered rose with indicator and emergency release	8.79	0.20	4.59	–	9.05	nr	**13.64**
50 Ø × 3 mm Modric anodized aluminium circular covered rose with heavy turn, 5–8 mm spindle	15.43	0.20	4.59	–	15.89	nr	**20.48**
Budget lock escutcheon – satin stainless steel 316	8.40	0.15	3.45	–	8.65	nr	**12.10**
50 Ø × 3 mm Stainless steel circular covered rose for profile cylinder	6.83	0.15	3.45	–	7.03	nr	**10.48**
50 Ø × 3 mm Stainless steel circular covered rose with indicator and emergency release	9.21	0.20	4.59	–	9.49	nr	**14.08**
50 Ø × 3 mm Stainless steel circular covered rose with heavy turn, 5–8 mm spindle	18.98	0.20	4.59	–	19.55	nr	**24.14**

24 DOORS, SHUTTERS AND HATCHES

Item	PC £	Labour hours	Labour £	Plant £	Material £	Unit	Total rate £
24.07 ASSOCIATED IRONMONGERY – cont							
Allgood Ltd or other equal – cont							
Door closers and furniture – cont							
330 × 76 × 1.6 mm Modric anodized aluminium push plate	3.36	0.20	4.59	–	3.46	nr	**8.05**
330 × 76 × 1.6 mm Stainless steel push plate	7.78	0.20	4.59	–	8.01	nr	**12.60**
800 × 150 × 1.5 mm Modric anodized aluminium kicking plate, drilled and countersunk with screws	8.28	0.35	8.04	–	8.53	nr	**16.57**
900 × 150 × 1.5 mm Modric anodized aluminium kicking plate, drilled and countersunk with screws	9.31	0.35	8.04	–	9.59	nr	**17.63**
1000 × 150 × 1.5 mm Modric anodized aluminium kicking plate, drilled & countersunk with screws	10.33	0.35	8.04	–	10.64	nr	**18.68**
800 × 150 × 1.5 mm Stainless steel kicking plate, drilled and countersunk with screws	14.65	0.35	8.04	–	15.09	nr	**23.13**
900 × 150 × 1.5 mm Stainless steel kicking plate, drilled and countersunk with screws	16.48	0.35	8.04	–	16.97	nr	**25.01**
1000 × 150 × 1.5 mm Stainless steel kicking plate, drilled & countersunk with screws	18.31	0.35	8.04	–	18.86	nr	**26.90**
610 × 70 × 19 mm Ø Modric anodized aluminium grab handle bolt through fixing for doors 10 to 55 mm thick	29.93	0.55	12.64	–	30.83	nr	**43.47**
400 × 19 mm Ø Stainless steel D line straight pull handle with M8 threaded holes, fixing centres 300 mm	46.31	0.45	10.34	–	52.77	nr	**63.11**
Hinges							
100 × 75 × 3 mm Stainless steel triple knuckle concealed twin Newtonbearings, button tipped butt hinges, jig drilled for metal doors/frames, complete with M6x12MT 'undercut' machine screws, stainless steel 316 CE marked to EN1935	18.39	0.35	8.04	–	18.94	pair	**26.98**
100 × 100 × 3 mm Stainless steel triple knuckle concealed twin Newton bearings, button tipped hinges, jig drilled, stainless steel grade 316 CE marked to EN1935	32.64	0.35	8.04	–	33.62	pair	**41.66**
Latches							
Modric anodized aluminium round cylinder for rim night latch, 2 keyed satin nickel plated	25.84	0.55	12.64	–	26.62	nr	**39.26**
93 × 75 mm Cylinder rim non-deadlocking night latch case only 60 mm backset	22.14	0.55	12.64	–	22.80	nr	**35.44**
71 series mortice latch, case only, low friction latchbolt, griptight follower, heavy spring for levers. Radius forend and sq strike. CE marked to BS EN12209 3/X/8/1/0G/-/B/02/0	16.74	1.05	24.13	–	17.24	nr	**41.37**
Modric anodized aluminium latch configuration with quadaxial assembly	30.23	1.05	24.13	–	31.14	nr	**55.27**
Modric anodized aluminium Nightlatch configuration with quadaxial assembly and single cylinder	56.07	1.05	24.13	–	57.75	nr	**81.88**

24 DOORS, SHUTTERS AND HATCHES

Item	PC £	Labour hours	Labour £	Plant £	Material £	Unit	Total rate £
Locks							
44 mm case Bright zinc plated steel mortice budget lock with slotted strike plate 33 mm backset	28.41	1.05	24.13	–	29.26	nr	**53.39**
76 × 58 mm b/s Stainless steel cubicle mortice deadlock with 8 mm follower	13.46	1.05	24.13	–	13.86	nr	**37.99**
'A' length European profile double cylinder lock, 2 keyed satin nickel plated	26.50	1.05	24.13	–	27.30	nr	**51.43**
'A' length European profile cylinder and large turn, 2 keyed satin nickel plated	30.11	1.05	24.13	–	31.01	nr	**55.14**
'A' length European profile cylinder and large turn, 2 keyed under master key, satin nickel plated	27.71	1.05	24.13	–	28.54	nr	**52.67**
'A' length European profile single cylinder, 2 keyed satin nickel plated	20.49	1.05	24.13	–	21.10	nr	**45.23**
'A' length European profile single cylinder, 2 keyed under master key, satin nickel plated	20.49	1.05	24.13	–	21.10	nr	**45.23**
93 × 60 mm b/s 71 series profile cylinder mortice deadlock, case only. Single throw 22 mm deadbolt. Radius forend and square strike. CE marked to BS EN12209 3/X/8/1/0/G/4/B/A/0/0	16.74	1.05	24.13	–	17.24	nr	**41.37**
92 × 60 mm b/s 71 series bathroom lock, case only, low friction latchbolt, griptight follower, heavy spring for levers, twin 8 mm followers at 78 mm centres. Radius forend and square strike. CE marked to BS EN12209 3/X/8/0/0/G-/B/0/2/0	19.83	1.05	24.13	–	20.42	nr	**44.55**
93 × 60 mm b/s 71 series profile cylinder mortice lock, case only, low friction latchbolt, griptight follower. Heavy spring for levers, 22 mm throw deadbolt, cylinder withdraws bolt bolts. Radius forend and square strike. CE marked to BS EN12209 3/X/8/1/0G/4/B/A2/0	19.83	1.05	24.13	–	20.42	nr	**44.55**
92 × 60 mm b/s71 series profile cylinder emergency lock, case only. Low friction latchbolt, griptight follower, heavy spring for lever, single throw 22 mm deadbolt, lever can withdraw both bolts. Radius forend and strike	65.25	1.05	24.13	–	67.21	nr	**91.34**
Modric anodized aluminium lock configuration with quadaxial assembly and cylinder with turn	86.38	1.05	24.13	–	88.97	nr	**113.10**
Sundries							
76 mm Ø Modric anodized aluminium circular sex symbol male	4.69	0.10	2.30	–	4.83	nr	**7.13**
76 mm Ø Modric anodized aluminium circular symbol fire door keep locked	4.69	0.10	2.30	–	4.83	nr	**7.13**
76 mm Ø Modric anodized aluminium circular symbol fire door keep shut	4.69	0.15	3.45	–	4.83	nr	**8.28**

24 DOORS, SHUTTERS AND HATCHES

Item	PC £	Labour hours	Labour £	Plant £	Material £	Unit	Total rate £
24.07 ASSOCIATED IRONMONGERY – cont							
Allgood Ltd or other equal – cont							
Sundries – cont							
38 × 47 mm Ø Modric anodized aluminium heavy circular floor door stop with cover	9.48	0.15	3.45	–	9.76	nr	**13.21**
38 × 47 mm Ø Stainless steel heavy circular floor door stop with cover	16.93	0.15	3.45	–	17.44	nr	**20.89**
63 × 19 mm Ø Modric anodized aluminium Circular heavy duty skirting buffer with thief-resistant insert	6.20	0.15	3.45	–	6.39	nr	**9.84**
102 × 25 mm Ø Stainless steel circular heavy duty skirting buffer with thief-resistant insert	10.74	0.15	3.45	–	11.06	nr	**14.51**
152 mm Cabin hook satin chrome on brass	22.46	0.20	4.59	–	23.13	nr	**27.72**
14 mm Ø × 145 × 94 mm Toilet roll holder, length 145 mm, colour white, satin stainless steel 316	62.52	0.20	4.59	–	66.08	nr	**70.67**
Towel rail with bushes, fixing centres 450 mm, satin stainless steel 316	84.97	0.35	8.04	–	92.59	nr	**100.63**
Toilet brush holder with toilet brush, with bushes, satin stainless steel 316	131.94	0.25	5.75	–	139.28	nr	**145.03**
Bathline 850 mm lift up support rail	199.85	0.75	17.24	–	205.85	set	**223.09**
Bathline 600 × 95 × 35 mm support rail with concealed fixing roses	102.44	0.50	11.49	–	105.51	set	**117.00**
Bathline 400 × 250 × 35 mm backrest rail with concealed fixing roses	102.44	0.50	11.49	–	105.51	set	**117.00**
ASSA ABLOY Ltd typical door ironmongery sets							
These are standard doorsets compilations which ASSA ABLOY most commonly supply							
Classroom doorset (fire-rated)	–	2.00	45.97	–	420.38	nr	**466.35**
Ironmongery schedule and cost							
UNION PowerLOAD 603 bushed bearing butt hinges (1.5 pair) with a fixed in and radius corners, 100 mm × 88 mm, satin stainless steel, grade 13 BS EN 1935	9.51	–	–	–	–	nr	–
ASSA ABLOY DC700 A overhead cam motion backcheck closure with slide arm size 2–6. Full cover standard silver	162.35	–	–	–	–	nr	–
UNION keyULTRA euro profile key and turn cylinder, turn/key, 35 mm/35 mm, satin chrome, BS EN 1303 Security Grade 2 when fitted with security escutcheon (sold seperately)	38.74	–	–	–	–	nr	–
Union 2S21 euro-profile sashlock, 72 mm centresm, 55 mm backset, satin stainless steel, Grade 3 category of use	11.78	–	–	–	–	nr	–
Union 1000 01 style, 19 mm dia. lever on 8 mm round rose, sprung, satin stainless steel grade 304, bolt through connections	8.13	–	–	–	–	pair	–
7012F ASSA finger guard 2.05 m push side, silver	96.33	–	–	–	–	nr	–
UNION 1000EE Euro Escutcheon, 54 mm dia. × 8 mm projection, satin stainless steel	0.65	–	–	–	–	nr	–

24 DOORS, SHUTTERS AND HATCHES

Item	PC £	Labour hours	Labour £	Plant £	Material £	Unit	Total rate £
Kick Plate 900 mm × 150 mm, satin stainless steel softened & finished edges with radius corners	17.97	–	–	–	–	nr	–
'Fire Door Keep Shut' 75 mm dia. sign, satin stainless steel	5.00	–	–	–	–	nr	–
UNION DS100 Floor Mounted Door Stop, half moon design, 45 mm dia. × 24 mm high, satin stainless steel	2.44	–	–	–	–	nr	–
Non-locking office doorset (fire-rated)	–	2.00	45.97	–	267.00	nr	**312.97**
Ironmongery schedule and cost							
UNION PowerLOAD 603 bushed bearing butt hinges (1.5 pair) with a fixed pin and radius corners, 100 mm × 88 mm, satin stainless steel, grade 13 BS EN 1935	28.54	–	–	–	–	nr	–
ASSA ABLOY DC700 A overhead cam motion backcheck closure with slide arm size 2–6. Full cover standard silver	162.35	–	–	–	–	nr	–
UNION keyULTRA euro profile key and turn cylinder, turn/key, 35 mm/35 mm, satin chrome, BS EN 1303 Security Grade 2 when fitted with security escutcheon (sold seperately)	38.74	–	–	–	–	nr	–
UNION 2C23 DIN mortice latch, 55 mm backset, satin stainless steel, radius forend, Grade 3 Category of Use, BS EN 12209, CE marked	11.83	–	–	–	–	nr	–
Union 1000 01 style, 19 mm dia. lever on 8 mm round rose, sprung, satin stainless steel grade 304, bolt through connections	8.13	–	–	–	–	pair	–
Kick Plate 900 mm × 150 mm, satin stainless steel softened & finished edges with radius corners	17.97	–	–	–	–	nr	–
'Fire Door Keep Shut' 75 mm dia. sign, satin stainless steel	5.00	–	–	–	–	nr	–
UNION DS100 Floor Mounted Door Stop, half moon design, 45 mm dia. × 24 mm high, satin stainless steel	2.44	–	–	–	–	nr	–
Office/Store locking doorset (fire-rated)	–	2.00	45.97	–	308.19	nr	**354.16**
Ironmongery schedule and cost							
UNION PowerLOAD 603 bushed bearing butt hinges (1.5 pair) with a fixed pin and radius corners, 100 mm × 88 mm, satin stainless steel, grade 13 BS EN 1935	28.54	–	–	–	–	nr	–
ASSA ABLOY DC700 A overhead cam motion backcheck closure with slide arm size 2–6. Full cover standard silver	162.35	–	–	–	–	nr	–
UNION keyULTRA euro profile key and turn cylinder, turn/key, 35 mm/35 mm, satin chrome, BS EN 1303 Security Grade 2 when fitted with security escutcheon (sold seperately)	38.74	–	–	–	–	nr	–

24 DOORS, SHUTTERS AND HATCHES

Item	PC £	Labour hours	Labour £	Plant £	Material £	Unit	Total rate £
24.07 ASSOCIATED IRONMONGERY – cont							
ASSA ABLOY Ltd typical door ironmongery sets – cont							
Ironmongery schedule and cost – cont							
Union 2S21 euro-profile sashlock, 72 mm centresm, 55 mm backset, satin stainless steel, Grade 3 category of use	11.78	–	–	–	–	nr	–
Union 1000 01 style, 19 mm dia. lever on 8 mm round rose, sprung, satin stainless steel grade 304, bolt through connections	8.13	–	–	–	–	pair	–
Kick Plate 200 mm × 1.5 mm, satin stainless steel softened & finished edges with radius corners	17.97	–	–	–	–	nr	–
'Fire Door Keep Shut' 75 mm dia. sign, satin stainless steel	5.00	–	–	–	–	nr	–
UNION DS100 Floor Mounted Door Stop, half moon design, 45 mm dia. × 24 mm high, satin stainless steel	2.44	–	–	–	–	nr	–
Common room locking doorset (fire-rated)	–	2.50	57.46	–	429.62	nr	**487.08**
Ironmongery schedule and cost							
UNION PowerLOAD 603 bushed bearing butt hinges (1.5 pair) with a fixed pin and radius corners, 100 mm × 88 mm, satin stainless steel, Grade 13 BS EN 1935	28.54	–	–	–	–	nr	–
ASSA ABLOY DC700 A overhead cam motion backcheck closure with slide arm size 2–6. Full cover standard silver	162.35	–	–	–	–	nr	–
UNION keyULTRA euro profile key and turn cylinder, turn/key (Large), 35 mm/35 mm, 6pin, satin chrome, BS EN 1303 Security Grade 2 when fitted with security escutcheon (sold seperately)	51.35	–	–	–	–	nr	–
UNION 2C22 DIN euro-profile deadlock, 55 mm backset, satin stainless steel, radius forend, Grade 3 Category of Use, BS EN 12209, CE marked	17.22	–	–	–	–	nr	–
UNION 01 style pull handle 19 mm dia. × 425 mm centres, satin stainless steel, Grade 304, bolt through fixing	5.18	–	–	–	–	nr	–
Kick Plate 200 mm × 1.5 mm, satin stainless steel softened & finished edges with radius corners	17.97	–	–	–	–	nr	–
Push Plate 450 mm × 100 mm × 1.5 mm satin stainless steel, softened & finished edges with radius corners	2.03	–	–	–	–	nr	–

24 DOORS, SHUTTERS AND HATCHES

Item	PC £	Labour hours	Labour £	Plant £	Material £	Unit	Total rate £
7012F ASSA finger guard 2.05 m push side, silver	96.33	–	–	–	–	nr	–
UNION 1000EE Euro Escutcheon, 54 mm dia. × 8 mm projection, satin stainless steel	0.65	–	–	–	–	nr	–
UNION DS100 Floor Mounted Door Stop, Half Moon Design, 45 mm Dia. × 24 mm high, Satin Stainless Steel	2.44	–	–	–	–	nr	–
'Fire Door Keep Shut' 75 mm dia. sign, satin stainless steel	5.00	–	–	–	–	nr	–
PULL or PUSH signage, 75 mm dia., satin stainless steel	2.22	–	–	–	–	nr	–
Maintenance/Plant room doorset (fire-rated)	–	1.75	40.22	–	294.71	nr	**334.93**
Ironmongery schedule and cost							
1½ pairs Union Powerload 603 brushed bearing butt hinge with a fixed pin; 100 mm × 88 mm; satin stainless steel grade 13 BS EN 1935	28.54	–	–	–	–	nr	–
ASSA ABLOY DC700 A overhead cam motion backcheck closure with slide arm size 2–6. Full cover standard silver	162.35	–	–	–	–	nr	–
UNION 2C22 DIN euro-profile deadlock, 55 mm backset, satin stainless steel, radius forend, Grade 3 Category of Use, BS EN 12209, CE marked	17.22	–	–	–	–	nr	–
UNION keyULTRA euro profile single cylinder, 35 mm/35 mm, satin chrome, BS EN 1303 Security Grade 2 when fitted with security escutcheon	51.35	–	–	–	–	nr	–
Kick Plate 900 mm × 150 mm, satin stainless steel softened & finished edges With Radius Corner	17.97	–	–	–	–	nr	–
'Fire Door Keep Locked' 75 mm Dia. Sign, Satin Stainless Steel	5.00	–	–	–	–	nr	–
Euro Rim Cylinder Pull, stainless steel	0.61	–	–	–	–	nr	–
UNION 1000EE Euro Escutcheon, 54 mm dia. × 8 mm projection, satin stainless steel	0.65	–	–	–	–	nr	–
UNION DS100 Floor Mounted Door Stop, half moon design, 45 mm Dia. × 24 mm high, satin stainless steel	2.44	–	–	–	–	nr	–
Standard bathroom doorset (unisex)	–	1.78	41.02	–	234.64	nr	**275.66**
Ironmongery schedule and cost							
1½ pairs Union Powerload 603 brushed bearing butt hinge with a fixed pin; 100 mm × 88 mm; satin stainless steel Grade 13 BS EN 1935	28.54	–	–	–	–	nr	–
ASSA ABLOY DC700 A overhead cam motion backcheck closure with slide arm size 2–6. Full cover standard silver	162.35	–	–	–	–	nr	–
UNION 2C27 DIN bathroom lock, 78 mm centres, 55 mm backset, satin stainless steel, radius forend, Grade 3 Category of Use, BS EN 12209, CE marked	12.24	–	–	–	–	nr	–

24 DOORS, SHUTTERS AND HATCHES

Item	PC £	Labour hours	Labour £	Plant £	Material £	Unit	Total rate £
24.07 ASSOCIATED IRONMONGERY – cont							
ASSA ABLOY Ltd typical door ironmongery sets – cont							
Ironmongery schedule and cost – cont							
Union 1000 01 style, 19 mm dia. lever on 8 mm round rose, sprung, satin stainless steel grade 304, bolt through connections	8.13	–	–	–	–	pair	–
UNION 1000ER emergency release, 54 mm dia. × 8 mm projection, satin stainless steel	3.90	–	–	–	–	nr	–
UNION 1000T standard turn, 54 mm dia. × 8 mm projection, satin stainless steel	5.80	–	–	–	–	nr	–
UNION DS100 Floor Mounted Door Stop, half moon design, 45 mm Dia. × 24 mm high, satin stainless steel	2.44	–	–	–	–	nr	–
UNION US200 Unisex Symbol, 75 mm dia., satin stainless steel, Grade 304	2.22	–	–	–	–	nr	–
UNION HC100 hat & coat hook with a buffer, satin stainless steel	2.19	–	–	–	–	nr	–
Accessible toilet doorset	–	2.55	58.61	–	101.66	nr	**160.27**
Ironmongery schedule and cost							
1½ pairs Union Powerload 603 brushed bearing butt hinge with a fixed pin; 100 mm × 88 mm; satin stainless steel Grade 13 BS EN 1935	28.54	–	–	–	–	nr	–
UNION 2C27 DIN bathroom lock, 78 mm centres, 55 mm backset, satin stainless steel, radius forend, Grade 3 Category of Use, BS EN 12209, CE marked	12.24	–	–	–	–	nr	–
Union 1000 01 style, 19 mm dia. lever on 8 mm round rose, sprung, satin stainless steel grade 304, bolt through connections	8.13	–	–	–	–	pair	–
Kick plates 900 mm × 150 mm, stainless steel	17.97	–	–	–	–	nr	–
UNION WS200 Disabled Symbol, 75 mm dia., satin stainless steel, Grade 304	2.22	–	–	–	–	nr	–
UNION 1000ER emergency release, 54 mm dia. × 8 mm projection, satin stainless steel	3.90	–	–	–	–	nr	–
UNION 1000LT Large Turn, 54 mm dia. × 8 mm projection, satin stainless steel	3.11	–	–	–	–	nr	–
UNION DS100 Floor Mounted Door Stop, half moon design, 45 mm dia. × 24 mm high, satin stainless steel	2.44	–	–	–	–	nr	–
UNION HC100 hat & coat hook with a buffer, satin stainless steel	2.19	–	–	–	–	nr	–

24 DOORS, SHUTTERS AND HATCHES

Item	PC £	Labour hours	Labour £	Plant £	Material £	Unit	Total rate £
Single lobby/corridor doorset (fire-rated)	–	2.25	51.72	–	357.67	nr	**409.39**
Ironmongery schedule and cost							
UNION PowerLOAD 603 bushed bearing butt hinges (1.5 pair) with a fixed pin and radius corners, 100 mm × 88 mm, satin stainless steel, Grade 13 BS EN 1935	28.54	–	–	–	–	nr	–
UNION 01 style pull handle 19 mm dia. × 425 mm centres, satin stainless steel, Grade 304, bolt through fixing	5.18	–	–	–	–	nr	–
ASSA ABLOY DC700 A overhead cam motion backcheck closure with slide arm size 2–6. Full cover standard silver	162.35	–	–	–	–	nr	–
7012F ASSA finger guard 2.05 m push side, silver	96.33	–	–	–	–	nr	–
Kick Plate 900 mm × 150 mm, satin stainless steel softened & finished edges with radius corners	17.97	–	–	–	–	nr	–
'Fire Door Keep Shut' 75 mm dia. sign, satin stainless steel	5.00	–	–	–	–	nr	–
PULL or PUSH signage, 75 mm dia., satin stainless steel	2.22	–	–	–	–	nr	–
UNION DS100 Floor Mounted Door Stop, half moon design, 45 mm dia. × 24 mm high, satin stainless steel	2.44	–	–	–	–	nr	–
Double lobby/corridor doorset (fire-rated)	–	4.00	91.94	–	738.75	nr	**830.69**
Ironmongery schedule and cost							
UNION PowerLOAD 603 bushed bearing butt hinges (3 pair) with a fixed pin and radius corners, 100 mm × 88 mm, satin stainless steel, Grade 13 BS EN 1935	57.07	–	–	–	–	nr	–
UNION 8899 Electro Mag closer hold open/swing free, fixed strength size 4, painted silver, app 1, 61 and 66, BS EN 1155, CE Marked, Certifire approved.Supplied with a curve cover c/w matching arms	173.72	–	–	–	–	nr	–
UNION 01 style pull handle 19 mm dia. × 425 mm centres, satin stainless steel, Grade 304, bolt through fixing	5.18	–	–	–	–	nr	–
Push Plate 450 mm × 100 mm × 1.5 mm satin stainless steel, softened & finished edges with radius corners	2.03	–	–	–	–	nr	–
7012F ASSA finger guard 2.05 m push side, silver	96.33	–	–	–	–	nr	–
Kick Plate 900 mm × 150 mm, satin stainless steel softened & finished edges with radius corners	17.97	–	–	–	–	nr	–
'Fire Door Keep Shut' 75 mm dia. sign, satin stainless steel	5.00	–	–	–	–	nr	–
PULL or PUSH signage, 75 mm dia., satin stainless steel	2.22	–	–	–	–	nr	–
UNION DS100 Floor Mounted Door Stop, half moon design, 45 mm dia. × 24 mm high, satin stainless steel	2.44	–	–	–	–	nr	–

24 DOORS, SHUTTERS AND HATCHES

Item	PC £	Labour hours	Labour £	Plant £	Material £	Unit	Total rate £
24.07 ASSOCIATED IRONMONGERY – cont							
External rebated double fire escape doorset	–	4.00	91.94	–	1618.98	nr	**1710.92**
Ironmongery schedule and cost							
UNION PowerLOAD 603 bushed bearing butt hinges (1.5 pair) with a fixed pin and radius corners, 100 mm × 88 mm, satin stainless steel, Grade 13 BS EN 1935	28.54	–	–	–	–	nr	–
UNION satin stainless steel, Ball Bearing Dog Bolt Hinge	144.44	–	–	–	–	nr	–
ASSA ABLOY DC700 A overhead cam motion backcheck closure with slide arm size 2–6. Full cover standard silver	162.35	–	–	–	–	nr	–
UNION 883T panic exit double rebated doorset, 3 point locking with pullman latches, touch bar design, 900 mm maximum door width, silver, EN1125, CE marked	204.01	–	–	–	–	nr	–
UNION 885 outside access device (OAD), lever handle to suit 880 series only, silver, uses a UNION 2X50 5 pin euro profile cylinder to be ordered separately	49.74	–	–	–	–	nr	–
UNION keyULTRA euro profile single cylinder, 35 mm/35 mm, satin chrome, BS EN 1303 Security Grade 2 when fitted with security escutcheon	51.35	–	–	–	–	nr	–
Generic door door limiting stays	43.90	–	–	–	–	nr	–
Kick Plate 900 mm × 150 mm, satin stainless steel softened & finished edges with radius corners	17.97	–	–	–	–	nr	–
'Fire Door Keep Shut' 75 mm dia. sign, satin stainless steel	5.00	–	–	–	–	nr	–
SP(102)FA1201 – Aluminium	29.50	–	–	–	–	nr	–
Double aluminium entrance doorset	–	5.00	114.92	–	743.68	nr	**858.60**
Ironmongery schedule and cost							
Adams Rite SENTINAL Radius Weatherstrip Faceplate, satin anodized aluminium	5.91	–	–	–	–	nr	–
Adams Rite 4700 Series Faceplate Radius w/s, satin anodized aluminium	4.94	–	–	–	–	nr	–
Adams Rite SENTINEL 6 Deadlock, 30 mm Backset BS EN 12209	16.09	–	–	–	–	nr	–
ADAMS RITE SENTINEL Armoured Strike Plate	5.64	–	–	–	–	nr	–
ADAMS RITE SENTINEL Radius Armoured Trim Strike, satin anodized aluminium	10.88	–	–	–	–	nr	–
ADAMS RITE 4596 Euro Paddle Handle, Pull to Left 45 mm Door Satin Anodized Aluminium	82.77	–	–	–	–	nr	–
ADAMS RITE 4750 Deadlatch EuroProfile 28 mm backset body only – left hand	43.33	–	–	–	–	nr	–

24 DOORS, SHUTTERS AND HATCHES

Item	PC £	Labour hours	Labour £	Plant £	Material £	Unit	Total rate £
ADAMS RITE 2 Point Latch Push to Left, satin anodized aluminium	160.80	–	–	–	–	nr	–
ADAMS RITE Transom ARC–51N EN3 Spring Closer HO side load 70 mm pivot	79.70	–	–	–	–	nr	–
ADAMS RITE 7108 12vDC Fail Safe Escape Electric Release, Satin Anodized Aluminium. Specifer to confirm depth of section as could clash with mechanism of 4781.	188.67	–	–	–	–	nr	–
ADAMS RITE Sentinal Euro Profile Security Escutcheon	43.58	–	–	–	–	nr	–
External toilet lobby doorset (fire-rated)	–	2.50	57.46	–	431.25	nr	**488.71**
Ironmongery schedule and cost							
1½ pairs Union Powerload 603 brushed bearing butt hinge with a fixed pin; 100 mm × 88 mm; satin stainless steel Grade 13 BS EN 1935	28.54	–	–	–	–	nr	–
ASSA ABLOY DC700 A overhead cam motion backcheck closure with slide arm size 2–6. Full cover standard silver	162.35	–	–	–	–	nr	–
UNION 2C22 DIN euro-profile deadlock, 55 mm backset, satin stainless steel, radius forend, Grade 3 Category of Use, BS EN 12209, CE marked	17.22	–	–	–	–	nr	–
UNION 01 style pull handle 19 mm dia. × 425 mm centres, satin stainless steel, Grade 304, bolt through fixing	5.18	–	–	–	–	nr	–
UNION keyULTRA euro profile single cylinder, 35 mm/35 mm, satin chrome, BS EN 1303 Security Grade 2 when fitted with security escutcheon	51.35	–	–	–	–	nr	–
Push Plate 450 mm × 100 mm × 1.5 mm satin stainless steel, softened & finished edges with radius corners	2.03	–	–	–	–	nr	–
Kick Plate 900 mm × 150 mm, satin stainless steel softened & finished edges with radius corner	17.97	–	–	–	–	nr	–
7012F ASSA finger guard 2.05 m push side, silver	96.33	–	–	–	–	nr	–
'Fire Door Keep Locked' 75 mm dia. sign, satin stainless steel	5.00	–	–	–	–	nr	–
UNION 1000EE Euro Escutcheon, 54 mm dia. × 8 mm projection, satin stainless steel	0.65	–	–	–	–	nr	–
UNION DS100 Floor Mounted Door Stop, half moon design, 45 mm dia. × 24 mm high, satin stainless steel	2.44	–	–	–	–	nr	–
PULL or PUSH signage, 75 mm dia., satin stainless steel	2.22	–	–	–	–	nr	–
UNION US200 Unisex Symbol, 75 mm dia., satin stainless steel, Grade 304	2.22	–	–	–	–	nr	–

Prices for Measured Works

24 DOORS, SHUTTERS AND HATCHES

Item	PC £	Labour hours	Labour £	Plant £	Material £	Unit	Total rate £
24.07 ASSOCIATED IRONMONGERY – cont							
Internal toilet lobby doorset (non fire-rated)	–	2.00	45.97	–	248.15	nr	**294.12**
Ironmongery schedule and cost							
1½ pairs Union Powerload 603 brushed bearing butt hinge with a fixed pin; 100 mm × 88 mm; satin stainless steel Grade 13 BS EN 1935	28.54	–	–	–	–	nr	–
ASSA ABLOY DC700 A overhead cam motion backcheck closure with slide arm size 2–6. Full cover standard silver	162.35	–	–	–	–	nr	–
UNION 01 style pull handle 19 mm dia. × 425 mm centres, satin stainless steel, Grade 304, bolt through fixing	5.18	–	–	–	–	nr	–
Push Plate 450 mm × 100 mm × 1.5 mm satin stainless steel, softened & finished edges with radius corners	2.03	–	–	–	–	nr	–
Kick Plate 900 mm × 150 mm, satin stainless steel softened & finished edges with radius corners	17.97	–	–	–	–	nr	–
UNION DS100 Floor Mounted Door Stop, half moon design, 45 mm Dia. × 24 mm high, satin stainless steel	2.44	–	–	–	–	nr	–
PULL or PUSH signage, 75 mm dia., satin stainless steel	2.22	–	–	–	–	nr	–
Double classroom store doorset (non fire-rated)	–	1.50	34.47	–	361.43	nr	**395.90**
Ironmongery schedule and cost							
UNION PowerLOAD 603 bushed bearing butt hinges (3 pair) with a fixed pin and radius corners, 100 mm × 88 mm, satin stainless steel, Grade 13 BS EN 1935	28.54	–	–	–	–	nr	–
Union 2S21 euro-profile sashlock, 72 mm centresm, 55 mm backset, satin stainless steel, Grade 3 category of use	11.78	–	–	–	–	nr	–
UNION keyULTRA euro profile key and turn cylinder, turn/key (Large), 35 mm/35 mm, 6pin, satin chrome, BS EN 1303 Security Grade 2 when fitted with security escutcheon (sold seperately)	51.35	–	–	–	–	nr	–
Union 1000 01 style, 19 mm dia. lever on 8 mm round rose, sprung, satin stainless steel grade 304, bolt through connections	8.13	–	–	–	–	pair	–
UNION 1000EE Euro Escutcheon, 54 mm dia. × 8 mm projection, Satin Stainless Steel	0.65	–	–	–	–	nr	–
Kick Plate 900 mm × 150 mm, satin stainless steel softened & finished edges with radius corners	17.97	–	–	–	–	nr	–
UNION 8060 aluminium lever action flush bolt 200 mm satin stainless steel satin chrome	7.72	–	–	–	–	pair	–

24 DOORS, SHUTTERS AND HATCHES

Item	PC £	Labour hours	Labour £	Plant £	Material £	Unit	Total rate £
Sliding door gear; Hillaldam Coburn Ltd or other equal and approved; Commercial/Light industrial; for top hung timber/metal doors, weight not exceeding 365 kg							
Sliding door gear							
bottom guide; fixed to concrete in groove	28.40	0.46	10.57	–	29.25	m	**39.82**
top track	38.60	0.23	5.28	–	39.76	m	**45.04**
detachable locking bar	48.30	0.31	7.13	–	49.75	nr	**56.88**
hangers; timber doors	77.92	0.46	10.57	–	80.26	nr	**90.83**
hangers; metal doors	49.79	0.46	10.57	–	51.28	nr	**61.85**
head brackets; open, soffit fixing; screwing to timber	9.93	0.32	7.35	–	10.26	nr	**17.61**
head brackets; open, side fixing; bolting to masonry	10.42	0.46	10.57	–	11.73	nr	**22.30**
door guide to timber door	8.73	0.23	5.28	–	8.99	nr	**14.27**
door stop; rubber buffers; to masonry	38.50	0.69	15.86	–	39.66	nr	**55.52**
drop bolt; screwing to timber	33.67	0.46	10.57	–	34.68	nr	**45.25**
bow handle; to timber	13.35	0.23	5.28	–	13.75	nr	**19.03**
Sundries							
20 mm rubber door stop; plugged and screwed to concrete	2.10	0.09	2.07	–	2.16	nr	**4.23**

25 STAIRS, WALKWAYS AND BALUSTRADES

Item	PC £	Labour hours	Labour £	Plant £	Material £	Unit	Total rate £
25.01 STAIRS							
Standard staircases; wrought softwood (Parana pine)							
Stairs; 25 mm thick treads with rounded nosings; 9 mm thick plywood risers; 32 mm thick strings; bullnose bottom tread; 50 mm × 75 mm hardwood handrail; 32 mm square plain balusters; 100 mm square plain newel posts							
straight flight; 838 mm wide; 2676 mm going; 2600 mm rise; with two newel posts	–	6.48	148.94	–	426.89	nr	**575.83**
straight flight with turn; 838 mm wide; 2676 mm going; 2600 mm rise; with two newel posts; three top treads winding	–	6.48	148.94	–	556.26	nr	**705.20**
dogleg staircase; 838 mm wide; 2676 mm going; 2600 mm rise; with two newel posts; quarter space landing third riser from top	–	6.48	148.94	–	521.65	nr	**670.59**
dogleg staircase; 838 mm wide; 2676 mm going; 2600 mm rise; with two newel posts; half space landing third riser from top	–	7.40	170.08	–	641.13	nr	**811.21**
Hardwood staircases; purpose-made; assembled at works							
Fixing only complete staircase including landings, balustrades, etc.							
plugging and screwing to brickwork or blockwork	–	13.88	319.02	–	3.02	nr	**322.04**
Spiral staircases, balustrades and handrails; mild steel; galvanized and polyester powder coated							
Staircase							
2080 mm dia. × 3695 mm high; 18 nr treads; 16 mm dia. intermediate balusters; 1040 mm × 1350 mm landing unit with matching balustrade both sides; fixing with 16 mm dia. resin anchors to masonry at landing and with 12 mm dia. expanding bolts to concrete at base	–	–	–	–	–	nr	**15000.00**
The following are supply only prices for purpose-made staircase components in selected Sapele supplied as part of an assembled staircase and may be used to arrive at a guide price for a complete hardwood staircase							
Board landings; cross-tongued joints; 100 mm × 50 mm sawn softwood bearers							
25 mm thick	–	–	–	–	99.46	m²	**99.46**
32 mm thick	–	–	–	–	111.79	m²	**111.79**

25 STAIRS, WALKWAYS AND BALUSTRADES

Item	PC £	Labour hours	Labour £	Plant £	Material £	Unit	Total rate £
Treads; cross-tongued joints and risers; rounded nosings; tongued, grooved, glued and blocked together; one 175 mm × 50 mm sawn softwood carriage							
25 mm treads; 19 mm risers	–	–	–	–	199.89	m²	**199.89**
ends; quadrant	–	–	–	–	60.88	nr	**60.88**
ends; housed to hardwood	–	–	–	–	1.13	nr	**1.13**
32 mm treads; 25 mm risers	–	–	–	–	207.07	m²	**207.07**
ends; quadrant	–	–	–	–	78.26	nr	**78.26**
ends; housed to hardwood	–	–	–	–	1.13	nr	**1.13**
Winders; cross-tongued joints and risers in one width; rounded nosings; tongued, grooved glued and blocked together; one 175 mm × 50 mm sawn softwood carriage							
25 mm treads; 19 mm risers	–	–	–	–	277.85	m²	**277.85**
32 mm treads; 25 mm risers	–	–	–	–	284.38	m²	**284.38**
wide ends; housed to hardwood	–	–	–	–	2.25	nr	**2.25**
narrow ends; housed to hardwood	–	–	–	–	1.69	nr	**1.69**
Closed strings; in one width; 230 mm wide; rounded twice							
32 mm thick	–	–	–	–	36.60	m	**36.60**
38 mm thick	–	–	–	–	39.86	m	**39.86**
50 mm thick	–	–	–	–	44.40	m	**44.40**
Closed strings; cross-tongued joints; 280 mm wide; once rounded							
32 mm thick	–	–	–	–	47.51	m	**47.51**
extra for short ramp	–	–	–	–	24.41	nr	**24.41**
38 mm thick	–	–	–	–	51.82	m	**51.82**
extra for short ramp	–	–	–	–	27.75	nr	**27.75**
50 mm thick	–	–	–	–	57.82	m	**57.82**
extra for short ramp	–	–	–	–	34.34	nr	**34.34**
The following labours are irrespective of timber width							
ends; fitted	–	–	–	–	1.44	nr	**1.44**
ends; framed	–	–	–	–	8.51	nr	**8.51**
extra for tongued heading joint	–	–	–	–	4.20	nr	**4.20**
Closed strings; ramped; crossed tongued joints 280 mm wide; once rounded							
32 mm thick	–	–	–	–	47.51	m	**47.51**
44 mm thick	–	–	–	–	51.82	m	**51.82**
57 mm thick	–	–	–	–	57.82	m	**57.82**
Apron linings; in one width 230 mm wide							
19 mm thick	–	–	–	–	12.63	m	**12.63**
25 mm thick	–	–	–	–	14.88	m	**14.88**

25 STAIRS, WALKWAYS AND BALUSTRADES

Item	PC £	Labour hours	Labour £	Plant £	Material £	Unit	Total rate £
25.01 STAIRS – cont							
The following are supply only prices for purpose-made staircase components in selected American Oak; supplied as part of an assembled staircase							
Board landings; cross-tongued joints;							
100 mm × 50 mm sawn softwood bearers							
25 mm thick	–	–	–	–	146.72	m²	**146.72**
32 mm thick	–	–	–	–	177.17	m²	**177.17**
Treads; cross-tongued joints and risers; rounded nosings; tongued, grooved, glued and blocked together; one 175 mm × 50 mm sawn softwood carriage							
25 mm treads; 19 mm risers	–	–	–	–	243.56	m²	**243.56**
ends; quadrant	–	–	–	–	121.67	nr	**121.67**
ends; housed to hardwood	–	–	–	–	1.48	nr	**1.48**
32 mm treads; 25 mm risers	–	–	–	–	279.09	m²	**279.09**
ends; quadrant	–	–	–	–	149.77	nr	**149.77**
ends; housed to hardwood	–	–	–	–	1.48	nr	**1.48**
Winders; cross-tongued joints and risers in one width; rounded nosings; tongued, grooved glued and blocked together; one 175 mm × 50 mm sawn softwood carriage							
25 mm treads; 19 mm risers	–	–	–	–	308.75	m²	**308.75**
32 mm treads; 25 mm risers	–	–	–	–	336.88	m²	**336.88**
wide ends; housed to hardwood	–	–	–	–	3.01	nr	**3.01**
narrow ends; housed to hardwood	–	–	–	–	2.26	nr	**2.26**
Closed strings; in one width; 230 mm wide; rounded twice							
32 mm thick	–	–	–	–	57.54	m	**57.54**
44 mm thick	–	–	–	–	66.41	m	**66.41**
57 mm thick	–	–	–	–	91.18	m	**91.18**
Closed strings; cross-tongued joints; 280 mm wide; once rounded							
32 mm thick	–	–	–	–	73.11	m	**73.11**
extra for short ramp	–	–	–	–	41.81	nr	**41.81**
38 mm thick	–	–	–	–	84.73	m	**84.73**
extra for short ramp	–	–	–	–	47.62	nr	**47.62**
50 mm thick	–	–	–	–	116.18	m	**116.18**
extra for short ramp	–	–	–	–	63.34	nr	**63.34**
Closed strings; ramped; crossed tongued joints 280 mm wide; once rounded							
32 mm thick	–	–	–	–	84.09	m	**84.09**
44 mm thick	–	–	–	–	97.45	m	**97.45**
57 mm thick	–	–	–	–	133.60	m	**133.60**
Apron linings; in one width 230 mm wide							
19 mm thick	–	–	–	–	19.59	m	**19.59**
25 mm thick	–	–	–	–	24.01	m	**24.01**

25 STAIRS, WALKWAYS AND BALUSTRADES

Item	PC £	Labour hours	Labour £	Plant £	Material £	Unit	Total rate £
Handrails; rounded							
40 mm × 50 mm	–	–	–	–	15.87	m	**15.87**
50 mm × 75 mm	–	–	–	–	20.34	m	**20.34**
57 mm × 87 mm	–	–	–	–	30.56	m	**30.56**
69 mm × 100 mm	–	–	–	–	41.14	m	**41.14**
Handrails; moulded							
40 mm × 50 mm	–	–	–	–	17.43	m	**17.43**
50 mm × 75 mm	–	–	–	–	21.89	m	**21.89**
57 mm × 87 mm	–	–	–	–	32.09	m	**32.09**
69 mm × 100 mm	–	–	–	–	42.66	m	**42.66**
Add to above for							
grooved once	–	–	–	–	0.79	m	**0.79**
ends; framed	–	–	–	–	7.87	nr	**7.87**
ends; framed on rake	–	–	–	–	9.97	nr	**9.97**
Heading joints to handrail; mitred or raked							
overall size not exceeding 50 mm × 75 mm	–	–	–	–	41.99	nr	**41.99**
overall size not exceeding 69 mm × 100 mm	–	–	–	–	49.84	nr	**49.84**
Knee piece to handrail; mitred or raked							
overall size not exceeding 69 mm × 100 mm	–	–	–	–	89.22	nr	**89.22**
Balusters; stiffeners							
25 mm × 25 mm	–	–	–	–	3.64	m	**3.64**
32 mm × 32 mm	–	–	–	–	4.57	m	**4.57**
44 mm × 44 mm	–	–	–	–	7.22	m	**7.22**
ends; housed	–	–	–	–	1.82	nr	**1.82**
Sub-rails							
32 mm × 63 mm	–	–	–	–	9.77	m	**9.77**
ends; framed joint to newel	–	–	–	–	7.87	nr	**7.87**
Knee rails							
32 mm × 140 mm	–	–	–	–	16.98	m	**16.98**
ends; framed joint to newel	–	–	–	–	7.87	nr	**7.87**
Newel posts							
44 mm × 94 mm; half newel	–	–	–	–	12.74	m	**12.74**
69 mm × 69 mm	–	–	–	–	21.71	m	**21.71**
94 mm × 94 mm	–	–	–	–	54.17	m	**54.17**
Newel caps; splayed on four sides							
62.50 mm × 125 mm × 50 mm	–	–	–	–	10.93	nr	**10.93**
100 mm × 100 mm × 50 mm	–	–	–	–	11.55	nr	**11.55**
125 mm × 125 mm × 50 mm	–	–	–	–	12.69	nr	**12.69**
25.02 WALKWAYS							
Flooring; metalwork							
Chequer plate flooring; galvanized mild steel; over 300 mm wide; bolted to steel supports							
6 mm thick	–	–	–	–	–	m²	**308.44**
8 mm thick	–	–	–	–	–	m²	**320.30**
Open mesh flooring; galvanized; over 300 mm wide; bolted to steel supports							
8 mm thick	–	–	–	–	–	m²	**308.44**

25 STAIRS, WALKWAYS AND BALUSTRADES

Item	PC £	Labour hours	Labour £	Plant £	Material £	Unit	Total rate £
25.02 WALKWAYS – cont							
Surface treatment							
At works							
galvanizing	–	–	–	–	–	tonne	**409.28**
shotblasting	–	–	–	–	–	m²	**5.34**
touch up primer and one coat of two pack epoxy zinc phosphate or chromate primer	–	–	–	–	–	m²	**10.68**
25.03 BALUSTRADES							
Standard balustrades; wrought softwood							
Landing balustrade; 50 mm × 75 mm hardwood handrail; 32 mm square plain balusters; one end of handrail jointed to newel post; other end built into wall; balusters housed in at bottom (newel post and mortices both not included)							
3.00 m long	–	3.70	85.04	–	107.12	nr	**192.16**
Steel							
Balustrades; galvanized mild steel CHS posts and top rail, with one infill rail							
1100 mm high	–	–	–	–	–	m	**249.12**
Balustrades; painted mild steel flat bar posts and CHS top rail, with 3 nr stainless steel infills							
1100 mm high	–	–	–	–	–	m	**355.90**
bends	–	–	–	–	–	nr	**32.03**
ends	–	–	–	–	–	nr	**21.35**
Stainless steel							
Balustrades; stainless steel flat bar posts and circular handrail, with 3 nr stainless steel infills							
1100 mm high	–	–	–	–	–	m	**427.07**
Balustrades; stainless steel 50 mm Ø posts and circular handrail, with 10 mm thick toughened glass infill panels							
1100 mm high	–	–	–	–	–	m	**711.79**
Glass							
Balustrades; laminated glass; with stainless steel cap channel to top and including all necessary support fixings							
1100 mm high	–	–	–	–	–	m	**711.79**
Softwood handrails							
Handrails; rounded							
40 mm × 50 mm	–	–	–	–	13.72	m	**13.72**
50 mm × 75 mm	–	–	–	–	16.54	m	**16.54**
57 mm × 87 mm	–	–	–	–	19.42	m	**19.42**
69 mm × 100 mm	–	–	–	–	24.12	m	**24.12**

25 STAIRS, WALKWAYS AND BALUSTRADES

Item	PC £	Labour hours	Labour £	Plant £	Material £	Unit	Total rate £
Handrails; moulded							
40 mm × 50 mm	–	–	–	–	15.24	m	**15.24**
50 mm × 75 mm	–	–	–	–	18.08	m	**18.08**
57 mm × 87 mm	–	–	–	–	20.95	m	**20.95**
69 mm × 100 mm	–	–	–	–	25.66	m	**25.66**
Add to above for							
grooved once	–	–	–	–	0.68	m	**0.68**
ends; framed	–	–	–	–	6.41	nr	**6.41**
ends; framed on rake	–	–	–	–	7.87	nr	**7.87**
Heading joints to handrail; mitred or raked							
overall size not exceeding 50 mm × 75 mm	–	–	–	–	31.49	nr	**31.49**
overall size not exceeding 69 mm × 100 mm	–	–	–	–	39.37	nr	**39.37**
Knee piece to handrail; mitred or raked							
overall size not exceeding 69 mm × 100 mm	–	–	–	–	83.97	nr	**83.97**
Balusters; stiffeners							
25 mm × 25 mm	–	–	–	–	3.52	m	**3.52**
32 mm × 32 mm	–	–	–	–	3.99	m	**3.99**
44 mm × 44 mm	–	–	–	–	5.23	m	**5.23**
ends; housed	–	–	–	–	1.58	nr	**1.58**
Sub-rails							
32 mm × 63 mm	–	–	–	–	8.10	m	**8.10**
ends; framed joint to newel	–	–	–	–	6.82	nr	**6.82**
Knee rails							
32 mm × 140 mm	–	–	–	–	13.41	m	**13.41**
ends; framed joint to newel	–	–	–	–	6.82	nr	**6.82**
Newel posts							
44 mm × 94 mm; half newel	–	–	–	–	9.31	m	**9.31**
69 mm × 69 mm	–	–	–	–	10.10	m	**10.10**
94 mm × 94 mm	–	–	–	–	20.66	m	**20.66**
Newel caps; splayed on four sides							
62.50 mm × 125 mm × 50 mm	–	–	–	–	9.88	nr	**9.88**
100 mm × 100 mm × 50 mm	–	–	–	–	10.06	nr	**10.06**
125 mm × 125 mm × 50 mm	–	–	–	–	10.57	nr	**10.57**
Metal handrails							
Wallrails; painted mild steel CHS wall rail; with wall rose bracket							
42 mm dia.	–	–	–	–	–	m	**100.84**
Wallrails; stainless steel circular wall rail; with wall rose bracket							
42 mm dia.	–	–	–	–	–	m	**130.49**

25 STAIRS, WALKWAYS AND BALUSTRADES

Item	PC £	Labour hours	Labour £	Plant £	Material £	Unit	Total rate £
25.04 LADDERS							
Loft ladders; fixing with screws to timber lining (not included)							
Loft ladders							
Youngman Easiway 3 section aluminium ladder; 2.3 m to 3.0 m ceiling height	–	0.93	21.37	–	96.64	nr	**118.01**
Youngman Eco folding 3 section timber ladder; 2.8 m ceiling height	–	2.00	45.97	–	164.88	nr	**210.85**
Youngman Spacemeaker aluminium sliding ladder; 2.6 m ceiling height	–	1.25	28.73	–	56.73	nr	**85.46**
Youngman Deluxe 2 section aluminium ladder; 3.25 m ceiling height; spring assisted	–	2.50	57.46	–	206.08	nr	**263.54**
Access ladders; mild steel							
Ladders							
400 mm wide; 3850 mm long (overall); 12 mm dia. rungs; 65 mm × 15 mm strings; 50 mm × 5 mm safety hoops; fixing with expanded bolts; to masonry; mortices; welded fabrication	–	–	–	–	–	nr	**2200.00**

27 GLAZING

Item	PC £	Labour hours	Labour £	Plant £	Material £	Unit	Total rate £
27.01 SUPPLY ONLY RATES FOR GLASS							
SUPPLY ONLY RATES							
Ordinary transluscent/patterned glass							
3 mm	–	–	–	–	35.56	m²	**35.56**
4 mm	–	–	–	–	37.74	m²	**37.74**
5 mm	–	–	–	–	45.88	m²	**45.88**
6 mm	–	–	–	–	50.37	m²	**50.37**
Obscured ground sheet glass – patterned							
4 mm white	–	–	–	–	53.34	m²	**53.34**
6 mm white	–	–	–	–	58.71	m²	**58.71**
Rough cast							
6 mm	–	–	–	–	50.00	m²	**50.00**
Ordinary Georgian wired							
7 mm cast	–	–	–	–	50.99	m²	**50.99**
6 mm polish	–	–	–	–	78.42	m²	**78.42**
Cetuff toughened; float							
4 mm	–	–	–	–	41.76	m²	**41.76**
5 mm	–	–	–	–	55.36	m²	**55.36**
6 mm	–	–	–	–	60.96	m²	**60.96**
10 mm	–	–	–	–	100.03	m²	**100.03**
Clear laminated; safety							
4.40 mm	–	–	–	–	49.78	m²	**49.78**
6.40 mm	–	–	–	–	59.43	m²	**59.43**
27.02 GENERAL GLAZING							
SUPPLY AND FIX PRICES							
NOTE: The following measured rates are provided by a glazing contractor and assume in excess of 500 m², within 20 miles of the suppliers branch.							
Standard plain glass; BS EN 14449; clear float; panes area 0.15 m²–4.00 m²							
3 mm thick; glazed with							
screwed beads	–	–	–	–	–	m²	**55.24**
4 mm thick; glazed with							
screwed beads	–	–	–	–	–	m²	**58.60**
5 mm thick; glazed with							
screwed beads	–	–	–	–	–	m²	**71.37**
6 mm thick; glazed with							
screwed beads	–	–	–	–	–	m²	**78.18**
Standard plain glass; BS EN 14449; obscure patterned; panes area 0.15 m²–4.00 m²							
4 mm thick; glazed with							
screwed beads	–	–	–	–	–	m²	**82.86**

27 GLAZING

Item	PC £	Labour hours	Labour £	Plant £	Material £	Unit	Total rate £
27.02 GENERAL GLAZING – cont							
Standard plain glass – cont							
6 mm thick; glazed with							
screwed beads	–	–	–	–	–	m²	91.18
Standard plain glass; BS EN 14449; rough cast; panes area 0.15 m²–4.00 m²							
6 mm thick; glazed with							
screwed beads	–	–	–	–	–	m²	74.98
Standard plain glass; BS EN 14449; Georgian wired cast; panes area 0.15 m²–4.00 m²							
7 mm thick; glazed with							
screwed beads	–	–	–	–	–	m²	76.42
extra for lining up wired glass	–	–	–	–	–	m²	3.95
Standard plain glass; BS EN 14449; Georgian wired polished; panes area 0.15 m²–4.00 m²							
6 mm thick; glazed with							
screwed beads	–	–	–	–	–	m²	103.37
extra for lining up wired glass	–	–	–	–	–	m²	3.95
Factory-made double hermetically sealed units; to wood or metal with screwed or clipped beads							
Two panes; BS EN 14449; clear float glass; 4 mm thick; 6 mm air space							
0.35 m²–2.00 m²	–	–	–	–	–	m²	151.95
Two panes; BS 952; clear float glass; 6 mm thick; 6 mm air space							
0.35 m²–2.0 m²	–	–	–	–	–	m²	176.96
2.00 m²–4.00 m²	–	–	–	–	–	m²	266.09
Factory-made double hermetically sealed units; with inner pane of Pilkington's K low emissivity coated glass; to wood or metal with screwed or clipped beads							
Two panes; BS EN 14449; clear float glass; 4 mm thick; 6 mm air space							
0.35 m²–2.00 m²	–	–	–	–	–	m²	184.65
Two panes; BS EN 14449; clear float glass; 6 mm thick; 6 mm air space							
0.35 m²–2.0 m²	–	–	–	–	–	m²	215.21
2.00 m²–4.00 m²	–	–	–	–	–	m²	323.67

27 GLAZING

Item	PC £	Labour hours	Labour £	Plant £	Material £	Unit	Total rate £
Factory-made triple hermetically sealed units; with inner pane of Pilkington's K low emissivity coated glass; to wood or metal with screwed or clipped beads							
Three panes; BS EN 14449; clear float glass; 4 mm thick; 6 mm air spaces							
0.35 m²–2.00 m²	–	–	–	–	–	m²	**297.60**
Three panes; BS EN 14449; clear float glass; 6 mm thick; 6 mm air spaces							
0.35 m²–2.0 m²	–	–	–	–	–	m²	**346.44**
2.00 m²–4.00 m²	–	–	–	–	–	m²	**521.14**
27.03 FIRE RESISTING GLAZING							
Special glass; BS EN 14449; Pyran half-hour fire-resisting glass or other equal or approved							
6.50 mm thick rectangular panes; glazed with screwed hardwood beads and Sealmaster Fireglaze intumescent compound or other equal and approved to rebated frame							
300 mm × 400 mm pane	–	0.74	19.60	–	66.41	nr	**86.01**
400 mm × 800 mm pane	–	0.92	24.37	–	170.85	nr	**195.22**
500 mm × 1400 mm pane	–	1.48	39.21	–	362.66	nr	**401.87**
600 mm × 1800 mm pane	–	1.86	49.28	–	582.75	nr	**632.03**
Special glass; BS EN 14449; Pyrostop one-hour fire-resisting glass or other equal and approved							
15 mm thick regular panes; glazed with screwed hardwood beads and Sealmaster Fireglaze intumescent liner and compound or other equal and approved both sides							
300 mm × 400 mm pane	–	2.22	58.81	–	122.45	nr	**181.26**
400 mm × 800 mm pane	–	2.78	73.65	–	249.45	nr	**323.10**
500 mm × 1400 mm pane	–	3.70	98.01	–	519.96	nr	**617.97**
600 mm × 1800 mm pane	–	4.62	122.39	–	780.54	nr	**902.93**
27.04 TOUGHENED AND LAMINATED GLAZING							
Special glass; BS EN 14449; toughened clear float; panes area 0.15 m²–4.00 m²							
4 mm thick; glazed with							
screwed beads	–	–	–	–	–	m²	**58.75**
5 mm thick; glazed with							
screwed beads	–	–	–	–	–	m²	**78.04**
6 mm thick; glazed with							
screwed beads	–	–	–	–	–	m²	**85.85**
10 mm thick; glazed with							
screwed beads	–	–	–	–	–	m²	**142.46**

27 GLAZING

Item	PC £	Labour hours	Labour £	Plant £	Material £	Unit	Total rate £
27.04 TOUGHENED AND LAMINATED GLAZING – cont							
Special glass; BS EN 14449; clear laminated safety glass; panes area 0.15 m² –4.00 m²							
4.40 mm thick; glazed with							
screwed beads	–	–	–	–	–	m²	**78.81**
6.40 mm thick; glazed with							
screwed beads	–	–	–	–	–	m²	**94.07**
Special glass; BS EN 14449; clear laminated security glass							
7.50 mm thick regular panes; glazed with screwed hardwood beads and Intergens intumescent strip							
300 mm × 400 mm pane	–	0.74	19.98	–	40.62	nr	**60.60**
400 mm × 800 mm pane	–	0.92	24.84	–	102.77	nr	**127.61**
500 mm × 1400 mm pane	–	1.48	39.97	–	219.20	nr	**259.17**
600 mm × 1800 mm pane	–	1.92	51.85	–	345.13	nr	**396.98**
27.05 MIRRORS AND LOUVRES							
Mirror panels; BS EN 14449; silvered; insulation backing							
4 mm thick float; fixing with adhesive							
1000 mm × 1000 mm	–	–	–	–	–	nr	**60.43**
1000 mm × 2000 mm	–	–	–	–	–	nr	**120.98**
1000 mm × 4000 mm	–	–	–	–	–	nr	**440.95**
Glass louvres; BS EN 14449; with long edges ground or smooth							
6 mm thick float							
150 mm wide	–	–	–	–	–	m	**30.12**
7 mm thick Georgian wired cast							
150 mm wide	–	–	–	–	–	m	**47.41**
6 mm thick Georgian wire polished							
150 mm wide	–	–	–	–	–	m	**66.61**
27.06 ADDITIONAL LABOURS							
Drill holes in glass							
Drill holes 6 mm to 15 mm dia. to glass thickness:							
not exceeding 6 mm thick	–	–	–	–	4.03	nr	**4.03**
not exceeding 10 mm thick	–	–	–	–	5.18	nr	**5.18**
not exceeding 12 mm thick	–	–	–	–	6.43	nr	**6.43**
not exceeding 19 mm thick	–	–	–	–	8.10	nr	**8.10**
not exceeding 25 mm thick	–	–	–	–	10.05	nr	**10.05**
Drill holes 16 mm to 38 mm dia. to glass thickness:							
not exceeding 6 mm thick	–	–	–	–	5.79	nr	**5.79**
not exceeding 10 mm thick	–	–	–	–	7.65	nr	**7.65**
not exceeding 12 mm thick	–	–	–	–	9.14	nr	**9.14**
not exceeding 19 mm thick	–	–	–	–	11.52	nr	**11.52**
not exceeding 25 mm thick	–	–	–	–	14.24	nr	**14.24**

27 GLAZING

Item	PC £	Labour hours	Labour £	Plant £	Material £	Unit	Total rate £
Drill holes over 38 mm dia. to glass thickness:							
not exceeding 6 mm thick	–	–	–	–	11.52	nr	**11.52**
not exceeding 10 mm thick	–	–	–	–	13.88	nr	**13.88**
not exceeding 12 mm thick	–	–	–	–	16.44	nr	**16.44**
not exceeding 19 mm thick	–	–	–	–	20.23	nr	**20.23**
not exceeding 25 mm thick	–	–	–	–	25.25	nr	**25.25**
Other works to glass							
Curved cutting to glass							
to 4 mm thick panes	–	–	–	–	5.54	m	**5.54**
to 6 mm thick panes	–	–	–	–	5.54	m	**5.54**
to 6 mm thick wired panes	–	–	–	–	8.45	m	**8.45**
Intumescant paste to glazed panels for die doors;							
per side treated	–	–	–	–	11.70	m	**11.70**
Imitation washleather/black velvet bedding to							
edge of glass	–	–	–	–	2.02	m	**2.02**

28 FLOOR, WALL, CEILING AND ROOF FINISHES

Item	PC £	Labour hours	Labour £	Plant £	Material £	Unit	Total rate £
28.01 FLOORS; SCREEDS (CEMENT: SAND; CONCRETE; GRANOLITHIC)							
Cement and sand (1:3) screeds; steel trowelled							
Work to floors; one coat level; to concrete base; screeded; over 600 mm wide							
25 mm thick	–	0.41	10.00	0.40	0.58	m²	**10.98**
50 mm thick	–	0.52	12.68	0.40	1.84	m²	**14.92**
75 mm thick	–	0.66	16.10	0.40	2.76	m²	**19.26**
100 mm thick	–	0.70	17.07	0.40	3.69	m²	**21.16**
Add to the above for work to falls and crossfalls and to slopes							
not exceeding 15° from horizontal	–	0.02	0.48	–	–	m²	**0.48**
over 15° from horizontal	–	0.09	2.19	–	–	m²	**2.19**
water repellent additive incorporated in the mix	–	0.02	0.48	–	6.80	m²	**7.28**
oil repellent additive incorporated in the mix	–	0.07	1.71	–	4.71	m²	**6.42**
Fine concrete (1: 4–5) levelling screeds; steel trowelled							
Work to floors; one coat; level; to concrete base; over 600 mm wide							
50 mm thick	–	0.52	12.68	–	5.04	m²	**17.72**
75 mm thick	–	0.66	16.10	–	7.56	m²	**23.66**
extra over last for isolation joint to perimeter	–	0.20	4.88	–	2.96	m	**7.84**
Early drying floor screed; RMC Mortars Readyscreed; or other equal and approved; steel trowelled							
Work to floors; one coat; level; to concrete base; over 600 mm wide							
100 mm thick	–	0.70	17.07	–	11.93	m²	**29.00**
100 mm thick with galvanized chicken wire anti-crack reinforcement	–	0.71	17.31	–	14.00	m²	**31.31**
Granolithic paving; cement and granite chippings 5 to dust (1:1:2); steel trowelled							
Work to floors; one coat; level; laid on concrete while green; bonded; over 600 mm wide							
25 mm thick	–	0.75	18.29	0.40	4.02	m²	**22.71**
38 mm thick	–	0.80	19.51	0.40	6.42	m²	**26.33**
Work to floors; two coat; laid on hacked concrete with slurry; over 600 mm wide							
50 mm thick	–	1.20	29.26	0.40	8.01	m²	**37.67**
75 mm thick	–	1.50	36.58	0.40	12.82	m²	**49.80**
Work to landings; one coat; level; laid on concrete while green; bonded; over 600 mm wide							
25 mm thick	–	0.80	19.51	0.40	4.02	m²	**23.93**
38 mm thick	–	1.00	24.39	0.40	6.42	m²	**31.21**

28 FLOOR, WALL, CEILING AND ROOF FINISHES

Item	PC £	Labour hours	Labour £	Plant £	Material £	Unit	Total rate £
Work to landings; two coat; laid on hacked concrete with slurry; over 600 mm wide							
50 mm thick	–	2.00	48.77	0.40	8.01	m²	**57.18**
75 mm thick	–	2.50	60.97	0.40	12.82	m²	**74.19**
Add to the above over 600 mm wide for							
liquid hardening additive incorporated in the mix	–	0.04	0.98	–	1.03	m²	**2.01**
oil-repellent additive incorporated in the mix	–	0.07	1.71	–	4.71	m²	**6.42**
25 mm work to treads; one coat; to concrete base							
225 mm wide	–	0.83	20.24	–	8.65	m	**28.89**
275 mm wide	–	0.83	20.24	–	9.69	m	**29.93**
returned end	–	0.17	4.14	–	–	nr	**4.14**
13 mm skirtings; rounded top edge and coved bottom junction; to brickwork or blockwork base							
75 mm wide on face	–	0.51	12.43	–	0.41	m	**12.84**
150 mm wide on face	–	0.69	16.83	–	7.61	m	**24.44**
ends; fair	–	0.04	0.98	–	–	nr	**0.98**
angles	–	0.06	1.46	–	–	nr	**1.46**
13 mm outer margin to stairs; to follow profile of and with rounded nosing to treads and risers; fair edge and arris at bottom, to concrete base							
75 mm wide	–	0.83	20.24	–	4.15	m	**24.39**
angles	–	0.06	1.46	–	–	nr	**1.46**
13 mm wall string to stairs; fair edge and arris on top; coved bottom junction with treads and risers; to brickwork or blockwork base							
275 mm (extreme) wide	–	0.74	18.05	–	7.26	m	**25.31**
ends	–	0.04	0.98	–	–	nr	**0.98**
angles	–	0.06	1.46	–	–	nr	**1.46**
ramps	–	0.07	1.71	–	–	nr	**1.71**
ramped and wreathed corners	–	0.09	2.19	–	–	nr	**2.19**
13 mm outer string to stairs; rounded nosing on top at junction with treads and risers; fair edge and arris at bottom; to concrete base							
300 mm (extreme) wide	–	0.74	18.05	–	8.99	m	**27.04**
ends	–	0.04	0.98	–	–	nr	**0.98**
angles	–	0.06	1.46	–	–	nr	**1.46**
ramps	–	0.07	1.71	–	–	nr	**1.71**
ramps and wreathed corners	–	0.09	2.19	–	–	nr	**2.19**
19 mm thick skirtings; rounded top edge and coved bottom junction; to brickwork or blockwork base							
75 mm wide on face	–	0.51	12.43	–	7.61	m	**20.04**
150 mm wide on face	–	0.69	16.83	–	11.76	m	**28.59**
ends; fair	–	0.04	0.98	–	–	nr	**0.98**
angles	–	0.06	1.46	–	–	nr	**1.46**
19 mm riser; one rounded nosing; to concrete base							
150 mm high; plain	–	0.83	20.24	–	6.57	m	**26.81**
150 mm high; undercut	–	0.83	20.24	–	6.57	m	**26.81**
180 mm high; plain	–	0.83	20.24	–	8.99	m	**29.23**
180 mm high; undercut	–	0.83	20.24	–	8.99	m	**29.23**

28 FLOOR, WALL, CEILING AND ROOF FINISHES

Item	PC £	Labour hours	Labour £	Plant £	Material £	Unit	Total rate £
28.02 FLOORS; SCREEDS (RESIN)							
Resins can be difficult to price as there may be site-specific variables, such as substrate condition and moisture content. There are numerous variables of thicknesses and surface textures, but we have provided cost guides for the most commonly used products. The installed cost will reduce by amounts increasing to approx. 10% as the overall area increases above 200 m² toward 4,000 m² and rise significantly as the area reduces toward 20 m²							
Latex self-levelling floor screeds; steel trowelled							
Work to floors; level; to concrete base; over 600 mm wide							
3 mm thick; one coat	–	–	–	–	–	m²	**8.24**
5 mm thick; two coats	–	–	–	–	–	m²	**10.30**
Resin flooring; Altro epoxy and polyurethane resin flooring system, level, to concrete; sand cement screed or securely bonded 25 mm plywood							
Work to floors; level; to concrete base; over 600 mm wide and over 300 m² total area							
Altrocoat water-based epoxy coating; 140 micron thick for two coats	–	0.45	6.64	–	7.40	m²	**14.04**
AltroTect solvent free high build epoxy coating; 350 micron thick for two coats	–	0.45	6.64	–	9.00	m²	**15.64**
Altroseal UVR chemically resistant polyurethane coating; 140 micron thick for two coats	–	0.45	6.64	–	8.54	m²	**15.18**
AltroFlow EP, epoxy self-smoothing resin screed; 2 mm thick including primer	–	0.90	13.30	–	20.44	m²	**33.74**
Altroflow PU, polyurethane self-smoothing resin screed; 2 mm thickness including primer	–	0.90	13.30	–	13.51	m²	**26.81**
Altroflow PU, polyurethane self-smoothing resin screed; 4 mm thickness including primer	–	0.90	13.30	–	22.69	m²	**35.99**
Altro Flexiflow, flexible polyurethame self-smoothing resin screed; 2 mm thickness including primer	–	0.90	13.30	–	23.54	m²	**36.84**
Altro TB screed, trowel applied multicoloured epoxy screed; 3 mm thickness including primer	–	1.20	17.73	–	24.49	m²	**42.22**
AltroScreed Quartz EP, trowel applied multicoloured epoxy screed; 4 mm thickness including primer	–	1.20	17.73	–	26.42	m²	**44.15**
Altro MultiScreed, trowel applied flecked epoxy screed; 4 mm thickness including primer	–	1.20	17.73	–	24.29	m²	**42.02**
AltroGrip EP, multilayer, textured, slip-resistant, plain coloured; 4 mm thickness including primer	–	1.20	17.73	–	25.06	m²	**42.79**

28 FLOOR, WALL, CEILING AND ROOF FINISHES

Item	PC £	Labour hours	Labour £	Plant £	Material £	Unit	Total rate £
AltroGrip PU, multilayer, textured, slip-resistant, plain coloured; 4 mm thickness including primer	–	1.20	17.73	–	20.17	m²	37.90
Altrocrete PU Excel, heavy duty polyurethane screed; 6 mm thickness including primer	–	1.20	17.73	–	27.26	m²	44.99
Altrocrete PU Excel, heavy duty polyurethane screed; 8 mm thickness including primer	–	1.20	17.73	–	31.33	m²	49.06
Perimeter site-formed resin cove, 100 mm high, 40 mm radius	–	–	–	–	27.18	m	27.18
Isocrete K screeds or other equal; steel trowelled							
Work to floors; level; to concrete base; over 600 mm wide							
35 mm thick; plus polymer bonder coat	–	–	–	–	–	m²	16.81
40 mm thick	–	–	–	–	–	m²	15.52
45 mm thick	–	–	–	–	–	m²	16.40
50 mm thick	–	–	–	–	–	m²	17.28
Work to floors; to falls or cross-falls; to concrete base; over 600 mm wide							
55 mm (average) thick	–	–	–	–	–	m²	18.15
60 mm (average) thick	–	–	–	–	–	m²	19.05
65 mm (average) thick	–	–	–	–	–	m²	19.92
75 mm (average) thick	–	–	–	–	–	m²	21.69
90 mm (average) thick	–	–	–	–	–	m²	24.36
Isocrete K screeds; quick drying; or other equal and approved; steel trowelled							
Work to floors; level or to floors n.e. 15° frojm the horizontal; to concrete base; over 600 mm wide							
55 mm thick	–	–	–	–	–	m²	23.53
75 mm thick	–	–	–	–	–	m²	29.39
Isocrete pumpable Self-Level Plus screeds; or other equal and approved; protected with Corex type polythene; knifed off prior to laying floor finish; flat smooth finish							
Work to floors; level or to floors n.e. 15° frojm the horizontal; to concrete base; over 600 mm wide							
20 mm thick	–	–	–	–	–	m²	28.65
50 mm thick	–	–	–	–	–	m²	38.22
28.03 FLOORS; MASTIC ASPHALT							
Mastic asphalt flooring; black							
20 mm thick; one coat coverings; felt isolating membrane; to concrete base; flat							
over 300 mm wide	–	–	–	–	–	m²	21.24
225 mm–300 mm wide	–	–	–	–	–	m²	39.45
150 mm–225 mm wide	–	–	–	–	–	m²	43.35
not exceeding 150 mm wide	–	–	–	–	–	m²	53.05

28 FLOOR, WALL, CEILING AND ROOF FINISHES

Item	PC £	Labour hours	Labour £	Plant £	Material £	Unit	Total rate £
28.03 FLOORS; MASTIC ASPHALT – cont							
Mastic asphalt flooring – cont							
25 mm thick; one coat coverings; felt isolating membrane; to concrete base; flat							
over 300 mm wide	–	–	–	–	–	m²	24.65
225 mm–300 mm wide	–	–	–	–	–	m²	42.08
150 mm–225 mm wide	–	–	–	–	–	m²	45.87
not exceeding 150 mm wide	–	–	–	–	–	m²	55.62
20 mm three coat skirtings to brickwork base							
not exceeding 150 mm girth	–	–	–	–	–	m	21.70
150 mm–225 mm girth	–	–	–	–	–	m	26.53
225 mm–300 mm girth	–	–	–	–	–	m	31.35
Mastic asphalt flooring; acid-resisting; black							
20 mm thick; one coat coverings; felt isolating membrane; to concrete base flat							
over 300 mm wide	–	–	–	–	–	m²	24.87
225 mm–300 mm wide	–	–	–	–	–	m²	45.50
150 mm–225 mm wide	–	–	–	–	–	m²	46.99
not exceeding 150 mm wide	–	–	–	–	–	m²	56.70
25 mm thick; one coat coverings; felt isolating membrane; to concrete base; flat							
over 300 mm wide	–	–	–	–	–	m²	29.39
225 mm–300 mm wide	–	–	–	–	–	m²	46.77
150 mm–225 mm wide	–	–	–	–	–	m²	50.62
not exceeding 150 mm wide	–	–	–	–	–	m²	60.36
20 mm thick; three coat skirtings to brickwork base							
not exceeding 150 mm girth	–	–	–	–	–	m	21.93
150 mm–225 mm girth	–	–	–	–	–	m	25.54
225 mm–300 mm girth	–	–	–	–	–	m	28.99
Mastic asphalt flooring; red							
20 mm thick; one coat coverings; felt isolating membrane; to concrete base; flat							
over 300 mm wide	–	–	–	–	–	m²	34.82
225 mm–300 mm wide	–	–	–	–	–	m²	57.50
150 mm–225 mm wide	–	–	–	–	–	m²	62.12
not exceeding 150 mm wide	–	–	–	–	–	m²	74.30
20 mm thick; three coat skirtings to brickwork base							
not exceeding 150 mm girth	–	–	–	–	–	m	27.32
150 mm–225 mm girth	–	–	–	–	–	m	34.82

28 FLOOR, WALL, CEILING AND ROOF FINISHES

Item	PC £	Labour hours	Labour £	Plant £	Material £	Unit	Total rate £
28.04 FLOORS; EDGE FIXED CARPETING; CARPET TILES							
Fitted carpeting; Wilton wool/nylon or other equal and approved; 80/20 velvet pile; heavy domestic plain							
Work to floors							
over 600 mm wide	38.00	0.37	6.12	–	43.05	m²	**49.17**
Work to treads and risers							
not exceeding 600 mm wide	22.80	0.74	12.23	–	25.83	m	**38.06**
Fitted carpet; Forbo Flooring; Flocked flooring							
Work to floors							
Flotex classic textile; 2.0 m wide roll	24.73	0.45	10.34	–	27.73	m²	**38.07**
Flotex HD textile; 2.0 m wide roll	26.47	0.45	10.34	–	29.65	m²	**39.99**
Carpet tiles; Forbo Flooring; 500 mm × 500 mm carpet tiles; adhesive applied to subfloor							
Work to floors:							
Teviot; low level loop pile; 2.50 mm thick	15.28	0.45	10.34	–	17.26	m²	**27.60**
Barcode; low level loop pile; 2.50 mm thick	20.36	0.45	10.34	–	22.89	m²	**33.23**
Create Space 2; random lay, batchless	21.74	0.45	10.34	–	24.42	m²	**34.76**
Arran; textured loop; 4.00 mm thick	29.24	0.45	10.34	–	32.72	m²	**43.06**
Alignment; textured cut and loop; 4.00 mm thick	27.68	0.45	10.34	–	30.98	m²	**41.32**
Fitted carpeting; Gradus broadloom polypropylene or other equal and approved							
Work to floors over 600 mm wide							
Bodega/Pacific/Predator ranges, loop pile	15.08	0.37	6.12	–	17.08	m²	**23.20**
Genus/Volnay ranges, cut pile	25.47	0.33	5.51	–	28.86	m²	**34.37**
Work to treads and risers							
not exceeding 600 mm wide	15.28	0.74	12.23	–	17.31	m	**29.54**
Underlay to carpeting							
Work to floors							
over 600 mm wide	3.04	0.07	1.15	–	3.29	m²	**4.44**
raking cutting	–	0.07	0.98	–	–	m	**0.98**
Carpet tiles to cement and sand base							
Work to floors over 600 mm wide							
Heuga 530 heavy duty loop pile	23.90	0.28	6.83	–	25.84	m²	**32.67**
Gradus Adventure range; bitumen backing; tufted pattern loop pile	35.28	0.28	6.83	–	38.15	m²	**44.98**
Latour/Predator ranges; bitumen backing; loop pile	20.48	0.28	6.83	–	22.15	m²	**28.98**
Sundries							
Carpet gripper fixed to floor; standard edging							
22 mm wide	–	0.04	0.56	–	0.40	m	**0.96**

28 FLOOR, WALL, CEILING AND ROOF FINISHES

Item	PC £	Labour hours	Labour £	Plant £	Material £	Unit	Total rate £
28.04 FLOORS; EDGE FIXED CARPETING; CARPET TILES – cont							
Stair nosings; aluminium; Gradus or equivalent							
Medium duty hard aluminium alloy stair tread nosings; plugged and screwed in concrete							
56 mm × 32 mm; ref AS11	10.25	0.23	4.00	–	10.89	m	**14.89**
84 mm × 32 mm; ref AS12	14.20	0.28	4.87	–	15.07	m	**19.94**
Heavy duty aluminium alloy stair tread nosings; plugged and screwed to concrete							
48 mm × 38 mm; ref HE1	11.97	0.28	4.87	–	12.71	m	**17.58**
82 mm × 38 mm; ref HE2	16.61	0.32	5.56	–	17.60	m	**23.16**
Door entrance mats							
Entrance mat systems; aluminium wiper bar; laying in position; 18 mm thick							
Gradus Topguard; 900 mm × 550 mm; single wipers	156.74	0.46	6.43	–	161.44	nr	**167.87**
Gradus Topguard; 1200 mm × 750 mm; single wipers	282.13	0.46	6.43	–	290.59	nr	**297.02**
Gradus Topguard; 2400 mm × 1200 mm; single wipers	902.82	0.93	13.00	–	929.90	nr	**942.90**
Gradus Topguard; 900 mm × 550 mm; double wipers	180.58	0.46	6.43	–	186.00	nr	**192.43**
Gradus Topguard; 1200 mm × 750 mm; double wipers	325.05	0.46	6.43	–	334.80	nr	**341.23**
Gradus Topguard; 2400 mm × 1200 mm; double wipers	1040.16	0.93	13.00	–	1071.36	nr	**1084.36**
Entrance mat systems; brass wiper bar; laying in position; 18 mm thick							
Gradus Topguard; 900 mm × 550 mm; single wipers	265.93	0.46	6.43	–	273.91	nr	**280.34**
Gradus Topguard; 1200 mm × 750 mm; single wipers	478.68	0.46	6.43	–	493.04	nr	**499.47**
Gradus Topguard; 2400 mm × 1200 mm; single wipers	1531.76	0.93	13.00	–	1577.71	nr	**1590.71**
Gradus Topguard; 900 mm × 550 mm; double wipers	272.60	0.46	6.43	–	280.78	nr	**287.21**
Gradus Topguard; 1200 mm × 750 mm; double wipers	490.68	0.46	6.43	–	505.40	nr	**511.83**
Gradus Topguard; 2400 mm × 1200 mm; double wipers	1570.17	0.93	13.00	–	1617.28	nr	**1630.28**
Other systems							
Coral Classic textile secondary and circulation matting system; 2.0 m wide roll	–	0.46	6.43	–	41.36	m²	**47.79**
Coral Brush Activ secondary and circulation matting system; 2.0 m wide roll	–	0.46	6.43	–	46.82	m²	**53.25**
Coral Duo secondary matting system; 2.0 m wide roll	–	0.46	6.43	–	46.82	m²	**53.25**

28 FLOOR, WALL, CEILING AND ROOF FINISHES

Item	PC £	Labour hours	Labour £	Plant £	Material £	Unit	Total rate £
Matwells							
Polished stainless steel matwell; angle rim with lugs; bedding in screed							
Stainless steel frame bed in screed							
1100 mm × 1400 mm	146.37	0.93	13.00	–	150.76	nr	**163.76**
Stainless steel frame bed in screed							
500 mm × 1300 mm	105.39	1.00	13.98	–	108.55	nr	**122.53**
Stainless steel frame bed in screed							
1000 mm × 1600 mm	152.22	2.00	27.95	–	156.79	nr	**184.74**
Stainless steel frame bed in screed							
1100 mm × 1400 mm	146.37	1.20	16.77	–	150.76	nr	**167.53**
Stainless steel frame bed in screed							
1800 mm × 3000 mm	281.03	1.50	20.97	–	289.46	nr	**310.43**
Polished aluminium matwell; angle rim with lugs brazed on; bedding in screed							
900 mm × 550 mm; constructed with							
25 mm × 25 mm × 3 mm angle	22.92	0.93	13.00	–	23.61	nr	**36.61**
1200 mm × 750 mm; constructed with							
34 mm × 26 mm × 6 mm angle	37.00	1.00	13.98	–	38.11	nr	**52.09**
2400 mm × 1200 mm; constructed with							
50 mm × 50 mm × 6 mm angle	79.80	1.50	20.97	–	82.19	nr	**103.16**
25 mm × 31 mm × 6 mm thick polished brass matwell; comprising angle rim with lugs brazed on; bedding in screed							
900 mm × 550 mm	74.74	0.93	13.00	–	76.98	nr	**89.98**
1200 mm × 750 mm	100.51	1.00	13.98	–	103.53	nr	**117.51**
2400 mm x1200 mm	185.56	1.50	20.97	–	191.13	nr	**212.10**
28.05 FLOORS; GRANITE; MARBLE; SLATE; TERRAZZO							
Clay floor quarries; level bedding 10 mm thick and jointing in cement and sand (1:3); butt joints; straight both ways; flush pointing with grout; to cement and sand base							
Work to floors; over 600 mm wide							
150 mm × 150 mm × 12 mm thick; heatherbrown	27.89	0.74	18.05	–	45.61	m²	**63.66**
150 mm × 150 mm × 12 mm thick; red	27.89	0.74	18.05	–	45.61	m²	**63.66**
200 mm × 200 mm × 12 mm thick; heatherbrown	28.00	0.60	14.64	–	45.73	m²	**60.37**
Works to floors; in staircase areas or plant rooms							
150 mm × 150 mm × 12 mm thick; heatherbrown	27.89	0.83	20.24	–	45.61	m²	**65.85**
150 mm × 150 mm × 12 mm thick; red	29.28	0.83	20.24	–	45.61	m²	**65.85**
200 mm × 200 mm × 12 mm thick; heatherbrown	28.00	0.69	16.83	–	45.73	m²	**62.56**
Work to floors; not exceeding 600 mm wide							
150 mm × 150 mm × 12 mm thick; heatherbrown	–	0.37	9.02	–	10.80	m	**19.82**
150 mm × 150 mm × 12 mm thick; red	–	0.37	9.02	–	10.80	m	**19.82**
200 mm × 200 mm × 12 mm thick; heatherbrown	–	0.31	7.56	–	10.63	m	**18.19**

28 FLOOR, WALL, CEILING AND ROOF FINISHES

Item	PC £	Labour hours	Labour £	Plant £	Material £	Unit	Total rate £
28.05 FLOORS; GRANITE; MARBLE; SLATE; TERRAZZO – cont							
Clay floor quarries – cont							
Work to floors – cont							
fair square cutting against flush edges of existing finishes	–	0.11	1.81	–	2.47	m	**4.28**
raking cutting	–	0.19	3.18	–	2.78	m	**5.96**
cutting around pipes; not exceeding 0.30 m girth	–	0.14	2.43	–	–	nr	**2.43**
extra for cutting and fitting into recessed manhole cover 600 mm × 600 mm	–	0.93	16.18	–	–	nr	**16.18**
Work to sills; round edge tiles							
150 mm × 150 mm × 12 mm thick; interior; heatherbrown or red	–	0.31	7.56	–	11.28	m	**18.84**
150 mm × 150 mm × 12 mm thick; exterior; heatherbrown or red	–	0.32	7.81	–	14.52	m	**22.33**
fitted end	–	0.14	2.43	–	–	nr	**2.43**
Coved skirtings; 150 mm high; rounded top edge							
150 mm × 150 mm × 12 mm thick; ref. CBTR; heatherbrown or red	–	0.23	5.61	–	28.40	m	**34.01**
ends	–	0.04	0.70	–	–	nr	**0.70**
angles	–	0.14	2.43	–	6.06	nr	**8.49**
50 mm × 50 mm × 5.50 mm thick slip-resistant mosaic floor tiles, in 300 mm × 300 mm sheets, fixing with adhesive; butt joints; straight both ways; flush pointing with white grout; to cement and sand base							
Work to floors							
over 600 mm wide	87.50	1.76	42.92	–	92.33	m²	**135.25**
not exceeding 600 mm wide	–	1.38	33.65	–	56.06	m	**89.71**
Dakota mahogany granite cladding; polished finish; jointed and pointed in coloured mortar (1:2:8)							
20 mm work to floors; level; to cement and sand base							
over 600 mm wide	–	–	–	–	–	m²	**400.15**
20 mm × 300 mm treads; plain nosings	–	–	–	–	–	m	**228.66**
raking, cutting	–	–	–	–	–	m	**40.02**
polished edges	–	–	–	–	–	m	**514.48**
birdsmouth	–	–	–	–	–	m	**51.44**

28 FLOOR, WALL, CEILING AND ROOF FINISHES

Item	PC £	Labour hours	Labour £	Plant £	Material £	Unit	Total rate £
Riven Welsh slate floor tiles; level; bedding 10 mm thick and jointing in cement and sand (1:3); butt joints; straight both ways; flush pointing with coloured mortar; to cement and sand base							
Work to floors; over 600 mm wide							
regular size and patter; 12 mm–15 mm thick	35.00	0.56	13.66	–	59.74	m²	**73.40**
Work to floors; not exceeding 600 mm wide							
regular size and patter; 12 mm–15 mm thick	–	0.56	13.66	–	36.15	m	**49.81**
Riven Chinese or Spanish slate floor tiles; level; bedding 10 mm thick and jointing in cement and sand (1:3); butt joints; straight both ways; flush pointing with coloured mortar; to cement and sand base							
Work to floors; over 600 mm wide							
regular size and patter; 12 mm–15 mm thick	20.00	0.56	13.66	–	42.75	m²	**56.41**
Work to floors; not exceeding 600 mm wide							
regular size and patter; 12 mm–15 mm thick	–	0.56	13.66	–	36.15	m	**49.81**
Roman Travertine marble cladding; polished finish; jointed and pointed in coloured mortar (1:2:8)							
20 mm thick work to floors; level; to cement and sand base							
over 600 mm wide	–	–	–	–	–	m²	**262.95**
20 mm × 300 mm treads; plain nosings	–	–	–	–	–	m	**157.78**
raking cutting	–	–	–	–	–	m	**30.87**
polished edges	–	–	–	–	–	m	**28.59**
birdsmouth	–	–	–	–	–	m	**57.16**
Terrazzo tiles; BS EN 13748; aggregate size random ground grouted and polished to 80's grit finish; standard colour range; 3 mm joints symmetrical layout; bedding in 42 mm cement semi-dry mix (1:4); grouting with neat matching cement							
300 mm × 300 mm × 28 mm (nominal) Terrazzo tile units; hydraulically pressed, mechanically vibrated, steam cured; to floors on concrete base (not included); sealed with penetrating case hardener or other equal and approved; 2 coats applied immediately after final polishing							
plain; laid level	–	–	–	–	–	m²	**51.28**
plain; to slopes exceeding 15° from horizontal	–	–	–	–	–	m²	**62.51**
to small areas/toilets	–	–	–	–	–	m²	**117.38**
Accessories							
plastic division strips; 6 mm × 38 mm; set into floor tiling above crack inducing joints, to the nearest full tile module	–	–	–	–	–	m	**3.53**

28 FLOOR, WALL, CEILING AND ROOF FINISHES

Item	PC £	Labour hours	Labour £	Plant £	Material £	Unit	Total rate £
28.05 FLOORS; GRANITE; MARBLE; SLATE; TERRAZZO – cont							
Specially made terrazzo precast units; BS EN 13748–1; aggregate size random; standard colour range; 3 mm joints; grouting with neat matching cement							
Standard tread and riser square combined terrazzo units (with riser cast down) or other equal and approved; 280 mm wide; 150 mm high; 40 mm thick; machine-made; vibrated and fully machine polished; incorporating 1 nr Ferodo anti-slip insert ref. OT40D or other equal and approved cast-in during manufacture; one end polished only							
fixed with cement: sand (1:4) mortar on prepared backgrounds (not included); grouted in neat tinted cement; wiped clean on completion of fixing	–	–	–	–	–	m	268.55
Standard tread square terrazzo units or other equal and approved; 40 mm thick; 280 mm wide; factory polished; incorporating 1 nr Ferodo anti-slip insert ref. OT40D or other equal and approved							
fixed with cement: sand (1:4) mortar on prepared backgrounds (not included); grouted in neat tinted cement; wiped clean on completion of fixing	–	–	–	–	–	m	161.58
extra over for 55 × 55 mm contrasting colour to step nosing	–	–	–	–	–	m	56.47
Standard riser square terrazzo units or other equal and approved; 40 mm thick; 150 mm high; factory polished							
fixed with cemnt: sand (1:4) mortar on prepared backgrounds (not included); grouted in neat tinted cement; wiped clean on completion of fixing	–	–	–	–	–	m	100.27
Standard coved terrazzo skirting units or other equal and approved; 904 mm long; 150 mm high; nominal finish; 23 mm thick; with square top edge							
fixed with cement: sand (1:4) mortar on prepared backgrounds (by others); grouted in neat tinted cement; wiped clean on completion of fixing	–	–	–	–	–	m	94.18
extra over for special internal/external angle pieces to match	–	–	–	–	–	m	26.75
extra over for special polished ends	–	–	–	–	–	nr	8.82

28 FLOOR, WALL, CEILING AND ROOF FINISHES

Item	PC £	Labour hours	Labour £	Plant £	Material £	Unit	Total rate £
28.06 FLOORS; CORK; RUBBER; VINYL							
Linoleum sheet; Forbo Flooring; laid level; fixing with adhesive; welded joints; to cement and sand base							
Work to floors; over 600 mm wide							
Marmoleum Real marbled sheet, 2.0 m wide × 2.50 mm thick	15.80	0.37	9.02	–	17.44	m²	**26.46**
Marmoleum Modular marbled tile 333 mm × 333 mm × 3.20 mm thick	21.07	0.45	10.97	–	23.13	m²	**34.10**
Decibel 17db marbled acoustic sheet, 2.0 m wide; 3.50 mm thick	21.07	0.45	10.97	–	23.13	m²	**34.10**
High performance vinyl sheet; Forbo Flooring Eternal; laid level with welded seams; fixing with adhesive; to cement and sand base							
Work to floors; over 600 mm wide							
2.0 m wide roll × 2.00 mm thick	15.95	0.35	8.54	–	17.59	m²	**26.13**
Slip-resistant vinyl sheet; Forbo Flooring Surestep; laid level with welded seams; fixing with adhesive; to cement and sand base							
Work to floors; over 600 mm wide							
2.0 m wide roll × 2.00 mm thick	15.95	0.37	9.02	–	17.59	m²	**26.61**
Anti-static vinyl tiles; Forbo Flooring Colorex SD; laid level; fixing with adhesive; to cement and sand base							
Work to floors; over 600 mm wide							
615 mm × 615 mm tiles × 2.00 mm thick	22.28	0.50	12.20	–	24.44	m²	**36.64**
Acoustic vinyl sheet; Forbo Flooring Sarlon Traffic 19 db; level with welded seams; fixing with adhesive; to cement and sand base							
Work to floors; over 600 mm wide							
2.0 m wide roll × 3.40 mm thick	17.02	0.37	9.02	–	18.75	m²	**27.77**
Vinyl sheet; Altro range; with welded seams; level; fixing with adhesive where appropriate; to cement and sand base							
Work to floors; over 600 mm wide and over 200 m² total area							
2.00 mm thick; Walkway 20 and Walkway Plus 20	12.08	0.56	13.66	–	14.76	m²	**28.42**
2.00 mm thick; Marine 20	19.84	0.55	13.41	–	23.64	m²	**37.05**
2.00 mm thick; Aquarius	19.84	0.55	13.41	–	23.64	m²	**37.05**
2.00 mm thick; Suprema II	21.30	0.55	13.41	–	24.98	m²	**38.39**
2.00 mm thick; Wood Safety	15.99	0.55	13.41	–	19.48	m²	**32.89**
2.20 mm thick; Xpresslay and Xpresslay Plus(NB No adhesive required)	14.60	0.44	10.73	–	15.79	m²	**26.52**
2.50 mm thick; Classic 25	21.82	0.60	14.64	–	25.30	m²	**39.94**

28 FLOOR, WALL, CEILING AND ROOF FINISHES

Item	PC £	Labour hours	Labour £	Plant £	Material £	Unit	Total rate £
28.06 FLOORS; CORK; RUBBER; VINYL – cont							
Vinyl sheet – cont							
Work to floors – cont							
2.50 mm thick; Designer 25	21.82	0.60	14.64	–	25.30	m²	**39.94**
2.50 mm thick; Unity 25	23.70	0.60	14.64	–	27.33	m²	**41.97**
3.50 mm thick; Stronghold 30	25.91	0.65	15.85	–	30.23	m²	**46.08**
Homogeneous Vinyl sheet; Marleyflor Plus or other equal; level; with welded seams; fixing with adhesive; level; to cement and sand base							
Work to floors; over 600 mm wide							
2.00 mm thick	7.00	0.42	10.24	–	8.00	m²	**18.24**
100 mm high skirtings	–	0.11	2.68	–	1.96	m	**4.64**
Safety sheet; Marleyflor Granite Multisafe or other equal; level; with welded seams; fixing with adhesive; level; to cement and sand base							
Work to floors; over 600 mm wide							
2.00 mm thick	13.90	0.42	10.24	–	15.47	m²	**25.71**
Vinyl sheet; Marleyflor Omnisports or other equal; level; with welded seams; fixing with adhesive; level; to cement and sand base							
Work to floors; over 600 mm wide							
7.65 mm thick; Pro	27.69	0.90	21.95	–	30.37	m²	**52.32**
8.75 mm thick; Competition	32.46	1.00	24.39	–	35.53	m²	**59.92**
Vinyl semi-flexible tiles; Marleyflor Homogeneous tiles range; level; fixing with adhesive; butt joints; straight both ways; to cement and sand base							
Work to floors; over 600 mm wide							
300 mm × 300 mm × 2.00 mm thick; Vylon Plus	5.92	0.23	5.61	–	6.84	m²	**12.45**
500 mm × 500 mm × 2.00 mm thick; Marleyflor Plus	7.27	0.20	4.88	–	8.30	m²	**13.18**
Vinyl tiles; Polyflex Plus; level; fixing with adhesive; butt joints; straight both ways; to cement and sand base							
Work to floors; over 600 mm wide							
300 mm × 300 mm × 2.00 mm thick	6.12	0.23	5.61	–	7.06	m²	**12.67**
Vinyl tiles; Polyflex Camaro; level; fixing with adhesive; butt joints; straight both ways; to cement and sand base							
Work to floors; over 600 mm wide							
300 mm × 300 mm × 2.00 mm thick	13.78	0.30	7.31	–	15.34	m²	**22.65**

28 FLOOR, WALL, CEILING AND ROOF FINISHES

Item	PC £	Labour hours	Labour £	Plant £	Material £	Unit	Total rate £
Vinyl tiles; Polyflor XL; level; fixing with adhesive; butt joints; straight both ways; to cement and sand base							
Work to floors; over 600 mm wide							
300 mm × 300 mm × 2.00 mm thick	7.64	0.32	7.81	–	8.69	m²	**16.50**
Vinyl sheet; Polyflor XL; level; fixing with adhesive; butt joints; straight both ways; to cement and sand base							
Work to floors; over 600 mm wide							
2.00 mm thick	5.25	0.28	6.71	–	6.11	m²	**12.82**
Vinyl sheet; Polysafe Standard; level; fixing with adhesive; welded seams; to cement and sand base							
Work to floors; over 600 mm wide							
2.00 mm thick	9.63	0.28	6.71	–	10.85	m²	**17.56**
2.50 mm thick	14.20	0.28	6.71	–	15.79	m²	**22.50**
Vinyl sheet; Polysafe Hydro; level; fixing with adhesive; welded seams; to cement and sand base							
Work to floors; over 600 mm wide							
2.00 mm thick	12.45	0.28	6.71	–	13.89	m²	**20.60**
Luxury mineral vinyl tiles; Marley I D Naturelle; level; fixing with adhesive; butt joints; straight both ways; to cement and sand base							
Work to floors; over 600 mm wide							
330 mm × 330 mm × 2.00 mm thick	9.71	0.23	5.61	–	10.94	m²	**16.55**
Acoustic vinyl tiles; Marley Tapiflex 243; level; fixing with adhesive; butt joints; straight both ways; to cement and sand base							
Work to floors; over 600 mm wide							
500 mm × 500 mm × 2.00 mm thick	13.11	0.20	4.88	–	14.62	m²	**19.50**
Linoleum tiles; Marley Veneto XF; level; fixing with adhesive; butt joints; straight both ways; to cement and sand base							
Work to floors; over 600 mm wide							
500 mm × 500 mm × 2.50 mm thick	15.54	0.20	4.88	–	17.24	m²	**22.12**
Cork tiles; natural finish, unsealed; level; fixing with adhesive; butt joints; straight both ways; to cement and sand base							
Work to floors; over 600 mm wide							
300 mm × 300 mm × 8.00 mm thick; contract quality	22.42	0.40	9.75	–	28.43	m²	**38.18**
300 mm × 300 mm × 6.00 mm thick; high density	18.13	0.40	9.75	–	23.57	m²	**33.32**
300 mm × 300 mm × 4.00 mm thick; domestic	15.79	0.37	9.02	–	20.93	m²	**29.95**
300 mm × 300 mm × 3.20 mm thick; domestic	12.85	0.37	9.02	–	17.58	m²	**26.60**

28 FLOOR, WALL, CEILING AND ROOF FINISHES

Item	PC £	Labour hours	Labour £	Plant £	Material £	Unit	Total rate £
28.06 FLOORS; CORK; RUBBER; VINYL – cont							
Cork tiles; PVC surfaced for heaviest wear areas; level; fixing with adhesive; butt joints; straight both ways; to cement and sand base							
600 mm × 300 mm × 3.20 mm thick	36.15	0.37	9.02	–	40.26	m²	**49.28**
Rubber studded tiles; Altro Mondopave; level; fixing with adhesive; butt joints; straight to cement and sand base							
Work to floors; over 600 mm wide							
500 mm × 500 mm × 2.50 mm thick; type MRB; black	25.03	0.60	14.64	–	29.43	m²	**44.07**
500 mm × 500 mm × 4.00 mm thick; type MRB; black	29.36	0.60	14.64	–	34.11	m²	**48.75**
Work to landings; over 600 mm wide							
500 mm × 500 mm × 4.00 mm thick; type MRB; black	29.36	0.74	18.05	–	37.26	m²	**55.31**
Work to stairs							
tread; 275 mm wide	–	0.46	11.22	–	11.10	m	**22.32**
riser; 180 mm wide	–	0.56	13.66	–	7.77	m	**21.43**
Sundry floor sheeting underlays							
For floor finishings; over 600 mm wide							
building paper to BS 1521; class A; 75 mm lap (laying only)	–	0.05	0.70	–	–	m²	**0.70**
3.20 mm thick hardboard	–	0.19	4.64	–	1.66	m²	**6.30**
6.00 mm thick plywood	–	0.28	6.83	–	6.34	m²	**13.17**
Skirtings; plastic; Gradus or equivalent							
Set-in skirtings							
100 mm high; ref. SI100	–	0.11	2.68	–	2.63	m	**5.31**
150 mm high; ref. SI150	–	0.22	5.37	–	4.27	m	**9.64**
Set-on skirtings							
100 mm high; ref. SO100	–	0.22	5.37	–	2.13	m	**7.50**
150 mm high; ref. SO150	–	0.40	9.75	–	4.02	m	**13.77**
Stair nosings; aluminium; Gradus or equivalent							
Medium duty hard aluminium alloy stair tread nosings; plugged and screwed in concrete							
56 mm × 32 mm; ref AS11	10.50	0.23	4.00	–	10.89	m	**14.89**
84 mm × 32 mm; ref AS12	14.56	0.28	4.87	–	15.07	m	**19.94**
Heavy duty aluminium alloy stair tread nosings; plugged and screwed to concrete							
48 mm × 38 mm; ref HE1	12.27	0.28	4.87	–	12.71	m	**17.58**
82 mm × 38 mm; ref HE2	17.02	0.32	5.56	–	17.60	m	**23.16**

28 FLOOR, WALL, CEILING AND ROOF FINISHES

Item	PC £	Labour hours	Labour £	Plant £	Material £	Unit	Total rate £
28.07 FLOORS; WOOD BLOCK							
Wood blocks; Havwoods or other equal and approved; 25 mm thick; level; laid to herringbone pattern with 2 block border; fixing with adhesive; to cement: sand base; sanded and sealed							
Work to floors; over 300 mm wide							
Merbau	–	–	–	–	–	m²	132.23
Iroko	–	–	–	–	–	m²	132.23
American Oak	–	–	–	–	–	m²	138.58
European Oak	–	–	–	–	–	m²	144.93
Add to wood block flooring over 300 mm wide for							
buff; one coat seal	–	–	–	–	–	m²	5.08
buff; two coats seal	–	–	–	–	–	m²	8.26
sand; three coats for seal or oil	–	–	–	–	–	m²	22.88
28.08 FLOORS; RAISED ACCESS FLOORS							
Raised flooring system; laid on or fixed to concrete floor							
Full access system; 150 mm high overall; pedestal supports							
PSA light grade; steel finish	–	–	–	–	–	m²	37.25
PSA medium grade; steel finish	–	–	–	–	–	m²	37.25
PSA heavy grade; steel finish	–	–	–	–	–	m²	53.56
Extra for							
factory applied needlepunch carpet	–	–	–	–	–	m²	17.46
factory applied anti-static vinyl	–	–	–	–	–	m²	29.10
factory applied black PVC edge strips	–	–	–	–	–	m	5.29
ramps; 3.00 m × 1.40 m (no finish)	–	–	–	–	–	nr	820.00
steps (no finish)	–	–	–	–	–	m	51.25
forming cut-out for electrical boxes	–	–	–	–	–	nr	4.66
supply and lay protection to raised floor; 2440 × 1220 mm polypropylene sheets with taped joints	–	–	–	–	–	m²	2.03
28.09 WALLS; PLASTERED; RENDERED; ROUGHCAST							
Plaster; one coat Thistle board finish or other equal; steel trowelled							
3 mm thick work to walls or ceilings; one coat; to plasterboard base							
over 600 mm wide	–	0.25	6.10	–	0.06	m²	6.16
over 600 mm wide; in staircase areas or plant rooms	–	0.35	8.54	–	0.06	m²	8.60
not exceeding 600 mm wide	–	0.15	3.66	–	0.04	m	3.70

Prices for Measured Works

28 FLOOR, WALL, CEILING AND ROOF FINISHES

Item	PC £	Labour hours	Labour £	Plant £	Material £	Unit	Total rate £
28.09 WALLS; PLASTERED; RENDERED; ROUGHCAST – cont							
Plaster; one coat Thistle board finish, 3 mm work to walls or ceilings; one coat on and including gypsum plasterboard; fixing with nails; 3 mm joints filled with plaster scrim; to softwood base; plain grade baseboard or lath with rounded edges							
9.50 mm thick boards to walls							
over 600 mm wide	–	0.97	16.02	–	5.01	m²	**21.03**
not exceeding 600 mm wide	–	0.37	6.44	–	1.50	m	**7.94**
9.50 mm thick boards to walls; in staircase areas or plant rooms							
over 600 mm wide	–	1.06	17.58	–	5.01	m²	**22.59**
not exceeding 600 mm wide	–	0.46	8.00	–	1.50	m	**9.50**
9.50 mm thick boards to isolated columns							
over 600 mm wide	–	1.06	17.58	–	5.01	m²	**22.59**
not exceeding 600 mm wide	–	0.56	9.74	–	1.50	m	**11.24**
12.50 mm thick boards to walls; in staircase areas or plant rooms							
over 600 mm wide	–	1.12	18.63	–	5.01	m²	**23.64**
not exceeding 600 mm wide	–	0.50	8.70	–	1.50	m	**10.20**
12.50 mm thick boards to isolated columns							
over 600 mm wide	–	1.12	18.63	–	5.01	m²	**23.64**
not exceeding 600 mm wide	–	0.59	10.27	–	1.50	m	**11.77**
Cement and sand (1:3) beds and backings							
10 mm thick work to walls; one coat; to brickwork or blockwork base							
over 600 mm wide	–	0.60	14.64	0.40	1.14	m²	**16.18**
not exceeding 600 mm wide	–	0.35	8.54	0.20	0.65	m	**9.39**
13 mm thick; work to walls; two coats; to brickwork or blockwork base							
over 600 mm wide	–	0.80	19.51	0.40	1.17	m²	**21.08**
not exceeding 600 mm wide	–	0.40	9.75	0.20	0.65	m	**10.60**
15 mm thick work to walls; two coats; to brickwork or blockwork base							
over 600 mm wide	–	0.90	21.95	0.40	1.26	m²	**23.61**
not exceeding 600 mm wide	–	0.40	9.75	0.20	0.76	m	**10.71**
Cement and sand (1:3); steel trowelled							
13 mm thick work to walls; two coats; to brickwork or blockwork base							
over 600 mm wide	–	0.80	19.51	0.40	1.17	m²	**21.08**
not exceeding 600 mm wide	–	0.40	9.75	0.20	0.65	m	**10.60**
16 mm thick work to walls; two coats; to brickwork or blockwork base							
over 600 mm wide	–	0.90	21.95	0.40	1.26	m²	**23.61**
not exceeding 600 mm wide	–	0.40	9.75	0.20	0.77	m	**10.72**

28 FLOOR, WALL, CEILING AND ROOF FINISHES

Item	PC £	Labour hours	Labour £	Plant £	Material £	Unit	Total rate £
19 mm thick work to walls; two coats; to brickwork or blockwork base							
over 600 mm wide	–	0.95	23.16	0.40	1.40	m²	**24.96**
not exceeding 600 mm wide	–	0.40	9.75	0.20	1.03	m	**10.98**
ADD to above							
over 600 mm wide in water repellent cement	–	0.10	2.44	–	1.04	m²	**3.48**
finishing coat in colour cement	–	0.35	8.54	0.20	1.04	m²	**9.78**
Cement-lime-sand (1:2:9); steel trowelled							
19 mm thick work to walls; two coats; to brickwork or blockwork base							
over 600 mm wide	–	0.95	23.16	0.40	1.12	m²	**24.68**
not exceeding 600 mm wide	–	0.40	9.75	0.20	4.44	m	**14.39**
Cement-lime-sand (1:1:6); steel trowelled							
13 mm thick work to walls; two coats; to brickwork or blockwork base							
over 600 mm wide	–	0.95	23.16	0.40	1.29	m²	**24.85**
not exceeding 600 mm wide	–	0.40	9.75	0.20	4.44	m	**14.39**
Add to the above over 600 mm wide for							
water repellent cement	–	0.10	2.44	–	1.04	m²	**3.48**
Plaster; first 11 mm coat of Thistle Hardwall plaster; second 2 mm finishing coat of Thistle Multi-Finish plaster; steel trowelled							
13 mm thick work to walls; two coats; to brickwork or blockwork base							
over 600 mm wide	–	0.50	12.20	–	1.71	m²	**13.91**
over 600 mm wide; in staircase areas, plant rooms and similar restricted areas	–	0.60	14.64	–	1.71	m²	**16.35**
not exceeding 600 mm wide	–	0.25	6.10	–	1.02	m	**7.12**
13 mm thick work to isolated brickwork or blockwork columns; two coats							
over 600 mm wide	–	1.00	24.39	–	1.71	m²	**26.10**
not exceeding 600 mm wide	–	0.50	12.20	–	1.02	m	**13.22**
Plaster; first 8 mm or 11 mm coat of Thistle Bonding plaster; second 2 mm finishing coat of Thistle Multi-Finish plaster; steel trowelled finish							
13 mm thick work to walls; two coats; to concrete base							
over 600 mm wide	–	0.50	12.20	–	1.72	m²	**13.92**
over 600 mm wide; in staircase areas, plant rooms and similar restricted areas	–	0.60	14.64	–	1.72	m²	**16.36**
not exceeding 600 mm wide	–	0.25	6.10	–	1.02	m	**7.12**
13 mm thick work to isolated piers or columns; two coats; to concrete base							
over 600 mm wide	–	1.00	24.39	–	1.72	m²	**26.11**
not exceeding 600 mm wide	–	0.50	12.20	–	1.03	m	**13.23**

28 FLOOR, WALL, CEILING AND ROOF FINISHES

Item	PC £	Labour hours	Labour £	Plant £	Material £	Unit	Total rate £
28.09 WALLS; PLASTERED; RENDERED; ROUGHCAST – cont							
Plaster – cont							
13 mm thick work to vertical face of metal lathing arch former							
not exceeding 0.50 m² per side	–	0.75	18.29	–	0.03	nr	**18.32**
0.50 m²–1 m² per side	–	1.00	24.39	–	0.05	nr	**24.44**
Plaster; first coat of Limelite renovating plaster; finishing coat of Limelite finishing plaster; or other equal							
13 mm thick work to walls; two coats; to brickwork or blockwork base							
over 600 mm wide	–	0.90	21.95	–	2.34	m²	**24.29**
over 600 mm wide; in staircase areas, plant rooms and similar restricted areas	–	1.00	24.39	–	2.34	m²	**26.73**
not exceeding 600 mm wide	–	0.45	10.97	–	1.33	m	**12.30**
Dubbing out existing walls with plaster; average 6 mm thick							
over 600 mm wide	–	0.25	6.10	–	0.07	m²	**6.17**
not exceeding 600 mm wide	–	0.15	3.66	–	0.04	m	**3.70**
Dubbing out existing walls with undercoat plaster; average 12 mm thick							
over 600 mm wide	–	0.50	12.20	–	0.12	m²	**12.32**
not exceeding 600 mm wide	–	0.25	6.10	–	0.07	m	**6.17**
Plaster; first coat of Thistle X-ray plaster or other equal; finishing coat of Thistle X-ray finishing plaster or other equal and approved; steel trowelled							
17 mm thick work to walls; two coats; to brickwork or blockwork base							
over 600 mm wide	–	–	–	–	–	m²	**75.35**
over 600 mm wide; in staircase areas or plant rooms	–	–	–	–	–	m²	**80.95**
not exceeding 600 mm wide	–	–	–	–	–	m	**30.12**
17 mm thick work to isolated columns; two coats							
over 600 mm wide	–	–	–	–	–	m²	**122.22**
not exceeding 600 mm wide	–	–	–	–	–	m	**48.86**
Plaster; one coat Thistle projection spray applied plaster or other equal; steel trowelled							
13 mm thick work to walls; one coat; to brickwork or blockwork base							
over 600 mm wide	–	0.15	3.66	1.10	5.86	m²	**10.62**
over 600 mm wide; in staircase areas or plant rooms	–	0.20	4.88	1.25	5.86	m²	**11.99**
not exceeding 600 mm wide	–	0.10	2.44	0.69	3.72	m	**6.85**

28 FLOOR, WALL, CEILING AND ROOF FINISHES

Item	PC £	Labour hours	Labour £	Plant £	Material £	Unit	Total rate £
Cemrend self-coloured render or other equal; **one coat; to brickwork or blockwork base** 20 mm thick work to walls; to brickwork or blockwork base							
over 600 mm wide	–	–	–	–	–	m²	**41.60**
not exceeding 600 mm wide	–	–	–	–	–	m	**26.78**
Tyrolean decorative rendering or similar; 13 mm **thick first coat of cement-lime-sand (1:1:6);** **finishing three coats of Cullamix or other equal** **and approved; applied with approved hand** **operated machine external** To walls; four coats; to brickwork or blockwork base							
over 600 mm wide	–	–	–	–	–	m²	**38.60**
not exceeding 600 mm wide	–	–	–	–	–	m	**19.28**
Drydash (pebbledash) finish of chippings on and **including cement-lime-sand (1:2:9) backing** 18 mm thick work to walls; two coats; to brickwork or blockwork base. Over 600 mm wide areas							
Glenarm White chippings	2.37	1.45	35.36	0.40	5.45	m²	**41.21**
Derbyshire Spar/Red Granite chippings	3.55	1.45	35.36	0.40	6.65	m²	**42.41**
Black and White, Red and White Spar chippings	3.99	1.45	35.36	0.40	7.11	m²	**42.87**
Calcined Flint, Snowdrop chippings	4.72	1.45	35.36	0.40	7.87	m²	**43.63**
Barleycorn, Durite chippings	5.10	1.45	35.36	0.40	8.25	m²	**44.01**
Beige Marble, Buff Quartz chippings	6.07	1.45	35.36	0.40	9.26	m²	**45.02**
Accessories Expamet render beads or other equal and approved; white PVC nosings; to brickwork or blockwork base							
external stop bead; 20 mm thick	–	0.07	1.22	–	3.62	m	**4.84**
Expamet render beads or other equal and approved; stainless steel; to brickwork or blockwork base							
stop bead; 20 mm thick	–	0.07	1.22	–	2.96	m	**4.18**
stop bead; 20 mm thick	–	0.07	1.22	–	2.96	m	**4.18**
Expamet plaster beads or other equal and approved; to brickwork or blockwork base							
angle bead; 13 mm thick	–	0.08	1.39	–	0.59	m	**1.98**
architrave bead; 13 mm thick	–	0.10	1.74	–	3.54	m	**5.28**
stop bead; 10 mm thick	–	0.07	1.22	–	0.78	m	**2.00**
stop beads; 13 mm thick	–	0.07	1.22	–	0.78	m	**2.00**
movement bead; 13 mm thick	–	0.09	1.57	–	12.03	m	**13.60**
Expamet plaster beads or other equal and approved; stainless steel; to brickwork or blockwork base							
angle bead; 20 mm thick	–	0.08	1.39	–	0.77	m	**2.16**
stop bead; 10 mm thick	–	0.07	1.22	–	2.96	m	**4.18**
stop bead; 13 mm thick	–	0.07	1.22	–	2.96	m	**4.18**
Expamet thin coat plaster beads or other equal and approved; galvanized steel; to timber base							
angle bead; 3 mm thick	–	0.07	1.22	–	0.59	m	**1.81**
stop bead; 3 mm thick	–	0.06	1.04	–	0.56	m	**1.60**
stop bead; 6 mm thick	–	0.06	1.04	–	0.56	m	**1.60**

28 FLOOR, WALL, CEILING AND ROOF FINISHES

Item	PC £	Labour hours	Labour £	Plant £	Material £	Unit	Total rate £
28.10 WALLS; INSULATED RENDER							
Sto Therm Classic M-system insulation render							
70 mm EPS insulation fixed with adhesive to SFS structure (measured separately) with horizontal PVC intermediate track and vertical T-spines; with glassfibre mesh reinforcement embedded in Sto Armat Classic Basecoat Render and Stolit K 1.5 Decorative Topcoat Render (white)							
over 300 mm wide	–	–	–	–	–	m²	69.44
70 mm EPS insulation mechanically fixed to SFS structure (measured separately) with horizontal PVC intermediate track and vertical T-spines; with glassfibre mesh reinforcement embedded in Sto Armat Classic Basecoat Render and Stolit K 1.5 Decorative Topcoat Render (white)							
over 300 mm wide	–	–	–	–	–	m²	76.41
rendered heads and reveals not exceeding 100 mm wide; including angle beads	–	–	–	–	–	m	19.19
Extra for							
aluminium starter track at base of insulated render system	–	–	–	–	–	m	11.68
external angle with PVC mesh angle bead	–	–	–	–	–	m	5.13
internal angle with Sto Armor angle	–	–	–	–	–	m	5.13
render stop bead	–	–	–	–	–	m	5.13
Sto seal tape to all vertical abutments	–	–	–	–	–	m	4.77
Sto Armor mat HD mesh reinforcement to areas prone to physical damage (e.g. 1800 mm high adjoining floor level)							
over 300 mm wide	–	–	–	–	–	m²	15.36
28.11 WALLS; CORK; CERAMIC TILING; MARBLE							
Glazed ceramic wall tiles; fixing with adhesive; butt joints; straight both ways; flush pointing with white grout; to plaster base. Ranges and prices vary enormously for tiles are toonumerous to schedule here. Prices are for good quality tiles							
Work to walls; over 600 mm wide							
152 mm × 152 mm × 5.50 mm thick; white	16.00	0.56	13.66	–	19.11	m²	32.77
152 mm × 152 mm × 5.50 mm thick; light colours	20.00	0.56	13.66	–	23.43	m²	37.09
152 mm × 152 mm × 5.50 mm thick; dark colours	25.00	0.56	13.66	–	28.84	m²	42.50
200 mm × 100 mm × 6.50 mm thick; white and light colours	15.00	0.56	13.66	–	18.03	m²	31.69
250 mm × 200 mm × 7.00 mm thick; white and light colours	20.00	0.56	13.66	–	23.43	m²	37.09

28 FLOOR, WALL, CEILING AND ROOF FINISHES

Item	PC £	Labour hours	Labour £	Plant £	Material £	Unit	Total rate £
Work to walls; in staircase areas or plant rooms							
152 mm × 152 mm × 5.50 mm thick; white	16.00	0.62	15.12	–	19.11	m²	**34.23**
Work to walls; not exceeding 600 mm wide							
152 mm × 152 mm × 5.50 mm thick; white	9.60	0.56	13.66	–	11.41	m	**25.07**
152 mm × 152 mm × 5.50 mm thick; light colours	12.00	0.56	13.66	–	14.58	m	**28.24**
152 mm × 152 mm × 5.50 mm thick; dark colours	15.00	0.56	13.66	–	17.06	m	**30.72**
200 mm × 100 mm × 6.50 mm thick; white and light colours	9.00	0.56	13.66	–	10.76	m	**24.42**
250 mm × 200 mm × 7.00 mm thick; white and light colours	12.00	0.46	11.22	–	14.01	m	**25.23**
cutting around pipes; not exceeding 0.30 m girth	–	0.09	1.57	–	–	nr	**1.57**
Work to sills; 150 mm wide; rounded edge tiles							
152 mm × 152 mm × 5.50 mm thick; white	2.40	0.23	5.61	–	2.85	m	**8.46**
fitted end	–	0.09	1.57	–	–	nr	**1.57**
198 mm × 64.50 mm × 6 mm thick wall tiles; fixing with adhesive; butt joints; straight both ways; flush pointing with white grout; to plaster base							
Work to walls							
over 600 mm wide	28.00	1.67	40.73	–	33.53	m²	**74.26**
not exceeding 600 mm wide	16.80	1.30	31.70	–	20.06	m	**51.76**
20 mm × 20 mm × 5.50 mm thick glazed mosaic wall tiles; fixing with adhesive; butt joints; straight both ways; flush pointing with white grout; to plaster base							
Work to walls							
over 600 mm wide	30.00	1.76	42.92	–	35.90	m²	**78.82**
not exceeding 600 mm wide	18.00	1.38	33.65	–	22.38	m	**56.03**
Dakota mahogany granite cladding; polished finish; jointed and pointed in coloured mortar (1:2:8)							
20 mm thick work to walls; to cement and sand base							
over 600 mm wide	–	–	–	–	–	m²	**411.58**
not exceeding 600 mm wide	–	–	–	–	–	m	**182.92**
40 mm thick work to walls; to cement and sand base							
over 600 mm wide	–	–	–	–	–	m²	**685.97**
not exceeding 600 mm wide	–	–	–	–	–	m	**308.69**
Roman Travertine marble cladding; polished finish; jointed and pointed in coloured mortar (1:2:8)							
20 mm thick work to walls; to cement and sand base							
over 600 mm wide	–	–	–	–	–	m²	**312.11**
not exceeding 600 mm wide	–	–	–	–	–	m	**142.92**
40 mm thick work to walls; to cement and sand base							
over 600 mm wide	–	–	–	–	–	m²	**429.87**
not exceeding 600 mm wide	–	–	–	–	–	m	**194.36**

28 FLOOR, WALL, CEILING AND ROOF FINISHES

Item	PC £	Labour hours	Labour £	Plant £	Material £	Unit	Total rate £
28.11 WALLS; CORK; CERAMIC TILING; MARBLE – cont							
Cork tiles; fixing with adhesive; butt joints; straight both ways; toplasterd walls							
3 mm thick							
3.00 mm thick; pinboard 2 tone	5.63	0.45	10.97	–	9.40	m²	**20.37**
3.00 mm thick; standard natural finish	9.90	0.40	9.75	–	14.24	m²	**23.99**
4 mm thick							
4.00 mm thick; character natural finish	8.82	0.40	9.75	–	13.03	m²	**22.78**
28.12 WALLS; WATERPROOF PANELLING							
Waterproof panelling system with moisture-resistant MDF core and high pressure laminate backing and decorative laminate face							
Supply and fit Showerwall 10.5 mm thick panels complete with aluminium trims, sealants and adhesive							
White gloss finish fixed to substrate with adhesive	–	0.25	5.75	–	63.54	m²	**69.29**
Perganmon Marble gloss finish fixed to substrate with adhesive	–	0.25	5.75	–	70.11	m²	**75.86**
28.13 CEILINGS; PLASTERED; RENDERED							
Plaster; first 8 mm or 11 mm coat of Thistle Bonding plaster; second 2 mm finishing coat of Thistle Multi-Finish plaster; steel trowelled finish							
10 mm thick work to ceilings; two coats; to concrete base							
over 600 mm wide	–	0.50	12.20	–	0.15	m²	**12.35**
over 600 wide; 3.50 m–5.00 m high	–	0.75	18.29	–	0.15	m²	**18.44**
over 600 mm wide; in staircase areas, plant rooms and similar restricted areas	–	0.75	18.29	–	0.15	m²	**18.44**
not exceeding 600 mm wide	–	0.33	8.12	–	0.08	m	**8.20**
10 mm thick work to isolated beams; two coats; to concrete base							
over 600 mm wide	–	1.10	26.82	–	0.15	m²	**26.97**
over 600 mm wide; 3.50 m–5.00 m high	–	1.20	29.26	–	0.15	m²	**29.41**
not exceeding 600 mm wide	–	0.50	12.20	–	0.08	m	**12.28**
Plaster; first 11 mm coat of Thistle Bonding plaster; second coat 2 mm finishing coat of Thistle Multi-Finish plaster; steel trowelled							
13 mm thick work to ceilings; to metal lathing base							
over 600 mm wide	–	0.50	12.20	–	0.15	m²	**12.35**
over 600 mm wide; in staircase areas, plant rooms and similar restricted areas	–	0.75	18.29	–	0.15	m²	**18.44**
not exceeding 600 mm wide	–	0.33	8.12	–	0.08	m	**8.20**

28 FLOOR, WALL, CEILING AND ROOF FINISHES

Item	PC £	Labour hours	Labour £	Plant £	Material £	Unit	Total rate £
Plaster; one coat Thistle board finish or similar; steel trowelled 3 mm work to ceilings; one coat on and including gypsum plasterboard; fixing with nails; 3 mm joints filled with plaster and jute scrim cloth; to softwood base; plain grade baseboard or lath with rounded edges							
9.50 mm thick boards to ceilings							
over 600 mm wide	–	0.89	14.63	–	5.01	m²	**19.64**
over 600 mm wide; 3.50 m–5.00 m high	–	1.03	17.07	–	5.01	m²	**22.08**
not exceeding 600 mm wide	–	0.43	7.48	–	1.50	m	**8.98**
9.50 mm thick boards to ceilings; in staircase areas or plant rooms							
over 600 mm wide	–	0.98	16.19	–	5.01	m²	**21.20**
not exceeding 600 mm wide	–	0.47	8.18	–	1.50	m	**9.68**
9.50 mm thick boards to isolated beams							
over 600 mm wide	–	1.05	17.41	–	5.01	m²	**22.42**
not exceeding 600 mm wide	–	0.50	8.70	–	1.50	m	**10.20**
12.50 mm thick boards to ceilings							
over 600 mm wide	–	0.95	15.68	–	5.01	m²	**20.69**
over 600 mm wide; 3.50 m–5.00 m high	–	1.06	17.58	–	5.01	m²	**22.59**
not exceeding 600 mm wide	–	0.45	7.83	–	1.50	m	**9.33**
12.50 mm thick boards to ceilings; in staircase areas or plant rooms							
over 600 mm wide	–	1.06	17.58	–	5.01	m²	**22.59**
not exceeding 600 mm wide	–	0.51	8.87	–	1.50	m	**10.37**
12.50 mm thick boards to isolated beams							
over 600 mm wide	–	1.15	19.15	–	5.01	m²	**24.16**
not exceeding 600 mm wide	–	0.56	9.74	–	1.50	m	**11.24**
Plaster; one coat Thistle board finish or other equal; steel trowelled							
3 mm thick work to walls or ceilings; one coat; to plasterboard base							
over 600 mm wide	–	0.25	6.10	–	0.06	m²	**6.16**
over 600 mm wide; in staircase areas, plant rooms and similar restricted areas	–	0.35	8.54	–	0.06	m²	**8.60**
not exceeding 600 mm wide	–	0.15	3.66	–	0.04	m	**3.70**
Cement-lime-sand (1:1:6); steel trowelled							
19 mm thick work to ceilings; three coats; to metal lathing base							
over 600 mm wide	–	0.95	23.16	0.40	1.12	m²	**24.68**
not exceeding 600 mm wide	–	0.40	9.75	0.20	4.44	m	**14.39**
Gyproc cove or other equal and approved; fixing with adhesive; filling and pointing joints with plaster							
Cove							
125 mm girth	–	0.19	3.31	–	1.23	m	**4.54**
angles	–	0.03	0.53	–	0.66	nr	**1.19**

28 FLOOR, WALL, CEILING AND ROOF FINISHES

Item	PC £	Labour hours	Labour £	Plant £	Material £	Unit	Total rate £
28.14 METAL MESH LATHING FOR PLASTERED COATINGS							
Arch frames							
Preformed galvanized expanded steel semi-circular arch frames; Expamet or other equal; to suit walls up to 230 mm thick							
for 760 mm opening; ref ESC 30	58.80	0.46	7.22	–	60.56	nr	**67.78**
for 920 mm opening; ref ESC 36	67.90	0.46	7.22	–	69.94	nr	**77.16**
for 1220 mm opening; ref ESC 48	82.23	0.46	7.22	–	84.70	nr	**91.92**
Lathing; Expamet BB expanded metal lathing or other equal; 50 mm laps							
6 mm thick mesh linings to ceilings; fixing with staples; to softwood base; over 300 mm wide							
ref BB263; 0.500 mm thick	10.05	0.56	8.79	–	10.35	m²	**19.14**
ref BB264; 0.675 mm thick	14.08	0.56	8.79	–	14.50	m²	**23.29**
6 mm thick mesh linings to ceilings; fixing with wire; to steelwork; over 300 mm wide							
ref BB263; 0.500 mm thick	–	0.59	9.27	–	10.35	m²	**19.62**
ref BB264; 0.675 mm thick	–	0.59	9.27	–	14.50	m²	**23.77**
6 mm thick mesh linings to ceilings; fixing with wire; to steelwork; not exceeding 300 mm wide							
ref BB263; 0.500 mm thick	–	0.37	5.79	–	10.35	m²	**16.14**
ref BB264; 0.675 mm thick	–	0.37	5.79	–	14.50	m²	**20.29**
raking cutting	–	0.19	3.31	–	–	m	**3.31**
cutting and fitting around pipes; not exceeding 0.30 m girth	–	0.28	4.87	–	–	nr	**4.87**
Lathing; Expamet Riblath or Spreedpro or other equal stiffened expanded metal lathing; 50 mm laps; 2.50 m × 0.60 m sheets							
rigid rib mesh lining to walls; fixing with nails; to softwood base; over 300 mm wide							
Riblath ref 269; galvanized steel	16.62	0.46	7.22	–	17.25	m²	**24.47**
Riblath ref 267; stanless steel	23.89	0.46	7.22	–	24.74	m²	**31.96**
rigid rib mesh lining to walls; fixing with nails; to softwood base; not exceeding 300 mm wide							
Riblath ref 269; galvanized steel	–	0.28	4.39	–	5.22	m	**9.61**
Riblath ref 267; stanless steel	–	0.28	4.39	–	7.47	m	**11.86**
rigid rib mesh lining to ceilings; fixing with wire; to steelwork; over 300 mm wide							
Riblath ref 269; galvanized steel	–	0.59	9.27	–	17.71	m²	**26.98**
Riblath ref 267; stanless steel	–	0.59	9.27	–	25.19	m²	**34.46**

28 FLOOR, WALL, CEILING AND ROOF FINISHES

Item	PC £	Labour hours	Labour £	Plant £	Material £	Unit	Total rate £
28.15 SPRAYED MINERAL FIBRE COATINGS							
Prepare and apply by spray Mandolite CP2 fire protection or other equal and approved on structural steel/metalwork							
16 mm thick (one hour) fire protection							
to walls and columns	–	–	–	–	–	m²	**12.01**
to ceilings and beams	–	–	–	–	–	m²	**13.25**
to isolated metalwork	–	–	–	–	–	m²	**26.41**
22 mm thick (one and a half hour) fire protection							
to walls and columns	–	–	–	–	–	m²	**13.97**
to ceilings and beams	–	–	–	–	–	m²	**15.49**
to isolated metalwork	–	–	–	–	–	m²	**30.99**
28 mm thick (two hour) fire protection							
to walls and columns	–	–	–	–	–	m²	**16.38**
to ceilings and beams	–	–	–	–	–	m²	**17.89**
to isolated metalwork	–	–	–	–	–	m²	**35.76**
52 mm thick (four hour) fire protection							
to walls and columns	–	–	–	–	–	m²	**24.76**
to ceilings and beams	–	–	–	–	–	m²	**27.57**
to isolated metalwork	–	–	–	–	–	m²	**54.87**
Prepare and apply by spray; cementitious Pyrok WF26 render or other equal and approved; on expanded metal lathing (not included)							
15 mm thick							
to ceilings and beams	–	–	–	–	–	m²	**38.01**
28.16 TIMBER FLOOR AND WALL LININGS							
Wrought softwood							
Boarding to internal walls; tongued and grooved and V-jointed							
12 mm × 100 mm boards	–	0.74	17.01	–	17.24	m²	**34.25**
16 mm × 100 mm boards	–	0.74	17.01	–	18.65	m²	**35.66**
19 mm × 100 mm boards	–	0.74	17.01	–	21.14	m²	**38.15**
19 mm × 125 mm boards	–	0.69	15.86	–	19.29	m²	**35.15**
19 mm × 125 mm boards; chevron pattern	–	1.11	25.51	–	19.29	m²	**44.80**
25 mm × 125 mm boards	–	0.69	15.86	–	16.59	m²	**32.45**
12 mm × 100 mm boards; knotty pine	–	0.74	17.01	–	10.61	m²	**27.62**
Wood strip; 22 mm thick; Junckers All in Beech Sylva Sport Premium pretreated or other equal and approved; tongued and grooved joints; on bearers etc.; level fixing to cement and sand base							
Strip flooring; over 300 mm wide							
on 45 × 45 mm blue bat bearers	–	–	–	–	–	m²	**75.18**
on 10 mm Pro Foam	–	–	–	–	–	m²	**85.87**
on Uno bat 50 mm bearers	–	–	–	–	–	m²	**78.64**

28 FLOOR, WALL, CEILING AND ROOF FINISHES

Item	PC £	Labour hours	Labour £	Plant £	Material £	Unit	Total rate £
28.16 TIMBER FLOOR AND WALL LININGS – cont							
Wood strip – cont							
Strip flooring – cont							
on New Era levelling system	–	–	–	–	–	m²	**81.20**
on Uno bat 62 mm bearers	–	–	–	–	–	m²	**79.92**
on Duo bat 110 mm bearers	–	–	–	–	–	m²	**134.13**
Wood strip; 22 mm thick; Junckers pretreated or other equal and approved flooring systems; tongued and grooved joints; on bearers etc.; level fixing to cement and sand base							
Strip flooring; over 300 mm wide							
Sylva Squash Beech untreated on blue bat bearers	–	–	–	–	–	m²	**78.82**
Classic Beech clip system	–	–	–	–	–	m²	**104.43**
Harmoni Oak clip system	–	–	–	–	–	m²	**109.35**
Classic Beech on blue bat bearers	–	–	–	–	–	m²	**105.54**
Harmoni Oak on blue bat bearers	–	–	–	–	–	m²	**108.07**
Unfinished wood strip; 22 mm thick; Havwoods or other equal and approved; tongued and grooved joints; secret fixed; laid on semi-sprung bearers; fixing to cement and sand base; sanded and sealed							
Strip flooring; over 300 mm wide							
Prime Iroko	–	–	–	–	–	m²	**101.71**
Prime Maple	–	–	–	–	–	m²	**101.71**
American Oak	–	–	–	–	–	m²	**111.49**

29 DECORATION

Item	PC £	Labour hours	Labour £	Plant £	Material £	Unit	Total rate £
29.01 SUPPLY ONLY RATES FOR PAINTS							
BASIC PAINT PRICES – MATERIAL ONLY							
Paints							
matt emulsion	–	–	–	–	17.25	5 litre	**17.25**
gloss	–	–	–	–	29.96	5 litre	**29.96**
Weathershield gloss	–	–	–	–	57.43	5 litre	**57.43**
Weathershield undercoat	–	–	–	–	71.43	5 litre	**71.43**
Sandtex masonry paint – brilliant white	–	–	–	–	16.48	5 litre	**16.48**
Sandtex masonry paint – coloured	–	–	–	–	29.29	5 litre	**29.29**
Primer/undercoats							
oil-based undercoat	–	–	–	–	26.96	5 litre	**26.96**
acrylic	–	–	–	–	31.08	5 litre	**31.08**
red oxide	–	–	–	–	33.89	5 litre	**33.89**
water-based	–	–	–	–	27.71	5 litre	**27.71**
zinc phosphate	–	–	–	–	52.56	5 litre	**52.56**
MDF primer	–	–	–	–	63.72	5 litre	**63.72**
knotting solution	–	–	–	–	100.80	5 litre	**100.80**
Special paints							
solar reflective aluminium	–	–	–	–	54.70	5 litre	**54.70**
anti-graffiti	–	–	–	–	184.07	5 litre	**184.07**
bituminous emulsion	–	–	–	–	19.60	5 litre	**19.60**
Hammerite	–	–	–	–	68.55	5 litre	**68.55**
fire retardant – undercoat	–	–	–	–	86.00	5 litre	**86.00**
fire retardent – top coat	–	–	–	–	74.18	5 litre	**74.18**
Stains and preservatives							
Cuprinaol – clear	–	–	–	–	43.15	5 litre	**43.15**
Cuprinaol – boiled linseed oil	–	–	–	–	47.98	5 litre	**47.98**
Sadolin – Extra	–	–	–	–	83.61	5 litre	**83.61**
Sadolin – New Base	–	–	–	–	28.90	5 litre	**28.90**
Sikkens – Cetol HLS	–	–	–	–	68.96	5 litre	**68.96**
Sikkens – Cetol TS	–	–	–	–	99.21	5 litre	**99.21**
Sikkens – Cetol Filter 7	–	–	–	–	104.93	5 litre	**104.93**
Protim Solignum – Architectural	–	–	–	–	77.25	5 litre	**77.25**
Protim Solignum – Green	–	–	–	–	77.25	5 litre	**77.25**
Protim Solignum – Cedar	–	–	–	–	77.25	5 litre	**77.25**
Varnishes							
polyurethane	–	–	–	–	50.09	5 litre	**50.09**

29 DECORATION

Item	PC £	Labour hours	Labour £	Plant £	Material £	Unit	Total rate £
29.02 PAINTING AND CLEAR FINISHES – EXTERNAL							
NOTE: The following prices include for preparing surfaces. Painting woodwork also includes for knotting prior to applying the priming coat and for all stopping of nail holes etc.							
Two coats of cement paint, Sandtex Matt or other equal and approved							
Brick or block walls							
over 300 mm girth	–	0.26	4.52	–	2.11	m²	**6.63**
Cement render or concrete walls							
over 300 mm girth	–	0.23	4.00	–	1.39	m²	**5.39**
Roughcast walls							
over 300 mm girth	–	0.40	6.96	–	1.39	m²	**8.35**
One coat sealer and two coats of external grade emulsion paint, Dulux Weathershield or other equal and approved							
Brick or block walls							
over 300 mm girth	–	0.43	7.48	–	10.00	m²	**17.48**
Cement render or concrete walls							
over 300 mm girth	–	0.35	6.09	–	6.66	m²	**12.75**
Concrete soffits							
over 300 mm girth	–	0.40	6.96	–	6.66	m²	**13.62**
One coat sealer (applied by brush) and two coats of external grade emulsion paint, Dulux Weathershield or other equal and approved (spray applied)							
Roughcast							
over 300 mm girth	–	0.29	5.05	–	13.55	m²	**18.60**
One coat sealer and two coats of anti-graffiti paint (spray applied)							
Brick or block walls							
over 300 mm girth	–	0.01	0.09	0.48	5.33	m²	**5.90**
Cement render or concrete walls							
over 300 mm girth	–	0.01	0.09	0.48	6.43	m²	**7.00**
2.5 mm of Vandalene or similar anti-climb paint (spray applied)							
General surfaces							
over 300 mm girth	–	0.01	0.09	0.48	5.25	m²	**5.82**
Two coats solar reflective aluminium paint; on bituminous roofing							
General surfaces							
over 300 mm girth	–	0.44	7.65	–	18.50	m²	**26.15**

29 DECORATION

Item	PC £	Labour hours	Labour £	Plant £	Material £	Unit	Total rate £
Touch up primer; two undercoats and one finishing coat of gloss oil paint; on wood surfaces							
General surfaces							
over 300 mm girth	–	0.35	6.09	–	2.71	m²	**8.80**
isolated surfaces not exceeding 300 mm girth	–	0.15	2.61	–	0.73	m	**3.34**
isolated areas not exceeding 1.00 m²; irrespective of girth	–	0.27	4.70	–	1.46	nr	**6.16**
Glazed windows and screens							
panes; area not exceeding 0.10 m²	–	0.59	10.27	–	2.40	m²	**12.67**
panes; area 0.10 m²–0.50 m²	–	0.59	10.27	–	2.03	m²	**12.30**
panes; area 0.50 m²–1.00 m²	–	0.47	8.18	–	1.77	m²	**9.95**
panes; area over 1.00 m²	–	0.35	6.09	–	1.46	m²	**7.55**
Glazed windows and screens; multicoloured work							
panes; area not exceeding 0.10 m²	–	0.68	11.83	–	2.40	m²	**14.23**
panes; area 0.10 m²–0.50 m²	–	0.55	9.57	–	2.08	m²	**11.65**
panes; area 0.50 m²–1.00 m²	–	0.47	8.18	–	1.77	m²	**9.95**
panes; area over 1.00 m²	–	0.41	7.13	–	1.46	m²	**8.59**
Knot; one coat primer; two undercoats and one finishing coat of gloss oil paint; on wood surfaces							
General surfaces							
over 300 mm girth	–	0.46	8.00	–	2.93	m²	**10.93**
isolated surfaces not exceeding 300 mm girth	–	0.19	3.31	–	4.77	m	**8.08**
isolated areas not exceeding 1.00 m²; irrespective of girth	–	0.35	6.09	–	1.98	nr	**8.07**
Glazed windows and screens							
panes; area not exceeding 0.10 m²	–	0.78	13.57	–	3.27	m²	**16.84**
panes; area 0.10 m²–0.50 m²	–	0.62	10.78	–	2.90	m²	**13.68**
panes; area 0.50 m²–1.00 m²	–	0.55	9.57	–	2.24	m²	**11.81**
panes; area over 1.00 m²	–	0.46	8.00	–	1.57	m²	**9.57**
Glazed windows and screens; multicoloured work							
panes; area not exceeding 0.10 m²	–	0.89	15.48	–	3.27	m²	**18.75**
panes; area 0.10 m²–0.50 m²	–	0.72	12.52	–	2.93	m²	**15.45**
panes; area 0.50 m²–1.00 m²	–	0.64	11.13	–	2.24	m²	**13.37**
panes; area over 1.00 m²	–	0.54	9.39	–	1.57	m²	**10.96**
Touch up primer; two undercoats and one finishing coat of gloss oil paint; on iron or steel surfaces							
General surfaces							
over 300 mm girth	–	0.35	6.09	–	2.18	m²	**8.27**
isolated surfaces not exceeding 300 mm girth	–	0.14	2.43	–	0.60	m	**3.03**
isolated areas not exceeding 1.00 m²; irrespective of girth	–	0.26	4.52	–	1.24	nr	**5.76**
Glazed windows and screens							
panes; area not exceeding 0.10 m²	–	0.59	10.27	–	2.25	m²	**12.52**
panes; area 0.10 m²–0.50 m²	–	0.47	8.18	–	1.95	m²	**10.13**
panes; area 0.50 m²–1.00 m²	–	0.41	7.13	–	1.65	m²	**8.78**
panes; area over 1.00 m²	–	0.35	6.09	–	1.34	m²	**7.43**

29 DECORATION

Item	PC £	Labour hours	Labour £	Plant £	Material £	Unit	Total rate £
29.02 PAINTING AND CLEAR FINISHES – EXTERNAL – cont							
Touch up primer – cont							
Structural steelwork							
over 300 mm girth	–	0.40	6.96	–	2.29	m²	**9.25**
Members of roof trusses							
over 300 mm girth	–	0.54	9.39	–	2.60	m²	**11.99**
Ornamental railings and the like; each side measured overall							
over 300 mm girth	–	0.60	10.43	–	2.68	m²	**13.11**
Eaves gutters							
over 300 mm girth	–	0.64	11.13	–	2.97	m²	**14.10**
not exceeding 300 mm girth	–	0.25	4.35	–	1.24	m	**5.59**
Pipes or conduits							
over 300 mm girth	–	0.54	9.39	–	2.97	m²	**12.36**
not exceeding 300 mm girth	–	0.21	3.66	–	1.00	m	**4.66**
One coat primer; two undercoats and one finishing coat of gloss oil paint; on iron or steel surfaces							
General surfaces							
over 300 mm girth	–	0.43	7.48	–	2.52	m²	**10.00**
isolated surfaces not exceeding 300 mm girth	–	0.18	3.13	–	0.66	m	**3.79**
isolated areas not exceeding 1.00 m²; irrespective of girth	–	0.32	5.56	–	1.32	nr	**6.88**
Glazed windows and screens							
panes; area not exceeding 0.10 m²	–	0.71	12.35	–	2.34	m²	**14.69**
panes; area 0.10 m²–0.50 m²	–	0.56	9.74	–	2.04	m²	**11.78**
panes; area 0.50 m²–1.00 m²	–	0.50	8.70	–	1.73	m²	**10.43**
panes; area over 1.00 m²	–	0.43	7.48	–	1.32	m²	**8.80**
Structural steelwork							
over 300 mm girth	–	0.48	8.35	–	2.64	m²	**10.99**
Members of roof trusses							
over 300 mm girth	–	0.64	11.13	–	2.94	m²	**14.07**
Ornamental railings and the like; each side measured overall							
over 300 mm girth	–	0.72	12.52	–	2.94	m²	**15.46**
Eaves gutters							
over 300 mm girth	–	0.76	13.23	–	3.37	m²	**16.60**
not exceeding 300 mm girth	–	0.31	5.40	–	1.16	m	**6.56**
Pipes or conduits							
over 300 mm girth	–	0.64	11.13	–	3.37	m²	**14.50**
not exceeding 300 mm girth	–	0.25	4.35	–	4.84	m	**9.19**
One coat of Andrews Hammerite paint or other equal and approved; on iron or steel surfaces							
General surfaces							
over 300 mm girth	–	0.15	2.61	–	2.15	m²	**4.76**
isolated surfaces not exceeding 300 mm girth	–	0.08	1.39	–	0.67	m	**2.06**
isolated areas not exceeding 1.00 m²; irrespective of girth	–	0.11	1.92	–	1.24	nr	**3.16**

29 DECORATION

Item	PC £	Labour hours	Labour £	Plant £	Material £	Unit	Total rate £
Glazed windows and screens							
panes; area not exceeding 0.10 m²	–	0.25	4.35	–	1.61	m²	**5.96**
panes; area 0.10 m²–0.50 m²	–	0.19	3.31	–	1.80	m²	**5.11**
panes; area 0.50 m²–1.00 m²	–	0.18	3.13	–	1.63	m²	**4.76**
panes; area over 1.00 m²	–	0.15	2.61	–	1.63	m²	**4.24**
Structural steelwork							
over 300 mm girth	–	0.17	2.96	–	1.98	m²	**4.94**
Members of roof trusses							
over 300 mm girth	–	0.23	4.00	–	2.15	m²	**6.15**
Ornamental railings and the like; each side measured overall							
over 300 mm girth	–	0.26	4.52	–	2.15	m²	**6.67**
Eaves gutters							
over 300 mm girth	–	0.27	4.70	–	2.33	m²	**7.03**
not exceeding 300 mm girth	–	0.08	1.39	–	1.13	m	**2.52**
Pipes or conduits							
over 300 mm girth	–	0.26	4.52	–	1.98	m²	**6.50**
not exceeding 300 mm girth	–	0.08	1.39	–	0.93	m	**2.32**
Two coats of creosote; on wood surfaces							
General surfaces							
over 300 mm girth	–	0.16	2.78	–	0.56	m²	**3.34**
isolated surfaces not exceeding 300 mm girth	–	0.01	0.09	–	0.34	m	**0.43**
Two coats of Solignum wood preservative or other equal and approved; on wood surfaces							
General surfaces							
over 300 mm girth	–	0.14	2.43	–	4.01	m²	**6.44**
isolated surfaces not exceeding 300 mm girth	–	0.05	0.87	–	1.16	m	**2.03**
Three coats of polyurethane; on wood surfaces							
General surfaces							
over 300 mm girth	–	0.29	5.05	–	3.13	m²	**8.18**
isolated surfaces not exceeding 300 mm girth	–	0.11	1.92	–	1.56	m	**3.48**
isolated areas not exceeding 1.00 m²; irrespective of girth	–	0.21	3.66	–	1.80	nr	**5.46**
Two coats of New Base primer or other equal and approved; and two coats of Extra or other equal and approved; Sadolin Ltd; pigmented; on wood surfaces							
General surfaces							
over 300 mm girth	–	0.43	7.48	–	5.51	m²	**12.99**
isolated surfaces not exceeding 300 mm girth	–	0.13	2.27	–	0.96	m	**3.23**
Glazed windows and screens							
panes; area not exceeding 0.10 m²	–	0.71	12.35	–	3.93	m²	**16.28**
panes; area 0.10 m²–0.50 m²	–	0.57	9.92	–	3.71	m²	**13.63**
panes; area 0.50 m²–1.00 m²	–	0.50	8.70	–	3.49	m²	**12.19**
panes; area over 1.00 m²	–	0.43	7.48	–	2.81	m²	**10.29**

Prices for Measured Works

29 DECORATION

Item	PC £	Labour hours	Labour £	Plant £	Material £	Unit	Total rate £
29.02 PAINTING AND CLEAR FINISHES – EXTERNAL – cont							
Two coats Sikkens Cetol Filter 7 exterior stain or other equal and approved; on wood surfaces							
General surfaces							
over 300 mm girth	–	0.20	3.48	–	5.27	m²	**8.75**
isolated surfaces not exceeding 300 mm girth	–	0.09	1.57	–	1.83	m	**3.40**
isolated areas not exceeding 1.00 m²; irrespective of girth	–	0.14	2.43	–	2.69	nr	**5.12**
29.03 PAINTING AND CLEAR FINISHES – INTERNAL							
NOTE: The following prices include for preparing surfaces. Painting woodwork also includes for knotting prior to applying the priming coat and for all stopping of nail holes etc.							
Touch up primer; two undercoats and one finishing coat of gloss oil paint; on wood surfaces							
General surfaces							
over 300 mm girth	–	0.35	6.09	–	2.71	m²	**8.80**
isolated surfaces not exceeding 300 mm girth	–	0.15	2.61	–	0.73	m	**3.34**
isolated areas not exceeding 1.00 m²; irrespective of girth	–	0.27	4.70	–	1.46	nr	**6.16**
One coat primer; on wood surfaces before fixing							
General surfaces							
over 300 mm girth	–	0.08	1.39	–	1.21	m²	**2.60**
isolated surfaces not exceeding 300 mm girth	–	0.02	0.35	–	0.44	m	**0.79**
isolated areas not exceeding 1.00 m²; irrespective of girth	–	0.06	1.04	–	0.45	nr	**1.49**
One coat polyurethane sealer; on wood surfaces before fixing							
General surfaces							
over 300 mm girth	–	0.10	1.74	–	1.08	m²	**2.82**
isolated surfaces not exceeding 300 mm girth	–	0.03	0.53	–	0.39	m	**0.92**
isolated areas not exceeding 1.00 m²; irrespective of girth	–	0.08	1.39	–	0.50	nr	**1.89**
One coat of Sikkens Cetol HLS stain or other equal and approved; on wood surfaces before fixing							
General surfaces							
over 300 mm girth	–	0.11	1.92	–	1.30	m²	**3.22**
isolated surfaces not exceeding 300 mm girth	–	0.03	0.53	–	0.93	m	**1.46**
isolated areas not exceeding 1.00 m²; irrespective of girth	–	0.08	1.39	–	0.62	nr	**2.01**

29 DECORATION

Item	PC £	Labour hours	Labour £	Plant £	Material £	Unit	Total rate £
One coat of Sikkens Cetol TS interior stain or other equal and approved; on wood surfaces before fixing							
General surfaces							
over 300 mm girth	–	0.11	1.92	–	1.79	m²	**3.71**
isolated surfaces not exceeding 300 mm girth	–	0.03	0.53	–	0.84	m	**1.37**
isolated areas not exceeding 1.00 m²; irrespective of girth	–	0.08	1.39	–	0.87	nr	**2.26**
One coat Cuprinol clear wood preservative or other equal and approved; on wood surfaces before fixing							
General surfaces							
over 300 mm girth	–	0.08	1.39	–	1.06	m²	**2.45**
isolated surfaces not exceeding 300 mm girth	–	0.02	0.35	–	0.38	m	**0.73**
isolated areas not exceeding 1.00 m²; irrespective of girth	–	0.05	0.87	–	0.49	nr	**1.36**
One coat HCC Protective Coatings Ltd Permacor urethane alkyd gloss finishing coat or other equal and approved; on previously primed steelwork							
Members of roof trusses							
over 300 mm girth	–	0.15	2.61	–	0.59	m²	**3.20**
Two coats emulsion paint							
Brick or block walls							
over 300 mm girth	–	0.21	3.66	–	0.99	m²	**4.65**
Cement render or concrete							
over 300 mm girth	–	0.20	3.48	–	0.89	m²	**4.37**
isolated surfaces not exceeding 300 mm girth	–	0.10	1.74	–	0.31	m	**2.05**
Plaster walls or plaster/plasterboard ceilings							
over 300 mm girth	–	0.18	3.13	–	0.87	m²	**4.00**
over 300 mm girth; in multi colours	–	0.24	4.17	–	1.02	m²	**5.19**
over 300 mm girth; in staircase areas	–	0.21	3.66	–	0.98	m²	**4.64**
cutting in edges on flush surfaces	–	0.08	1.39	–	–	m	**1.39**
Plaster/plasterboard ceilings							
over 300 mm girth; 3.50 m–5.00 m high	–	0.21	3.66	–	0.88	m²	**4.54**
One mist and two coats emulsion paint							
Brick or block walls							
over 300 mm girth	–	0.19	3.31	–	1.29	m²	**4.60**
Cement render or concrete							
over 300 mm girth	–	0.19	3.31	–	1.19	m²	**4.50**
Plaster walls or plaster/plasterboard ceilings							
over 300 mm girth	–	0.18	3.13	–	1.19	m²	**4.32**
over 300 mm girth; in multi colours	–	0.25	4.35	–	1.22	m²	**5.57**
over 300 mm girth; in staircase areas	–	0.21	3.66	–	1.19	m²	**4.85**
cutting in edges on flush surfaces	–	0.09	1.57	–	–	m	**1.57**
Plaster/plasterboard ceilings							
over 300 mm girth; 3.50 m–5.00 m high	–	0.21	3.66	–	1.19	m²	**4.85**

Prices for Measured Works

29 DECORATION

Item	PC £	Labour hours	Labour £	Plant £	Material £	Unit	Total rate £
29.03 PAINTING AND CLEAR FINISHES – INTERNAL – cont							
One mist Supermatt; one full Supermatt and one full coat of quick drying Acrylic Eggshell							
Brick or block walls							
over 300 mm girth	–	0.19	3.31	–	2.04	m²	**5.35**
Cement render or concrete							
over 300 mm girth	–	0.19	3.31	–	1.88	m²	**5.19**
Plaster walls or plaster/plasterboard ceilings							
over 300 mm girth	–	0.18	3.13	–	1.88	m²	**5.01**
over 300 mm girth; in multi colours	–	0.25	4.35	–	1.88	m²	**6.23**
over 300 mm girth; in staircase areas	–	0.21	3.66	–	1.88	m²	**5.54**
cutting in edges on flush surfaces	–	0.09	1.57	–	–	m	**1.57**
Plaster/plasterboard ceilings							
over 300 mm girth; 3.50 m–5.00 m high	–	0.21	3.66	–	1.88	m²	**5.54**
Clean and prepare surfaces, one coat Keim dilution, one coat primer and two coats of Keim Ecosil paint							
Brick or block walls							
over 300 mm girth	–	0.30	5.22	–	5.81	m²	**11.03**
Cement render or concrete							
over 300 mm girth	–	0.30	5.22	–	5.09	m²	**10.31**
Plaster walls or plaster/plasterboard ceilings							
over 300 mm girth	–	0.30	5.22	–	5.09	m²	**10.31**
over 300 mm girth; in staircase areas	–	0.30	5.22	–	5.09	m²	**10.31**
cutting in edges on flush surfaces	–	0.09	1.57	–	–	m	**1.57**
Plaster/plasterboard ceilings							
over 300 mm girth; 3.50 m–5.00 m high	–	0.25	4.35	–	5.29	m²	**9.64**
One coat Tretol No 10 Sealer or other equal and approved; two coats Tretol sprayed Supercover Spraytone emulsion paint or other equal and approved							
Plaster walls or plaster/plasterboard ceilings							
over 300 mm girth	–	0.25	6.10	–	0.26	m²	**6.36**
Textured plastic; Artex or other equal and approved finish							
Plasterboard ceilings							
over 300 mm girth	–	0.19	3.31	–	3.67	m²	**6.98**
Concrete walls or ceilings							
over 300 mm girth	–	0.23	4.00	–	3.42	m²	**7.42**

29 DECORATION

Item	PC £	Labour hours	Labour £	Plant £	Material £	Unit	Total rate £
Touch up primer; one undercoat and one finishing coat of gloss oil paint; on wood surfaces							
General surfaces							
over 300 mm girth	–	0.20	3.48	–	2.63	m²	**6.11**
isolated surfaces not exceeding 300 mm girth	–	0.08	1.39	–	0.92	m	**2.31**
isolated areas not exceeding 1.00 m²; irrespective of girth	–	0.18	3.13	–	1.43	nr	**4.56**
Glazed windows and screens							
panes; area not exceeding 0.10 m²	–	0.38	6.61	–	2.13	m²	**8.74**
panes; area 0.10 m²–0.50 m²	–	0.31	5.40	–	1.63	m²	**7.03**
panes; area 0.50 m²–1.00 m²	–	0.26	4.52	–	1.31	m²	**5.83**
panes; area over 1.00 m²	–	0.23	4.00	–	1.12	m²	**5.12**
Knot; one coat primer; stop; one undercoat and one finishing coat of gloss oil paint; on wood surfaces							
General surfaces							
over 300 mm girth	–	0.33	5.74	–	2.70	m²	**8.44**
isolated surfaces not exceeding 300 mm girth	–	0.13	2.27	–	0.92	m	**3.19**
isolated areas not exceeding 0.50 m²; irrespective of girth	–	0.25	4.35	–	1.78	nr	**6.13**
Glazed windows and screens							
panes; area not exceeding 0.10 m²	–	0.56	9.74	–	2.70	m²	**12.44**
panes; area 0.10 m²–0.50 m²	–	0.45	7.83	–	2.26	m²	**10.09**
panes; area 0.50 m²–1.00 m²	–	0.40	6.96	–	2.26	m²	**9.22**
panes; area over 1.00 m²	–	0.33	5.74	–	1.67	m²	**7.41**
One coat primer; one undercoat and one finishing coat of gloss oil paint							
Plaster surfaces							
over 300 mm girth	–	0.30	5.22	–	3.44	m²	**8.66**
One coat primer; two undercoats and one finishing coat of gloss oil paint							
Plaster surfaces							
over 300 mm girth	–	0.40	6.96	–	4.55	m²	**11.51**
One coat primer; two undercoats and one finishing coat of eggshell paint							
Plaster surfaces							
over 300 mm girth	–	0.40	6.96	–	4.90	m²	**11.86**
Touch up primer; one undercoat and one finishing coat of gloss paint; on iron or steel surfaces							
General surfaces							
over 300 mm girth	–	0.23	4.00	–	2.02	m²	**6.02**
isolated surfaces not exceeding 300 mm girth	–	0.09	1.57	–	0.70	m	**2.27**
isolated areas not exceeding 0.50 m²; irrespective of girth	–	0.18	3.13	–	1.12	nr	**4.25**

29 DECORATION

Item	PC £	Labour hours	Labour £	Plant £	Material £	Unit	Total rate £
29.03 PAINTING AND CLEAR FINISHES – INTERNAL – cont							
Touch up primer – cont							
Glazed windows and screens							
panes; area not exceeding 0.10 m²	–	0.38	6.61	–	2.10	m²	**8.71**
panes; area 0.10 m²–0.50 m²	–	0.31	5.40	–	1.64	m²	**7.04**
panes; area 0.50 m²–1.00 m²	–	0.26	4.52	–	1.27	m²	**5.79**
panes; area over 1.00 m²	–	0.23	4.00	–	1.07	m²	**5.07**
Structural steelwork							
over 300 mm girth	–	0.25	4.35	–	2.13	m²	**6.48**
Members of roof trusses							
over 300 mm girth	–	0.34	5.91	–	2.42	m²	**8.33**
Ornamental railings and the like; each side measured overall							
over 300 mm girth	–	0.40	6.96	–	2.67	m²	**9.63**
Iron or steel radiators							
over 300 mm girth	–	0.23	4.00	–	2.22	m²	**6.22**
Pipes or conduits							
over 300 mm girth	–	0.34	5.91	–	2.33	m²	**8.24**
not exceeding 300 mm girth	–	0.13	2.27	–	0.78	m	**3.05**
One coat primer; one undercoat and one finishing coat of gloss oil paint; on iron or steel surfaces							
General surfaces							
over 300 mm girth	–	0.30	5.22	–	2.04	m²	**7.26**
isolated surfaces not exceeding 300 mm girth	–	0.12	2.09	–	1.17	m	**3.26**
isolated areas not exceeding 0.50 m²; irrespective of girth	–	0.23	4.00	–	1.98	nr	**5.98**
Glazed windows and screens							
panes; area not exceeding 0.10 m²	–	0.50	8.70	–	3.21	m²	**11.91**
panes; area 0.10 m²–0.50 m²	–	0.40	6.96	–	2.58	m²	**9.54**
panes; area 0.50 m²–1.00 m²	–	0.34	5.91	–	2.20	m²	**8.11**
panes; area over 1.00 m²	–	0.30	5.22	–	1.98	m²	**7.20**
Structural steelwork							
over 300 mm girth	–	0.33	5.74	–	3.14	m²	**8.88**
Members of roof trusses							
over 300 mm girth	–	0.45	7.83	–	3.33	m²	**11.16**
Ornamental railings and the like; each side measured overall							
over 300 mm girth	–	0.51	8.87	–	3.94	m²	**12.81**
Iron or steel radiators							
over 300 mm girth	–	0.30	5.22	–	3.33	m²	**8.55**
Pipes or conduits							
over 300 mm girth	–	0.45	7.83	–	3.33	m²	**11.16**
not exceeding 300 mm girth	–	0.18	3.13	–	1.10	m	**4.23**

29 DECORATION

Item	PC £	Labour hours	Labour £	Plant £	Material £	Unit	Total rate £
Two coats of bituminous paint; on iron or steel surfaces							
General surfaces							
over 300 mm girth	–	0.23	4.00	–	1.03	m²	**5.03**
Inside of galvanized steel cistern							
over 300 mm girth	–	0.34	5.91	–	1.23	m²	**7.14**
Two coats bituminous paint; first coat blinded with clean sand prior to second coat; on concrete surfaces							
General surfaces							
over 300 mm girth	–	0.79	13.74	–	3.04	m²	**16.78**
Mordant solution; one coat HCC Protective Coatings Ltd Permacor Alkyd MIO or other equal and approved; one coat Permatex Epoxy Gloss finishing coat or other equal and approved on galvanized steelwork							
Structural steelwork							
over 300 mm girth	–	0.44	7.65	–	3.88	m²	**11.53**
One coat HCC Protective Coatings Ltd Epoxy Zinc Primer or other equal and approved; two coats Permacor Alkyd MIO or other equal and approved; one coat Permacor Epoxy Gloss finishing coat or other equal and approved on steelwork							
Structural steelwork							
over 300 mm girth	–	0.63	10.96	–	7.30	m²	**18.26**
Steel protection; HCC Protective Coatings Ltd Unitherm or other equal and approved; two coats to steelwork							
Structural steelwork							
over 300 mm girth	–	0.99	17.22	–	2.61	m²	**19.83**
Two coats of epoxy anti-slip floor paint; on screeded concrete surfaces							
General surfaces							
over 300 mm girth	–	0.25	4.35	–	17.14	m²	**21.49**
Nitoflor Lithurin floor hardener and dust proofer or other equal and approved; Fosroc Expandite Ltd; two coats; on concrete surfaces							
General surfaces							
over 300 mm girth	–	0.24	3.36	–	0.47	m²	**3.83**

29 DECORATION

Item	PC £	Labour hours	Labour £	Plant £	Material £	Unit	Total rate £
29.03 PAINTING AND CLEAR FINISHES – INTERNAL – cont							
Two coats of boiled linseed oil; on hardwood surfaces							
General surfaces							
over 300 mm girth	–	0.18	3.13	–	3.79	m²	**6.92**
isolated surfaces not exceeding 300 mm girth	–	0.07	1.22	–	1.22	m	**2.44**
isolated areas not exceeding 0.50 m²; irrespective of girth	–	0.13	2.27	–	2.20	nr	**4.47**
Two coats polyurethane varnish; on wood surfaces							
General surfaces							
over 300 mm girth	–	0.18	3.13	–	1.92	m²	**5.05**
isolated surfaces not exceeding 300 mm girth	–	0.07	1.22	–	0.71	m	**1.93**
isolated areas not exceeding 0.50 m²; irrespective of girth	–	0.13	2.27	–	0.26	nr	**2.53**
Three coats polyurethane varnish; on wood surfaces							
General surfaces							
over 300 mm girth	–	0.26	4.52	–	2.87	m²	**7.39**
isolated surfaces not exceeding 300 mm girth	–	0.10	1.74	–	0.92	m	**2.66**
isolated areas not exceeding 0.50 m²; irrespective of girth	–	0.19	3.31	–	1.62	nr	**4.93**
One undercoat; and one finishing coat; of Albi clear flame retardant surface coating or other equal and approved; on wood surfaces							
General surfaces							
over 300 mm girth	–	0.34	5.91	–	6.38	m²	**12.29**
isolated surfaces not exceeding 300 mm girth	–	0.14	2.43	–	2.21	m	**4.64**
isolated areas not exceeding 0.50 m²; irrespective of girth	–	0.19	3.31	–	4.85	nr	**8.16**
Two undercoats; and one finishing coat; of Albi clear flame retardant surface coating or other equal and approved; on wood surfaces							
General surfaces							
over 300 mm girth	–	0.40	6.96	–	8.06	m²	**15.02**
isolated surfaces not exceeding 300 mm girth	–	0.20	3.48	–	3.24	m	**6.72**
isolated areas not exceeding 0.50 m²; irrespective of girth	–	0.33	5.74	–	4.31	nr	**10.05**
Seal and wax polish; dull gloss finish on wood surfaces							
General surfaces							
over 300 mm girth	–	0.50	12.20	–	2.47	m²	**14.67**
isolated surfaces not exceeding 300 mm girth	–	0.25	6.10	–	0.47	m	**6.57**
isolated areas not exceeding 0.50 m²; irrespective of girth	–	0.35	8.54	–	1.17	nr	**9.71**

29 DECORATION

Item	PC £	Labour hours	Labour £	Plant £	Material £	Unit	Total rate £
One coat of Sadolin Extra or other equal and approved; clear or pigmented; one further coat of Holdex clear interior silk matt lacquer or similar							
General surfaces							
over 300 mm girth	–	0.25	4.35	–	6.83	m²	**11.18**
isolated surfaces not exceeding 300 mm girth	–	0.10	1.74	–	3.19	m	**4.93**
isolated areas not exceeding 0.50 m²; irrespective of girth	–	0.20	3.48	–	3.32	nr	**6.80**
Glazed windows and screens							
panes; area not exceeding 0.10 m²	–	0.42	7.30	–	3.89	m²	**11.19**
panes; area 0.10 m²–0.50 m²	–	0.33	5.74	–	3.64	m²	**9.38**
panes; area 0.50 m²–1.00 m²	–	0.29	5.05	–	3.37	m²	**8.42**
panes; area over 1.00 m²	–	0.25	4.35	–	3.19	m²	**7.54**
Two coats of Sadolin Extra or other equal and approved; clear or pigmented; two further coats of PV67 clear interior silk matt lacquer or similar							
General surfaces							
over 300 mm girth	–	0.40	6.96	–	12.55	m²	**19.51**
isolated surfaces not exceeding 300 mm girth	–	0.16	2.78	–	6.27	m	**9.05**
isolated areas not exceeding 0.50 m²; irrespective of girth	–	0.30	5.22	–	7.16	nr	**12.38**
Glazed windows and screens							
panes; area not exceeding 0.10 m²	–	0.66	11.48	–	7.68	m²	**19.16**
panes; area 0.10 m²–0.50 m²	–	0.52	9.04	–	7.16	m²	**16.20**
panes; area 0.50 m²–1.00 m²	–	0.45	7.83	–	6.62	m²	**14.45**
panes; area over 1.00 m²	–	0.40	6.96	–	6.27	m²	**13.23**
Two coats of Sikkens Cetol TS interior stain or other equal and approved; on wood surfaces							
General surfaces							
over 300 mm girth	–	0.19	3.31	–	3.16	m²	**6.47**
isolated surfaces not exceeding 300 mm girth	–	0.08	1.39	–	1.13	m	**2.52**
isolated areas not exceeding 0.50 m²; irrespective of girth	–	0.13	2.27	–	1.74	nr	**4.01**
Body in and wax polish; dull gloss finish; on hardwood surfaces							
General surfaces							
over 300 mm girth	–	0.60	14.64	–	1.17	m²	**15.81**
isolated surfaces not exceeding 300 mm girth	–	0.30	7.31	–	0.56	m	**7.87**
isolated areas not exceeding 0.50 m²; irrespective of girth	–	0.40	9.75	–	0.56	nr	**10.31**
Stain; body in and wax polish; dull gloss finish; on hardwood surfaces							
General surfaces							
over 300 mm girth	–	0.75	18.29	–	1.17	m²	**19.46**
isolated surfaces not exceeding 300 mm girth	–	0.30	7.31	–	1.17	m	**8.48**
isolated areas not exceeding 0.50 m²; irrespective of girth	–	0.50	12.20	–	1.17	nr	**13.37**

29 DECORATION

Item	PC £	Labour hours	Labour £	Plant £	Material £	Unit	Total rate £
29.03 PAINTING AND CLEAR FINISHES – INTERNAL – cont							
Seal; two coats of synthetic resin lacquer; decorative flatted finish; wire down, wax and burnish; on wood surfaces							
General surfaces							
over 300 mm girth	–	1.00	24.39	–	1.14	m²	**25.53**
isolated surfaces not exceeding 300 mm girth	–	0.40	9.75	–	0.94	m	**10.69**
isolated areas not exceeding 0.50 m²; irrespective of girth	–	0.65	15.85	–	0.94	nr	**16.79**
Stain; body in and fully French polish; full gloss finish; on hardwood surfaces							
General surfaces							
over 300 mm girth	–	1.10	26.82	–	2.00	m²	**28.82**
isolated surfaces not exceeding 300 mm girth	–	0.50	12.20	–	1.12	m	**13.32**
isolated areas not exceeding 0.50 m²; irrespective of girth	–	0.75	18.29	–	1.94	nr	**20.23**
Stain; fill grain and fully French polish; full gloss finish; on hardwood surfaces							
General surfaces							
over 300 mm girth	–	1.60	39.02	–	3.38	m²	**42.40**
isolated surfaces not exceeding 300 mm girth	–	0.60	14.64	–	1.12	m	**15.76**
isolated areas not exceeding 0.50 m²; irrespective of girth	–	1.00	24.39	–	1.94	nr	**26.33**
Stain black; body in and fully French polish; ebonized finish; on hardwood surfaces							
General surfaces							
over 300 mm girth	–	1.75	42.67	–	3.78	m²	**46.45**
isolated surfaces not exceeding 300 mm girth	–	0.80	19.51	–	1.12	m	**20.63**
isolated areas not exceeding 0.50 m²; irrespective of girth	–	1.30	31.70	–	1.94	nr	**33.64**
29.04 DECORATIVE PAPERS OR FABRICS							
Lining paper; and hanging							
Plaster walls or columns							
over 1.00 m² (PC £ per roll)	2.41	0.19	3.31	–	0.34	m²	**3.65**
Plaster ceilings or beams							
over 1.00 m² girth (PC £ per roll)	2.41	0.23	4.00	–	0.34	m²	**4.34**

29 DECORATION

Item	PC £	Labour hours	Labour £	Plant £	Material £	Unit	**Total rate £**
Decorative paper-backed vinyl wallpaper; and hanging. Because of the enormous range of papers available we have simply allowed a prime cost amount to purchase the paper							
Plaster walls or columns							
over 1.00 m² girth (PC £ per roll)	25.00	0.23	4.00	–	4.38	m²	**8.38**
PVC Wall lining; Altro Whiterock; or other equal and approved; fixed directly to plastered brick or blockwork							
Work to walls over 300 mm wide; over 100 m² total area							
Standard white	35.58	1.25	21.74	–	38.48	m²	**60.22**
Satins	45.01	1.25	21.74	–	48.68	m²	**70.42**
Chameleon	69.60	1.25	21.74	–	75.27	m²	**97.01**

30 SUSPENDED CEILINGS

Item	PC £	Labour hours	Labour £	Plant £	Material £	Unit	Total rate £
30.01 SUSPENDED CEILINGS							
Gyproc M/F suspended ceiling system or other equal approved; hangers plugged and screwed to concrete soffit, 900 mm × 1800 mm × 12.50 mm tapered edge wallboard infill; joints filled with joint filler and taped to receive direct decoration							
Lining to ceilings; hangers 150 mm–500 mm long							
over 600 mm wide	–	–	–	–	–	m²	35.22
not exceeding 600 mm wide in isolated strips	–	–	–	–	–	m	35.70
Edge treatments							
20 × 20 mm SAS perimeter shadow gap; screwed to plasterboard	–	–	–	–	–	m	6.42
20 × 20 mm SAS shadow gap around 450 mm dia. column; including 15 × 44 mm batten plugged and screwed to concrete	–	–	–	–	–	nr	76.15
Vertical bulkhead; including additional hangers							
over 600 mm wide	–	–	–	–	–	m²	44.47
not exceeding 600 mm wide in isolated strips	–	–	–	–	–	m	43.86
Rockfon, or other equal suspended concealed ceiling system; 400 mm to 600 mm long hangers plugged and screwed to concrete soffit.							
Lining to ceilings; 600 mm × 600 mm × 20 mm Sonar suspended ceiling tiles							
over 600 mm wide	–	–	–	–	–	m²	45.08
not exceeding 600 mm wide in isolated strips	–	–	–	–	–	m	26.58
edge trim; shadow-line trim	–	–	–	–	–	m	5.27
Vertical bulkhead, as upstand to rooflight well; including additional hangers; perimeter trim							
300 mm × 600 mm wide	–	–	–	–	–	m	49.35
Ecophon, or other equal; Z demountable suspended concealed ceiling system; 400 mm long hangers plugged and screwed to concrete soffit.							
Lining to ceilings; 600 mm × 600 mm × 20 mm Gedina ET15 suspended ceiling tiles							
over 600 mm wide	–	–	–	–	–	m²	39.12
not exceeding 600 mm wide in isolated strips	–	–	–	–	–	m	24.67
edge trim; shadow-line trim	–	–	–	–	–	m	4.87
Vertical bulkhead, as upstand to rooflight well; including additional hangers; perimeter trim							
300 mm × 600 mm wide	–	–	–	–	–	m	45.42
Lining to ceilings; 600 mm × 600 mm × 20 mm Hygiene Performance washable suspended ceiling tiles							
over 600 mm wide	–	–	–	–	–	m²	52.24
not exceeding 600 mm wide in isolated strips	–	–	–	–	–	m	42.83
edge trim; shadow-line trim	–	–	–	–	–	m	6.88

30 SUSPENDED CEILINGS

Item	PC £	Labour hours	Labour £	Plant £	Material £	Unit	Total rate £
Vertical bulkhead, as upstand to rooflight well; including additional hangers; perimeter trim							
not exceeding 600 mm wide in isolated strips	–	–	–	–	–	m	**47.15**
Lining to ceilings; 1200 mm × 1200 mm × 20 mm Focus DG suspended ceiling tiles							
over 600 mm wide	–	–	–	–	–	m²	**47.39**
not exceeding 600 mm wide in isolated strips	–	–	–	–	–	m	**28.09**
edge trim; shadow-line trim	–	–	–	–	–	m	**4.87**
Vertical bulkhead, as upstand to rooflight well; including additional hangers; perimeter trim							
300 mm × 600 mm wide	–	–	–	–	–	m	**47.85**
Z demountable suspended ceiling system or other equal; hangers plugged and screwed to concrete soffit, 600 mm × 600 mm × 19 mm Echostop glass reinforced fibrous plaster lightweight plain bevelled edge tiles							
Lining to ceilings; hangers 150 mm–500 mm long							
over 600 mm wide	–	–	–	–	–	m²	**96.83**
not exceeding 600 mm wide in isolated strips	–	–	–	–	–	m	**69.05**
Concealed galvanized steel suspension system; hangers plugged and screwed to concrete soffit, Burgess white stove enamelled perforated mild steel tiles 600 mm × 600 mm							
Lining to ceilings; hangers average 150 mm–500 mm long							
over 600 mm wide	–	–	–	–	–	m²	**46.55**
not exceeding 600 mm wide in isolated strips	–	–	–	–	–	m	**40.82**
Semi-concealed galvanized steel Trulok suspension system or other equal; hangers plugged and screwed to concrete; Armstrong Microlook Ultima Plain 600 mm × 600 mm × 18 mm mineral ceiling tiles							
Linings to ceilings; hangers 500 mm–1000 mm long							
over 600 mm wide	–	–	–	–	–	m²	**37.77**
over 600 mm wide; 3.50 m–5.00 m high	–	–	–	–	–	m²	**43.17**
over 600 mm wide; in staircase areas or plant rooms	–	–	–	–	–	m²	**43.17**
not exceeding 600 mm wide in isolated strips	–	–	–	–	–	m	**37.77**
cutting and fitting around modular downlighter including yoke	–	–	–	–	–	nr	**17.61**
24 mm × 19 mm white finished angle edge trim	–	–	–	–	–	m	**4.36**
Vertical bulkhead; including additional hangers							
over 600 mm wide	–	–	–	–	–	m²	**54.01**
not exceeding 600 mm wide in isolated strips	–	–	–	–	–	m	**48.70**

30 SUSPENDED CEILINGS

Item	PC £	Labour hours	Labour £	Plant £	Material £	Unit	Total rate £
30.01 SUSPENDED CEILINGS – cont							
Metal; SAS system 330; EMAC suspension system; 100 mm Omega C profiles at 1500 mm centres filled in with 1400 mm × 250 mm perforated metal tiles with 18 mm thick × 80 kg/ m³ density foil wrapped tissue-faced acoustic pad adhered above; ceiling to achieve 40dB with 0.7 absorption coefficient							
Linings to ceilings; hangers 500 mm–700 mm long							
over 600 mm wide	–	–	–	–	–	m²	52.26
not exceeding 600 mm wide in isolated strips	–	–	–	–	–	m	26.67
cutting and reinforcing to receive a recessed light maximum 1300 mm × 500 mm.	–	–	–	–	–	nr	14.93
edge trim; to perimeter	–	–	–	–	–	m	13.33
edge trim around 450 mm dia. column;	–	–	–	–	–	nr	51.19
Galvanized steel suspension system; hangers plugged and screwed to concrete soffit, Luxalon stove enamelled aluminium linear panel ceiling, type 80B or other equal and approved, complete with mineral insulation							
Linings to ceilings; hangers 500 mm–700 mm long							
over 600 mm wide	–	–	–	–	–	m²	87.88
not exceeding 600 mm wide in isolated strips	–	–	–	–	–	m	42.60
Armstrong suspended ceiling; assume large rooms over 250 m²							
Mineral fibre							
mineral fibre; basic range; Cortega, Tatra, Academy; exposed grid	–	–	–	–	–	m²	22.54
mineral fibre; medium quality; Corline; exposed grid	–	–	–	–	–	m²	32.70
mineral fibre; medium quality; Corline; concealed grid	–	–	–	–	–	m²	48.71
mineral fibre; medium quality; Corline; silhouette grid	–	–	–	–	–	m²	60.89
mineral fibre; Specific; Bioguard; exposed grid	–	–	–	–	–	m²	27.06
mineral fibre; Specific; Bioguard; clean room grid	–	–	–	–	–	m²	42.84
mineral fibre; Specific; Hygiene, Cleanroom; exposed grid	–	–	–	–	–	m²	58.63
Open cell							
metal open cell; Cellio Global White; exposed grid	–	–	–	–	–	m²	55.25
metal open cell; Cellio Black; exposed grid	–	–	–	–	–	m²	66.53
Wood veneer							
wood veneer board; Microlook 8; plain; exposed grid	–	–	–	–	–	m²	72.16
wood veneer; plain; exposed grid	–	–	–	–	–	m²	99.22
wood veneers; perforated; exposed grid	–	–	–	–	–	m²	171.38
wood veneers; plain; concealed grid	–	–	–	–	–	m²	163.49
wood veneers; perforated; concealed grid	–	–	–	–	–	m²	225.51

30 SUSPENDED CEILINGS

Item	PC £	Labour hours	Labour £	Plant £	Material £	Unit	Total rate £
Acoustic Panel linings; Troldekt Ultrafine 1200 mm × 600 mm × 25 mm on and including 50 mm × 70 mm timber framework at 600 mm centres both ways and supported from the roof with 50 mm × 50 mm hangers 400 mm–600 mm long							
Ceilings							
Allow for ceiling tiles to be clipped to prevent wind uplift when external doors are opened	–	–	–	–	–	m²	**118.71**
Perimeter edge detail with 50 mm × 50 mm support batten	–	–	–	–	–	m	**7.52**
30.02 SOLID SUSPENDED CEILINGS							
Gypsum plasterboard; BS EN 520; plain grade tapered edge wallboard; fixing on dabs or with nails; joints left open to receive Artex finish or other equal and approved; to softwood base							
9.50 mm board to ceilings							
over 600 mm wide	–	0.23	5.28	–	3.44	m²	**8.72**
9.50 mm board to beams							
girth not exceeding 600 mm	–	0.28	6.44	–	2.07	m²	**8.51**
girth 600 mm–1200 mm	–	0.37	8.51	–	4.13	m²	**12.64**
12.50 mm board to ceilings							
over 300 mm wide	–	0.31	7.13	–	3.45	m²	**10.58**
12.50 mm board to beams							
girth not exceeding 600 mm	–	0.28	6.44	–	2.07	m²	**8.51**
girth 600 mm–1200 mm	–	0.37	8.51	–	4.14	m²	**12.65**
Gypsum plasterboard; fixing on dabs or with nails; joints filled with joint tape and filler to receive direct decoration; to softwood base (measured elsewhere)							
Plain grade tapered edge wallboard							
9.50 mm board to ceilings							
over 600 mm wide	–	0.29	6.81	–	4.75	m²	**11.56**
9.50 mm board to faces of beams – 3 nr							
not exceeding 600 mm total girth	–	0.46	10.86	–	4.71	m	**15.57**
600 mm–1200 mm total girth	–	0.67	15.75	–	7.09	m	**22.84**
1200 mm–1800 mm total girth	–	0.75	17.65	–	9.47	m	**27.12**
12.50 mm board to ceilings							
over 600 mm wide	–	0.31	7.26	–	6.66	m²	**13.92**
12.50 mm board to faces of beams – 3 nr							
not exceeding 600 mm total girth	–	0.46	10.86	–	5.85	m	**16.71**
600 mm–1200 mm total girth	–	0.67	15.75	–	9.37	m	**25.12**
1200 mm–1800 mm total girth	–	0.75	17.65	–	12.91	m	**30.56**
external angle; with joint tape bedded and covered with Jointex or other equal and approved	–	0.11	2.68	–	0.58	m	**3.26**

30 SUSPENDED CEILINGS

Item	PC £	Labour hours	Labour £	Plant £	Material £	Unit	Total rate £
30.02 SOLID SUSPENDED CEILINGS – cont							
Tapered edge wallboard TEN							
12.50 mm board to ceilings							
over 600 mm wide	–	0.41	9.70	–	4.92	m²	**14.62**
12.50 mm board to faces of beams – 3 nr							
not exceeding 600 mm total girth	–	0.56	13.29	–	4.81	m	**18.10**
600 mm–1200 mm total girth	–	1.02	24.29	–	7.29	m	**31.58**
1200 mm–1800 mm total girth	–	1.30	31.08	–	9.76	m	**40.84**
external angle; with joint tape bedded and							
covered with Jointex or other equal and approved	–	0.11	2.68	–	0.58	m	**3.26**
Tapered edge plank							
19 mm plank to ceilings							
over 600 mm wide	–	0.43	10.17	–	8.19	m²	**18.36**
19 mm plank to faces of beams – 3 nr							
not exceeding 600 mm total girth	–	0.60	14.21	–	6.77	m	**20.98**
600 mm–1200 mm total girth	–	1.06	25.20	–	11.21	m	**36.41**
1200 mm–1800 mm total girth	–	1.34	31.99	–	15.65	m	**47.64**
ThermaLine Plus Board							
27 mm board to ceilings							
over 600 mm wide	–	0.46	10.57	–	15.02	m²	**25.59**
27 mm board to faces of beams – 3 nr							
not exceeding 600 mm total girth	–	0.56	12.88	–	10.87	m	**23.75**
600 mm–1200 mm total girth	–	1.06	24.36	–	19.41	m	**43.77**
1200 mm–1800 mm total girth	–	1.43	32.87	–	27.95	m	**60.82**
48 mm board to ceilings							
over 600 mm wide	–	0.49	11.26	–	20.49	m²	**31.75**
48 mm board to faces of beams – 3 nr							
not exceeding 600 mm total girth	–	0.58	13.33	–	14.17	m	**27.50**
600 mm–1200 mm total girth	–	1.17	26.89	–	26.00	m	**52.89**
1200 mm–1800 mm total girth	–	1.57	36.08	–	37.82	m	**73.90**
ThermaLine Super boards							
50 mm board to ceilings							
over 600 mm wide	–	0.49	11.26	–	27.44	m²	**38.70**
50 mm board to faces of beams – 3 nr							
not exceeding 600 mm total girth	–	0.58	13.33	–	18.34	m	**31.67**
600 mm–1200 mm total girth	–	1.17	26.89	–	34.34	m	**61.23**
1200 mm–1800 mm total girth	–	1.57	36.08	–	50.34	m	**86.42**
60 mm board to ceilings							
over 600 mm wide	–	0.49	11.26	–	38.14	m²	**49.40**
60 mm board to faces of beams – 3 nr							
not exceeding 600 mm total girth	–	0.58	13.33	–	24.76	m	**38.09**
600 mm–1200 mm total girth	–	1.17	26.89	–	47.17	m	**74.06**
1200 mm–1800 mm total girth	–	1.57	36.08	–	69.59	m	**105.67**

31 INSULATION, FIRE STOPPING AND FIRE PROTECTION

Item	PC £	Labour hours	Labour £	Plant £	Material £	Unit	Total rate £
31.01 INSULATION							
Altrernative insulation supply only prices,							
delivered in full loads							
Crown FrameTherm Roll 40 (Thermal conductivity							
0.040 W/mK)							
90 mm thick	–	–	–	–	1.32	m²	**1.32**
140 mm thick	–	–	–	–	3.10	m²	**3.10**
Crown FrameTherm Roll 35 (Thermal conductivity							
0.035 W/mK)							
90 mm thick	–	–	–	–	5.87	m²	**5.87**
140 mm thick	–	–	–	–	8.02	m²	**8.02**
Crown Factoryclad 40 (Thermal conductivity							
0.040 W/mK)							
180 mm thick	–	–	–	–	3.09	m²	**3.09**
100 mm thick	–	–	–	–	1.46	m²	**1.46**
Thermafleece EcoRoll (0.039 W/mK); 75% sheep's							
wool, 15% recycled polyester and 10% polyester							
binder with a high recycled content							
50 mm thick	–	–	–	–	4.21	m²	**4.21**
75 mm thick	–	–	–	–	6.27	m²	**6.27**
100 mm thick	–	–	–	–	8.20	m²	**8.20**
140 mm thick	–	–	–	–	11.60	m²	**11.60**
Thermafleece TF35 high density wool insulating							
batts (0.035 W/mK); 60% British wool, 30% recycled							
polyester and 10% polyester binder with a high							
recycled content							
50 mm thick	–	–	–	–	6.03	m²	**6.03**
75 mm thick	–	–	–	–	9.37	m²	**9.37**
SUPPLY AND FIX PRICES							
Sisalkraft building papers/vapour barriers or							
other equal and approved							
Building paper; 150 mm laps; fixed to softwood							
Moistop grade 728 (class A1F)	–	0.08	1.32	–	1.48	m²	**2.80**
Vapour barrier/reflective insulation 150 mm laps;							
fixed to softwood							
Insulex grade 714; single sided	–	0.08	1.32	–	1.75	m²	**3.07**
Insulation board underlays							
Vapour barrier							
reinforced; metal lined	–	–	–	–	–	m²	**15.50**
Rockwool; Duorock flat insulation board							
140 mm thick (0.25 U-value)	–	–	–	–	–	m²	**42.46**
Kingspan Thermaroof TR21 zero OPD urethene							
insulation board							
50 mm thick	–	–	–	–	–	m²	**26.09**
90 mm thick	–	–	–	–	–	m²	**44.96**
100 mm thick	–	–	–	–	–	m²	**49.94**

31 INSULATION, FIRE STOPPING AND FIRE PROTECTION

Item	PC £	Labour hours	Labour £	Plant £	Material £	Unit	Total rate £
31.01 INSULATION – cont							
Insulation board underlays – cont							
Wood fibre boards; impregnated; density 220–350 kg/m^3							
12.70 mm thick	–	–	–	–	–	m^2	**7.04**
Mat or quilt insulation							
Glass fibre roll; Crown Loft Roll 44 (Thermal conductivity 0.044 W/mK) or other equal; laid loose							
100 mm thick	1.38	0.09	1.48	–	1.42	m^2	**2.90**
150 mm thick	2.08	0.10	1.65	–	2.14	m^2	**3.79**
200 mm thick	2.89	0.11	1.81	–	2.98	m^2	**4.79**
Glass fibre quilt; Isover Modular roll (Thermal conductivity 0.043 W/mK) or other equal and approved; laid loose							
100 mm thick	1.32	0.09	1.48	–	1.36	m^2	**2.84**
150 mm thick	1.99	0.10	1.65	–	2.05	m^2	**3.70**
200 mm thick	2.96	0.15	2.48	–	3.05	m^2	**5.53**
Mineral fibre quilt; Isover Acoustic Partition Roll (APR 1200) or other equal; pinned vertically to softwood							
25 mm thick	1.13	0.08	1.32	–	1.43	m^2	**2.75**
50 mm thick	1.86	0.09	1.48	–	2.18	m^2	**3.66**
75 mm thick	3.11	0.10	1.65	–	3.47	m^2	**5.12**
100 mm thick	4.03	0.15	2.48	–	4.42	m^2	**6.90**
Crown Dritherm Cavity Slab 37 (Thermal conductivity 0.037 W/mK) glass fibre batt or other equal; as full or partial cavity fill; including cutting and fitting around wall ties and retaining discs							
50 mm thick	2.17	0.12	1.99	–	2.50	m^2	**4.49**
75 mm thick	3.70	0.13	2.15	–	4.08	m^2	**6.23**
100 mm thick	4.44	0.14	2.32	–	4.83	m^2	**7.15**
Crown Dritherm Cavity Slab 34 (Thermal conductivity 0.034 W/mK) glass fibre batt or other equal; as full or partial cavity fill; including cutting and fitting around wall ties and retaining discs							
65 mm thick	2.38	0.12	1.99	–	2.71	m^2	**4.70**
75 mm thick	2.75	0.13	2.15	–	3.09	m^2	**5.24**
85 mm thick	4.17	0.13	2.15	–	4.56	m^2	**6.71**
100 mm thick	4.17	0.14	2.32	–	4.56	m^2	**6.88**
Crown Dritherm Cavity Slab 32 (Thermal conductivity 0.032 W/mK) glass fibre batt or other equal; as full or partial cavity fill; including cutting and fitting around wall ties and retaining discs							
65 mm thick	3.09	0.12	1.99	–	3.44	m^2	**5.43**
75 mm thick	3.60	0.13	2.15	–	3.98	m^2	**6.13**
85 mm thick	4.08	0.13	2.15	–	4.47	m^2	**6.62**
100 mm thick	4.70	0.14	2.32	–	5.11	m^2	**7.43**

31 INSULATION, FIRE STOPPING AND FIRE PROTECTION

Item	PC £	Labour hours	Labour £	Plant £	Material £	Unit	Total rate £
Crown Frametherm Roll 40 (Thermal conductivity 0.040 W/mK) glass fibre semi-rigid or rigid batt or other equal; pinned vertically in timber frame construction							
90 mm thick	3.84	0.14	2.32	–	3.96	m²	**6.28**
140 mm thick	5.58	0.16	2.65	–	5.75	m²	**8.40**
Crown Rafter Roll 32 (Thermal conductivity 0.032 W/mK) glass fibre flanged building roll; pinned vertically or to slope between timber framing							
50 mm thick	3.72	0.13	2.15	–	3.83	m²	**5.98**
75 mm thick	5.32	0.14	2.32	–	5.48	m²	**7.80**
100 mm thick	6.85	0.15	2.48	–	7.06	m²	**9.54**
Board or slab insulation							
Kay-Cel (Thermal conductivity 0.033 W/mK) expanded polystyrene board standard grade SD/N or other equal; fixed with adhesive							
25 mm thick	–	0.14	3.21	–	2.54	m²	**5.75**
50 mm thick	–	0.16	3.68	–	4.18	m²	**7.86**
75 mm thick	–	0.18	4.14	–	5.80	m²	**9.94**
100 mm thick	–	0.19	4.37	–	7.43	m²	**11.80**
KIngspan Kooltherm K8 (Thermal conductivity 0.022 W/mK) Premium performance rigid thermoset modified resin insulant manufactured with a blowing agent that has zero Ozone Depletion Potential (ODP) and low Global Warming Potential (GWP) or other equal; as partial cavity fill; including cutting and fitting around wall ties and retaining discs							
40 mm thick	–	0.15	3.45	–	12.71	m²	**16.16**
50 mm thick	–	0.15	3.45	–	12.71	m²	**16.16**
60 mm thick	–	0.15	3.45	–	15.24	m²	**18.69**
75 mm thick	–	0.15	3.45	–	15.07	m²	**18.52**
Kingspan Thermawall TW50 (Thermal conductivity 0.022 W/mK) High performance rigid thermoset polyisocyanurate (PIR) insulant or other equal and approved; as partial cavity fill; including cutting and fitting around wall ties and retaining discs							
25 mm thick	–	0.17	3.90	–	8.61	m²	**12.51**
50 mm thick	–	0.17	3.90	–	11.26	m²	**15.16**
75 mm thick	–	0.18	4.14	–	16.39	m²	**20.53**
100 mm thick	–	0.19	4.37	–	21.23	m²	**25.60**
Kingspan Thermafloor TF70 (Thermal conductivity 0.022 W/mK) high performance rigid urethane floor insulation for solid concrete and suspended ground floors							
50 mm thick	–	0.17	4.02	–	8.29	m²	**12.31**
75 mm thick	–	0.17	4.02	–	12.28	m²	**16.30**
100 mm thick	–	0.17	4.02	–	15.35	m²	**19.37**
125 mm thick	–	0.17	4.02	–	21.28	m²	**25.30**
150 mm thick	–	0.20	4.59	–	24.92	m²	**29.51**

31 INSULATION, FIRE STOPPING AND FIRE PROTECTION

Item	PC £	Labour hours	Labour £	Plant £	Material £	Unit	Total rate £
31.01 INSULATION – cont							
Board or slab insulation – cont							
Styrofoam Floormate 500 (Thermal conductivity 0.033 W/mK) extruded polystyrene foam or other equal and approved							
50 mm thick	–	0.46	10.57	–	8.79	m²	**19.36**
75 mm thick	–	0.46	10.57	–	9.81	m²	**20.38**
100 mm thick	–	0.46	10.57	–	13.17	m²	**23.74**
Dow Chemicals Styrofoam SP or other equal and approved; cold bridging insulation fixed with adhesive to brick, block or concrete base							
Insulation to walls							
50 mm thick	–	0.33	7.58	–	11.38	m²	**18.96**
75 mm thick	–	0.35	8.04	–	14.89	m²	**22.93**
Insulation to isolated columns							
50 mm thick	–	0.41	9.42	–	11.38	m²	**20.80**
75 mm thick	–	0.43	9.89	–	14.89	m²	**24.78**
Insulation to ceilings							
50 mm thick	–	0.36	8.27	–	11.38	m²	**19.65**
75 mm thick	–	0.39	8.96	–	14.89	m²	**23.85**
Insulation to isolated beams							
50 mm thick	–	0.43	9.89	–	11.38	m²	**21.27**
75 mm thick	–	0.46	10.57	–	14.89	m²	**25.46**
Pittsburgh Company Foamglas cellular glass insulation with PC 56 component adhesive as required							
Flat roof insulation; Foamglas T3+; thermal conductivity 0.036 W/mK; compressive strength > 500 kPa size 600 mm × 450 mm slabs							
80 mm thick slab with adhered joints	36.64	1.20	26.32	–	45.81	m²	**72.13**
100 mm thick slab with adhered joints	45.79	1.20	26.32	–	55.25	m²	**81.57**
150 mm thick slab with adhered joints	68.69	1.30	28.51	–	78.84	m²	**107.35**
180 mm thick slab with adhered joints	82.43	1.40	30.70	–	92.99	m²	**123.69**
Flat roof insulation; Foamglas T4+; thermal conductivity 0.041 W/mK; compressive strength > 600 kPa size 600 mm × 450 mm slabs							
80 mm thick slab with adhered joints	36.64	1.20	26.32	–	45.81	m²	**72.13**
100 mm thick slab with adhered joints	45.79	1.20	26.32	–	55.25	m²	**81.57**
150 mm thick slab with adhered joints	68.69	1.30	28.51	–	78.84	m²	**107.35**
180 mm thick slab with adhered joints	82.43	1.40	30.70	–	92.99	m²	**123.69**
Cavity wall insulation slabs; Foamglas W+F; thermal conductivity 0.038 W/mK; 600 mm × 450 mm slabs							
80 mm thick slab with adhered joints	36.70	1.25	27.42	–	45.88	m²	**73.30**
100 mm thick slab with adhered joints	45.87	1.25	27.42	–	55.32	m²	**82.74**
150 mm thick slab with adhered joints	68.81	1.25	27.42	–	78.95	m²	**106.37**

31 INSULATION, FIRE STOPPING AND FIRE PROTECTION

Item	PC £	Labour hours	Labour £	Plant £	Material £	Unit	Total rate £
Sheepswool mix							
Thermafleece TF35 high density wool insulating batts (0.035 W/mK); 60% British wool, 30% recycled polyester and 10% polyester binder with a high recycled content							
50 mm thick	–	0.35	8.04	–	6.03	m²	**14.07**
75 mm thick	–	0.38	8.62	–	9.37	m²	**17.99**
Thermafleece EcoRoll (0.039 W/mK); 75% sheep's wool, 15% recycled polyester and 10% polyester binder with a high recycled content							
50 mm thick	–	0.15	2.48	–	4.21	m²	**6.69**
75 mm thick	–	0.14	2.32	–	6.27	m²	**8.59**
100 mm thick	–	0.15	2.48	–	8.20	m²	**10.68**
140 mm thick	–	0.17	2.89	–	11.60	m²	**14.49**
Tapered insulation board underlays							
Tapered insulation £/m² prices can vary dramatically depending upon the factors which determine the scheme layout; these primarily being gutter/outlet locations and the length of fall involved. The following guide assumes a U-value of 0.18W/m² K as a benchmark.							
As the required insulation value will vary from project to project the required U-value should be determined by calculating the buildings energy consumption at the design stage. Due to tapered insulation scheme prices varying by project, the following prices are indicative. Please contact a specialist for a project specific quotation. U-Value must be calculated in accordance with BSENISO 6946:2007 Annex C							
Tapered PIR (Polyisocyanurate) boards; bedded in hot bitumen							
effective thickness achieving 0.18 W/m² K	30.30	–	–	–	–	m²	**66.64**
minimum thickness achieving 0.18 W/m² K	36.05	–	–	–	–	m²	**73.99**
Tapered PIR (Polyisocyanurate) board; mechanically fastened							
effective thickness achieving 0.18 W/m² K	30.30	–	–	–	–	m²	**69.56**
Tapered Aspire; Hybrid EPS/PIR boards; bedded in hot bitumen							
effective thickness achieving 0.18 W/m² K	28.21	–	–	–	–	m²	**66.64**
minimum thickness achieving 0.18 W/m² K	33.96	–	–	–	–	m²	**73.99**
Tapered PIR (Polyisocyanurate) board; mechanically fastened							
effective thickness achieving 0.18 W/m² K	28.21	–	–	–	–	m²	**69.56**
minimum thickness achieving 0.18 W/m² K	33.96	–	–	–	–	m²	**76.91**
Tapered Rockwool boards; bedded in hot bitumen							
effective thickness achieving 0.18 W/m² K	54.13	–	–	–	–	m²	**98.04**
minimum thickness achieving 0.18 W/m² K	66.88	–	–	–	–	m²	**111.24**

31 INSULATION, FIRE STOPPING AND FIRE PROTECTION

Item	PC £	Labour hours	Labour £	Plant £	Material £	Unit	Total rate £
31.01 INSULATION – cont							
Tapered insulation board underlays – cont							
Tapered Rockwool boards; mechanically fastened							
effective thickness achieving 0.18 W/m^2 K	54.13	–	–	–	–	m^2	**100.97**
minimum thickness achieving 0.18 W/m^2 K	66.88	–	–	–	–	m^2	**109.54**
Tapered EPS (Expanded polystyrene) boards; bedded in hot bitumen							
effective thickness achieving 0.18 W/m^2 K	24.77	–	–	–	–	m^2	**59.38**
minimum thickness achieving 0.18 W/m^2 K	30.30	–	–	–	–	m^2	**66.64**
Tapered EPS (Expanded polystyrene) boards; mechanicaly fastened							
effective thickness achieving 0.18 W/m^2 K	24.77	–	–	–	–	m^2	**62.21**
minimum thickness achieving 0.18 W/m^2 K	30.30	–	–	–	–	m^2	**69.55**
Insulation board overlays							
Dow Roofmate SL extruded polystyrene foam boards or other equal and approved; Thermal conductivity – 0.028 W/mK							
50 mm thick	–	–	–	–	–	m^2	**17.27**
140 mm thick	–	–	–	–	–	m^2	**30.15**
160 mm thick	–	–	–	–	–	m^2	**32.66**
Dow Roofmate LG extruded polystyrene foam boards or other equal and approved; Thermal conductivity – 0.028 W/mK							
80 mm thick	–	–	–	–	–	m^2	**61.51**
100 mm thick	–	–	–	–	–	m^2	**65.92**
120 mm thick	–	–	–	–	–	m^2	**70.40**
31.02 FIRE STOPPING							
Fire stopping							
Cape Firecheck channel; intumescent coatings on cut mitres; fixing with brass cups and screws							
19 mm × 44 mm or 19 mm × 50 mm	–	0.56	12.88	–	21.92	m	**34.80**
Sealmaster intumescent fire and smoke seals or other equal and approved; pinned into groove in timber							
type N30; for single leaf half hour door	–	0.28	6.44	–	2.88	m	**9.32**
type N60; for single leaf one hour door	–	0.31	7.13	–	3.76	m	**10.89**
type IMN or IMP; for meeting or pivot stiles of pair of one hour doors; per stile	–	0.31	7.13	–	3.76	m	**10.89**
intumescent plugs in timber; including boring	–	0.09	2.07	–	0.42	nr	**2.49**
Rockwool fire stops or other equal and approved; between top of brick/block wall and concrete soffit							
30 mm deep × 100 mm wide	–	0.07	1.61	–	4.91	m	**6.52**
30 mm deep × 150 mm wide	–	0.09	2.07	–	7.45	m	**9.52**
30 mm deep × 200 mm wide	–	0.11	2.52	–	9.96	m	**12.48**
60 mm deep × 100 mm wide	–	0.08	1.84	–	6.47	m	**8.31**
60 mm deep × 150 mm wide	–	0.10	2.30	–	9.67	m	**11.97**
60 mm deep × 200 mm wide	–	0.12	2.76	–	13.01	m	**15.77**
90 mm deep × 100 mm wide	–	0.10	2.30	–	10.33	m	**12.63**

31 INSULATION, FIRE STOPPING AND FIRE PROTECTION

Item	PC £	Labour hours	Labour £	Plant £	Material £	Unit	**Total rate £**
90 mm deep × 150 mm wide	–	0.12	2.76	–	15.44	m	**18.20**
90 mm deep × 200 mm wide	–	0.14	3.21	–	20.65	m	**23.86**
Fire protection compound							
Quelfire QF4, fire protection compound or other							
equal and approved; filling around pipes, ducts and							
the like; including all necessary formwork							
300 mm × 300 mm × 250 mm; pipes – 2	–	0.93	17.43	–	16.52	nr	**33.95**
500 mm × 500 mm × 250 mm; pipes – 2	–	1.16	21.17	–	49.57	nr	**70.74**
Fire barriers							
Rockwool fire barrier or other equal and approved;							
between top of suspended ceiling and concrete soffit							
one 50 mm layer × 900 mm wide; half hour	–	0.56	12.88	–	28.30	m²	**41.18**
two 50 mm layers × 900 mm wide; one hour	–	0.83	19.08	–	56.32	m²	**75.40**
three 50 mm layers × 900 mm wide; two hour	–	1.10	25.29	–	80.74	m²	**106.03**
Corofil C144 fire barrier to edge of slab; fixed with							
non-flammable contact adhesive							
to suit void 30 mm wide × 100 mm deep; one hour	–	–	–	–	–	m	**15.65**
Lamatherm fire barrier or other equal and approved;							
to void below raised access floors							
75 mm thick × 300 mm high; half hour	–	0.17	3.90	–	10.00	m	**13.90**
75 mm thick × 600 mm high; half hour	–	0.17	3.90	–	21.91	m	**25.81**
90 mm thick × 300 mm high; half hour	–	0.17	3.90	–	14.04	m	**17.94**
90 mm thick × 600 mm high; half hour	–	0.17	3.90	–	29.23	m	**33.13**

32 FURNITURE, FITTINGS AND EQUIPMENT

Item	PC £	Labour hours	Labour £	Plant £	Material £	Unit	Total rate £
32.01 GENERAL FIXTURES, FURNISHES AND EQUIPMENT							
SUPPLY ONLY PRICES							
NOTE: The fixing of general fixtures will vary considerably dependent upon the size of the fixture and the method of fixing employed. Prices for fixing like sized kitchen fittings may be suitable for certain fixtures, although adjustment to those rates will almost invariably be necessary.							
The following supply only prices are for purpose-made fittings components in various materials supplied as part of an assembled fitting and therefore may be used to arrive at a guide price for a complete fitting.							
Fitting components; medium density fibreboard							
Backs, fronts, sides or divisions; over 300 mm wide							
12 mm thick	–	–	–	–	23.36	m²	**23.36**
18 mm thick	–	–	–	–	24.75	m²	**24.75**
25 mm thick	–	–	–	–	27.56	m²	**27.56**
Shelves or worktops; over 300 mm wide							
18 mm thick	–	–	–	–	24.75	m²	**24.75**
25 mm thick	–	–	–	–	27.56	m²	**27.56**
Flush doors; lipped on four edges							
450 mm × 750 mm × 18 mm	–	–	–	–	35.85	nr	**35.85**
450 mm × 750 mm × 25 mm	–	–	–	–	36.54	nr	**36.54**
600 mm × 900 mm × 18 mm	–	–	–	–	42.32	nr	**42.32**
600 mm × 900 mm × 25 mm	–	–	–	–	43.42	nr	**43.42**
Fitting components; moisture-resistant medium density fibreboard							
Backs, fronts, sides or divisions; over 300 mm wide							
12 mm thick	–	–	–	–	26.15	m²	**26.15**
18 mm thick	–	–	–	–	28.95	m²	**28.95**
25 mm thick	–	–	–	–	31.73	m²	**31.73**
Shelves or worktops; over 300 mm wide							
18 mm thick	–	–	–	–	28.95	m²	**28.95**
25 mm thick	–	–	–	–	31.73	m²	**31.73**
Flush doors; lipped on four edges							
450 mm × 750 mm × 18 mm	–	–	–	–	36.54	nr	**36.54**
450 mm × 750 mm × 25 mm	–	–	–	–	37.56	nr	**37.56**
600 mm × 900 mm × 18 mm	–	–	–	–	43.42	nr	**43.42**
600 mm × 900 mm × 25 mm	–	–	–	–	45.10	nr	**45.10**
Fitting components; medium density fibreboard; melamine faced both sides							
Backs, fronts, sides or divisions; over 300 mm wide							
12 mm thick	–	–	–	–	30.82	m²	**30.82**
18 mm thick	–	–	–	–	34.32	m²	**34.32**

32 FURNITURE, FITTINGS AND EQUIPMENT

Item	PC £	Labour hours	Labour £	Plant £	Material £	Unit	Total rate £
Shelves or worktops; over 300 mm wide							
18 mm thick	–	–	–	–	34.32	m²	**34.32**
Flush doors; lipped on four edges							
450 mm × 750 mm × 18 mm	–	–	–	–	25.65	nr	**25.65**
600 mm × 900 mm × 25 mm	–	–	–	–	32.53	nr	**32.53**
Fitting components; medium density fibreboard; formica faced both sides							
Backs, fronts, sides or divisions; over 300 mm wide							
12 mm thick	–	–	–	–	93.85	m²	**93.85**
18 mm thick	–	–	–	–	97.59	m²	**97.59**
Shelves or worktops; over 300 mm wide							
18 mm thick	–	–	–	–	97.59	m²	**97.59**
Flush doors; lipped on four edges							
450 mm × 750 mm × 18 mm	–	–	–	–	51.68	nr	**51.68**
600 mm × 900 mm × 25 mm	–	–	–	–	52.80	nr	**52.80**
Fitting components; wrought softwood							
Backs, fronts, sides or divisions; cross-tongued joints; over 300 mm wide							
25 mm thick	–	–	–	–	44.86	m²	**44.86**
Shelves or worktops; cross-tongued joints; over 300 mm wide							
25 mm thick	–	–	–	–	44.86	m²	**44.86**
Bearers							
19 mm × 38 mm	–	–	–	–	2.35	m	**2.35**
25 mm × 50 mm	–	–	–	–	2.60	m	**2.60**
44 mm × 44 mm	–	–	–	–	2.77	m	**2.77**
44 mm × 75 mm	–	–	–	–	3.19	m	**3.19**
Bearers; framed; to backs, fronts or sides							
19 mm × 38 mm	–	–	–	–	5.38	m	**5.38**
25 mm × 50 mm	–	–	–	–	5.81	m	**5.81**
50 mm × 50 mm	–	–	–	–	7.51	m	**7.51**
50 mm × 75 mm	–	–	–	–	8.61	m	**8.61**
Add 5% to the above material prices for selected softwood staining							
Fitting components; selected Sapele							
Backs, fronts, sides or divisions; cross-tongued joints; over 300 mm wide							
25 mm thick	–	–	–	–	70.60	m²	**70.60**
Shelves or worktops; cross-tongued joints; over 300 mm wide							
25 mm thick	–	–	–	–	70.60	m²	**70.60**
Bearers							
19 mm × 38 mm	–	–	–	–	3.53	m	**3.53**
25 mm × 50 mm	–	–	–	–	4.41	m	**4.41**
50 mm × 50 mm	–	–	–	–	4.94	m	**4.94**
50 mm × 75 mm	–	–	–	–	6.45	m	**6.45**

32 FURNITURE, FITTINGS AND EQUIPMENT

Item	PC £	Labour hours	Labour £	Plant £	Material £	Unit	Total rate £
32.01 GENERAL FIXTURES, FURNISHES AND EQUIPMENT – cont							
Fitting components – cont							
Bearers; framed; to backs, fronts or sides							
19 mm × 38 mm	–	–	–	–	6.84	m	**6.84**
25 mm × 50 mm	–	–	–	–	7.58	m	**7.58**
50 mm × 50 mm	–	–	–	–	10.17	m	**10.17**
50 mm × 75 mm	–	–	–	–	12.74	m	**12.74**
Fitting components; Iroko							
Backs, fronts, sides or divisions; cross-tongued joints; over 300 mm wide							
25 mm thick	–	–	–	–	77.82	m²	**77.82**
Shelves or worktops; cross-tongued joints; over 300 mm wide							
25 mm thick	–	–	–	–	77.82	m²	**77.82**
Draining boards; cross-tongued joints; over 300 mm wide							
25 mm thick	–	–	–	–	97.49	m²	**97.49**
stopped flutes	–	–	–	–	5.05	m	**5.05**
grooves; cross-grain	–	–	–	–	0.75	m	**0.75**
Bearers							
19 mm × 38 mm	–	–	–	–	3.86	m	**3.86**
25 mm × 50 mm	–	–	–	–	4.93	m	**4.93**
50 mm × 50 mm	–	–	–	–	5.61	m	**5.61**
50 mm × 75 mm	–	–	–	–	7.41	m	**7.41**
Bearers; framed; to backs, fronts or sides							
19 mm × 38 mm	–	–	–	–	7.08	m	**7.08**
25 mm × 50 mm	–	–	–	–	7.87	m	**7.87**
50 mm × 50 mm	–	–	–	–	10.64	m	**10.64**
50 mm × 75 mm	–	–	–	–	13.80	m	**13.80**
Lockers and cupboards; Welconstruct Distribution or other equal and approved							
Standard clothes lockers; steel body and door within reinforced 19G frame, powder coated finish, cam locks							
1 compartment; placing in position							
300 mm × 300 mm × 1800 mm	–	0.23	3.21	–	59.25	nr	**62.46**
380 mm × 380 mm × 1800 mm	–	0.23	3.21	–	84.80	nr	**88.01**
450 mm × 450 mm × 1800 mm	–	0.28	3.91	–	86.57	nr	**90.48**
Compartment lockers; steel body and door within reinforced 19G frame, powder coated finish, cam locks							
2 compartments; placing in position							
300 mm × 300 mm × 1800 mm	–	0.23	3.21	–	68.52	nr	**71.73**
380 mm × 380 mm × 1800 mm	–	0.23	3.21	–	82.62	nr	**85.83**
450 mm × 450 mm × 1800 mm	–	0.28	3.91	–	89.59	nr	**93.50**

32 FURNITURE, FITTINGS AND EQUIPMENT

Item	PC £	Labour hours	Labour £	Plant £	Material £	Unit	Total rate £
4 compartments; placing in position							
300 mm × 300 mm × 1800 mm	–	0.23	3.21	–	80.57	nr	**83.78**
380 mm × 380 mm × 1800 mm	–	0.23	3.21	–	96.72	nr	**99.93**
450 mm × 450 mm × 1800 mm	–	0.28	3.91	–	96.72	nr	**100.63**
Timber clothes lockers; veneered MDF finish, routed door, cam locks							
1 compartment; placing in position							
380 mm × 380 mm × 1830 mm	–	0.28	3.91	–	211.02	nr	**214.93**
4 compartments; placing in position							
380 mm × 380 mm × 1830 mm	–	0.28	3.91	–	314.53	nr	**318.44**
Vandal-resistant lockers							
1030 high mm × 370 mm × 560 mm; one compartment	–	0.23	3.21	–	223.72	nr	**226.93**
1930 mm × 370 mm × 560 mm; two compartments	–	0.23	3.21	–	347.01	nr	**350.22**
850 mm × 740 mm × 560 mm; 2 high × 2 wide	–	0.23	3.21	–	455.55	nr	**458.76**
1930 mm × 740 mm × 560 mm; 5 high × 2 wide	–	0.23	3.21	–	1031.37	nr	**1034.58**
Shelving support systems; The Welconstruct Company or other equal and approved standard duty; maximum bayload of 2000 kg							
Shelving support systems; steel body; stove enamelled finish; assembling							
open initial bay; 5 shelves; placing in position							
1000 mm × 300 mm × 1850 mm	–	0.69	11.40	–	164.20	nr	**175.60**
1000 mm × 600 mm × 1850 mm	–	0.69	11.40	–	208.18	nr	**219.58**
open extension bay; 5 shelves; placing in position							
1000 mm × 300 mm × 1850 mm	–	0.83	13.71	–	106.64	nr	**120.35**
1000 mm × 600 mm × 1850 mm	–	0.83	13.71	–	146.41	nr	**160.12**
closed initial bay; 5 shelves; placing in position							
1000 mm × 300 mm × 1850 mm	–	0.69	11.40	–	230.19	nr	**241.59**
1000 mm × 600 mm × 1850 mm	–	0.69	11.40	–	298.10	nr	**309.50**
closed extension bay; 5 shelves; placing in position							
1000 mm × 300 mm × 1850 mm	–	0.83	13.71	–	171.04	nr	**184.75**
1000 mm × 600 mm × 1850 mm	–	0.83	13.71	–	222.84	nr	**236.55**
extra for pair of doors; fixing in position							
1000 mm × 1850 mm	–	0.75	12.39	–	323.52	nr	**335.91**
Cloakroom racks; The Welconstruct Company or other equal and approved							
Cloakroom racks; 40 mm × 40 mm square tube framing, polyester powder coated finish; beech slatted seats and rails to one side only; placing in position							
1675 mm × 325 mm × 1500 mm; 5 nr coat hooks	–	0.30	4.95	–	368.47	nr	**373.42**
1825 mm × 325 mm × 1500 mm; 15 nr coat hangers	–	0.30	4.95	–	422.23	nr	**427.18**
Extra for							
shoe baskets	–	–	–	–	82.99	nr	**82.99**
mesh bottom shelf	–	–	–	–	57.95	nr	**57.95**

32 FURNITURE, FITTINGS AND EQUIPMENT

Item	PC £	Labour hours	Labour £	Plant £	Material £	Unit	Total rate £
32.01 GENERAL FIXTURES, FURNISHES AND EQUIPMENT – cont							
Cloakroom racks – cont							
Cloakroom racks; 40 mm × 40 mm square tube framing, polyester powder coated finish; beech slatted seats and rails to both sides; placing in position							
1675 mm × 600 mm × 1500 mm; 10 nr coat hooks	–	0.40	6.61	–	505.31	nr	**511.92**
1825 mm × 600 mm × 1500 mm; 30 nr coat hangers	–	0.40	6.61	–	526.80	nr	**533.41**
Extra for							
shoe baskets	–	–	–	–	103.69	nr	**103.69**
mesh bottom shelf	–	–	–	–	70.47	nr	**70.47**
6 mm thick rectangular glass mirrors; silver backed; fixed with chromium plated domed headed screws; to background requiring plugging							
Mirror with polished edges							
365 mm × 254 mm	12.17	0.74	12.88	–	12.84	nr	**25.72**
400 mm × 300 mm	15.87	0.74	12.88	–	16.67	nr	**29.55**
560 mm × 380 mm	27.50	0.83	14.44	–	28.63	nr	**43.07**
640 mm × 460 mm	35.94	0.93	16.18	–	37.33	nr	**53.51**
Mirror with bevelled edges							
365 mm × 254 mm	21.68	0.74	12.88	–	22.64	nr	**35.52**
400 mm × 300 mm	25.39	0.74	12.88	–	26.47	nr	**39.35**
560 mm × 380 mm	42.31	0.83	14.44	–	43.89	nr	**58.33**
640 mm × 460 mm	52.88	0.93	16.18	–	54.78	nr	**70.96**
Internal blinds							
Roller blinds; Luxaflex EOS type 10 roller; Compact Fabric; plain type material; 1219 mm drop; fixing with screws							
1016 mm wide	50.33	0.93	13.00	–	51.84	nr	**64.84**
2031 mm wide	74.30	1.45	20.27	–	76.53	nr	**96.80**
2843 mm wide	92.28	1.97	27.53	–	95.05	nr	**122.58**
Roller blinds; Luxaflex EOS type 10 roller; Compact Fabric; fire-resisting material; 1219 mm drop; fixing with screws							
1016 mm wide	65.92	0.93	13.00	–	67.90	nr	**80.90**
2031 mm wide	98.27	1.45	20.27	–	101.22	nr	**121.49**
2843 mm wide	124.64	1.97	27.53	–	128.38	nr	**155.91**
Roller blinds; Luxaflex EOS type 10 roller; Light-resistant; blackout material; 1219 mm drop; fixing with screws							
1016 mm wide	85.09	0.93	13.00	–	87.64	nr	**100.64**
2031 mm wide	142.61	1.45	20.27	–	146.89	nr	**167.16**
2843 mm wide	192.95	1.97	27.53	–	198.74	nr	**226.27**

32 FURNITURE, FITTINGS AND EQUIPMENT

Item	PC £	Labour hours	Labour £	Plant £	Material £	Unit	Total rate £
Roller blinds; Luxaflex Lite-master Crank Op; 100% blackout; 1219 mm drop; fixing with screws							
1016 mm wide	237.29	1.96	27.40	–	244.41	nr	**271.81**
2031 mm wide	317.58	2.75	38.44	–	327.11	nr	**365.55**
2843 mm wide	409.87	3.53	49.34	–	422.17	nr	**471.51**
Vertical louvre blinds; 89 mm wide louvres; Luxaflex EOS type; Florida Fabric; 1219 mm drop; fixing with screws							
1016 mm wide	68.31	0.82	11.46	–	70.36	nr	**81.82**
2031 mm wide	104.27	1.30	18.17	–	107.40	nr	**125.57**
3046 mm wide	142.61	1.77	24.74	–	146.89	nr	**171.63**
Vertical louvre blinds; 127 mm wide louvres; Luxaflex EOS type; Florida Fabric; 1219 mm drop; fixing with screws							
1016 mm wide	57.52	0.88	12.30	–	59.25	nr	**71.55**
2031 mm wide	87.50	1.35	18.87	–	90.13	nr	**109.00**
3046 mm wide	117.45	1.81	25.30	–	120.97	nr	**146.27**
Venetian blinds; 80 mm wide louvres; Levolux 480 type; solid slat; stock colour; manual hand crank; 2700 mm drop; standard brackets fixed to suitable grounds							
875 mm wide	–	–	–	–	–	nr	**225.00**
1150 mm wide	–	–	–	–	–	nr	**255.00**
1500 mm wide	–	–	–	–	–	nr	**300.00**
2500 mm wide	–	–	–	–	–	nr	**400.00**
3000 mm wide	–	–	–	–	–	nr	**430.00**
3500 mm wide	–	–	–	–	–	nr	**490.00**
4000 mm wide	–	–	–	–	–	nr	**600.00**
Door entrance mats							
Entrance mats; single aluminium wiper bar; laying in position; 12 mm thick							
Nuway Tuftiguard Plain 12 mm depth; 500 mm × 1300 mm	–	0.50	6.99	–	235.79	nr	**242.78**
Nuway Tuftiguard Plain 12 mm depth; 1000 mm × 1600 mm	–	0.65	9.08	–	538.94	nr	**548.02**
Nuway Tuftiguard Plain 12 mm depth; 1800 mm × 3000 mm	–	1.25	17.47	–	1818.90	nr	**1836.37**
Nuway Tuftiguard Classic 12 mm depth; 500 mm × 1300 mm	–	0.50	6.99	–	259.38	nr	**266.37**
Nuway Tuftiguard Classic 12 mm depth; 1000 mm × 1600 mm	–	0.65	9.08	–	592.87	nr	**601.95**
Nuway Tuftiguard Classic 12 mm depth; 1800 mm × 3000 mm	–	1.25	17.47	–	2000.94	nr	**2018.41**
Entrance mats; double aluminium wiper bar; laying in position; 18 mm thick							
Gradus Topguard; 900 mm × 550 mm	–	0.46	6.43	–	161.44	nr	**167.87**
Gradus Topguard; 1200 mm × 750 mm	–	0.46	6.43	–	290.59	nr	**297.02**
Gradus Topguard; 2400 mm × 1200 mm	–	0.93	13.00	–	929.90	nr	**942.90**

32 FURNITURE, FITTINGS AND EQUIPMENT

Item	PC £	Labour hours	Labour £	Plant £	Material £	Unit	Total rate £
32.01 GENERAL FIXTURES, FURNISHES AND EQUIPMENT – cont							
Door entrance mats – cont							
Entrance mats – cont							
Nuway Tuftiguard Plain 17 mm depth;							
500 mm × 1300 mm	–	0.55	7.68	–	285.78	nr	**293.46**
Nuway Tuftiguard Plain 17 mm depth;							
1000 mm × 1600 mm	–	0.72	10.14	–	653.23	nr	**663.37**
Nuway Tuftiguard Plain 17 mm depth;							
1800 mm × 3000 mm	–	1.38	19.22	–	2204.63	nr	**2223.85**
Nuway Tuftiguard Classic 17 mm depth;							
500 mm × 1300 mm	–	0.55	7.68	–	314.75	nr	**322.43**
Nuway Tuftiguard Classic 17 mm depth;							
1000 mm × 1600 mm	–	0.72	10.14	–	719.41	nr	**729.55**
Nuway Tuftiguard Classic 17 mm depth;							
1800 mm × 3000 mm	–	1.38	19.22	–	2428.02	nr	**2447.24**
Coral Classic textile secondary and circulation matting system for moisture removal; 2.0 m wide roll	–	0.46	6.43	–	42.37	m²	**48.80**
Coral Brush Activ secondary and circulation matting system for soil and moisture removal; 2.0 m wide roll	–	0.46	6.43	–	47.83	m²	**54.26**
Coral Duo secondary matting system ribbed; 2.0 m wide roll	–	0.46	6.43	–	47.83	m²	**54.26**
Matwells							
Polished stainless steel matwell; angle rim with lugs; bedding in screed							
Stainless steel frame bed in screed							
500 mm × 1300 mm	105.39	0.93	13.00	–	108.55	nr	**121.55**
Stainless steel frame bed in screed							
1000 mm × 1600 mm	93.68	0.93	13.00	–	96.49	nr	**109.49**
Stainless steel frame bed in screed							
1100 mm × 1400 mm	146.37	1.00	13.98	–	150.76	nr	**164.74**
Stainless steel frame bed in screed							
1800 mm × 3000 mm	281.03	2.00	27.95	–	289.46	nr	**317.41**
Polished aluminium matwell; angle rim with lugs brazed on; bedding in screed							
900 mm × 550 mm; constructed with 25 × 25 × 3 mm angle	22.92	0.93	13.00	–	23.61	nr	**36.61**
1200 mm × 750 mm; constructed with 34 × 26 × 6 mm angle	37.00	1.00	13.98	–	38.11	nr	**52.09**
2400 mm × 1200 mm; constructed with 50 × 50 × 6 mm angle	79.80	1.50	20.97	–	82.19	nr	**103.16**

32 FURNITURE, FITTINGS AND EQUIPMENT

Item	PC £	Labour hours	Labour £	Plant £	Material £	Unit	Total rate £
Polished brass matwell; comprising angle rim with lugs brazed on; bedding in screed							
900 mm × 550 mm; constructed with							
25 × 25 × 5 mm angle	74.74	0.93	13.00	–	76.98	nr	**89.98**
1200 mm × 750 mm; constructed with							
38 × 38 × 6 mm angle	100.51	1.00	13.98	–	103.53	nr	**117.51**
2400 mm x1200 mm; constructed with							
38 × 38 × 6 mm angle	185.56	1.50	20.97	–	191.13	nr	**212.10**
32.02 KITCHEN FITTINGS							
NOTE: Kitchen fittings vary considerably. PC supply prices for reasonable quantities for a moderately priced range of kitchen fittings have been shown.							
Kitchen units: supplying and fixing to backgrounds requiring plugging; including any pre-assembly							
Wall units							
300 mm × 300 mm × 720 mm	115.00	1.11	19.31	–	118.60	nr	**137.91**
500 mm × 300 mm × 720 mm	140.00	1.16	20.18	–	144.35	nr	**164.53**
600 mm × 300 mm × 720 mm	150.00	1.30	22.62	–	154.65	nr	**177.27**
800 mm × 300 mm × 720 mm	230.00	1.48	25.75	–	237.05	nr	**262.80**
Floor units with drawers							
500 mm × 600 mm × 870 mm	210.00	1.16	20.18	–	216.45	nr	**236.63**
600 mm × 600 mm × 870 mm	230.00	1.30	22.62	–	237.05	nr	**259.67**
1000 mm × 600 mm × 870 mm	345.00	1.57	27.32	–	355.50	nr	**382.82**
Sink units (excluding sink top)							
1000 mm × 600 mm × 870 mm	315.00	1.48	25.75	–	324.60	nr	**350.35**
Kitchen worktop: laminated plastics worktops; single rolled edge; prices include for fixing							
38 mm thick; 600 mm wide	59.50	0.37	6.44	–	61.34	m	**67.78**
extra for forming hole for inset sink	–	0.69	12.00	–	–	nr	**12.00**
extra for jointing strip at corner intersection of worktops	–	0.14	2.43	–	12.36	nr	**14.79**
extra for butt and scribe joint at corner intersection of worktops	–	4.16	72.37	–	–	nr	**72.37**

32 FURNITURE, FITTINGS AND EQUIPMENT

Item	PC £	Labour hours	Labour £	Plant £	Material £	Unit	Total rate £
32.03 SANITARY APPLIANCES AND FITTINGS							
Sinks; Armitage Shanks or equal							
Sinks; white glazed fireclay; BS 6465; pointing all round with silicone sealant							
Belfast sink; 46 cm × 38 cm × 21 cm; pair of Nuastyle 21 basin taps with dual indices, chrome handle; wall mounts 38 mm slotted waste, chain and plug, screw stay; pair of 40.5 cm aluminium alloy build-in brackets with 35.5 cm studs	328.78	2.78	73.65	–	431.75	nr	**505.40**
Belfast sink; 61 cm × 38 cm × 21 cm; pair of Nuastyle 21 basin taps with dual indices, chrome handle; wall mounts; 38 mm slotted waste, chain and plug, screw stay; pair of 40.5 cm aluminium alloy build-in brackets with 35.5 cm studs	391.31	2.78	73.65	–	496.47	nr	**570.12**
Belfast sink; 76 cm × 38 cm × 21 cm; pair of Nuastyle 21 basin taps with dual indices, chrome handle; wall mounts; 38 mm slotted waste, chain and plug, screw stay; pair of 40.5 cm aluminium alloy build-in brackets with 35.5 cm studs	549.57	2.78	73.65	–	659.78	nr	**733.43**
Lavatory basins; Armitage Shanks or equal							
Basins; white vitreous china; BS 6465 Part 3; pointing all round with silicone sealant							
Portman 21 40 cm basin; with overflow, chain hole and two tapholes; pair of Nuastyle 21 basin taps with dual indices; slotted basin waste with plastic plug, chain waste and plug; 32 × 75 mm seal plastic standard bottle trap; pair of Portman concealed brackets with waste support; Isovalve 15 mm plastic servicing valve with outlet for copper	154.39	2.13	56.42	–	219.35	nr	**275.77**
Portman 21 50 cm basin; with overflow, chain hole and two tapholes; pair of Nuastyle 21 basin taps with dual indices; slotted basin waste with plastic plug, chain waste and plug; 32 × 75 mm seal plastic standard bottle trap; pair of Portman concealed brackets with waste support; Isovalve 15 mm plastic servicing valve with outlet for copper	188.46	2.13	56.42	–	254.65	nr	**311.07**
Portman 21 60 cm basin; with overflow, chain hole and two tapholes; pair of Nuastyle 21 basin taps with dual indices; slotted basin waste with plastic plug, chain waste and plug; 32 × 75 mm seal plastic standard bottle trap; pair of Portman concealed brackets with waste support; Isovalve 15 mm plastic servicing valve with outlet for copper	247.92	2.13	56.42	–	316.10	nr	**372.52**

32 FURNITURE, FITTINGS AND EQUIPMENT

Item	PC £	Labour hours	Labour £	Plant £	Material £	Unit	Total rate £
Tiffany 50 cm pedestal basin; with two tapholes; Millenia STD dual control one taphole standard basin mixer with pop-up waste; pair of Millenia STD handles; Full pedestal; Isovalve 15 mm plastic servicing valve with outlet for copper	283.94	2.31	61.19	–	321.27	nr	**382.46**
Tiffany or similar 55 cm pedestal basin; with two tapholes; Millenia STD dual control one taphole standard basin mixer with pop-up waste; pair of Millenia STD handles; Full pedestal; Isovalve 15 mm plastic servicing valve with outlet for copper	278.00	2.31	61.19	–	315.15	nr	**376.34**
Tiffany 60 cm pedestal basin; with two tapholes; Millenia STD dual control one taphole standard basin mixer with pop-up waste; pair of Millenia STD handles; Full pedestal; Isovalve 15 mm plastic servicing valve with outlet for copper	289.94	2.31	61.19	–	331.08	nr	**392.27**
Montana 51 cm pedestal basin; with one taphole; Millenia STD dual control one taphole standard basin mixer with pop-up waste; pair of Millenia STD handles; Full pedestal; Isovalve 15 mm plastic servicing valve with outlet for copper	268.67	2.31	61.19	–	305.44	nr	**366.63**
Montana 58 cm pedestal basin; with one taphole; Millenia STD dual control one taphole standard basin mixer with pop-up waste; pair of Millenia STD handles; Full pedestal; Isovalve 15 mm plastic servicing valve with outlet for copper	274.94	2.31	61.19	–	312.04	nr	**373.23**
Drinking fountains; Armitage Shanks or equal							
White vitreous china fountains; pointing all round with silicone selant							
Aqualon wall mounted drinking fountain; Aqualon self-closing valve with fittings and plastic waste; 32 × 75 mm seal plastic standard bottle trap	462.35	2.31	61.19	–	490.12	nr	**551.31**
Polished stainless steel fountains; pointing all round with silicone selant							
Purita wall mounted drinking fountain with self-closing valve and fittings; 32 mm unslotted basin strainer waste	366.08	2.31	61.19	–	377.77	nr	**438.96**
Purita pedestal mounted drinking fountain 90 cm high with self-closing valve and fittings; 32 mm unslotted basin strainer waste	942.84	2.78	73.65	–	973.42	nr	**1047.07**

32 FURNITURE, FITTINGS AND EQUIPMENT

Item	PC £	Labour hours	Labour £	Plant £	Material £	Unit	Total rate £
32.03 SANITARY APPLIANCES AND FITTINGS – cont							
Baths; Armitage Shanks or equal							
Pointing all round with silicone selant							
Sandringham acrylic rectangular bath with chrome plated grips and two tapholes; standard pair of standard bath taps with chrome handles; bath chain waste with plastic plug and overflow; cast brass P trap with plain outlet and overflow connection; 170 cm long × 70 cm wide; white or coloured	228.99	3.50	92.72	–	235.86	nr	**328.58**
Nisa lowline heavy gauge steel rectangular bath with chrome plated grips and two tapholes; standard pair of standard bath taps with chrome handles; bath chain waste with plastic plug and overflow; cast brass P trap with plain outlet and overflow connection; 170 cm long × 70 cm wide; white or coloured	581.29	3.50	92.72	–	598.73	nr	**691.45**
Water closets; Armitage Shanks or equal							
White vitreous china pans and cisterns; pointing all round base with silicone sealant							
Wentworth close coupled washdown closet pan with horizontal outlet; Orion 3 plastic toilet seat and cover; Panketa pan connector 14° finned; Universal close coupled bottom inlet cistern with syphon	263.63	3.05	80.80	–	282.98	nr	**363.78**
Tiffany back to wall washdown closet pan with horizontal outlet; Saturn plastic toilet seat and cover; Panketa pan connector 14° finned; Conceala 2 6 litre low level side inlet cistern with syphon and lever	346.89	3.05	80.80	–	368.74	nr	**449.54**
Extra for; Panketa pan connector 90° finned	–	–	–	–	2.67	nr	**2.67**
Tiffany close coupled washdown closet pan with horizontal outlet; Saturn plastic toilet seat and cover; Panketa pan connector 14° finned; Tiffany 7½ litre close coupled cistern with dual flush valve	358.35	3.05	80.80	–	380.54	nr	**461.34**
Extra for; Panketa pan connector 90° finned	–	–	–	–	2.67	nr	**2.67**
Cameo close coupled washdown closet pan with horizontal outlet; Accolade/Cameo plastic toilet seat and cover; Panketa pan connector 14° finned; Cameo 6 litre close coupled cistern with dual flush valve	446.89	3.05	80.80	–	471.74	nr	**552.54**
Extra for; Panketa pan connector 90° finned	–	–	–	–	2.67	nr	**2.67**

32 FURNITURE, FITTINGS AND EQUIPMENT

Item	PC £	Labour hours	Labour £	Plant £	Material £	Unit	Total rate £
Wall urinals; Armitage Shanks or equal							
White vitreous china bowls and cisterns; pointing all round with silicone sealant							
Single Sanura 40 cm urinal bowl; Sanura top inlet spreader; pair of wall hangers for urinal bowl; 38 mm plastic domed waste; 38 × 75 mm seal plastic standard bottle trap; Conceala 4½ litre capacity auto cistern and cover; Sanura concealed flushpipe for single urinal bowl; screwing	310.22	3.70	98.01	–	354.62	nr	**452.63**
Single Sanura 40 cm urinal bowl; Sanura top inlet spreader; pair of wall hangers for urinal bowl; 38 mm plastic domed waste; 38 × 75 mm seal plastic standard bottle trap; Mura 4½ litre capacity auto cistern and cover; Sanura/Mura exposed flushpipe for single urinal bowl	353.44	3.70	98.01	–	399.14	nr	**497.15**
Single Sanura 50 cm urinal bowl; Sanura top inlet spreader; pair of wall hangers for urinal bowl; 38 mm plastic domed waste; 38 × 75 mm seal plastic standard bottle trap; Conceala 4½ litre capacity auto cistern and cover; Sanura concealed flushpipe for single urinal bowl	423.60	3.70	98.01	–	471.55	nr	**569.56**
Single Sanura 50 cm urinal bowl; Sanura top inlet spreader; pair of wall hangers for urinal bowl; 38 mm plastic domed waste; 38 × 75 mm seal plastic standard bottle trap; Mura 4½ litre capacity auto cistern and cover; Sanura/Mura exposed flushpipe for single urinal bowl	466.82	3.70	98.01	–	516.07	nr	**614.08**
Range of 2 nr Sanura 40 cm urinal bowls; Sanura top inlet spreader; pairs of wall hangers for urinal bowls; 38 mm plastic domed wastes; 38 × 75 mm seal plastic standard bottle traps; Conceala 9 litre capacity auto cistern and cover; Sanura concealed flushpipe for range of 2 nr urinal bowls	522.31	6.95	184.11	–	608.17	nr	**792.28**
Range of 2 nr Sanura 50 cm urinal bowls; Sanura top inlet spreader; pairs of wall hangers for urinal bowls; 38 mm plastic domed wastes; 38 × 75 mm seal plastic standard bottle traps; Conceala 9 litre capacity auto cistern and cover; Sanura concealed flushpipe for range of 2 nr urinal bowls	749.07	6.95	184.11	–	842.05	nr	**1026.16**
Range of 3 nr Sanura 40 cm urinal bowls; Sanura top inlet spreader; pairs of wall hangers for urinal bowls; 38 mm plastic domed wastes; 38 × 75 mm seal plastic standard bottle traps; Conceala 9 litre capacity auto cistern and cover; Sanura concealed flushpipe for range of 3 nr urinal bowls	727.16	10.15	268.89	–	854.26	nr	**1123.15**

32 FURNITURE, FITTINGS AND EQUIPMENT

Item	PC £	Labour hours	Labour £	Plant £	Material £	Unit	Total rate £
32.03 SANITARY APPLIANCES AND FITTINGS – cont							
Wall urinals – cont							
White vitreous china bowls and cisterns – cont							
Range of 3 nr Sanura 50 cm urinal bowls; Sanura top inlet spreader; pairs of wall hangers for urinal bowls; 38 mm plastic domed wastes; 38 × 75 mm seal plastic standard bottle traps; Conceala 9 litre capacity auto cistern and cover; Sanura concealed flushpipe for range of 3 nr urinal bowls	1067.30	10.15	268.89	–	1204.61	nr	**1473.50**
Range of 4 nr Sanura 40 cm urinal bowls; Sanura top inlet spreader; pairs of wall hangers for urinal bowls; 38 mm plastic domed wastes; 38 × 75 mm seal plastic standard bottle traps; Conceala 9 litre capacity auto cistern and cover; Sanura concealed flushpipe for range of 4 nr urinal bowls; screwing	939.49	13.40	354.99	–	1108.06	nr	**1463.05**
Range of 4 nr Sanura 50 cm urinal bowls; Sanura top inlet spreader; pairs of wall hangers for urinal bowls; 38 mm plastic domed wastes; 38 × 75 mm seal plastic standard bottle traps; Conceala 9 litre capacity auto cistern and cover; Sanura concealed flushpipe for range of 4 nr urinal bowls	1393.01	13.40	354.99	–	1575.80	nr	**1930.79**
Range of 5 nr Sanura 40 cm urinal bowls; Sanura top inlet spreader; pairs of wall hangers for urinal bowls; 38 mm plastic domed wastes; 38 × 75 mm seal plastic standard bottle traps; Conceala 9 litre capacity auto cistern and cover; Sanura concealed flushpipe for range of 5 nr urinal bowls	1144.43	16.65	441.09	–	1354.24	nr	**1795.33**
Range of 5 nr Sanura 50 cm urinal bowls; Sanura top inlet spreader; pairs of wall hangers for urinal bowls; 38 mm plastic domed wastes; 38 × 75 mm seal plastic standard bottle traps; Conceala 9 litre capacity auto cistern and cover; Sanura concealed flushpipe for range of 5 nr urinal bowls	1711.33	16.65	441.09	–	1938.92	nr	**2380.01**
White vitreous china division panels; pointing all round with silicone sealant							
Urinal division with screw and hanger	97.24	0.70	18.54	–	101.85	nr	**120.39**
Bidets; Armitage Shanks or equal							
Tiffany back to wall bidet with one taphole; vitreous china; chromium plated pop-up waste and mixer tap with hand wheels; 58 cm × 39 cm; white or coloured	516.10	3.50	92.72	–	531.58	nr	**624.30**
Shower tray and fittings							
Simplicity shower tray; acrylic; with outlet and grated waste; chain and plug; bedding and pointing in waterproof cement mortar							
760 mm × 760 mm; white or coloured	81.90	3.00	79.47	–	84.36	nr	**163.83**

32 FURNITURE, FITTINGS AND EQUIPMENT

Item	PC £	Labour hours	Labour £	Plant £	Material £	Unit	Total rate £
Shower fitting; riser pipe with mixing valve and shower rose; chromium plated; plugging and screwing mixing valve and pipe bracket							
15 mm dia. riser pipe; 127 mm dia. shower rose	440.68	5.00	132.46	–	453.90	nr	**586.36**
Corner fitting shower enclosure							
Bliss flat top hinged door with front panel and clear glass side panel	867.61	3.00	41.93	–	893.64	nr	**935.57**
Miscellaneous fittings; Magrini Ltd or equal							
Vertical nappy changing unit							
ref KBCS; screwing	–	0.60	10.43	–	211.36	nr	**221.79**
Horizontal nappy changing unit							
ref KBHS; screwing	–	0.60	10.43	–	211.36	nr	**221.79**
Stay Safe baby seat							
ref MX33; screwing to wall	–	0.55	9.57	–	79.26	nr	**88.83**
Miscellaneous fittings; Pressalit Ltd or equal; stainless steel							
Grab rails							
300 mm long ref RT100000; screw fix to wall	–	0.50	8.70	–	47.23	nr	**55.93**
450 mm long ref RT101000; screw fix to wall	–	0.50	8.70	–	54.60	nr	**63.30**
600 mm long ref RT102000; screw fix to wall	–	0.50	8.70	–	62.63	nr	**71.33**
800 mm long ref RT103000; screw fix to wall	–	0.50	8.70	–	70.43	nr	**79.13**
1000 mm long ref RT104000; screw fix to wall	–	0.50	8.70	–	81.19	nr	**89.89**
Angled grab rails							
900 mm long, angled 135° ref RT110000; screw fix to wall	–	0.50	8.70	–	102.07	nr	**110.77**
1300 mm long, angled 90° ref RT119000; screw fix to wall	–	0.75	13.05	–	159.87	nr	**172.92**
Hinged grab rails							
600 mm long ref R3016000; screw fix to wall	–	0.35	6.09	–	165.78	nr	**171.87**
600 mm long with spring counter balance ref RF016000; screw fix to wall	–	0.35	6.09	–	231.36	nr	**237.45**
850 mm long ref R3010000; screw fix to wall	–	0.35	6.09	–	201.30	nr	**207.39**
850 mm long with spring counter balance ref RF010000; screw fix to wall	–	0.35	6.09	–	248.31	nr	**254.40**
Shower seat; wall mounted; with padded seat and back	–	1.50	26.10	–	245.72	nr	**271.82**
32.04 NOTICES AND SIGNS							
Plain script; in gloss oil paint; on painted or varnished surfaces							
Capital letters; lower case letters or numerals							
per coat; per 25 mm high	–	0.09	1.57	–	–	nr	**1.57**

33 DRAINAGE ABOVE GROUND

Item	PC £	Labour hours	Labour £	Plant £	Material £	Unit	Total rate £
33.01 RAINWATER INSTALLATIONS							
Aluminium							
Pipes and fittings; ears cast on; polyester powder coated finish							
63 mm dia. pipes; plugged and screwed	23.18	0.34	6.76	–	25.54	m	**32.30**
extra for							
fittings with one end	–	0.20	3.98	–	22.45	nr	**26.43**
fittings with two ends	–	0.39	7.76	–	29.08	nr	**36.84**
fittings with three ends	–	0.56	11.13	–	31.84	nr	**42.97**
shoe	22.83	0.20	3.98	–	24.36	nr	**28.34**
bend	30.04	0.39	7.76	–	30.98	nr	**38.74**
single branch	30.04	0.56	11.13	–	31.84	nr	**42.97**
offset 228 projection	37.09	0.39	8.96	–	39.05	nr	**48.01**
offset 304 projection	41.33	0.39	7.76	–	43.42	nr	**51.18**
access pipe	45.78	0.39	7.76	–	47.18	nr	**54.94**
connection to clay pipes; cement and sand (1:2) joint	–	0.14	2.78	–	0.15	nr	**2.93**
75 mm dia. pipes; plugged and screwed	27.01	0.37	7.35	–	29.50	m	**36.85**
extra for							
shoe	23.51	0.23	4.57	–	25.06	nr	**29.63**
bend	31.03	0.42	8.34	–	32.02	nr	**40.36**
single branch	31.03	0.60	11.93	–	32.87	nr	**44.80**
offset 228 projection	40.99	0.42	8.34	–	43.07	nr	**51.41**
offset 304 projection	45.35	0.42	8.34	–	47.57	nr	**55.91**
access pipe	50.03	0.42	8.34	–	51.55	nr	**59.89**
connection to clay pipes; cement and sand (1:2) joint	–	0.16	3.18	–	0.15	nr	**3.33**
100 mm dia. pipes; plugged and screwed	45.97	0.42	8.34	–	49.03	m	**57.37**
extra for							
shoe	25.90	0.26	5.17	–	27.54	nr	**32.71**
bend	34.42	0.46	9.15	–	35.52	nr	**44.67**
single branch	34.42	0.69	13.71	–	36.39	nr	**50.10**
offset 228 projection	47.43	0.46	9.15	–	48.89	nr	**58.04**
offset 304 projection	52.65	0.46	9.15	–	54.26	nr	**63.41**
access pipe	59.29	0.46	9.15	–	61.10	nr	**70.25**
connection to clay pipes; cement and sand (1:2) joint	–	0.19	3.78	–	0.15	nr	**3.93**
Roof outlets; circular aluminium; with flat or domed grating; joint to pipe							
50 mm dia.	73.82	0.56	12.41	–	76.03	nr	**88.44**
75 mm dia.	96.75	0.60	13.30	–	99.65	nr	**112.95**
100 mm dia.	126.00	0.65	14.40	–	129.78	nr	**144.18**
150 mm dia.	161.42	0.69	15.29	–	166.26	nr	**181.55**
Roof outlets; d-shaped; balcony; with flat or domed grating; joint to pipe							
50 mm dia.	73.82	0.56	12.41	–	76.03	nr	**88.44**
75 mm dia.	97.97	0.60	13.30	–	100.91	nr	**114.21**
100 mm dia.	137.42	0.65	14.40	–	141.54	nr	**155.94**

33 DRAINAGE ABOVE GROUND

Item	PC £	Labour hours	Labour £	Plant £	Material £	Unit	Total rate £
PVC balloon grating							
110 mm dia.	7.65	0.06	1.33	–	7.88	nr	**9.21**
63 mm dia.	4.80	0.06	1.33	–	4.94	nr	**6.27**
Aluminium gutters and fittings; polyester powder coated finish							
100 mm half round gutters; on brackets; screwed to timber	26.97	0.32	7.35	–	34.60	m	**41.95**
extra for							
stop end	7.13	0.15	3.45	–	13.77	nr	**17.22**
running outlet	15.79	0.31	7.13	–	17.94	nr	**25.07**
stop end outlet	14.07	0.15	3.45	–	20.92	nr	**24.37**
angle	14.61	0.31	7.13	–	16.74	nr	**23.87**
113 mm half round gutters; on brackets; screwed to timber	28.23	0.32	7.35	–	35.97	m	**43.32**
extra for							
stop end	7.50	0.15	3.45	–	14.21	nr	**17.66**
running outlet	17.23	0.31	7.13	–	19.43	nr	**26.56**
stop end outlet	16.14	0.15	3.45	–	23.05	nr	**26.50**
angle	16.47	0.31	7.13	–	13.70	nr	**20.83**
125 mm half round gutters; on brackets; screwed to timber	31.72	0.37	8.51	–	43.23	m	**51.74**
extra for							
stop end	9.13	0.17	3.90	–	19.23	nr	**23.13**
running outlet	18.67	0.32	7.35	–	20.91	nr	**28.26**
stop end outlet	17.12	0.17	3.90	–	27.46	nr	**31.36**
angle	18.28	0.32	7.35	–	20.52	nr	**27.87**
100 mm ogee gutters; on brackets; screwed to timber	33.63	0.34	7.82	–	44.57	m	**52.39**
extra for							
stop end	7.53	0.16	3.68	–	9.16	nr	**12.84**
running outlet	18.53	0.32	7.35	–	19.66	nr	**27.01**
stop end outlet	18.53	0.16	3.68	–	28.24	nr	**31.92**
angle	15.64	0.32	7.35	–	10.82	nr	**18.17**
112 mm ogee gutters; on brackets; screwed to timber	37.40	0.39	8.96	–	49.04	m	**58.00**
extra for							
stop end	8.04	0.16	3.68	–	9.68	nr	**13.36**
running outlet	18.77	0.32	7.35	–	19.90	nr	**27.25**
stop end outlet	18.77	0.16	3.68	–	29.00	nr	**32.68**
angle	18.62	0.32	7.35	–	19.78	nr	**27.13**
125 mm ogee gutters; on brackets; screwed to timber	41.31	0.39	8.96	–	53.93	m	**62.89**
extra for							
stop end	8.81	0.18	4.14	–	10.48	nr	**14.62**
running outlet	20.50	0.34	7.82	–	21.69	nr	**29.51**
stop end outlet	20.50	0.18	4.14	–	31.59	nr	**35.73**
angle	21.71	0.34	7.82	–	15.73	nr	**23.55**

33 DRAINAGE ABOVE GROUND

Item	PC £	Labour hours	Labour £	Plant £	Material £	Unit	Total rate £
33.01 RAINWATER INSTALLATIONS – cont							
Cast iron; primed finish only							
Pipes and fittings; ears cast on; joints							
65 mm pipes; primed; nailed to masonry	44.19	0.48	9.54	–	46.04	m	**55.58**
extra for							
shoe	37.85	0.30	5.96	–	36.95	nr	**42.91**
bend	23.18	0.53	10.54	–	21.84	nr	**32.38**
single branch	48.65	0.67	13.32	–	47.63	nr	**60.95**
offset 225 mm projection	41.28	0.53	10.54	–	37.75	nr	**48.29**
offset 305 mm projection	48.39	0.53	10.54	–	44.16	nr	**54.70**
connection to clay pipes; cement and sand (1:2) joint	–	0.14	2.78	–	0.16	nr	**2.94**
75 mm pipes; primed; nailed to masonry	44.19	0.51	10.14	–	46.33	m	**56.47**
extra for							
shoe	37.85	0.32	6.36	–	37.44	nr	**43.80**
bend	28.13	1.11	23.93	–	27.42	nr	**51.35**
single branch	53.64	0.69	13.71	–	53.76	nr	**67.47**
offset 225 mm projection	41.28	0.56	11.13	–	38.24	nr	**49.37**
offset 305 mm projection	50.76	0.56	11.13	–	47.09	nr	**58.22**
connection to clay pipes; cement and sand (1:2) joint	–	0.16	3.18	–	0.16	nr	**3.34**
100 mm pipes; primed; nailed to masonry	59.35	0.56	11.13	–	62.26	m	**73.39**
extra for							
shoe	50.27	0.37	7.35	–	49.89	nr	**57.24**
bend	39.73	0.60	11.93	–	39.05	nr	**50.98**
single branch	62.50	0.74	14.71	–	62.85	nr	**77.56**
offset 225 mm projection	81.04	0.60	11.93	–	77.93	nr	**89.86**
offset 305 mm projection	82.62	0.60	11.93	–	78.33	nr	**90.26**
connection to clay pipes; cement and sand (1:2) joint	–	0.19	3.78	–	0.15	nr	**3.93**
100 mm × 75 mm rectangular pipes; primed; nailing to masonry	127.44	0.56	11.13	–	132.40	m	**143.53**
extra for							
shoe	151.45	0.37	7.35	–	148.51	nr	**155.86**
offset 225 mm projection	194.68	0.37	7.35	–	185.15	nr	**192.50**
offset 305 mm projection	224.33	0.37	7.35	–	213.07	nr	**220.42**
connection to clay pipes; cement and sand (1:2) joint	–	0.19	3.78	–	0.15	nr	**3.93**
Rainwater head; 225 mm × 125 mm × 125 mm, rectangular; for pipes							
65 mm dia.	116.34	0.53	10.54	–	121.43	nr	**131.97**
75 mm dia.	116.34	0.56	11.13	–	121.93	nr	**133.06**
Rainwater head; 280 mm × 130 mm × 130 mm, rectangular; for pipes							
100 mm dia.	160.64	0.60	11.93	–	168.47	nr	**180.40**
Rainwater head; flat back, octagonal; for pipes							
65 mm dia.	83.62	0.53	10.54	–	87.73	nr	**98.27**
75 mm dia.	83.62	0.56	11.13	–	88.22	nr	**99.35**

33 DRAINAGE ABOVE GROUND

Item	PC £	Labour hours	Labour £	Plant £	Material £	Unit	Total rate £
Gutters and fittings; primed finish only							
100 mm half round gutters; primed; on brackets;							
screwed to timber	21.61	0.37	8.51	–	29.47	m	**37.98**
extra for							
stop end	5.21	0.16	3.68	–	8.52	nr	**12.20**
running outlet	15.05	0.32	7.35	–	14.46	nr	**21.81**
angle	15.47	0.32	7.35	–	17.66	nr	**25.01**
115 mm half round gutters; primed; on brackets;							
screwed to timber	22.53	0.37	8.51	–	30.53	m	**39.04**
extra for							
stop end	6.71	0.16	3.68	–	10.06	nr	**13.74**
running outlet	16.42	0.32	7.35	–	15.87	nr	**23.22**
angle	16.65	0.32	7.35	–	18.78	nr	**26.13**
125 mm half round gutters; primed; on brackets;							
screwed to timber	26.37	0.42	9.65	–	34.56	m	**44.21**
extra for							
stop end	6.71	0.19	4.37	–	10.21	nr	**14.58**
running outlet	18.72	0.37	8.51	–	17.99	nr	**26.50**
angle	18.72	0.37	8.51	–	20.31	nr	**28.82**
150 mm half round gutters; primed; on brackets;							
screwed to timber	41.47	0.46	10.57	–	50.57	m	**61.14**
extra for							
stop end	9.34	0.20	4.59	–	16.58	nr	**21.17**
running outlet	32.45	0.42	9.65	–	30.96	nr	**40.61**
angle	34.24	0.42	9.65	–	34.96	nr	**44.61**
100 mm ogee gutters; primed; on brackets;							
screwed to timber	24.10	0.39	8.96	–	32.17	m	**41.13**
extra for							
stop end	5.31	0.17	3.90	–	11.51	nr	**15.41**
running outlet	16.42	0.34	7.82	–	15.82	nr	**23.64**
angle	16.12	0.34	7.82	–	18.42	nr	**26.24**
115 mm ogee gutters; primed; on brackets;							
screwed to timber	26.51	0.39	8.96	–	34.70	m	**43.66**
extra for							
stop end	6.88	0.17	3.90	–	13.19	nr	**17.09**
running outlet	17.47	0.34	7.82	–	16.70	nr	**24.52**
angle	17.47	0.34	7.82	–	19.39	nr	**27.21**
125 mm ogee gutters; primed; on brackets;							
screwed to timber	27.82	0.43	9.89	–	36.78	m	**46.67**
extra for							
stop end	6.88	0.19	4.37	–	13.94	nr	**18.31**
running outlet	21.45	0.39	8.96	–	20.83	nr	**29.79**
angle	19.06	0.39	8.96	–	21.60	nr	**30.56**
Cast iron; pre-painted finish; black standard							
Pipes and fittings; ears cast on; joints							
65 mm pipes; primed; nailed to masonry	52.16	0.48	9.54	–	54.25	m	**63.79**
extra for							
shoe	43.74	0.30	5.96	–	42.35	nr	**48.31**
bend	28.71	0.53	10.54	–	26.87	nr	**37.41**
single branch	54.27	0.67	13.32	–	52.43	nr	**65.75**

33 DRAINAGE ABOVE GROUND

Item	PC £	Labour hours	Labour £	Plant £	Material £	Unit	Total rate £
33.01 RAINWATER INSTALLATIONS – cont							
Cast iron – cont							
Pipes and fittings – cont							
extra for – cont							
offset 225 mm projection	47.35	0.53	10.54	–	42.85	nr	**53.39**
offset 305 mm projection	54.59	0.53	10.54	–	49.22	nr	**59.76**
connection to clay pipes; cement and sand (1:2) joint	–	0.14	2.78	–	0.16	nr	**2.94**
75 mm pipes; primed; nailed to masonry	52.16	0.51	10.14	–	54.54	m	**64.68**
extra for							
shoe	43.74	0.32	6.36	–	42.85	nr	**49.21**
bend	33.82	1.11	23.93	–	32.63	nr	**56.56**
single branch	59.29	0.69	13.71	–	58.59	nr	**72.30**
offset 225 mm projection	47.35	0.56	11.13	–	43.34	nr	**54.47**
offset 305 mm projection	57.01	0.56	11.13	–	52.21	nr	**63.34**
connection to clay pipes; cement and sand (1:2) joint	–	0.16	3.18	–	0.16	nr	**3.34**
100 mm pipes; primed; nailed to masonry	73.30	0.56	11.13	–	76.64	m	**87.77**
extra for							
shoe	55.59	0.37	7.35	–	54.23	nr	**61.58**
bend	45.74	0.60	11.93	–	44.08	nr	**56.01**
single branch	69.42	0.74	14.71	–	68.25	nr	**82.96**
offset 225 mm projection	88.15	0.60	11.93	–	83.23	nr	**95.16**
offset 305 mm projection	88.15	0.60	11.93	–	81.73	nr	**93.66**
connection to clay pipes; cement and sand (1:2) joint	–	0.19	3.78	–	0.15	nr	**3.93**
100 mm × 75 mm rectangular pipes; primed; nailing to masonry	127.44	0.56	11.13	–	132.40	m	**143.53**
extra for							
shoe	151.45	0.37	7.35	–	148.51	nr	**155.86**
offset 225 mm projection	194.68	0.37	7.35	–	185.15	nr	**192.50**
offset 305 mm projection	224.33	0.37	7.35	–	213.07	nr	**220.42**
connection to clay pipes; cement and sand (1:2) joint	–	0.19	3.78	–	0.15	nr	**3.93**
Rainwater head; 225 mm × 125 mm × 125 mm, rectangular; for pipes							
65 mm dia.	128.03	0.53	10.54	–	133.47	nr	**144.01**
75 mm dia.	128.03	0.56	11.13	–	133.96	nr	**145.09**
Rainwater head; 280 mm × 130 mm × 130 mm, rectangular; for pipes							
100 mm dia.	172.73	0.60	11.93	–	180.93	nr	**192.86**
Rainwater head; flat back, octagonal; for pipes							
65 mm dia.	91.02	0.53	10.54	–	95.35	nr	**105.89**
75 mm dia.	91.02	0.56	11.13	–	95.84	nr	**106.97**
Gutters and fittings; pre-painted finish; black standard							
100 mm half round gutters; pre-painted; on brackets; screwed to timber	28.71	0.37	8.51	–	38.17	m	**46.68**

33 DRAINAGE ABOVE GROUND

Item	PC £	Labour hours	Labour £	Plant £	Material £	Unit	Total rate £
extra for							
stop end	7.92	0.16	3.68	–	11.94	nr	**15.62**
running outlet	18.00	0.32	7.35	–	16.92	nr	**24.27**
angle	18.38	0.32	7.35	–	20.68	nr	**28.03**
115 mm half round gutters; pre-painted; on brackets; screwed to timber	29.51	0.37	8.51	–	39.66	m	**48.17**
extra for							
stop end	9.21	0.16	3.68	–	13.51	nr	**17.19**
running outlet	19.26	0.32	7.35	–	18.22	nr	**25.57**
angle	18.81	0.32	7.35	–	21.54	nr	**28.89**
125 mm half round gutters; pre-painted; on brackets; screwed to timber	9.67	0.42	9.65	–	19.77	m	**29.42**
extra for							
stop end	9.21	0.19	4.37	–	13.87	nr	**18.24**
running outlet	21.45	0.37	8.51	–	22.18	nr	**30.69**
angle	–	0.37	8.51	–	23.91	nr	**32.42**
150 mm half round gutters; pre-painted; on brackets; screwed to timber	50.40	0.46	10.57	–	61.04	m	**71.61**
extra for							
stop end	13.14	0.20	4.59	–	21.64	nr	**26.23**
running outlet	45.71	0.42	9.65	–	43.89	nr	**53.54**
angle	54.23	0.42	9.65	–	55.15	nr	**64.80**
100 mm ogee gutters; pre-painted; on brackets; screwed to timber	32.79	0.39	8.96	–	42.45	m	**51.41**
extra for							
stop end	7.99	0.17	3.90	–	15.47	nr	**19.37**
running outlet	19.14	0.34	7.82	–	17.90	nr	**25.72**
angle	18.94	0.34	7.82	–	21.00	nr	**28.82**
115 mm ogee gutters; pre-painted; on brackets; screwed to timber	34.08	0.39	8.96	–	44.36	m	**53.32**
extra for							
stop end	9.28	0.17	3.90	–	17.36	nr	**21.26**
running outlet	20.14	0.34	7.82	–	18.82	nr	**26.64**
angle	20.07	0.34	7.82	–	22.43	nr	**30.25**
125 mm ogee gutters; pre-painted; on brackets; screwed to timber	35.50	0.43	9.89	–	46.94	m	**56.83**
extra for							
stop end	10.44	0.19	4.37	–	19.63	nr	**24.00**
running outlet	21.70	0.39	8.96	–	21.53	nr	**30.49**
angle	21.45	0.39	8.96	–	27.03	nr	**35.99**
Steel							
3 mm thick magnesium coated galvanized pressed steel gutters and fittings; joggle joints; including bracket and stiffeners							
200 mm × 100 mm (400 mm girth) box gutter; screwed to timber	–	0.60	11.93	–	16.84	m	**28.77**
extra for							
stop end	–	0.32	6.36	–	5.15	nr	**11.51**
running outlet	–	0.65	12.92	–	29.82	nr	**42.74**
stop end outlet	–	0.32	6.36	–	38.18	nr	**44.54**

33 DRAINAGE ABOVE GROUND

Item	PC £	Labour hours	Labour £	Plant £	Material £	Unit	Total rate £
33.01 RAINWATER INSTALLATIONS – cont							
Steel – cont							
3 mm thick magnesium coated galvanized pressed steel gutters and fittings – cont							
angle	–	0.65	12.92	–	32.56	nr	**45.48**
381 mm boundary wall gutters (900 mm girth);							
bent twice; screwed to timber	–	0.60	11.93	–	18.85	m	**30.78**
extra for							
stop end	–	0.37	7.35	–	5.69	nr	**13.04**
running outlet	–	0.65	12.92	–	39.12	nr	**52.04**
stop end outlet	–	0.32	6.36	–	69.88	nr	**76.24**
angle	–	0.65	12.92	–	45.19	nr	**58.11**
457 mm boundary wall gutters (1200 mm girth);							
bent twice; screwed to timber	–	0.69	13.71	–	78.23	m	**91.94**
extra for							
stop end	–	0.37	7.35	–	27.25	nr	**34.60**
running outlet	–	0.74	14.71	–	55.83	nr	**70.54**
stop end outlet	–	0.37	7.35	–	56.95	nr	**64.30**
angle	–	0.74	14.71	–	59.86	nr	**74.57**
750 mm girth; valley gutters; Kalzip Membrane lined composite gutter system	–	–	–	–	–	m	**104.33**
extra for							
stop end	–	–	–	–	–	nr	**78.24**
running outlet	–	–	–	–	–	nr	**99.11**
uPVC							
External rainwater pipes and fittings; slip-in joints							
50 mm pipes; fixing with pipe or socket brackets;							
plugged and screwed	11.70	0.28	5.56	–	16.16	m	**21.72**
extra for							
shoe	6.86	0.19	3.78	–	8.67	nr	**12.45**
bend	8.05	0.28	5.56	–	9.90	nr	**15.46**
two bends to form offset 229 mm projection	16.10	0.28	5.56	–	16.15	nr	**21.71**
connection to clay pipes; cement and sand (1:2) joint	–	0.12	2.39	–	0.16	nr	**2.55**
68 mm pipes; fixing with pipe or socket brackets;							
plugged and screwed	9.05	0.31	6.16	–	14.24	m	**20.40**
extra for							
shoe	6.86	0.20	3.98	–	9.42	nr	**13.40**
bend	8.81	0.31	6.16	–	11.43	nr	**17.59**
single branch	17.71	0.41	8.15	–	20.60	nr	**28.75**
two bends to form offset 229 mm projection	17.63	0.31	6.16	–	18.93	nr	**25.09**
loose drain connector; cement and sand (1:2) joint	–	0.14	2.78	–	21.33	nr	**24.11**
110 mm pipes; fixing with pipe or socket brackets;							
plugged and screwed	19.60	0.33	6.56	–	30.74	m	**37.30**
extra for							
shoe	22.61	0.22	4.38	–	26.15	nr	**30.53**
bend	33.38	0.33	6.56	–	37.24	nr	**43.80**
single branch	50.42	0.44	8.74	–	54.80	nr	**63.54**

33 DRAINAGE ABOVE GROUND

Item	PC £	Labour hours	Labour £	Plant £	Material £	Unit	Total rate £
two bends to form offset 229 mm projection	66.76	0.33	6.56	–	68.80	nr	**75.36**
loose drain connector; cement and sand (1:2) joint	–	0.32	6.36	–	22.04	nr	**28.40**
65 mm square pipes; fixing with pipe or socket brackets; plugged and screwed	7.87	0.31	6.16	–	12.22	m	**18.38**
extra for							
shoe	5.76	0.20	3.98	–	7.91	nr	**11.89**
bend	8.81	0.31	6.16	–	11.05	nr	**17.21**
single branch	17.71	0.41	8.15	–	20.22	nr	**28.37**
two bends to form offset 229 mm projection	17.63	0.31	6.16	–	19.16	nr	**25.32**
drain connector; square to round; cement and sand (1:2) joint	–	0.32	6.36	–	9.04	nr	**15.40**
Rainwater head; rectangular; for pipes							
50 mm dia.	37.44	0.42	8.34	–	41.83	nr	**50.17**
68 mm dia.	30.24	0.43	8.55	–	35.91	nr	**44.46**
110 mm dia.	64.76	0.51	10.14	–	72.47	nr	**82.61**
65 mm square	25.30	0.43	8.55	–	30.05	nr	**38.60**
Gutters and fittings							
76 mm half round gutters; on brackets screwed to timber	8.89	0.28	6.44	–	13.27	m	**19.71**
extra for							
stop end	3.17	0.12	2.76	–	4.17	nr	**6.93**
running outlet	8.89	0.23	5.28	–	8.43	nr	**13.71**
stop end outlet	8.87	0.12	2.76	–	9.31	nr	**12.07**
angle	8.89	0.23	5.28	–	10.21	nr	**15.49**
112 mm half round gutters; on brackets screwed to timber	9.00	0.31	7.13	–	15.61	m	**22.74**
extra for							
stop end	5.23	0.12	2.76	–	6.97	nr	**9.73**
running outlet	10.24	0.26	5.97	–	9.81	nr	**15.78**
stop end outlet	10.24	0.12	2.76	–	11.39	nr	**14.15**
angle	11.44	0.26	5.97	–	14.20	nr	**20.17**
170 mm half round gutters; on brackets; screwed to timber	19.54	0.31	7.13	–	30.41	m	**37.54**
extra for							
stop end	8.54	0.15	3.45	–	11.86	nr	**15.31**
running outlet	19.11	0.29	6.66	–	18.07	nr	**24.73**
stop end outlet	18.16	0.15	3.45	–	20.16	nr	**23.61**
angle	20.30	0.29	6.66	–	25.42	nr	**32.08**
114 mm rectangular gutters; on brackets; screwed to timber	9.26	0.31	7.13	–	17.49	m	**24.62**
extra for							
stop end	5.23	0.12	2.76	–	6.97	nr	**9.73**
running outlet	10.24	0.29	6.66	–	9.78	nr	**16.44**
stop end outlet	10.24	0.12	2.76	–	11.37	nr	**14.13**
angle	11.44	0.26	5.97	–	14.18	nr	**20.15**

33 DRAINAGE ABOVE GROUND

Item	PC £	Labour hours	Labour £	Plant £	Material £	Unit	Total rate £
33.01 RAINWATER INSTALLATIONS – cont							
Kingspan Insulated Gutter							
Pre-laminated insulated membrane gutters							
eaves; 1250 mm girth	–	–	–	–	–	m	**121.59**
valley; 1250 mm girth	–	–	–	–	–	m	**121.59**
weir overflows	–	–	–	–	–	nr	**202.66**
stop ends	–	–	–	–	–	nr	**249.95**
T-sections	–	–	–	–	–	nr	**405.31**
33.02 FOUL DRAINAGE INSTALLATIONS							
Cast iron							
Cast iron Timesaver pipes and fittings or other equal and approved; BS 416							
50 mm pipes; primed; 3 m lengths; fixing with expanding bolts; to masonry	28.28	0.51	10.14	–	41.63	m	**51.77**
extra for							
fittings with two ends	–	0.51	10.16	–	36.28	nr	**46.44**
fittings with three ends	–	0.69	13.71	–	61.46	nr	**75.17**
bends; short radius	25.18	0.51	10.14	–	36.28	nr	**46.42**
access bends; short radius	62.07	0.51	10.14	–	74.27	nr	**84.41**
boss; 38 BSP	52.14	0.51	10.14	–	63.31	nr	**73.45**
single branch	37.88	0.69	13.71	–	62.77	nr	**76.48**
isolated Timesaver coupling joint	14.28	0.28	5.56	–	14.71	nr	**20.27**
connection to clay pipes; cement and sand (1:2) joint	–	0.12	2.39	–	0.15	nr	**2.54**
75 mm pipes; primed; 3 m lengths; fixing with standard brackets; plugged and screwed to masonry	31.63	0.51	10.14	–	51.56	m	**61.70**
extra for							
bends; short radius	28.49	0.55	10.93	–	40.72	nr	**51.65**
access bends; short radius	67.33	0.51	10.14	–	80.72	nr	**90.86**
boss; 38 BSP	52.14	0.55	10.93	–	65.08	nr	**76.01**
single branch	42.89	0.79	15.70	–	70.01	nr	**85.71**
double branch	63.68	1.02	20.27	–	107.69	nr	**127.96**
offset 115 mm projection	40.87	0.55	10.93	–	49.72	nr	**60.65**
offset 150 mm projection	48.00	0.55	10.93	–	56.09	nr	**67.02**
access pipe, round door	60.59	0.55	10.93	–	69.55	nr	**80.48**
isolated Timesaver coupling joint	15.79	0.32	6.36	–	16.26	nr	**22.62**
connection to clay pipes; cement and sand (1:2) joint	–	0.14	2.78	–	0.15	nr	**2.93**
100 mm pipes; primed; 3 m lengths; fixing with standard brackets; plugged and screwed to masonry	38.24	0.55	10.93	–	71.72	m	**82.65**
extra for							
WC connector	42.12	0.55	10.93	–	50.62	nr	**61.55**
bends; short radius	34.86	0.62	12.32	–	51.22	nr	**63.54**
access bends; short radius	73.72	0.62	12.32	–	91.25	nr	**103.57**

33 DRAINAGE ABOVE GROUND

Item	PC £	Labour hours	Labour £	Plant £	Material £	Unit	Total rate £
boss; 38 BSP	62.24	0.62	12.32	–	79.42	nr	**91.74**
single branch	53.90	0.93	18.49	–	88.70	nr	**107.19**
double branch	66.65	1.20	23.85	–	123.06	nr	**146.91**
offset 225 mm projection	59.15	0.62	12.32	–	71.12	nr	**83.44**
offset 300 mm projection	56.43	0.62	12.32	–	67.15	nr	**79.47**
access pipe, round door	63.68	0.62	12.32	–	75.00	nr	**87.32**
roof connector; for asphalt	60.20	0.62	12.32	–	75.95	nr	**88.27**
isolated Timesaver coupling joint	20.61	0.39	7.76	–	21.23	nr	**28.99**
transitional clayware socket; cement and sand (1:2) joint	41.00	0.37	7.35	–	63.59	nr	**70.94**
150 mm pipes; primed; 3 m lengths; fixing with standard brackets; plugged and screwed to masonry	79.84	0.69	13.71	–	145.16	m	**158.87**
extra for							
bends; short radius	62.24	0.77	15.31	–	94.14	nr	**109.45**
access bends; short radius	104.70	0.77	15.31	–	137.88	nr	**153.19**
boss; 38 BSP	101.60	0.77	15.31	–	132.62	nr	**147.93**
single branch	133.56	1.11	22.06	–	200.11	nr	**222.17**
double branch	187.67	1.48	29.42	–	294.53	nr	**323.95**
access pipe, round door	105.94	0.77	15.31	–	122.37	nr	**137.68**
isolated Timesaver coupling joint	–	0.46	9.15	–	42.37	nr	**51.52**
transitional clayware socket; cement and sand (1:2) joint	71.80	0.48	9.54	–	116.47	nr	**126.01**
Cast iron Ensign lightweight pipes and fittings or other equal and approved; BS EN 877							
50 mm pipes; primed; 3 m lengths; fixing with standard brackets; plugged and screwed to masonry	18.73	0.31	6.46	–	99.75	m	**106.21**
extra for							
bends; short radius	14.56	0.27	5.65	–	23.67	nr	**29.32**
single branch	23.36	0.33	6.87	–	41.42	nr	**48.29**
access pipe	38.82	0.27	5.27	–	48.67	nr	**53.94**
70 mm pipes; primed; 3 m lengths; fixing with standard brackets; plugged and screwed to masonry	21.67	0.34	7.13	–	103.06	m	**110.19**
extra for							
bends; short radius	16.39	0.30	6.19	–	26.44	nr	**32.63**
single branch	24.68	0.37	7.67	–	44.54	nr	**52.21**
access pipe	41.07	0.30	6.19	–	51.86	nr	**58.05**
100 mm pipes; primed; 3 m lengths; fixing with standard brackets; plugged and screwed to masonry	25.78	0.37	7.67	–	109.07	m	**116.74**
extra for							
bends; short radius	19.41	0.32	6.73	–	32.41	nr	**39.14**
single branch	31.43	0.39	8.08	–	57.23	nr	**65.31**
double branch	45.22	0.46	9.56	–	83.85	nr	**93.41**
access pipe	45.14	0.32	6.73	–	58.92	nr	**65.65**
connector (transitional)	41.09	0.21	4.32	–	54.74	nr	**59.06**
reducer	26.35	0.32	6.73	–	39.57	nr	**46.30**

33 DRAINAGE ABOVE GROUND

Item	PC £	Labour hours	Labour £	Plant £	Material £	Unit	Total rate £
33.02 FOUL DRAINAGE INSTALLATIONS – cont							
uPVC							
muPVC waste pipes and fittings; BS EN 1329; solvent welded joints							
32 mm pipes; fixing with pipe clips; plugged and screwed	2.02	0.23	4.57	–	3.05	m	**7.62**
extra for							
fittings with one end	–	0.16	3.18	–	1.57	nr	**4.75**
fittings with two ends	–	0.23	4.57	–	1.68	nr	**6.25**
fittings with three ends	–	0.31	6.16	–	2.21	nr	**8.37**
access plug	1.14	0.16	3.18	–	1.57	nr	**4.75**
straight coupling	1.23	0.16	3.18	–	1.65	nr	**4.83**
expansion coupling	2.17	0.23	4.57	–	2.62	nr	**7.19**
male iron to muPVC coupling	2.19	0.35	6.95	–	2.45	nr	**9.40**
sweep bend	1.25	0.23	4.57	–	1.68	nr	**6.25**
spigot/socket bend	–	0.23	4.57	–	2.49	nr	**7.06**
sweep tee	1.69	0.31	6.16	–	2.21	nr	**8.37**
40 mm pipes; fixing with pipe clips; plugged and screwed	2.50	0.28	5.56	–	3.61	m	**9.17**
extra for							
fittings with one end	–	0.18	3.57	–	1.57	nr	**5.14**
fittings with two ends	–	0.28	5.56	–	1.82	nr	**7.38**
fittings with three ends	–	0.37	7.35	–	2.68	nr	**10.03**
fittings with four ends	5.18	0.49	9.74	–	6.00	nr	**15.74**
access plug	1.14	0.18	3.57	–	1.57	nr	**5.14**
straight coupling	1.22	0.19	3.78	–	1.65	nr	**5.43**
expansion coupling	2.61	0.28	5.56	–	3.07	nr	**8.63**
male iron to muPVC coupling	2.19	0.35	6.95	–	2.45	nr	**9.40**
level invert taper	1.54	0.28	5.56	–	1.98	nr	**7.54**
sweep bend	1.40	0.28	5.56	–	1.82	nr	**7.38**
spigot/socket bend	2.35	0.28	5.56	–	2.80	nr	**8.36**
sweep tee	2.13	0.37	7.35	–	2.68	nr	**10.03**
sweep cross	5.18	0.49	9.74	–	6.00	nr	**15.74**
50 mm pipes; fixing with pipe clips; plugged and screwed	3.77	0.32	6.36	–	5.81	m	**12.17**
extra for							
fittings with one end	–	0.19	3.78	–	2.08	nr	**5.86**
fittings with two ends	–	0.32	6.36	–	2.90	nr	**9.26**
fittings with three ends	–	0.43	8.55	–	4.75	nr	**13.30**
fittings with four ends	–	0.57	11.33	–	6.26	nr	**17.59**
access plug	1.65	0.19	3.78	–	2.08	nr	**5.86**
straight coupling	2.25	0.21	4.17	–	2.70	nr	**6.87**
expansion coupling	3.53	0.32	6.36	–	4.02	nr	**10.38**
male iron to muPVC coupling	3.17	0.42	8.34	–	3.45	nr	**11.79**
level invert taper	1.91	0.32	6.36	–	2.36	nr	**8.72**
sweep bend	2.45	0.32	6.36	–	2.90	nr	**9.26**
spigot/socket bend	3.35	0.32	6.36	–	3.83	nr	**10.19**
sweep tee	2.13	0.37	7.35	–	2.68	nr	**10.03**
sweep cross	5.42	0.57	11.33	–	6.26	nr	**17.59**

33 DRAINAGE ABOVE GROUND

Item	PC £	Labour hours	Labour £	Plant £	Material £	Unit	Total rate £
uPVC overflow pipes and fittings; solvent welded joints							
19 mm pipes; fixing with pipe clips; plugged and screwed	0.65	0.20	3.98	–	1.97	m	**5.95**
extra for							
splay cut end	–	0.01	0.20	–	–	nr	**0.20**
fittings with one end	–	0.16	3.18	–	2.28	nr	**5.46**
fittings with two ends	–	0.16	3.18	–	2.65	nr	**5.83**
fittings with three ends	–	0.20	3.98	–	2.96	nr	**6.94**
straight connector	2.02	0.16	3.18	–	2.28	nr	**5.46**
female iron to uPVC coupling	–	0.19	3.78	–	3.29	nr	**7.07**
bend	2.38	0.16	3.18	–	2.65	nr	**5.83**
bent tank connector	3.80	0.19	3.78	–	4.01	nr	**7.79**
uPVC pipes and fittings; with solvent welded joints (unless otherwise described)							
82 mm pipes; fixing with holderbats; plugged and screwed	12.27	0.37	7.35	–	17.24	m	**24.59**
extra for							
socket plug	7.08	0.19	3.78	–	8.67	nr	**12.45**
slip coupling; push fit	12.13	0.34	6.76	–	12.49	nr	**19.25**
expansion coupling	9.92	0.37	7.35	–	11.60	nr	**18.95**
sweep bend	16.51	0.37	7.35	–	18.39	nr	**25.74**
boss connector	6.84	0.25	4.96	–	8.43	nr	**13.39**
single branch	17.45	0.49	9.74	–	20.18	nr	**29.92**
access door	16.64	0.56	11.13	–	17.83	nr	**28.96**
110 mm pipes; fixing with holderbats; plugged and screwed	10.66	0.41	8.15	–	15.57	m	**23.72**
extra for							
socket plug	8.59	0.20	3.98	–	10.64	nr	**14.62**
slip coupling; push fit	11.35	0.37	7.35	–	11.69	nr	**19.04**
expansion coupling	10.14	0.41	8.15	–	12.24	nr	**20.39**
WC connector	13.83	0.27	5.37	–	15.20	nr	**20.57**
sweep bend	15.00	0.41	8.15	–	17.24	nr	**25.39**
WC connecting bend	22.69	0.27	5.37	–	24.34	nr	**29.71**
access bend	40.55	0.43	8.55	–	43.56	nr	**52.11**
boss connector	6.84	0.27	5.37	–	8.84	nr	**14.21**
single branch	19.34	0.54	10.73	–	22.68	nr	**33.41**
single branch with access	33.09	0.56	11.13	–	36.85	nr	**47.98**
double branch	47.78	0.68	13.51	–	52.94	nr	**66.45**
WC manifold	18.98	0.27	5.37	–	22.32	nr	**27.69**
access door	–	0.56	11.13	–	17.83	nr	**28.96**
access pipe connector	31.07	0.46	9.15	–	33.79	nr	**42.94**
connection to clay pipes; caulking ring and cement and sand (1:2) joint	–	0.39	7.76	–	13.13	nr	**20.89**
160 mm pipes; fixing with holderbats; plugged and screwed	33.17	0.46	9.15	–	45.27	m	**54.42**
extra for							
socket plug	15.77	0.23	4.57	–	20.25	nr	**24.82**
slip coupling; push fit	29.81	0.42	8.34	–	30.70	nr	**39.04**
expansion coupling	30.53	0.46	9.15	–	35.45	nr	**44.60**

33 DRAINAGE ABOVE GROUND

Item	PC £	Labour hours	Labour £	Plant £	Material £	Unit	Total rate £
33.02 FOUL DRAINAGE INSTALLATIONS – cont							
uPVC – cont							
uPVC pipes and fittings – cont							
extra for – cont							
sweep bend	39.99	0.46	9.15	–	45.20	nr	**54.35**
boss connector	9.66	0.31	6.16	–	13.96	nr	**20.12**
single branch	41.05	0.61	12.12	–	48.22	nr	**60.34**
double branch	86.33	0.77	15.31	–	96.80	nr	**112.11**
access door	29.70	0.56	11.13	–	31.29	nr	**42.42**
access pipe connector	31.07	0.46	9.15	–	33.79	nr	**42.94**
Weathering apron; for pipe							
82 mm dia.	3.51	0.31	6.16	–	4.31	nr	**10.47**
110 mm dia.	4.04	0.35	6.95	–	5.13	nr	**12.08**
160 mm dia.	12.15	0.39	7.76	–	14.45	nr	**22.21**
Weathering slate; for pipe							
110 mm dia.	42.93	0.83	16.50	–	45.19	nr	**61.69**
Vent cowl; for pipe							
82 mm dia.	3.51	0.31	6.16	–	4.31	nr	**10.47**
110 mm dia.	3.57	0.31	6.16	–	4.65	nr	**10.81**
160 mm dia.	9.32	0.31	6.16	–	11.54	nr	**17.70**
Polypropylene							
Polypropylene (PP) waste pipes and fittings; BS EN 1451; push fit O – ring joints							
32 mm pipes; fixing with pipe clips; plugged and screwed	1.88	0.20	3.98	–	2.78	m	**6.76**
extra for							
fittings with one end	–	0.15	2.98	–	1.58	nr	**4.56**
fittings with two ends	–	0.20	3.98	–	1.61	nr	**5.59**
fittings with three ends	–	0.28	5.56	–	2.77	nr	**8.33**
access plug	1.53	0.15	2.98	–	1.58	nr	**4.56**
double socket	1.18	0.14	2.78	–	1.22	nr	**4.00**
male iron to PP coupling	2.12	0.26	5.17	–	2.18	nr	**7.35**
sweep bend	1.46	0.20	3.98	–	1.50	nr	**5.48**
spigot bend	2.12	0.23	4.57	–	2.18	nr	**6.75**
40 mm pipes; fixing with pipe clips; plugged and screwed	2.32	0.20	3.98	–	3.24	m	**7.22**
extra for							
fittings with one end	–	0.18	3.57	–	1.66	nr	**5.23**
fittings with two ends	–	0.28	5.56	–	1.91	nr	**7.47**
fittings with three ends	–	0.37	7.35	–	2.93	nr	**10.28**
access plug	1.61	0.18	3.57	–	1.66	nr	**5.23**
double socket	1.18	0.19	3.78	–	1.22	nr	**5.00**
universal connector	2.28	0.23	4.57	–	2.35	nr	**6.92**
sweep bend	1.64	0.28	5.56	–	1.69	nr	**7.25**
spigot bend	2.06	0.28	5.56	–	2.12	nr	**7.68**
reducer 40 mm–32 mm	1.10	0.28	5.56	–	1.13	nr	**6.69**
50 mm pipes; fixing with pipe clips; plugged and screwed	2.98	0.32	6.36	–	4.76	m	**11.12**

33 DRAINAGE ABOVE GROUND

Item	PC £	Labour hours	Labour £	Plant £	Material £	Unit	**Total rate £**
extra for							
fittings with one end	–	0.19	3.78	–	2.94	nr	**6.72**
fittings with two ends	–	0.32	6.36	–	3.15	nr	**9.51**
fittings with three ends	–	0.43	8.55	–	4.36	nr	**12.91**
access plug	2.85	0.19	3.78	–	2.94	nr	**6.72**
double socket	2.41	0.21	4.17	–	2.48	nr	**6.65**
sweep bend	3.13	0.32	6.36	–	3.22	nr	**9.58**
spigot bend	2.63	0.32	6.36	–	2.71	nr	**9.07**
reducer 50 mm–40 mm	1.89	0.32	6.36	–	1.95	nr	**8.31**
Polypropylene ancillaries; screwed joint to waste fitting							
Tubular S trap; bath; shallow seal							
40 mm dia.	7.01	0.51	10.14	–	7.22	nr	**17.36**
P trap; two piece; 76 mm seal							
32 mm dia.	4.74	0.35	6.95	–	4.88	nr	**11.83**
40 mm dia.	5.47	0.42	8.34	–	5.63	nr	**13.97**
S trap; two piece; 76 mm seal							
32 mm dia.	6.00	0.35	6.95	–	6.18	nr	**13.13**
40 mm dia.	7.01	0.42	8.34	–	7.22	nr	**15.56**
P bottle trap; 76 mm seal							
32 dia.	5.29	0.35	6.95	–	5.45	nr	**12.40**
40 dia.	6.30	0.42	8.34	–	6.49	nr	**14.83**

34 DRAINAGE BELOW GROUND

Item	PC £	Labour hours	Labour £	Plant £	Material £	Unit	Total rate £
34.01 URBAN AND LANDSCAPE DRAINAGE							
Slot and grate drainage							
ACO RoadDrain one piece channel drainage system for medium to heavy duty highway and distribution yards							
100 F900 units, 500 mm long	–	0.46	6.43	–	217.34	m	**223.77**
200 F900 units, 500 mm long	–	0.48	6.71	–	278.74	m	**285.45**
ACO MultiDrain Monoblock PD100D one piece channel drainage system for pedestrian and medium duty vehicle applications made from high performance recycled materials							
PD100 units, 500 mm long with integral heelguard inlets	–	0.45	6.29	–	155.04	m	**161.33**
ACO MultiDrain M100D polymer concrete channel drainage system; galvanized steel edge trim; nominal bore 100 mm; type of fall constant; bedding and haunching with in situ concrete (not included)							
Slotted grating							
galvanized steel grating, load class A15 (pedestrian areas)	–	0.46	6.43	–	107.93	m	**114.36**
galvanized steel grating, load class C250 (cars and light vans)	–	0.46	6.43	–	148.67	m	**155.10**
ductile iron grating, load class D400 (driving lanes of roads)	–	0.46	6.43	–	155.03	m	**161.46**
Heelguard resin composite grating, load class C250 (cars and light vans)	–	0.46	6.43	–	146.72	m	**153.15**
end caps	–	0.09	1.26	–	8.55	nr	**9.81**
sump unit	–	1.39	19.43	–	148.01	nr	**167.44**
ACO universal gully	–	1.50	20.97	–	883.85	nr	**904.82**
ACO MultiDrain M150D polymer concrete channel drainage system; galvanized steel edge trim; nominal bore 150 mm; type of fall constant; bedding and haunching with in situ concrete (not included)							
Slotted grating							
galvanized steel grating, load class A15 (pedestrian areas)	–	0.46	6.43	–	128.88	m	**135.31**
galvanized steel grating, load class C250 (cars and light vans)	–	0.46	6.43	–	132.13	m	**138.56**
sump unit	–	1.45	20.27	–	229.32	nr	**249.59**
ACO MultiDrain M200D polymer concrete channel drainage system; galvanized steel edge trim; nominal bore 200 mm; type of fall constant; bedding and haunching with in situ concrete (not included)							
Slotted rrating; galvanized steel grating, load class A15 (pedestrian areas)	–	0.46	6.43	–	168.30	m	**174.73**
sump unit	–	1.50	20.97	–	288.32	nr	**309.29**

34 DRAINAGE BELOW GROUND

Item	PC £	Labour hours	Labour £	Plant £	Material £	Unit	Total rate £
ACO S Range polymer concrete channel drainage system; bolted ductile iron grating, load class F900 (airfields); bedding and haunching with in situ concrete (not included)							
S100 F900 channel and grate	–	1.00	13.98	–	163.83	m	**177.81**
S150 F900 channel and grate	–	1.00	13.98	–	176.19	m	**190.17**
S200 F900 channel and grate	–	1.10	15.38	–	186.84	m	**202.22**
S300 F900 channel and grate	–	1.20	16.77	–	290.89	m	**307.66**
end caps	–	0.09	1.26	–	23.64	nr	**24.90**
sump unit	–	1.50	20.97	–	299.75	nr	**320.72**
ACO Qmax large capacity slot drainage channel with MDPE body and hot dipped galvanized steel edge rail, up to load class F900; bedding and haunching with in situ concrete (not included)							
ACO Qmax 225	–	0.75	10.49	–	101.23	m	**111.72**
ACO Qmax 350	–	1.00	13.98	–	133.24	m	**147.22**
ACO Qmax 900	–	1.50	20.97	–	254.33	m	**275.30**
shallow access chamber	–	1.50	20.97	–	287.21	nr	**308.18**
deep access chamber	–	2.00	27.95	–	685.25	nr	**713.20**
ACO Kerbdrain one-piece polymer concrete combined drainage system, load class D400; bedding and haunching in in situ concrete (not included). Manufactured from recycled and recyclable material							
KerbDrain KD305	–	0.50	6.99	–	97.28	m	**104.27**
KerbDrain KD480	–	0.65	9.08	–	105.65	m	**114.73**
KerbDrain KD305 drop kerb (left drop, one centre stone and right drop) total length 2745 mm	–	2.00	27.95	–	196.84	nr	**224.79**
KerbDrain KD305 mitre unit	–	0.25	3.49	–	106.09	nr	**109.58**
KerbDrain KD end cap	–	0.09	1.26	–	58.29	nr	**59.55**
KerbDrain KD610 shallow gully assembly	–	1.50	20.97	–	836.89	nr	**857.86**
Interconnecting drainage channel; Birco-lite or other equal; Marshalls Plc; galvanized steel grating ref 8041; bedding and haunching in in situ concrete (not included)							
100 mm wide							
laid level or to falls	–	0.46	6.43	–	50.66	m	**57.09**
100 mm dia. trapped outlet unit	–	1.39	19.43	–	111.48	nr	**130.91**
end caps	–	0.09	1.26	–	6.31	nr	**7.57**
Oil separators (polyethylene single chamber design)							
Supply and install only oil separator complete with lockable cover (excavations, filling etc. measured elsewhere)							
1000 litre ACO Q-ceptor by-pass oil separators NSB3 Class 1 (discharge concentrations of less than 5 mg/litre of oil) for discharges to surface water drains	–	2.00	27.95	61.13	1098.87	nr	**1187.95**
1000 litre ACO Q-ceptor full retention oil separators NS3 Class 1	–	2.25	31.45	69.86	1193.05	nr	**1294.36**

Prices for Measured Works

34 DRAINAGE BELOW GROUND

Item	PC £	Labour hours	Labour £	Plant £	Material £	Unit	Total rate £
34.02 TRENCHES							
Excavating trenches in firm soil; by machine; grading bottoms; earthwork support; filling with excavated material and compacting; disposal of surplus soil; spreading on site average 50 m from excavations							
Pipes not exceeding 200 mm nominal size							
average depth of trench 0.50 m	–	0.28	3.94	1.97	–	m	**5.91**
average depth of trench 0.75 m	–	0.37	5.21	2.91	–	m	**8.12**
average depth of trench 1.00 m	–	0.79	11.13	4.15	1.48	m	**16.76**
average depth of trench 1.25 m	–	1.16	16.35	4.37	2.04	m	**22.76**
average depth of trench 1.50 m	–	1.48	20.86	4.87	2.41	m	**28.14**
average depth of trench 1.75 m	–	1.85	26.07	5.10	2.97	m	**34.14**
average depth of trench 2.00 m	–	2.13	30.01	5.82	3.34	m	**39.17**
average depth of trench 2.25 m	–	2.64	37.20	7.06	4.45	m	**48.71**
average depth of trench 2.50 m	–	3.10	43.68	8.22	5.20	m	**57.10**
average depth of trench 2.75 m	–	3.42	48.19	9.24	5.76	m	**63.19**
average depth of trench 3.00 m	–	3.75	52.85	10.19	6.31	m	**69.35**
average depth of trench 3.25 m	–	4.07	57.35	10.70	6.87	m	**74.92**
average depth of trench 3.50 m	–	4.35	61.30	11.13	7.43	m	**79.86**
Pipes 225 mm nominal size							
average depth of trench 0.50 m	–	0.28	3.94	1.97	–	m	**5.91**
average depth of trench 0.75 m	–	0.37	5.21	2.91	–	m	**8.12**
average depth of trench 1.00 m	–	0.79	11.13	4.15	1.48	m	**16.76**
average depth of trench 1.25 m	–	1.16	16.35	4.37	2.04	m	**22.76**
average depth of trench 1.50 m	–	1.48	20.86	4.87	2.41	m	**28.14**
average depth of trench 1.75 m	–	1.85	26.07	5.10	2.97	m	**34.14**
average depth of trench 2.00 m	–	2.13	30.01	5.82	3.34	m	**39.17**
average depth of trench 2.25 m	–	2.64	37.20	7.06	4.45	m	**48.71**
average depth of trench 2.50 m	–	3.10	43.68	8.22	5.20	m	**57.10**
average depth of trench 2.75 m	–	3.42	48.19	9.24	5.76	m	**63.19**
average depth of trench 3.00 m	–	3.75	52.85	10.19	6.31	m	**69.35**
average depth of trench 3.25 m	–	4.07	57.35	10.70	6.87	m	**74.92**
average depth of trench 3.50 m	–	4.35	61.30	11.13	7.43	m	**79.86**
Pipes 300 mm nominal size							
average depth of trench 0.75 m	–	0.44	6.20	3.64	–	m	**9.84**
average depth of trench 1.00 m	–	0.93	13.10	4.15	1.48	m	**18.73**
average depth of trench 1.25 m	–	1.25	17.61	4.58	2.04	m	**24.23**
average depth of trench 1.50 m	–	1.62	22.82	5.10	2.41	m	**30.33**
average depth of trench 1.75 m	–	1.85	26.07	5.31	2.97	m	**34.35**
average depth of trench 2.00 m	–	2.13	30.01	6.55	3.34	m	**39.90**
average depth of trench 2.25 m	–	2.64	37.20	7.50	4.45	m	**49.15**
average depth of trench 2.50 m	–	3.10	43.68	8.52	5.20	m	**57.40**
average depth of trench 2.75 m	–	3.42	48.19	9.46	5.76	m	**63.41**
average depth of trench 3.00 m	–	3.75	52.85	10.40	6.31	m	**69.56**
average depth of trench 3.25 m	–	4.07	57.35	11.42	6.87	m	**75.64**
average depth of trench 3.50 m	–	4.35	61.30	11.64	7.43	m	**80.37**

34 DRAINAGE BELOW GROUND

Item	PC £	Labour hours	Labour £	Plant £	Material £	Unit	Total rate £
Pipes 375 mm nominal size							
average depth of trench 0.75 m	–	0.46	6.48	4.37	–	m	**10.85**
average depth of trench 1.00 m	–	0.97	13.67	4.87	1.48	m	**20.02**
average depth of trench 1.25 m	–	1.34	18.88	5.82	2.04	m	**26.74**
average depth of trench 1.50 m	–	1.71	24.10	6.04	2.41	m	**32.55**
average depth of trench 1.75 m	–	1.99	28.05	6.55	2.97	m	**37.57**
average depth of trench 2.00 m	–	2.27	31.99	6.77	3.34	m	**42.10**
average depth of trench 2.25 m	–	2.82	39.74	8.22	4.45	m	**52.41**
average depth of trench 2.50 m	–	3.38	47.63	9.46	5.20	m	**62.29**
average depth of trench 2.75 m	–	3.70	52.14	10.19	5.76	m	**68.09**
average depth of trench 3.00 m	–	4.02	56.65	10.92	6.31	m	**73.88**
average depth of trench 3.25 m	–	4.35	61.30	11.87	6.87	m	**80.04**
average depth of trench 3.50 m	–	4.67	65.81	12.59	7.43	m	**85.83**
Pipes 450 mm nominal size							
average depth of trench 0.75 m	–	0.51	7.19	4.37	–	m	**11.56**
average depth of trench 1.00 m	–	1.02	14.38	5.31	1.48	m	**21.17**
average depth of trench 1.25 m	–	1.48	20.86	6.33	2.04	m	**29.23**
average depth of trench 1.50 m	–	1.85	26.07	6.77	2.41	m	**35.25**
average depth of trench 1.75 m	–	2.13	30.01	7.06	2.97	m	**40.04**
average depth of trench 2.00 m	–	2.45	34.53	7.50	3.34	m	**45.37**
average depth of trench 2.25 m	–	3.05	42.98	8.73	4.45	m	**56.16**
average depth of trench 2.50 m	–	3.61	50.87	10.19	5.20	m	**66.26**
average depth of trench 2.75 m	–	3.98	56.08	11.13	5.76	m	**72.97**
average depth of trench 3.00 m	–	4.26	60.03	12.15	6.31	m	**78.49**
average depth of trench 3.25 m	–	4.63	65.25	13.32	6.87	m	**85.44**
average depth of trench 3.50 m	–	5.00	70.46	14.55	7.43	m	**92.44**
Pipes 600 mm nominal size							
average depth of trench 1.00 m	–	1.11	15.65	5.82	1.48	m	**22.95**
average depth of trench 1.25 m	–	1.57	22.12	6.77	2.04	m	**30.93**
average depth of trench 1.50 m	–	2.04	28.75	7.79	2.41	m	**38.95**
average depth of trench 1.75 m	–	2.31	32.55	7.79	2.97	m	**43.31**
average depth of trench 2.00 m	–	2.73	38.47	8.52	3.34	m	**50.33**
average depth of trench 2.25 m	–	3.28	46.23	10.19	4.45	m	**60.87**
average depth of trench 2.50 m	–	3.89	54.82	11.87	5.20	m	**71.89**
average depth of trench 2.75 m	–	4.30	60.59	13.32	5.76	m	**79.67**
average depth of trench 3.00 m	–	4.72	66.52	14.55	6.31	m	**87.38**
average depth of trench 3.25 m	–	5.09	71.73	15.50	6.87	m	**94.10**
average depth of trench 3.50 m	–	5.46	76.94	16.23	7.43	m	**100.60**
Pipes 900 mm nominal size							
average depth of trench 1.25 m	–	1.90	26.77	8.22	2.04	m	**37.03**
average depth of trench 1.50 m	–	2.41	33.96	9.24	2.41	m	**45.61**
average depth of trench 1.75 m	–	2.78	39.17	9.46	2.97	m	**51.60**
average depth of trench 2.00 m	–	3.10	43.68	10.92	3.34	m	**57.94**
average depth of trench 2.25 m	–	3.84	54.12	12.89	4.45	m	**71.46**
average depth of trench 2.50 m	–	4.53	63.84	14.77	5.20	m	**83.81**
average depth of trench 2.75 m	–	5.00	70.46	16.23	5.76	m	**92.45**
average depth of trench 3.00 m	–	5.46	76.94	17.69	6.31	m	**100.94**
average depth of trench 3.25 m	–	5.92	83.42	19.14	6.87	m	**109.43**
average depth of trench 3.50 m	–	6.38	89.91	20.37	7.43	m	**117.71**

34 DRAINAGE BELOW GROUND

Item	PC £	Labour hours	Labour £	Plant £	Material £	Unit	Total rate £
34.02 TRENCHES – cont							
Excavating trenches – cont							
Pipes 1200 mm nominal size							
average depth of trench 1.50 m	–	2.73	38.47	9.97	2.41	m	**50.85**
average depth of trench 1.75 m	–	3.19	44.95	11.42	2.97	m	**59.34**
average depth of trench 2.00 m	–	3.56	50.17	13.10	3.34	m	**66.61**
average depth of trench 2.25 m	–	4.35	61.30	15.50	4.45	m	**81.25**
average depth of trench 2.50 m	–	5.18	73.00	17.69	5.20	m	**95.89**
average depth of trench 2.75 m	–	5.69	80.19	19.65	5.76	m	**105.60**
average depth of trench 3.00 m	–	6.20	87.37	21.32	6.31	m	**115.00**
average depth of trench 3.25 m	–	6.75	95.12	23.07	6.87	m	**125.06**
average depth of trench 3.50 m	–	7.26	102.31	24.74	7.43	m	**134.48**
Extra over excavating trenches; irrespective of depth; breaking out existing materials							
brick	–	1.80	25.37	10.78	–	m³	**36.15**
concrete	–	2.54	35.79	14.88	–	m³	**50.67**
reinforced concrete	–	3.61	50.87	21.48	–	m³	**72.35**
Extra over excavating trenches; irrespective of depth; breaking out existing hard pavings; 75 mm thick							
tarmacadam	–	0.19	2.68	1.09	–	m²	**3.77**
Extra over excavating trenches; irrsepective of depth; breaking out existing hard pavings; 150 mm thick							
concrete	–	0.37	5.21	2.41	–	m²	**7.62**
tarmacadam and hardcore	–	0.28	3.94	1.34	–	m²	**5.28**
Excavating trenches; by hand; grading bottoms; earthwork support; filling with excavated material and compacting; disposal of surplus soil on site; spreading on site average 25 m from excavations							
Pipes not exceeding 200 mm nominal size							
average depth of trench 0.50 m	–	0.93	13.10	–	–	m	**13.10**
average depth of trench 0.75 m	–	1.39	19.59	–	–	m	**19.59**
average depth of trench 1.00 m	–	2.04	28.75	–	1.48	m	**30.23**
average depth of trench 1.25 m	–	2.87	40.45	–	2.04	m	**42.49**
average depth of trench 1.50 m	–	3.93	55.38	–	2.48	m	**57.86**
average depth of trench 1.75 m	–	5.18	73.00	–	2.97	m	**75.97**
average depth of trench 2.00 m	–	5.92	83.42	–	3.34	m	**86.76**
average depth of trench 2.25 m	–	7.40	104.28	–	4.45	m	**108.73**
average depth of trench 2.50 m	–	8.88	125.13	–	5.20	m	**130.33**
average depth of trench 2.75 m	–	9.76	137.54	–	5.76	m	**143.30**
average depth of trench 3.00 m	–	10.64	149.94	–	6.31	m	**156.25**
average depth of trench 3.25 m	–	11.52	162.34	–	6.87	m	**169.21**
average depth of trench 3.50 m	–	12.40	174.74	–	7.43	m	**182.17**
Pipes 225 mm nominal size							
average depth of trench 0.50 m	–	0.93	13.10	–	–	m	**13.10**
average depth of trench 0.75 m	–	1.39	19.59	–	–	m	**19.59**
average depth of trench 1.00 m	–	2.04	28.75	–	1.48	m	**30.23**

34 DRAINAGE BELOW GROUND

Item	PC £	Labour hours	Labour £	Plant £	Material £	Unit	Total rate £
average depth of trench 1.25 m	–	2.87	40.45	–	2.04	m	**42.49**
average depth of trench 1.50 m	–	3.93	55.38	–	2.48	m	**57.86**
average depth of trench 1.75 m	–	5.18	73.00	–	2.97	m	**75.97**
average depth of trench 2.00 m	–	5.92	83.42	–	3.34	m	**86.76**
average depth of trench 2.25 m	–	7.40	104.28	–	4.45	m	**108.73**
average depth of trench 2.50 m	–	8.88	125.13	–	5.20	m	**130.33**
average depth of trench 2.75 m	–	9.76	137.54	–	5.76	m	**143.30**
average depth of trench 3.00 m	–	10.64	149.94	–	6.31	m	**156.25**
average depth of trench 3.25 m	–	11.52	162.34	–	6.87	m	**169.21**
average depth of trench 3.50 m	–	12.40	174.74	–	7.43	m	**182.17**
Pipes 300 mm nominal size							
average depth of trench 0.75 m	–	1.62	22.82	–	–	m	**22.82**
average depth of trench 1.00 m	–	2.36	33.26	–	1.48	m	**34.74**
average depth of trench 1.25 m	–	3.33	46.93	–	2.04	m	**48.97**
average depth of trench 1.50 m	–	4.44	62.57	–	2.48	m	**65.05**
average depth of trench 1.75 m	–	5.18	73.00	–	2.97	m	**75.97**
average depth of trench 2.00 m	–	5.92	83.42	–	3.34	m	**86.76**
average depth of trench 2.25 m	–	7.40	104.28	–	4.45	m	**108.73**
average depth of trench 2.50 m	–	8.88	125.13	–	5.20	m	**130.33**
average depth of trench 2.75 m	–	9.76	137.54	–	5.76	m	**143.30**
average depth of trench 3.00 m	–	10.64	149.94	–	6.31	m	**156.25**
average depth of trench 3.25 m	–	11.52	162.34	–	6.87	m	**169.21**
average depth of trench 3.50 m	–	12.40	174.74	–	7.43	m	**182.17**
Pipes 375 mm nominal size							
average depth of trench 0.75 m	–	1.80	25.37	–	–	m	**25.37**
average depth of trench 1.00 m	–	2.64	37.20	–	1.48	m	**38.68**
average depth of trench 1.25 m	–	3.70	52.14	–	2.04	m	**54.18**
average depth of trench 1.50 m	–	4.93	69.47	–	2.48	m	**71.95**
average depth of trench 1.75 m	–	5.74	80.89	–	2.97	m	**83.86**
average depth of trench 2.00 m	–	6.57	92.59	–	3.34	m	**95.93**
average depth of trench 2.25 m	–	8.23	115.98	–	4.45	m	**120.43**
average depth of trench 2.50 m	–	9.90	139.51	–	5.20	m	**144.71**
average depth of trench 2.75 m	–	10.87	153.18	–	5.76	m	**158.94**
average depth of trench 3.00 m	–	11.84	166.85	–	6.31	m	**173.16**
average depth of trench 3.25 m	–	12.86	181.22	–	6.87	m	**188.09**
average depth of trench 3.50 m	–	13.88	195.60	–	7.43	m	**203.03**
Pipes 450 mm nominal size							
average depth of trench 0.75 m	–	2.04	28.75	–	–	m	**28.75**
average depth of trench 1.00 m	–	2.94	41.43	–	1.48	m	**42.91**
average depth of trench 1.25 m	–	4.13	58.20	–	2.04	m	**60.24**
average depth of trench 1.50 m	–	5.41	76.24	–	2.48	m	**78.72**
average depth of trench 1.75 m	–	6.31	88.92	–	2.97	m	**91.89**
average depth of trench 2.00 m	–	7.22	101.74	–	3.34	m	**105.08**
average depth of trench 2.25 m	–	9.05	127.53	–	4.45	m	**131.98**
average depth of trench 2.50 m	–	10.87	153.18	–	5.20	m	**158.38**
average depth of trench 2.75 m	–	11.96	168.54	–	5.76	m	**174.30**
average depth of trench 3.00 m	–	13.04	183.76	–	6.31	m	**190.07**
average depth of trench 3.25 m	–	14.11	198.84	–	6.87	m	**205.71**
average depth of trench 3.50 m	–	15.17	213.78	–	7.43	m	**221.21**

34 DRAINAGE BELOW GROUND

Item	PC £	Labour hours	Labour £	Plant £	Material £	Unit	Total rate £
34.02 TRENCHES – cont							
Excavating trenches – cont							
Pipes 600 mm nominal size							
average depth of trench 1.00 m	–	3.24	45.66	–	1.48	m	**47.14**
average depth of trench 1.25 m	–	4.63	65.25	–	2.04	m	**67.29**
average depth of trench 1.50 m	–	6.20	87.37	–	2.48	m	**89.85**
average depth of trench 1.75 m	–	7.17	101.04	–	2.97	m	**104.01**
average depth of trench 2.00 m	–	8.19	115.41	–	3.34	m	**118.75**
average depth of trench 2.25 m	–	9.20	129.65	–	4.45	m	**134.10**
average depth of trench 2.50 m	–	11.56	162.90	–	5.20	m	**168.10**
average depth of trench 2.75 m	–	12.35	174.04	–	5.76	m	**179.80**
average depth of trench 3.00 m	–	14.80	208.56	–	6.31	m	**214.87**
average depth of trench 3.25 m	–	16.03	225.89	–	6.87	m	**232.76**
average depth of trench 3.50 m	–	17.25	243.09	–	7.43	m	**250.52**
Pipes 900 mm nominal size							
average depth of trench 1.25 m	–	5.78	81.45	–	2.04	m	**83.49**
average depth of trench 1.50 m	–	7.63	107.52	–	2.48	m	**110.00**
average depth of trench 1.75 m	–	8.88	125.13	–	2.97	m	**128.10**
average depth of trench 2.00 m	–	10.13	142.75	–	3.34	m	**146.09**
average depth of trench 2.25 m	–	12.72	179.25	–	4.45	m	**183.70**
average depth of trench 2.50 m	–	15.31	215.74	–	5.20	m	**220.94**
average depth of trench 2.75 m	–	16.84	237.31	–	5.76	m	**243.07**
average depth of trench 3.00 m	–	18.32	258.17	–	6.31	m	**264.48**
average depth of trench 3.25 m	–	19.84	279.58	–	6.87	m	**286.45**
average depth of trench 3.50 m	–	21.37	301.14	–	7.43	m	**308.57**
Pipes 1200 mm nominal size							
average depth of trench 1.50 m	–	9.11	128.38	–	2.48	m	**130.86**
average depth of trench 1.75 m	–	10.59	149.24	–	2.97	m	**152.21**
average depth of trench 2.00 m	–	12.12	170.79	–	3.34	m	**174.13**
average depth of trench 2.25 m	–	15.20	214.20	–	4.45	m	**218.65**
average depth of trench 2.50 m	–	18.27	257.46	–	5.20	m	**262.66**
average depth of trench 2.75 m	–	20.07	282.83	–	5.76	m	**288.59**
average depth of trench 3.00 m	–	21.88	308.33	–	6.31	m	**314.64**
average depth of trench 3.25 m	–	23.66	333.41	–	6.87	m	**340.28**
average depth of trench 3.50 m	–	25.44	358.50	–	7.43	m	**365.93**
Extra over excavating trenches irrespective of depth; breaking out existing materials using compressor and breaker							
brick	–	2.78	39.17	8.83	–	m³	**48.00**
concrete	–	4.16	58.62	14.71	–	m³	**73.33**
reinforced concrete	–	5.55	78.21	20.61	–	m³	**98.82**
150 mm concrete	–	0.65	9.16	2.07	–	m²	**11.23**
75 mm thick tarmacadam	–	0.37	5.21	1.18	–	m²	**6.39**
150 mm thick tarmacadam and hardcore	–	0.46	6.48	1.46	–	m²	**7.94**

34 DRAINAGE BELOW GROUND

Item	PC £	Labour hours	Labour £	Plant £	Material £	Unit	Total rate £
34.03 BEDS AND FILLINGS							
Sand filling							
Beds; to receive pitch fibre pipes							
600 mm × 50 mm thick	–	0.07	0.99	0.67	1.36	m	**3.02**
700 mm × 50 mm thick	–	0.09	1.27	0.67	1.59	m	**3.53**
800 mm × 50 mm thick	–	0.11	1.55	0.67	1.81	m	**4.03**
Granular (shingle) filling							
Beds; 100 mm thick; to pipes							
100 mm nominal size	–	0.09	1.27	0.67	3.58	m	**5.52**
150 mm nominal size	–	0.09	1.27	0.67	4.18	m	**6.12**
225 mm nominal size	–	0.11	1.55	0.67	4.78	m	**7.00**
300 mm nominal size	–	0.13	1.83	0.67	5.38	m	**7.88**
375 mm nominal size	–	0.15	2.11	0.67	5.97	m	**8.75**
450 mm nominal size	–	0.17	2.40	0.67	6.57	m	**9.64**
600 mm nominal size	–	0.19	2.68	0.67	7.17	m	**10.52**
Beds; 150 mm thick; to pipes							
100 mm nominal size	–	0.13	1.83	0.67	5.38	m	**7.88**
150 mm nominal size	–	0.15	2.11	0.67	5.97	m	**8.75**
225 mm nominal size	–	0.17	2.40	0.67	6.57	m	**9.64**
300 mm nominal size	–	0.19	2.68	0.67	7.17	m	**10.52**
375 mm nominal size	–	0.22	3.10	0.67	8.96	m	**12.73**
450 mm nominal size	–	0.24	3.38	0.67	9.56	m	**13.61**
600 mm nominal size	–	0.28	3.94	0.67	11.35	m	**15.96**
Beds and benchings; beds 100 mm thick; to pipes							
100 nominal size	–	0.21	2.96	0.67	6.57	m	**10.20**
150 nominal size	–	0.23	3.24	0.67	6.57	m	**10.48**
225 nominal size	–	0.28	3.94	0.67	8.96	m	**13.57**
300 nominal size	–	0.32	4.51	0.67	10.16	m	**15.34**
375 nominal size	–	0.42	5.92	0.67	13.74	m	**20.33**
450 nominal size	–	0.48	6.77	0.67	15.52	m	**22.96**
600 nominal size	–	0.62	8.73	0.67	20.31	m	**29.71**
Beds and benchings; beds 150 mm thick; to pipes							
100 nominal size	–	0.23	3.24	0.67	7.17	m	**11.08**
150 nominal size	–	0.26	3.67	0.67	7.77	m	**12.11**
225 nominal size	–	0.32	4.51	0.67	10.75	m	**15.93**
300 nominal size	–	0.42	5.92	0.67	13.14	m	**19.73**
375 nominal size	–	0.48	6.77	0.67	15.52	m	**22.96**
450 nominal size	–	0.57	8.03	0.67	18.52	m	**27.22**
600 nominal size	–	0.68	9.58	0.67	23.90	m	**34.15**
Beds and coverings; 100 mm thick; to pipes							
100 nominal size	–	0.33	4.65	0.67	8.96	m	**14.28**
150 nominal size	–	0.42	5.92	0.67	10.75	m	**17.34**
225 nominal size	–	0.56	7.89	0.67	14.92	m	**23.48**
300 nominal size	–	0.67	9.45	0.67	17.92	m	**28.04**
375 nominal size	–	0.80	11.28	0.67	21.50	m	**33.45**
450 nominal size	–	0.94	13.25	0.67	25.68	m	**39.60**
600 nominal size	–	1.22	17.19	0.67	32.86	m	**50.72**

34 DRAINAGE BELOW GROUND

Item	PC £	Labour hours	Labour £	Plant £	Material £	Unit	Total rate £
34.03 BEDS AND FILLINGS – cont							
Granular (shingle) filling – cont							
Beds and coverings; 150 mm thick; to pipes							
100 nominal size	–	0.50	7.05	0.67	13.14	m	**20.86**
150 nominal size	–	0.56	7.89	0.67	14.92	m	**23.48**
225 nominal size	–	0.72	10.15	0.67	19.12	m	**29.94**
300 nominal size	–	0.86	12.12	0.67	22.70	m	**35.49**
375 nominal size	–	1.00	14.09	0.67	26.88	m	**41.64**
450 nominal size	–	1.19	16.77	0.67	32.26	m	**49.70**
600 nominal size	–	1.44	20.29	0.67	38.82	m	**59.78**
Plain in situ ready mixed designated concrete; C10–40 mm aggregate							
Beds; 100 mm thick; to pipes							
100 mm nominal size	–	0.17	2.81	0.67	5.06	m	**8.54**
150 mm nominal size	–	0.17	2.81	0.67	5.06	m	**8.54**
225 mm nominal size	–	0.20	3.31	0.67	6.07	m	**10.05**
300 mm nominal size	–	0.23	3.80	0.67	7.08	m	**11.55**
375 mm nominal size	–	0.27	4.46	0.67	8.09	m	**13.22**
450 mm nominal size	–	0.30	4.95	0.67	9.09	m	**14.71**
600 mm nominal size	–	0.33	5.45	0.67	10.10	m	**16.22**
900 mm nominal size	–	0.40	6.61	0.67	12.12	m	**19.40**
1200 mm nominal size	–	0.54	8.92	0.67	16.17	m	**25.76**
Beds; 150 mm thick; to pipes							
100 mm nominal size	–	0.23	3.80	0.67	7.08	m	**11.55**
150 mm nominal size	–	0.27	4.46	0.67	8.09	m	**13.22**
225 mm nominal size	–	0.30	4.95	0.67	9.09	m	**14.71**
300 mm nominal size	–	0.33	5.45	0.67	10.10	m	**16.22**
375 mm nominal size	–	0.40	6.61	0.67	12.12	m	**19.40**
450 mm nominal size	–	0.43	7.11	0.67	13.14	m	**20.92**
600 mm nominal size	–	0.50	8.26	0.67	15.16	m	**24.09**
900 mm nominal size	–	0.63	10.41	0.67	19.20	m	**30.28**
1200 mm nominal size	–	0.77	12.72	0.67	23.25	m	**36.64**
Beds and benchings; beds 100 mm thick; to pipes							
100 mm nominal size	–	0.33	5.45	0.67	9.09	m	**15.21**
150 mm nominal size	–	0.38	6.28	0.67	10.10	m	**17.05**
225 mm nominal size	–	0.45	7.44	0.67	12.12	m	**20.23**
300 mm nominal size	–	0.53	8.76	0.67	14.14	m	**23.57**
375 mm nominal size	–	0.68	11.24	0.67	18.19	m	**30.10**
450 mm nominal size	–	0.80	13.21	0.67	21.22	m	**35.10**
600 mm nominal size	–	1.02	16.85	0.67	27.28	m	**44.80**
900 mm nominal size	–	1.65	27.26	0.67	44.45	m	**72.38**
1200 mm nominal size	–	2.44	40.31	0.67	65.67	m	**106.65**
Beds and benchings; beds 150 mm thick; to pipes							
100 mm nominal size	–	0.38	6.28	0.67	10.10	m	**17.05**
150 mm nominal size	–	0.42	6.94	0.67	11.11	m	**18.72**
225 mm nominal size	–	0.53	8.76	0.67	14.14	m	**23.57**
300 mm nominal size	–	0.68	11.24	0.67	18.19	m	**30.10**
375 mm nominal size	–	0.80	13.21	0.67	21.22	m	**35.10**
450 mm nominal size	–	0.94	15.53	0.67	25.26	m	**41.46**
600 mm nominal size	–	1.20	19.83	0.67	32.33	m	**52.83**

34 DRAINAGE BELOW GROUND

Item	PC £	Labour hours	Labour £	Plant £	Material £	Unit	Total rate £
900 mm nominal size	–	1.91	31.56	0.67	51.53	m	**83.76**
1200 mm nominal size	–	2.70	44.61	0.67	72.75	m	**118.03**
Beds and coverings; 100 mm thick; to pipes							
100 mm nominal size	–	0.50	8.26	0.67	12.12	m	**21.05**
150 mm nominal size	–	0.58	9.58	0.67	14.14	m	**24.39**
225 mm nominal size	–	0.83	13.71	0.67	20.21	m	**34.59**
300 mm nominal size	–	1.00	16.52	0.67	24.25	m	**41.44**
375 mm nominal size	–	1.21	19.99	0.67	29.29	m	**49.95**
450 mm nominal size	–	1.42	23.46	0.67	34.35	m	**58.48**
600 mm nominal size	–	1.83	30.24	0.67	44.45	m	**75.36**
900 mm nominal size	–	2.79	46.10	0.67	67.70	m	**114.47**
1200 mm nominal size	–	3.83	63.28	0.67	92.96	m	**156.91**
Beds and coverings; 150 mm thick; to pipes							
100 mm nominal size	–	0.75	12.39	0.67	18.19	m	**31.25**
150 mm nominal size	–	0.83	13.71	0.67	20.21	m	**34.59**
225 mm nominal size	–	1.08	17.85	0.67	26.27	m	**44.79**
300 mm nominal size	–	1.30	21.48	0.67	31.31	m	**53.46**
375 mm nominal size	–	1.50	24.78	0.67	36.37	m	**61.82**
450 mm nominal size	–	1.79	29.58	0.67	43.45	m	**73.70**
600 mm nominal size	–	2.16	35.69	0.67	52.54	m	**88.90**
900 mm nominal size	–	3.54	58.49	0.67	85.88	m	**145.04**
1200 mm nominal size	–	5.00	82.62	0.67	121.24	m	**204.53**
Plain in situ ready mixed designated concrete; C20–40 mm aggregate							
Beds; 100 mm thick; to pipes							
100 mm nominal size	–	0.17	2.81	0.67	5.24	m	**8.72**
150 mm nominal size	–	0.17	2.81	0.67	5.24	m	**8.72**
225 mm nominal size	–	0.20	3.31	0.67	6.28	m	**10.26**
300 mm nominal size	–	0.23	3.80	0.67	7.33	m	**11.80**
375 mm nominal size	–	0.27	4.46	0.67	8.37	m	**13.50**
450 mm nominal size	–	0.30	4.95	0.67	9.43	m	**15.05**
600 mm nominal size	–	0.33	5.45	0.67	10.48	m	**16.60**
900 mm nominal size	–	0.40	6.61	0.67	12.57	m	**19.85**
1200 mm nominal size	–	0.54	8.92	0.67	16.76	m	**26.35**
Beds; 150 mm thick; to pipes							
100 mm nominal size	–	0.23	3.80	0.67	7.33	m	**11.80**
150 mm nominal size	–	0.27	4.46	0.67	8.37	m	**13.50**
225 mm nominal size	–	0.30	4.95	0.67	9.43	m	**15.05**
300 mm nominal size	–	0.33	5.45	0.67	10.48	m	**16.60**
375 mm nominal size	–	0.40	6.61	0.67	12.57	m	**19.85**
450 mm nominal size	–	0.43	7.11	0.67	13.62	m	**21.40**
600 mm nominal size	–	0.50	8.26	0.67	15.72	m	**24.65**
900 mm nominal size	–	0.63	10.41	0.67	19.90	m	**30.98**
1200 mm nominal size	–	0.77	12.72	0.67	24.09	m	**37.48**
Beds and benchings; beds 100 mm thick; to pipes							
100 mm nominal size	–	0.33	5.45	0.67	9.43	m	**15.55**
150 mm nominal size	–	0.38	6.28	0.67	10.48	m	**17.43**
225 mm nominal size	–	0.45	7.44	0.67	12.57	m	**20.68**

34 DRAINAGE BELOW GROUND

Item	PC £	Labour hours	Labour £	Plant £	Material £	Unit	Total rate £
34.03 BEDS AND FILLINGS – cont							
Plain in situ ready mixed designated concrete – cont							
Beds and benchings – cont							
300 mm nominal size	–	0.53	8.76	0.67	14.66	m	**24.09**
375 mm nominal size	–	0.68	11.24	0.67	18.85	m	**30.76**
450 mm nominal size	–	0.80	13.21	0.67	21.99	m	**35.87**
600 mm nominal size	–	1.02	16.85	0.67	28.28	m	**45.80**
900 mm nominal size	–	1.65	27.26	0.67	46.08	m	**74.01**
1200 mm nominal size	–	2.44	40.31	0.67	68.06	m	**109.04**
Beds and benchings; beds 150 mm thick; to pipes							
100 mm nominal size	–	0.38	6.28	0.67	10.48	m	**17.43**
150 mm nominal size	–	0.42	6.94	0.67	11.52	m	**19.13**
225 mm nominal size	–	0.53	8.76	0.67	14.66	m	**24.09**
300 mm nominal size	–	0.68	11.24	0.67	18.85	m	**30.76**
375 mm nominal size	–	0.80	13.21	0.67	21.99	m	**35.87**
450 mm nominal size	–	0.94	15.53	0.67	26.17	m	**42.37**
600 mm nominal size	–	1.20	19.83	0.67	33.52	m	**54.02**
900 mm nominal size	–	1.91	31.56	0.67	53.42	m	**85.65**
1200 mm nominal size	–	2.70	44.61	0.67	75.40	m	**120.68**
Beds and coverings; 100 mm thick; to pipes							
100 mm nominal size	–	0.50	8.26	0.67	12.57	m	**21.50**
150 mm nominal size	–	0.58	9.58	0.67	14.66	m	**24.91**
225 mm nominal size	–	0.83	13.71	0.67	20.94	m	**35.32**
300 mm nominal size	–	1.00	16.52	0.67	25.13	m	**42.32**
375 mm nominal size	–	1.21	19.99	0.67	30.36	m	**51.02**
450 mm nominal size	–	1.42	23.46	0.67	35.61	m	**59.74**
600 mm nominal size	–	1.83	30.24	0.67	46.08	m	**76.99**
900 mm nominal size	–	2.79	46.10	0.67	70.17	m	**116.94**
1200 mm nominal size	–	3.83	63.28	0.67	96.35	m	**160.30**
Beds and coverings; 150 mm thick; to pipes							
100 mm nominal size	–	0.75	12.39	0.67	18.85	m	**31.91**
150 mm nominal size	–	0.83	13.71	0.67	20.94	m	**35.32**
225 mm nominal size	–	1.08	17.85	0.67	27.23	m	**45.75**
300 mm nominal size	–	1.30	21.48	0.67	32.46	m	**54.61**
375 mm nominal size	–	1.50	24.78	0.67	37.70	m	**63.15**
450 mm nominal size	–	1.79	29.58	0.67	45.02	m	**75.27**
600 mm nominal size	–	2.16	35.69	0.67	54.46	m	**90.82**
900 mm nominal size	–	3.54	58.49	0.67	89.01	m	**148.17**
1200 mm nominal size	–	5.00	82.62	0.67	125.67	m	**208.96**

34.04 PIPES, FITTINGS AND ACCESSORIES
NOTE: The following items unless otherwise described include for all appropriate joints/couplings in the running length. The prices for gullies and rainwater shoes, etc. include for appropriate joints to pipes and for setting on and surrounding accessory with site mixed in situ concrete 10.00 N/mm^2–40 mm aggregate

34 DRAINAGE BELOW GROUND

Item	PC £	Labour hours	Labour £	Plant £	Material £	Unit	Total rate £
Cast Iron							
Timesaver drain pipes and fittings or other equal; coated; with mechanical coupling joints							
100 mm dia. pipes							
laid straight	48.36	0.46	6.43	–	63.42	m	**69.85**
in shorter runs not exceeding 3 m long	48.36	0.63	8.81	–	98.95	m	**107.76**
bend; medium radius	59.41	0.56	7.83	–	86.05	nr	**93.88**
bend; medium radius with access	165.07	0.56	7.83	–	194.88	nr	**202.71**
bend; long radius	98.18	0.56	7.83	–	123.36	nr	**131.19**
rest bend	68.14	0.56	7.83	–	92.43	nr	**100.26**
single branch	78.80	0.69	9.64	–	133.76	nr	**143.40**
single branch; with access	181.80	0.79	11.04	–	239.85	nr	**250.89**
double branch	134.00	0.88	12.30	–	219.12	nr	**231.42**
isolated Timesaver joint	31.75	0.32	4.47	–	32.70	nr	**37.17**
transitional pipe; for WC	42.12	0.46	6.43	–	76.09	nr	**82.52**
150 mm dia. pipes							
laid straight	89.52	0.56	7.83	–	110.28	m	**118.11**
in shorter runs not exceeding 3 m long	89.52	0.76	10.62	–	162.25	m	**172.87**
bend; medium radius	136.69	0.65	9.08	–	165.89	nr	**174.97**
bend; medium radius with access	289.84	0.65	9.08	–	323.63	nr	**332.71**
bend; long radius	186.57	0.65	9.08	–	212.43	nr	**221.51**
rest bend	190.90	0.56	7.83	–	218.58	nr	**226.41**
single branch	170.20	0.79	11.04	–	179.11	nr	**190.15**
single 100 mm branch; with access	355.60	0.79	11.04	–	425.69	nr	**436.73**
taper pipe 150 mm to 100 mm	77.43	0.65	9.08	–	100.01	nr	**109.09**
isolated Timesaver joint	38.47	0.39	5.45	–	39.62	nr	**45.07**
Accessories in Timesaver cast iron or equal; with mechanical coupling joints							
Gully fittings; comprising low invert gully trap and round hopper							
100 mm outlet	76.88	0.88	12.30	–	119.11	nr	**131.41**
150 mm outlet	191.32	1.20	16.77	–	247.89	nr	**264.66**
Gulley inlet for bellmouth; circular plain grate							
100 mm nominal size; 200 mm plain grate	80.08	0.42	5.87	–	127.61	nr	**133.48**
100 mm nominal size; 100 mm horizontal inlet; 200 mm plain grate	107.59	0.42	5.87	–	156.65	nr	**162.52**
100 mm nominal size; 100 mm vertical inlet; 200 mm plain grate	107.59	0.42	5.87	–	156.65	nr	**162.52**
Yard gully (Deans); trapped; galvanized sediment pan; 300 mm round heavy grate							
100 mm outlet	519.75	2.68	37.46	–	627.33	nr	**664.79**
Yard gully (garage); trapless; galvanized sediment pan; 300 mm round heavy grate							
100 mm outlet	571.38	2.50	34.94	–	642.23	nr	**677.17**
Yard gully (garage); trapped; with rodding eye, galvanized perforated sediment pan; stopper; 405 mm round heavy grate							
100 mm outlet	1077.87	2.50	34.94	–	1293.75	nr	**1328.69**

34 DRAINAGE BELOW GROUND

Item	PC £	Labour hours	Labour £	Plant £	Material £	Unit	Total rate £
34.04 PIPES, FITTINGS AND ACCESSORIES – cont							
Accessories in Timesaver cast iron or equal – cont							
Grease trap; internal access; galvanized perforated bucket; lid and frame							
100 mm outlet; 20 gallon capacity	1080.17	3.70	51.72	–	1193.70	nr	**1245.42**
Ensign lightweight drain pipes and fittings or other equal; ductile iron couplings							
100 mm dia. pipes							
laid straight	34.51	0.19	3.90	–	41.39	m	**45.29**
bend; long radius	53.40	0.19	3.90	–	72.54	nr	**76.44**
single branch	36.92	0.23	4.84	–	73.11	nr	**77.95**
150 mm dia. pipes							
laid straight	67.07	0.22	4.57	–	81.02	m	**85.59**
bend; medium radius	160.15	0.22	4.57	–	200.79	nr	**205.36**
single branch	86.78	0.28	5.79	–	161.03	nr	**166.82**
Clay							
Extra strength vitrified clay pipes and fittings; Hepworth Supersleve or other equal; plain ends with push fit polypropylene flexible couplings							
100 mm dia. pipes (Supersleve House Drain)							
laid straight	6.25	0.19	2.66	–	11.67	m	**14.33**
bend	9.28	0.19	2.66	–	18.09	nr	**20.75**
rest bend	16.44	0.19	2.66	–	25.46	nr	**28.12**
rodding point	52.71	0.19	2.66	–	62.17	nr	**64.83**
socket adaptor	12.13	0.16	2.24	–	17.08	nr	**19.32**
saddle	19.69	0.69	9.64	–	25.51	nr	**35.15**
single junction	20.06	0.23	3.21	–	33.77	nr	**36.98**
single access junction with alloy frame and lid							
(pedestrian areas only)	70.62	0.25	3.49	–	127.17	nr	**130.66**
150 mm dia. pipes							
laid straight	12.65	0.23	3.21	–	22.19	m	**25.40**
bend	18.45	0.22	3.08	–	33.41	nr	**36.49**
access bend	10.15	0.22	3.08	–	131.70	nr	**134.78**
90° bend	18.45	0.22	3.08	–	33.41	nr	**36.49**
taper pipe	22.83	0.22	3.08	–	34.00	nr	**37.08**
rodding point	66.51	0.22	3.08	–	81.62	nr	**84.70**
socket adaptor	18.71	0.19	2.66	–	27.13	nr	**29.79**
saddle	28.24	0.83	11.60	–	38.24	nr	**49.84**
single junction	24.68	0.28	3.91	–	47.69	nr	**51.60**
single access junction (pedestrian areas only)	101.22	0.28	3.91	–	167.86	nr	**171.77**
225 mm dia. pipes							
laid straight	36.91	0.23	3.21	–	38.02	m	**41.23**
bend	77.85	0.22	3.08	–	103.93	nr	**107.01**
90° bend	77.85	0.22	3.08	–	103.93	nr	**107.01**
taper pipe	71.22	0.22	3.08	–	79.52	nr	**82.60**
socket adaptor	42.65	0.19	2.66	–	57.70	nr	**60.36**

34 DRAINAGE BELOW GROUND

Item	PC £	Labour hours	Labour £	Plant £	Material £	Unit	Total rate £
300 mm dia. pipes includiing EPDM coupling							
laid straight	56.23	0.23	3.21	–	94.02	m	**97.23**
bend	141.52	0.24	3.36	–	200.57	nr	**203.93**
90° bend	141.52	0.24	3.36	–	200.57	nr	**203.93**
taper pipe	71.22	0.24	3.36	–	92.07	nr	**95.43**
socket adaptor	63.81	0.20	2.79	–	96.03	nr	**98.82**
Accessories in vitrified clay; set in concrete; with polypropylene coupling joints to pipes							
Rodding point; with square aluminium plate							
100 mm nominal size	44.96	0.46	6.43	–	53.56	nr	**59.99**
Gully fittings; comprising low back trap and square hopper; 150 mm × 150 mm square gully grid							
100 mm nominal size	46.27	0.79	11.04	–	60.13	nr	**71.17**
Gully fittings; comprising low back trap and square hopper with back inlet; 150 mm × 150 mm square gully grid							
100 mm nominal size	63.11	0.85	11.88	–	77.48	nr	**89.36**
Accessories in vitrified clay; set in concrete; with cement and sand (1:2) joints to pipes							
Yard gully; 225 mm dia.; including domestic duty grate and frame (up to 1 tonne) and combined filter and silk bucket							
100 mm outlet	194.45	2.50	34.94	–	200.90	nr	**235.84**
100 mm outlet; 100 mm back inlet	335.13	2.70	37.74	–	345.79	nr	**383.53**
150 mm outlet	194.45	3.50	48.91	–	200.90	nr	**249.81**
150 mm outlet; 150 mm back inlet	340.65	3.70	51.72	–	351.49	nr	**403.21**
Yard gully; 225 mm dia.; including medium duty grate and frame (up to 5 tonnes) and combined filter and silk bucket							
100 mm outlet	253.53	2.50	34.94	–	261.74	nr	**296.68**
100 mm outlet; 100 mm back inlet	270.69	2.70	37.74	–	279.43	nr	**317.17**
150 mm outlet	272.85	3.50	48.91	–	281.64	nr	**330.55**
150 mm outlet; 150 mm back inlet	276.19	3.70	51.72	–	285.09	nr	**336.81**
Road gully; trapped with rodding eye and stopper (grate not included)							
300 mm × 600 mm × 100 mm outlet	137.35	3.05	42.63	–	164.85	nr	**207.48**
300 mm × 600 mm × 150 mm outlet	140.66	3.05	42.63	–	168.26	nr	**210.89**
400 mm × 750 mm × 150 mm outlet	163.13	3.70	51.72	–	203.53	nr	**255.25**
450 mm × 900 mm × 150 mm outlet	220.71	4.65	64.99	–	269.91	nr	**334.90**
Grease trap; with internal access; galvanized perforated bucket; lid and frame							
600 mm × 450 mm × 600 mm deep; 100 mm outlet	1124.51	3.89	54.37	–	1193.15	nr	**1247.52**
Interceptor; trapped with inspection arm; lever locking stopper; chain and staple; cement and sand (1:2) joints to pipes; building in, and cutting and fitting brickwork around							
100 mm outlet; 100 mm inlet	109.82	3.70	51.72	–	113.71	nr	**165.43**
150 mm outlet; 150 mm inlet	158.51	4.16	58.14	–	163.86	nr	**222.00**
225 mm outlet; 225 mm inlet	494.18	4.63	64.71	–	509.69	nr	**574.40**

34 DRAINAGE BELOW GROUND

Item	PC £	Labour hours	Labour £	Plant £	Material £	Unit	Total rate £
34.04 PIPES, FITTINGS AND ACCESSORIES – cont							
Accessories in polypropylene; cover set in concrete; with coupling joints to pipes							
Inspection chamber; 5 nr 100 mm inlets; cast iron cover and frame							
475 mm dia. × 595 mm deep	319.21	2.13	29.77	–	337.89	nr	**367.66**
475 mm dia. × 175 mm deep raising piece	22.83	0.25	3.49	–	23.51	nr	**27.00**
Accessories; grates and covers							
Aluminium alloy gully grids; set in position							
120 mm × 120 mm	5.52	0.09	1.26	–	5.69	nr	**6.95**
150 mm × 150 mm	5.27	0.09	1.26	–	5.43	nr	**6.69**
225 mm × 225 mm	16.41	0.09	1.26	–	16.90	nr	**18.16**
100 mm dia.	5.52	0.09	1.26	–	5.69	nr	**6.95**
150 mm dia.	8.43	0.09	1.26	–	8.68	nr	**9.94**
225 mm dia.	18.19	0.09	1.26	–	18.74	nr	**20.00**
Polypropylene access covers and frames; supplied by Manhole Covers Ltd or other equal and approved; to suit PPIC inspection chambers; bedding and pointing in frame.							
450 mm dia.; class A15	45.65	1.30	18.17	–	49.64	nr	**67.81**
450 mm dia.; class B125; kite-marked	34.70	1.30	18.17	–	38.36	nr	**56.53**
Ductile iron heavy duty road grate and frame; supplied by Manhole Covers Ltd or other equal and approved; bedding and pointing in cement and sand (1:3); one course half brick thick wall in semi-engineering bricks in cement mortar (1:3)							
225 mm × 225 mm × 80 mm hinged and dished road grate and frame; class C250	30.45	2.25	31.45	–	36.27	nr	**67.72**
300 mm × 300 mm × 80 mm hinged and dished road grate and frame; class C250	50.31	2.25	31.45	–	56.73	nr	**88.18**
420 mm × 420 mm × 75 mm hinged road grate and frame; class C250; kite-marked	62.22	2.25	31.45	–	69.00	nr	**100.45**
445 mm × 445 mm × 75 mm double triangular road grate and frame; class C250; kite-marked	66.19	2.25	31.45	–	73.09	nr	**104.54**
435 mm × 435 mm × 100 mm pedestrian mesh road grate and frame; class D400	67.53	2.25	31.45	–	74.47	nr	**105.92**
440 mm × 400 mm × 150 mm hinged road grate and frame; class D400; kite-marked	87.39	2.25	31.45	–	94.92	nr	**126.37**
Concrete							
Vibrated concrete pipes and fittings; with flexible joints; BS 5911 Part 1; trench 2.00 m deep							
300 mm dia. pipes							
Class M; laid straight	23.21	0.65	9.08	8.73	23.91	m	**41.72**
bend; <= 45°	–	0.65	9.08	–	239.05	nr	**248.13**
bend; > 45°	–	0.65	9.08	–	358.58	nr	**367.66**
junction; 300 mm × 100 mm	–	0.46	6.43	–	125.51	nr	**131.94**

34 DRAINAGE BELOW GROUND

Item	PC £	Labour hours	Labour £	Plant £	Material £	Unit	Total rate £
450 mm dia. pipes							
Class H; laid straight	36.27	1.02	14.26	8.73	37.36	m	**60.35**
bend; <= 45°	–	1.02	14.26	–	373.57	nr	**387.83**
bend; > 45°	–	1.02	14.26	–	560.36	nr	**574.62**
junction; 450 mm × 150 mm	–	0.65	9.08	–	196.12	nr	**205.20**
525 mm dia. pipes							
Class H; laid straight	44.52	1.48	20.68	8.73	45.86	m	**75.27**
bend; <= 45°	–	1.48	20.68	8.73	458.55	nr	**487.96**
bend; > 45°	–	1.48	20.68	8.73	687.81	nr	**717.22**
junction; 600 mm × 150 mm	–	0.83	11.60	8.73	320.98	nr	**341.31**
900 mm dia. pipes							
Class H; laid straight	147.15	2.59	36.20	8.73	151.56	m	**196.49**
bend; <= 45°	–	2.59	36.20	8.73	1515.60	nr	**1560.53**
bend; > 45°	–	2.59	36.20	8.73	2273.41	nr	**2318.34**
junction; 900 mm × 150 mm	–	1.02	14.26	8.73	682.02	nr	**705.01**
1200 mm dia. pipes							
Class H; laid straight	253.54	3.70	51.72	8.73	261.15	m	**321.60**
bend; <= 45°	–	3.70	51.72	8.73	2611.49	nr	**2671.94**
bend; > 45°	–	3.70	51.72	8.73	3917.24	nr	**3977.69**
junction; 1200 mm × 150 mm	–	1.48	20.68	8.73	1175.18	nr	**1204.59**
Accessories in precast concrete; top set in with rodding eye and stopper; cement and sand (1:2) joint to pipe							
Concrete road gully; trapped with rodding eye and stopper; cement and sand (1:2) joint to pipe							
450 mm dia. × 900 mm deep; 100 mm or 150 mm outlet	63.22	4.39	61.36	–	92.09	nr	**153.45**
450 mm dia. × 1050 mm deep; 100 mm or 150 mm outlet	68.81	4.39	61.36	–	97.85	nr	**159.21**
Gully adapter type 1501	–	–	–	–	5.80	nr	**5.80**
uPVC							
Osmadrain uPVC pipes and fittings or other equal and approved; with ring seal joints							
110 mm dia. pipes							
laid straight	6.88	0.17	2.38	–	8.46	m	**10.84**
bend; short radius	20.90	0.15	2.10	–	21.10	nr	**23.20**
bend; long radius	39.59	0.15	2.10	–	38.65	nr	**40.75**
spigot/socket bend	17.66	0.15	2.10	–	25.43	nr	**27.53**
socket plug	9.15	0.04	0.56	–	9.42	nr	**9.98**
adjustable double socket bend	25.01	0.15	2.10	–	32.58	nr	**34.68**
adaptor to clay	27.10	0.09	1.26	–	27.63	nr	**28.89**
single junction	2.44	0.21	2.94	–	0.39	nr	**3.33**
sealed access junction	64.48	0.19	2.66	–	64.29	nr	**66.95**
slip coupler	12.09	0.09	1.26	–	12.45	nr	**13.71**
160 mm dia. pipes							
laid straight	15.78	0.21	2.94	–	18.78	m	**21.72**
bend; short radius	49.80	0.18	2.51	–	50.32	nr	**52.83**
spigot/socket bend	45.07	0.18	2.51	–	59.33	nr	**61.84**

34 DRAINAGE BELOW GROUND

Item	PC £	Labour hours	Labour £	Plant £	Material £	Unit	Total rate £
34.04 PIPES, FITTINGS AND ACCESSORIES – cont							
Osmadrain uPVC pipes and fittings or other equal and approved – cont							
160 mm dia. pipes – cont							
socket plug	19.63	0.07	0.98	–	20.22	nr	**21.20**
adaptor to clay	58.89	0.12	1.68	–	59.82	nr	**61.50**
level invert taper	24.10	0.18	2.51	–	36.75	nr	**39.26**
single junction	81.38	0.24	3.36	–	83.82	nr	**87.18**
slip coupler	29.37	0.11	1.53	–	30.25	nr	**31.78**
uPVC Osma Ultra-Rib ribbed pipes and fittings or other equal and approved; WIS approval; with sealed ring push-fit joints							
150 mm dia. pipes							
laid straight	8.93	0.19	2.66	–	9.20	m	**11.86**
bend; short radius	25.78	0.17	2.38	–	26.01	nr	**28.39**
adaptor to 160 mm dia. uPVC	34.31	0.10	1.40	–	34.24	nr	**35.64**
adaptor to clay	42.35	0.10	1.40	–	43.06	nr	**44.46**
level invert taper	12.60	0.18	2.51	–	11.33	nr	**13.84**
single junction	51.90	0.22	3.08	–	50.70	nr	**53.78**
225 mm dia. pipes							
laid straight	23.12	0.22	3.08	–	23.81	m	**26.89**
bend; short radius	121.28	0.20	2.79	–	123.49	nr	**126.28**
level invert taper	21.11	0.20	2.79	–	17.46	nr	**20.25**
single junction	180.03	0.27	3.77	–	178.28	nr	**182.05**
300 mm dia. pipes							
laid straight	32.80	0.32	4.47	–	33.78	m	**38.25**
bend; short radius	191.03	0.29	4.06	–	194.73	nr	**198.79**
level invert taper	68.52	0.29	4.06	–	64.50	nr	**68.56**
single junction	415.97	0.37	5.17	–	418.31	nr	**423.48**
Accessories in uPVC; with ring seal joints to pipes (unless otherwise described)							
Rodding eye							
square top, sealed with coupling	63.63	0.43	6.02	–	70.60	nr	**76.62**
Universal gully fitting; comprising gully trap, plain hopper							
150 mm × 150 mm grate	37.20	0.93	13.00	–	45.39	nr	**58.39**
Bottle gully; square comprising gully with rotating							
217 mm × 217 mm grate	68.71	0.78	10.90	–	77.85	nr	**88.75**
Shallow inspection chamber; 250 mm dia.; 600 mm deep; sealed cover and frame (1.0 tonne loading)							
4 nr 110 mm outlets/inlets	134.11	1.28	17.89	–	163.39	nr	**181.28**
Universal inspection chamber; 450 mm dia.; 4 nr 110 mm branch outlet; single seal cast iron cover and frame (12.5 tonnes loading)							
500 mm deep	272.86	1.35	18.87	–	306.30	nr	**325.17**
730 mm deep	314.66	1.60	22.36	–	354.41	nr	**376.77**
960 mm deep	356.47	1.85	25.85	–	402.52	nr	**428.37**

34 DRAINAGE BELOW GROUND

Item	PC £	Labour hours	Labour £	Plant £	Material £	Unit	Total rate £
Equal manhole base; 500 mm dia.							
6 nr 110 mm branch outlets	156.76	1.21	16.91	–	176.62	nr	**193.53**
Unequal manhole base; 500 mm dia.							
2 nr 160 mm, 4 nr 110 mm branch outlets	168.08	1.21	16.91	–	188.28	nr	**205.19**
Kerb to gullies; class B engineering bricks on edge to three sides in cement mortar (1:3) rendering in cement mortar (1:3) to top and two sides and skirting to brickwork 230 mm high; dishing in cement mortar (1:3) to gully; steel trowelled							
230 mm × 230 mm internally	–	1.39	19.43	–	2.78	nr	**22.21**

34.05 LAND DRAINAGE

Excavating; by hand; grading bottoms; earthwork support; filling to within 150 mm of surface with gravel rejects; remainder filled with excavated material and compacting; disposal of surplus soil on site; spreading on site average 50 m

Item	PC £	Labour hours	Labour £	Plant £	Material £	Unit	Total rate £
Pipes not exceeding 200 nominal size							
average depth of trench 0.75 m	–	1.57	22.12	–	14.69	m	**36.81**
average depth of trench 1.00 m	–	2.08	29.31	–	24.03	m	**53.34**
average depth of trench 1.25 m	–	2.91	41.00	–	30.39	m	**71.39**
average depth of trench 1.50 m	–	5.00	70.46	–	37.23	m	**107.69**
average depth of trench 1.75 m	–	5.92	83.42	–	43.59	m	**127.01**
average depth of trench 2.00 m	–	6.85	96.53	–	50.44	m	**146.97**
Disposal; load lorry by machine							
Excavated material							
inactive waste off site; to tip not exceeding 13 km (using lorries); including Landfill Tax	–	0.10	1.41	14.95	17.09	m³	**33.45**
active non-hazardous waste off site; to tip not exceeding 13 km (using lorries); including Landfill Tax	–	0.06	1.04	14.95	246.61	m³	**262.60**
Vitrified clay perforated subsoil pipes; BS 65; Hepworth Hepline or other equal and approved							
Pipes; laid straight							
100 mm dia.	10.50	0.20	2.79	–	10.82	m	**13.61**
150 mm dia.	18.04	0.25	3.49	–	18.58	m	**22.07**
225 mm dia.	40.36	0.33	4.61	–	41.57	m	**46.18**
Concrete canvas cement impregnated cloth lining to form ditch linings; holding ponds; slope protection and similar							
Type CC8, 8 mm thick and sprayed with water	–	0.10	1.65	0.10	53.54	m²	**55.29**

34 DRAINAGE BELOW GROUND

Item	PC £	Labour hours	Labour £	Plant £	Material £	Unit	Total rate £
34.06 MANHOLES AND SOAKAWAYS							
Excavating; by machine							
Manholes							
maximum depth not exceeding 1.00 m	–	0.19	2.68	5.10	–	m³	**7.78**
maximum depth not exceeding 2.00 m	–	0.21	2.96	5.60	–	m³	**8.56**
maximum depth not exceeding 4.00 m	–	0.25	3.52	6.55	–	m³	**10.07**
Excavating; by hand							
Manholes							
maximum depth not exceeding 1.00 m	–	3.05	42.98	–	–	m³	**42.98**
maximum depth not exceeding 2.00 m	–	3.61	50.87	–	–	m³	**50.87**
maximum depth not exceeding 4.00 m	–	4.63	65.25	–	–	m³	**65.25**
Earthwork support (average risk prices)							
Maximum depth not exceeding 1.00 m							
distance between opposing faces not exceeding 2.00 m	–	0.14	1.98	–	7.57	m²	**9.55**
Maximum depth not exceeding 2.00 m							
distance between opposing faces not exceeding 2.00 m	–	0.18	2.53	–	14.76	m²	**17.29**
Maximum depth not exceeding 4.00 m							
distance between opposing faces not exceeding 2.00 m	–	0.22	3.10	–	21.96	m²	**25.06**
Disposal; by machine							
Excavated material							
off site; to tip not exceeding 13 km (using lorries) including Landfill Tax based on inactive waste	–	0.10	1.41	14.95	17.09	m³	**33.45**
active non-hazardous waste off site; to tip not exceeding 13 km (using lorries); including Landfill Tax	–	0.06	1.04	14.95	246.61	m³	**262.60**
on site; depositing on site in spoil heaps; average 50 m distance	–	0.14	1.98	4.55	–	m³	**6.53**
Disposal; by hand							
Excavated material							
on site; depositing on site in spoil heaps; average 50 m distance	–	1.20	16.91	–	–	m³	**16.91**
Filling to excavations; by machine							
Average thickness not exceeding 0.25 m							
arising excavations	–	0.14	1.98	2.40	–	m³	**4.38**
Filling to excavations; by hand							
Average thickness not exceeding 0.25 m							
arising from excavations	–	0.93	13.10	–	–	m³	**13.10**

34 DRAINAGE BELOW GROUND

Item	PC £	Labour hours	Labour £	Plant £	Material £	Unit	Total rate £
Plain in situ ready mixed designated concrete; C10–40 mm aggregate							
Beds							
thickness not exceeding 150 mm	103.00	2.78	45.94	–	106.09	m³	**152.03**
thickness 150 mm–450 mm	–	2.08	34.37	–	106.09	m³	**140.46**
thickness exceeding 450 mm	–	1.76	29.08	–	106.09	m³	**135.17**
Plain in situ ready mixed designated concrete; C20–20 mm aggregate							
Beds							
thickness not exceeding 150 mm	106.76	2.78	45.94	–	109.96	m³	**155.90**
thickness 150 mm–450 mm	–	2.08	34.37	–	109.96	m³	**144.33**
thickness exceeding 450 mm	–	1.76	29.08	–	109.96	m³	**139.04**
Plain in situ ready mixed designated concrete; C25–20 mm aggregate; (small quantities)							
Benching in bottoms							
150 mm–450 mm average thickness	108.31	8.33	148.45	–	111.56	m³	**260.01**
Reinforced in situ ready mixed designated concrete; C20–20 mm aggregate; (small quantities)							
Isolated cover slabs							
thickness not exceeding 150 mm	101.67	6.48	107.07	–	104.72	m³	**211.79**
Reinforcement; fabric to BS 4449; lapped; in beds or suspended slabs							
Ref D98 (1.54 kg/m²)							
400 mm minimum laps	1.51	0.11	2.26	–	1.56	m²	**3.82**
Ref A142 (2.22 kg/m²)							
400 mm minimum laps	1.91	0.11	2.26	–	1.97	m²	**4.23**
Ref A193 (3.02 kg/m²)							
400 mm minimum laps	2.59	0.11	2.26	–	2.67	m²	**4.93**
Formwork; basic finish							
Soffits of isolated cover slabs							
horizontal	–	2.64	51.63	–	12.16	m²	**63.79**
Edges of isolated cover slabs							
height not exceeding 250 mm	–	0.78	15.25	–	3.55	m	**18.80**
Precast concrete circular manhole rings; bedding, jointing and pointing in cement mortar (1:3) on prepared bed							
Chamber or shaft rings; plain							
900 mm dia.	77.24	5.09	71.14	–	81.10	m	**152.24**
1050 mm dia.	81.94	6.01	84.01	–	87.49	m	**171.50**
1200 mm dia.	99.86	6.94	97.01	–	107.49	m	**204.50**
Chamber or shaft rings; reinforced							
1350 mm dia.	156.72	7.86	109.86	43.66	167.60	m	**321.12**
1500 mm dia.	184.28	8.79	122.86	43.66	199.08	m	**365.60**
1800 mm dia.	265.00	11.10	155.15	78.60	286.86	m	**520.61**

34 DRAINAGE BELOW GROUND

Item	PC £	Labour hours	Labour £	Plant £	Material £	Unit	Total rate £
34.06 MANHOLES AND SOAKAWAYS – cont							
Precast concrete circular manhole rings – cont							
Chamber or shaft rings – cont							
2100 mm dia.	533.58	13.88	194.00	87.32	568.13	m	**849.45**
step irons built in	8.69	0.14	1.96	–	8.95	nr	**10.91**
integrated ladder system 150 mm projection							
polypropylene encapusulated steps and rails	10.05	0.50	6.99	–	10.35	step	**17.34**
Reducing slabs							
1200 mm dia.	141.97	5.55	77.57	43.66	149.32	nr	**270.55**
1350 mm dia.	227.63	8.79	122.86	43.66	240.64	nr	**407.16**
1500 mm dia.	259.45	10.18	142.28	43.66	274.96	nr	**460.90**
1800 mm dia.	408.35	12.95	181.00	78.60	432.96	nr	**692.56**
2100 mm dia.	866.50	12.95	181.00	78.60	904.86	nr	**1164.46**
Heavy duty cover slabs; to suit rings							
900 mm dia.	84.81	2.78	38.85	17.47	88.90	nr	**145.22**
1050 mm dia.	89.95	3.24	45.29	17.47	94.50	nr	**157.26**
1200 mm dia.	109.22	3.70	51.72	20.08	115.59	nr	**187.39**
1350 mm dia.	175.09	4.16	58.14	21.84	184.98	nr	**264.96**
1500 mm dia.	199.57	4.63	64.71	36.68	211.74	nr	**313.13**
1800 mm dia.	314.11	5.55	77.57	41.92	332.10	nr	**451.59**
2100 mm dia.	666.54	6.48	90.57	43.66	698.12	nr	**832.35**
Precast concrete circular manhole; CPM Perfect Manhole or similar; complete with preformed benching and outlets to base; elastomeric seal to joints to rings; single steps as required							
1200 mm with up to four outlets 100 mm or 150 mm outlets; effective internal depth:							
1250 mm deep	–	5.00	73.31	34.93	1388.44	nr	**1496.68**
1500 mm deep	–	5.00	73.31	34.93	1386.60	nr	**1494.84**
1800 mm deep	–	5.00	73.31	34.93	1353.17	nr	**1461.41**
2000 mm deep	–	5.00	73.31	34.93	1364.46	nr	**1472.70**
2500 mm deep	–	4.50	62.89	43.66	1346.61	nr	**1453.16**
1500 mm dia. with up to four outlets; effective internal depth:							
2000 mm deep; upto 450 mm pipe	–	4.50	62.89	43.66	1868.12	nr	**1974.67**
2500 mm deep; up to 450 mm pipe	–	4.75	66.39	48.90	2439.52	nr	**2554.81**
3000 mm deep; up to 450 mm pipe	–	4.75	66.39	48.90	2636.56	nr	**2751.85**
3500 mm deep; up to 450 mm pipe	–	5.00	69.89	52.40	2689.76	nr	**2812.05**
4000 mm deep; up to 450 mm pipe	–	5.00	69.89	52.40	2869.32	nr	**2991.61**
4500 mm deep; up to 450 mm pipe	–	5.00	69.89	52.40	2886.82	nr	**3009.11**
4000 mm deep; 600 mm pipe	–	5.00	69.89	52.40	3277.05	nr	**3399.34**
4500 mm deep; 600 mm pipe	–	5.00	69.89	52.40	3300.84	nr	**3423.13**
Common bricks; in cement mortar (1:3)							
Walls to manholes							
one brick thick	380.00	2.22	48.69	–	69.85	m²	**118.54**
one and a half brick thick	–	3.24	71.06	–	104.79	m²	**175.85**
Projections of footings							
two brick thick	–	4.53	99.34	–	139.72	m²	**239.06**

34 DRAINAGE BELOW GROUND

Item	PC £	Labour hours	Labour £	Plant £	Material £	Unit	Total rate £
Class A engineering bricks; in cement mortar (1:3)							
Walls to manholes							
one brick thick (PC £ per 1000)	513.70	2.50	54.83	–	91.18	m²	**146.01**
one and a half brick thick	–	3.61	79.18	–	95.81	m²	**174.99**
Projections of footings							
two brick thick	–	5.09	111.63	–	182.35	m²	**293.98**
Class B engineering bricks; in cement mortar (1:3)							
Walls to manholes							
one brick thick (PC £ per 1000)	438.10	2.50	54.83	–	79.12	m²	**133.95**
one and a half brick thick	–	3.61	79.18	–	118.69	m²	**197.87**
Projections of footings							
two brick thick	–	5.09	111.63	–	158.25	m²	**269.88**
Brickwork sundries							
Extra over for fair face; flush smooth pointing							
manhole walls	–	0.19	4.17	–	–	m²	**4.17**
Building ends of pipes into brickwork; making good fair face or rendering							
not exceeding 55 mm nominal size	–	0.09	1.98	–	–	nr	**1.98**
55 mm–110 mm nominal size	–	0.14	3.07	–	–	nr	**3.07**
over 110 mm nominal size	–	0.19	4.17	–	–	nr	**4.17**
Step irons; BS 1247; malleable; galvanized; building into joints							
general purpose pattern	–	0.14	3.07	–	8.95	nr	**12.02**
Cement and sand (1:3) in situ finishings; steel trowelled							
13 mm work to manhole walls; one coat; to							
brickwork base over 300 wide	–	0.65	14.26	–	3.09	m²	**17.35**
Cast iron inspection chambers; with bolted flat covers; BS 437; bedded in cement mortar (1:3); with mechanical coupling joints							
100 mm × 100 mm							
one branch either side	373.02	1.40	19.57	–	385.76	nr	**405.33**
two branches either side	705.54	2.00	27.95	–	728.25	nr	**756.20**
150 mm × 100 mm							
one branch either side	461.85	1.55	21.66	–	477.25	nr	**498.91**
two branches either side	896.35	2.15	30.06	–	926.33	nr	**956.39**
150 mm × 150 mm							
one branch either side	572.07	1.80	25.16	–	592.32	nr	**617.48**
two branches either side	1104.23	2.60	36.34	–	1140.45	nr	**1176.79**

34 DRAINAGE BELOW GROUND

Item	PC £	Labour hours	Labour £	Plant £	Material £	Unit	Total rate £
34.06 MANHOLES AND SOAKAWAYS – cont							
Coated cast or ductile iron access covers and frames; to BS EN124; supplied by Manhole Covers Ltd or other equal; bedding frame in cement and sand (1:3); cover in grease and sand							
Light duty; cast iron; rectangular single seal solid top							
450 mm × 450 mm; class A15	48.29	1.50	20.97	–	52.35	nr	**73.32**
600 mm × 450 mm; class A15	48.29	1.50	20.97	–	52.53	nr	**73.50**
600 mm × 600 mm; class A15	74.62	1.50	20.97	–	79.83	nr	**100.80**
750 mm × 600 mm; class A15	146.62	1.50	20.97	–	153.99	nr	**174.96**
Medium duty; ductile iron; rectangular single seal solid top							
450 mm × 450 mm × 40 mm; class C250; kite-marked	89.55	2.00	27.95	–	95.21	nr	**123.16**
600 mm × 450 mm × 40 mm; slide-out; class C250; kite-marked	92.18	2.00	27.95	–	97.92	nr	**125.87**
600 mm × 600 mm × 40 mm; slide-out; class C250; kite-marked	111.50	2.00	27.95	–	117.81	nr	**145.76**
760 mm × 600 mm × 40 mm; slide-out; class C250; kite-marked	165.49	2.00	27.95	–	173.43	nr	**201.38**
Heavy duty; ductile iron; solid top							
450 mm × 450 mm × 75 mm; single seal; class C250; kite-marked	131.07	2.50	34.94	–	137.97	nr	**172.91**
600 mm × 450 mm × 75 mm; single seal; class C250; kite-marked	145.63	2.50	34.94	–	152.98	nr	**187.92**
600 mm × 600 mm × 75 mm; single seal; class C250; kite-marked	166.81	2.50	34.94	–	174.79	nr	**209.73**
450 mm × 450 mm × 100 mm; double triangular; class D400; kite-marked	119.15	2.50	34.94	–	125.70	nr	**160.64**
600 mm × 450 mm × 100 mm; double triangular; class D400; kite-marked	156.22	2.50	34.94	–	163.87	nr	**198.81**
600 mm × 600 mm × 100 mm; double triangular; class D400; kite-marked	95.33	2.50	34.94	–	101.17	nr	**136.11**
750 mm × 600 mm × 100 mm; double triangular; class D400; kite-marked	243.61	2.50	34.94	–	253.88	nr	**288.82**
1220 mm × 675 mm × 100 mm; double triangular; class D400; kite-marked	271.40	3.50	48.91	–	282.52	nr	**331.43**
British Standard best quality vitrified clay channels; bedding and jointing in cement and sand (1:2)							
Half section straight							
100 mm dia. × 1 m long	13.63	0.74	10.34	–	14.04	nr	**24.38**
150 mm dia. × 1 m long	22.68	0.93	13.00	–	23.36	nr	**36.36**
225 mm dia. × 1 m long	50.92	1.20	16.77	–	52.45	nr	**69.22**
300 mm dia. × 1 m long	104.53	1.48	20.68	–	107.67	nr	**128.35**
Half section bend							
100 mm dia.	15.34	0.56	7.83	–	15.80	nr	**23.63**
150 mm dia.	25.28	0.69	9.64	–	26.04	nr	**35.68**
225 mm dia.	84.33	0.93	13.00	–	86.86	nr	**99.86**

34 DRAINAGE BELOW GROUND

Item	PC £	Labour hours	Labour £	Plant £	Material £	Unit	Total rate £
Taper straight							
150 mm–100 mm dia.	63.72	0.65	9.08	–	65.63	nr	**74.71**
300 mm–225 mm dia.	279.73	0.83	11.60	–	288.12	nr	**299.72**
Taper bend							
150 mm–100 mm dia.	97.02	0.83	11.60	–	99.93	nr	**111.53**
225 mm–150 mm dia.	278.01	1.06	14.81	–	286.35	nr	**301.16**
Three quarter section branch bend							
100 mm dia.	34.55	0.46	6.43	–	35.59	nr	**42.02**
150 mm dia.	58.00	0.69	9.64	–	59.74	nr	**69.38**
Glass fibre septic tank; Klargester or other equal and approved; fixing lockable manhole cover and frame; placing in position (excavations, fill etc. measured elsewhere)							
2800 litre capacity; depth to invert							
1000 mm deep	–	2.45	34.25	69.86	912.31	nr	**1016.42**
1500 mm deep	–	2.73	38.16	69.86	966.68	nr	**1074.70**
3800 litre capacity; depth to invert							
1000 mm deep	–	2.64	36.89	87.32	1165.66	nr	**1289.87**
1500 mm deep	–	2.91	40.67	96.06	1220.03	nr	**1356.76**
4600 litre capacity; depth to invert							
1000 mm deep	–	2.75	38.44	87.32	1336.38	nr	**1462.14**
1500 mm deep	–	3.20	44.72	96.06	1390.75	nr	**1531.53**
SOAKAWAYS							
Soakaway crates, lightweight modular water storage cells for stormwater attenuation with geotextile membrane wrapping							
Polystorm Heavy Duty (60 tonne) crates							
1000 mm × 500 mm × 400 mm deep	–	1.25	17.47	–	152.60	m³	**170.07**
Polystorm Lite Duty (20 tonne) crates							
1000 mm × 500 mm × 400 mm deep	–	1.25	17.47	–	151.56	m³	**169.03**
Brett Martin Heavy Duty (60 tonne) crates							
1200 mm × 600 mm × 420 mm deep	–	1.25	17.47	–	93.81	m³	**111.28**
Brett Martin Light Duty (20 tonne) crates							
1200 mm × 600 mm × 420 mm deep	–	1.25	17.47	–	86.48	m³	**103.95**
Rainwater harvesting							
FP McCann Ltd Easi-Rain domestic and commercial rainwater harvesting tank system designed to reduce demands on the mains network by providing a sustainable water collection and distribution system ideal for non-potable uses such as toilet flushing, washing and landscape or garden watering.							
Excluding excavations							
1200 litre tank	2077.77	1.50	20.97	20.19	2140.10	nr	**2181.26**
1500 litre tank	2085.05	1.75	24.46	24.23	2147.60	nr	**2196.29**
3000 litre tank	2560.01	1.80	25.16	28.26	2636.81	nr	**2690.23**
5000 litre tank	2589.77	1.90	26.55	32.30	2667.46	nr	**2726.31**
7500 litre tank	2698.92	2.00	27.95	32.30	2779.89	nr	**2840.14**

35 SITE WORK

Item	PC £	Labour hours	Labour £	Plant £	Material £	Unit	Total rate £
35.01 KERBS, EDGINGS AND CHANNELS							
Excavating; by machine							
Excavating trenches; to receive kerb foundations; average size							
300 mm × 100 mm	–	0.02	0.28	0.36	–	m	**0.64**
450 mm × 150 mm	–	0.02	0.28	0.73	–	m	**1.01**
600 mm × 200 mm	–	0.03	0.45	1.02	–	m	**1.47**
Excavating curved trenches; to receive kerb foundations; average size							
300 mm × 100 mm	–	0.01	0.15	0.59	–	m	**0.74**
450 mm × 150 mm	–	0.03	0.42	0.88	–	m	**1.30**
600 mm × 200 mm	–	0.04	0.57	1.09	–	m	**1.66**
Excavating; by hand							
Excavating trenches; to receive kerb foundations; average size							
150 mm × 50 mm	–	0.02	0.28	–	–	m	**0.28**
200 mm × 75 mm	–	0.06	0.84	–	–	m	**0.84**
250 mm × 100 mm	–	0.10	1.41	–	–	m	**1.41**
300 mm × 100 mm	–	0.13	1.83	–	–	m	**1.83**
Excavating curved trenches; to receive kerb foundations; average size							
150 mm × 50 mm	–	0.03	0.42	–	–	m	**0.42**
200 mm × 75 mm	–	0.07	0.99	–	–	m	**0.99**
250 mm × 100 mm	–	0.11	1.55	–	–	m	**1.55**
300 mm × 100 mm	–	0.14	1.98	–	–	m	**1.98**
Plain in situ ready mixed designated concrete; C7.5–40 mm aggregate; poured on or against earth or unblinded hardcore							
Foundations	101.93	1.16	19.17	–	104.99	m³	**124.16**
Blinding beds							
thickness not exceeding 150 mm	101.93	1.71	28.25	–	104.99	m³	**133.24**
Plain in situ ready mixed designated concrete; C10–40 mm aggregate; poured on or against earth or unblinded hardcore							
Foundations	103.00	1.16	19.17	–	106.09	m³	**125.26**
Blinding beds							
thickness not exceeding 150 mm	103.00	1.71	28.25	–	106.09	m³	**134.34**
Plain in situ ready mixed designated concrete; C20–20 mm aggregate; poured on or against earth or unblinded hardcore							
Foundations	106.76	1.16	19.17	–	109.96	m³	**129.13**
Blinding beds							
thickness not exceeding 150 mm	106.76	1.71	28.25	–	109.96	m³	**138.21**

35 SITE WORK

Item	PC £	Labour hours	Labour £	Plant £	Material £	Unit	Total rate £
Filling to make up levels; by machine							
Average thickness not exceeding 0.25 m							
obtained off site; hardcore	–	0.28	3.94	1.21	42.76	m³	**47.91**
obtained off site; granular fill type one	–	0.28	3.94	1.21	45.18	m³	**50.33**
obtained off site; granular fill type two	–	0.28	3.94	1.21	45.18	m³	**50.33**
Average thickness exceeding 0.25 m							
obtained off site; hardcore	–	0.24	3.38	1.10	36.65	m³	**41.13**
obtained off site; granular fill type one	–	0.24	3.38	1.10	45.18	m³	**49.66**
obtained off site; granular fill type two	–	0.24	3.38	1.10	45.18	m³	**49.66**
Filling to make up levels; by hand							
Average thickness not exceeding 0.25 m							
obtained off site; hardcore	–	0.61	8.60	2.08	42.76	m³	**53.44**
obtained off site; sand	–	0.71	10.00	2.43	55.68	m³	**68.11**
Average thickness exceeding 0.25 m							
obtained off site; hardcore	–	0.51	7.19	1.73	36.65	m³	**45.57**
obtained off site; sand	–	0.60	8.46	2.05	55.68	m³	**66.19**
Surface treatments							
Compacting							
filling; blinding with sand	–	0.04	0.57	0.04	2.89	m²	**3.50**
Precast concrete kerbs, channels, edgings, etc.; BS 340; bedded, jointed and pointed in cement mortar (1:3); including haunching up one side with in situ ready mix designated concrete C10– 40 mm aggregate; to concrete base							
Edgings; straight; square edge							
50 mm × 150 mm	–	0.23	5.61	–	5.17	m	**10.78**
50 mm × 200 mm	–	0.23	5.61	–	7.25	m	**12.86**
50 mm × 255 mm	–	0.23	5.61	–	8.53	m	**14.14**
Kerbs; straight							
125 mm × 255 mm; half battered	–	0.31	7.56	–	13.25	m	**20.81**
125 mm × 255 mm; half battered drop kerb	–	0.31	7.56	–	18.82	m	**26.38**
150 mm × 305 mm; half battered	–	0.31	7.56	–	24.37	m	**31.93**
150 mm × 305 mm; half battered drop kerb	–	0.31	7.56	–	44.26	m	**51.82**
Kerbs; curved							
125 mm × 255 mm; half battered	–	0.46	11.22	–	13.19	m	**24.41**
150 mm × 305 mm; half battered	–	0.46	11.22	–	18.48	m	**29.70**
Channels; 255 × 125 mm							
straight	–	0.31	7.56	–	15.11	m	**22.67**
Quadrants; half battered							
305 mm × 305 mm × 150 mm	–	0.32	7.81	–	22.68	nr	**30.49**
305 mm × 305 mm × 255 mm	–	0.32	7.81	–	26.48	nr	**34.29**
455 mm × 455 mm × 255 mm	–	0.37	9.02	–	28.72	nr	**37.74**

Prices for Measured Works

35 SITE WORK

Item	PC £	Labour hours	Labour £	Plant £	Material £	Unit	Total rate £
35.01 KERBS, EDGINGS AND CHANNELS – cont							
Natural stone kerbs; Silver Grey Granite bedded, jointed and pointed in cement mortar (1:3); including haunching up one side with in situ ready mix designated concrete C10–40 mm aggregate; to concrete base							
Kerbs; straight							
kerbs with granite setts							
100 mm × 100 mm × 200 mm, split all sides	10.03	0.35	8.54	–	19.03	m	**27.57**
kerbs with granite setts							
300 mm × 50 mm × 600 mm, sawn all sides, flame textured top and front face	50.35	0.35	8.54	–	64.73	m	**73.27**
kerbs with granite setts							
250 mm × 50 mm × 600 mm, sawn all sides, flame textured top and front face	50.35	0.35	8.54	–	64.73	m	**73.27**
Bullnose kerbs with granite setts							
150 mm × 150 mm × 600 mm, 10 mm rounding to top one long side, flame textured top and front face	19.80	0.50	12.20	–	30.10	m	**42.30**
Bullnose kerbs with granite setts							
300 mm × 150 mm × 600 mm, 10 mm rounding to top one long side, flame textured top and front face	39.00	0.50	12.20	–	51.86	m	**64.06**
transition Bullnose kerbs with granite setts							
300 mm to 150 mm width change across length × 150 mm × 600 mm, 10 mm rounding to top one long side, flame textured top and front face	10.48	0.03	0.73	–	19.54	nr	**20.27**
Kerbs; curved							
Bullnose kerbs with granite setts							
150 mm × 175 mm × 600 mm, sawn all sides, 10 mm rounding to top one long side, flame textured top and front face, curved on plan 1–12 m	33.91	0.50	12.20	–	46.09	m	**58.29**
Bullnose kerbs with granite setts							
300 mm × 150 mm × 600 mm, 10 mm rounding to top one long side, flame textured top and front face, curved on plan 1–12 m	46.62	0.50	12.20	–	60.49	m	**72.69**
35.02 IN SITU CONCRETE ROADS AND PAVINGS							
Reinforced in situ ready mixed designated concrete; C10–40 mm aggregate							
Roads; to hardcore base							
thickness not exceeding 150 mm	98.09	1.85	30.57	–	101.03	m³	**131.60**
thickness 150 mm–450 mm	98.09	1.30	21.48	–	101.03	m³	**122.51**
Reinforced in situ ready mixed designated concrete; C20–20 mm aggregate							
Roads; to hardcore base							
thickness not exceeding 150 mm	101.67	1.85	30.57	–	104.72	m³	**135.29**
thickness 150 mm–450 mm	101.67	1.30	21.48	–	104.72	m³	**126.20**

35 SITE WORK

Item	PC £	Labour hours	Labour £	Plant £	Material £	Unit	Total rate £
Reinforced in situ ready mixed designated concrete; C25–20 mm aggregate							
Roads; to hardcore base							
thickness ot exceeding 150 mm	108.31	1.85	30.57	–	111.56	m³	**142.13**
thickness 150 mm–450 mm	108.31	1.30	21.48	–	111.56	m³	**133.04**
Formwork; sides of foundations; basic finish							
Plain vertical							
height not exceeding 250 mm	–	0.39	7.63	–	3.73	m	**11.36**
height 250 mm–500 mm	–	0.57	11.14	–	6.55	m	**17.69**
height 500 mm–1.00 m	–	0.83	16.23	–	11.22	m	**27.45**
add to above for curved radius 6 m	–	0.03	0.59	–	0.30	m	**0.89**
Reinforcement; fabric; BS 4449; lapped; in roads, footpaths or pavings							
Ref A142 (2.22 kg/m²)							
400 mm minimum laps	1.91	0.14	2.87	–	1.97	m²	**4.84**
Ref A193 (3.02 kg/m²)							
400 mm minimum laps	–	0.14	2.87	–	2.67	m²	**5.54**
Formed joints; Fosroc Expandite Flexcell impregnated joint filler or other equal and approved							
Width not exceeding 150 mm							
12.50 mm thick	–	0.14	2.74	–	2.68	m	**5.42**
25 mm thick	–	0.19	3.72	–	3.69	m	**7.41**
Width 150–300 mm							
12.50 mm thick	–	0.19	3.72	–	4.84	m	**8.56**
25 mm thick	–	0.19	3.72	–	6.82	m	**10.54**
Width 300–450 mm							
12.50 mm thick	–	0.23	4.50	–	7.26	m	**11.76**
25 mm thick	–	0.23	4.50	–	10.22	m	**14.72**
Sealants; Fosroc Expandite Pliastic N2 hot poured rubberized bituminous compound or other equal and approved							
Width 25 mm							
25 mm depth	–	0.20	3.91	–	2.05	m	**5.96**
Concrete sundries							
Treating surfaces							
unset concrete; grading to cambers; tamping with a 75 mm thick steel shod tamper	–	0.23	3.80	–	–	m²	**3.80**
Sundries							
Line marking							
width not exceeding 300 mm; NB minimum charge usually applies	–	0.04	0.70	–	0.27	m	**0.97**

35 SITE WORK

Item	PC £	Labour hours	Labour £	Plant £	Material £	Unit	Total rate £
35.03 COATED MACADAM, ASPHALT ROADS AND PAVINGS							
NOTE: The prices for all bitumen macadam and hot rolled asphalt materials are for individual courses to roads and footpaths and need combining to arrive at complete specifications and costs for full construction. Costs include for work to falls, crossfalls or slopes not exceeding 15° from horizontal; for laying on prepared bases (prices not included) and for rolling with an appropriate roller. The following rates are based on black bitumen macadam. Red bitumen macadam rates are approximately 50% dearer. PSV is Polished Stone Value.							
Dense bitumen macadam base course; bitumen penetration 100/125							
Carriageway, hardshoulder and hardstrip							
100 mm thick; one coat; with 0/32 mm aggregate size	–	–	–	–	–	m²	22.07
200 mm thick; one coat; with 0/32 mm aggregate size	–	–	–	–	–	m²	38.86
extra for increase thickness in 10 mm increments	–	–	–	–	–	m²	1.58
Hot rolled asphalt base course							
Carriageway, hardshoulder and hardstrip							
150 mm thick; one coat; 60% 0/32 mm aggregate size; to column 2/5	–	–	–	–	–	m²	36.70
200 mm thick; one coat; 60% 0/32 mm aggregate size; to column 2/5	–	–	–	–	–	m²	48.81
extra for increase thickness in 10 mm increments	–	–	–	–	–	m²	2.02
Dense bitumen macadam binder course; bitumen penetration 100/125							
Carriageway, hardshoulder and hardstrip							
60 mm thick; one coat; with 0/32 mm aggregate size	–	–	–	–	–	m²	13.49
60 mm thick; one coat; with 0/32 mm aggregate size; to clause 6.5	–	–	–	–	–	m²	13.61
extra for increase thickness in 10 mm increments	–	–	–	–	–	m²	1.80
Hot rolled asphalt binder course							
Carriageway, hardshoulder and hardstrip							
40 mm thick; one coat; 50% 0/14 mm aggregate size; to column 2/2; 55 PSV	–	–	–	–	–	m²	12.88
60 mm thick; one coat; 50% 0/14 mm aggregate size; to column 2/2	–	–	–	–	–	m²	14.83
60 mm thick; one coat; 50% 0/20 mm aggregate size; to column 2/3	–	–	–	–	–	m²	14.57

35 SITE WORK

Item	PC £	Labour hours	Labour £	Plant £	Material £	Unit	Total rate £
60 mm thick; one coat; 60% 0/32 mm aggregate size; to column 2/5	–	–	–	–	–	m²	**13.83**
100 mm thick; one coat; 60% 0/32 mm aggregate size; to column 2/5	–	–	–	–	–	m²	**22.33**
extra for increase thickness in 10 mm increments	–	–	–	–	–	m²	**2.67**
Macadam surface course; bitumen penetration 100/125							
Carriageway, hardsholder and hardstrip							
30 mm thick; one coat; medium graded with 0/6 mm nominal aggregate binder	–	–	–	–	–	m²	**10.42**
40 mm thick; one coat; close graded with 0/14 mm nominal aggregate binder; to clause 7.3	–	–	–	–	–	m²	**9.55**
40 mm thick; one coat; close graded with 0/10 mm nominal aggregate binder; to clause 7.4	–	–	–	–	–	m²	**10.42**
extra over above items for increase/reduction in 10 mm thick increments	–	–	–	–	–	m²	**1.88**
extra over above items for coarse aggregate 60–64 PSV	–	–	–	–	–	m²	**1.89**
extra over above items for coarse aggregate 65–67 PSV	–	–	–	–	–	m²	**2.07**
extra over above items for coarse aggregate 68 PSV	–	–	–	–	–	m²	**2.75**
Hot rolled asphalt surface course; bitumen penetration 40/60							
Carriageway, hardsholder and hardstrip							
40 mm thick; one coat; 30% mix 0/10 mm aggregate size; to column 3/2; with 20 mm pre-coated chippings 60–64 PSV	–	–	–	–	–	m²	**12.87**
40 mm thick; one coat; 30% mix 0/10 mm aggregate size; to column 3/2; with 14 mm pre-coated chippings 60–64 PSV	–	–	–	–	–	m²	**12.96**
extra over above items for increase/reduction in 10 mm increments	–	–	–	–	–	m²	**2.15**
extra over above items for chippings with 65–67 PSV	–	–	–	–	–	m²	**0.13**
extra over above items for chippings with 68 PSV	–	–	–	–	–	m²	**0.19**
extra over above items for 6–10 KN High Traffic Flows	–	–	–	–	–	m²	**0.92**
Stone mastic asphalt surface course							
Carriageway, hardsholder and hardstrip							
35 mm thick; one coat; with 0/14 mm nominal aggregate size; 55 PSV	–	–	–	–	–	m²	**11.83**
35 mm thick; one coat; with 0/10 mm nominal aggregate size; 55 PSV	–	–	–	–	–	m²	**11.83**
extra for increase thickness in 10 mm increments	–	–	–	–	–	m²	**2.65**

35 SITE WORK

Item	PC £	Labour hours	Labour £	Plant £	Material £	Unit	Total rate £
35.03 COATED MACADAM, ASPHALT ROADS AND PAVINGS – cont							
Thin surface course with 60 PSV							
Carriageway, hardshoulder and hardstrip							
35 mm thick; one coat; with 0/10 mm nominal aggregate size	–	–	–	–	–	m²	**11.83**
extra for increase thickness in 10 mm increments	–	–	–	–	–	m²	**1.34**
extra over above items for coarse aggregate 60–64 PSV	–	–	–	–	–	m²	**0.32**
extra over above items for coarse aggregate 65–67 PSV	–	–	–	–	–	m²	**0.32**
extra over above items for coarse aggregate 68 PSV	–	–	–	–	–	m²	**0.61**
Regulating courses							
Carriageway, hardshoulder and hardstrip							
Dense Bitumen Macadam; bitumen penetration 100/125; with 0/20 mm nominal aggregate regulating course	–	–	–	–	–	tonne	**101.20**
Hot rolled asphalt; 50% 0/20 mm aggregate size	–	–	–	–	–	tonne	**111.90**
Stone mastic asphalt; 0/6 mm aggregate	–	–	–	–	–	tonne	**142.86**
Bitumen emulsion tack coats							
Carriageway, hardshoulder and hardstrip							
K1–40; applied 0.35–0.45 l/m²	–	–	–	–	–	m²	**0.19**
K1–70; applied 0.35–0.45 l/m²	–	–	–	–	–	m²	**0.31**
Sundries							
Line marking							
width not exceeding 300 mm; NB minimum charge usually applies	–	0.04	0.70	–	0.27	m	**0.97**
35.04 GRAVEL PAVINGS							
Two coat gravel paving; level and to falls; first layer course clinker aggregate and wearing layer fine gravel aggregate							
Pavings; over 300 mm wide							
50 mm thick	–	0.07	1.71	0.07	2.77	m²	**4.55**
63 mm thick	–	0.09	2.19	0.10	3.79	m²	**6.08**
Resin bonded gravel paving; level and to falls							
Pavings; over 300 mm wide							
50 mm thick	–	–	–	–	–	m²	**52.91**

35 SITE WORK

Item	PC £	Labour hours	Labour £	Plant £	Material £	Unit	Total rate £
35.05 SLAB, BRICK/BLOCK SETTS AND COBBLE PAVINGS							
Artificial stone paving; Charcon's Moordale Textured or other equal and approved; to falls or crossfalls; bedding 25 mm thick in cement mortar (1:3); staggered joints; jointing in coloured cement mortar (1:3), brushed in; to sand base							
Pavings; over 300 mm wide							
600 mm × 600 mm × 50 mm thick; natural	13.25	0.39	9.51	–	19.83	m²	**29.34**
Brick paviors; 215 mm × 103 mm × 65 mm rough stock bricks; to falls or crossfalls; bedding on 50 mm sharp sand; kiln dried sand to joints							
Pavings; over 300 mm wide; straight joints both ways							
bricks laid flat (PC £ per 1000)	450.00	1.50	32.90	–	21.74	m²	**54.64**
bricks laid on edge	–	2.50	54.83	–	31.70	m²	**86.53**
Pavings; over 300 mm wide; laid to herringbone pattern							
bricks laid flat bricks laid flat (PC £ per 1000)	450.00	0.93	20.39	–	21.74	m²	**42.13**
bricks laid on edge	–	1.30	28.51	–	31.70	m²	**60.21**
Add or deduct for variation of £10.00/1000 in PC of brick paviors							
bricks laid flat bricks laid flat (PC £ per 1000)	–	–	–	–	0.46	m²	**0.46**
bricks laid on edge	–	–	–	–	0.70	m²	**0.70**
River washed cobble paving; 50 mm–75 mm; to falls or crossfalls; bedding 13 mm thick in cement mortar (1:3); jointing to a height of two thirds of cobbles in dry mortar (1:3); tightly butted, washed and brushed; to concrete							
Pavings; over 300 mm wide							
regular (PC £ per tonne)	140.00	3.70	90.23	–	37.42	m²	**127.65**
laid to pattern	–	4.63	112.91	–	37.42	m²	**150.33**
Concrete paving flags; to falls or crossfalls; bedding 25 mm thick in cement and sand mortar (1:4); butt joints straight both ways; jointing in cement and sand (1:3); brushed in; to sand base							
Pavings; over 300 mm wide							
450 mm × 600 mm × 50 mm thick; grey	13.32	0.42	10.24	–	17.50	m²	**27.74**
450 mm × 600 mm × 60 mm thick; coloured	26.20	0.42	10.24	–	31.43	m²	**41.67**
600 mm × 600 mm × 50 mm thick; grey	12.00	0.39	9.51	–	16.10	m²	**25.61**
600 mm × 600 mm × 50 mm thick; coloured	21.86	0.39	9.51	–	26.79	m²	**36.30**
750 mm × 600 mm × 50 mm thick; grey	12.35	0.36	8.78	–	16.46	m²	**25.24**
750 mm × 600 mm × 50 mm thick; coloured	21.76	0.36	8.78	–	26.65	m²	**35.43**
900 mm × 600 mm × 50 mm thick; grey	10.76	0.33	8.04	–	14.72	m²	**22.76**
900 mm × 600 mm × 50 mm thick; coloured	19.96	0.33	8.04	–	24.67	m²	**32.71**

35 SITE WORK

Item	PC £	Labour hours	Labour £	Plant £	Material £	Unit	Total rate £
35.05 SLAB, BRICK/BLOCK SETTS AND COBBLE PAVINGS – cont							
Blister Tactile paving flags; to falls or crossfalls; bedding 25 mm thick in cement and sand mortar (1:4); butt joints straight both ways; jointing in cement and sand (1:3); brushed in; to sand base							
Pavings; over 300 mm wide							
400 mm × 400 mm × 50 mm thick; buff	72.00	0.45	10.97	–	80.96	m²	**91.93**
400 mm × 400 mm × 60 mm thick; buff	65.50	0.45	10.97	–	73.93	m²	**84.90**
450 mm × 450 mm × 50 mm thick; buff	55.00	0.45	10.97	–	62.57	m²	**73.54**
450 mm × 450 mm × 70 mm thick; buff	65.00	0.45	10.97	–	73.39	m²	**84.36**
Concrete rectangular paving blocks; to falls or crossfalls; herringbone pattern; on prepared base (not included here); bedding on 50 mm thick dry sharp sand; filling joints with sharp sand brushed in							
Pavings; Keyblock or other equal and approved; over 300 mm wide; straight joints both ways							
200 mm × 100 mm × 60 mm thick; grey	15.84	1.25	30.48	–	21.49	m²	**51.97**
200 mm × 100 mm × 60 mm thick; coloured	15.84	1.25	30.48	–	21.49	m²	**51.97**
200 mm × 100 mm × 60 mm thick; brindle	15.84	1.25	30.48	–	21.49	m²	**51.97**
200 mm × 100 mm × 80 mm thick; grey	20.00	1.35	32.92	–	26.35	m²	**59.27**
200 mm × 100 mm × 80 mm thick; coloured	20.00	1.35	32.92	–	26.35	m²	**59.27**
200 mm × 100 mm × 80 mm thick; brindle	20.00	1.35	32.92	–	26.35	m²	**59.27**
Pavings; Cemex or other equal; over 300 mm wide; laid to herringbone pattern							
200 mm × 100 mm × 60 mm thick; grey	15.84	1.35	32.92	–	21.49	m²	**54.41**
200 mm × 100 mm × 60 mm thick; coloured	15.84	1.35	32.92	–	21.49	m²	**54.41**
200 mm × 100 mm × 80 mm thick; grey	20.00	1.35	32.92	–	26.35	m²	**59.27**
200 mm × 100 mm × 80 mm thick; coloured	20.00	1.35	32.92	–	26.35	m²	**59.27**
Extra for two row boundary edging to herringbone pavings; 200 mm wide; including a 150 mm high in situ concrete mix C10–40 mm aggregate haunching to one side; blocks laid breaking joint							
200 mm × 100 mm × 60 mm; coloured	–	0.28	6.83	–	3.22	m	**10.05**
200 mm × 100 mm × 80 mm; coloured	–	0.28	6.83	–	3.53	m	**10.36**
Pavings; Europa or other equal; over 300 mm wide; straight joints both ways							
200 mm × 100 mm × 60 mm thick; grey	16.00	1.25	30.48	–	21.65	m²	**52.13**
200 mm × 100 mm × 60 mm thick; coloured	17.60	1.25	30.48	–	23.38	m²	**53.86**
200 mm × 100 mm × 80 mm thick; grey	19.20	1.35	32.92	–	25.48	m²	**58.40**
200 mm × 100 mm × 80 mm thick; coloured	20.80	1.35	32.92	–	27.21	m²	**60.13**
Pavings; Metropolitan or other equal; over 300 mm wide; straight joints both ways							
200 mm × 100 mm × 80 mm thick; grey	22.60	1.35	32.92	–	29.16	m²	**62.08**
200 mm × 100 mm × 80 mm thick; coloured	22.60	1.35	32.92	–	29.16	m²	**62.08**

35 SITE WORK

Item	PC £	Labour hours	Labour £	Plant £	Material £	Unit	Total rate £
Concrete rectangular paving blocks; to falls or crossfalls; 6 mm wide joints; symmetrical layout; bedding in 15 mm semi-dry cement mortar (1:4); jointing and pointing in cement and sand (1:4); on concrete base							
Pavings; Trafica or other equal and approved; over 300 mm wide							
400 mm × 400 mm × 65 mm; Saxon textured; natural	24.24	–	–	–	–	m²	–
400 mm × 400 mm × 65 mm; Saxon textured; buff	26.78	0.44	10.73	–	32.05	m²	**42.78**
400 mm × 400 mm × 65 mm; Perfecta; natural	31.03	0.44	10.73	–	35.05	m²	**45.78**
400 mm × 400 mm × 65 mm; Perfecta; buff	33.86	0.44	10.73	–	39.72	m²	**50.45**
450 mm × 450 mm × 70 mm; Saxon textured; natural	25.34	0.43	10.49	–	30.50	m²	**40.99**
450 mm × 450 mm × 70 mm; Saxon textured; buff	28.99	0.43	10.49	–	34.44	m²	**44.93**
450 mm × 450 mm × 70 mm; Perfecta; natural	27.29	0.43	10.49	–	32.60	m²	**43.09**
450 mm × 450 mm × 70 mm; Perfecta; buff	31.67	0.43	10.49	–	37.35	m²	**47.84**
Blue grey granite setts tactile pavings (blister or corduroy); 400 mm × 80 mm × 400 mm; standard B dressing; sandblasted top finish, tightly butted to falls or crossfalls; bedding 25 mm thick in cement mortar (1:3); filling joints with dry mortar (1:6); washed and brushed; on concrete base							
Pavings; over 300 mm wide							
straight joints	76.59	1.48	36.09	–	88.53	m²	**124.62**
York stone slab pavings; to falls or crossfalls; bedding 25 mm thick in cement: sand mortar (1:4); 5 mm wide joints; jointing in coloured cement mortar (1:3); brushed in; to sand base							
Pavings; over 300 mm wide							
50 mm thick; random rectangular pattern	86.67	0.69	15.13	–	95.59	m²	**110.72**
600 mm × 600 mm × 50 mm thick	82.53	0.39	8.55	–	91.23	m²	**99.78**
600 mm × 900 mm × 50 mm thick	82.53	0.33	7.24	–	91.23	m²	**98.47**
Granite setts; 200 mm × 100 mm × 100 mm; standard C dressing; tightly butted to falls or crossfalls; bedding 25 mm thick in cement mortar (1:3); filling joints with dry mortar (1:6); washed and brushed; on concrete base							
Pavings; over 300 mm wide							
straight joints (PC £ per tonne)	228.80	1.48	36.09	–	71.44	m²	**107.53**
laid to pattern	228.80	1.85	45.11	–	71.44	m²	**116.55**
Boundary edging							
two rows of granite setts as boundary edging; 200 mm wide; including a 150 mm high ready mixed designated concrete C10–40 mm aggregate; haunching to one side; blocks laid breaking joint	–	0.65	15.85	–	15.94	m	**31.79**

35 SITE WORK

Item	PC £	Labour hours	Labour £	Plant £	Material £	Unit	Total rate £
35.05 SLAB, BRICK/BLOCK SETTS AND COBBLE PAVINGS – cont							
Blue grey granite setts paving bands; 100 mm × 100 mm × 100 mm; standard C dressing; flame textured top, tightly butted to falls or crossfalls; bedding 25 mm thick in cement mortar (1:3); filling joints with dry mortar (1:6); washed and brushed; on concrete base							
Pavings; over 300 mm wide							
straight joints	53.27	1.48	36.09	–	63.91	m²	**100.00**
laid to pattern	53.27	1.85	45.11	–	63.91	m²	**109.02**
Boundary edging							
two rows of granite setts as boundary edging; 200 mm wide; including a 150 mm high ready mixed designated concrete C10–40 mm aggregate; haunching to one side; blocks laid breaking joint	–	0.65	15.85	–	14.71	m	**30.56**
Black Basalt setts paving bands; 100 mm × 100 mm × 100 mm; standard C dressing; flame textured top, tightly butted to falls or crossfalls; bedding 25 mm thick in cement mortar (1:3); filling joints with dry mortar (1:6); washed and brushed; on concrete base							
Pavings; over 300 mm wide							
straight joints	78.81	1.48	36.09	–	90.88	m²	**126.97**
laid to pattern	78.81	1.85	45.11	–	90.88	m²	**135.99**
Boundary edging							
two rows of granite setts as boundary edging; 200 mm wide; including a 150 mm high ready mixed designated concrete C10–40 mm aggregate; haunching to one side; blocks laid breaking joint	–	0.65	15.85	–	20.24	m	**36.09**
35.06 EXTERNAL UNDERGROUND SERVICE RUNS							
Excavating trenches; by machine; grading bottoms; earthwork support; filling with excavated material and compacting; disposal of surplus soil on site; spreading on site average 50 m from excavations							
Services not exceeding 200 mm nominal size							
average depth of run not exceeding 0.50 m	–	0.28	3.94	1.45	–	m	**5.39**
average depth of run not exceeding 0.75 m	–	0.37	5.21	2.40	–	m	**7.61**
average depth of run not exceeding 1.00 m	–	0.79	11.13	3.13	1.37	m	**15.63**
average depth of run not exceeding 1.25 m	–	1.16	16.35	4.15	2.04	m	**22.54**
average depth of run not exceeding 1.50 m	–	1.48	20.86	5.60	2.48	m	**28.94**
average depth of run not exceeding 1.75 m	–	1.85	26.07	7.28	3.05	m	**36.40**
average depth of run not exceeding 2.00 m	–	2.13	30.01	8.52	3.33	m	**41.86**

35 SITE WORK

Item	PC £	Labour hours	Labour £	Plant £	Material £	Unit	Total rate £
Excavating trenches; by hand; grading bottoms; earthwork support; filling with excavated material and compacting; disposal; of surplus soil on site; spreading on site average 50 m from excavations							
Services not exceeding 200 mm nominal size							
average depth of run not exceeding 0.50 m	–	0.93	13.10	–	–	m	**13.10**
average depth of run not exceeding 0.75 m	–	1.39	19.59	–	–	m	**19.59**
average depth of run not exceeding 1.00 m	–	2.04	28.75	–	1.48	m	**30.23**
average depth of run not exceeding 1.25 m	–	2.87	40.45	–	2.04	m	**42.49**
average depth of run not exceeding 1.50 m	–	3.93	55.38	–	2.48	m	**57.86**
average depth of run not exceeding 1.75 m	–	5.18	73.00	–	3.01	m	**76.01**
average depth of run not exceeding 2.00 m	–	5.92	83.42	–	3.31	m	**86.73**
Stopcock pits, valve chambers and the like; excavating; half brick thick walls in common bricks in cement mortar (1:3); on in situ concrete designated mix C20–20 mm aggregate bed; 100 mm thick							
Pits							
100 mm × 100 mm × 750 mm deep; internal holes for one small pipe; polypropylene hinged box cover; bedding in cement mortar (1:3)	–	3.89	85.31	–	55.91	nr	**141.22**
Communication boxes							
FP McCann Ltd precast reinforced concrete BT approved communication boxes cable and junction protection box. Medium duty cast iron cover							
Excavations and bedding excluded							
C2 communication box 1200 mm long × 600 mm wide × 895 mm high	259.35	0.50	6.99	0.80	365.68	nr	**373.47**
DP communication box 910 mm long × 890 mm wide × 1000 mm high	323.40	0.50	6.99	0.80	431.65	nr	**439.44**
J4 communication box 910 mm long × 440 mm wide × 800 mm high	115.50	0.50	6.99	0.80	217.52	nr	**225.31**
MCX communication box 1300 mm long × 850 mm wide × 900 mm high	339.15	0.60	8.38	0.80	521.42	nr	**530.60**
MCX communication box riser piece 1300 mm long × 850 mm wide × 300 mm high	98.70	0.45	6.29	0.40	101.66	nr	**108.35**

Prices for Measured Works

36 FENCING

Item	PC £	Labour hours	Labour £	Plant £	Material £	Unit	Total rate £
36.01 FENCING							
NOTE: The prices for all fencing include for setting posts in position, to a depth of 0.60 m for fences not exceeding 1.40 m high and of 0.76 m for fences over 1.40 m high. The prices allow for excavating post holes; filling to within 150 mm of ground level with concrete and all necessary backfilling.							
For more examples please consult *Spon's External Works and Landscaping Book*							
Fencing; strained wire							
Strained wire fencing; BS 1722 Part 3; 4 mm dia. galvanized mild steel plain wire threaded through posts and strained with eye bolts							
900 mm high; three line; concrete posts at 2.75 m centres	–	–	–	–	–	m	20.22
end concrete straining post; one strut	–	–	–	–	–	nr	49.16
angle concrete straining post; two struts	–	–	–	–	–	nr	57.10
1.07 m high; six line; concrete posts at 2.75 m centres	–	–	–	–	–	m	21.04
end concrete straining post; one strut	–	–	–	–	–	nr	55.29
angle concrete straining post; two struts	–	–	–	–	–	nr	63.23
1.20 m high; six line; concrete posts at 2.75 m centres	–	–	–	–	–	m	21.16
end concrete straining post; one strut	–	–	–	–	–	nr	56.86
angle concrete straining post; two struts	–	–	–	–	–	nr	64.77
1.40 m high; eight line; concrete posts at 2.75 m centres	–	–	–	–	–	m	21.73
end concrete straining post; one strut	–	–	–	–	–	nr	58.09
angle concrete straining post; two struts	–	–	–	–	–	nr	66.00
Chain link fencing; BS 1722 Part 1; 3 mm dia. galvanized mild steel wire; 50 mm mesh; galvanized mild steel tying and line wire; three line wires threaded through posts and strained with eye bolts and winding brackets							
900 mm high; galvanized mild steel angle posts at 3.00 m centres	–	–	–	–	–	m	28.45
end steel straining post; one strut	–	–	–	–	–	nr	81.96
angle steel straining post; two struts	–	–	–	–	–	nr	94.54
900 mm high; concrete posts at 3.00 m centres	–	–	–	–	–	m	20.61
end concrete straining post; one strut	–	–	–	–	–	nr	44.03
angle concrete straining post; two struts	–	–	–	–	–	nr	51.97
1.20 m high; galvanized mild steel angle posts at 3.00 m centres	–	–	–	–	–	m	20.98
end steel straining post; one strut	–	–	–	–	–	nr	87.46
angle steel straining post; two struts	–	–	–	–	–	nr	112.02
1.20 m high; concrete posts at 3.00 m centres	–	–	–	–	–	m	20.09
end concrete straining post; one strut	–	–	–	–	–	nr	50.40
angle concrete straining post; two struts	–	–	–	–	–	nr	59.54

36 FENCING

Item	PC £	Labour hours	Labour £	Plant £	Material £	Unit	Total rate £
1.80 m high; galvanized mild steel angle posts at 3.00 m centres	–	–	–	–	–	m	23.57
end steel straining post; one strut	–	–	–	–	–	nr	88.96
angle steel straining post; two struts	–	–	–	–	–	nr	110.66
1.80 m high; concrete posts at 3.00 m centres	–	–	–	–	–	m	27.23
end concrete straining post; one strut	–	–	–	–	–	nr	70.48
angle concrete straining post; two struts	–	–	–	–	–	nr	83.14
Gates							
Pair of gates and gate posts; gates to match galvanized chain link fencing, with angle framing, braces, etc., complete with hinges, locking bar, lock and bolts; two 100 mm × 100 mm angle section gate posts; each with one strut							
2.44 m × 0.90 m	–	–	–	–	–	nr	691.20
2.44 m × 1.20 m	–	–	–	–	–	nr	713.33
2.44 m × 1.80 m	–	–	–	–	–	nr	769.50
Fencing; chain link							
Chain link fencing; BS 1722 Part 1; 3 mm dia. plastic coated mild steel wire; 50 mm mesh; plastic coated mild steel tying and line wire; three line wires threaded through posts and strained with eye bolts and winding brackets							
900 mm high; galvanized mild steel angle posts at 3.00 m centres	–	–	–	–	–	m	26.08
end steel straining post; one strut	–	–	–	–	–	nr	72.28
angle steel straining post; two struts	–	–	–	–	–	nr	80.46
900 mm high; concrete posts at 3.00 m centres	–	–	–	–	–	m	19.45
end concrete straining post; one strut	–	–	–	–	–	nr	44.03
angle concrete straining post; two struts	–	–	–	–	–	nr	51.97
1.20 m high; galvanized mild steel angle posts at 3.00 m centres	–	–	–	–	–	m	19.26
end steel straining post; one strut	–	–	–	–	–	nr	75.82
angle steel straining post; two struts	–	–	–	–	–	nr	81.00
1.20 m high; concrete posts at 3.00 m centres	–	–	–	–	–	m	19.71
end concrete straining post; one strut	–	–	–	–	–	nr	50.40
angle concrete straining post; two struts	–	–	–	–	–	nr	59.54
1.80 m high; galvanized mild steel angle posts at 3.00 m centres	–	–	–	–	–	m	21.86
end steel straining post; one strut	–	–	–	–	–	nr	75.04
angle steel straining post; two struts	–	–	–	–	–	nr	89.62
1.80 m high; concrete posts at 3.00 mm centres	–	–	–	–	–	m	25.13
end concrete straining post; one strut	–	–	–	–	–	nr	70.48
angle concrete straining post; two struts	–	–	–	–	–	nr	83.14

36 FENCING

Item	PC £	Labour hours	Labour £	Plant £	Material £	Unit	Total rate £
36.01 FENCING – cont							
Fencing – cont							
Chain link fencing for tennis courts; BS 1722 Part 13; 2.5 dia. galvanized mild wire; 45 mm mesh; line and tying wires threaded through 45 mm × 45 mm × 5 mm galvanized mild steel angle standards, posts and struts; 60 mm × 60 mm × 6 mm straining posts and gate posts							
fencing to tennis court							
36.00 m × 18.00 m × 2.745 m high; standards at 3.00 m centres	–	–	–	–	–	nr	2762.01
fencing to tennis court 36.00 m × 18.00 m × 3.66 m high; standards at 2.50 mm centres	–	–	–	–	–	nr	4231.15
Gates							
Pair of gates and gate posts; gates to match plastic chain link fencing; with angle framing, braces, etc. complete with hinges, locking bar, lock and bolts; two 100 mm × 100 mm angle section gate posts; each with one strut							
2.44 m × 0.90 m	–	–	–	–	–	nr	604.39
2.44 m × 1.20 m	–	–	–	–	–	nr	620.14
2.44 m × 1.80 m	–	–	–	–	–	nr	667.92
Fencing; timber							
Cleft chestnut pale fencing; BS 1722 Part 4; pales spaced 51 mm apart; on two lines of galvanized wire; 64 mm dia. posts; 76 mm × 51 mm struts							
900 mm high; posts at 2.50 m centres	–	–	–	–	–	m	11.76
straining post; one strut	–	–	–	–	–	nr	31.14
corner straining post; two struts	–	–	–	–	–	nr	31.14
1.05 m high; posts at 2.50 m centres	–	–	–	–	–	m	13.18
straining post; one strut	–	–	–	–	–	nr	31.38
corner straining post; two struts	–	–	–	–	–	nr	31.38
Closeboarded fencing; BS 1722 Part 5; 76 mm × 38 mm softwood rails; 89 mm × 19 mm softwood pales lapped 13 mm; 152 mm × 25 mm softwood gravel boards; all softwood treated; posts at 3.00 m centres							
Fencing; two rail; concrete posts							
height 1.00 m	–	–	–	–	–	m	35.73
height 1.20 m	–	–	–	–	–	m	36.13
Fencing; three rail; concrete posts							
height 1.40 m	–	–	–	–	–	m	39.67
height 1.60 m	–	–	–	–	–	m	39.67
height 1.80 m	–	–	–	–	–	m	41.13

36 FENCING

Item	PC £	Labour hours	Labour £	Plant £	Material £	Unit	Total rate £
Fencing; concrete							
Precast concrete slab fencing;							
305 mm × 38 mm × 1753 mm slabs; fitted into twice grooved concrete posts at 1.83 m centres							
height 1.50 m	–	–	–	–	–	m	**70.51**
height 1.80 m	–	–	–	–	–	m	**76.39**
Fencing; security							
Mild steel unclimbable fencing; in rivetted panels 2440 mm long; 44 mm × 13 mm flat section top and bottom rails; two 44 mm × 19 mm flat section standards; one with foot plate; and 38 mm × 13 mm raking stay with foot plate; 20 mm dia. pointed verticals at 120 mm centres; two 44 mm × 19 mm supports 760 mm long with ragged ends to bottom rail; the whole bolted together; coated with red oxide primer; setting standards and stays in ground at 2440 mm centres and supports at 815 mm centres							
height 1.67 m	–	–	–	–	–	m	**131.64**
height 2.13 m	–	–	–	–	–	m	**152.79**
Pair of gates and gate posts, to match mild steel unclimbable fencing; with flat section framing, braces, etc., complete with locking bar, lock, handles, drop bolt, gate stop and holding back catches; two 102 mm × 102 mm hollow section gate posts with cap and foot plates							
2.44 m × 1.67 m	–	–	–	–	–	nr	**1134.18**
2.44 m × 2.13 m	–	–	–	–	–	nr	**1316.35**
4.88 m × 1.67 m	–	–	–	–	–	nr	**1762.98**
4.88 m × 2.13 m	–	–	–	–	–	nr	**2221.36**
PVC coated, galvanized mild steel high security fencing; Sentinal Sterling fencing or other equal and approved; 50 mm × 50 mm mesh; 3/3.50 mm gauge wire; barbed edge – 1; Sentinal Bi-steel colour coated posts or other equal and approved at 2.44 m centres							
1.80 m	–	0.93	13.10	–	46.93	m	**60.03**
2.10 m	–	1.16	16.35	–	51.84	m	**68.19**

37 SOFT LANDSCAPING

Item	PC £	Labour hours	Labour £	Plant £	Material £	Unit	Total rate £
37.01 SEEDING AND TURFING							
For more examples and depth of information please consult *Spon's External Works and Landscape Price Book*							
Top soil							
By machine							
Selected from spoil heaps not exceeding 50 m; grading; prepared for turfing or seeding; to general surfaces							
average 100 mm thick	–	–	0.06	1.47	–	m²	**1.53**
average 150 mm thick	–	0.01	0.08	1.97	–	m²	**2.05**
average 200 mm thick	–	0.01	0.11	3.44	–	m²	**3.55**
Selected from spoil heaps; grading; prepared for turfing or seeding; to cuttings or embankments							
average 100 mm thick	–	–	0.07	1.67	–	m²	**1.74**
average 150 mm thick	–	0.01	0.09	2.27	–	m²	**2.36**
average 200 mm thick	–	0.01	0.12	3.93	–	m²	**4.05**
By hand							
Selected from spoil heaps not exceeding 20 m; grading; prepared for turfing or seeding; to general surfaces							
average 100 mm thick	–	0.40	5.63	–	–	m²	**5.63**
average 150 mm thick	–	0.50	7.05	–	–	m²	**7.05**
average 200 mm thick	–	0.60	8.46	–	–	m²	**8.46**
Selected from spoil heaps; grading; prepared for turfing or seeding; to cuttings or embankments							
average 100 mm thick	–	0.51	7.13	–	–	m²	**7.13**
average 150 mm thick	–	0.63	8.91	–	–	m²	**8.91**
average 200 mm thick	–	0.76	10.69	–	–	m²	**10.69**
Imported top soil, planting quality							
By machine							
Grading; prepared for turfing or seeding; to general surfaces							
average 100 mm thick	–	–	0.06	1.47	3.11	m²	**4.64**
average 150 mm thick	–	0.01	0.08	1.97	4.68	m²	**6.73**
average 200 mm thick	–	0.01	0.11	3.44	5.71	m²	**9.26**
Grading; preparing for turfing or seeding; to cuttings or embankments							
average 100 mm thick	–	–	0.07	1.67	3.81	m²	**5.55**
average 150 mm thick	–	0.01	0.09	2.27	6.23	m²	**8.59**
average 200 mm thick	–	0.01	0.12	3.93	7.61	m²	**11.66**
By hand							
Grading; prepared for turfing or seeding; to general surfaces							
average 100 mm thick	–	0.51	7.13	–	3.81	m²	**10.94**
average 150 mm thick	–	0.63	8.91	–	6.23	m²	**15.14**
average 200 mm thick	–	0.76	10.69	–	7.61	m²	**18.30**

37 SOFT LANDSCAPING

Item	PC £	Labour hours	Labour £	Plant £	Material £	Unit	Total rate £
Fertilizer							
Fertilizer 0.07 kg/m^2; raking in							
general surfaces (PC £ per 25 kg)	28.35	0.03	0.42	–	0.09	m^2	**0.51**
Selected grass seed							
Grass seed; sowing at a rate of 40 g/m^2 two							
applications; raking in							
general surfaces (PC £ per kg)	2.94	0.17	2.40	–	0.16	m^2	**2.56**
cuttings or embankments	–	0.20	2.82	–	0.16	m^2	**2.98**
Turfing							
Imported turf; cultivated							
general surfaces	2.64	0.10	1.41	–	2.72	m^2	**4.13**
cuttings or embankments; shallow	2.64	0.15	2.11	–	2.72	m^2	**4.83**
cuttings or embankments; steep; pegged	2.64	0.28	3.94	–	2.72	m^2	**6.66**
Preserved turf from stack on site; lay only							
general surfaces	–	0.19	2.68	–	–	m^2	**2.68**
cuttings or embankments; shallow	–	0.20	2.82	–	–	m^2	**2.82**
cuttings or embankments; steep; pegged	–	0.28	3.94	–	–	m^2	**3.94**
37.02 PLANTING							
Hedge plants							
instant hedge planting; delivered in 2.5 m lengths;							
planting and backfilling with excavted material;							
trench not included; hedge height; hedge 1.0 m							
high × 300 mm wide	–	0.75	10.57	5.68	136.27	nr	**152.52**
Tree planting							
semi-mature or mature tree, 40–45 cm girth	630.00	8.00	112.73	85.14	648.90	nr	**846.77**
semi-mature or mature tree, 80–90 cm girth	5250.00	20.00	281.84	227.04	5407.50	nr	**5916.38**

39 ELECTRICAL SERVICES

Item	PC £	Labour hours	Labour £	Plant £	Material £	Unit	Total rate £
39.01 LIGHTNING PROTECTION							
For more examples and depth of information please consult *Spon's Mechanical and Electrical Services Price Book*							
Lightning protection equipment							
Copper strip roof or down conductors fixed with bracket or saddle clips							
20 mm × 3 mm flat section	–	–	–	–	–	m	**39.27**
25 mm × 6 mm flat section	–	–	–	–	–	m	**64.47**
Joints in tapes	–	–	–	–	–	nr	**16.96**
Bonding connections to roof and structural metalwork	–	–	–	–	–	nr	**80.63**
Testing points	–	–	–	–	–	nr	**72.57**
Earth electrodes							
16 mm dia. driven copper electrodes in 1220 mm long sectional lengths; 2440 mm long overall	–	–	–	–	–	nr	**212.13**

41 BUILDER'S WORK IN CONNECTION WITH SERVICES

Item	PC £	Labour hours	Labour £	Plant £	Material £	Unit	Total rate £
41.01 HOLES, CHASES FOR SERVICES							
Electrical installations; cutting away for and making good after electrician; including cutting or leaving all holes, notches, mortices, sinkings and chases, in both the structure and its coverings, for the following electrical points							
Exposed installation							
lighting points	–	0.28	4.95	–	–	nr	**4.95**
socket outlet points	–	0.46	8.51	–	–	nr	**8.51**
fitting outlet points	–	0.46	8.51	–	–	nr	**8.51**
equipment points or control gear points	–	0.65	12.21	–	–	nr	**12.21**
Concealed installation							
lighting points	–	0.37	6.74	–	–	nr	**6.74**
socket outlet points	–	0.65	12.21	–	–	nr	**12.21**
fitting outlet points	–	0.65	12.21	–	–	nr	**12.21**
equipment points or control gear points	–	0.93	17.16	–	–	nr	**17.16**
Builders' work for other services installations							
Cutting chases in brickwork							
for one pipe; not exceeding 55 mm nominal size; vertical	–	0.37	5.17	–	–	m	**5.17**
for one pipe; 55 mm–110 mm nominal size; vertical	–	0.65	9.08	–	–	m	**9.08**
Cutting and pinning to brickwork or blockwork; ends of supports							
for pipes not exceeding 55 mm nominal size	–	0.19	4.17	–	–	nr	**4.17**
for cast iron pipes 55 mm–110 mm nominal size	–	0.31	6.80	–	–	nr	**6.80**
Cutting or forming holes for pipes or the like; not exceeding 55 mm nominal size; making good							
reinforced concrete; not exceeding 100 mm deep	–	0.75	12.39	0.76	–	nr	**13.15**
reinforced concrete; 100 mm–200 mm deep	–	1.15	19.00	1.17	–	nr	**20.17**
reinforced concrete; 200 mm–300 mm deep	–	1.50	24.78	1.52	–	nr	**26.30**
half brick thick	–	0.31	5.12	–	–	nr	**5.12**
one brick thick	–	0.51	8.43	–	–	nr	**8.43**
one and a half brick thick	–	0.83	13.71	–	–	nr	**13.71**
100 mm blockwork	–	0.28	4.62	–	–	nr	**4.62**
140 mm blockwork	–	0.37	6.12	–	–	nr	**6.12**
215 mm blockwork	–	0.46	7.60	–	–	nr	**7.60**
plasterboard partition or suspended ceiling	–	0.35	5.78	–	–	nr	**5.78**
Cutting or forming holes for pipes or the like; 55 mm–110 mm nominal size; making good							
reinforced concrete; not exceeding 100 mm deep	–	1.15	19.00	1.17	–	nr	**20.17**
reinforced concrete; 100 mm–200 mm deep	–	1.75	28.91	1.78	–	nr	**30.69**
reinforced concrete; 200 mm–300 mm deep	–	2.25	37.17	2.29	–	nr	**39.46**
half brick thick	–	0.37	6.12	–	–	nr	**6.12**
one brick thick	–	0.65	10.74	–	–	nr	**10.74**
one and a half brick thick	–	1.02	16.85	–	–	nr	**16.85**
100 mm blockwork	–	0.32	5.28	–	–	nr	**5.28**
140 mm blockwork	–	0.46	7.60	–	–	nr	**7.60**
215 mm blockwork	–	0.56	9.25	–	–	nr	**9.25**
plasterboard partition or suspended ceiling	–	0.40	6.61	–	–	nr	**6.61**

41 BUILDER'S WORK IN CONNECTION WITH SERVICES

Item	PC £	Labour hours	Labour £	Plant £	Material £	Unit	Total rate £
41.01 HOLES, CHASES FOR SERVICES – cont							
Builders' work for other services installations – cont							
Cutting or forming holes for pipes or the like; over 110 mm nominal size; making good							
reinforced concrete; not exceeding 100 mm deep	–	1.15	19.00	1.17	–	nr	**20.17**
reinforced concrete; 100 mm–200 mm deep	–	1.75	28.91	1.78	–	nr	**30.69**
reinforced concrete; 200 mm–300 mm deep	–	2.25	37.17	2.29	–	nr	**39.46**
half brick thick	–	0.46	7.60	–	–	nr	**7.60**
one brick thick	–	0.79	13.05	–	–	nr	**13.05**
one and a half brick thick	–	1.25	20.65	–	–	nr	**20.65**
100 mm blockwork	–	0.42	6.94	–	–	nr	**6.94**
140 mm blockwork	–	0.56	9.25	–	–	nr	**9.25**
215 mm blockwork	–	0.69	11.40	–	–	nr	**11.40**
plasterboard partition or suspended ceiling	–	0.45	7.44	–	–	nr	**7.44**
Add for making good fair face or facings one side							
pipe; not exceeding 55 mm nominal size	–	0.07	1.53	–	–	nr	**1.53**
pipe; 55 mm–110 mm nominal size	–	0.09	1.98	–	–	nr	**1.98**
pipe; over 110 mm nominal size	–	0.11	2.41	–	–	nr	**2.41**
Add for fixing sleeve (supply not included)							
for pipe; small	–	0.14	3.07	–	–	nr	**3.07**
for pipe; large	–	0.19	4.17	–	–	nr	**4.17**
for pipe; extra large	–	0.28	6.14	–	–	nr	**6.14**
Add for supplying and fixing two hour intumescent sleeve							
for 55 mm uPVC pipe	–	0.25	4.13	–	7.12	nr	**11.25**
for 110 mm uPVC pipe	–	0.28	4.62	–	7.74	nr	**12.36**
for 200 mm uPVC pipe	–	0.30	4.95	–	50.63	nr	**55.58**
Cutting or forming holes for ducts; girth not exceeding 1.00 m; making good							
half brick thick	–	0.56	9.25	–	–	nr	**9.25**
one brick thick	–	0.93	15.37	–	–	nr	**15.37**
one and a half brick thick	–	1.48	24.45	–	–	nr	**24.45**
100 mm blockwork	–	0.46	7.60	–	–	nr	**7.60**
140 mm blockwork	–	0.65	10.74	–	–	nr	**10.74**
215 mm blockwork	–	0.83	13.71	–	–	nr	**13.71**
plasterboard partition or suspended ceiling	–	0.65	10.74	–	–	nr	**10.74**
Cutting or forming holes for ducts; girth 1.00 m–2.00 m; making good							
half brick thick	–	0.65	10.74	–	–	nr	**10.74**
one brick thick	–	1.11	18.34	–	–	nr	**18.34**
one and a half brick thick	–	1.76	29.08	–	–	nr	**29.08**
100 mm blockwork	–	0.56	9.25	–	–	nr	**9.25**
140 mm blockwork	–	0.74	12.23	–	–	nr	**12.23**
215 mm blockwork	–	0.93	15.37	–	–	nr	**15.37**
plasterboard partition or suspended ceiling	–	0.75	12.39	–	–	nr	**12.39**
Cutting or forming holes for ducts; girth 2.00 m–3.00 m; making good							
half brick thick	–	1.02	16.85	–	–	nr	**16.85**
one brick thick	–	1.76	29.08	–	–	nr	**29.08**
one and a half brick thick	–	2.78	45.94	–	–	nr	**45.94**

41 BUILDER'S WORK IN CONNECTION WITH SERVICES

Item	PC £	Labour hours	Labour £	Plant £	Material £	Unit	Total rate £
100 mm blockwork	–	0.88	14.54	–	–	nr	**14.54**
140 mm blockwork	–	1.20	19.83	–	–	nr	**19.83**
215 mm blockwork	–	1.53	25.28	–	–	nr	**25.28**
plasterboard partition or suspended ceiling	–	1.00	16.52	–	–	nr	**16.52**
Cutting or forming holes for ducts; girth 3.00 m–4.00 m; making good							
half brick thick	–	1.39	22.97	–	–	nr	**22.97**
one brick thick	–	2.31	38.17	–	–	nr	**38.17**
one and a half brick thick	–	3.70	61.14	–	–	nr	**61.14**
100 mm blockwork	–	1.02	16.85	–	–	nr	**16.85**
140 mm blockwork	–	1.39	22.97	–	–	nr	**22.97**
215 mm blockwork	–	1.76	29.08	–	–	nr	**29.08**
plasterboard partition or suspended ceiling	–	1.25	20.65	–	–	nr	**20.65**
Mortices in brickwork							
for expansion bolt	–	0.19	3.14	–	–	nr	**3.14**
for 20 mm dia. bolt; 75 mm deep	–	0.14	2.32	–	–	nr	**2.32**
for 20 mm dia. bolt; 150 mm deep	–	0.23	3.80	–	–	nr	**3.80**
Mortices in brickwork; grouting with cement mortar (1:1)							
75 mm × 75 mm × 200 mm deep	–	0.28	4.62	–	0.18	nr	**4.80**
75 mm × 75 mm × 300 mm deep	–	0.37	6.12	–	0.27	nr	**6.39**
Holes in softwood for pipes, bars, cables and the like							
12 mm thick	–	0.03	0.69	–	–	nr	**0.69**
25 mm thick	–	0.05	1.15	–	–	nr	**1.15**
50 mm thick	–	0.09	2.07	–	–	nr	**2.07**
100 mm thick	–	0.14	3.21	–	–	nr	**3.21**
Holes in hardwood for pipes, bars, cables and the like							
12 mm thick	–	0.05	1.15	–	–	nr	**1.15**
25 mm thick	–	0.08	1.84	–	–	nr	**1.84**
50 mm thick	–	0.14	3.21	–	–	nr	**3.21**
100 mm thick	–	0.20	4.59	–	–	nr	**4.59**
Diamond drilling for cutting holes and mortices in masonry or concrete							
Cutting holes and mortices in brickwork; per 25 mm depth							
25 mm dia.	–	–	–	–	–	nr	**2.01**
32 mm dia.	–	–	–	–	–	nr	**1.62**
52 mm dia.	–	–	–	–	–	nr	**1.94**
78 mm dia.	–	–	–	–	–	nr	**2.13**
107 mm dia.	–	–	–	–	–	nr	**2.24**
127 mm dia.	–	–	–	–	–	nr	**2.77**
152 mm dia.	–	–	–	–	–	nr	**3.25**
200 mm dia.	–	–	–	–	–	nr	**4.19**
250 mm dia.	–	–	–	–	–	nr	**6.32**
300 mm dia.	–	–	–	–	–	nr	**8.39**
Diamond chasing; per 25 × 25 mm section							
in facing or common brickwork	–	–	–	–	–	m	**3.81**
in semi-engineering brickwork	–	–	–	–	–	m	**7.63**
in engineering brickwork	–	–	–	–	–	m	**10.64**

41 BUILDER'S WORK IN CONNECTION WITH SERVICES

Item	PC £	Labour hours	Labour £	Plant £	Material £	Unit	Total rate £
41.01 HOLES, CHASES FOR SERVICES – cont							
Diamond drilling for cutting holes and mortices in masonry or concrete – cont							
Diamond chasing – cont							
in lightweight blockwork	–	–	–	–	–	m	**2.99**
in heavyweight blockwork	–	–	–	–	–	m	**6.00**
in render/screed	–	–	–	–	–	m	**11.81**
Forming boxes; 100 × 100 mm; per 25 mm depth							
in facing or common brickwork	–	–	–	–	–	nr	**1.52**
in semi-engineering brickwork	–	–	–	–	–	nr	**3.06**
in engineering brickwork	–	–	–	–	–	nr	**4.25**
in lightweight blockwork	–	–	–	–	–	nr	**1.21**
in heavyweight blockwork	–	–	–	–	–	nr	**2.41**
in render/screed	–	–	–	–	–	nr	**4.73**
Other items							
diamond track mount or ring sawing brickwork	–	–	–	–	–	m	**7.51**
diamond floor sawing asphalte	–	–	–	–	–	m	**1.26**
stitch drilling 107 mm dia. hole in brickwork	–	–	–	–	–	nr	**1.62**
41.02 INTERNAL FLOOR DUCTS FOR SERVICES							
Screed floor ducting; with side flanges; laid within floor screed; galvanized mild steel							
Floor ducting							
100 mm wide × 50 mm deep	11.82	0.19	4.37	–	12.17	m	**16.54**
bend	–	0.09	2.07	–	18.85	nr	**20.92**
tee section	–	0.09	2.07	–	18.85	nr	**20.92**
connector/stop end	–	0.09	2.07	–	2.17	nr	**4.24**
ply cover 15 mm/16 mm thick WBP exterior grade		0.09	2.07	–	2.70	m	**4.77**
100 mm wide × 70 mm deep	12.98	0.20	4.59	–	13.37	m	**17.96**
bend	–	0.09	2.07	–	18.85	nr	**20.92**
tee section	–	0.09	2.07	–	18.85	nr	**20.92**
connector/stop end	–	0.09	2.07	–	2.17	nr	**4.24**
ply cover 15 mm/16 mm thick WBP exterior grade	–	0.09	2.07	–	2.70	m	**4.77**
200 mm wide × 50 mm deep	16.34	0.19	4.37	–	16.83	m	**21.20**
bend	–	0.09	2.07	–	20.98	nr	**23.05**
tee section	–	0.09	2.07	–	20.98	nr	**23.05**
connector/stop end	–	0.09	2.07	–	2.17	nr	**4.24**
ply cover 15 mm/16 mm thick WBP exterior grade	–	0.09	2.07	–	4.90	m	**6.97**

41 BUILDER'S WORK IN CONNECTION WITH SERVICES

Item	PC £	Labour hours	Labour £	Plant £	Material £	Unit	Total rate £
41.03 EXTERNAL SERVICES							
Service Runs							
Water main; all laid in trenches including excavation and backfill with type 1 hardcore. Surface finish not included							
MDPE pipe; 20 mm dia.	–	2.85	40.16	9.92	48.99	m	**99.07**
MDPE pipe; 25 mm dia.	–	2.85	40.16	9.92	49.11	m	**99.19**
MDPE pipe; 32 mm dia.	–	2.85	40.16	9.92	49.35	m	**99.43**
MDPE pipe; 50 mm dia.	–	2.85	40.16	9.92	51.02	m	**101.10**
Electric mains 25 mm 2 core armoured cable laid in service duct, including excavation and backfill with type 1 hardcore. Surface finish not included							
12.5 mm 2 core armoured cable	–	3.00	42.27	9.92	62.97	m	**115.16**
25 mm 3 core armoured cable	–	3.00	42.27	9.92	64.15	m	**116.34**
25 mm 5 core armoured cable	–	3.00	42.27	9.92	65.22	m	**117.41**

Understanding NEC4: Term Service Contract

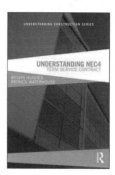

Kelvin Hughes and Patrick Waterhouse

With the release of the new NEC (formerly the New Engineering Contract), the use of the family of contracts continues to grow worldwide, as does the importance of understanding its clauses and nuances to everyone working in the built environment. Now in its fourth edition, this set of contracts is different to others in concept as well as format, so users may well find themselves needing a helping hand along the way.

Written in plain English, Understanding NEC4: Term Service Contract offers a practical guide to the use and management of the NEC4 Term Service Contract (TSC). The authors describe the full life of a contract, from the initial selection of options and contract formation through to the operations period and ultimately termination and dispute resolution. Although born of the same stable as the NEC4 construction contracts, the TSC is aimed at maintaining infrastructure and differs significantly from its siblings.

This is the first book to be written on the new contract and takes the reader through the important provisions including communications, planning, early warnings, compensation events and payments. It is ideal for clients, contractors and their advisors describing how to deploy the contract successfully.

March 2018: 234 x 156 mm: 230 pp
Pb: 978-0-8153-4836-8 : £47.99

To Order: Tel: +44 (0) 1235 400524 Fax: +44 (0) 1235 400525
or Post: Taylor and Francis Customer Services,
Bookpoint Ltd, Unit T1, 200 Milton Park, Abingdon, Oxon, OX14 4TA UK
Email: book.orders@tandf.co.uk

For a complete listing of all our titles visit:
www.tandf.co.uk

PART 5

Fees for Professional Services

This part contains the following sections:

The Streetwise Subbie, 4th Edition

Barry J Ashmore

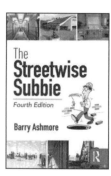

Very few books explore the problems which are particular to the relationship between Specialist Contractors and the Main Contractor, or Clients with whom they are in contract. Fewer still provide solutions in such a down to earth no-nonsense way as The Streetwise Subbie does.

The Streetwise Subbie is a highly regarded practical guide to contractual matters. Its original author Jack Russell was well known for his 'contractual terrier' column in Electrical Times. Now, Barry Ashmore has updated and revised this work by drawing on his 46 years of construction experience and professional expertise gained at the sharp end resolving disputes and solving contractual problems for Specialist Contractors.

Thousands of subbies have already benefited from the insights and the streetwise approach to avoiding or resolving contractual problems, and the clarity of thought and advice the book provides. The fourth edition features all the old favourites such as payment, delay and disruption, extension of time and the all-important checklists and site records. But it has now been brought bang up to date to reflect the importance of the 2011 revisions to the Construction Act and the emergence of adjudication as the pre-eminent means of dispute resolution.

It is an easy to read, practical, and essential guide, aimed at Specialist Contractors of all sizes and specialisations, be they sole traders, company directors, or any member of the subbie's team that has to handle the commercial and contractual aspects of the projects they undertake.

It's the kind of book that you keep handy, because it has so many answers that you can refer to it over and over again

May 2018: 234 x 156 mm: 218 pp
Pb: 978-1-138-30016-3 : £16.99

To Order: Tel: +44 (0) 1235 400524 Fax: +44 (0) 1235 400525
or Post: Taylor and Francis Customer Services,
Bookpoint Ltd, Unit T1, 200 Milton Park, Abingdon, Oxon, OX14 4TA UK
Email: book.orders@tandf.co.uk

For a complete listing of all our titles visit:
www.tandf.co.uk

QUANTITY SURVEYORS' FEES

Guidance on basic quantity surveying services is set out in the RICS Standard Form of Consultant's Appointment for Quantity Surveyors. Services are separated into core services for a variety of contracts and supplementary services and it is advisable to refer to this guidance. Copies can be obtained from the Royal Institution of Chartered Surveyors (RICS) at www.ricsbooks.com.

Quantity surveying services

Preparation:

- liaising with clients and the professional team
- advice on cost
- preparation of initial budget/cost plan/cash flow forecasts

Design:

- prepare and maintain cost plan
- advise design team on impact of design development on cost

Pre-construction:

- liaise with professional team
- advise on procurement strategy
- liaise with client's legal advisors on contract matters
- prepare tender documents
- define prospective tenderers
- obtain tenders/check tenders/prepare recommendation for client
- maintain and develop cost plan

Construction:

- visit the site
- prepare interim valuations
- advise on the cost of variations
- agree the cost of claims
- advise on contractual matters

Supplementary services may include:

- preparation of mechanical and electrical tender documentation
- preparation of cost analyses
- advice on insurance claims
- facilitate value management exercises
- prepare life cycle calculations/sustainability
- capital allowance advice/VAT
- attend adjudication/mediation proceedings

Fee guide

These do not represent minimum or maximum values.

	Mean	Lower Quartile (Not complicated projects)	Upper Quartile (Complicated/ Sophisticated project)
Quantity Surveying Services Benchmark	2.2%	1.9%	3.4%

The level of fees above, are expressed as a percentage of the contract value of £4,000,000 for a new build project and do not include VAT.

Fee levels vary depending on many factors, including, but not limited to the following; type of project; complexity; procurement route.

ARCHITECTS' FEES

RIBA Agreements are designed to be:

- in line with current working practices, legislative changes and procurement methods
- attractive to clients, architects and other consultants, with robust but fair terms
- a flexible system of components that can be assembled and customized to create tailored and bespoke contracts
- suitable for a wide range of projects and services
- based upon the updated RIBA Outline Plan of Work 2013
- available in electronic and printed formats

RIBA Agreements 2013	Suitable for
Standard Agreement	• a commission where detailed contract terms are necessary for a wide range of projects using most procurement methods • where the client is acting for business or commercial purposes • where the commission is for work to the client's home where the size or value of the Project merits use of the JCT Standard or Intermediate forms of building contract or similar and the terms have been negotiated with the client as a consumer
Concise Agreement	• for a commission where the concise contract terms are compatible with the complexity of the Project and the risks to each party • where the Client is acting for business or commercial purposes • where the commission is for work to the Client's home and the terms have been negotiated with the Client as a 'consumer'. A consumer is a natural person acting for purposes outside his trade, business or profession • where the building works, including extensions and alterations, will be carried out using forms of building contract, such as JCT Agreement for Minor Works or JCT Intermediate Form of Building Contract
Domestic Project Agreement	• the commission relates to work to the client's home, provided that they have elected to use these conditions in their own name, i.e. not as a limited company or other legal entity • the contract terms are compatible with the complexity of the project and the risks to each party and have been negotiated with the client as a consumer • the building works, including extensions and alterations, will be carried out using forms of building contract, such as the JCT Building Contract for a homeowner/ occupier, JCT Agreement for Minor Works or JCT Intermediate Form of Building Contract
Sub-consultant's Agreement	• a consultant wishes or perhaps is required by the client to appoint another consultant (thus, a sub-consultant) to perform part of the consultant's services • the contract terms are compatible with the (head) agreement between the consultant and the client, with the complexity of the project and the risks to each party
Electronic and print formats	• all the RIBA Agreements 2013 and their components are available as electronic files. A limited number of the conditions and core components are published in print. The electronic and printed versions may be used in combination
Electronic components	• conditions, notes and guides are available as locked PDFs and all other components, e.g. schedules and model letters, are available in Rich Text Format (RTF), which can be customized using most commonly used word processing software to meet project requirements or modified to match the house style of the practice

Each agreement comprises the selected Conditions of Appointment (i.e. Standard, Concise or Domestic), related components, and a schedule or schedules of Services.

Notes on use and completion and model letters for business clients and domestic clients are included with each pack.

For further information, readers are advised to log onto the RIBA Publications website at *www.ribabookshops.com/agreements*

ARCHITECTS' FEES

RIBA Plan of Work 2013

The table below shows how the new Plan of Work maps to the old RIBA Plan of Work 2007 and the CIC's work stages

RIBA Plan of Work 2013	RIBA Plan of Work 2007	CIC/BIM Task Group Coordinated work stages
0 Strategic definition	A Appraisal	0 Strategy
1 Preparation and brief	B Design Brief	1 Brief
2 Concept design	C Concept	2 Concept
3 Developed design	D Design development	3 Definition
4 Technical design	E Technical design	4 Design
	F Production information	
	G Tender documentation	
	H Tender Action	
5 Construction	J Mobilization	5 Build and commission
	K Construction to practical completion	
6 Handover and close out		6 Handover & close out
7 In use	L Post-practical completion	7 Operation and end of life

More details can be found at the RIBA website.

Appointment guidance

A guide, 'A Client's Guide to Engaging an Architect', is available from RIBA Bookshops at www.ribabookshops.com.

This guide includes an introduction to the services an Architect can be expected to provide, advice on the forms to use, linking the RIBA Plan of Work Stages with fees (which are a matter of negotiation) and classifying buildings according to three levels of complexity.

Generally, the more complex the building the higher the level of fee.

Example categories include:

- Simple: for buildings such as car parks, warehouses, factories and speculative retail schemes
- Average: for buildings such as offices, most retail outlets, general housing, schools etc.
- Complex: for multi-purpose developments, specialist buildings e.g. hospitals, research laboratories etc.

Procurement option can also influence the architects' fees.

ARCHITECTS' FEES

Fee guide

These do not represent minimum or maximum values.

Architectural Services Benchmark	Mean	Lower Quartile (Not complicated projects)	Upper Quartile (Complicated/Sophisticated project)
	4.5%	3.6%	7.0%

The level of fees above, are expressed as a percentage of the contract value of £4,000,000 for a new build project and do not include VAT.

Projects carried out under the traditional form of contracts tend to attract higher fees than for design and build contract led jobs.

Sectors project have less of an impact on fee levels, however health and education projects attract slightly higher fees than private housing and industrial.

CONSULTING ENGINEERS' FEES

CONDITIONS OF APPOINTMENT

A scale of professional charges for consulting engineering services is published by the Association for Consultancy and Engineering (ACE)

Copies of the document can be obtained direct from:

Association for Consultancy and Engineering
Alliance House
12 Caxton Street
London SW1H OQL
Tel 020 7222 6557
Fax: 020 7222 0750

Comparisons

Instead of the previous arrangement of having different agreements designed for each major discipline of engineering, the current agreements have been developed primarily to suit the different roles that Consulting Engineers may be required to perform, with variants of some of them for different disciplines. The agreements have been standardized as far as possible whilst retaining essential differences.

Greater attention is required than with previous agreements to ensure the documents are completed properly. This is because of the perceived need to allow for a wider choice of arrangements, particularly of methods of payment.

The agreements are not intended to be used as unsigned reference material with the details of an engagement being covered in an exchange of letters, although much of their content could be used as a basis for drafting such correspondence.

For 2009 the ACE has published a new suite of Agreements with a broader set of services, these are listed below.

Forms of Agreement

- ACE Agreement 1: Design
- ACE Agreement 2: Advise and Report
- ACE Agreement 3: Design and Construct
- ACE Agreement 4: Sub-Consultancy
- ACE Agreement 5: Homeowner
- ACE Agreement 6: Expert Witness (Sole Practitioner)
- ACE Agreement 7: Expert Witness (Firm)
- ACE Agreement 8: Adjudicator

To a number of the the above ACE Agreements Schedules of Services are appended and currently these are:

For use with ACE Agreement 1: Design

- ACE Schedule of Services – Part G(a):
  Civil and Structural Engineer – single consultant or non-lead consultant
- ACE Schedule of Services – Part G(b):
  Mechanical and Electrical Engineering (detailed design in buildings)
- ACE Schedule of Services – Part G(c):
  Mechanical and Electrical Engineering (performance design in buildings)
- ACE Schedule of Services – Part G(d):
  Civil and Structural Engineer – Lead consultant
- ACE Schedule of Services – Part G(e):
  Mechanical and Electrical Engineering Design in buildings – Lead consultant

CONSULTING ENGINEERS' FEES

For use with ACE Agreement 3: Design and Construct

- ACE Schedule of Services – Part G(f):
  Civil and Structural Engineer
- ACE Schedule of Services – Part G(g):
  Mechanical and Electrical Engineering (detailed design in buildings)
- ACE Schedule of Services – Part G(h):
  Mechanical and Electrical Engineering (performance design in buildings)

ACE Agreement 1: Design

Design for the appointment of a consultant by a client to undertake detailed design and/or specification of permanent works to be undertaken or installed by a contractor including any studies, appraisals, investigations, contract administration or construction monitoring leading to or resulting from such detailed design and/or specification.

ACE Agreement 2: Advise and Report

Advise and report for the appointment of a consultant by a client to provide any type of advisory, research, checking, reviewing, investigatory, monitoring, reporting or technical services in the built and natural environments where such services do not consist of detailed design or specification of permanent works to be constructed or installed by a contractor.

ACE Agreement 3: Design and Construct

For the appointment of a consultant by a contractor in circumstances where the contractor is to construct permanent works designed by the consultant.

ACE Agreement 4: Sub-consultancy

For the appointment of a sub-consultant by a consultant in circumstances where the consultant is appointed on the terms of an ACE Agreement by its client.

ACE Agreement 5: Homeowner

Model letter for the appointment of a consultant by a homeowner.

ACE Agreement 6: Expert Witness (Sole Practitioner)

For the appointment of an individual to act as an expert witness.

ACE Agreement 7: Expert Witness (Firm)

For the appointment of a firm to provide an expert witness.

ACE Agreement 8: Adjudicator

For the appointment of an adjudicator.

Fee Guide

Profession

These do not represent minimum and maximum values. Based on value of service not project.

Structural Engineering Services Benchmark	Mean	Lower Quartile	Upper Quartile
	2.5%	1.7%	2.9%

Services Engineering Benchmark	Mean	Lower Quartile	Upper Quartile
	2.0%	1.5%	2.6%

THE TOWN AND COUNTRY PLANNING FEES AND BUILDING REGULATION FEES

THE TOWN AND COUNTRY PLANNING APPLICATION FEES

Author's Note

This is only a small extract of typical the fees chargeable. Users should always obtain actual fees from the local authority concerned with the particular planning application.

All Outline Applications	
Not more than 2.5 hectares	£462 per 0.1 hectare
More than 2.5 hectares	£11,432 + £138 per 0.1 hectare up to a maximum of £150,000

There is a very useful online calculator which will give prices for planning applications at

http://www.planningportal.gov.uk/pins/FeeCalculatorStandalone

The fee is paid at the time of application. You should always check your fees with your local planning authority

THE BUILDING (LOCAL AUTHORITY CHARGES) REGULATIONS

CHARGE SCHEDULES

Fees vary from one authority to another, so always check with your Local Authority if you decide to use their officers to provide building control services.

The Construction Industry Council (CIC) has been designated by the government as a body for approving inspectors (AI). Individual and Corporate Approved Inspectors registered with CIC are qualified to undertake building control work in accordance with section 49 of the Building Act 1984 and regulation 4 of the Building (Approved Inspectors etc.) Regulations 1985, and the Building (Approved Inspectors etc.) Regulations 2010.

Approved Inspectors provide building control services on all types of construction projects. The Construction Industry Council (CIC) maintains a list of approved inspectors, see www.cic.org.uk.

Fees vary. Each Local Authority publishes a schedule of rates applicable to project values commonly up to £250,000. The fee for a project of £250,000 in value is approximately £1,500.

For higher value projects the fees are individually determined but are typically approximately 0.1% of the estimated project cost.

Estimator's Pocket Book, 2nd edition

Duncan Cartlidge

The *Estimator's Pocket Book, Second Edition* is a concise and practical reference covering the main pricing approaches, as well as useful information such as how to process sub-contractor quotations, tender settlement and adjudication. It is fully up to date with NRM2 throughout, features a look ahead to NRM3 and describes the implications of BIM for estimators.
It includes instructions on how to handle:

*   the NRM order of cost estimate;
*   unit-rate pricing for different trades;
*   pro-rata pricing and dayworks;
*   builders' quantities;
*   approximate quantities.

Worked examples show how each of these techniques should be carried out in clear, easy-to-follow steps. This is the indispensable estimating reference for all quantity surveyors, cost managers, project managers and anybody else with estimating responsibilities. Particular attention is given to NRM2, but the overall focus is on the core estimating skills needed in practice. Updates to this edition include a greater reference to BIM, an update on the current state of the construction industry as well as up-to-date wage rates, legislative changes and guidance notes.

Routledge Pocket Books

February 2019: 198 × 129 mm: 292pp
Pb: 978-1-138-36670-1 : £24.99

To Order: Tel: +44 (0) 1235 400524 Fax: +44 (0) 1235 400525
or Post: Taylor and Francis Customer Services,
Bookpoint Ltd, Unit T1, 200 Milton Park, Abingdon, Oxon, OX14 4TA UK
Email: book.orders@tandf.co.uk

For a complete listing of all our titles visit:
www.tandf.co.uk

PART 6

Daywork and Prime Cost

Quality Auditing in Construction Projects

Abdul Razzak Rumane

This book provides construction professionals, designers, contractors and quality auditors involved in construction projects with the auditing skills and processes required to improve construction quality and make their projects more competitive and economical.

The processes within the book focus on auditing compliance to ISO, corporate quality management systems, project specific quality management systems, contract management, regulatory authorities' requirements, safety, and environmental considerations. The book is divided into seven chapters and each chapter is divided into numbered sections covering auditing-related topics that have importance or relevance for understanding quality auditing concepts for construction projects.

No other book covers construction quality auditing in such detail and with this level of practical application. It is an essential guide for construction and quality professionals, but also for students and academics interested in learning about quality auditing in construction projects.

June 2019: 234 × 156 mm: 600pp
Hb: 978-0-8153-8531-8 : £100.00

To Order: Tel: +44 (0) 1235 400524 Fax: +44 (0) 1235 400525
or Post: Taylor and Francis Customer Services,
Bookpoint Ltd, Unit T1, 200 Milton Park, Abingdon, Oxon, OX14 4TA UK
Email: book.orders@tandf.co.uk

For a complete listing of all our titles visit:
www.tandf.co.uk

BUILDING INDUSTRY

When work is carried out which cannot be valued in any other way it is customary to assess the value on a cost basis with an allowance to cover overheads and profit. The basis of costing is a matter for agreement between the parties concerned, but definitions of prime cost for the building industry have been prepared and published jointly by the Royal Institution of Chartered Surveyors and the National Federation of Building Trades Employers (now the Construction Confederation) for the convenience of those who wish to use them. These documents are reproduced with the permission of the Royal Institution of Chartered Surveyors, which owns the copyright.

The daywork schedule published by the Civil Engineering Contractors Association is included in the *Architects' & Builders'* companion title, *Spon's Civil Engineering and Highway Works Price Book*.

For larger Prime Cost contracts the reader is referred to the form of contract issued by the Royal Institute of British Architects.

DEFINITION OF PRIME COST OF BUILDING WORKS OF A JOBBING OR MAINTENANCE CHARACTER (2010 EDITION)

This definition of Prime Cost is published by the Royal Institution of Chartered Surveyors and the National Federation of Building Trades Employers, for convenience and for use by people who choose to use it. Members of the National Federation of Building Trades Employers are not in any way debarred from defining Prime Cost and rendering their accounts for work carried out on that basis in any way they choose. Building owners are advised to reach agreement with contractors on the Definition of Prime Cost to be used prior to issuing instructions.

SECTION 1 – APPLICATION

1.1. This definition provides a basis for the valuation of work of a jobbing or maintenance character executed under such building contracts as provide for its use.

1.2. It is not applicable in any other circumstances, such as daywork executed under or incidental to a building contract.

SECTION 2 – COMPOSITION OF TOTAL CHARGES

2.1. The prime cost of jobbing work comprises the sum of the following costs:
(a) Labour as defined in Section 3.
(b) Materials and goods as defined in Section 4.
(c) Plant, consumable stores and services as defined in Section 5.
(d) Subcontracts as defined in Section 6.

2.2. Incidental costs, overhead and profit as defined in Section 7 and expressed as percentage adjustments are applicable to each of 2.1 (a)–(d).

SECTION 3 – LABOUR

3.1. Labour costs comprise all payments made to or in respect of all persons directly engaged upon the work, whether on or off the site, except those included in Section 7.

3.2. Such payments are based upon the standard wage rates, emoluments and expenses as laid down for the time being in the rules or decisions of the National Joint Council for the Building Industry and the terms of the Building and Civil Engineering Annual and Public Holiday Agreements applying to the works, or the rules of decisions or agreements of such other body as may relate to the class of labour concerned, at the time when and in the area where the work is executed, together with the Contractor's statutory obligations, including:
(a) Guaranteed minimum weekly earnings (e.g. Standard Basic Rate of Wages and Guaranteed Minimum Bonus Payment in the case of NJCBI rules).
(b) All other guaranteed minimum payments (unless included in Section 7).
(c) Payments in respect of incentive schemes or productivity agreements applicable to the works.
(d) Payments in respect of overtime normally worked; or necessitated by the particular circumstances of the work; or as otherwise agreed between the parties.
(e) Differential or extra payments in respect of skill, responsibility, discomfort or inconvenience.
(f) Tool allowance.
(g) Subsistence and periodic allowances.
(h) Fares, travelling and lodging allowances.
(i) Employer's contributions to annual holiday credits.

BUILDING INDUSTRY

(j)　Employer's contributions to death benefit schemes.

(k)　Any amounts which may become payable by the Contractor to or in respect of operatives arising from the operation of the rules referred to in 3.2 which are not provided for in 3.2 (a)–(k) or in Section 7.

(l)　Employer's National Insurance contributions and any contribution, levy or tax imposed by statute, payable by the Contractor in his capacity as employer.

Note:

Any payments normally made by the Contractor which are of a similar character to those described in 3.2 (a)–(c) but which are not within the terms of the rules and decisions referred to above are applicable subject to the prior agreement of the parties, as an alternative to 3.2 (a)–(c).

3.3.　The wages or salaries of supervisory staff, timekeepers, storekeepers, and the like, employed on or regularly visiting site, where the standard wage rates, etc., are not applicable, are those normally paid by the Contractor together with any incidental payments of a similar character to 3.2 (c)–(k).

3.4.　Where principals are working manually their time is chargeable, in respect of the trades practised, in accordance with 3.2.

SECTION 4 – MATERIALS AND GOODS

4.1.　The prime cost of materials and goods obtained by the Contractor from stockists or manufacturers is the invoice cost after deduction of all trade discounts but including cash discounts not exceeding 5 per cent, and includes the cost of delivery to site.

4.2.　The prime cost of materials and goods supplied from the Contractor's stock is based upon the current market prices plus any appropriate handling charges.

4.3.　The prime cost under 4.1 and 4.2 also includes any costs of:

(a)　non-returnable crates or other packaging.

(b)　returning crates and other packaging less any credit obtainable.

4.4.　Any value added tax which is treated, or is capable of being treated, as input tax (as defined in the Finance Act, 1972 or any re-enactment thereof) by the Contractor is excluded.

SECTION 5 – PLANT, CONSUMABLE STORES AND SERVICES

5.1.　The prime cost of plant and consumable stores as listed below is the cost at hire rates agreed between the parties or in the absence of prior agreement at rates not exceeding those normally applied in the locality at the time when the works are carried out, or on a use and waste basis where applicable:

(a)　Machinery in workshops.

(b)　Mechanical plant and power-operated tools.

(c)　Scaffolding and scaffold boards.

(d)　Non-mechanical plant excluding hand tools.

(e)　Transport including collection and disposal of rubbish.

(f)　Tarpaulins and dust sheets.

(g)　Temporary roadways, shoring, planking and strutting, hoarding, centring, formwork, temporary fans, partitions or the like.

(h)　Fuel and consumable stores for plant and power-operated tools unless included in 5.1 (a), (b), (d) or (e) above.

(i)　Fuel and equipment for drying out the works and fuel for testing mechanical services.

5.2.　The prime cost also includes the net cost incurred by the Contractor of the following services, excluding any such cost included under Sections 3, 4 or 7:

(a)　Charges for temporary water supply including the use of temporary plumbing and storage.

(b)　Charges for temporary electricity or other power and lighting including the use of temporary installations.

(c)　Charges arising from work carried out by local authorities or public undertakings.

(d)　Fees, royalties and similar charges.

(e)　Testing of materials.

(f)　The use of temporary buildings including rates and telephone and including heating and lighting not charged under (b) above.

BUILDING INDUSTRY

(g)   The use of canteens, sanitary accommodation, protective clothing and other provision for the welfare of persons engaged in the work in accordance with the current Working Rule Agreement and any Act of Parliament, statutory instrument, rule, order, regulation or bye-law.

(h)   The provision of safety measures necessary to comply with any Act of Parliament.

(i)   Premiums or charges for any performance bonds or insurances which are required by the Building Owner and which are not referred to elsewhere in this Definition.

SECTION 6 – SUBCONTRACTS

6.1.   The prime cost of work executed by subcontractors, whether nominated by the Building Owner or appointed by the Contractor, is the amount which is due from the Contractor to the subcontractors in accordance with the terms of the subcontracts after deduction of all discounts except any cash discount offered by any subcontractor to the Contractor not exceeding 2.5%.

SECTION 7 – INCIDENTAL COSTS, OVERHEADS AND PROFIT

7.1.   The percentage adjustments provided in the building contract, which are applicable to each of the totals of Sections 3–6, provide for the following:

(a)   Head Office charges.

(b)   Off-site staff including supervisory and other administrative staff in the Contractor's workshops and yard.

(c)   Payments in respect of public holidays.

(d)   Payments in respect of apprentices' study time.

(e)   Sick pay or insurance in respect thereof.

(f)   Third party employer's liability insurance.

(g)   Liability in respect of redundancy payments made to employees.

(h)   Use, repair and sharpening of non-mechanical hand tools.

(i)   Any variations to basic rates required by the Contractor in cases where the building contract provides for the use of a specified schedule of basic plant charges (to the extent that no other provision is made for such variation).

(j)   All other liabilities and obligations whatsoever not specifically referred to in this Section nor chargeable under any other section.

(k)   Profit.

BUILDING INDUSTRY

SPECIMEN ACCOUNT FORMAT

If this Definition of Prime Cost is followed the Contractor's account could be in the following format:

	£
Labour (as defined in Section 3) Add __ % (see Section 7)	
Materials and goods (as defined in Section 4) Add __ % (see Section 7)	£
Plant, consumable stores and services (as defined in Section 5) Add __ % (see Section 7)	£
Subcontracts (as defined in Section 6) Add __ % (see Section 7)	£
Total	£

VAT to be added if applicable.

Example Calculations of Prime Cost of Labour in Daywork

Example of calculation of typical standard hourly base rate (as defined in Section 3) for CIJC Building Craft operative and General Operative based upon assumed rates applicable June 2019.

		Craft Operative		General Operative	
		Rate £	Annual Cost £	Rate £	Annual Cost £
Basic Wages	52 weeks	495.18	25,749.52	371.67	19,326.84
CITB levy (0.5% of payroll)	0.5%		128.75		96.63
Pension and welfare benefit	52 weeks	12.86	668.72	12.86	668.72
Employer's National Insurance contribution *(13.8% after the first £155 per week)*	46.2 weeks		2,495.05		1,608.75
Annual labour cost:			29,042.07		21,700.95
Hourly base rate:			**16.19**		**12.10**

This is the prime cost of employment per person, which the employer has to meet even if there is no work for the employee to do

Note:

(1) Calculated following Definition of Prime Cost of Daywork carried out under a Building Contract, published by the Royal Institution of Chartered Surveyors and the Construction Confederation.
(2) Standard basic assumed rates effective from June 2019.
(3) Standard working hours per annum calculated as follows:

BUILDING INDUSTRY

52 weeks @ 39 hours =		2028
Less		
22 days holiday @ 39 hours =	163.8	
8 days public holidays @ 7.8 hours =	<u>62.4</u>	
	= −226.2	
Standard working hours per year =		1801.8

(4) All labour costs incurred by the contractor in his capacity as an employer, other than those contained in the hourly base rate, are to be taken into account under Section 6.

(5) The above example is for guidance only and does not form part of the Definition; all the basic costs are subject to re-examination according to the time when and in the area where the daywork is executed.

(6) NI payments are at not-contracted out rates applicable from April 2018.

(7) Basic rate and GMP number of weeks =
52 Weeks − 4.2 weeks annual holiday − 1.6 weeks public holiday = 46.2 weeks

BUILDING INDUSTRY

SCHEDULE OF BASIC PLANT CHARGES (1 JULY 2010 ISSUE)

This Schedule is published by the Royal Institution of Chartered Surveyors, Parliament Square, London, and is for use in connection with Dayworks under a Building Contract.

EXPLANATORY NOTES

1. The rates in the Schedule are intended to apply solely to daywork carried out under and incidental to a Building Contract. They are NOT intended to apply to:
   (i) Jobbing or any other work carried out as a main or separate contract; or
   (ii) Work carried out after the date of commencement of the Defects Liability Period.
2. The rates apply only to plant and machinery already on site, whether hired or owned by the Contractor.
3. The rates, unless otherwise stated, include the cost of fuel and power of every description, lubricating oils, grease, maintenance, sharpening of tools, replacement of spare parts, all consumable stores and for licences and insurances applicable to items of plant.
4. The rates, unless otherwise stated, do not include the costs of drivers and attendants.
5. The rates do not include for any possible discounts which may be given to the Contractor if the plant or machinery was hired.
6. The rates are base costs and may be subject to the overall adjustment for price movement, overheads and profit, quoted by the Contractor prior to the placing of the Contract.
7. The rates should be applied to the time during which the plant is actually engaged in daywork.
8. Whether or not plant is chargeable on daywork depends on the daywork agreement in use and the inclusion of an item of plant in this schedule does not necessarily indicate that the item is chargeable.
9. Rates for plant not included in the Schedule or which is not already on site and is specifically provided or hired for daywork shall be settled at prices which are reasonably related to the rates in the Schedule having regard to any overall adjustment quoted by the Contractor in the Conditions of Contract.

Item of plant	Size/Rating	Unit	Rate per Hour (£)
PUMPS			
Mobile Pumps			
Including pump hoses, values and strainers, etc.			
Diaphragm	50 mm dia.	Each	1.17
Diaphragm	76 mm dia.	Each	1.89
Diaphragm	102 mm dia.	Each	3.54
Submersible	50 mm dia.	Each	0.76
Submersible	76 mm dia.	Each	0.86
Submersible	102 mm dia.	Each	1.03
Induced Flow	50 mm dia.	Each	0.77
Induced Flow	76 mm dia.	Each	1.67
Centrifugal, self-priming	25 mm dia.	Each	1.30
Centrifugal, self-priming	50 mm dia.	Each	1.92
Centrifugal, self-priming	75 mm dia.	Each	2.74
Centrifugal, self-priming	102 mm dia.	Each	3.35
Centrifugal, self-priming	152 mm dia.	Each	4.27

BUILDING INDUSTRY

Item of plant	Size/Rating	Unit	Rate per Hour (£)
SCAFFOLDING, SHORING, FENCING			
Complete Scaffolding			
Mobile working towers, single width	2.0 m × 0.72 m base × 7.45 m high	Each	3.36
Mobile working towers, single width	2.0 m × 0.72 m base × 8.84 m high	Each	3.79
Mobile working towers, double width	2.0 m × 1.35 m × 7.45 m high	Each	3.79
Mobile working towers, double width	2.0 m × 1.35 m × 15.8 m high	Each	7.13
Chimney scaffold, single unit		Each	1.92
Chimney scaffold, twin unit		Each	3.59
Push along access platform	1.63 – 3.1 m	Each	5.00
Push along access platform	1.80 m × 0.70 m	Each	1.79
Trestles			
Trestle, adjustable	Any height	Pair	0.41
Trestle, painters	1.8 m high	Pair	0.31
Trestle, painters	2.4 m high	Pair	0.36
Shoring, Planking and Strutting			
Acrow adjustable prop	Sizes up to 4.9 m (open)	Each	0.06
Strong Boy support attachment		Each	0.22
Adjustable trench strut	Sizes up to 1.67 m (open)	Each	0.16
Trench sheet		Metre	0.03
Backhoe trench box	Base unit	Each	1.23
Backhoe trench box	Top unit	Each	0.87
Temporary Fencing			
Including block and coupler			
Site fencing steel grid panel	3.5 m × 2.0 m	Each	0.05
Anti-climb site steel grid fence panel	3.5 m × 2.0 m	Each	0.08
Solid panel Heras	2.0 m × 2.0 m	Each	0.09
Pedestrian gate		Each	0.36
Roadway gate		Each	0.60
LIFTING APPLIANCES AND CONVEYORS			
Cranes			
<u>Mobile Cranes</u>			
Rates are inclusive of drivers			
Lorry mounted, telescopic jib			
Two wheel drive	5 tonnes	Each	19.00
Two wheel drive	8 tonnes	Each	42.00
Two wheel drive	10 tonnes	Each	50.00

BUILDING INDUSTRY

Item of plant	Size/Rating		Unit	Rate per Hour (£)
LIFTING APPLIANCES AND CONVEYORS				
Two wheel drive	12 tonnes		Each	77.00
Two wheel drive	20 tonnes		Each	89.69
Four wheel drive	18 tonnes		Each	46.51
Four wheel drive	25 tonnes		Each	35.90
Four wheel drive	30 tonnes		Each	38.46
Four wheel drive	45 tonnes		Each	46.15
Four wheel drive	50 tonnes		Each	53.85
Four wheel drive	60 tonnes		Each	61.54
Four wheel drive	70 tonnes		Each	71.79
Static tower crane				
Rates inclusive of driver				
Note: Capacity equals maximum lift in tonnes times maximum radius at which it can be lifted				
	Capacity (Metre/tonnes) Up to	Height under hook above ground (m) Up to		
Tower crane	30	22	Each	22.23
Tower crane	40	22	Each	26.62
Tower crane	40	30	Each	33.33
Tower crane	50	22	Each	29.16
Tower crane	60	22	Each	35.90
Tower crane	60	36	Each	35.90
Tower crane	70	22	Each	41.03
Tower crane	80	22	Each	39.12
Tower crane	90	42	Each	37.18
Tower crane	110	36	Each	47.62
Tower crane	140	36	Each	55.77
Tower crane	170	36	Each	64.11
Tower crane	200	36	Each	71.95
Tower crane	250	36	Each	84.77
Tower crane with luffing jig	30	25	Each	22.23
Tower crane with luffing jig	40	30	Each	26.62
Tower crane with luffing jig	50	30	Each	29.16
Tower crane with luffing jig	60	36	Each	41.03
Tower crane with luffing jig	65	30	Each	33.13
Tower crane with luffing jig	80	22	Each	48.72
Tower crane with luffing jig	100	45	Each	48.72
Tower crane with luffing jig	125	30	Each	53.85

BUILDING INDUSTRY

Item of plant	Size/Rating		Unit	Rate per Hour (£)
LIFTING APPLIANCES AND CONVEYORS				
Tower crane with luffing jig	160	50	Each	53.85
Tower crane with luffing jig	200	50	Each	74.36
Tower crane with luffing jig	300	60	Each	100.00
Crane Equipment				
Muck tipping skip	Up to 200 litres		Each	0.67
Muck tipping skip	500 litres		Each	0.82
Muck tipping skip	750 litres		Each	1.08
Muck tipping skip	1000 litres		Each	1.28
Muck tipping skip	1500 litres		Each	1.41
Muck tipping skip	2000 litres		Each	1.67
Mortar skip	250 litres, plastic		Each	0.41
Mortar skip	350 litres, steel		Each	0.77
Boat skip	250 litres		Each	0.92
Boat skip	500 litres		Each	1.08
Boat skip	750 litres		Each	1.23
Boat skip	1000 litres		Each	1.38
Boat skip	1500 litres		Each	1.64
Boat skip	2000 litres		Each	1.90
Boat skip	3000 litres		Each	2.82
Boat skip	4000 litres		Each	3.23
Master flow skip	250 litres		Each	0.77
Master flow skip	500 litres		Each	1.03
Master flow skip	750 litres		Each	1.28
Master flow skip	1000 litres		Each	1.44
Master flow skip	1500 litres		Each	1.69
Master flow skip	2000 litres		Each	1.85
Grand master flow skip	500 litres		Each	1.28
Grand master flow skip	750 litres		Each	1.64
Grand master flow skip	1000 litres		Each	1.69
Grand master flow skip	1500 litres		Each	1.95
Grand master flow skip	2000 litres		Each	2.21
Cone flow skip	500 litres		Each	1.33
Cone flow skip	1000 litres		Each	1.69
Geared rollover skip	500 litres		Each	1.28
Geared rollover skip	750 litres		Each	1.64
Geared rollover skip	1000 litres		Each	1.69

BUILDING INDUSTRY

Item of plant	Size/Rating	Unit	Rate per Hour (£)
LIFTING APPLIANCES AND CONVEYORS			
Geared rollover skip	1500 litres	Each	1.95
Geared rollover skip	2000 litres	Each	2.21
Multi-skip, rope operated	200 mm outlet size, 500 litres	Each	1.49
Multi-skip, rope operated	200 mm outlet size, 750 litres	Each	1.64
Multi-skip, rope operated	200 mm outlet size, 1000 litres	Each	1.74
Multi-skip, rope operated	200 mm outlet size, 1500 litres	Each	2.00
Multi-skip, rope operated	200 mm outlet size, 2000 litres	Each	2.26
Multi-skip, man riding	200 mm outlet size, 1000 litres	Each	2.00
Multi-skip	4 point lifting frame	Each	0.90
Multi-skip	Chain brothers	Set	0.87
Crane accessories			
Multi-purpose crane forks	1.5 and 2 tonnes S.W.L.	Each	1.13
Self-levelling crane forks		Each	1.28
Man cage	1 man, 230 kg S.W.L.	Each	1.90
Man cage	2 man, 500 kg S.W.L.	Each	1.95
Man cage	4 man, 750 kg S.W.L.	Each	2.15
Man cage	8 man, 1000 kg S.W.L.	Each	3.33
Stretcher cage	500 kg, S.W.L.	Each	2.69
Goods carrying cage	1500 kg, S.W.L.	Each	1.33
Goods carrying cage	3000 kg, S.W.L.	Each	1.85
Builders' skip lifting cradle	12 tonnes, S.W.L.	Each	2.31
Board/pallet fork	1600 kg, S.W.L.	Each	1.90
Gas bottle carrier	500 kg, S.W.L.	Each	0.92
Hoists			
Scaffold hoist	200 kg	Each	2.46
Rack and pinion (goods only)	500 kg	Each	4.56
Rack and pinion (goods only)	1100 kg	Each	5.90
Rack and pinion (goods and passenger)	8 person, 80 kg	Each	7.44
Rack and pinion (goods and passenger)	14 person, 1400 kg	Each	8.72
Wheelbarrow chain sling		Each	1.67

BUILDING INDUSTRY

Item of plant	Size/Rating		Unit	Rate per Hour (£)
LIFTING APPLIANCES AND CONVEYORS				
Conveyors				
Belt conveyors				
Conveyor	8 m long × 450 mm wide		Each	5.90
Miniveyor, control box and loading hopper	3 m unit		Each	4.49
Other conveying equipment				
Wheelbarrow			Each	0.62
Hydraulic superlift			Each	4.56
Pavac slab lifter (tile hoist)			Each	4.49
High lift pallet truck			Each	3.08
	Payload	Maximum Lift		
Lifting Trucks				
Fork lift, two wheel drive	1100 kg	up to 3.0 m	Each	5.64
Fork lift, two wheel drive	2540 kg	up to 3.7 m	Each	5.64
Fork lift, four wheel drive	1524 kg	up to 6.0 m	Each	5.64
Fork lift, four wheel drive	2600 kg	up to 5.4 m	Each	7.44
Fork life, four wheel drive	4000 kg	up to 17 m	Each	10.77
Lifting Platforms				
Hydraulic platform (Cherry picker)	9 m		Each	4.62
Hydraulic platform (Cherry picker)	12 m		Each	7.56
Hydraulic platform (Cherry picker)	15 m		Each	10.13
Hydraulic platform (Cherry picker)	17 m		Each	15.63
Hydraulic platform (Cherry picker)	20 m		Each	18.13
Hydraulic platform (Cherry picker)	25.6 m		Each	32.38
Scissor lift	7.6 m, electric		Each	3.85
Scissor lift	7.8 m, electric		Each	5.13
Scissor lift	9.7 m, electric		Each	4.23
Scissor lift	10 m, diesel		Each	6.41
Telescopic handler	7 m, 2 tonnes		Each	5.13
Telescopic handler	13 m, 3 tonnes		Each	7.18
Lifting and Jacking Gear				
Pipe winch including gantry	1 tonne		Set	1.92
Pipe winch including gantry	3 tonnes		Set	3.21
Chain block	1 tonne		Each	0.35
Chain block	2 tonnes		Each	0.58
Chain block	5 tonnes		Each	1.14
Pull lift (Tirfor winch)	1 tonne		Each	0.64
Pull lift (Tirfor winch)	1.6 tonnes		Each	0.90

BUILDING INDUSTRY

Item of plant	Size/Rating	Unit	Rate per Hour (£)
LIFTING APPLIANCES AND CONVEYORS			
Pull lift (Tirfor winch)	3.2 tonnes	Each	1.15
Brother or chain slings, two legs	not exceeding 3.1 tonnes	Set	0.21
Brother or chain slings, two legs	not exceeding 4.25 tonnes	Set	0.31
Brother or chain slings, four legs	not exceeding 11.2 tonnes	Set	1.09
CONSTRUCTION VEHICLES			
Lorries			
Plated lorries (Rates are inclusive of driver)			
Platform lorry	7.5 tonnes	Each	16.21
Platform lorry	17 tonnes	Each	22.90
Platform lorry	24 tonnes	Each	30.68
Extra for lorry with crane attachment	up to 2.5 tonnes	Each	3.25
Extra for lorry with crane attachment	up to 5 tonnes	Each	6.00
Extra for lorry with crane attachment	up to 7.5 tonnes	Each	9.10
Tipper Lorries			
(Rates are inclusive of driver)			
Tipper lorry	up to 11 tonnes	Each	15.78
Tipper lorry	up to 17 tonnes	Each	23.95
Tipper lorry	up to 25 tonnes	Each	31.35
Tipper lorry	up to 31 tonnes	Each	37.79
Dumpers			
Site use only (excl. tax, insurance			
and extra cost of DERV etc. when			
operating on highway)	Makers Capacity		
Two wheel drive	1 tonne	Each	1.71
Four wheel drive	2 tonnes	Each	2.43
Four wheel drive	3 tonnes	Each	2.44
Four wheel drive	5 tonnes	Each	3.08
Four wheel drive	6 tonnes	Each	3.85
Four wheel drive	9 tonnes	Each	5.65
Tracked	0.5 tonnes	Each	3.33
Tracked	1.5 tonnes	Each	4.23
Tracked	3.0 tonnes	Each	8.33
Tracked	6.0 tonnes	Each	16.03
Dumper Trucks *(Rates are inclusive of drivers)*			
Dumper truck	up to 15 tonnes	Each	28.56

BUILDING INDUSTRY

Item of plant	Size/Rating	Unit	Rate per Hour (£)
CONSTRUCTION VEHICLES			
Dumper truck	up to 17 tonnes	Each	32.82
Dumper truck	up to 23 tonnes	Each	54.64
Dumper truck	up to 30 tonnes	Each	63.50
Dumper truck	up to 35 tonnes	Each	73.02
Dumper truck	up to 40 tonnes	Each	87.84
Dumper truck	up to 50 tonnes	Each	133.44
Tractors			
Agricultural type			
Wheeled, rubber-clad tyred	up to 40 kW	Each	8.63
Wheeled, rubber-clad tyred	up to 90 kW	Each	25.31
Wheeled, rubber-clad tyred	up to 140 kW	Each	36.49
Crawler tractors			
With bull or angle dozer	up to 70 kW	Each	29.38
With bull or angle dozer	up to 85 kW	Each	38.63
With bull or angle dozer	up to 100 kW	Each	52.59
With bull or angle dozer	up to 115 kW	Each	55.85
With bull or angle dozer	up to 135 kW	Each	60.43
With bull or angle dozer	up to 185 kW	Each	76.44
With bull or angle dozer	up to 200 kW	Each	96.43
With bull or angle dozer	up to 250 kW	Each	117.68
With bull or angle dozer	up to 350 kW	Each	160.03
With bull or angle dozer	up to 450 kW	Each	219.86
With loading shovel	0.8 m³	Each	26.92
With loading shovel	1.0 m³	Each	32.59
With loading shovel	1.2 m³	Each	37.53
With loading shovel	1.4 m³	Each	42.89
With loading shovel	1.8 m³	Each	52.22
With loading shovel	2.0 m³	Each	57.22
With loading shovel	2.1 m³	Each	60.12
With loading shovel	3.5 m³	Each	87.26
Light Vans			
VW Caddivan or the like		Each	5.26
VW Transport transit or the like	1.0 tonnes	Each	6.03
Luton Box Van or the like	1.8 tonnes	Each	9.87

Daywork and Prime Cost

BUILDING INDUSTRY

Item of plant	Size/Rating	Unit	Rate per Hour (£)
CONSTRUCTION VEHICLES			
Water/Fuel Storage			
Mobile water container	110 litres	Each	0.62
Water bowser	1100 litres	Each	0.72
Water bowser	3000 litres	Each	0.87
Mobile fuel container	110 litres	Each	0.62
Fuel bowser	1100 litres	Each	1.23
Fuel bowser	3000 litres	Each	1.87
EXCAVATIONS AND LOADERS			
Excavators			
Wheeled, hydraulic	up to 11 tonnes	Each	25.86
Wheeled, hydraulic	up to 14 tonnes	Each	30.82
Wheeled, hydraulic	up to 16 tonnes	Each	34.50
Wheeled, hydraulic	up to 21 tonnes	Each	39.10
Wheeled, hydraulic	up to 25 tonnes	Each	43.81
Wheeled, hydraulic	up to 30 tonnes	Each	55.30
Crawler, hydraulic	up to 11 tonnes	Each	25.86
Crawler, hydraulic	up to 14 tonnes	Each	30.82
Crawler, hydraulic	up to 17 tonnes	Each	34.50
Crawler, hydraulic	up to 23 tonnes	Each	39.10
Crawler, hydraulic	up to 30 tonnes	Each	43.81
Crawler, hydraulic	up to 35 tonnes	Each	55.30
Crawler, hydraulic	up to 38 tonnes	Each	71.73
Crawler, hydraulic	up to 55 tonnes	Each	95.63
Mini excavator	1000/1500 kg	Each	4.87
Mini excavator	2150/2400 kg	Each	6.67
Mini excavator	2700/3500 kg	Each	7.31
Mini excavator	3500/4500 kg	Each	8.21
Mini excavator	4500/6000 kg	Each	9.23
Mini excavator	7000 kg	Each	14.10
Micro excavator	725 mm wide	Each	5.13
Loaders			
Shovel loader	0.4 m³	Each	7.69
Shovel loader	1.57 m³	Each	8.97
Shovel loader, four wheel drive	1.7 m³	Each	4.83
Shovel loader, four wheel drive	2.3 m³	Each	4.38

BUILDING INDUSTRY

Item of plant	Size/Rating	Unit	Rate per Hour (£)
EXCAVATIONS AND LOADERS			
Shovel loader, four wheel drive	3.3 m³	Each	5.06
Skid steer loader wheeled	300/400 kg payload	Each	7.31
Skid steer loader wheeled	625 kg payload	Each	7.67
Tracked skip loader	650 kg	Each	4.42
Excavator Loaders			
Wheeled tractor type with black-hoe excavator			
Four wheel drive			
Four wheel drive, 2 wheel steer	6 tonnes	Each	6.41
Four wheel drive, 2 wheel steer	8 tonnes	Each	8.59
Attachments			
Breakers for excavator		Each	8.72
Breakers for mini excavator		Each	1.75
Breakers for back-hoe excavator/loader		Each	5.13
COMPACTION EQUIPMENT			
Rollers			
Vibrating roller	368–420 kg	Each	1.43
Single roller	533 kg	Each	1.94
Single roller	750 kg	Each	3.43
Twin roller	up to 650 kg	Each	6.03
Twin roller	up to 950 kg	Each	6.62
Twin roller with seat end steering wheel	up to 1400 kg	Each	7.68
Twin roller with seat end steering wheel	up to 2500 kg	Each	10.61
Pavement roller	3–4 tonnes dead weight	Each	6.00
Pavement roller	4–6 tonnes	Each	6.86
Pavement roller	6–10 tonnes	Each	7.17
Pavement roller	10–13 tonnes	Each	19.86
Rammers			
Tamper rammer 2 stroke-petrol	225 mm–275 mm	Each	1.52
Soil Compactors			
Plate compactor	75 mm–mm	Each	1.53
Plate compactor rubber pad	375 mm–1400 mm	Each	1.53
Plate compactor reversible plate-petrol	400 mm	Each	2.44

BUILDING INDUSTRY

Item of plant	Size/Rating	Unit	Rate per Hour (£)
CONCRETE EQUIPMENT			
Concrete/Mortar Mixers			
Open drum without hopper	0.09/0.06 m³	Each	0.61
Open drum without hopper	0.12/0.09 m³	Each	1.22
Open drum without hopper	0.15/0.10 m³	Each	0.72
Concrete/Mortar Transport Equipment			
Concrete pump incl. hose, valve and couplers			
Lorry mounted concrete pump	24 m max. distance	Each	50.00
Lorry mounted concrete pump	34 m max. distance	Each	66.00
Lorry mounted concrete pump	42 m max. distance	Each	91.50
Concrete Equipment			
Vibrator, poker, petrol type	up to 75 mm dia.	Each	0.69
Air vibrator (*excluding compressor and hose*)	up to 75 mm dia.	Each	0.64
Extra poker heads	25/36/60 mm dia.	Each	0.76
Vibrating screed unit with beam	5.00 m	Each	2.48
Vibrating screed unit with adjustable beam	3.00–5.00 m	Each	3.54
Power float	725 mm–900 mm	Each	2.56
Power float finishing pan		Each	0.62
Floor grinder	660 × 1016 mm, 110 V electric	Each	4.31
Floor plane	450 × 1100 mm	Each	4.31
TESTING EQUIPMENT			
Pipe Testing Equipment			
Pressure testing pump, electric		Set	2.19
Pressure test pump		Set	0.80
SITE ACCOMODATION AND TEMPORARY SERVICES			
Heating equipment			
Space heater – propane	80,000 Btu/hr	Each	1.03
Space heater – propane/electric	125,000 Btu/hr	Each	2.09
Space heater – propane/electric	250,000 Btu/hr	Each	2.33
Space heater – propane	125,000 Btu/hr	Each	1.54
Space heater – propane	260,000 Btu/hr	Each	1.88
Cabinet heater		Each	0.82
Cabinet heater, catalytic		Each	0.57

BUILDING INDUSTRY

Item of plant	Size/Rating	Unit	Rate per Hour (£)
SITE ACCOMODATION AND TEMPORARY SERVICES			
Electric halogen heater		Each	1.27
Ceramic heater	3 kW	Each	0.99
Fan heater	3 kW	Each	0.66
Cooling fan		Each	1.92
Mobile cooling unit, small		Each	3.60
Mobile cooling unit, large		Each	4.98
Air conditioning unit		Each	2.81
Site Lighting and Equipment			
Tripod floodlight	500 W	Each	0.48
Tripod floodlight	1000 W	Each	0.62
Towable floodlight	4 × 100 W	Each	3.85
Hand held floodlight	500 W	Each	0.51
Rechargeable light		Each	0.41
Inspection light		Each	0.37
Plasterer's light		Each	0.65
Lighting mast		Each	2.87
Festoon light string	25 m	Each	0.55
Site Electrical Equipment			
Extension leads	240 V/14 m	Each	0.26
Extension leads	110 V/14 m	Each	0.36
Cable reel	25 m 110 V/240 V	Each	0.46
Cable reel	50 m 110 V/240 V	Each	0.88
4 way junction box	110 V	Each	0.56
Power Generating Units			
Generator – petrol	2 kVA	Each	1.23
Generator – silenced petrol	2 kVA	Each	2.87
Generator – petrol	3 kVA	Each	1.47
Generator – diesel	5 kVA	Each	2.44
Generator – silenced diesel	10 kVA	Each	1.90
Generator – silenced diesel	15 kVA	Each	2.26
Generator – silenced diesel	30 kVA	Each	3.33
Generator – silenced diesel	50 kVA	Each	4.10
Generator – silenced diesel	75 kVA	Each	4.62
Generator – silenced diesel	100 kVA	Each	5.64
Generator – silenced diesel	150 kVA	Each	7.18
Generator – silenced diesel	200 kVA	Each	9.74

BUILDING INDUSTRY

Item of plant	Size/Rating	Unit	Rate per Hour (£)
SITE ACCOMODATION AND TEMPORARY SERVICES			
Generator – silenced diesel	250 kVA	Each	11.28
Generator – silenced diesel	350 kVA	Each	14.36
Generator – silenced diesel	500 kVA	Each	15.38
Tail adaptor	240 V	Each	0.10
Transformers			
Transformer	3 kVA	Each	0.32
Transformer	5 kVA	Each	1.23
Transformer	7.5 kVA	Each	0.59
Transformer	10 kVA	Each	2.00
Rubbish Collection and Disposal			
Equipment			
Rubbish chutes			
Standard plastic module	1 m section	Each	0.15
Steel liner insert		Each	0.30
Steel top hopper		Each	0.22
Plastic side entry hopper		Each	0.22
Plastic side entry hopper liner		Each	0.22
Dust Extraction Plant			
Dust extraction unit, light duty		Each	2.97
Dust extraction unit, heavy duty		Each	2.97
SITE EQUIPMENT – Welding Equipment			
Arc-(Electric) complete with leads			
Welder generator – petrol	200 amp	Each	3.53
Welder generator – diesel	300/350 amp	Each	3.78
Welder generator – diesel	4000 amp	Each	7.92
Extra welding lead sets		Each	0.69
Gas-Oxy welder			
Welding and cutting set (including oxygen and acetylene, excluding underwater equipment and thermic boring)			
Small		Each	2.24
Large		Each	3.75
Lead burning gun		Each	0.50
Mig welder		Each	1.38
Fume extractor		Each	2.46

BUILDING INDUSTRY

Item of plant	Size/Rating	Unit	Rate per Hour (£)
SITE EQUIPMENT – Welding Equipment			
Road Works Equipment			
Traffic lights, mains/generator	2-way	Set	10.94
Traffic lights, mains/generator	3-way	Set	11.56
Traffic lights, mains/generator	4-way	Set	12.19
Flashing light		Each	0.10
Road safety cone	450 mm	Each	0.08
Safety cone	750 mm	Each	0.10
Safety barrier plank	1.25 m	Each	0.13
Safety barrier plank	2 m	Each	0.15
Safety barrier plank post		Each	0.13
Safety barrier plank post base		Each	0.10
Safety four gate barrier	1 m each gate	Set	0.77
Guard barrier	2 m	Each	0.19
Road sign	750 mm	Each	0.23
Road sign	900 mm	Each	0.31
Road sign	1200 mm	Each	0.42
Speed ramp/cable protection	500 mm section	Each	0.14
Hose ramp open top	3 m section	Each	0.07
DPC Equipment			
Damp-proofing injection machine		Each	2.56
Cleaning Equipment			
Vacuum cleaner (industrial wet) single motor		Each	1.08
Vacuum cleaner (industrial wet) twin motor	30 litre capacity	Each	1.79
Vacuum cleaner (industrial wet) twin motor	70 litre capacity	Each	2.21
Steam cleaner	Diesel/electric 1 phase	Each	3.33
Steam cleaner	Diesel/electric 3 phase	Each	3.85
Pressure washer, light duty electric	1450 PSI	Each	0.72
Pressure washer, heavy duty, diesel	2500 PSI	Each	1.33
Pressure washer, heavy duty, diesel	4000 PSI	Each	2.18
Cold pressure washer, electric		Each	2.39
Hot pressure washer, petrol		Each	4.19
Hot pressure washer, electric		Each	5.13
Cold pressure washer, petrol		Each	2.92

Daywork and Prime Cost

BUILDING INDUSTRY

Item of plant	Size/Rating	Unit	Rate per Hour (£)
SITE EQUIPMENT – Welding Equipment			
Sandblast attachment to last washer		Each	1.23
Drain cleaning attachment to last washer		Each	1.03
Surface Preparation Equipment			
Rotavator	5 h.p.	Each	2.46
Rotavator	9 h.p.	Each	5.00
Scabbler, up to three heads		Each	1.53
Scabbler, pole		Each	2.68
Scrabbler, multi-headed floor		Each	3.89
Floor preparation machine		Each	1.05
Compressors and Equipment			
<u>Portable compressors</u>			
Compressor – electric	4 cfm	Each	1.36
Compressor – electric	8 cfm lightweight	Each	1.31
Compressor – electric	8 cfm	Each	1.36
Compressor – electric	14 cfm	Each	1.56
Compressor – petrol	24 cfm	Each	2.15
Compressor – electric	25 cfm	Each	2.10
Compressor – electric	30 cfm	Each	2.36
Compressor – diesel	100 cfm	Each	2.56
Compressor – diesel	250 cfm	Each	5.54
Compressor – diesel	400 cfm	Each	8.72
<u>Mobile compressors</u>			
Lorry mounted compressor			
(machine plus lorry only)	up to 3 m³	Each	41.47
(machine plus lorry only)	up to 5 m³	Each	48.94
Tractor mounted compressor			
(machine plus rubber tyred tractor)	Up to 4 m³	Each	21.03
<u>Accessories (pneumatic tools) *(with and including up to 15 m of air hose)*</u>			
Demolition pick, medium duty		Each	0.90
Demolition pick, heavy duty		Each	1.03
Breakers (with six steels) light	up to 150 kg	Each	1.19
Breakers (with six steels) medium	295 kg	Each	1.24
Breakers (with six steels) heavy	386 kg	Each	1.44
Rock drill (for use with compressor) hand held		Each	1.18

BUILDING INDUSTRY

Item of plant	Size/Rating	Unit	Rate per Hour (£)
SITE EQUIPMENT – Welding Equipment			
Additional hoses	15 m	Each	0.09
Breakers			
Demolition hammer drill, heavy duty, electric		Each	1.54
Road breaker, electric		Each	2.41
Road breaker, 2 stroke, petrol		Each	4.06
Hydraulic breaker unit, light duty, petrol		Each	3.06
Hydraulic breaker unit, heavy duty, petrol		Each	3.46
Hydraulic breaker unit, heavy duty, diesel		Each	4.62
Quarrying and Tooling Equipment			
Block and stone splitter, hydraulic	600 mm × 600 mm	Each	1.90
Block and stone splitter, manual		Each	1.64
Steel Reinforcement Equipment			
Bar bending machine – manual	up to 13 mm dia. rods	Each	1.03
Bar bending machine – manual	up to 20 mm dia. rods	Each	1.41
Bar shearing machine – electric	up to 38 mm dia. rods	Each	3.08
Bar shearing machine – electric	up to 40 mm dia. rods	Each	4.62
Bar cropper machine – electric	up to 13 mm dia. rods	Each	2.05
Bar cropper machine – electric	up to 20 mm dia. rods	Each	2.56
Bar cropper machine – electric	up to 40 mm dia. rods	Each	4.62
Bar cropper machine – 3 phase	up to 40 mm dia. rods	Each	4.62
Dehumidifiers			
110/240 V Water	68 litres extraction per 24 hours	Each	2.46
110/240 V Water	90 litres extraction per 24 hours	Each	3.38
SMALL TOOLS			
Saws			
Masonry bench saw	350 mm–500 mm dia.	Each	1.13
Floor saw	125 mm max. cut	Each	1.15
Floor saw	150 mm max. cut	Each	3.83
Floor saw, reversible	350 mm max. cut	Each	3.32
Wall saw, electric		Each	2.05
Chop/cut off saw, electric	350 mm dia.	Each	1.79
Circular saw, electric	230 mm dia.	Each	0.72
Tyrannosaw		Each	1.74
Reciprocating saw		Each	0.79

BUILDING INDUSTRY

Item of plant	Size/Rating	Unit	Rate per Hour (£)
SMALL TOOLS			
Door trimmer		Each	1.17
Stone saw	300 mm	Each	1.44
Chainsaw, petrol	500 mm	Each	3.92
Full chainsaw safety kit		Each	0.41
Worktop jig		Each	1.08
Pipe Work Equipment			
Pipe bender	15 mm–22 mm	Each	0.92
Pipe bender, hydraulic	50 mm	Each	1.76
Pipe bender, electric	50 mm–150 mm dia.	Each	2.19
Pipe cutter, hydraulic		Each	0.46
Tripod pipe vice		Set	0.75
Ratchet threader	12 mm–32 mm	Each	0.93
Pipe threading machine, electric	12 mm–75 mm	Each	3.07
Pipe threading machine, electric	12 mm–100 mm	Each	4.93
Impact wrench, electric		Each	1.33
Hand-Held Drills and Equipment			
Impact or hammer drill	up to 25 mm dia.	Each	1.03
Impact or hammer drill	35 mm dia.	Each	1.29
Dry diamond core cutter		Each	0.99
Angle head drill		Each	0.90
Stirrer, mixer drill		Each	1.13
Paint, Insulation Application Equipment			
Airless spray unit		Each	4.13
Portaspray unit		Each	1.16
HPVL turbine spray unit		Each	2.23
Compressor and spray gun		Each	1.91
Other Handtools			
Staple gun		Each	0.96
Air nail gun	110 V	Each	1.01
Cartridge hammer		Each	1.08
Tongue and groove nailer complete with mallet		Each	1.59
Diamond wall chasing machine		Each	2.63
Masonry chain saw	300 mm	Each	5.49
Floor grinder		Each	3.99
Floor plane		Each	1.79

BUILDING INDUSTRY

Item of plant	Size/Rating	Unit	Rate per Hour (£)
SMALL TOOLS			
Diamond concrete planer		Each	1.93
Autofeed screwdriver, electric		Each	1.38
Laminate trimmer		Each	0.91
Biscuit jointer		Each	1.49
Random orbital sander		Each	0.97
Floor sander		Each	1.54
Palm, delta, flap or belt sander		Each	0.75
Disc cutter, electric	300 mm	Each	1.49
Disc cutter, 2 stroke petrol	300 mm	Each	1.24
Dust suppressor for petrol disc cutter		Each	0.51
Cutter cart for petrol disc cutter		Each	1.21
Grinder, angle or cutter	up to 225 mm	Each	0.50
Grinder, angle or cutter	300 mm	Each	1.41
Mortar raking tool attachment		Each	0.19
Floor/polisher scrubber	325 mm	Each	1.76
Floor tile stripper		Each	2.44
Wallpaper stripper, electric		Each	0.81
Hot air paint stripper		Each	0.50
Electric diamond tile cutter	All sizes	Each	2.42
Hand tile cutter		Each	0.82
Electric needle gun		Each	1.29
Needle chipping gun		Each	1.85
Pedestrian floor sweeper	250 mm dia.	Each	0.82
Pedestrian floor sweeper	Petrol	Each	2.20
Diamond tile saw		Each	1.84
Blow lamp equipment and glass		Set	0.50

The Client Role in Successful Construction Projects

Jason Challender *et al.*

The Client Role in Successful Construction Projects is a practical guide for clients on how to initiate, procure and manage construction projects and developments. This book is written from the perspective of the client initiating a construction project as part of a business venture and differs from most available construction literature which can externalise the client as a risk to be managed by the design team. The book provides a practical framework for new and novice clients undertaking construction, giving them a voice and enabling them to:

- Understand the challenges that they and the project are likely to face.
- Communicate and interact effectively with key stakeholders and professionals within the industry.
- Understand in straightforward terms where they can have a positive impact on the project.
- Put in place a client-side due diligence process.
- Reduce their institutional risk and the risk of project failure.
- Discover how their standard models are able to co-exist and even transfer to a common client-side procedure for managing a construction project.

Written by clients, for clients, this book is highly recommended not only for clients, but for construction industry professionals who want to develop their own skills and enhance their working relationship with their clients. A supporting website for the book will be available, which will give practical examples of the points illustrated in the book and practical advice from specialists in the field.

May 2019: 234 × 156 mm: 318pp
Pb: 978-1-138-05821-7 : £29.99

To Order: Tel: +44 (0) 1235 400524 Fax: +44 (0) 1235 400525
or Post: Taylor and Francis Customer Services,
Bookpoint Ltd, Unit T1, 200 Milton Park, Abingdon, Oxon, OX14 4TA UK
Email: book.orders@tandf.co.uk

For a complete listing of all our titles visit:
www.tandf.co.uk

PART 7

Useful Addresses for Further Information

ACOUSTICAL INVESTIGATION & RESEARCH
ORGANISATION LTD (AIRO)
Duxon's Turn
Maylands Avenue
Hemel Hempstead
Hertfordshire
HP2 4SB
Tel: 01442 247 146
Fax: 01442 256 749
Website: www.airo.co.uk

Aluminium Federation Ltd (ALFED)
National Metalforming Centre
47 Birmingham Road
West Bromwich
West Midlands
B70 6PY
Tel: 0121 601 6361
Fax: 0870 138 9714
Website: www.alfed.org.uk

AMERICAN HARDWOOD EXPORT COUNCIL (AHEC)
3 St Michaels Alley
London
EC3 V 9DS
Tel: 020 7626 4111
Fax: 020 7626 4222
Website: www.ahec-europe.org

ANCIENT MONUMENTS SOCIETY (AMS)
Saint Ann's Vestry Hall
2 Church Entry
London
EC4 V 5HB
Tel: 020 7236 3934
Fax: 020 7329 3677
Website: www.ancientmonumentssociety.org.uk

APA – THE ENGINEERED WOOD ASSOCIATION
Claridge House
29 Barnes, High Street
London
SW13 9LW
Tel: 0845 123 3721
Fax: 020 8282 1660
Website: http: //www.apa-europe.org/

ARBORICULTURAL ASSOCIATION
The Moult House
Stround Green
Standish
Stonehouse
Gloucestershire
GL10 3DL
Tel: 01242 522152
Fax: 01242 577766
Website: www.trees.org.uk

ARCHITECTURAL ADVISORY SERVICE CENTRE
(POWDER/ANODIC METAL FINISHES)
Barn One
Barn Road
Longwick
Buckinghamshire
HP27 9RW
Tel: 01844 342 425
Fax: 01844 274 781
Website: http: //www.aasc.org.uk/

ARCHITECTURAL ASSOCIATION (AA)
34–36 Bedford Square
London
WC1B 3ES
Tel: 020 7887 4000
Fax: 020 7414 0782
Website: http: //www.aaschool.ac.uk/

ARCHITECTURAL CLADDING ASSOCIATION (ACA)
60 Charles Street
Leicester
Leicestershire
LE1 1FB
Tel: 0116 253 6161
Fax: 0116 251 4568
Website: http: //www.architectural-cladding-association.org.uk/

ASBESTOS INFORMATION CENTRE (AIC)
ARCA House,
237 Branston Road,
Burton upon Trent,
Staffordshire
DE14 3BT
Tel: 01283 531 126
Fax: 01283 568 228
Website: http: //www.aic.org.uk

ASSOCIATION FOR CONSULTANCY AND ENGINEERING
Alliance House
12 Caxton Street
London
SW1H 0QL
Tel: 020 7222 6557
Fax: 020 7990 9202
Website: www.acenet.co.uk

ASSOCIATION OF INTERIOR SPECIALISTS
Olton Bridge
245 Warwick Road
Solihull
West Midlands
B92 7 AH
Tel: 0121 707 0077
Fax: 0121 706 1949
Website: www.ais-interiors.org.uk

ASSOCIATION OF LOADING AND ELEVATING EQUIPMENT MANUFACTURERS
Airport House,
Purley Way,
Croydon,
Surrey
CR0 0XY
Tel: 020 8253 4501
Fax: 020 8253 4510
Website: www.alem.org.uk

ASSOCIATION OF PROJECT MANAGEMENT
IBIS House
Summerleys Road
Princes Risborough
Bucks
HP27 9LE
Tel: 0845 458 1944
Fax: 01494 528 937
Website: www.apm.org.uk

BOX CULVERT ASSOCIATION (BCA)
60 Charles Street
Leicester
Leicestershire
LE1 1FB
Tel: 0116 253 6161
Fax: 0116 251 4568
Website: www.boxculvert.org.uk

BRITISH ADHESIVES AND SEALANTS ASSOCIATION
5 Alderson Road
Worksop
Notts
S80 1UZ
Tel: 01909 480 888
Fax: 01909 473 834

BRITISH AGGREGATE CONSTRUCTION MATERIALS INDUSTRIES LTD (BACMI)
156 Buckingham Palace Road
London
SW1 W 9TR
Tel: 020 7730 8194

BRITISH APPROVALS FOR FIRE EQUIPMENT (BAFE)
Bridges 2
The Fire College
London Road
Moreton in Marsh
Gloucestershire
GL56 0RH
Tel: 0844 335 0897
Fax: 01608 653359
Website: www.bafe.org.uk

BRITISH APPROVALS SERVICE FOR CABLES (BASEC)
23 Presley Way
Crownhill
Milton Keynes
Buckinghamshire
MK8 0ES
Tel: 01908 267 300
Fax: 01908 267 255
Website: www.basec.org.uk

BRITISH ARCHITECTURAL LIBRARY (BAL)
Royal Institute of British Architects
66 Portland Place
London
W1B 1AD
Tel: 020 7580 5533
Fax: 020 7631 1802
Website: www.architecture.com

BRITISH ASSOCIATION OF LANDSCAPE INDUSTRIES (BALI)
Landscape House
National Agricultural Centre
Stoneleigh Park
Warwickshire
CV8 2LG
Tel: 024 7669 0333
Fax: 024 7669 0077
Website: www.bali.co.uk

BRITISH ASSOCIATION OF REINFORCEMENT
Riverside House
4 Meadows Business Park
Station Approach
Camberley
Surrey
GU17 9AB
Tel: 07802 747031
Website: www.uk-bar.org

BRITISH BATHROOM COUNCIL
(BATHROOM MANUFACTURERS ASSOCIATION)
Federation House
Station Road
Stoke-on-Trent
Staffordshire
ST4 2RT
Tel: 01782 747 123
Fax: 01782 747 161
Website: www.bathroom-assciation.org

BRITISH BOARD OF AGREMENT (BBA)
PO Box 195
Bucknalls Lane
Garston
Watford
Hertfordshire
WD25 9BA
Tel: 01923 665 300
Fax: 01923 665 301
Website: www.bbacerts.co.uk

BRITISH CABLES ASSOCIATION (BCA)
37a Walton Road
East Molesey
Surrey
KT8 0DH
Tel: 020 8941 4079
Fax: 020 8783 0104
Website: www.bcauk.org

BRITISH CARPET MANUFACTURERS ASSOCIATION LTD (BCMA)
PO Box 1155
MCF Complex
60 New Road
Kidderminster
Worcestershire
DY10 1AQ
Tel: 01562 755 568
Fax: 01562 865 4055
Website: www.carpetfoundation.com

BRITISH CEMENT ASSOCIATION (BCA)
CENTRE FOR CONCRETE INFORMATION
Century House
Telford Avenue
Crowthorne
Berkshire
RG45 6YS
Tel: 01344 466 007
Fax: 01344 466 008
Website: www.cementindustry.co.uk

BRITISH CERAMIC CONFEDERATION (BCC)
Federation House
Station Road
Stoke-on-Trent
Staffordshire
ST4 2SA
Tel: 01782 744 631
Fax: 01782 744 102
Website: www.ceramfed.co.uk

BRITISH CERAMIC RESEARCH LTD (BCR)
Queens Road
Penkhull
Stoke-on-Trent
Staffordshire
ST4 7LQ
Tel: 01782 764 444
Fax: 01782 412 331
Website: www.ceram.co.uk

BRITISH CERAMIC TILE COUNCIL (BCTC TILE
ASSOCIATION)
Federation house
Station Road
Stoke On Trent
ST4 2RT
Tel: 01782 747 147
Fax: 01782 747 161
Website: www.tpb.org.uk/

BRITISH COMBUSTION EQUIPMENT MANUFACTURERS
ASSOCIATION (BCEMA)
58 London Road
Leicester
LE2 0QD
Tel: 0116 275 7111
Fax: 0116 275 7222
Website: bcema.co.uk

BRITISH CONSTRUCTIONAL STEELWORK
ASSOCIATION LTD (BCSA)
4 Whitehall Court
Westminster
London
SW1A 2ES
Tel: 020 7839 8566
Fax: 020 7976 1634
Website: www.steelconstruction.org

BRITISH CONTRACT FURNISHING ASSOCIATION (BCFA)
Suite 2/4
The Business Design Centre
52 Upper Street
Islington Green
London
N1 0QH
Tel: 020 7226 6641
Fax: 020 7288 6190
Website: thebcfa.com

BRITISH ELECTROTECHNICAL APPROVALS BOARD
(BEAB)
1 Station View
Guildford
Surrey
GU1 4JY
Tel: 01483 455 466
Fax: 01483 455 477
Website: www.beab.co.uk

BRITISH FIRE PROTECTION SYSTEMS ASSOCIATION
LTD (BFPSA)
Neville House
55 Eden Street
KT1 1BW
Tel: 0208 549 5855
Fax: 0208 547 1564

BRITISH FURNITURE MANUFACTURERS FEDERATION
LTD (BFM Ltd)
30 Harcourt Street
London
W1H 2AA
Tel: 020 7724 0851
Fax: 020 7723 0622
Website: www.bfm.org.uk

BRITISH GEOLOGICAL SURVEY (BGS)
Keyworth Headquarters
Kingsley Drive
Dunham Centre
Nottingham
Nottinghamshire
NG12 5GG
Tel: 0115 936 3100
Fax: 0115 936 3200
Website: www.thebgs.co.uk

BRITISH INSTITUTE OF ARCHITECTURAL
TECHNOLOGISTS (BIAT)
397 City Road
London
EC1V 1NH
Tel: 020 7278 2206
Fax: 020 7837 3194
Website: www.biat.org.uk

BRITISH LAMINATE FABRICATORS ASSOCIATION
PO Box 775
Broseley Wood
TF7 9FG
Tel: 0845 056 8496
Website: www.blfa.co.uk

BRITISH LIBRARY BIBLIOGRAPHIC SERVICE AND
DOCUMENT SUPPLY
Boston Spa
Wetherby
West Yorkshire
LS23 7BQ
Tel: 01937 546 548
Fax: 01937 546 586
Website: www.bl.uk

BRITISH LIBRARY ENVIRONMENTAL INFORMATION
SERVICE
96 Euston Road
London
NW1 2DB
Tel: 020 7412 7000
Website: www.bl.uk/environment

BRITISH NON-FERROUS METALS FEDERATION
Broadway House
60 Calthorpe Road
Edgbaston
Birmingham
West Midlands
B15 1TN
Tel: 0121 456 6110
Fax: 0121 456 2274

BRITISH PLASTICS FEDERATION (BPF)
Plastics & Rubber Advisory Service
6 Bath Place
Rivington Street
London
EC2A 3JE
Tel: 020 7457 5000
Fax: 020 7457 5020
Website: www.bpf.co.uk

BRITISH PRECAST CONCRETE FEDERATION LTD
The Old Rectory
Main Street
Glenfield
LE3 8DG
Tel: 0116 232 5170
Website: www.britishprecast.org

BRITISH PROPERTY FEDERATION (BPF)
1 Warwick Row
London
SW1E 5ER
Tel: 020 7828 0111
Fax: 020 7824 3442
Website: www.bpf.org.uk

BRITISH RUBBER MANUFACTURERS' ASSOCIATION
LTD (BRMA)
6 Bath Place
Rivington Street
London
EC2A 3JE
Tel: 020 7457 5040
Fax: 020 7972 9008
Website: www.brma.co.uk

BRITISH STAINLESS STEEL ASSOCIATION
Broomgrove
59 Clarkehouse Road
Sheffield
S10 2LE
Tel: 0114 267 1260
Fax: 0114 266 1252
Website: www.bssa.org.uk

BRITISH STANDARDS INSTITUTION (BSI)
389 Chiswick High Road
London
W4 4 AL
Tel: 020 8996 9001
Fax: 020 8996 7001
Website: www.bsigroup.com

BRITISH WATER
1 Queen Anne's Gate
London
SW1H 9BT
Tel: 020 7957 4554
Fax: 020 7957 4565
Website: www.britishwater.co.uk

BRITISH WOOD PRESERVING & DAMP-PROOFING
ASSOCIATION (BWPDA)
6 Office Village
Romford Road
London
E15 4ED
Tel: 020 8519 2588
Fax: 020 8519 3444
Website: www.bwpda.co.uk

BRITISH WOODWORKING FEDERATION
55 Tufton Street
London
SW1 3QL
Tel: 0870 458 6939
Fax: 0870 458 6949
Website: www.bwf.org.uk

BUILDERS MERCHANTS FEDERATION
Soho Square
London
W1D 3HL
Tel: 020 7439 1753
Fax: 020 7734 2766
Website: www.bmf.org.uk

BUILDING & ENGINEERING SERVICES ASSOCIATION
ESCA House
34 Palace Court
Bayswater
London
W2 4JG
Tel: 020 7313 4900
Fax: 020 7727 9268
Website: www.hvca.org.uk

BUILDING CENTRE
The Building Centre
26 Store Street
London
WC1E 7BT
Tel: 020 7692 4000
Fax: 020 7580 9641
Website: www.buildingcentre.co.uk

BUILDING COST INFORMATION SERVICE LTD (BCIS)
Royal Institution of Chartered Surveyors
12 Great George Street
London
SW1P 3 AD
Tel: 020 7695 1500
Fax: 020 7695 1501
Website: www.bcis.co.uk

BUILDING EMPLOYERS CONFEDERATION (BEC)
55 Tufton Street
Westminster
London
SW1P 3QL
Tel: 0870 898 9090
Fax: 0870 898 9095
Website:
www.thecc.org.uk

BUILDING MAINTENANCE INFORMATION (BMI)
Royal Institution of Chartered Surveyors
12 Great George Street
London
SW1P 3 AD
Tel: 020 7695 1500
Fax: 020 7695 1501
Website: www.bcis.co.uk

BUILDING RESEARCH ESTABLISHMENT (BRE)
BRE Garston
Watford
WD5 9XX
Tel: 01923 664 000
Website: www.bre.co.uk

BUILDING RESEARCH ESTABLISHMENT: SCOTLAND
(BRE)
Kelvin Road
East Kilbride
Glasgow
G75 0RZ
Tel: 01355 576 200
Fax: 01355 241 895
Website: www.bre.co.uk

BUILDING SERVICES RESEARCH AND INFORMATION
ASSOCIATION LTD
Old Bracknell Lane West
Bracknell
Berkshire
RG12 7 AH
Tel: 01344 465 600
Fax: 01344 465 626
Website: www.bsria.co.uk

CASTINGS TECHNOLOGY INTERNATIONAL
7 East Bank Road
Sheffield
S2 3PT
Tel: 0114 272 8647
Fax: 0114 273 0854
Website: www.castingsdev.com

CATERING EQUIPMENT MANUFACTURERS
ASSOCIATION (CEMA)
Carlyle House
235 Vauxhall Bridge Road
London
SW1 V 1EJ
Tel: 020 7233 7724
Fax: 020 7828 0667

CEMENT ADMIXTURES ASSOCIATION
38 Tilehouse
Green Lane
Knowle
West Midlands
B93 9EY
Tel: + Fax: 01564 776 362
Website: www.admixtures.org.uk

CHARTERED INSTITUTE OF ARBITRATORS
12 Bloomsbury Square
London
WC1 A 2LP
Tel: 0207 421 7444
Fax: 0207 404 4023
Website: www.ciarb.org

CHARTERED INSTITUTE OF ARCHITECTURAL
TECHNOLOGISTS
397 City Road
London
EC1 V 1NH
Tel: 020 7278 2206
Fax: 020 7837 3194
Website: www.ciat.org.uk

CHARTERED INSTITUTE OF BUILDING (CIOB)
1 Arlington Square
Downshire Way
Bracknell
RG12 1WA
Tel: 01344 630 700
Fax: 01344 306 430
Website: www.ciob.org.uk

CHARTERED INSTITUTE OF WASTES MANAGEMENT
9 Saxon Court
St Peter's Gardens
Northampton
NN1 1SX
Tel: 01604 620 426
Fax: 01604 621 339
Website: www.iwm.co.uk

CHARTERED INSTITUTION OF BUILDING
SERVICES ENGINEERS (CIBSE)
Delta House
222 Balham High Road
London
SW12 9BS
Tel: 020 8675 5211
Fax: 020 8675 5449
Website: www.cibse.org

CIVIL ENGINEERING CONTRACTORS ASSOCIATION
1 Birdcage Walk
London
SW1H 9JJ
Tel: 020 7340 0450
Website: www.ceca.co.uk

CLAY PIPE DEVELOPMENT ASSOCIATION (CPDA)
Copsham House
53 Broad Street
Chesham
HP5 3EA
Tel: 01494 791 456
Fax: 01494 792 378
Website: www.cpda.co.uk

CLAY ROOF TILE COUNCIL
Federation House
Station Road
Stoke-on-Trent
Staffordshire
ST4 2SA
Tel: 01782 744 631
Fax: 01782 744 102
Website: www.clayroof.co.uk

COLD ROLLED SECTIONS ASSOCIATION (CRSA)
National Metal Forming Centre
47 Birmingham Road
West Bromwich
West Midlands
B70 6PY
Tel: 0121 601 6350
Fax: 0121 601 6373
Website: www.crsauk.com

COMMONWEALTH ASSOCIATION OF ARCHITECTS
(CAA)
PO BOX 508
Edgware
HA8 9XZ
Tel: 020 8951 0550
Website: www.comarchitect.org

CONCRETE BRIDGE DEVELOPMENT GROUP
Riverside House
4 Meadows Business Park
Station Approach
Blackwater, Camberley
Surrey
GU17 9AB
Tel: 01276 33777
Fax: 01276 38899
Website: www.concrete.org.uk

CONCRETE PIPE ASSOCIATION (CPA)
Main Street
Glenfield
LE3 8DG
Tel: 0116 232 5170
Website: www.concretepipes.co.uk

CONCRETE REPAIR ASSOCIATION (CRA)
Association House
235 Ash Road
Aldershot
Hampshire
GU12 4DD
Tel: 01252 321 302
Fax: 01252 333 901
Website: www.concreterepair.org.uk

CONCRETE SOCIETY ADVISORY SERVICE
Riverside House
4 Meadows Business Park
Station Approach
Blackwater, Camberley
Surrey
GU17 9AB
Tel: 01276 607 140
Fax: 01276 607 141
Website: www.concrete.org.uk

CONFEDERATION OF BRITISH INDUSTRY (CBI)
Centre Point
103 New Oxford Street
London
WC1A 1DU
Tel: 020 7379 7400
Fax: 020 7240 1578
Website: www.cbi.org.uk

CONSTRUCT – CONCRETE STRUCTURES GROUP LTD
Riverside House
4 Meadows Business Park
Station Approach
Blackwater, Camberley
Surrey
GU17 9AB
Tel: 01276 38444
Fax: 01276 38899
Website: www.construct.org.uk

CONSTRUCTION EMPLOYERS FEDERATION LTD (CEF)
143 Malone Road
Belfast
Northern Ireland
BT9 6SU
Tel: 028 9087 7143
Fax: 028 9087 7155
Website: www.cefni.co.uk

CONSTRUCTION INDUSTRY JOINT COUNCIL (CIJC)
55 Tufton Street
London
SW1P 3QL
Tel: 0870 898 9090
Fax: 0870 898 9095
Website: www.nscc.org.uk

CONSTRUCTION INDUSTRY RESEARCH AND INFORMATION ASSOCIATION
Classic House
174–180 Old Street
London
EC1V 9BP
Tel: 020 7549 3300
Fax: 020 7253 0523
Website: www.ciria.org.uk

CONSTRUCTION PLANT-HIRE ASSOCIATION (CPA)
27–28 Newbury Street
London
EC1A 7HU
Tel: 020 7796 3366
Website: www.cpa.uk.net

CONTRACT FLOORING ASSOCIATION (CFA)
4c Saint Mary's Place
The Lace Market
Nottingham
Nottinghamshire
NG1 1PH
Tel: 0115 941 1126
Fax: 0115 941 2238
Website: www.cfa.org.uk

CONTRACTORS MECHANICAL PLANT ENGINEERS (CMPE)
43 Portsmouth Road
Horndeam
Waterlooville
Hampshire
PO8 9LN
Tel: 023 925 70011
Fax: 023 925 70022
Website: www.cmpe.co.uk/

COPPER DEVELOPMENT ASSOCIATION
Verulam Industrial Estate
224 London Road
Saint Albans
Hertfordshire
AL1 1AQ
Tel: 01727 731 200
Fax: 01727 731 216
Website: www.cda.org.uk

CORUS RESEARCH DEVELOPMENT AND TECHNOLOGY
Swinden Technology Centre
Moorgate
Rotherham
South Yorkshire
S60 3AR
Tel: 01709 825 335
Fax: 01709 825 464
Website: www.corusgroup.com/en/technology/research_and_development/

COUNCIL FOR ALUMINIUM IN BUILDING (CAB)
Bank House
Bond's Mill
Stonehouse
Gloucestershire
GL10 3RF
Tel: 01453 828851
Fax: 01453 828861
Website: www.c-a-b.org.uk

DEPARTMENT FOR BUSINESS, INNOVATIN AND SKILLS
1 Victoria Street
London
SW1H 0ET
Tel: 020 7215 5000
Website: www.gov.uk/government/organizations/department-for-business-innovation-skills

DEPARTMENT FOR TRANSPORT
Great Minister House
33 Horseferry Road,
London
SW1 P 4DR
Tel: 0300 330 3000
Website: www.dft.gov.uk

DOORS & HARDWARE FEDERATION
42 Heath Street
Tamworth
Staffordshire
B79 7JH
Tel: 01827 52337
Fax: 01827 310 827
Website: www.abhm.org.uk

DRY STONE WALLING ASSOCIATION OF GREAT
BRITAIN (DSWA)
Westmorland County Showground
Lane Fram
Crooklands, Milnthorpe
Cumbria
LA7 7NH
Tel: 01539 567 953
Website: www.dswa.org.uk

ELECTRICAL CONTRACTORS ASSOCIATION (ECA)
ESCA House
34 Palace Court
Bayswater
London
W2 4HY
Tel: 020 7313 4800
Fax: 020 7221 7344
Website: www.eca.co.uk

ELECTRICAL CONTRACTORS ASSOCIATION OF
SCOTLAND (SELECT)
The Walled Gardens
Bush Estate
Midlothian
Scotland
EH26 0SB
Tel: 0131 445 5577
Fax: 0131 445 5548
Website: www.select.org.uk

ELECTRICAL INSTALLATION EQUIPMENT
MANUFACTURERS ASSOCIATION LTD (EIEMA)
Beama Installation Ltd
Westminster Tower
3 Albert Embankment
London
SE1 7SL
Tel: 020 7793 3000
Fax: 020 7793 3003
Website: www.eiema.org.uk

EUROPEAN LIQUID ROOFING ASSOCIATION (ELRA)
Fields House
Gower Road
Haywards Heath
West Sussex
RH16 4PL
Tel: 01444 417 458
Fax: 01444 415 616
Website: www.elra.org.uk

FEDERATION OF ENVIRONMENTAL TRADE
ASSOCIATIONS
2 Waltham Court
Milley Lane
Hare Hatch
Reading
RG10 9TH
Tel: 0118 940 3416
Fax: 0118 940 6258
Website: www.feta.co.uk

FEDERATION OF MANUFACTURERS OF
CONSTRUCTION EQUIPMENT & CRANES
Ambassador House
Brigstock Road
Thornton Heath
Surrey
CR7 7JG
Tel: 020 8665 5727
Fax: 020 8665 6447
Website: www.coneq.org.uk

FEDERATION OF MASTER BUILDERS
Gordon Fisher House
14–15 Great James Street
London
WC1N 3DP
Tel: 020 7242 7583
Fax: 020 7404 0296
Website: www.fmb.org.uk/

FEDERATION OF PILING SPECIALISTS
Forum Court
83 Coppers Cope Road
Beckenham
Kent
BR3 1NR
Tel: 020 8663 0947
Fax: 020 8663 0949
Website: www.fps.org.uk

FEDERATION OF PLASTERING & DRYWALL
CONTRACTORS
Construction House
56–64 Leonard Street
London
EC2A 4JX
Tel: 020 7608 5092
Fax: 020 7608 5081
Website: www.fpdc.org

FENCING CONTRACTORS ASSOCIATION
Warren Road
Trellech
Monmouthshire
NP5 4PQ
Tel: 07000 560 722
Fax: 01600 860 614
Website: www.fencingcontractors.org

FINNISH PLYWOOD INTERNATIONAL
PO Box 99
Welwyn Garden City
Herts
AL6 0HS
Tel: 01438 798 746
Fax: 01438 798 305

FLAT ROOFING ALLIANCE
Fields House
Gower Road
Haywards Heath
West Sussex
RH16 4PL
Tel: 01444 440 027
Fax: 01444 415 616
Website: www.fra.org.uk

FURNITURE INDUSTRY RESEARCH ASSOCIATION
(FIRA INTERNATIONAL LTD)
Maxwell Road
Stevenage
Hertfordshire
SG1 2EW
Tel: 01438 777 700
Fax: 01438 777 800
Website: www.fira.co.uk

GLASS & GLAZING FEDERATION (GGF)
44–48 Borough High Street
London
SE1 1XB
Tel: 020 7403 7177
Website: www.ggf.org.uk

HOUSING CORPORATION HEADQUARTERS
Maple House
149 Tottenham Court Road
London
W1N 7BN
Tel: 0845 230 7000
Fax: 0207 393 2111
Website: www.housingcorp.gov.uk

ICOM Energy Association
Camden House
Warwick Road
Kenilworth
Warwickshire
CV8 1TH
Tel: 01926 513748
Fax: 01926 21 855017
Website: www.icome.org.uk

INSTITUTE OF ACOUSTICS
77 A Saint Peter' Street
Saint Albans
Hertfordshire
AL1 3BN
Tel: 01727 848 195
Fax: 01727 850 553
Website: www.ioa.org.uk

INSTITUTE OF ASPHALT TECHNOLOGY
Paper Mews Place
290 High Street
Dorking
Surrey
RH4 1QT
Tel: 01306 742 792
Fax: 01306 888 902
Website: www.instofasphalt.org

INSTITUTE OF MAINTENANCE AND BUILDING
MANAGEMENT
Keets House
30 East Street
Farnham
Surrey
GU9 7SW
Tel: 01252 710 994
Fax: 01252 737 741
Website: www.imbm.org.uk

INSTITUTE OF MATERIALS
Headquarters
1 Carlton House Terrace
London
SW1Y 5 AF
Tel: 020 7451 7300
Fax: 020 7839 1702
Website: www.materials.org.uk

INSTITUTE OF PLUMBING
64 Station Lane
Hornchurch
Essex
RM12 6NB
Tel: 01708 472 791
Fax: 01708 448 987
Website: www.plumbers.org.uk

INSTITUTE OF WOOD SCIENCE
Stocking Lane
Hughenden Valley
High Wycombe
Buckinghamshire
HP14 4NU
Tel: 01494 565 374
Fax: 01494 565 395
Website: www.iwsc.org.uk

INSTITUTION OF CIVIL ENGINEERS (ICE)
1 Great George Street
London
SW1P 3 AA
Tel: 020 7222 7722
Fax: 020 7222 7500
Website: www.ice.org.uk

INSTITUTION OF ENGINEERING AND TECHNOLOGY
The Institution of Engineering and Technology
Michael Faraday House
Stevenage
Herts
SG1 2 AY
Tel: 01438 313 311
Fax: 01438 765 526
Website: www.theiet.org

INSTITUTION OF MECHANICAL ENGINEERS
1 Birdcage Walk
London
SW1H 9JJ
Tel: 020 7222 7899
Fax: 020 7222 4557
Website: www.imeche.org

INSTITUTION OF STRUCTURAL ENGINEERS (ISE)
11 Upper Belgrave Street
London
SW1X 8BH
Tel: 020 7235 4535
Fax: 020 7235 4294
Website: www.istructe.org

INTERNATIONAL LEAD ASSOCIATION
17 A Welbeck Way
London
W1G 9YJ
Tel: 020 7499 8422
Fax: 020 7493 1555
Website: www.ldaint.org

INTERPAVE (THE PRECAST CONCRETE PAVING
& KERB ASSOCIATION)
60 Charles Street
Leicester
Leicestershire
LE1 1FB
Tel: 0116 253 6161
Fax: 0116 251 4568
Website: www.paving.org.uk/

JOINT CONTRACTS TRIBUNAL LTD
9 Cavendish Place
London
W1G 0GD
Tel: 020 7630 8650
Fax: 020 7630 8670
Website: www.jctltd.co.uk

KITCHEN SPECIALISTS ASSOCIATION
12 TopBarn Business Centre
Holt Heath
Worcester
Worcestershire
WR6 6NH
Tel: 01905 621 787
Fax: 01905 621 887
Website: www.kbsa.co.uk

LIGHTING ASSOCIATION LTD
Stafford Park 7
Telford
Shropshire
TF3 3BQ
Tel: 01952 290 905
Fax: 01952 290 906
Website: www.lightingassociation.com/

MASTIC ASPHALT COUNCIL LTD
PO BOX 77
Hastings
Kent
TN35 4WL
Tel: 01424 814 400
Fax: 01424 814 446
Website: www.masticasphaltcouncil.co.uk

METAL CLADDING & ROOFING MANUFACTURERS
ASSOCIATION
18 Mere Farm Road
Prenton
Wirral
Cheshire
CH43 9TT
Tel: 0151 652 3846
Fax: 0151 653 4080
Website: www.mcrma.co.uk

METAL GUTTER MANUFACTURERS ASSOCIATION
106 Ruskin Avenue
Rogerstone
Newport
South Wales
NP10 0BD
Tel: 01633 891584
Website: www.mgma.co.uk

MET OFFICE
Fitzroy Road
Exeter
Devon
EX1 3PB
Tel: 0870 900 0100
Fax: 0870 900 5050
Website: www.metoffice.gov.uk

NATIONAL AACCESS AND SCAFFOLDING
CONFEDERATION
NASC
4th Floor
12 Bridewell Place
London
EC4 V 6 AP
Tel: 020 7822 7400
Website: www.nasc.org.uk

NATIONAL ASSOCIATION OF STEEL STOCKHOLDERS
The Citadel
190 Corporation Street
Birmingham
B4 6QD
Tel: 0121 200 2288
Fax: 0121 236 7444
Website: www.nass.org.uk

NATIONAL HOUSE-BUILDING COUNCIL (NHBC)
NHBC
NHBC House
Davy Avenue
Knowlhill
Milton Keynes
MK5 8FP
Tel: 0844 633 1000
Website: www.nhbc.co.uk

NATURAL SLATE QUARRIES ASSOCIATION
26 Store Street
London
WC1E 7BT
Tel: 020 7323 3770
Fax: 020 7323 0307
Website: http://slateassociation.org

NHS ESTATES
Departments of Health
1 Trevelyan Square
Boar Lane
Leeds
West Yorkshire
LS1 6 AE
Tel: 0113 254 7000
Fax: 0113 254 7299
Website: http://www.buyingsolutions.gov.uk/healthcms/

ORDNANCE SURVEY
Romsey Road
Southampton
SO16 4GU
Tel: 08456 050 504
Fax: 02380 792 615
Website: www.ordnancesurvey.co.uk

PAINTING AND DECORATING ASSOCIATION
32 Coton Road
Nuneaton
Warwickshire
CV11 5TW
Tel: 01203 353 776
Fax: 01203 354 4513
Website: www.paintingdecoratingassociation.co.uk

PIPELINE INDUSTRIES GUILD
14–15 Belgrave Square
London
SW1X 8PS
Tel: 020 7235 7938
Fax: 020 7235 0074
Website: www.pipeguild.co.uk

PLASTIC PIPES GROUP
c/o British Plastics Federation
6 Bath Place
Rivington Street
London
EC2 A 3JE
Tel: 020 7457 5024
Website: www.plasticpipesgroup.com

PRECAST FLOORING FEDERATION
60 Charles Street
Leicester
Leicestershire
LE1 1FB
Tel: 0116 253 6161
Fax: 0116 251 4568
Website: www.pff.org.uk

PRESTRESSED CONCRETE ASSOCIATION
60 Charles Street
Leicester
Leicestershire
LE1 1FB
Tel: 0116 253 6161
Fax: 0116 251 4568
Website: www.britishprecast.org

PROPERTY CONSULTANTS SOCIETY LTD
Basement Office
1 Surrey Street
Arundel
West Sussex
BN18 9DT
Tel: 01903 883 787
Fax: 01903 889 590
Website: www.p-c-s.org.uk

QUARRY PRODUCTS ASSOCIATION
Gillingham House
38–44 Gillingham Street
London
SW1 V1HU
Tel: 020 7963 8000
Fax: 020 7963 8001
Website: www.qpa.org

READY-MIXED CONCRETE BUREAU
Century House
Telford Avenue
Crowthorne
Berkshire
RG45 6YS
Tel: 01344 725 732
Fax: 01344 774 976
Website: www.rcb.org.uk

REINFORCED CONCRETE COUNCIL
Riverside House
4 Meadows Business Park
Station Approach
Camberley
Berkshire
GU17 9AB
Tel: 01276 607140
Fax: 01276 607141
Website: www.rcc-info.org.uk

ROYAL INCORPORATION OF ARCHITECTS IN
SCOTLAND (RIAS)
15 Rutland Square
Edinburgh
Scotland
EH1 2BE
Tel: 0131 229 7545
Fax: 0131 228 2188
Website: www.rias.org.uk

ROYAL INSTITUTE OF BRITISH ARCHITECTS (RIBA)
66 Portland Place
London
W1B 1AD
Tel: 020 7580 5533
Fax: 020 7255 1541
Website: www.architecture.com

ROYAL INSTITUTION OF CHARTERED SURVEYORS
(RICS)
12 Great George Street
Parliament Square
London
SW1P 3AD
Tel: 020 7222 7000
Fax: 020 7222 9430
Website: www.rics.org

ROYAL TOWN PLANNING INSTITUTE (RTPI)
41 Botolph Lane
London
EC3R 8DL
Tel: 020 7929 9494
Fax: 020 7323 1582
Website: www.rtpi.org.uk/

RURAL DESIGN AND BUILDING ASSOCIATION
ATSS House
Station Road East
Stowmarket
Suffolk
IP14 1RQ
Tel: 01449 676 049
Fax: 01449 770 028
Website: www.rdba.org.uk

SCOTTISH BUILDING EMPLOYERS FEDERATION
Carron Grange
Carron Grange Avenue
Stenhousemuir
Scotland
FK5 3BQ
Tel: 01324 555 550
Fax: 01324 555 551
Website: www.scottish-building.co.uk

SCOTTISH HOMES- Community Scotland
Thistle House
91 Haymarket Terrace
Edinburgh
Scotland
EH12 5HE
Tel: 0131 313 0044
Fax: 0131 313 2680
Website: www.scotland.gov.uk

SCOTTISH NATURAL HERITAGE
Communications Directorate
12 Hope Terrace
Edinburgh
EH9 2AS
Tel: 0131 447 4784
Fax: 0131 446 2277
Website: www.snh.org.uk

SINGLE PLY ROOFING ASSOCIATION
177 Bagnall Road
Basford
Nottinghamshire
NG6 8SJ
Tel: 0115 914 4445
Fax: 0115 974 9827
Website: www.spra.co.uk/

SMOKE CONTROL ASSOCIATION
2 Waltham Court
Milley Lane, Hare Hatch
Reading
Berkshire
RG10 9TH
Tel: 0118 940 3416
Fax: 0118 940 6258
Website: www.feta.co.uk

SOCIETY FOR THE PROTECTION OF ANCIENT
BUILDINGS (SPAB)
37 Spital Square
London E1 6DY
Tel: 020 7377 1644
Fax: 020 7247 5296
Website: www.spab.org.uk

SOCIETY OF GLASS TECHNOLOGY
Don Valley House
Saville Street East
Sheffield
South Yorkshire
S4 7UQ
Tel: 0114 263 4455
Fax: 0114 263 4411
Website: www.sgt.org

SOIL SURVEY AND LAND RESEARCH INSTITUTE
Cranfield University
Silsoe Campus
Bedford
Bedfordshire
MK45 4DT
Tel: 01525 863 000
Fax: 01525 863 253
Website: www.cranfield.ac.uk/sslrc

SOLAR ENERGY SOCIETY
c/o School of Engineering
Oxford Brookes University
Gipsy Lane
Headington, Oxford
OX3 0BP
Tel: 01865 741 111
Fax: 01525 863 253
Website: www.uk-ises.org/

SPON'S PRICE BOOK AUTHORS
AECOM
MidCity Place
71 High Holborn
Holborn
London
WC1V 6QS
Tel: 020 7061 7000
Website: www.aecom.com

SPORT ENGLAND
3rd Floor, Victoria House
Bloomsbury Square
London
WC1B 4SE
Tel: 0845 850 8508
Fax: 020 7383 5740
Website: www.sportengland.org

SPORT SCOTLAND
Caledonia House
South Gyle
Edinburgh
Scotland
EH12 9DQ
Tel: 0131 317 7200
Fax: 0131 317 7202
Website: www.sportscotland.org.uk

SPORTS COUNCIL FOR WALES
Welsh Institute of Sport
Sophia Gardens
Cardiff
CF11 9SW
Tel: 0845 045 0904
Fax: 0845 846 0014
Website: www.sports-council-wales.co.uk

SPORTS TURF RESEARCH INSTITUTE (STRI)
Saint Ives Estate
Bingley
West Yorkshire
BD16 1AU
Tel: 01274 565 131
Fax: 01274 561 891
Website: www.stri.co.uk

SPRAYED CONCRETE ASSOCIATION
Association House
235 Ash Road
Aldershot
Hampshire
GU12 4DD
Tel: 01252 321 302
Fax: 01252 333 901
Website: www.sca.org.uk

STEEL CONSTRUCTION INSTITUTE
Silwood Park
Ascot
Berkshire
SL5 7QN
Tel: 01344 636 505
Fax: 01344 636 570
Website: www.steel-sci.org

STEEL WINDOW ASSOCIATION
The Building Centre
26 Store Street
London
WC1E 7BT
Tel: 020 7637 3571
Fax: 020 7637 3572
Website: www.steel-window-association.co.uk

STONE FEDERATION GREAT BRITAIN
Channel Business Centre
Ingles Manor
Castle Hill Avenue
Folkestone
Kent
CT20 2RD
Tel: 01303 856123
Fax: 01303 856117
Website: www.stone-federationgb.org.uk

SUSPENDED ACCESS EQUIPMENT
MANUFACTURERS ASSOCIATION
56–54 Leonard Street
London
EC2A 4JX
Tel: 020 7608 5098
Fax: 020 7636 5984
Website: www.saema.org

SWIMMING POOL & ALLIED TRADES ASSOCIATION
(SPATA)
Spata House
1a Junction Road
Andover
Hampshire
SP10 3QT
Tel: 01264 356210
Fax: 01264 332628
Website: www.spata.co.uk

THE PIPELINE INDUSTRIES GUILD
F150 First Floor
Cherwell Business Village
Southam Road
Banbury
OX16 2SP
Tel: 020 7235 7938
Fax: 020 7235 0074
Website: www.pipeline.com

THERMAL INSULATION CONTRACTORS ASSOCIATION
Tica House
Allington Way
Yarm Road Business Park
Darlington
County Durham
DL1 4QB
Tel: 01325 466 704
Fax: 01325 487 691
Website: www.tica-acad.co.uk

TIMBER RESEARCH & DEVELOPMENT ASSOCIATION
(TRADA)
Stocking Lane
Hughenden Valley
High Wycombe
Buckinghamshire
HP14 4ND
Tel: 01494 569 600
Fax: 01494 565 487
Website www.trada.co.uk

TIMBER TRADE FEDERATION
4th Floor
Clareville House
26–27 Oxenden Street
London
SW1Y 4EL
Tel: 020 7839 1891
Fax: 020 7930 0094
Website: www.ttf.co.uk

TOWN & COUNTRY PLANNING ASSOCIATION (TCPA)
17 Carlton House Terrace
London
SW1Y 5AS
Tel: 020 7930 8903
Fax: 020 7930 3280
Website: tcpa.org.uk

TREE COUNCIL
71 Newcomen Street
London
SE1 1WT
Tel: 020 7407 9992
Fax: 020 7407 9908
Website: www.treecouncil.org.uk

TRUSSED RAFTER ASSOCIATION
31 Station Road
Sutton
Retford
Nottinghamshire
DN22 8PZ
Tel: 01777 869 281
Fax: 01777 869 281
Website: www.tra.org.uk

TWI
Granta Park
Great Abington
Cambridge
Cambridgeshire
CB1 6AL
Tel: 01223 899 000
Fax: 01223 892 588
Website: www.twi.co.uk

UNDERFLOOR HEATING MANUFACTURERS'
ASSOCIATION
Belhaven House
67 Walton Road
East Moseley
Surrey
KT8 0DB
Tel: 020 8941 7177
Fax: 020 8941 815
Website: www.uhma.org.uk

VERMICULITE INFORMATION SERVICE
1 A Guildford Business Park
Guildford
Surrey
GU2 8XG
Tel: 01483 242 100
Fax: 01483 242 101

WALLCOVERING MANUFACTURERS ASSOCIATION
James House
Bridge Street
Leatherhead
Surrey
KT22 7EP
Tel: 01372 360 660
Fax: 01372 376 069

WATERHEATER MANUFACTURERS ASSOCIATION
Waterheater Manufacturers Association
c/o Andrews Waterheaters
Wednesbury One
Black Country New Road
Wednesbury
WS10 7NZ
Tel: 07775 754456
Fax: 0161 456 7106
Website: www.waterheating.fsnet.co.uk/wma.htm

WATER RESEARCH CENTRE
Henley Road
Medmenham
Marlow
Buckinghamshire
SL7 2HD
Tel: 01491 636 500
Fax: 01491 636 501
Website: www.wrcplc.co.uk

WATER UK
1 Queen Anne's Gate
London
SW1H 9BT
Tel: 020 7344 1844
Fax: 020 7344 1866
Website: www.water.org.uk/

WELDING MANUFACTURERS' ASSOCIATION
Westminster Tower
3 Albert Embankment
London
SE1 7SL
Tel: 020 7793 3041
Fax: 020 7582 8020
Website: www.wma.uk.com

WOOD PANEL INDUSTRIES FEDERATION
Grantham
Lincolnshire
NG31 6LR
Tel: 01476 563 707
Fax: 01476 579 314
Website: www.wpif.org.uk

WRAP
The Old Academy
21 Horse Fair
Banbury
OX16 0 AH
Tel: 01295 819 900
Website: www.wrap.org.uk

ZINC DEVELOPMENT ASSOCIATION
42 Weymouth Street
London
W1N 3LQ
Tel: 020 7499 6636
Fax: 020 7493 135
Website: www.zincinfocentre.org

BIM and Quantity Surveying

S Pittard *et al.*

The sudden arrival of Building Information Modelling (BIM) as a key part of the building industry is redefining the roles and working practices of its stakeholders. Many clients, designers, contractors, quantity surveyors, and building managers are still finding their feet in an industry where BIM compliance can bring great rewards.

This guide is designed to help quantity surveying practitioners and students understand what BIM means for them, and how they should prepare to work successfully on BIM compliant projects. The case studies show how firms at the forefront of this technology have integrated core quantity surveying responsibilities like cost estimating, tendering, and development appraisal into high profile BIM projects. In addition to this, the implications for project management, facilities management, contract administration and dispute resolution are also explored through case studies, making this a highly valuable guide for those in a range of construction project management roles.

Featuring a chapter describing how the role of the quantity surveyor is likely to permanently shift as a result of this development, as well as descriptions of tools used, this covers both the organisational and practical aspects of a crucial topic.

December 2015: 234 x 156 mm: 258 pp
Pbk: 978-0-415-87043-6; £30.99

To Order: Tel: +44 (0) 1235 400524 Fax: +44 (0) 1235 400525
or Post: Taylor and Francis Customer Services,
Bookpoint Ltd, Unit T1, 200 Milton Park, Abingdon, Oxon, OX14 4TA UK
Email: book.orders@tandf.co.uk

For a complete listing of all our titles visit:
www.tandf.co.uk

PART 8

Tables and Memoranda

This part contains the following sections:

Project Risk and Opportunity Management

Agnar Johansen *et al.*

Effective risk and opportunity management is key to the successful delivery of any major engineering and construction project. This book looks at how all those involved can manage risk and capitalise on the opportunities that uncertainty present. The authors of this book highlight that uncertainties should be managed rather than avoided. This book will look at simple projects with a small team, to megaprojects where some hundreds of people are involved, and the consequences of delays or unforeseen costs. However, while the obvious risks can be planned for, the authors argue that it is often the opportunities in these situations that can have unexploited potential.

This book is about opportunity management seen from the owner's perspective. It will be an invaluable resource for those studying Engineering both undergraduate and postgraduate and set out ways in which projects should be managed from planning to completion. This book is also a great tool for those working in project management and the construction industry. While there are many books that demonstrate effective construction management, this book is the first of its kind to emphasise that there is opportunity in uncertainty, and possibility in the unexpected.

April 2019: 234 × 156 mm: 162pp
Pb: 978-1-138-36582-7 : £44.99

CONVERSION TABLES

Length	Unit	Conversion factors			
Millimetre	mm	1 in	= 25.4 mm	1 mm	= 0.0394 in
Centimetre	cm	1 in	= 2.54 cm	1 cm	= 0.3937 in
Metre	m	1 ft	= 0.3048 m	1 m	= 3.2808 ft
		1 yd	= 0.9144 m		= 1.0936 yd
Kilometre	km	1 mile	= 1.6093 km	1 km	= 0.6214 mile

Note:
1 cm	= 10 mm	1 ft	= 12 in
1 m	= 1 000 mm	1 yd	= 3 ft
1 km	= 1 000 m	1 mile	= 1 760 yd

Area	Unit	Conversion factors			
Square Millimetre	mm^2	1 in^2	= 645.2 mm^2	1 mm^2	= 0.0016 in^2
Square Centimetre	cm^2	1 in^2	= 6.4516 cm^2	1 cm^2	= 1.1550 in^2
Square Metre	m^2	1 ft^2	= 0.0929 m^2	1 m^2	= 10.764 ft^2
		1 yd^2	= 0.8361 m^2	1 m^2	= 1.1960 yd^2
Square Kilometre	km^2	1 mile^2	= 2.590 km^2	1 km^2	= 0.3861 mile^2

Note:
1 cm^2	= 100 mm^2	1 ft^2	= 144 in^2
1 m^2	= 10 000 cm^2	1 yd^2	= 9 ft^2
1 km^2	= 100 hectares	1 acre	= 4 840 yd^2
		1 mile^2	= 640 acres

Volume	Unit	Conversion factors			
Cubic Centimetre	cm^3	1 cm^3	= 0.0610 in^3	1 in^3	= 16.387 cm^3
Cubic Decimetre	dm^3	1 dm^3	= 0.0353 ft^3	1 ft^3	= 28.329 dm^3
Cubic Metre	m^3	1 m^3	= 35.3147 ft^3	1 ft^3	= 0.0283 m^3
		1 m^3	= 1.3080 yd^3	1 yd^3	= 0.7646 m^3
Litre	l	1 l	= 1.76 pint	1 pint	= 0.5683 l
			= 2.113 US pt		= 0.4733 US l

Note:
1 dm^3	= 1 000 cm^3	1 ft^3	= 1 728 in^3	1 pint	= 20 fl oz
1 m^3	= 1 000 dm^3	1 yd^3	= 27 ft^3	1 gal	= 8 pints
1 l	= 1 dm^3				

Neither the Centimetre nor Decimetre are SI units, and as such their use, particularly that of the Decimetre, is not widespread outside educational circles.

Mass	Unit	Conversion factors			
Milligram	mg	1 mg	= 0.0154 grain	1 grain	= 64.935 mg
Gram	g	1 g	= 0.0353 oz	1 oz	= 28.35 g
Kilogram	kg	1 kg	= 2.2046 lb	1 lb	= 0.4536 kg
Tonne	t	1 t	= 0.9842 ton	1 ton	= 1.016 t

Note:
1 g	= 1000 mg	1 oz	= 437.5 grains	1 cwt	= 112 lb
1 kg	= 1000 g	1 lb	= 16 oz	1 ton	= 20 cwt
1 t	= 1000 kg	1 stone	= 14 lb		

Force	Unit	Conversion factors			
Newton	N	1 lbf	= 4.448 N	1 kgf	= 9.807 N
Kilonewton	kN	1 lbf	= 0.004448 kN	1 ton f	= 9.964 kN
Meganewton	MN	100 tonf	= 0.9964 MN		

CONVERSION TABLES

Pressure and stress	Unit	Conversion factors	
Kilonewton per square metre	kN/m^2	1 lbf/in^2	= 6.895 kN/m^2
		1 bar	= 100 kN/m^2
Meganewton per square metre	MN/m^2	1 tonf/ft^2	= 107.3 kN/m^2 = 0.1073 MN/m^2
		1 kgf/cm^2	= 98.07 kN/m^2
		1 lbf/ft^2	= 0.04788 kN/m^2

Coefficient of consolidation (Cv) or swelling	Unit	Conversion factors	
Square metre per year	m^2/year	1 cm^2/s	= 3 154 m^2/year
		1 ft^2/year	= 0.0929 m^2/year

Coefficient of permeability	Unit	Conversion factors	
Metre per second	m/s	1 cm/s	= 0.01 m/s
Metre per year	m/year	1 ft/year	= 0.3048 m/year
			= 0.9651 × (10)^8 m/s

Temperature	Unit	Conversion factors	
Degree Celsius	°C	°C = 5/9 × (°F − 32)	°F = (9 × °C)/ 5 + 32

CONVERSION TABLES

SPEED CONVERSION

km/h	m/min	mph	fpm
1	16.7	0.6	54.7
2	33.3	1.2	109.4
3	50.0	1.9	164.0
4	66.7	2.5	218.7
5	83.3	3.1	273.4
6	100.0	3.7	328.1
7	116.7	4.3	382.8
8	133.3	5.0	437.4
9	150.0	5.6	492.1
10	166.7	6.2	546.8
11	183.3	6.8	601.5
12	200.0	7.5	656.2
13	216.7	8.1	710.8
14	233.3	8.7	765.5
15	250.0	9.3	820.2
16	266.7	9.9	874.9
17	283.3	10.6	929.6
18	300.0	11.2	984.3
19	316.7	11.8	1038.9
20	333.3	12.4	1093.6
21	350.0	13.0	1148.3
22	366.7	13.7	1203.0
23	383.3	14.3	1257.7
24	400.0	14.9	1312.3
25	416.7	15.5	1367.0
26	433.3	16.2	1421.7
27	450.0	16.8	1476.4
28	466.7	17.4	1531.1
29	483.3	18.0	1585.7
30	500.0	18.6	1640.4
31	516.7	19.3	1695.1
32	533.3	19.9	1749.8
33	550.0	20.5	1804.5
34	566.7	21.1	1859.1
35	583.3	21.7	1913.8
36	600.0	22.4	1968.5
37	616.7	23.0	2023.2
38	633.3	23.6	2077.9
39	650.0	24.2	2132.5
40	666.7	24.9	2187.2

Tables and Memoranda

CONVERSION TABLES

km/h	m/min	mph	fpm
41	683.3	25.5	2241.9
42	700.0	26.1	2296.6
43	716.7	26.7	2351.3
44	733.3	27.3	2405.9
45	750.0	28.0	2460.6
46	766.7	28.6	2515.3
47	783.3	29.2	2570.0
48	800.0	29.8	2624.7
49	816.7	30.4	2679.4
50	833.3	31.1	2734.0

GEOMETRY

Two dimensional figures

Figure	Diagram of figure	Surface area	Perimeter
Square		a^2	$4a$
Rectangle		ab	$2(a+b)$
Triangle		$\frac{1}{2}ch$	$a+b+c$
Circle		πr^2 $\frac{1}{4}\pi d^2$ where $2r = d$	$2\pi r$ πd
Parallelogram		ah	$2(a+b)$
Trapezium		$\frac{1}{2}h(a+b)$	$a+b+c+d$
Ellipse		Approximately πab	$\pi(a+b)$
Hexagon		$2.6 \times a^2$	

GEOMETRY

Figure	Diagram of figure	Surface area	Perimeter
Octagon		$4.83 \times a^2$	$6a$
Sector of a circle		$\frac{1}{2}rb$ or $\frac{q}{360}\pi r^2$ note b = angle $\frac{q}{360} \times \pi 2r$	
Segment of a circle		$S - T$ where S = area of sector, T = area of triangle	
Bellmouth		$\frac{3}{14} \times r^2$	

GEOMETRY

Three dimensional figures

Figure	Diagram of figure	Surface area	Volume
Cube		$6a^2$	a^3
Cuboid/ rectangular block		$2(ab + ac + bc)$	abc
Prism/ triangular block		$bd + hc + dc + ad$	$\frac{1}{2}hcd$
Cylinder		$2\pi r^2 + 2\pi h$	$\pi r^2 h$
Sphere		$4\pi r^2$	$\frac{4}{3}\pi r^3$
Segment of sphere		$2\pi Rh$	$\frac{1}{6}\pi h(3r^2 + h^2)$ $\frac{1}{3}\pi h^2(3R - H)$
Pyramid		$(a + b)l + ab$	$\frac{1}{3}abh$

GEOMETRY

Figure	Diagram of figure	Surface area	Volume
Frustum of a pyramid		$l(a + b + c + d) + \sqrt{(ab + cd)}$ [rectangular figure only]	$\frac{h}{3}(ab + cd + \sqrt{abcd})$
Cone		πrl (excluding base) $\pi rl + \pi r^2$ (including base)	$\frac{1}{3}\pi r^2 h$ $\frac{1}{12}\pi d^2 h$
Frustum of a cone		$\pi r^2 + \pi R^2 + \pi l(R + r)$	$\frac{1}{3}\pi(R^2 + Rr + r^2)$

FORMULAE

Formulae

Formula	Description
Pythagoras Theorem	$A^2 = B^2 + C^2$ where A is the hypotenuse of a right-angled triangle and B and C are the two adjacent sides

Simpsons Rule	The Area is divided into an even number of strips of equal width, and therefore has an odd number of ordinates at the division points
	
	$$\text{area} = \frac{S(A + 2B + 4C)}{3}$$
	
	where S = common interval (strip width)
	              A = sum of first and last ordinates
	              B = sum of remaining odd ordinates
	              C = sum of the even ordinates
	
	The Volume can be calculated by the same formula, but by substituting the area of each coordinate rather than its length

Trapezoidal Rule	A given trench is divided into two equal sections, giving three ordinates, the first, the middle and the last
	
	$$\text{volume} = \frac{S \times (A + B + 2C)}{2}$$
	
	where S = width of the strips
	              A = area of the first section
	              B = area of the last section
	              C = area of the rest of the sections

Prismoidal Rule	A given trench is divided into two equal sections, giving three ordinates, the first, the middle and the last
	
	$$\text{volume} = \frac{L \times (A + 4B + C)}{6}$$
	
	where L = total length of trench
	              A = area of the first section
	              B = area of the middle section
	              C = area of the last section

TYPICAL THERMAL CONDUCTIVITY OF BUILDING MATERIALS

(Always check manufacturer's details – variation will occur depending on product and nature of materials)

	Thermal conductivity (W/mK)		Thermal conductivity (W/mK)
Acoustic plasterboard	0.25	Oriented strand board	0.13
Aerated concrete slab (500 kg/m^3)	0.16	Outer leaf brick	0.77
Aluminium	237	Plasterboard	0.22
Asphalt (1700 kg/m^3)	0.5	Plaster dense (1300 kg/m^3)	0.5
Bitumen-impregnated fibreboard	0.05	Plaster lightweight (600 kg/m^3)	0.16
Blocks (standard grade 600 kg/m^3)	0.15	Plywood (950 kg/m^3)	0.16
Blocks (solar grade 460 kg/m^3)	0.11	Prefabricated timber wall panels (check manufacturer)	0.12
Brickwork (outer leaf 1700 kg/m^3)	0.84	Screed (1200 kg/m^3)	0.41
Brickwork (inner leaf 1700 kg/m^3)	0.62	Stone chippings (1800 kg/m^3)	0.96
Dense aggregate concrete block 1800 kg/m^3 (exposed)	1.21	Tile hanging (1900 kg/m^3)	0.84
Dense aggregate concrete block 1800 kg/m^3 (protected)	1.13	Timber (650 kg/m^3)	0.14
Calcium silicate board (600 kg/m^3)	0.17	Timber flooring (650 kg/m^3)	0.14
Concrete general	1.28	Timber rafters	0.13
Concrete (heavyweight 2300 kg/m^3)	1.63	Timber roof or floor joists	0.13
Concrete (dense 2100 kg/m^3 typical floor)	1.4	Roof tile (1900 kg/m^3)	0.84
Concrete (dense 2000 kg/m^3 typical floor)	1.13	Timber blocks (650 kg/m^3)	0.14
Concrete (medium 1400 kg/m^3)	0.51	Cellular glass	0.045
Concrete (lightweight 1200 kg/m^3)	0.38	Expanded polystyrene	0.034
Concrete (lightweight 600 kg/m^3)	0.19	Expanded polystyrene slab (25 kg/m^3)	0.035
Concrete slab (aerated 500 kg/m^3)	0.16	Extruded polystyrene	0.035
Copper	390	Glass mineral wool	0.04
External render sand/cement finish	1	Mineral quilt (12 kg/m^3)	0.04
External render (1300 kg/m^3)	0.5	Mineral wool slab (25 kg/m^3)	0.035
Felt – Bitumen layers (1700 kg/m^3)	0.5	Phenolic foam	0.022
Fibreboard (300 kg/m^3)	0.06	Polyisocyanurate	0.025
Glass	0.93	Polyurethane	0.025
Marble	3	Rigid polyurethane	0.025
Metal tray used in wriggly tin concrete floors (7800 kg/m^3)	50	Rock mineral wool	0.038
Mortar (1750 kg/m^3)	0.8		

EARTHWORK

Weights of Typical Materials Handled by Excavators

The weight of the material is that of the state in its natural bed and includes moisture.
Adjustments should be made to allow for loose or compacted states

Material	Mass (kg/m^3)	Mass (lb/cu yd)
Ashes, dry	610	1028
Ashes, wet	810	1365
Basalt, broken	1954	3293
Basalt, solid	2933	4943
Bauxite, crushed	1281	2159
Borax, fine	849	1431
Caliche	1440	2427
Cement, clinker	1415	2385
Chalk, fine	1221	2058
Chalk, solid	2406	4055
Cinders, coal, ash	641	1080
Cinders, furnace	913	1538
Clay, compacted	1746	2942
Clay, dry	1073	1808
Clay, wet	1602	2700
Coal, anthracite, solid	1506	2538
Coal, bituminous	1351	2277
Coke	610	1028
Dolomite, lumpy	1522	2565
Dolomite, solid	2886	4864
Earth, dense	2002	3374
Earth, dry, loam	1249	2105
Earth, Fullers, raw	673	1134
Earth, moist	1442	2430
Earth, wet	1602	2700
Felsite	2495	4205
Fieldspar, solid	2613	4404
Fluorite	3093	5213
Gabbro	3093	5213
Gneiss	2696	4544
Granite	2690	4534
Gravel, dry ¼ to 2 inch	1682	2835
Gravel, dry, loose	1522	2565
Gravel, wet ¼ to 2 inch	2002	3374
Gypsum, broken	1450	2444
Gypsum, solid	2787	4697
Hardcore (consolidated)	1928	3249
Lignite, dry	801	1350
Limestone, broken	1554	2619
Limestone, solid	2596	4375
Magnesite, magnesium ore	2993	5044
Marble	2679	4515
Marl, wet	2216	3735
Mica, broken	1602	2700
Mica, solid	2883	4859
Peat, dry	400	674
Peat, moist	700	1179

EARTHWORK

Material	Mass (kg/m³)	Mass (lb/cu yd)
Peat, wet	1121	1889
Potash	1281	2159
Pumice, stone	640	1078
Quarry waste	1438	2423
Quartz sand	1201	2024
Quartz, solid	2584	4355
Rhyolite	2400	4045
Sand and gravel, dry	1650	2781
Sand and gravel, wet	2020	3404
Sand, dry	1602	2700
Sand, wet	1831	3086
Sandstone, solid	2412	4065
Shale, solid	2637	4444
Slag, broken	2114	3563
Slag, furnace granulated	961	1619
Slate, broken	1370	2309
Slate, solid	2667	4495
Snow, compacted	481	810
Snow, freshly fallen	160	269
Taconite	2803	4724
Trachyte	2400	4045
Trap rock, solid	2791	4704
Turf	400	674
Water	1000	1685

Transport Capacities

Type of vehicle	Capacity of vehicle	
	Payload	Heaped capacity
Wheelbarrow	150	0.10
1 tonne dumper	1250	1.00
2.5 tonne dumper	4000	2.50
Articulated dump truck (Volvo A20 6 × 4)	18500	11.00
Articulated dump truck (Volvo A35 6 × 6)	32000	19.00
Large capacity rear dumper (Euclid R35)	35000	22.00
Large capacity rear dumper (Euclid R85)	85000	50.00

EARTHWORK

Machine Volumes for Excavating and Filling

Machine type	Cycles per minute	Volume per minute (m³)
1.5 tonne excavator	1	0.04
	2	0.08
	3	0.12
3 tonne excavator	1	0.13
	2	0.26
	3	0.39
5 tonne excavator	1	0.28
	2	0.56
	3	0.84
7 tonne excavator	1	0.28
	2	0.56
	3	0.84
21 tonne excavator	1	1.21
	2	2.42
	3	3.63
Backhoe loader JCB3CX excavator Rear bucket capacity 0.28 m³	1	0.28
	2	0.56
	3	0.84
Backhoe loader JCB3CX loading Front bucket capacity 1.00 m³	1	1.00
	2	2.00

Machine Volumes for Excavating and Filling

Machine type	Loads per hour	Volume per hour (m³)
1 tonne high tip skip loader Volume 0.485 m³	5	2.43
	7	3.40
	10	4.85
3 tonne dumper Max volume 2.40 m³ Available volume 1.9 m³	4	7.60
	5	9.50
	7	13.30
	10	19.00
6 tonne dumper Max volume 3.40 m³ Available volume 3.77 m³	4	15.08
	5	18.85
	7	26.39
	10	37.70

EARTHWORK

Bulkage of Soils (after excavation)

Type of soil	Approximate bulking of 1 m³ after excavation
Vegetable soil and loam	25–30%
Soft clay	30–40%
Stiff clay	10–20%
Gravel	20–25%
Sand	40–50%
Chalk	40–50%
Rock, weathered	30–40%
Rock, unweathered	50–60%

Shrinkage of Materials (on being deposited)

Type of soil	Approximate bulking of 1 m³ after excavation
Clay	10%
Gravel	8%
Gravel and sand	9%
Loam and light sandy soils	12%
Loose vegetable soils	15%

Voids in Material Used as Subbases or Beddings

Material	m³ of voids/m³
Alluvium	0.37
River grit	0.29
Quarry sand	0.24
Shingle	0.37
Gravel	0.39
Broken stone	0.45
Broken bricks	0.42

Angles of Repose

Type of soil		Degrees
Clay	– dry	30
	– damp, well drained	45
	– wet	15–20
Earth	– dry	30
	– damp	45
Gravel	– moist	48
Sand	– dry or moist	35
	– wet	25
Loam		40

EARTHWORK

Slopes and Angles

Ratio of base to height	Angle in degrees
5:1	11
4:1	14
3:1	18
2:1	27
1½:1	34
1:1	45
1:1½	56
1:2	63
1:3	72
1:4	76
1:5	79

Grades (in Degrees and Percents)

Degrees	Percent	Degrees	Percent
1	1.8	24	44.5
2	3.5	25	46.6
3	5.2	26	48.8
4	7.0	27	51.0
5	8.8	28	53.2
6	10.5	29	55.4
7	12.3	30	57.7
8	14.0	31	60.0
9	15.8	32	62.5
10	17.6	33	64.9
11	19.4	34	67.4
12	21.3	35	70.0
13	23.1	36	72.7
14	24.9	37	75.4
15	26.8	38	78.1
16	28.7	39	81.0
17	30.6	40	83.9
18	32.5	41	86.9
19	34.4	42	90.0
20	36.4	43	93.3
21	38.4	44	96.6
22	40.4	45	100.0
23	42.4		

EARTHWORK

Bearing Powers

Ground conditions		Bearing power		
		kg/m^2	lb/in^2	Metric t/m^2
Rock,	broken	483	70	50
	solid	2415	350	240
Clay,	dry or hard	380	55	40
	medium dry	190	27	20
	soft or wet	100	14	10
Gravel,	cemented	760	110	80
Sand,	compacted	380	55	40
	clean dry	190	27	20
Swamp and alluvial soils		48	7	5

Earthwork Support

Maximum depth of excavation in various soils without the use of earthwork support

Ground conditions	Feet (ft)	Metres (m)
Compact soil	12	3.66
Drained loam	6	1.83
Dry sand	1	0.3
Gravelly earth	2	0.61
Ordinary earth	3	0.91
Stiff clay	10	3.05

It is important to note that the above table should only be used as a guide. Each case must be taken on its merits and, as the limited distances given above are approached, careful watch must be kept for the slightest signs of caving in

CONCRETE WORK

Weights of Concrete and Concrete Elements

Type of material		kg/m³	lb/cu ft
Ordinary concrete (dense aggregates)			
Non-reinforced plain or mass concrete			
Nominal weight		2305	144
Aggregate	– limestone	2162 to 2407	135 to 150
	– gravel	2244 to 2407	140 to 150
	– broken brick	2000 (av)	125 (av)
	– other crushed stone	2326 to 2489	145 to 155
Reinforced concrete			
Nominal weight		2407	150
Reinforcement	– 1%	2305 to 2468	144 to 154
	– 2%	2356 to 2519	147 to 157
	– 4%	2448 to 2703	153 to 163
Special concretes			
Heavy concrete			
Aggregates	– barytes, magnetite	3210 (min)	200 (min)
	– steel shot, punchings	5280	330
Lean mixes			
Dry-lean (gravel aggregate)		2244	140
Soil-cement (normal mix)		1601	100

CONCRETE WORK

Type of material		kg/m² per mm thick	lb/sq ft per inch thick
Ordinary concrete (dense aggregates)			
Solid slabs (floors, walls etc.)			
Thickness:	75 mm or 3 in	184	37.5
	100 mm or 4 in	245	50
	150 mm or 6 in	378	75
	250 mm or 10 in	612	125
	300 mm or 12 in	734	150
Ribbed slabs			
Thickness:	125 mm or 5 in	204	42
	150 mm or 6 in	219	45
	225 mm or 9 in	281	57
	300 mm or 12 in	342	70
Special concretes			
Finishes etc.			
	Rendering, screed etc.		
	Granolithic, terrazzo	1928 to 2401	10 to 12.5
	Glass-block (hollow) concrete	1734 (approx)	9 (approx)
Prestressed concrete		Weights as for reinforced concrete (upper limits)	
Air-entrained concrete		Weights as for plain or reinforced concrete	

CONCRETE WORK

Average Weight of Aggregates

Materials	Voids %	Weight kg/m³
Sand	39	1660
Gravel 10–20 mm	45	1440
Gravel 35–75 mm	42	1555
Crushed stone	50	1330
Crushed granite (over 15 mm)	50	1345
(n.e. 15 mm)	47	1440
'All-in' ballast	32	1800–2000

Material	kg/m³	lb/cu yd
Vermiculite (aggregate)	64–80	108–135
All-in aggregate	1999	125

Applications and Mix Design

Site mixed concrete

Recommended mix	Class of work suitable for	Cement (kg)	Sand (kg)	Coarse aggregate (kg)	Nr 25 kg bags cement per m³ of combined aggregate
1:3:6	Roughest type of mass concrete such as footings, road haunching over 300 mm thick	208	905	1509	8.30
1:2.5:5	Mass concrete of better class than 1:3:6 such as bases for machinery, walls below ground etc.	249	881	1474	10.00
1:2:4	Most ordinary uses of concrete, such as mass walls above ground, road slabs etc. and general reinforced concrete work	304	889	1431	12.20
1:1.5:3	Watertight floors, pavements and walls, tanks, pits, steps, paths, surface of 2 course roads, reinforced concrete where extra strength is required	371	801	1336	14.90
1:1:2	Works of thin section such as fence posts and small precast work	511	720	1206	20.40

CONCRETE WORK

Ready mixed concrete

Application	Designated concrete	Standardized prescribed concrete	Recommended consistence (nominal slump class)
Foundations			
Mass concrete fill or blinding	GEN 1	ST2	S3
Strip footings	GEN 1	ST2	S3
Mass concrete foundations			
Single storey buildings	GEN 1	ST2	S3
Double storey buildings	GEN 3	ST4	S3
Trench fill foundations			
Single storey buildings	GEN 1	ST2	S4
Double storey buildings	GEN 3	ST4	S4
General applications			
Kerb bedding and haunching	GEN 0	ST1	S1
Drainage works – immediate support	GEN 1	ST2	S1
Other drainage works	GEN 1	ST2	S3
Oversite below suspended slabs	GEN 1	ST2	S3
Floors			
Garage and house floors with no embedded steel	GEN 3	ST4	S2
Wearing surface: Light foot and trolley traffic	RC30	ST4	S2
Wearing surface: General industrial	RC40	N/A	S2
Wearing surface: Heavy industrial	RC50	N/A	S2
Paving			
House drives, domestic parking and external parking	PAV 1	N/A	S2
Heavy-duty external paving	PAV 2	N/A	S2

CONCRETE WORK

Prescribed Mixes for Ordinary Structural Concrete

Weights of cement and total dry aggregates in kg to produce approximately one cubic metre of fully compacted concrete together with the percentages by weight of fine aggregate in total dry aggregates

Conc. grade	Nominal max size of aggregate (mm)	40		20		14		10	
	Workability	**Med.**	**High**	**Med.**	**High**	**Med.**	**High**	**Med.**	**High**
	Limits to slump that may be expected (mm)	**50–100**	**100–150**	**25–75**	**75–125**	**10–50**	**50–100**	**10–25**	**25–50**
7	Cement (kg)	180	200	210	230	–	–	–	–
	Total aggregate (kg)	1950	1850	1900	1800	–	–	–	–
	Fine aggregate (%)	30–45	30–45	35–50	35–50	–	–	–	–
10	Cement (kg)	210	230	240	260	–	–	–	–
	Total aggregate (kg)	1900	1850	1850	1800	–	–	–	–
	Fine aggregate (%)	30–45	30–45	35–50	35–50	–	–	–	–
15	Cement (kg)	250	270	280	310	–	–	–	–
	Total aggregate (kg)	1850	1800	1800	1750	–	–	–	–
	Fine aggregate (%)	30–45	30–45	35–50	35–50	–	–	–	–
20	Cement (kg)	300	320	320	350	340	380	360	410
	Total aggregate (kg)	1850	1750	1800	1750	1750	1700	1750	1650
	Sand								
	Zone 1 (%)	35	40	40	45	45	50	50	55
	Zone 2 (%)	30	35	35	40	40	45	45	50
	Zone 3 (%)	30	30	30	35	35	40	40	45
25	Cement (kg)	340	360	360	390	380	420	400	450
	Total aggregate (kg)	1800	1750	1750	1700	1700	1650	1700	1600
	Sand								
	Zone 1 (%)	35	40	40	45	45	50	50	55
	Zone 2 (%)	30	35	35	40	40	45	45	50
	Zone 3 (%)	30	30	30	35	35	40	40	45
30	Cement (kg)	370	390	400	430	430	470	460	510
	Total aggregate (kg)	1750	1700	1700	1650	1700	1600	1650	1550
	Sand								
	Zone 1 (%)	35	40	40	45	45	50	50	55
	Zone 2 (%)	30	35	35	40	40	45	45	50
	Zone 3 (%)	30	30	30	35	35	40	40	45

REINFORCEMENT

Weights of Bar Reinforcement

Nominal sizes (mm)	Cross-sectional area (mm²)	Mass (kg/m)	Length of bar (m/tonne)
6	28.27	0.222	4505
8	50.27	0.395	2534
10	78.54	0.617	1622
12	113.10	0.888	1126
16	201.06	1.578	634
20	314.16	2.466	405
25	490.87	3.853	260
32	804.25	6.313	158
40	1265.64	9.865	101
50	1963.50	15.413	65

Weights of Bars (at specific spacings)

Weights of metric bars in kilogrammes per square metre

Size (mm)	Spacing of bars in millimetres									
	75	100	125	150	175	200	225	250	275	300
6	2.96	2.220	1.776	1.480	1.27	1.110	0.99	0.89	0.81	0.74
8	5.26	3.95	3.16	2.63	2.26	1.97	1.75	1.58	1.44	1.32
10	8.22	6.17	4.93	4.11	3.52	3.08	2.74	2.47	2.24	2.06
12	11.84	8.88	7.10	5.92	5.07	4.44	3.95	3.55	3.23	2.96
16	21.04	15.78	12.63	10.52	9.02	7.89	7.02	6.31	5.74	5.26
20	32.88	24.66	19.73	16.44	14.09	12.33	10.96	9.87	8.97	8.22
25	51.38	38.53	30.83	25.69	22.02	19.27	17.13	15.41	14.01	12.84
32	84.18	63.13	50.51	42.09	36.08	31.57	28.06	25.25	22.96	21.04
40	131.53	98.65	78.92	65.76	56.37	49.32	43.84	39.46	35.87	32.88
50	205.51	154.13	123.31	102.76	88.08	77.07	68.50	61.65	56.05	51.38

Basic weight of steelwork taken as 7850 kg/m³
Basic weight of bar reinforcement per metre run = 0.00785 kg/mm²
The value of π has been taken as 3.141592654

Fabric Reinforcement

Preferred range of designated fabric types and stock sheet sizes

Fabric reference	Longitudinal wires			Cross wires			
	Nominal wire size (mm)	Pitch (mm)	Area (mm²/m)	Nominal wire size (mm)	Pitch (mm)	Area (mm²/m)	Mass (kg/m²)
Square mesh							
A393	10	200	393	10	200	393	6.16
A252	8	200	252	8	200	252	3.95
A193	7	200	193	7	200	193	3.02
A142	6	200	142	6	200	142	2.22
A98	5	200	98	5	200	98	1.54
Structural mesh							
B1131	12	100	1131	8	200	252	10.90
B785	10	100	785	8	200	252	8.14
B503	8	100	503	8	200	252	5.93
B385	7	100	385	7	200	193	4.53
B283	6	100	283	7	200	193	3.73
B196	5	100	196	7	200	193	3.05
Long mesh							
C785	10	100	785	6	400	70.8	6.72
C636	9	100	636	6	400	70.8	5.55
C503	8	100	503	5	400	49.0	4.34
C385	7	100	385	5	400	49.0	3.41
C283	6	100	283	5	400	49.0	2.61
Wrapping mesh							
D98	5	200	98	5	200	98	1.54
D49	2.5	100	49	2.5	100	49	0.77

Stock sheet size 4.8 m × 2.4 m, Area 11.52 m²

Average weight kg/m³ of steelwork reinforcement in concrete for various building elements

Substructure	kg/m³ concrete	Substructure	kg/m³ concrete
Pile caps	110–150	Plate slab	150–220
Tie beams	130–170	Cant slab	145–210
Ground beams	230–330	Ribbed floors	130–200
Bases	125–180	Topping to block floor	30–40
Footings	100–150	Columns	210–310
Retaining walls	150–210	Beams	250–350
Raft	60–70	Stairs	130–170
Slabs – one way	120–200	Walls – normal	40–100
Slabs – two way	110–220	Walls – wind	70–125

Note: For exposed elements add the following %:
Walls 50%, Beams 100%, Columns 15%

FORMWORK

Formwork Stripping Times – Normal Curing Periods

Conditions under which concrete is maturing	Minimum periods of protection for different types of cement					
	Number of days (where the average surface temperature of the concrete exceeds 10°C during the whole period)			Equivalent maturity (degree hours) calculated as the age of the concrete in hours multiplied by the number of degrees Celsius by which the average surface temperature of the concrete exceeds 10°C		
	Other	SRPC	OPC or RHPC	Other	SRPC	OPC or RHPC
1. Hot weather or drying winds	7	4	3	3500	2000	1500
2. Conditions not covered by 1	4	3	2	2000	1500	1000

KEY
OPC – Ordinary Portland Cement
RHPC – Rapid-hardening Portland Cement
SRPC – Sulphate-resisting Portland Cement

Minimum Period before Striking Formwork

	Minimum period before striking		
	Surface temperature of concrete		
	16°C	17°C	t°C (0–25)
Vertical formwork to columns, walls and large beams	12 hours	18 hours	300 hours t+10
Soffit formwork to slabs	4 days	6 days	100 days t+10
Props to slabs	10 days	15 days	250 days t+10
Soffit formwork to beams	9 days	14 days	230 days t+10
Props to beams	14 days	21 days	360 days t+10

MASONRY

MASONRY

Number of Bricks Required for Various Types of Work per m² of Walling

Description	Brick size	
	215 × 102.5 × 50 mm	215 × 102.5 × 65 mm
Half brick thick		
Stretcher bond	74	59
English bond	108	86
English garden wall bond	90	72
Flemish bond	96	79
Flemish garden wall bond	83	66
One brick thick and cavity wall of two half brick skins		
Stretcher bond	148	119

Quantities of Bricks and Mortar Required per m² of Walling

	Unit	No of bricks required	Mortar required (cubic metres)		
Standard bricks			No frogs	Single frogs	Double frogs
Brick size 215 × 102.5 × 50 mm					
half brick wall (103 mm)	m²	72	0.022	0.027	0.032
2 × half brick cavity wall (270 mm)	m²	144	0.044	0.054	0.064
one brick wall (215 mm)	m²	144	0.052	0.064	0.076
one and a half brick wall (322 mm)	m²	216	0.073	0.091	0.108
Mass brickwork	m³	576	0.347	0.413	0.480
Brick size 215 × 102.5 × 65 mm					
half brick wall (103 mm)	m²	58	0.019	0.022	0.026
2 × half brick cavity wall (270 mm)	m²	116	0.038	0.045	0.055
one brick wall (215 mm)	m²	116	0.046	0.055	0.064
one and a half brick wall (322 mm)	m²	174	0.063	0.074	0.088
Mass brickwork	m³	464	0.307	0.360	0.413
Metric modular bricks			Perforated		
Brick size 200 × 100 × 75 mm					
90 mm thick	m²	67	0.016	0.019	
190 mm thick	m²	133	0.042	0.048	
290 mm thick	m²	200	0.068	0.078	
Brick size 200 × 100 × 100 mm					
90 mm thick	m²	50	0.013	0.016	
190 mm thick	m²	100	0.036	0.041	
290 mm thick	m²	150	0.059	0.067	
Brick size 300 × 100 × 75 mm					
90 mm thick	m²	33	–	0.015	
Brick size 300 × 100 × 100 mm					
90 mm thick	m²	44	0.015	0.018	

Note: Assuming 10 mm thick joints

MASONRY

Mortar Required per m² Blockwork (9.88 blocks/m²)

Wall thickness	75	90	100	125	140	190	215
Mortar m³/m²	0.005	0.006	0.007	0.008	0.009	0.013	0.014

Mortar Group	Cement: lime: sand	Masonry cement: sand	Cement: sand with plasticizer
1	1:0–0.25:3		
2	1:0.5:4–4.5	1:2.5-3.5	1:3–4
3	1:1:5–6	1:4–5	1:5–6
4	1:2:8–9	1:5.5–6.5	1:7–8
5	1:3:10–12	1:6.5–7	1:8

Group 1: strong inflexible mortar
Group 5: weak but flexible

All mixes within a group are of approximately similar strength
Frost resistance increases with the use of plasticizers
Cement: lime: sand mixes give the strongest bond and greatest resistance to rain penetration
Masonry cement equals ordinary Portland cement plus a fine neutral mineral filler and an air entraining agent

Calcium Silicate Bricks

Type	Strength	Location
Class 2 crushing strength	14.0 N/mm²	not suitable for walls
Class 3	20.5 N/mm²	walls above dpc
Class 4	27.5 N/mm²	cappings and copings
Class 5	34.5 N/mm²	retaining walls
Class 6	41.5 N/mm²	walls below ground
Class 7	48.5 N/mm²	walls below ground

The Class 7 calcium silicate bricks are therefore equal in strength to Class B bricks
Calcium silicate bricks are not suitable for DPCs

Durability of Bricks	
FL	Frost resistant with low salt content
FN	Frost resistant with normal salt content
ML	Moderately frost resistant with low salt content
MN	Moderately frost resistant with normal salt content

MASONRY

Brickwork Dimensions

No. of horizontal bricks	Dimensions (mm)	No. of vertical courses	Height of vertical courses (mm)
½	112.5	1	75
1	225.0	2	150
1½	337.5	3	225
2	450.0	4	300
2½	562.5	5	375
3	675.0	6	450
3½	787.5	7	525
4	900.0	8	600
4½	1012.5	9	675
5	1125.0	10	750
5½	1237.5	11	825
6	1350.0	12	900
6½	1462.5	13	975
7	1575.0	14	1050
7½	1687.5	15	1125
8	1800.0	16	1200
8½	1912.5	17	1275
9	2025.0	18	1350
9½	2137.5	19	1425
10	2250.0	20	1500
20	4500.0	24	1575
40	9000.0	28	2100
50	11250.0	32	2400
60	13500.0	36	2700
75	16875.0	40	3000

TIMBER

Weights of Timber

Material	kg/m³	lb/cu ft
General	806 (avg)	50 (avg)
Douglas fir	479	30
Yellow pine, spruce	479	30
Pitch pine	673	42
Larch, elm	561	35
Oak (English)	724 to 959	45 to 60
Teak	643 to 877	40 to 55
Jarrah	959	60
Greenheart	1040 to 1204	65 to 75
Quebracho	1285	80
Material	**kg/m² per mm thickness**	**lb/sq ft per inch thickness**
Wooden boarding and blocks		
Softwood	0.48	2.5
Hardwood	0.76	4
Hardboard	1.06	5.5
Chipboard	0.76	4
Plywood	0.62	3.25
Blockboard	0.48	2.5
Fibreboard	0.29	1.5
Wood-wool	0.58	3
Plasterboard	0.96	5
Weather boarding	0.35	1.8

TIMBER

Conversion Tables (for timber only)

Inches	Millimetres	Feet	Metres
1	25	1	0.300
2	50	2	0.600
3	75	3	0.900
4	100	4	1.200
5	125	5	1.500
6	150	6	1.800
7	175	7	2.100
8	200	8	2.400
9	225	9	2.700
10	250	10	3.000
11	275	11	3.300
12	300	12	3.600
13	325	13	3.900
14	350	14	4.200
15	375	15	4.500
16	400	16	4.800
17	425	17	5.100
18	450	18	5.400
19	475	19	5.700
20	500	20	6.000
21	525	21	6.300
22	550	22	6.600
23	575	23	6.900
24	600	24	7.200

Planed Softwood
The finished end section size of planed timber is usually 3/16" less than the original size from which it is produced. This however varies slightly depending upon availability of material and origin of the species used.

Standards (timber) to cubic metres and cubic metres to standards (timber)

Cubic metres	Cubic metres standards	Standards
4.672	1	0.214
9.344	2	0.428
14.017	3	0.642
18.689	4	0.856
23.361	5	1.070
28.033	6	1.284
32.706	7	1.498
37.378	8	1.712
42.050	9	1.926
46.722	10	2.140
93.445	20	4.281
140.167	30	6.421
186.890	40	8.561
233.612	50	10.702
280.335	60	12.842
327.057	70	14.982
373.779	80	17.122

TIMBER

1 cu metre = 35.3148 cu ft = 0.21403 std

1 cu ft = 0.028317 cu metres

1 std = 4.67227 cu metres

Basic sizes of sawn softwood available (cross-sectional areas)

Thickness (mm)	Width (mm)								
	75	100	125	150	175	200	225	250	300
16	X	X	X	X					
19	X	X	X	X					
22	X	X	X	X					
25	X	X	X	X	X	X	X	X	X
32	X	X	X	X	X	X	X	X	X
36	X	X	X	X					
38	X	X	X	X	X	X	X		
44	X	X	X	X	X	X	X	X	X
47*	X	X	X	X	X	X	X	X	X
50	X	X	X	X	X	X	X	X	X
63	X	X	X	X	X	X	X		
75	X	X	X	X	X	X	X	X	
100		X		X		X		X	X
150				X		X			X
200						X			
250								X	
300									X

* This range of widths for 47 mm thickness will usually be found to be available in construction quality only

Note: The smaller sizes below 100 mm thick and 250 mm width are normally but not exclusively of European origin. Sizes beyond this are usually of North and South American origin

Basic lengths of sawn softwood available (metres)

1.80	2.10	3.00	4.20	5.10	6.00	7.20
	2.40	3.30	4.50	5.40	6.30	
	2.70	3.60	4.80	5.70	6.60	
		3.90			6.90	

Note: Lengths of 6.00 m and over will generally only be available from North American species and may have to be recut from larger sizes

TIMBER

Reductions from basic size to finished size by planning of two opposed faces

Purpose	Reductions from basic sizes for timber			
	15–35 mm	36–100 mm	101–150 mm	over 150 mm
a) Constructional timber	3 mm	3 mm	5 mm	6 mm
b) Matching interlocking boards	4 mm	4 mm	6 mm	6 mm
c) Wood trim not specified in BS 584	5 mm	7 mm	7 mm	9 mm
d) Joinery and cabinet work	7 mm	9 mm	11 mm	13 mm

Note: The reduction of width or depth is overall the extreme size and is exclusive of any reduction of the face by the machining of a tongue or lap joints

Maximum Spans for Various Roof Trusses

Maximum permissible spans for rafters for Fink trussed rafters

Basic size	Actual size	Pitch (degrees)								
(mm)	(mm)	15 (m)	17.5 (m)	20 (m)	22.5 (m)	25 (m)	27.5 (m)	30 (m)	32.5 (m)	35 (m)
38 × 75	35 × 72	6.03	6.16	6.29	6.41	6.51	6.60	6.70	6.80	6.90
38 × 100	35 × 97	7.48	7.67	7.83	7.97	8.10	8.22	8.34	8.47	8.61
38 × 125	35 × 120	8.80	9.00	9.20	9.37	9.54	9.68	9.82	9.98	10.16
44 × 75	41 × 72	6.45	6.59	6.71	6.83	6.93	7.03	7.14	7.24	7.35
44 × 100	41 × 97	8.05	8.23	8.40	8.55	8.68	8.81	8.93	9.09	9.22
44 × 125	41 × 120	9.38	9.60	9.81	9.99	10.15	10.31	10.45	10.64	10.81
50 × 75	47 × 72	6.87	7.01	7.13	7.25	7.35	7.45	7.53	7.67	7.78
50 × 100	47 × 97	8.62	8.80	8.97	9.12	9.25	9.38	9.50	9.66	9.80
50 × 125	47 × 120	10.01	10.24	10.44	10.62	10.77	10.94	11.00	11.00	11.00

TIMBER

Sizes of Internal and External Doorsets

Description	Internal size (mm)	Permissible deviation	External size (mm)	Permissible deviation
Coordinating dimension: height of door leaf height sets	2100		2100	
Coordinating dimension: height of ceiling height set	2300 2350 2400 2700 3000		2300 2350 2400 2700 3000	
Coordinating dimension: width of all doorsets S = Single leaf set D = Double leaf set	600 S 700 S 800 S&D 900 S&D 1000 S&D 1200 D 1500 D 1800 D 2100 D		900 S 1000 S 1200 D 1800 D 2100 D	
Work size: height of door leaf height set	2090	± 2.0	2095	± 2.0
Work size: height of ceiling height set	2285 2335 2385 2685 2985	± 2.0	2295 2345 2395 2695 2995	± 2.0
Work size: width of all doorsets S = Single leaf set D = Double leaf set	590 S 690 S 790 S&D 890 S&D 990 S&D 1190 D 1490 D 1790 D 2090 D	± 2.0	895 S 995 S 1195 D 1495 D 1795 D 2095 D	± 2.0
Width of door leaf in single leaf sets F = Flush leaf P = Panel leaf	526 F 626 F 726 F&P 826 F&P 926 F&P	± 1.5	806 F&P 906 F&P	± 1.5
Width of door leaf in double leaf sets F = Flush leaf P = Panel leaf	362 F 412 F 426 F 562 F&P 712 F&P 826 F&P 1012 F&P	± 1.5	552 F&P 702 F&P 852 F&P 1002 F&P	± 1.5
Door leaf height for all doorsets	2040	± 1.5	1994	± 1.5

ROOFING

Total Roof Loadings for Various Types of Tiles/Slates

	Roof load (slope) kg/m^2		
	Slate/Tile	Roofing underlay and battens^2	Total dead load kg/m
Asbestos cement slate (600 × 300)	21.50	3.14	24.64
Clay tile — interlocking	67.00	5.50	72.50
plain	43.50	2.87	46.37
Concrete tile — interlocking	47.20	2.69	49.89
plain	78.20	5.50	83.70
Natural slate (18" × 10")	35.40	3.40	38.80
	Roof load (plan) kg/m^2		
Asbestos cement slate (600 × 300)	28.45	76.50	104.95
Clay tile — interlocking	53.54	76.50	130.04
plain	83.71	76.50	60.21
Concrete tile — interlocking	57.60	76.50	134.10
plain	96.64	76.50	173.14

ROOFING

Tiling Data

Product		Lap (mm)	Gauge of battens	No. slates per m²	Battens (m/m²)	Weight as laid (kg/m²)
CEMENT SLATES						
Eternit slates	600 × 300 mm	100	250	13.4	4.00	19.50
(Duracem)		90	255	13.1	3.92	19.20
		80	260	12.9	3.85	19.00
		70	265	12.7	3.77	18.60
	600 × 350 mm	100	250	11.5	4.00	19.50
		90	255	11.2	3.92	19.20
	500 × 250 mm	100	200	20.0	5.00	20.00
		90	205	19.5	4.88	19.50
		80	210	19.1	4.76	19.00
		70	215	18.6	4.65	18.60
	400 × 200 mm	90	155	32.3	6.45	20.80
		80	160	31.3	6.25	20.20
		70	165	30.3	6.06	19.60
CONCRETE TILES/SLATES						
Redland Roofing						
Stonewold slate	430 × 380 mm	75	355	8.2	2.82	51.20
Double Roman tile	418 × 330 mm	75	355	8.2	2.91	45.50
Grovebury pantile	418 × 332 mm	75	343	9.7	2.91	47.90
Norfolk pantile	381 × 227 mm	75	306	16.3	3.26	44.01
		100	281	17.8	3.56	48.06
Renown interlocking tile	418 × 330 mm	75	343	9.7	2.91	46.40
'49' tile	381 × 227 mm	75	306	16.3	3.26	44.80
		100	281	17.8	3.56	48.95
Plain, vertical tiling	265 × 165 mm	35	115	52.7	8.70	62.20
Marley Roofing						
Bold roll tile	420 × 330 mm	75	344	9.7	2.90	47.00
		100	–	10.5	3.20	51.00
Modern roof tile	420 × 330 mm	75	338	10.2	3.00	54.00
		100	–	11.0	3.20	58.00
Ludlow major	420 × 330 mm	75	338	10.2	3.00	45.00
		100	–	11.0	3.20	49.00
Ludlow plus	387 × 229 mm	75	305	16.1	3.30	47.00
		100	–	17.5	3.60	51.00
Mendip tile	420 × 330 mm	75	338	10.2	3.00	47.00
		100	–	11.0	3.20	51.00
Wessex	413 × 330 mm	75	338	10.2	3.00	54.00
		100	–	11.0	3.20	58.00
Plain tile	267 × 165 mm	65	100	60.0	10.00	76.00
		75	95	64.0	10.50	81.00
		85	90	68.0	11.30	86.00
Plain vertical tiles (feature)	267 × 165 mm	35	110	53.0	8.70	67.00
		34	115	56.0	9.10	71.00

ROOFING

Slate Nails, Quantity per Kilogram

Length	Type			
	Plain wire	**Galvanized wire**	**Copper nail**	**Zinc nail**
28.5 mm	325	305	325	415
34.4 mm	286	256	254	292
50.8 mm	242	224	194	200

Metal Sheet Coverings

Thicknesses and weights of sheet metal coverings									
Lead to BS 1178									
BS Code No	3	4	5	6	7	8			
Colour code	Green	Blue	Red	Black	White	Orange			
Thickness (mm)	1.25	1.80	2.24	2.50	3.15	3.55			
Density kg/m^2	14.18	20.41	25.40	30.05	35.72	40.26			
Copper to BS 2870									
Thickness (mm)		0.60	0.70						
Bay width									
Roll (mm)		500	650						
Seam (mm)		525	600						
Standard width to form bay	600	750							
Normal length of sheet	1.80	1.80							
Zinc to BS 849									
Zinc Gauge (Nr)	9	10	11	12	13	14	15	16	
Thickness (mm)	0.43	0.48	0.56	0.64	0.71	0.79	0.91	1.04	
Density (kg/m^2)	3.1	3.2	3.8	4.3	4.8	5.3	6.2	7.0	
Aluminium to BS 4868									
Thickness (mm)	0.5	0.6	0.7	0.8	0.9	1.0	1.2		
Density (kg/m^2)	12.8	15.4	17.9	20.5	23.0	25.6	30.7		

ROOFING

Type of felt	Nominal mass per unit area (kg/10 m)	Nominal mass per unit area of fibre base (g/m²)	Nominal length of roll (m)
Class 1			
1B fine granule	14	220	10 or 20
surfaced bitumen	18	330	10 or 20
	25	470	10
1E mineral surfaced bitumen	38	470	10
1F reinforced bitumen	15	160 (fibre)	15
		110 (hessian)	
1F reinforced bitumen, aluminium faced	13	160 (fibre)	15
		110 (hessian)	
Class 2			
2B fine granule surfaced bitumen asbestos	18	500	10 or 20
2E mineral surfaced bitumen asbestos	38	600	10
Class 3			
3B fine granule surfaced bitumen glass fibre	18	60	20
3E mineral surfaced bitumen glass fibre	28	60	10
3E venting base layer bitumen glass fibre	32	60*	10
3H venting base layer bitumen glass fibre	17	60*	20

* Excluding effect of perforations

GLAZING

Nominal thickness (mm)	Tolerance on thickness (mm)	Approximate weight (kg/m²)	Normal maximum size (mm)
Float and polished plate glass			
3	+ 0.2	7.50	2140 × 1220
4	+ 0.2	10.00	2760 × 1220
5	+ 0.2	12.50	3180 × 2100
6	+ 0.2	15.00	4600 × 3180
10	+ 0.3	25.00)	6000 × 3300
12	+ 0.3	30.00)	
15	+ 0.5	37.50	3050 × 3000
19	+ 1.0	47.50)	3000 × 2900
25	+ 1.0	63.50)	
Clear sheet glass			
2 *	+ 0.2	5.00	1920 × 1220
3	+ 0.3	7.50	2130 × 1320
4	+ 0.3	10.00	2760 × 1220
5 *	+ 0.3	12.50)	2130 × 2400
6 *	+ 0.3	15.00)	
Cast glass			
3	+ 0.4 / − 0.2	6.00)	2140 × 1280
4	+ 0.5	7.50)	
5	+ 0.5	9.50	2140 × 1320
6	+ 0.5	11.50)	3700 × 1280
10	+ 0.8	21.50)	
Wired glass (Cast wired glass)			
6	+ 0.3 / − 0.7	−))	3700 × 1840
7	+ 0.7	−)	
(Polished wire glass)			
6	+ 1.0	−	330 × 1830

* The 5 mm and 6 mm thickness are known as *thick drawn sheet*. Although 2 mm sheet glass is available it is not recommended for general glazing purposes

METAL

Weights of Metals

Material	kg/m³	lb/cu ft
Metals, steel construction, etc.		
Iron		
– cast	7207	450
– wrought	7687	480
– ore – general	2407	150
– (crushed) Swedish	3682	230
Steel	7854	490
Copper		
– cast	8731	545
– wrought	8945	558
Brass	8497	530
Bronze	8945	558
Aluminium	2774	173
Lead	11322	707
Zinc (rolled)	7140	446

	g/mm² per metre	lb/sq ft per foot
Steel bars	7.85	3.4

Structural steelwork	Net weight of member @ 7854 kg/m³
riveted	+ 10% for cleats, rivets, bolts, etc.
welded	+ 1.25% to 2.5% for welds, etc.
Rolled sections	
beams	+ 2.5%
stanchions	+ 5% (extra for caps and bases)
Plate	
web girders	+ 10% for rivets or welds, stiffeners, etc.

	kg/m	lb/ft
Steel stairs: industrial type		
1 m or 3 ft wide	84	56
Steel tubes		
50 mm or 2 in bore	5 to 6	3 to 4
Gas piping		
20 mm or ¾ in	2	1¼

METAL

Universal Beams BS 4: Part 1: 2005

Designation	Mass (kg/m)	Depth of section (mm)	Width of section (mm)	Thickness		Surface area (m²/m)
				Web (mm)	Flange (mm)	
1016 × 305 × 487	487.0	1036.1	308.5	30.0	54.1	3.20
1016 × 305 × 438	438.0	1025.9	305.4	26.9	49.0	3.17
1016 × 305 × 393	393.0	1016.0	303.0	24.4	43.9	3.15
1016 × 305 × 349	349.0	1008.1	302.0	21.1	40.0	3.13
1016 × 305 × 314	314.0	1000.0	300.0	19.1	35.9	3.11
1016 × 305 × 272	272.0	990.1	300.0	16.5	31.0	3.10
1016 × 305 × 249	249.0	980.2	300.0	16.5	26.0	3.08
1016 × 305 × 222	222.0	970.3	300.0	16.0	21.1	3.06
914 × 419 × 388	388.0	921.0	420.5	21.4	36.6	3.44
914 × 419 × 343	343.3	911.8	418.5	19.4	32.0	3.42
914 × 305 × 289	289.1	926.6	307.7	19.5	32.0	3.01
914 × 305 × 253	253.4	918.4	305.5	17.3	27.9	2.99
914 × 305 × 224	224.2	910.4	304.1	15.9	23.9	2.97
914 × 305 × 201	200.9	903.0	303.3	15.1	20.2	2.96
838 × 292 × 226	226.5	850.9	293.8	16.1	26.8	2.81
838 × 292 × 194	193.8	840.7	292.4	14.7	21.7	2.79
838 × 292 × 176	175.9	834.9	291.7	14.0	18.8	2.78
762 × 267 × 197	196.8	769.8	268.0	15.6	25.4	2.55
762 × 267 × 173	173.0	762.2	266.7	14.3	21.6	2.53
762 × 267 × 147	146.9	754.0	265.2	12.8	17.5	2.51
762 × 267 × 134	133.9	750.0	264.4	12.0	15.5	2.51
686 × 254 × 170	170.2	692.9	255.8	14.5	23.7	2.35
686 × 254 × 152	152.4	687.5	254.5	13.2	21.0	2.34
686 × 254 × 140	140.1	383.5	253.7	12.4	19.0	2.33
686 × 254 × 125	125.2	677.9	253.0	11.7	16.2	2.32
610 × 305 × 238	238.1	635.8	311.4	18.4	31.4	2.45
610 × 305 × 179	179.0	620.2	307.1	14.1	23.6	2.41
610 × 305 × 149	149.1	612.4	304.8	11.8	19.7	2.39
610 × 229 × 140	139.9	617.2	230.2	13.1	22.1	2.11
610 × 229 × 125	125.1	612.2	229.0	11.9	19.6	2.09
610 × 229 × 113	113.0	607.6	228.2	11.1	17.3	2.08
610 × 229 × 101	101.2	602.6	227.6	10.5	14.8	2.07
533 × 210 × 122	122.0	544.5	211.9	12.7	21.3	1.89
533 × 210 × 109	109.0	539.5	210.8	11.6	18.8	1.88
533 × 210 × 101	101.0	536.7	210.0	10.8	17.4	1.87
533 × 210 × 92	92.1	533.1	209.3	10.1	15.6	1.86
533 × 210 × 82	82.2	528.3	208.8	9.6	13.2	1.85
457 × 191 × 98	98.3	467.2	192.8	11.4	19.6	1.67
457 × 191 × 89	89.3	463.4	191.9	10.5	17.7	1.66
457 × 191 × 82	82.0	460.0	191.3	9.9	16.0	1.65
457 × 191 × 74	74.3	457.0	190.4	9.0	14.5	1.64
457 × 191 × 67	67.1	453.4	189.9	8.5	12.7	1.63
457 × 152 × 82	82.1	465.8	155.3	10.5	18.9	1.51
457 × 152 × 74	74.2	462.0	154.4	9.6	17.0	1.50
457 × 152 × 67	67.2	458.0	153.8	9.0	15.0	1.50
457 × 152 × 60	59.8	454.6	152.9	8.1	13.3	1.50
457 × 152 × 52	52.3	449.8	152.4	7.6	10.9	1.48
406 × 178 × 74	74.2	412.8	179.5	9.5	16.0	1.51
406 × 178 × 67	67.1	409.4	178.8	8.8	14.3	1.50
406 × 178 × 60	60.1	406.4	177.9	7.9	12.8	1.49

Tables and Memoranda

METAL

Designation	Mass (kg/m)	Depth of section (mm)	Width of section (mm)	Thickness		Surface area (m²/m)
				Web (mm)	Flange (mm)	
406 × 178 × 50	54.1	402.6	177.7	7.7	10.9	1.48
406 × 140 × 46	46.0	403.2	142.2	6.8	11.2	1.34
406 × 140 × 39	39.0	398.0	141.8	6.4	8.6	1.33
356 × 171 × 67	67.1	363.4	173.2	9.1	15.7	1.38
356 × 171 × 57	57.0	358.0	172.2	8.1	13.0	1.37
356 × 171 × 51	51.0	355.0	171.5	7.4	11.5	1.36
356 × 171 × 45	45.0	351.4	171.1	7.0	9.7	1.36
356 × 127 × 39	39.1	353.4	126.0	6.6	10.7	1.18
356 × 127 × 33	33.1	349.0	125.4	6.0	8.5	1.17
305 × 165 × 54	54.0	310.4	166.9	7.9	13.7	1.26
305 × 165 × 46	46.1	306.6	165.7	6.7	11.8	1.25
305 × 165 × 40	40.3	303.4	165.0	6.0	10.2	1.24
305 × 127 × 48	48.1	311.0	125.3	9.0	14.0	1.09
305 × 127 × 42	41.9	307.2	124.3	8.0	12.1	1.08
305 × 127 × 37	37.0	304.4	123.3	7.1	10.7	1.07
305 × 102 × 33	32.8	312.7	102.4	6.6	10.8	1.01
305 × 102 × 28	28.2	308.7	101.8	6.0	8.8	1.00
305 × 102 × 25	24.8	305.1	101.6	5.8	7.0	0.992
254 × 146 × 43	43.0	259.6	147.3	7.2	12.7	1.08
254 × 146 × 37	37.0	256.0	146.4	6.3	10.9	1.07
254 × 146 × 31	31.1	251.4	146.1	6.0	8.6	1.06
254 × 102 × 28	28.3	260.4	102.2	6.3	10.0	0.904
254 × 102 × 25	25.2	257.2	101.9	6.0	8.4	0.897
254 × 102 × 22	22.0	254.0	101.6	5.7	6.8	0.890
203 × 133 × 30	30.0	206.8	133.9	6.4	9.6	0.923
203 × 133 × 25	25.1	203.2	133.2	5.7	7.8	0.915
203 × 102 × 23	23.1	203.2	101.8	5.4	9.3	0.790
178 × 102 × 19	19.0	177.8	101.2	4.8	7.9	0.738
152 × 89 × 16	16.0	152.4	88.7	4.5	7.7	0.638
127 × 76 × 13	13.0	127.0	76.0	4.0	7.6	0.537

METAL

Universal Columns BS 4: Part 1: 2005

Designation	Mass	Depth of section	Width of section	Thickness		Surface area
	(kg/m)	(mm)	(mm)	Web (mm)	Flange (mm)	(m²/m)
356 × 406 × 634	633.9	474.7	424.0	47.6	77.0	2.52
356 × 406 × 551	551.0	455.6	418.5	42.1	67.5	2.47
356 × 406 × 467	467.0	436.6	412.2	35.8	58.0	2.42
356 × 406 × 393	393.0	419.0	407.0	30.6	49.2	2.38
356 × 406 × 340	339.9	406.4	403.0	26.6	42.9	2.35
356 × 406 × 287	287.1	393.6	399.0	22.6	36.5	2.31
356 × 406 × 235	235.1	381.0	384.8	18.4	30.2	2.28
356 × 368 × 202	201.9	374.6	374.7	16.5	27.0	2.19
356 × 368 × 177	177.0	368.2	372.6	14.4	23.8	2.17
356 × 368 × 153	152.9	362.0	370.5	12.3	20.7	2.16
356 × 368 × 129	129.0	355.6	368.6	10.4	17.5	2.14
305 × 305 × 283	282.9	365.3	322.2	26.8	44.1	1.94
305 × 305 × 240	240.0	352.5	318.4	23.0	37.7	1.91
305 × 305 × 198	198.1	339.9	314.5	19.1	31.4	1.87
305 × 305 × 158	158.1	327.1	311.2	15.8	25.0	1.84
305 × 305 × 137	136.9	320.5	309.2	13.8	21.7	1.82
305 × 305 × 118	117.9	314.5	307.4	12.0	18.7	1.81
305 × 305 × 97	96.9	307.9	305.3	9.9	15.4	1.79
254 × 254 × 167	167.1	289.1	265.2	19.2	31.7	1.58
254 × 254 × 132	132.0	276.3	261.3	15.3	25.3	1.55
254 × 254 × 107	107.1	266.7	258.8	12.8	20.5	1.52
254 × 254 × 89	88.9	260.3	256.3	10.3	17.3	1.50
254 × 254 × 73	73.1	254.1	254.6	8.6	14.2	1.49
203 × 203 × 86	86.1	222.2	209.1	12.7	20.5	1.24
203 × 203 × 71	71.0	215.8	206.4	10.0	17.3	1.22
203 × 203 × 60	60.0	209.6	205.8	9.4	14.2	1.21
203 × 203 × 52	52.0	206.2	204.3	7.9	12.5	1.20
203 × 203 × 46	46.1	203.2	203.6	7.2	11.0	1.19
152 × 152 × 37	37.0	161.8	154.4	8.0	11.5	0.912
152 × 152 × 30	30.0	157.6	152.9	6.5	9.4	0.901
152 × 152 × 23	23.0	152.4	152.2	5.8	6.8	0.889

METAL

Joists BS 4: Part 1: 2005 (retained for reference, Corus have ceased manufacture in UK)

Designation	Mass (kg/m)	Depth of section (mm)	Width of section (mm)	Thickness		Surface area (m²/m)
				Web (mm)	Flange (mm)	
254 × 203 × 82	82.0	254.0	203.2	10.2	19.9	1.210
203 × 152 × 52	52.3	203.2	152.4	8.9	16.5	0.932
152 × 127 × 37	37.3	152.4	127.0	10.4	13.2	0.737
127 × 114 × 29	29.3	127.0	114.3	10.2	11.5	0.646
127 × 114 × 27	26.9	127.0	114.3	7.4	11.4	0.650
102 × 102 × 23	23.0	101.6	101.6	9.5	10.3	0.549
102 × 44 × 7	7.5	101.6	44.5	4.3	6.1	0.350
89 × 89 × 19	19.5	88.9	88.9	9.5	9.9	0.476
76 × 76 × 13	12.8	76.2	76.2	5.1	8.4	0.411

Parallel Flange Channels

Designation	Mass (kg/m)	Depth of section (mm)	Width of section (mm)	Thickness		Surface area (m²/m)
				Web (mm)	Flange (mm)	
430 × 100 × 64	64.4	430	100	11.0	19.0	1.23
380 × 100 × 54	54.0	380	100	9.5	17.5	1.13
300 × 100 × 46	45.5	300	100	9.0	16.5	0.969
300 × 90 × 41	41.4	300	90	9.0	15.5	0.932
260 × 90 × 35	34.8	260	90	8.0	14.0	0.854
260 × 75 × 28	27.6	260	75	7.0	12.0	0.79
230 × 90 × 32	32.2	230	90	7.5	14.0	0.795
230 × 75 × 26	25.7	230	75	6.5	12.5	0.737
200 × 90 × 30	29.7	200	90	7.0	14.0	0.736
200 × 75 × 23	23.4	200	75	6.0	12.5	0.678
180 × 90 × 26	26.1	180	90	6.5	12.5	0.697
180 × 75 × 20	20.3	180	75	6.0	10.5	0.638
150 × 90 × 24	23.9	150	90	6.5	12.0	0.637
150 × 75 × 18	17.9	150	75	5.5	10.0	0.579
125 × 65 × 15	14.8	125	65	5.5	9.5	0.489
100 × 50 × 10	10.2	100	50	5.0	8.5	0.382

METAL

Equal Angles BS EN 10056-1

Designation	Mass (kg/m)	Surface area (m²/m)
200 × 200 × 24	71.1	0.790
200 × 200 × 20	59.9	0.790
200 × 200 × 18	54.2	0.790
200 × 200 × 16	48.5	0.790
150 × 150 × 18	40.1	0.59
150 × 150 × 15	33.8	0.59
150 × 150 × 12	27.3	0.59
150 × 150 × 10	23.0	0.59
120 × 120 × 15	26.6	0.47
120 × 120 × 12	21.6	0.47
120 × 120 × 10	18.2	0.47
120 × 120 × 8	14.7	0.47
100 × 100 × 15	21.9	0.39
100 × 100 × 12	17.8	0.39
100 × 100 × 10	15.0	0.39
100 × 100 × 8	12.2	0.39
90 × 90 × 12	15.9	0.35
90 × 90 × 10	13.4	0.35
90 × 90 × 8	10.9	0.35
90 × 90 × 7	9.61	0.35
90 × 90 × 6	8.30	0.35

Unequal Angles BS EN 10056-1

Designation	Mass (kg/m)	Surface area (m²/m)
200 × 150 × 18	47.1	0.69
200 × 150 × 15	39.6	0.69
200 × 150 × 12	32.0	0.69
200 × 100 × 15	33.7	0.59
200 × 100 × 12	27.3	0.59
200 × 100 × 10	23.0	0.59
150 × 90 × 15	26.6	0.47
150 × 90 × 12	21.6	0.47
150 × 90 × 10	18.2	0.47
150 × 75 × 15	24.8	0.44
150 × 75 × 12	20.2	0.44
150 × 75 × 10	17.0	0.44
125 × 75 × 12	17.8	0.40
125 × 75 × 10	15.0	0.40
125 × 75 × 8	12.2	0.40
100 × 75 × 12	15.4	0.34
100 × 75 × 10	13.0	0.34
100 × 75 × 8	10.6	0.34
100 × 65 × 10	12.3	0.32
100 × 65 × 8	9.94	0.32
100 × 65 × 7	8.77	0.32

METAL

Structural Tees Split from Universal Beams BS 4: Part 1: 2005

Designation	Mass (kg/m)	Surface area (m²/m)
305 × 305 × 90	89.5	1.22
305 × 305 × 75	74.6	1.22
254 × 343 × 63	62.6	1.19
229 × 305 × 70	69.9	1.07
229 × 305 × 63	62.5	1.07
229 × 305 × 57	56.5	1.07
229 × 305 × 51	50.6	1.07
210 × 267 × 61	61.0	0.95
210 × 267 × 55	54.5	0.95
210 × 267 × 51	50.5	0.95
210 × 267 × 46	46.1	0.95
210 × 267 × 41	41.1	0.95
191 × 229 × 49	49.2	0.84
191 × 229 × 45	44.6	0.84
191 × 229 × 41	41.0	0.84
191 × 229 × 37	37.1	0.84
191 × 229 × 34	33.6	0.84
152 × 229 × 41	41.0	0.76
152 × 229 × 37	37.1	0.76
152 × 229 × 34	33.6	0.76
152 × 229 × 30	29.9	0.76
152 × 229 × 26	26.2	0.76

Universal Bearing Piles BS 4: Part 1: 2005

Designation	Mass (kg/m)	Depth of Section (mm)	Width of section (mm)	Thickness Web (mm)	Flange (mm)
356 × 368 × 174	173.9	361.4	378.5	20.3	20.4
356 × 368 × 152	152.0	356.4	376.0	17.8	17.9
356 × 368 × 133	133.0	352.0	373.8	15.6	15.7
356 × 368 × 109	108.9	346.4	371.0	12.8	12.9
305 × 305 × 223	222.9	337.9	325.7	30.3	30.4
305 × 305 × 186	186.0	328.3	320.9	25.5	25.6
305 × 305 × 149	149.1	318.5	316.0	20.6	20.7
305 × 305 × 126	126.1	312.3	312.9	17.5	17.6
305 × 305 × 110	110.0	307.9	310.7	15.3	15.4
305 × 305 × 95	94.9	303.7	308.7	13.3	13.3
305 × 305 × 88	88.0	301.7	307.8	12.4	12.3
305 × 305 × 79	78.9	299.3	306.4	11.0	11.1
254 × 254 × 85	85.1	254.3	260.4	14.4	14.3
254 × 254 × 71	71.0	249.7	258.0	12.0	12.0
254 × 254 × 63	63.0	247.1	256.6	10.6	10.7
203 × 203 × 54	53.9	204.0	207.7	11.3	11.4
203 × 203 × 45	44.9	200.2	205.9	9.5	9.5

METAL

Hot Formed Square Hollow Sections EN 10210 S275J2H & S355J2H

Size (mm)	Wall thickness (mm)	Mass (kg/m)	Superficial area (m²/m)
40 × 40	2.5	2.89	0.154
	3.0	3.41	0.152
	3.2	3.61	0.152
	3.6	4.01	0.151
	4.0	4.39	0.150
	5.0	5.28	0.147
50 × 50	2.5	3.68	0.194
	3.0	4.35	0.192
	3.2	4.62	0.192
	3.6	5.14	0.191
	4.0	5.64	0.190
	5.0	6.85	0.187
	6.0	7.99	0.185
	6.3	8.31	0.184
60 × 60	3.0	5.29	0.232
	3.2	5.62	0.232
	3.6	6.27	0.231
	4.0	6.90	0.230
	5.0	8.42	0.227
	6.0	9.87	0.225
	6.3	10.30	0.224
	8.0	12.50	0.219
70 × 70	3.0	6.24	0.272
	3.2	6.63	0.272
	3.6	7.40	0.271
	4.0	8.15	0.270
	5.0	9.99	0.267
	6.0	11.80	0.265
	6.3	12.30	0.264
	8.0	15.00	0.259
80 × 80	3.2	7.63	0.312
	3.6	8.53	0.311
	4.0	9.41	0.310
	5.0	11.60	0.307
	6.0	13.60	0.305
	6.3	14.20	0.304
	8.0	17.50	0.299
90 × 90	3.6	9.66	0.351
	4.0	10.70	0.350
	5.0	13.10	0.347
	6.0	15.50	0.345
	6.3	16.20	0.344
	8.0	20.10	0.339
100 × 100	3.6	10.80	0.391
	4.0	11.90	0.390
	5.0	14.70	0.387
	6.0	17.40	0.385
	6.3	18.20	0.384
	8.0	22.60	0.379
	10.0	27.40	0.374
120 × 120	4.0	14.40	0.470
	5.0	17.80	0.467
	6.0	21.20	0.465

METAL

Size (mm)	Wall thickness (mm)	Mass (kg/m)	Superficial area (m²/m)
	6.3	22.20	0.464
	8.0	27.60	0.459
	10.0	33.70	0.454
	12.0	39.50	0.449
	12.5	40.90	0.448
140 × 140	5.0	21.00	0.547
	6.0	24.90	0.545
	6.3	26.10	0.544
	8.0	32.60	0.539
	10.0	40.00	0.534
	12.0	47.00	0.529
	12.5	48.70	0.528
150 × 150	5.0	22.60	0.587
	6.0	26.80	0.585
	6.3	28.10	0.584
	8.0	35.10	0.579
	10.0	43.10	0.574
	12.0	50.80	0.569
	12.5	52.70	0.568
Hot formed from seamless hollow	16.0	65.2	0.559
160 × 160	5.0	24.10	0.627
	6.0	28.70	0.625
	6.3	30.10	0.624
	8.0	37.60	0.619
	10.0	46.30	0.614
	12.0	54.60	0.609
	12.5	56.60	0.608
	16.0	70.20	0.599
180 × 180	5.0	27.30	0.707
	6.0	32.50	0.705
	6.3	34.00	0.704
	8.0	42.70	0.699
	10.0	52.50	0.694
	12.0	62.10	0.689
	12.5	64.40	0.688
	16.0	80.20	0.679
200 × 200	5.0	30.40	0.787
	6.0	36.20	0.785
	6.3	38.00	0.784
	8.0	47.70	0.779
	10.0	58.80	0.774
	12.0	69.60	0.769
	12.5	72.30	0.768
	16.0	90.30	0.759
250 × 250	5.0	38.30	0.987
	6.0	45.70	0.985
	6.3	47.90	0.984
	8.0	60.30	0.979
	10.0	74.50	0.974
	12.0	88.50	0.969
	12.5	91.90	0.968
	16.0	115.00	0.959

METAL

Size (mm)	Wall thickness (mm)	Mass (kg/m)	Superficial area (m²/m)
300 × 300	6.0	55.10	1.18
	6.3	57.80	1.18
	8.0	72.80	1.18
	10.0	90.20	1.17
	12.0	107.00	1.17
	12.5	112.00	1.17
	16.0	141.00	1.16
350 × 350	8.0	85.40	1.38
	10.0	106.00	1.37
	12.0	126.00	1.37
	12.5	131.00	1.37
	16.0	166.00	1.36
400 × 400	8.0	97.90	1.58
	10.0	122.00	1.57
	12.0	145.00	1.57
	12.5	151.00	1.57
	16.0	191.00	1.56
(Grade S355J2H only)	20.00*	235.00	1.55

Note: * SAW process

METAL

Hot Formed Square Hollow Sections JUMBO RHS: JIS G3136

Size (mm)	Wall thickness (mm)	Mass (kg/m)	Superficial area (m²/m)
350 × 350	19.0	190.00	1.33
	22.0	217.00	1.32
	25.0	242.00	1.31
400 × 400	22.0	251.00	1.52
	25.0	282.00	1.51
450 × 450	12.0	162.00	1.76
	16.0	213.00	1.75
	19.0	250.00	1.73
	22.0	286.00	1.72
	25.0	321.00	1.71
	28.0 *	355.00	1.70
	32.0 *	399.00	1.69
500 × 500	12.0	181.00	1.96
	16.0	238.00	1.95
	19.0	280.00	1.93
	22.0	320.00	1.92
	25.0	360.00	1.91
	28.0 *	399.00	1.90
	32.0 *	450.00	1.89
	36.0 *	498.00	1.88
550 × 550	16.0	263.00	2.15
	19.0	309.00	2.13
	22.0	355.00	2.12
	25.0	399.00	2.11
	28.0 *	443.00	2.10
	32.0 *	500.00	2.09
	36.0 *	555.00	2.08
	40.0 *	608.00	2.06
600 × 600	25.0 *	439.00	2.31
	28.0 *	487.00	2.30
	32.0 *	550.00	2.29
	36.0 *	611.00	2.28
	40.0 *	671.00	2.26
700 × 700	25.0 *	517.00	2.71
	28.0 *	575.00	2.70
	32.0 *	651.00	2.69
	36.0 *	724.00	2.68
	40.0 *	797.00	2.68

Note: * SAW process

METAL

Hot Formed Rectangular Hollow Sections: EN10210 S275J2h & S355J2H

Size (mm)	Wall thickness (mm)	Mass (kg/m)	Superficial area (m²/m)
50 × 30	2.5	2.89	0.154
	3.0	3.41	0.152
	3.2	3.61	0.152
	3.6	4.01	0.151
	4.0	4.39	0.150
	5.0	5.28	0.147
60 × 40	2.5	3.68	0.194
	3.0	4.35	0.192
	3.2	4.62	0.192
	3.6	5.14	0.191
	4.0	5.64	0.190
	5.0	6.85	0.187
	6.0	7.99	0.185
	6.3	8.31	0.184
80 × 40	3.0	5.29	0.232
	3.2	5.62	0.232
	3.6	6.27	0.231
	4.0	6.90	0.230
	5.0	8.42	0.227
	6.0	9.87	0.225
	6.3	10.30	0.224
	8.0	12.50	0.219
76.2 × 50.8	3.0	5.62	0.246
	3.2	5.97	0.246
	3.6	6.66	0.245
	4.0	7.34	0.244
	5.0	8.97	0.241
	6.0	10.50	0.239
	6.3	11.00	0.238
	8.0	13.40	0.233
90 × 50	3.0	6.24	0.272
	3.2	6.63	0.272
	3.6	7.40	0.271
	4.0	8.15	0.270
	5.0	9.99	0.267
	6.0	11.80	0.265
	6.3	12.30	0.264
	8.0	15.00	0.259
100 × 50	3.0	6.71	0.292
	3.2	7.13	0.292
	3.6	7.96	0.291
	4.0	8.78	0.290
	5.0	10.80	0.287
	6.0	12.70	0.285
	6.3	13.30	0.284
	8.0	16.30	0.279

METAL

Size (mm)	Wall thickness (mm)	Mass (kg/m)	Superficial area (m²/m)
100 × 60	3.0	7.18	0.312
	3.2	7.63	0.312
	3.6	8.53	0.311
	4.0	9.41	0.310
	5.0	11.60	0.307
	6.0	13.60	0.305
	6.3	14.20	0.304
	8.0	17.50	0.299
120 × 60	3.6	9.70	0.351
	4.0	10.70	0.350
	5.0	13.10	0.347
	6.0	15.50	0.345
	6.3	16.20	0.344
	8.0	20.10	0.339
120 × 80	3.6	10.80	0.391
	4.0	11.90	0.390
	5.0	14.70	0.387
	6.0	17.40	0.385
	6.3	18.20	0.384
	8.0	22.60	0.379
	10.0	27.40	0.374
150 × 100	4.0	15.10	0.490
	5.0	18.60	0.487
	6.0	22.10	0.485
	6.3	23.10	0.484
	8.0	28.90	0.479
	10.0	35.30	0.474
	12.0	41.40	0.469
	12.5	42.80	0.468
160 × 80	4.0	14.40	0.470
	5.0	17.80	0.467
	6.0	21.20	0.465
	6.3	22.20	0.464
	8.0	27.60	0.459
	10.0	33.70	0.454
	12.0	39.50	0.449
	12.5	40.90	0.448
200 × 100	5.0	22.60	0.587
	6.0	26.80	0.585
	6.3	28.10	0.584
	8.0	35.10	0.579
	10.0	43.10	0.574
	12.0	50.80	0.569
	12.5	52.70	0.568
	16.0	65.20	0.559
250 × 150	5.0	30.40	0.787
	6.0	36.20	0.785
	6.3	38.00	0.784
	8.0	47.70	0.779
	10.0	58.80	0.774
	12.0	69.60	0.769
	12.5	72.30	0.768
	16.0	90.30	0.759

METAL

Size (mm)	Wall thickness (mm)	Mass (kg/m)	Superficial area (m²/m)
300 × 200	5.0	38.30	0.987
	6.0	45.70	0.985
	6.3	47.90	0.984
	8.0	60.30	0.979
	10.0	74.50	0.974
	12.0	88.50	0.969
	12.5	91.90	0.968
	16.0	115.00	0.959
400 × 200	6.0	55.10	1.18
	6.3	57.80	1.18
	8.0	72.80	1.18
	10.0	90.20	1.17
	12.0	107.00	1.17
	12.5	112.00	1.17
	16.0	141.00	1.16
450 × 250	8.0	85.40	1.38
	10.0	106.00	1.37
	12.0	126.00	1.37
	12.5	131.00	1.37
	16.0	166.00	1.36
500 × 300	8.0	98.00	1.58
	10.0	122.00	1.57
	12.0	145.00	1.57
	12.5	151.00	1.57
	16.0	191.00	1.56
	20.0	235.00	1.55

METAL

Hot Formed Circular Hollow Sections EN 10210 S275J2H & S355J2H

Outside diameter (mm)	Wall thickness (mm)	Mass (kg/m)	Superficial area (m²/m)
21.3	3.2	1.43	0.067
26.9	3.2	1.87	0.085
33.7	3.0	2.27	0.106
	3.2	2.41	0.106
	3.6	2.67	0.106
	4.0	2.93	0.106
42.4	3.0	2.91	0.133
	3.2	3.09	0.133
	3.6	3.44	0.133
	4.0	3.79	0.133
48.3	2.5	2.82	0.152
	3.0	3.35	0.152
	3.2	3.56	0.152
	3.6	3.97	0.152
	4.0	4.37	0.152
	5.0	5.34	0.152
60.3	2.5	3.56	0.189
	3.0	4.24	0.189
	3.2	4.51	0.189
	3.6	5.03	0.189
	4.0	5.55	0.189
	5.0	6.82	0.189
76.1	2.5	4.54	0.239
	3.0	5.41	0.239
	3.2	5.75	0.239
	3.6	6.44	0.239
	4.0	7.11	0.239
	5.0	8.77	0.239
	6.0	10.40	0.239
	6.3	10.80	0.239
88.9	2.5	5.33	0.279
	3.0	6.36	0.279
	3.2	6.76	0.27
	3.6	7.57	0.279
	4.0	8.38	0.279
	5.0	10.30	0.279
	6.0	12.30	0.279
	6.3	12.80	0.279
114.3	3.0	8.23	0.359
	3.2	8.77	0.359
	3.6	9.83	0.359
	4.0	10.09	0.359
	5.0	13.50	0.359
	6.0	16.00	0.359
	6.3	16.80	0.359

METAL

Outside diameter (mm)	Wall thickness (mm)	Mass (kg/m)	Superficial area (m²/m)
139.7	3.2	10.80	0.439
	3.6	12.10	0.439
	4.0	13.40	0.439
	5.0	16.60	0.439
	6.0	19.80	0.439
	6.3	20.70	0.439
	8.0	26.00	0.439
	10.0	32.00	0.439
168.3	3.2	13.00	0.529
	3.6	14.60	0.529
	4.0	16.20	0.529
	5.0	20.10	0.529
	6.0	24.00	0.529
	6.3	25.20	0.529
	8.0	31.60	0.529
	10.0	39.00	0.529
	12.0	46.30	0.529
	12.5	48.00	0.529
193.7	5.0	23.30	0.609
	6.0	27.80	0.609
	6.3	29.10	0.609
	8.0	36.60	0.609
	10.0	45.30	0.609
	12.0	53.80	0.609
	12.5	55.90	0.609
219.1	5.0	26.40	0.688
	6.0	31.50	0.688
	6.3	33.10	0.688
	8.0	41.60	0.688
	10.0	51.60	0.688
	12.0	61.30	0.688
	12.5	63.70	0.688
	16.0	80.10	0.688
244.5	5.0	29.50	0.768
	6.0	35.30	0.768
	6.3	37.00	0.768
	8.0	46.70	0.768
	10.0	57.80	0.768
	12.0	68.80	0.768
	12.5	71.50	0.768
	16.0	90.20	0.768
273.0	5.0	33.00	0.858
	6.0	39.50	0.858
	6.3	41.40	0.858
	8.0	52.30	0.858
	10.0	64.90	0.858
	12.0	77.20	0.858
	12.5	80.30	0.858
	16.0	101.00	0.858

METAL

Outside diameter (mm)	Wall thickness (mm)	Mass (kg/m)	Superficial area (m²/m)
323.9	5.0	39.30	1.02
	6.0	47.00	1.02
	6.3	49.30	1.02
	8.0	62.30	1.02
	10.0	77.40	1.02
	12.0	92.30	1.02
	12.5	96.00	1.02
	16.0	121.00	1.02
355.6	6.3	54.30	1.12
	8.0	68.60	1.12
	10.0	85.30	1.12
	12.0	102.00	1.12
	12.5	106.00	1.12
	16.0	134.00	1.12
406.4	6.3	62.20	1.28
	8.0	79.60	1.28
	10.0	97.80	1.28
	12.0	117.00	1.28
	12.5	121.00	1.28
	16.0	154.00	1.28
457.0	6.3	70.00	1.44
	8.0	88.60	1.44
	10.0	110.00	1.44
	12.0	132.00	1.44
	12.5	137.00	1.44
	16.0	174.00	1.44
508.0	6.3	77.90	1.60
	8.0	98.60	1.60
	10.0	123.00	1.60
	12.0	147.00	1.60
	12.5	153.00	1.60
	16.0	194.00	1.60

METAL

Spacing of Holes in Angles

Nominal leg length (mm)	Spacing of holes						Maximum diameter of bolt or rivet		
	A	B	C	D	E	F	A	B and C	D, E and F
200		75	75	55	55	55		30	20
150		55	55					20	
125		45	60					20	
120									
100	55						24		
90	50						24		
80	45						20		
75	45						20		
70	40						20		
65	35						20		
60	35						16		
50	28						12		
45	25								
40	23								
30	20								
25	15								

KERBS, PAVING, ETC.

KERBS/EDGINGS/CHANNELS

Precast Concrete Kerbs to BS 7263

Straight kerb units: length from 450 to 915 mm

150 mm high × 125 mm thick		
bullnosed	type BN	
half battered	type HB3	
255 mm high × 125 mm thick		
45° splayed	type SP	
half battered	type HB2	
305 mm high × 150 mm thick		
half battered	type HB1	
Quadrant kerb units		
150 mm high × 305 and 455 mm radius to match	type BN	type QBN
150 mm high × 305 and 455 mm radius to match	type HB2, HB3	type QHB
150 mm high × 305 and 455 mm radius to match	type SP	type QSP
255 mm high × 305 and 455 mm radius to match	type BN	type QBN
255 mm high × 305 and 455 mm radius to match	type HB2, HB3	type QHB
225 mm high × 305 and 455 mm radius to match	type SP	type QSP
Angle kerb units		
305 × 305 × 225 mm high × 125 mm thick		
bullnosed external angle	type XA	
splayed external angle to match type SP	type XA	
bullnosed internal angle	type IA	
splayed internal angle to match type SP	type IA	
Channels		
255 mm wide × 125 mm high flat	type CS1	
150 mm wide × 125 mm high flat type	CS2	
255 mm wide × 125 mm high dished	type CD	

KERBS, PAVING, ETC.

Transition kerb units			
from kerb type SP to HB	left handed	type TL	
	right handed	type TR	
from kerb type BN to HB	left handed	type DL1	
	right handed	type DR1	
from kerb type BN to SP	left handed	type DL2	
	right handed	type DR2	

Number of kerbs required per quarter circle (780 mm kerb lengths)

Radius (m)	Number in quarter circle
12	24
10	20
8	16
6	12
5	10
4	8
3	6
2	4
1	2

Precast Concrete Edgings

Round top type ER	Flat top type EF	Bullnosed top type EBN
150 × 50 mm	150 × 50 mm	150 × 50 mm
200 × 50 mm	200 × 50 mm	200 × 50 mm
250 × 50 mm	250 × 50 mm	250 × 50 mm

KERBS, PAVING, ETC.

BASES

Cement Bound Material for Bases and Subbases

CBM1:	very carefully graded aggregate from 37.5–75 mm, with a 7-day strength of 4.5 N/mm^2
CBM2:	same range of aggregate as CBM1 but with more tolerance in each size of aggregate with a 7-day strength of 7.0 N/mm^2
CBM3:	crushed natural aggregate or blast furnace slag, graded from 37.5–150 mm for 40 mm aggregate, and from 20–75 mm for 20 mm aggregate, with a 7-day strength of 10 N/mm^2
CBM4:	crushed natural aggregate or blast furnace slag, graded from 37.5–150 mm for 40 mm aggregate, and from 20–75 mm for 20 mm aggregate, with a 7-day strength of 15 N/mm^2

INTERLOCKING BRICK/BLOCK ROADS/PAVINGS

Sizes of Precast Concrete Paving Blocks

Type R blocks
200 × 100 × 60 mm
200 × 100 × 65 mm
200 × 100 × 80 mm
200 × 100 × 100 mm

Type S
Any shape within a 295 mm space

Sizes of clay brick pavers
200 × 100 × 50 mm
200 × 100 × 65 mm
210 × 105 × 50 mm
210 × 105 × 65 mm
215 × 102.5 × 50 mm
215 × 102.5 × 65 mm

Type PA: 3 kN
Footpaths and pedestrian areas, private driveways, car parks, light vehicle traffic and over-run

Type PB: 7 kN
Residential roads, lorry parks, factory yards, docks, petrol station forecourts, hardstandings, bus stations

KERBS, PAVING, ETC.

PAVING AND SURFACING

Weights and Sizes of Paving and Surfacing

Description of item	Size	Quantity per tonne
Paving 50 mm thick	900 × 600 mm	15
Paving 50 mm thick	750 × 600 mm	18
Paving 50 mm thick	600 × 600 mm	23
Paving 50 mm thick	450 × 600 mm	30
Paving 38 mm thick	600 × 600 mm	30
Path edging	914 × 50 × 150 mm	60
Kerb (including radius and tapers)	125 × 254 × 914 mm	15
Kerb (including radius and tapers)	125 × 150 × 914 mm	25
Square channel	125 × 254 × 914 mm	15
Dished channel	125 × 254 × 914 mm	15
Quadrants	300 × 300 × 254 mm	19
Quadrants	450 × 450 × 254 mm	12
Quadrants	300 × 300 × 150 mm	30
Internal angles	300 × 300 × 254 mm	30
Fluted pavement channel	255 × 75 × 914 mm	25
Corner stones	300 × 300 mm	80
Corner stones	360 × 360 mm	60
Cable covers	914 × 175 mm	55
Gulley kerbs	220 × 220 × 150 mm	60
Gulley kerbs	220 × 200 × 75 mm	120

KERBS, PAVING, ETC.

Weights and Sizes of Paving and Surfacing

Material	kg/m³	lb/cu yd
Tarmacadam	2306	3891
Macadam (waterbound)	2563	4325
Vermiculite (aggregate)	64–80	108–135
Terracotta	2114	3568
Cork – compressed	388	24
	kg/m²	lb/sq ft
Clay floor tiles, 12.7 mm	27.3	5.6
Pavement lights	122	25
Damp-proof course	5	1
	kg/m² per mm thickness	lb/sq ft per inch thickness
Paving slabs (stone)	2.3	12
Granite setts	2.88	15
Asphalt	2.30	12
Rubber flooring	1.68	9
Polyvinyl chloride	1.94 (avg)	10 (avg)

Coverage (m²) Per Cubic Metre of Materials Used as Subbases or Capping Layers

Consolidated thickness laid in (mm)	Square metre coverage		
	Gravel	Sand	Hardcore
50	15.80	16.50	–
75	10.50	11.00	–
100	7.92	8.20	7.42
125	6.34	6.60	5.90
150	5.28	5.50	4.95
175	–	–	4.23
200	–	–	3.71
225	–	–	3.30
300	–	–	2.47

KERBS, PAVING, ETC.

Approximate Rate of Spreads

Average thickness of course (mm)	Description	Approximate rate of spread			
		Open Textured		Dense, Medium & Fine Textured	
		(kg/m²)	(m²/t)	(kg/m²)	(m²/t)
35	14 mm open textured or dense wearing course	60–75	13–17	70–85	12–14
40	20 mm open textured or dense base course	70–85	12–14	80–100	10–12
45	20 mm open textured or dense base course	80–100	10–12	95–100	9–10
50	20 mm open textured or dense, or 28 mm dense base course	85–110	9–12	110–120	8–9
60	28 mm dense base course, 40 mm open textured of dense base course or 40 mm single course as base course		8–10	130–150	7–8
65	28 mm dense base course, 40 mm open textured or dense base course or 40 mm single course	100–135	7–10	140–160	6–7
75	40 mm single course, 40 mm open textured or dense base course, 40 mm dense roadbase	120–150	7–8	165–185	5–6
100	40 mm dense base course or roadbase	–	–	220–240	4–4.5

KERBS, PAVING, ETC.

Surface Dressing Roads: Coverage (m²) per Tonne of Material

Size in mm	Sand	Granite chips	Gravel	Limestone chips
Sand	168	–	–	–
3	–	148	152	165
6	–	130	133	144
9	–	111	114	123
13	–	85	87	95
19	–	68	71	78

Sizes of Flags

Reference	Nominal size (mm)	Thickness (mm)
A	600 × 450	50 and 63
B	600 × 600	50 and 63
C	600 × 750	50 and 63
D	600 × 900	50 and 63
E	450 × 450	50 and 70 chamfered top surface
F	400 × 400	50 and 65 chamfered top surface
G	300 × 300	50 and 60 chamfered top surface

Sizes of Natural Stone Setts

Width (mm)		Length (mm)		Depth (mm)
100	×	100	×	100
75	×	150 to 250	×	125
75	×	150 to 250	×	150
100	×	150 to 250	×	100
100	×	150 to 250	×	150

SEEDING/TURFING AND PLANTING

Topsoil Quality

Topsoil grade	Properties
Premium	Natural topsoil, high fertility, loamy texture, good soil structure, suitable for intensive cultivation.
General purpose	Natural or manufactured topsoil of lesser quality than Premium, suitable for agriculture or amenity landscape, may need fertilizer or soil structure improvement.
Economy	Selected subsoil, natural mineral deposit such as river silt or greensand. The grade comprises two subgrades; 'Low clay' and 'High clay' which is more liable to compaction in handling. This grade is suitable for low-production agricultural land and amenity woodland or conservation planting areas.

Forms of Trees

Standards:	Shall be clear with substantially straight stems. Grafted and budded trees shall have no more than a slight bend at the union. Standards shall be designated as Half, Extra light, Light, Standard, Selected standard, Heavy, and Extra heavy.
Sizes of Standards	
Heavy standard	12–14 cm girth × 3.50 to 5.00 m high
Extra Heavy standard	14–16 cm girth × 4.25 to 5.00 m high
Extra Heavy standard	16–18 cm girth × 4.25 to 6.00 m high
Extra Heavy standard	18–20 cm girth × 5.00 to 6.00 m high
Semi-mature trees:	Between 6.0 m and 12.0 m tall with a girth of 20 to 75 cm at 1.0 m above ground.
Feathered trees:	Shall have a defined upright central leader, with stem furnished with evenly spread and balanced lateral shoots down to or near the ground.
Whips:	Shall be without significant feather growth as determined by visual inspection.
Multi-stemmed trees:	Shall have two or more main stems at, near, above or below ground.

Seedlings grown from seed and not transplanted shall be specified when ordered for sale as:

1+0	one year old seedling
2+0	two year old seedling
1+1	one year seed bed, one year transplanted = two year old seedling
1+2	one year seed bed, two years transplanted = three year old seedling
2+1	two years seed bed, one year transplanted = three year old seedling
1u1	two years seed bed, undercut after 1 year = two year old seedling
2u2	four years seed bed, undercut after 2 years = four year old seedling

SEEDING/TURFING AND PLANTING

Cuttings

The age of cuttings (plants grown from shoots, stems, or roots of the mother plant) shall be specified when ordered for sale. The height of transplants and undercut seedlings/cuttings (which have been transplanted or undercut at least once) shall be stated in centimetres. The number of growing seasons before and after transplanting or undercutting shall be stated.

0 + 1	one year cutting
0 + 2	two year cutting
0 + 1 + 1	one year cutting bed, one year transplanted = two year old seedling
0 + 1 + 2	one year cutting bed, two years transplanted = three year old seedling

Grass Cutting Capacities in m² per hour

Speed mph	Width of cut in metres												
	0.5	0.7	1.0	1.2	1.5	1.7	2.0	2.0	2.1	2.5	2.8	3.0	3.4
1.0	724	1127	1529	1931	2334	2736	3138	3219	3380	4023	4506	4828	5472
1.5	1086	1690	2293	2897	3500	4104	4707	4828	5069	6035	6759	7242	8208
2.0	1448	2253	3058	3862	4667	5472	6276	6437	6759	8047	9012	9656	10944
2.5	1811	2816	3822	4828	5834	6840	7846	8047	8449	10058	11265	12070	13679
3.0	2173	3380	4587	5794	7001	8208	9415	9656	10139	12070	13518	14484	16415
3.5	2535	3943	5351	6759	8167	9576	10984	11265	11829	14082	15772	16898	19151
4.0	2897	4506	6115	7725	9334	10944	12553	12875	13518	16093	18025	19312	21887
4.5	3259	5069	6880	8690	10501	12311	14122	14484	15208	18105	20278	21726	24623
5.0	3621	5633	7644	9656	11668	13679	15691	16093	16898	20117	22531	24140	27359
5.5	3983	6196	8409	10622	12834	15047	17260	17703	18588	22128	24784	26554	30095
6.0	4345	6759	9173	11587	14001	16415	18829	19312	20278	24140	27037	28968	32831
6.5	4707	7322	9938	12553	15168	17783	20398	20921	21967	26152	29290	31382	35566
7.0	5069	7886	10702	13518	16335	19151	21967	22531	23657	28163	31543	33796	38302

Number of Plants per m² at the following offset spacings

(All plants equidistant horizontally and vertically)

Distance mm	Nr of Plants
100	114.43
200	28.22
250	18.17
300	12.35
400	6.72
500	4.29
600	3.04
700	2.16
750	1.88
900	1.26
1000	1.05
1200	0.68
1500	0.42
2000	0.23

SEEDING/TURFING AND PLANTING

Grass Clippings Wet: Based on 3.5 m³/tonne

Annual kg/100 m²		Average 20 cuts kg/100 m²			m²/tonne	m²/m³
32.0		1.6			61162.1	214067.3

Nr of cuts	22	20	18	16	12	4
kg/cut	1.45	1.60	1.78	2.00	2.67	8.00
Area capacity of 3 tonne vehicle per load						
m²	206250	187500	168750	150000	112500	37500
Load m³	**100 m² units/m³ of vehicle space**					
1	196.4	178.6	160.7	142.9	107.1	35.7
2	392.9	357.1	321.4	285.7	214.3	71.4
3	589.3	535.7	482.1	428.6	321.4	107.1
4	785.7	714.3	642.9	571.4	428.6	142.9
5	982.1	892.9	803.6	714.3	535.7	178.6

Transportation of Trees

To unload large trees a machine with the necessary lifting strength is required. The weight of the trees must therefore be known in advance. The following table gives a rough overview. The additional columns with root ball dimensions and the number of plants per trailer provide additional information, for example about preparing planting holes and calculating unloading times.

Girth in cm	Rootball diameter in cm	Ball height in cm	Weight in kg	Numbers of trees per trailer
16–18	50–60	40	150	100–120
18–20	60–70	40–50	200	80–100
20–25	60–70	40–50	270	50–70
25–30	80	50–60	350	50
30–35	90–100	60–70	500	12–18
35–40	100–110	60–70	650	10–15
40–45	110–120	60–70	850	8–12
45–50	110–120	60–70	1100	5–7
50–60	130–140	60–70	1600	1–3
60–70	150–160	60–70	2500	1
70–80	180–200	70	4000	1
80–90	200–220	70–80	5500	1
90–100	230–250	80–90	7500	1
100–120	250–270	80–90	9500	1

Data supplied by Lorenz von Ehren GmbH
The information in the table is approximate; deviations depend on soil type, genus and weather

FENCING AND GATES

Types of Preservative

Creosote (tar oil) can be 'factory' applied	by pressure to BS 144: pts 1&2 by immersion to BS 144: pt 1 by hot and cold open tank to BS 144: pts 1&2
Copper/chromium/arsenic (CCA)	by full cell process to BS 4072 pts 1&2
Organic solvent (OS)	by double vacuum (vacvac) to BS 5707 pts 1&3 by immersion to BS 5057 pts 1&3
Pentachlorophenol (PCP)	by heavy oil double vacuum to BS 5705 pts 2&3
Boron diffusion process (treated with disodium octaborate to BWPA Manual 1986)	

Note: Boron is used on green timber at source and the timber is supplied dry

Cleft Chestnut Pale Fences

Pales	Pale spacing	Wire lines	
900 mm	75 mm	2	temporary protection
1050 mm	75 or 100 mm	2	light protective fences
1200 mm	75 mm	3	perimeter fences
1350 mm	75 mm	3	perimeter fences
1500 mm	50 mm	3	narrow perimeter fences
1800 mm	50 mm	3	light security fences

Close-Boarded Fences

Close-boarded fences 1.05 to 1.8 m high
Type BCR (recessed) or BCM (morticed) with concrete posts 140 × 115 mm tapered and Type BW with timber posts

Palisade Fences

Wooden palisade fences
Type WPC with concrete posts 140 × 115 mm tapered and Type WPW with timber posts

For both types of fence:
Height of fence 1050 mm: two rails
Height of fence 1200 mm: two rails
Height of fence 1500 mm: three rails
Height of fence 1650 mm: three rails
Height of fence 1800 mm: three rails

FENCING AND GATES

Post and Rail Fences

Wooden post and rail fences Type MPR 11/3 morticed rails and Type SPR 11/3 nailed rails Height to top of rail 1100 mm Rails: three rails 87 mm, 38 mm Type MPR 11/4 morticed rails and Type SPR 11/4 nailed rails Height to top of rail 1100 mm Rails: four rails 87 mm, 38 mm Type MPR 13/4 morticed rails and Type SPR 13/4 nailed rails Height to top of rail 1300 mm Rail spacing 250 mm, 250 mm, and 225 mm from top Rails: four rails 87 mm, 38 mm

Steel Posts

Rolled steel angle iron posts for chain link fencing

Posts	Fence height	Strut	Straining post
1500 × 40 × 40 × 5 mm	900 mm	1500 × 40 × 40 × 5 mm	1500 × 50 × 50 × 6 mm
1800 × 40 × 40 × 5 mm	1200 mm	1800 × 40 × 40 × 5 mm	1800 × 50 × 50 × 6 mm
2000 × 45 × 45 × 5 mm	1400 mm	2000 × 45 × 45 × 5 mm	2000 × 60 × 60 × 6 mm
2600 × 45 × 45 × 5 mm	1800 mm	2600 × 45 × 45 × 5 mm	2600 × 60 × 60 × 6 mm
3000 × 50 × 50 × 6 mm	1800 mm	2600 × 45 × 45 × 5 mm	3000 × 60 × 60 × 6 mm
with arms			

Concrete Posts

Concrete posts for chain link fencing

Posts and straining posts	Fence height	Strut
1570 mm 100 × 100 mm	900 mm	1500 mm × 75 × 75 mm
1870 mm 125 × 125 mm	1200 mm	1830 mm × 100 × 75 mm
2070 mm 125 × 125 mm	1400 mm	1980 mm × 100 × 75 mm
2620 mm 125 × 125 mm	1800 mm	2590 mm × 100 × 85 mm
3040 mm 125 × 125 mm	1800 mm	2590 mm × 100 × 85 mm (with arms)

FENCING AND GATES

Rolled Steel Angle Posts

Rolled steel angle posts for rectangular wire mesh (field) fencing

Posts	Fence height	Strut	Straining post
1200 × 40 × 40 × 5 mm	600 mm	1200 × 75 × 75 mm	1350 × 100 × 100 mm
1400 × 40 × 40 × 5 mm	800 mm	1400 × 75 × 75 mm	1550 × 100 × 100 mm
1500 × 40 × 40 × 5 mm	900 mm	1500 × 75 × 75 mm	1650 × 100 × 100 mm
1600 × 40 × 40 × 5 mm	1000 mm	1600 × 75 × 75 mm	1750 × 100 × 100 mm
1750 × 40 × 40 × 5 mm	1150 mm	1750 × 75 × 100 mm	1900 × 125 × 125 mm

Concrete Posts

Concrete posts for rectangular wire mesh (field) fencing

Posts	Fence height	Strut	Straining post
1270 × 100 × 100 mm	600 mm	1200 × 75 × 75 mm	1420 × 100 × 100 mm
1470 × 100 × 100 mm	800 mm	1350 × 75 × 75 mm	1620 × 100 × 100 mm
1570 × 100 × 100 mm	900 mm	1500 × 75 × 75 mm	1720 × 100 × 100 mm
1670 × 100 × 100 mm	600 mm	1650 × 75 × 75 mm	1820 × 100 × 100 mm
1820 × 125 × 125 mm	1150 mm	1830 × 75 × 100 mm	1970 × 125 × 125 mm

Cleft Chestnut Pale Fences

Timber Posts

Timber posts for wire mesh and hexagonal wire netting fences

Round timber for general fences

Posts	Fence height	Strut	Straining post
1300 × 65 mm dia.	600 mm	1200 × 80 mm dia.	1450 × 100 mm dia.
1500 × 65 mm dia.	800 mm	1400 × 80 mm dia.	1650 × 100 mm dia.
1600 × 65 mm dia.	900 mm	1500 × 80 mm dia.	1750 × 100 mm dia.
1700 × 65 mm dia.	1050 mm	1600 × 80 mm dia.	1850 × 100 mm dia.
1800 × 65 mm dia.	1150 mm	1750 × 80 mm dia.	2000 × 120 mm dia.

Squared timber for general fences

Posts	Fence height	Strut	Straining post
1300 × 75 × 75 mm	600 mm	1200 × 75 × 75 mm	1450 × 100 × 100 mm
1500 × 75 × 75 mm	800 mm	1400 × 75 × 75 mm	1650 × 100 × 100 mm
1600 × 75 × 75 mm	900 mm	1500 × 75 × 75 mm	1750 × 100 × 100 mm
1700 × 75 × 75 mm	1050 mm	1600 × 75 × 75 mm	1850 × 100 × 100 mm
1800 × 75 × 75 mm	1150 mm	1750 × 75 × 75 mm	2000 × 125 × 100 mm

FENCING AND GATES

Steel Fences to BS 1722: Part 9: 1992

	Fence height	Top/bottom rails and flat posts	Vertical bars
Light	1000 mm	40 × 10 mm 450 mm in ground	12 mm dia. at 115 mm cs
	1200 mm	40 × 10 mm 550 mm in ground	12 mm dia. at 115 mm cs
	1400 mm	40 × 10 mm 550 mm in ground	12 mm dia. at 115 mm cs
Light	1000 mm	40 × 10 mm 450 mm in ground	16 mm dia. at 120 mm cs
	1200 mm	40 × 10 mm 550 mm in ground	16 mm dia. at 120 mm cs
	1400 mm	40 × 10 mm 550 mm in ground	16 mm dia. at 120 mm cs
Medium	1200 mm	50 × 10 mm 550 mm in ground	20 mm dia. at 125 mm cs
	1400 mm	50 × 10 mm 550 mm in ground	20 mm dia. at 125 mm cs
	1600 mm	50 × 10 mm 600 mm in ground	22 mm dia. at 145 mm cs
	1800 mm	50 × 10 mm 600 mm in ground	22 mm dia. at 145 mm cs
Heavy	1600 mm	50 × 10 mm 600 mm in ground	22 mm dia. at 145 mm cs
	1800 mm	50 × 10 mm 600 mm in ground	22 mm dia. at 145 mm cs
	2000 mm	50 × 10 mm 600 mm in ground	22 mm dia. at 145 mm cs
	2200 mm	50 × 10 mm 600 mm in ground	22 mm dia. at 145 mm cs

Notes: Mild steel fences: round or square verticals; flat standards and horizontals. Tops of vertical bars may be bow-top, blunt, or pointed. Round or square bar railings

Timber Field Gates to BS 3470: 1975

Gates made to this standard are designed to open one way only
All timber gates are 1100 mm high
Width over stiles 2400, 2700, 3000, 3300, 3600, and 4200 mm
Gates over 4200 mm should be made in two leaves

Steel Field Gates to BS 3470: 1975

All steel gates are 1100 mm high
Heavy duty: width over stiles 2400, 3000, 3600 and 4500 mm
Light duty: width over stiles 2400, 3000, and 3600 mm

FENCING AND GATES

Domestic Front Entrance Gates to BS 4092: Part 1: 1966

Metal gates:	Single gates are 900 mm high minimum, 900 mm, 1000 mm and 1100 mm wide

Domestic Front Entrance Gates to BS 4092: Part 2: 1966

Wooden gates:	All rails shall be tenoned into the stiles Single gates are 840 mm high minimum, 801 mm and 1020 mm wide Double gates are 840 mm high minimum, 2130, 2340 and 2640 mm wide

Timber Bridle Gates to BS 5709:1979 (Horse or Hunting Gates)

Gates open one way only Minimum width between posts Minimum height	 1525 mm 1100 mm

Timber Kissing Gates to BS 5709:1979

Minimum width	700 mm
Minimum height	1000 mm
Minimum distance between shutting posts	600 mm
Minimum clearance at mid-point	600 mm

Metal Kissing Gates to BS 5709:1979

Sizes are the same as those for timber kissing gates Maximum gaps between rails 120 mm

Categories of Pedestrian Guard Rail to BS 3049:1976

Class A for normal use Class B where vandalism is expected Class C where crowd pressure is likely

DRAINAGE

Width Required for Trenches for Various Diameters of Pipes

Pipe diameter (mm)	Trench n.e. 1.50 m deep	Trench over 1.50 m deep
n.e. 100 mm	450 mm	600 mm
100–150 mm	500 mm	650 mm
150–225 mm	600 mm	750 mm
225–300 mm	650 mm	800 mm
300–400 mm	750 mm	900 mm
400–450 mm	900 mm	1050 mm
450–600 mm	1100 mm	1300 mm

Weights and Dimensions – Vitrified Clay Pipes

Product	Nominal diameter (mm)	Effective length (mm)	BS 65 limits of tolerance min (mm)	max (mm)	Crushing strength (kN/m)	Weight (kg/pipe)	(kg/m)
Supersleve	100	1600	96	105	35.00	14.71	9.19
	150	1750	146	158	35.00	29.24	16.71
Hepsleve	225	1850	221	236	28.00	84.03	45.42
	300	2500	295	313	34.00	193.05	77.22
	150	1500	146	158	22.00	37.04	24.69
Hepseal	225	1750	221	236	28.00	85.47	48.84
	300	2500	295	313	34.00	204.08	81.63
	400	2500	394	414	44.00	357.14	142.86
	450	2500	444	464	44.00	454.55	181.63
	500	2500	494	514	48.00	555.56	222.22
	600	2500	591	615	57.00	796.23	307.69
	700	3000	689	719	67.00	1111.11	370.45
	800	3000	788	822	72.00	1351.35	450.45
Hepline	100	1600	95	107	22.00	14.71	9.19
	150	1750	145	160	22.00	29.24	16.71
	225	1850	219	239	28.00	84.03	45.42
	300	1850	292	317	34.00	142.86	77.22
Hepduct (conduit)	90	1500	–	–	28.00	12.05	8.03
	100	1600	–	–	28.00	14.71	9.19
	125	1750	–	–	28.00	20.73	11.84
	150	1750	–	–	28.00	29.24	16.71
	225	1850	–	–	28.00	84.03	45.42
	300	1850	–	–	34.00	142.86	77.22

DRAINAGE

Weights and Dimensions – Vitrified Clay Pipes

Nominal internal diameter (mm)	Nominal wall thickness (mm)	Approximate weight (kg/m)
150	25	45
225	29	71
300	32	122
375	35	162
450	38	191
600	48	317
750	54	454
900	60	616
1200	76	912
1500	89	1458
1800	102	1884
2100	127	2619

Wall thickness, weights and pipe lengths vary, depending on type of pipe required
The particulars shown above represent a selection of available diameters and are applicable to strength class 1 pipes with flexible rubber ring joints
Tubes with Ogee joints are also available

DRAINAGE

Weights and Dimensions – PVC-u Pipes

	Nominal size	Mean outside diameter (mm)		Wall thickness	Weight
		min	max	(mm)	(kg/m)
Standard pipes	82.4	82.4	82.7	3.2	1.2
	110.0	110.0	110.4	3.2	1.6
	160.0	160.0	160.6	4.1	3.0
	200.0	200.0	200.6	4.9	4.6
	250.0	250.0	250.7	6.1	7.2
Perforated pipes heavy grade	As above	As above	As above	As above	As above
thin wall	82.4	82.4	82.7	1.7	–
	110.0	110.0	110.4	2.2	–
	160.0	160.0	160.6	3.2	–

Width of Trenches Required for Various Diameters of Pipes

Pipe diameter (mm)	Trench n.e. 1.5 m deep (mm)	Trench over 1.5 m deep (mm)
n.e. 100	450	600
100–150	500	650
150–225	600	750
225–300	650	800
300–400	750	900
400–450	900	1050
450–600	1100	1300

DRAINAGE

DRAINAGE BELOW GROUND AND LAND DRAINAGE

Flow of Water Which Can Be Carried by Various Sizes of Pipe

Clay or concrete pipes

Pipe size	Gradient of pipeline							
	1:10	1:20	1:30	1:40	1:50	1:60	1:80	1:100
	Flow in litres per second							
DN 100 15.0	8.5	6.8	5.8	5.2	4.7	4.0	3.5	
DN 150 28.0	19.0	16.0	14.0	12.0	11.0	9.1	8.0	
DN 225 140.0	95.0	76.0	66.0	58.0	53.0	46.0	40.0	

Plastic pipes

Pipe size	Gradient of pipeline							
	1:10	1:20	1:30	1:40	1:50	1:60	1:80	1:100
	Flow in litres per second							
82.4 mm i/dia.	12.0	8.5	6.8	5.8	5.2	4.7	4.0	3.5
110 mm i/dia.	28.0	19.0	16.0	14.0	12.0	11.0	9.1	8.0
160 mm i/dia.	76.0	53.0	43.0	37.0	33.0	29.0	25.0	22.0
200 mm i/dia.	140.0	95.0	76.0	66.0	58.0	53.0	46.0	40.0

Vitrified (Perforated) Clay Pipes and Fittings to BS En 295-5 1994

Length not specified		
75 mm bore	**250 mm bore**	**600 mm bore**
100	300	700
125	350	800
150	400	1000
200	450	1200
225	500	

Precast Concrete Pipes: Prestressed Non-pressure Pipes and Fittings: Flexible Joints to BS 5911: Pt. 103: 1994

Rationalized metric nominal sizes: 450, 500	
Length:	500–1000 by 100 increments
	1000–2200 by 200 increments
	2200–2800 by 300 increments
Angles: length:	450–600 angles 45, 22.5,11.25°
	600 or more angles 22.5, 11.25°

DRAINAGE

Precast Concrete Pipes: Unreinforced and Circular Manholes and Soakaways to BS 5911: Pt. 200: 1994

Nominal sizes:	
Shafts:	675, 900 mm
Chambers:	900, 1050, 1200, 1350, 1500, 1800, 2100, 2400, 2700, 3000 mm
Large chambers:	To have either tapered reducing rings or a flat reducing slab in order to accept the standard cover
Ring depths:	1. 300–1200 mm by 300 mm increments except for bottom slab and rings below cover slab, these are by 150 mm increments
	2. 250–1000 mm by 250 mm increments except for bottom slab and rings below cover slab, these are by 125 mm increments
Access hole:	750 × 750 mm for DN 1050 chamber 1200 × 675 mm for DN 1350 chamber

Calculation of Soakaway Depth

The following formula determines the depth of concrete ring soakaway that would be required for draining given amounts of water.

$$h = \frac{4ar}{3\pi D^2}$$

h = depth of the chamber below the invert pipe
a = the area to be drained
r = the hourly rate of rainfall (50 mm per hour)
π = pi
D = internal diameter of the soakaway

This table shows the depth of chambers in each ring size which would be required to contain the volume of water specified. These allow a recommended storage capacity of ⅓ (one third of the hourly rainfall figure).

Table Showing Required Depth of Concrete Ring Chambers in Metres

Area m²	50	100	150	200	300	400	500
Ring size							
0.9	1.31	2.62	3.93	5.24	7.86	10.48	13.10
1.1	0.96	1.92	2.89	3.85	5.77	7.70	9.62
1.2	0.74	1.47	2.21	2.95	4.42	5.89	7.37
1.4	0.58	1.16	1.75	2.33	3.49	4.66	5.82
1.5	0.47	0.94	1.41	1.89	2.83	3.77	4.72
1.8	0.33	0.65	0.98	1.31	1.96	2.62	3.27
2.1	0.24	0.48	0.72	0.96	1.44	1.92	2.41
2.4	0.18	0.37	0.55	0.74	1.11	1.47	1.84
2.7	0.15	0.29	0.44	0.58	0.87	1.16	1.46
3.0	0.12	0.24	0.35	0.47	0.71	0.94	1.18

DRAINAGE

Precast Concrete Inspection Chambers and Gullies to BS 5911: Part 230: 1994

Nominal sizes:	375 diameter, 750, 900 mm deep
	450 diameter, 750, 900, 1050, 1200 mm deep
Depths:	from the top for trapped or untrapped units:
	centre of outlet 300 mm
	invert (bottom) of the outlet pipe 400 mm
Depth of water seal for trapped gullies:	
	85 mm, rodding eye int. dia. 100 mm
Cover slab:	65 mm min

Bedding Flexible Pipes: PVC-u Or Ductile Iron

Type 1 =	100 mm fill below pipe, 300 mm above pipe: single size material
Type 2 =	100 mm fill below pipe, 300 mm above pipe: single size or graded material
Type 3 =	100 mm fill below pipe, 75 mm above pipe with concrete protective slab over
Type 4 =	100 mm fill below pipe, fill laid level with top of pipe
Type 5 =	200 mm fill below pipe, fill laid level with top of pipe
Concrete =	25 mm sand blinding to bottom of trench, pipe supported on chocks, 100 mm concrete under the pipe, 150 mm concrete over the pipe

DRAINAGE

Bedding Rigid Pipes: Clay or Concrete
(for vitrified clay pipes the manufacturer should be consulted)

Class D:	Pipe laid on natural ground with cut-outs for joints, soil screened to remove stones over 40 mm and returned over pipe to 150 m min depth. Suitable for firm ground with trenches trimmed by hand.
Class N:	Pipe laid on 50 mm granular material of graded aggregate to Table 4 of BS 882, or 10 mm aggregate to Table 6 of BS 882, or as dug light soil (not clay) screened to remove stones over 10 mm. Suitable for machine dug trenches.
Class B:	As Class N, but with granular bedding extending half way up the pipe diameter.
Class F:	Pipe laid on 100 mm granular fill to BS 882 below pipe, minimum 150 mm granular fill above pipe: single size material. Suitable for machine dug trenches.
Class A:	Concrete 100 mm thick under the pipe extending half way up the pipe, backfilled with the appropriate class of fill. Used where there is only a very shallow fall to the drain. Class A bedding allows the pipes to be laid to an exact gradient.
Concrete surround:	25 mm sand blinding to bottom of trench, pipe supported on chocks, 100 mm concrete under the pipe, 150 mm concrete over the pipe. It is preferable to bed pipes under slabs or wall in granular material.

PIPED SUPPLY SYSTEMS

Identification of Service Tubes From Utility to Dwellings

Utility	Colour	Size	Depth
British Telecom	grey	54 mm od	450 mm
Electricity	black	38 mm od	450 mm
Gas	yellow	42 mm od rigid 60 mm od convoluted	450 mm
Water	may be blue	(normally untubed)	750 mm

ELECTRICAL SUPPLY/POWER/LIGHTING SYSTEMS

Electrical Insulation Class En 60.598 BS 4533

Class 1:	luminaires comply with class 1 (I) earthed electrical requirements
Class 2:	luminaires comply with class 2 (II) double insulated electrical requirements
Class 3:	luminaires comply with class 3 (III) electrical requirements

Protection to Light Fittings

BS EN 60529:1992 Classification for degrees of protection provided by enclosures.
(IP Code – International or ingress Protection)

1st characteristic: against ingress of solid foreign objects

The figure	2	indicates that fingers cannot enter
	3	that a 2.5 mm diameter probe cannot enter
	4	that a 1.0 mm diameter probe cannot enter
	5	the fitting is dust proof (no dust around live parts)
	6	the fitting is dust tight (no dust entry)

2nd characteristic: ingress of water with harmful effects

The figure	0	indicates unprotected
	1	vertically dripping water cannot enter
	2	water dripping 15° (tilt) cannot enter
	3	spraying water cannot enter
	4	splashing water cannot enter
	5	jetting water cannot enter
	6	powerful jetting water cannot enter
	7	proof against temporary immersion
	8	proof against continuous immersion

Optional additional codes:		A–D protects against access to hazardous parts
	H	high voltage apparatus
	M	fitting was in motion during water test
	S	fitting was static during water test
	W	protects against weather

Marking code arrangement:	(example) IPX5S = IP (International or Ingress Protection)
	X (denotes omission of first characteristic)
	5 = jetting
	S = static during water test

RAIL TRACKS

	kg/m of track	lb/ft of track
Standard gauge Bull-head rails, chairs, transverse timber (softwood) sleepers etc.	245	165
Main lines Flat-bottom rails, transverse prestressed concrete sleepers, etc.	418	280
Add for electric third rail	51	35
Add for crushed stone ballast	2600	1750
	kg/m²	**lb/sq ft**
Overall average weight – rails connections, sleepers, ballast, etc.	733	150
	kg/m of track	**lb/ft of track**
Bridge rails, longitudinal timber sleepers, etc.	112	75

RAIL TRACKS

Heavy Rails

British Standard Section No.	Rail height (mm)	Foot width (mm)	Head width (mm)	Min web thickness (mm)	Section weight (kg/m)
Flat Bottom Rails					
60 A	114.30	109.54	57.15	11.11	30.62
70 A	123.82	111.12	60.32	12.30	34.81
75 A	128.59	114.30	61.91	12.70	37.45
80 A	133.35	117.47	63.50	13.10	39.76
90 A	142.88	127.00	66.67	13.89	45.10
95 A	147.64	130.17	69.85	14.68	47.31
100 A	152.40	133.35	69.85	15.08	50.18
110 A	158.75	139.70	69.85	15.87	54.52
113 A	158.75	139.70	69.85	20.00	56.22
50 'O'	100.01	100.01	52.39	10.32	24.82
80 'O'	127.00	127.00	63.50	13.89	39.74
60R	114.30	109.54	57.15	11.11	29.85
75R	128.59	122.24	61.91	13.10	37.09
80R	133.35	127.00	63.50	13.49	39.72
90R	142.88	136.53	66.67	13.89	44.58
95R	147.64	141.29	68.26	14.29	47.21
100R	152.40	146.05	69.85	14.29	49.60
95N	147.64	139.70	69.85	13.89	47.27
Bull Head Rails					
95R BH	145.26	69.85	69.85	19.05	47.07

Light Rails

British Standard Section No.	Rail height (mm)	Foot width (mm)	Head width (mm)	Min web thickness (mm)	Section weight (kg/m)
Flat Bottom Rails					
20M	65.09	55.56	30.96	6.75	9.88
30M	75.41	69.85	38.10	9.13	14.79
35M	80.96	76.20	42.86	9.13	17.39
35R	85.73	82.55	44.45	8.33	17.40
40	88.11	80.57	45.64	12.3	19.89
Bridge Rails					
13	48.00	92	36.00	18.0	13.31
16	54.00	108	44.50	16.0	16.06
20	55.50	127	50.00	20.5	19.86
28	67.00	152	50.00	31.0	28.62
35	76.00	160	58.00	34.5	35.38
50	76.00	165	58.50	–	50.18
Crane Rails					
A65	75.00	175.00	65.00	38.0	43.10
A75	85.00	200.00	75.00	45.0	56.20
A100	95.00	200.00	100.00	60.0	74.30
A120	105.00	220.00	120.00	72.0	100.00
175CR	152.40	152.40	107.95	38.1	86.92

RAIL TRACKS

Fish Plates

British Standard Section No.	Overall plate length		Hole diameter	Finished weight per pair	
	4 Hole (mm)	6 Hole (mm)	(mm)	4 Hole (kg/pair)	6 Hole (kg/pair)
For British Standard Heavy Rails: Flat Bottom Rails					
60 A	406.40	609.60	20.64	9.87	14.76
70 A	406.40	609.60	22.22	11.15	16.65
75 A	406.40	–	23.81	11.82	17.73
80 A	406.40	609.60	23.81	13.15	19.72
90 A	457.20	685.80	25.40	17.49	26.23
100 A	508.00	–	pear	25.02	–
110 A (shallow)	507.00	–	27.00	30.11	54.64
113 A (heavy)	507.00	–	27.00	30.11	54.64
50 'O' (shallow)	406.40	–	–	6.68	10.14
80 'O' (shallow)	495.30	–	23.81	14.72	22.69
60R (shallow)	406.40	609.60	20.64	8.76	13.13
60R (angled)	406.40	609.60	20.64	11.27	16.90
75R (shallow)	406.40	–	23.81	10.94	16.42
75R (angled)	406.40	–	23.81	13.67	–
80R (shallow)	406.40	609.60	23.81	11.93	17.89
80R (angled)	406.40	609.60	23.81	14.90	22.33
For British Standard Heavy Rails: Bull head rails					
95R BH (shallow)	–	457.20	27.00	14.59	14.61
For British Standard Light Rails: Flat Bottom Rails					
30M	355.6	–	–	–	2.72
35M	355.6	–	–	–	2.83
40	355.6	–	–	3.76	–

FRACTIONS, DECIMALS AND MILLIMETRE EQUIVALENTS

Fractions	Decimals	(mm)		Fractions	Decimals	(mm)
1/64	0.015625	0.396875		33/64	0.515625	13.096875
1/32	0.03125	0.79375		17/32	0.53125	13.49375
3/64	0.046875	1.190625		35/64	0.546875	13.890625
1/16	0.0625	1.5875		9/16	0.5625	14.2875
5/64	0.078125	1.984375		37/64	0.578125	14.684375
3/32	0.09375	2.38125		19/32	0.59375	15.08125
7/64	0.109375	2.778125		39/64	0.609375	15.478125
1/8	0.125	3.175		5/8	0.625	15.875
9/64	0.140625	3.571875		41/64	0.640625	16.271875
5/32	0.15625	3.96875		21/32	0.65625	16.66875
11/64	0.171875	4.365625		43/64	0.671875	17.065625
3/16	0.1875	4.7625		11/16	0.6875	17.4625
13/64	0.203125	5.159375		45/64	0.703125	17.859375
7/32	0.21875	5.55625		23/32	0.71875	18.25625
15/64	0.234375	5.953125		47/64	0.734375	18.653125
1/4	0.25	6.35		3/4	0.75	19.05
17/64	0.265625	6.746875		49/64	0.765625	19.446875
9/32	0.28125	7.14375		25/32	0.78125	19.84375
19/64	0.296875	7.540625		51/64	0.796875	20.240625
5/16	0.3125	7.9375		13/16	0.8125	20.6375
21/64	0.328125	8.334375		53/64	0.828125	21.034375
11/32	0.34375	8.73125		27/32	0.84375	21.43125
23/64	0.359375	9.128125		55/64	0.859375	21.828125
3/8	0.375	9.525		7/8	0.875	22.225
25/64	0.390625	9.921875		57/64	0.890625	22.621875
13/32	0.40625	10.31875		29/32	0.90625	23.01875
27/64	0.421875	10.71563		59/64	0.921875	23.415625
7/16	0.4375	11.1125		15/16	0.9375	23.8125
29/64	0.453125	11.50938		61/64	0.953125	24.209375
15/32	0.46875	11.90625		31/32	0.96875	24.60625
31/64	0.484375	12.30313		63/64	0.984375	25.003125
1/2	0.5	12.7		1.0	1	25.4

IMPERIAL STANDARD WIRE GAUGE (SWG)

SWG No.	Diameter (inches)	Diameter (mm)	SWG No.	Diameter (inches)	Diameter (mm)
7/0	0.5	12.7	23	0.024	0.61
6/0	0.464	11.79	24	0.022	0.559
5/0	0.432	10.97	25	0.02	0.508
4/0	0.4	10.16	26	0.018	0.457
3/0	0.372	9.45	27	0.0164	0.417
2/0	0.348	8.84	28	0.0148	0.376
1/0	0.324	8.23	29	0.0136	0.345
1	0.3	7.62	30	0.0124	0.315
2	0.276	7.01	31	0.0116	0.295
3	0.252	6.4	32	0.0108	0.274
4	0.232	5.89	33	0.01	0.254
5	0.212	5.38	34	0.009	0.234
6	0.192	4.88	35	0.008	0.213
7	0.176	4.47	36	0.008	0.193
8	0.16	4.06	37	0.007	0.173
9	0.144	3.66	38	0.006	0.152
10	0.128	3.25	39	0.005	0.132
11	0.116	2.95	40	0.005	0.122
12	0.104	2.64	41	0.004	0.112
13	0.092	2.34	42	0.004	0.102
14	0.08	2.03	43	0.004	0.091
15	0.072	1.83	44	0.003	0.081
16	0.064	1.63	45	0.003	0.071
17	0.056	1.42	46	0.002	0.061
18	0.048	1.22	47	0.002	0.051
19	0.04	1.016	48	0.002	0.041
20	0.036	0.914	49	0.001	0.031
21	0.032	0.813	50	0.001	0.025
22	0.028	0.711			

PIPES, WATER, STORAGE, INSULATION

WATER PRESSURE DUE TO HEIGHT

Imperial

Head (Feet)	Pressure (lb/in²)		Head (Feet)	Pressure (lb/in²)
1	0.43		70	30.35
5	2.17		75	32.51
10	4.34		80	34.68
15	6.5		85	36.85
20	8.67		90	39.02
25	10.84		95	41.18
30	13.01		100	43.35
35	15.17		105	45.52
40	17.34		110	47.69
45	19.51		120	52.02
50	21.68		130	56.36
55	23.84		140	60.69
60	26.01		150	65.03
65	28.18			

Metric

Head (m)	Pressure (bar)		Head (m)	Pressure (bar)
0.5	0.049		18.0	1.766
1.0	0.098		19.0	1.864
1.5	0.147		20.0	1.962
2.0	0.196		21.0	2.06
3.0	0.294		22.0	2.158
4.0	0.392		23.0	2.256
5.0	0.491		24.0	2.354
6.0	0.589		25.0	2.453
7.0	0.687		26.0	2.551
8.0	0.785		27.0	2.649
9.0	0.883		28.0	2.747
10.0	0.981		29.0	2.845
11.0	1.079		30.0	2.943
12.0	1.177		32.5	3.188
13.0	1.275		35.0	3.434
14.0	1.373		37.5	3.679
15.0	1.472		40.0	3.924
16.0	1.57		42.5	4.169
17.0	1.668		45.0	4.415

1 bar	=	14.5038 lbf/in²
1 lbf/in²	=	0.06895 bar
1 metre	=	3.2808 ft or 39.3701 in
1 foot	=	0.3048 metres
1 in wg	=	2.5 mbar (249.1 N/m²)

PIPES, WATER, STORAGE, INSULATION

Dimensions and Weights of Copper Pipes to BSEN 1057, BSEN 12499, BSEN 14251

Outside diameter (mm)	Internal diameter (mm)	Weight per metre (kg)	Internal diameter (mm)	Weight per metre (kg)	Internal siameter (mm)	Weight per metre (kg)
	Formerly Table X		Formerly Table Y		Formerly Table Z	
6	4.80	0.0911	4.40	0.1170	5.00	0.0774
8	6.80	0.1246	6.40	0.1617	7.00	0.1054
10	8.80	0.1580	8.40	0.2064	9.00	0.1334
12	10.80	0.1914	10.40	0.2511	11.00	0.1612
15	13.60	0.2796	13.00	0.3923	14.00	0.2031
18	16.40	0.3852	16.00	0.4760	16.80	0.2918
22	20.22	0.5308	19.62	0.6974	20.82	0.3589
28	26.22	0.6814	25.62	0.8985	26.82	0.4594
35	32.63	1.1334	32.03	1.4085	33.63	0.6701
42	39.63	1.3675	39.03	1.6996	40.43	0.9216
54	51.63	1.7691	50.03	2.9052	52.23	1.3343
76.1	73.22	3.1287	72.22	4.1437	73.82	2.5131
108	105.12	4.4666	103.12	7.3745	105.72	3.5834
133	130.38	5.5151	–	–	130.38	5.5151
159	155.38	8.7795	–	–	156.38	6.6056

Dimensions of Stainless Steel Pipes to BS 4127

Outside siameter (mm)	Maximum outside siameter (mm)	Minimum outside diameter (mm)	Wall thickness (mm)	Working pressure (bar)
6	6.045	5.940	0.6	330
8	8.045	7.940	0.6	260
10	10.045	9.940	0.6	210
12	12.045	11.940	0.6	170
15	15.045	14.940	0.6	140
18	18.045	17.940	0.7	135
22	22.055	21.950	0.7	110
28	28.055	27.950	0.8	121
35	35.070	34.965	1.0	100
42	42.070	41.965	1.1	91
54	54.090	53.940	1.2	77

PIPES, WATER, STORAGE, INSULATION

Dimensions of Steel Pipes to BS 1387

Nominal Size	Approx. Outside Diameter	Outside diameter				Thickness		
		Light		Medium & Heavy		Light	Medium	Heavy
		Max	Min	Max	Min			
(mm)	(mm)	(mm)	(mm)	(mm)	(mm)	(mm)	(mm)	(mm)
6	10.20	10.10	9.70	10.40	9.80	1.80	2.00	2.65
8	13.50	13.60	13.20	13.90	13.30	1.80	2.35	2.90
10	17.20	17.10	16.70	17.40	16.80	1.80	2.35	2.90
15	21.30	21.40	21.00	21.70	21.10	2.00	2.65	3.25
20	26.90	26.90	26.40	27.20	26.60	2.35	2.65	3.25
25	33.70	33.80	33.20	34.20	33.40	2.65	3.25	4.05
32	42.40	42.50	41.90	42.90	42.10	2.65	3.25	4.05
40	48.30	48.40	47.80	48.80	48.00	2.90	3.25	4.05
50	60.30	60.20	59.60	60.80	59.80	2.90	3.65	4.50
65	76.10	76.00	75.20	76.60	75.40	3.25	3.65	4.50
80	88.90	88.70	87.90	89.50	88.10	3.25	4.05	4.85
100	114.30	113.90	113.00	114.90	113.30	3.65	4.50	5.40
125	139.70	–	–	140.60	138.70	–	4.85	5.40
150	165.1*	–	–	166.10	164.10	–	4.85	5.40

* 165.1 mm (6.5in) outside diameter is not generally recommended except where screwing to BS 21 is necessary
All dimensions are in accordance with ISO R65 except approximate outside diameters which are in accordance with ISO R64
Light quality is equivalent to ISO R65 Light Series II

Approximate Metres Per Tonne of Tubes to BS 1387

Nom. size	BLACK						GALVANIZED					
	Plain/screwed ends			Screwed & socketed			Plain/screwed ends			Screwed & socketed		
	L	M	H	L	M	H	L	M	H	L	M	H
(mm)	(m)	(m)	(m)	(m)	(m)	(m)	(m)	(m)	(m)	(m)	(m)	(m)
6	2765	2461	2030	2743	2443	2018	2604	2333	1948	2584	2317	1937
8	1936	1538	1300	1920	1527	1292	1826	1467	1254	1811	1458	1247
10	1483	1173	979	1471	1165	974	1400	1120	944	1386	1113	939
15	1050	817	688	1040	811	684	996	785	665	987	779	661
20	712	634	529	704	628	525	679	609	512	673	603	508
25	498	410	336	494	407	334	478	396	327	474	394	325
32	388	319	260	384	316	259	373	308	254	369	305	252
40	307	277	226	303	273	223	296	268	220	292	264	217
50	244	196	162	239	194	160	235	191	158	231	188	157
65	172	153	127	169	151	125	167	149	124	163	146	122
80	147	118	99	143	116	98	142	115	97	139	113	96
100	101	82	69	98	81	68	98	81	68	95	79	67
125	–	62	56	–	60	55	–	60	55	–	59	54
150	–	52	47	–	50	46	–	51	46	–	49	45

The figures for 'plain or screwed ends' apply also to tubes to BS 1775 of equivalent size and thickness
Key:
L – Light
M – Medium
H – Heavy

PIPES, WATER, STORAGE, INSULATION

Flange Dimension Chart to BS 4504 & BS 10

Normal Pressure Rating (PN 6) 6 Bar

Nom. size	Flange outside dia.	Table 6/2 Forged Welding Neck	Table 6/3 Plate Slip on	Table 6/4 Forged Bossed Screwed	Table 6/5 Forged Bossed Slip on	Table 6/8 Plate Blank	Raised face		Nr. bolt hole	Size of bolt
							Dia.	T'ness		
15	80	12	12	12	12	12	40	2	4	M10 × 40
20	90	14	14	14	14	14	50	2	4	M10 × 45
25	100	14	14	14	14	14	60	2	4	M10 × 45
32	120	14	16	14	14	14	70	2	4	M12 × 45
40	130	14	16	14	14	14	80	3	4	M12 × 45
50	140	14	16	14	14	14	90	3	4	M12 × 45
65	160	14	16	14	14	14	110	3	4	M12 × 45
80	190	16	18	16	16	16	128	3	4	M16 × 55
100	210	16	18	16	16	16	148	3	4	M16 × 55
125	240	18	20	18	18	18	178	3	8	M16 × 60
150	265	18	20	18	18	18	202	3	8	M16 × 60
200	320	20	22	–	20	20	258	3	8	M16 × 60
250	375	22	24	–	22	22	312	3	12	M16 × 65
300	440	22	24	–	22	22	365	4	12	M20 × 70

Normal Pressure Rating (PN 16) 16 Bar

Nom. size	Flange outside dia.	Table 6/2 Forged Welding Neck	Table 6/3 Plate Slip on	Table 6/4 Forged Bossed Screwed	Table 6/5 Forged Bossed Slip on	Table 6/8 Plate Blank	Raised face		Nr. bolt hole	Size of bolt
							Dia.	T'ness		
15	95	14	14	14	14	14	45	2	4	M12 × 45
20	105	16	16	16	16	16	58	2	4	M12 × 50
25	115	16	16	16	16	16	68	2	4	M12 × 50
32	140	16	16	16	16	16	78	2	4	M16 × 55
40	150	16	16	16	16	16	88	3	4	M16 × 55
50	165	18	18	18	18	18	102	3	4	M16 × 60
65	185	18	18	18	18	18	122	3	4	M16 × 60
80	200	20	20	20	20	20	138	3	8	M16 × 60
100	220	20	20	20	20	20	158	3	8	M16 × 65
125	250	22	22	22	22	22	188	3	8	M16 × 70
150	285	22	22	22	22	22	212	3	8	M20 × 70
200	340	24	24	–	24	24	268	3	12	M20 × 75
250	405	26	26	–	26	26	320	3	12	M24 × 90
300	460	28	28	–	28	28	378	4	12	M24 × 90

PIPES, WATER, STORAGE, INSULATION

Minimum Distances Between Supports/Fixings

Material	BS Nominal pipe size		Pipes – Vertical	Pipes – Horizontal on to low gradients
	(inch)	(mm)	Support distance in metres	Support distance in metres
Copper	0.50	15.00	1.90	1.30
	0.75	22.00	2.50	1.90
	1.00	28.00	2.50	1.90
	1.25	35.00	2.80	2.50
	1.50	42.00	2.80	2.50
	2.00	54.00	3.90	2.50
	2.50	67.00	3.90	2.80
	3.00	76.10	3.90	2.80
	4.00	108.00	3.90	2.80
	5.00	133.00	3.90	2.80
	6.00	159.00	3.90	2.80
muPVC	1.25	32.00	1.20	0.50
	1.50	40.00	1.20	0.50
	2.00	50.00	1.20	0.60
Polypropylene	1.25	32.00	1.20	0.50
	1.50	40.00	1.20	0.50
uPVC	–	82.40	1.20	0.50
	–	110.00	1.80	0.90
	–	160.00	1.80	1.20
Steel	0.50	15.00	2.40	1.80
	0.75	20.00	3.00	2.40
	1.00	25.00	3.00	2.40
	1.25	32.00	3.00	2.40
	1.50	40.00	3.70	2.40
	2.00	50.00	3.70	2.40
	2.50	65.00	4.60	3.00
	3.00	80.40	4.60	3.00
	4.00	100.00	4.60	3.00
	5.00	125.00	5.50	3.70
	6.00	150.00	5.50	4.50
	8.00	200.00	8.50	6.00
	10.00	250.00	9.00	6.50
	12.00	300.00	10.00	7.00
	16.00	400.00	10.00	8.25

PIPES, WATER, STORAGE, INSULATION

Litres of Water Storage Required Per Person Per Building Type

Type of building	Storage (litres)
Houses and flats (up to 4 bedrooms)	120/bedroom
Houses and flats (more than 4 bedrooms)	100/bedroom
Hostels	90/bed
Hotels	200/bed
Nurses homes and medical quarters	120/bed
Offices with canteen	45/person
Offices without canteen	40/person
Restaurants	7/meal
Boarding schools	90/person
Day schools – Primary	15/person
Day schools – Secondary	20/person

Recommended Air Conditioning Design Loads

Building type	Design loading
Computer rooms	500 W/m² of floor area
Restaurants	150 W/m² of floor area
Banks (main area)	100 W/m² of floor area
Supermarkets	25 W/m² of floor area
Large office block (exterior zone)	100 W/m² of floor area
Large office block (interior zone)	80 W/m² of floor area
Small office block (interior zone)	80 W/m² of floor area

PIPES, WATER, STORAGE, INSULATION

Capacity and Dimensions of Galvanized Mild Steel Cisterns – BS 417

Capacity (litres)	BS type (SCM)	Dimensions		
		Length (mm)	Width (mm)	Depth (mm)
18	45	457	305	305
36	70	610	305	371
54	90	610	406	371
68	110	610	432	432
86	135	610	457	482
114	180	686	508	508
159	230	736	559	559
191	270	762	584	610
227	320	914	610	584
264	360	914	660	610
327	450/1	1220	610	610
336	450/2	965	686	686
423	570	965	762	787
491	680	1090	864	736
709	910	1070	889	889

Capacity of Cold Water Polypropylene Storage Cisterns – BS 4213

Capacity (litres)	BS type (PC)	Maximum height (mm)
18	4	310
36	8	380
68	15	430
91	20	510
114	25	530
182	40	610
227	50	660
273	60	660
318	70	660
455	100	760

PIPES, WATER, STORAGE, INSULATION

Minimum Insulation Thickness to Protect Against Freezing for Domestic Cold Water Systems (8 Hour Evaluation Period)

Pipe size (mm)	Insulation thickness (mm)					
	Condition 1			Condition 2		
	λ = 0.020	λ = 0.030	λ = 0.040	λ = 0.020	λ = 0.030	λ = 0.040
Copper pipes						
15	11	20	34	12	23	41
22	6	9	13	6	10	15
28	4	6	9	4	7	10
35	3	5	7	4	5	7
42	3	4	5	8	4	6
54	2	3	4	2	3	4
76	2	2	3	2	2	3
Steel pipes						
15	9	15	24	10	18	29
20	6	9	13	6	10	15
25	4	7	9	5	7	10
32	3	5	6	3	5	7
40	3	4	5	3	4	6
50	2	3	4	2	3	4
65	2	2	3	2	3	3

Condition 1: water temperature 7°C; ambient temperature −6°C; evaluation period 8 h; permitted ice formation 50%; normal installation, i.e. inside the building and inside the envelope of the structural insulation
Condition 2: water temperature 2°C; ambient temperature −6°C; evaluation period 8 h; permitted ice formation 50%; extreme installation, i.e. inside the building but outside the envelope of the structural insulation
λ = thermal conductivity [W/(mK)]

Insulation Thickness for Chilled And Cold Water Supplies to Prevent Condensation

On a Low Emissivity Outer Surface (0.05, i.e. Bright Reinforced Aluminium Foil) with an Ambient Temperature of +25°C and a Relative Humidity of 80%

Steel pipe size (mm)	t = +10			t = +5			t = 0		
	Insulation thickness (mm)			Insulation thickness (mm)			Insulation thickness (mm)		
	λ = 0.030	λ = 0.040	λ = 0.050	λ = 0.030	λ = 0.040	λ = 0.050	λ = 0.030	λ = 0.040	λ = 0.050
15	16	20	25	22	28	34	28	36	43
25	18	24	29	25	32	39	32	41	50
50	22	28	34	30	39	47	38	49	60
100	26	34	41	36	47	57	46	60	73
150	29	38	46	40	52	64	51	67	82
250	33	43	53	46	60	74	59	77	94
Flat surfaces	39	52	65	56	75	93	73	97	122

t = temperature of contents (°C)
λ = thermal conductivity at mean temperature of insulation [W/(mK)]

PIPES, WATER, STORAGE, INSULATION

Insulation Thickness for Non-domestic Heating Installations to Control Heat Loss

Steel pipe size (mm)	t = 75			t = 100			t = 150		
	Insulation thickness (mm)			Insulation thickness (mm)			Insulation thickness (mm)		
	$\lambda = 0.030$	$\lambda = 0.040$	$\lambda = 0.050$	$\lambda = 0.030$	$\lambda = 0.040$	$\lambda = 0.050$	$\lambda = 0.030$	$\lambda = 0.040$	$\lambda = 0.050$
10	18	32	55	20	36	62	23	44	77
15	19	34	56	21	38	64	26	47	80
20	21	36	57	23	40	65	28	50	83
25	23	38	58	26	43	68	31	53	85
32	24	39	59	28	45	69	33	55	87
40	25	40	60	29	47	70	35	57	88
50	27	42	61	31	49	72	37	59	90
65	29	43	62	33	51	74	40	63	92
80	30	44	62	35	52	75	42	65	94
100	31	46	63	37	54	76	45	68	96
150	33	48	64	40	57	77	50	73	100
200	35	49	65	42	59	79	53	76	103
250	36	50	66	43	61	80	55	78	105

t = hot face temperature (°C)
λ = thermal conductivity at mean temperature of insulation [W/(mK)]

Index

Ebook Single-User Licence Agreement

We welcome you as a user of this Spon Price Book ebook and hope that you find it a useful and valuable tool. Please read this document carefully. **This is a legal agreement** between you (hereinafter referred to as the "Licensee") and Taylor and Francis Books Ltd. (the "Publisher"), which defines the terms under which you may use the Product. **By accessing and retrieving the access code on the label inside the front cover of this book you agree to these terms and conditions outlined herein. If you do not agree to these terms you must return the Product to your supplier intact, with the seal on the label unbroken and with the access code not accessed**.

1. **Definition of the Product**
    The product which is the subject of this Agreement, (the "Product") consists of online and offline access to the VitalSource ebook edition of *Spon's Mechanical & Electrical Services Price Book 2020*.

2. **Commencement and licence**
    2.1   This Agreement commences upon the breaking open of the document containing the access code by the Licensee (the "Commencement Date").
    2.2   This is a licence agreement (the "Agreement") for the use of the Product by the Licensee, and not an agreement for sale.
    2.3   The Publisher licenses the Licensee on a non-exclusive and non-transferable basis to use the Product on condition that the Licensee complies with this Agreement. The Licensee acknowledges that it is only permitted to use the Product in accordance with this Agreement.

3. **Multiple use**
    Use of the Product is not provided or allowed for more than one user or for a wide area network or consortium.

4. **Installation and Use**
    4.1   The Licensee may provide access to the Product for individual study in the following manner: The Licensee may install the Product on a secure local area network on a single site for use by one user.
    4.2   The Licensee shall be responsible for installing the Product and for the effectiveness of such installation.
    4.3   Text from the Product may be incorporated in a coursepack. Such use is only permissible with the express permission of the Publisher in writing and requires the payment of the appropriate fee as specified by the Publisher and signature of a separate licence agreement.
    4.4   The Product is a free addition to the book and the Publisher is under no obligation to provide any technical support.

5. **Permitted Activities**
    5.1   The Licensee shall be entitled to use the Product for its own internal purposes;
    5.2   The Licensee acknowledges that its rights to use the Product are strictly set out in this Agreement, and all other uses (whether expressly mentioned in Clause 6 below or not) are prohibited.

6. **Prohibited Activities**
    The following are prohibited without the express permission of the Publisher:
    6.1   The commercial exploitation of any part of the Product.
    6.2   The rental, loan, (free or for money or money's worth) or hire purchase of this product, save with the express consent of the Publisher.
    6.3   Any activity which raises the reasonable prospect of impeding the Publisher's ability or opportunities to market the Product.
    6.4   Any networking, physical or electronic distribution or dissemination of the product save as expressly permitted by this Agreement.
    6.5   Any reverse engineering, decompilation, disassembly or other alteration of the Product save in accordance with applicable national laws.
    6.6   The right to create any derivative product or service from the Product save as expressly provided for in this Agreement.
    6.7   Any alteration, amendment, modification or deletion from the Product, whether for the purposes of error correction or otherwise.

7. **General Responsibilities of the License**

7.1 The Licensee will take all reasonable steps to ensure that the Product is used in accordance with the terms and conditions of this Agreement.

7.2 The Licensee acknowledges that damages may not be a sufficient remedy for the Publisher in the event of breach of this Agreement by the Licensee, and that an injunction may be appropriate.

7.3 The Licensee undertakes to keep the Product safe and to use its best endeavours to ensure that the product does not fall into the hands of third parties, whether as a result of theft or otherwise.

7.4 Where information of a confidential nature relating to the product of the business affairs of the Publisher comes into the possession of the Licensee pursuant to this Agreement (or otherwise), the Licensee agrees to use such information solely for the purposes of this Agreement, and under no circumstances to disclose any element of the information to any third party save strictly as permitted under this Agreement. For the avoidance of doubt, the Licensee's obligations under this sub-clause 7.4 shall survive the termination of this Agreement.

8. **Warrant and Liability**

8.1 The Publisher warrants that it has the authority to enter into this agreement and that it has secured all rights and permissions necessary to enable the Licensee to use the Product in accordance with this Agreement.

8.2 The Publisher warrants that the Product as supplied on the Commencement Date shall be free of defects in materials and workmanship, and undertakes to replace any defective Product within 28 days of notice of such defect being received provided such notice is received within 30 days of such supply. As an alternative to replacement, the Publisher agrees fully to refund the Licensee in such circumstances, if the Licensee so requests, provided that the Licensee returns this copy of *Spon's Architects & Builders' Price Book 2020* to the Publisher. The provisions of this sub-clause 8.2 do not apply where the defect results from an accident or from misuse of the product by the Licensee.

8.3 Sub-clause 8.2 sets out the sole and exclusive remedy of the Licensee in relation to defects in the Product.

8.4 The Publisher and the Licensee acknowledge that the Publisher supplies the Product on an "as is" basis. The Publisher gives no warranties:

8.4.1 that the Product satisfies the individual requirements of the Licensee; or

8.4.2 that the Product is otherwise fit for the Licensee's purpose; or

8.4.3 that the Product is compatible with the Licensee's hardware equipment and software operating environment.

8.5 The Publisher hereby disclaims all warranties and conditions, express or implied, which are not stated above.

8.6 Nothing in this Clause 8 limits the Publisher's liability to the Licensee in the event of death or personal injury resulting from the Publisher's negligence.

8.7 The Publisher hereby excludes liability for loss of revenue, reputation, business, profits, or for indirect or consequential losses, irrespective of whether the Publisher was advised by the Licensee of the potential of such losses.

8.8 The Licensee acknowledges the merit of independently verifying the price book data prior to taking any decisions of material significance (commercial or otherwise) based on such data. It is agreed that the Publisher shall not be liable for any losses which result from the Licensee placing reliance on the data under any circumstances.

8.9 Subject to sub-clause 8.6 above, the Publisher's liability under this Agreement shall be limited to the purchase price.

9. **Intellectual Property Rights**

9.1 Nothing in this Agreement affects the ownership of copyright or other intellectual property rights in the Product.

9.2 The Licensee agrees to display the Publishers' copyright notice in the manner described in the Product.

9.3 The Licensee hereby agrees to abide by copyright and similar notice requirements required by the Publisher, details of which are as follows:

"© 2020 Taylor & Francis. All rights reserved. All materials in *Spon's Architects' & Builders' Price Book 2020* are copyright protected. All rights reserved. No such materials may be used, displayed, modified, adapted, distributed, transmitted, transferred, published or otherwise reproduced in any form or by any means now or hereafter developed other than strictly in accordance with the terms of the licence agreement enclosed with *Spon's Architects' & Builders' Price Book 2020*. However, text and images may be printed and copied for research and private study within the preset program limitations. Please note the copyright notice above, and that any text or images printed or copied must credit the source."

9.4 This Product contains material proprietary to and copyedited by the Publisher and others. Except for the licence granted herein, all rights, title and interest in the Product, in all languages, formats and media throughout the world, including copyrights therein, are and remain the property of the Publisher or other copyright holders identified in the Product.

10. Non-assignment

This Agreement and the licence contained within it may not be assigned to any other person or entity without the written consent of the Publisher.

11. Termination and Consequences of Termination.

11.1 The Publisher shall have the right to terminate this Agreement if:

11.1.1 the Licensee is in material breach of this Agreement and fails to remedy such breach (where capable of remedy) within 14 days of a written notice from the Publisher requiring it to do so; or

11.1.2 the Licensee becomes insolvent, becomes subject to receivership, liquidation or similar external administration; or

11.1.3 the Licensee ceases to operate in business.

11.2 The Licensee shall have the right to terminate this Agreement for any reason upon two month's written notice. The Licensee shall not be entitled to any refund for payments made under this Agreement prior to termination under this sub-clause 11.2.

11.3 Termination by either of the parties is without prejudice to any other rights or remedies under the general law to which they may be entitled, or which survive such termination (including rights of the Publisher under sub-clause 7.4 above).

11.4 Upon termination of this Agreement, or expiry of its terms, the Licensee must destroy all copies and any back up copies of the product or part thereof.

12. General

12.1 *Compliance with export provisions*
The Publisher hereby agrees to comply fully with all relevant export laws and regulations of the United Kingdom to ensure that the Product is not exported, directly or indirectly, in violation of English law.

12.2 *Force majeure*
The parties accept no responsibility for breaches of this Agreement occurring as a result of circumstances beyond their control.

12.3 *No waiver*
Any failure or delay by either party to exercise or enforce any right conferred by this Agreement shall not be deemed to be a waiver of such right.

12.4 *Entire agreement*
This Agreement represents the entire agreement between the Publisher and the Licensee concerning the Product. The terms of this Agreement supersede all prior purchase orders, written terms and conditions, written or verbal representations, advertising or statements relating in any way to the Product.

12.5 *Severability*
If any provision of this Agreement is found to be invalid or unenforceable by a court of law of competent jurisdiction, such a finding shall not affect the other provisions of this Agreement and all provisions of this Agreement unaffected by such a finding shall remain in full force and effect.

12.6 *Variations*
This agreement may only be varied in writing by means of variation signed in writing by both parties.

12.7 *Notices*
All notices to be delivered to: Spon's Price Books, Taylor & Francis Books Ltd., 3 Park Square, Milton Park, Abingdon, Oxfordshire, OX14 4RN, UK.

12.8 *Governing law*
This Agreement is governed by English law and the parties hereby agree that any dispute arising under this Agreement shall be subject to the jurisdiction of the English courts.

If you have any queries about the terms of this licence, please contact:

Spon's Price Books
Taylor & Francis Books Ltd.
3 Park Square, Milton Park, Abingdon, Oxfordshire, OX14 4RN
Tel: +44 (0) 20 7017 6000
www.sponpress.com

Spon Press
an imprint of Taylor & Francis

Ebook options

Your print copy comes with a free ebook on the VitalSource® Bookshelf platform. Further copies of the ebook are available from https://www.crcpress.com/search/results?kw=spon+2020
Pick your price book and select the ebook under the 'Select Format' option.

To buy Spon's ebooks for five or more users of in your organisation please contact:

Spon's Price Books
eBooks & Online Sales
Taylor & Francis Books Ltd.
3 Park Square, Milton Park, Oxfordshire, OX14 4RN
Tel: (020) 337 73480
onlinesales@informa.com